NATIONAL ACADEMIES *Sciences Engineering Medicine*

Accelerating Decarbonization in the United States

Technology, Policy, and Societal Dimensions

Committee on Accelerating Decarbonization
in the United States: Technology, Policy, and
Societal Dimensions

Board on Energy and Environmental Systems

Division on Engineering and Physical Sciences

Board on Atmospheric Sciences and Climate

Division on Earth and Life Studies

Board on Environmental Change and Society

Division of Behavioral and Social Sciences and
Education

Transportation Research Board

Consensus Study Report

NATIONAL ACADEMIES PRESS 500 Fifth Street, NW Washington, DC 20001

This activity was supported by the Alfred P. Sloan Foundation, Breakthrough Energy, Heising-Simons Foundation, Incite Labs, Quadrivium Foundation, and U.S. Energy Foundation, with support from the National Academy of Sciences Andrew W. Mellon Foundation Fund, National Academy of Sciences Arthur L. Day Fund, National Academy of Sciences Thomas Lincoln Casey Fund, and the National Academies Mitchell Fund. Any opinions, findings, conclusions, or recommendations expressed in this publication do not necessarily reflect the views of any organization or agency that provided support for the project.

International Standard Book Number-13: 978-0-309-68284-8
International Standard Book Number-10: 0-309-68284-3
Digital Object Identifier: https://doi.org/10.17226/25931
Library of Congress Control Number: 2024930545

This publication is available in limited quantities from:

Board on Energy and Environmental Systems
500 Fifth Street, NW
Washington, DC 20001
bees@nas.edu
http://www.sites.nationalacademies.org/DEPS/BEES

This publication is available from the National Academies Press, 500 Fifth Street, NW, Keck 360, Washington, DC 20001; (800) 624-6242 or (202) 334-3313; http://www.nap.edu.

Suggested citation: National Academies of Sciences, Engineering, and Medicine. 2024. *Accelerating Decarbonization in the United States: Technology, Policy, and Societal Dimensions*. Washington, DC: The National Academies Press. https://doi.org/10.17226/25931.

The **National Academy of Sciences** was established in 1863 by an Act of Congress, signed by President Lincoln, as a private, nongovernmental institution to advise the nation on issues related to science and technology. Members are elected by their peers for outstanding contributions to research. Dr. Marcia McNutt is president.

The **National Academy of Engineering** was established in 1964 under the charter of the National Academy of Sciences to bring the practices of engineering to advising the nation. Members are elected by their peers for extraordinary contributions to engineering. Dr. John L. Anderson is president.

The **National Academy of Medicine** (formerly the Institute of Medicine) was established in 1970 under the charter of the National Academy of Sciences to advise the nation on medical and health issues. Members are elected by their peers for distinguished contributions to medicine and health. Dr. Victor J. Dzau is president.

The three Academies work together as the **National Academies of Sciences, Engineering, and Medicine** to provide independent, objective analysis and advice to the nation and conduct other activities to solve complex problems and inform public policy decisions. The National Academies also encourage education and research, recognize outstanding contributions to knowledge, and increase public understanding in matters of science, engineering, and medicine.

Learn more about the National Academies of Sciences, Engineering, and Medicine at **www.nationalacademies.org**.

COMMITTEE ON ACCELERATING DECARBONIZATION IN THE UNITED STATES: TECHNOLOGY, POLICY, AND SOCIETAL DIMENSIONS

STEPHEN W. PACALA (NAS), Princeton University, *Chair*
DANIELLE DEANE-RYAN, The New School
ALEXANDRA "SANDY" FAZELI, National Association of State Energy Officials
KELLY SIMS GALLAGHER, Tufts University
JULIA H. HAGGERTY, Montana State University
CHRIS T. HENDRICKSON (NAE), Carnegie Mellon University
ADRIENNE L. HOLLIS,[1] National Wildlife Federation
JESSE JENKINS,[2] Princeton University
ROXANNE JOHNSON, BlueGreen Alliance
TIMOTHY C. LIEUWEN (NAE), Georgia Institute of Technology
VIVIAN E. LOFTNESS, Carnegie Mellon University
CARLOS E. MARTÍN, The Brookings Institution
MICHAEL A. MÉNDEZ, University of California, Irvine
CLARK A. MILLER, Arizona State University
JONATHAN A. PATZ (NAM), University of Wisconsin–Madison
KEITH PAUSTIAN, Colorado State University
WILLIAM "BILLY" PIZER, Resources for the Future
VARUN RAI,[3] The University of Texas at Austin
EDWARD "ED" RIGHTOR, Information Technology and Innovation Foundation (retired)
PATRICIA "PATY" ROMERO-LANKAO, University of Toronto Scarborough
DEVASHREE SAHA, World Resources Institute
ESTHER S. TAKEUCHI (NAE),[4] Stony Brook University
SUSAN F. TIERNEY, Analysis Group
WILLIAM "REED" WALKER, University of California, Berkeley

NOTE: See Appendix B, Disclosure of Unavoidable Conflicts of Interest.

---

[1] Resigned from the committee October 2022.
[2] Resigned from the committee March 2022.
[3] Resigned from the committee April 2022.
[4] Resigned from the committee April 2022.

Staff

K. JOHN HOLMES, Senior Director/Scholar, Board on Energy and Environmental Systems (BEES), *Study Director*
ELIZABETH ZEITLER, Associate Director, BEES
BRENT HEARD, Program Officer, BEES
CATHERINE WISE, Program Officer, BEES
KASIA KORNECKI, Program Officer, BEES
REBECCA DeBOER, Research Associate, BEES
KYRA HOWE, Research Assistant, BEES
JASMINE VICTORIA BRYANT, Research Assistant, BEES
KAIA RUSSELL, Program Assistant, BEES
IPPOLYTI DELATOLAS, Christine Mirzayan Science and Technology Policy Fellow, BEES
RAPHAEL APEANING, Christine Mirzayan Science and Technology Policy Fellow, BEES
KAVITHA CHINTAM, Christine Mirzayan Science and Technology Policy Fellow, BEES
STEPHEN GODWIN, Scholar, Transportation Research Board
CHANDRA MIDDLETON, Program Officer, Board on Environmental Change and Society
HANNAH STEWART, Associate Program Officer, Board on Environmental Change and Society
ALEX REICH, Program Officer, Board on Atmospheric Sciences and Climate
ELI NASS, Research Assistant, Division on Engineering and Physical Sciences
TOM THORNTON, Director, Board on Environmental Change and Society
ELIZABETH FINKLEMAN, Chief of Staff, National Academy of Medicine
DAVID BUTLER, J. Herbert Hollomon Scholar, National Academy of Engineering
JENELL WALSH-THOMAS, Program Officer, Board on Environmental Change and Society

Consultants

JENNIFER R. BRATBURD, University of Wisconsin–Madison
IPPOLYTI DELATOLAS, Massachusetts Institute of Technology

Reviewers

This Consensus Study Report was reviewed in draft form by individuals chosen for their diverse perspectives and technical expertise. The purpose of this independent review is to provide candid and critical comments that will assist the National Academies of Sciences, Engineering, and Medicine in making each published report as sound as possible and to ensure that it meets the institutional standards for quality, objectivity, evidence, and responsiveness to the study charge. The review comments and draft manuscript remain confidential to protect the integrity of the process.

We thank the following individuals for their review of this report:

JOSEPH ALDY, Harvard Kennedy School
DOUGLAS ARENT, National Renewable Energy Laboratory
RICHARD BIRDSEY, Woodwell Climate Research Center
DANA BOURLAND, The JPB Foundation
SANYA CARLEY, Indiana University
MIJIN CHA, University of California, Santa Cruz
JARED L. COHEN (NAE), Carnegie Mellon University
PAMELA EATON, Green West Strategies
THOMAS A. FANNING, Southern Company
HOWARD FRUMKIN (NAM), University of Washington and Trust for Public Land
DOUGLAS HOLLETT, Melroy-Hollett Technology Partners
TARA HUDIBURG, University of Idaho
THERESA KOTANCHEK (NAE), Evolved Analytics, LLC
JOHN LARSEN, Rhodium Group
ROBERT LITTERMAN, Kepos Capital
PAMELA MATSON (NAS), Stanford University
RONI NEFF, Johns Hopkins University
JOSEPH POWELL (NAE), ChemePD, LLC
JOSÉ G. SANTIESTEBAN (NAE), ExxonMobil Research and Engineering Company (retired)
EMILY SCHAPIRA, Philadelphia Energy Authority
JOSEPH L. SCHOFER, Northwestern University at Evanston
DREW SHINDELL (NAS), Duke University
CHRISTOPHER TESSUM, University of Illinois at Urbana-Champaign
MICHAEL VANDENBERGH, Vanderbilt University School of Law

Although the reviewers listed above provided many constructive comments and suggestions, they were not asked to endorse the findings or recommendations of this report nor did they see the final draft before its release. The review of this report was overseen by **CHERRY MURRAY (NAS/NAE),** University of Arizona at Tucson, and **DAVID ALLEN (NAE),** The University of Texas at Austin. They were responsible for making certain that an independent examination of this report was carried out in accordance with standards of the National Academies and that all review comments were carefully considered. Responsibility for the final content rests entirely with the authoring committee and the National Academies.

Contents

APPENDIXES

Preface

As the Committee on Accelerating Decarbonization in the United States: Technology, Policy, and Societal Dimensions issues its second and final report, it is worth reflecting on global events that have transpired since we began work in late 2019. Our first meeting in early March 2020 ended with the doors of the National Academies building on Fifth Street NW in Washington, DC, being shut behind us for a COVID-19 lockdown lasting about 2 years. Before our next in-person meeting, the world had weathered its first global pandemic in over a century. The United States was undergoing a reckoning of racial injustices in the aftermath of George Floyd's murder, and equity rose as a priority for organizations of all sizes. Russia had invaded Ukraine and demonstrated to everyone the strategic and economic disadvantages associated with a fossil fuel economy. The U.S. Congress had passed the most ambitious set of legislative climate and energy initiatives ever enacted in the United States. These events affected our work in large ways and small, from how we interacted as a committee to the scope and arc of our reports.

While the setting changed for this study, the motivation has not. The adverse impacts of climate change continue to grow, exposing ever-wider swathes of society to its destructive effects. Low, non-emitting, and negative emissions technologies continue to be deployed at ever-increasing scales and ever-lower prices across the globe. Governments, companies, and institutions across the globe and within the United States continue to adopt emissions reduction goals and develop plans to achieve zero net emissions of anthropogenic greenhouse gases, usually by midcentury.

It was within that context that our first report was released in February 2021. That report produced a technical blueprint and policy portfolio for the first 10 years of a just, prosperous, and equitable 30-year transition to net-zero U.S. emissions. The committee fully appreciated that public support for a decades-long transition could be maintained only by fairly distributing benefits and costs. Amid the expanding focus on climate policy that began with the new Biden administration, we briefed our first report widely and now find that many of its recommendations are either enacted or similar to those implemented in recent legislation.

In embarking on the work for our second report, we expanded the committee's expertise in energy justice, health, workforce, and the role of subnational actors. We also expanded the study to include non-CO_2 greenhouse gases, land use, and sectoral analyses. The committee undertook ambitious public information gathering, holding 14 webinars on wide-ranging issues, including leveraging financial systems for decarbonization; soil carbon offsets; government, nonprofit, and philanthropic perspectives on implementing a just and equitable energy transition; manufacturing and industrial decarbonization; public engagement strategies; and research and development priorities for the buildings sector. The committee gathered for its second in-person meeting on July 26, 2022, to hold a workshop: Pathways to an Equitable and Just Transition: Principles, Best Practices, and Inclusive Stakeholder Engagement.

Our meeting in July was especially significant because it was then that the committee concluded that major climate legislation was *not* forthcoming, and that it was time to move ahead with the knowledge that almost none of the federal actions recommended in the first report would be implemented. Famous last words. Within hours of that decision, it became clear that something historic was afoot in Congress. By the end of the summer, our plate was full with analyses of multiple pieces of federal legislation, executive orders, and regulatory actions aiming to put the country on track for 50 percent emissions reductions by 2030 and net zero by midcentury. This policy environment inspired the report we have today, one that focuses on filling gaps between the current policy portfolio and the goal of a fair, just, and equitable transition to net zero, and about how to overcome barriers to implementing this robust and unprecedented set of policies.

I wish to express my deepest gratitude to the study committee for its patience, perseverance, hard work, and equanimity as we struggled together to communicate across disciplines, to meet difficult deadlines, and to wrestle with a complex and continuously evolving set of policies that will affect everyone. I want to thank the governmental leaders who took bold action to combat a global crisis and especially those inside and outside government who now work every day to implement these actions. We offer our recommendations to these experts with humility, knowing that they have far better understanding of facts on the ground than we do, and that they continually expand and adapt their strategies. I want to thank everyone who provided input to the committee during our public sessions—a list running to hundreds of individuals. And last but by no means least, I wish to express my gratitude to the staff of the National Academies who devoted significant portions of their lives to this effort, with calm expertise and almost supernatural energy and efficiency. Special thanks to K. John Holmes for his tireless leadership and wisdom. Because of this large cast of participants, the committee has produced what we hope will be a useful and significant report, something that will serve both policy makers and the public as we work together to accelerate decarbonization in the United States.

Stephen W. Pacala, *Chair*
Committee on Accelerating Decarbonization
in the United States: Technology, Policy, and
Societal Dimensions

Executive Summary

The world is coalescing around the need to reduce greenhouse gas (GHG) emissions to limit the effects of anthropogenic climate change, with many nations setting goals of net-zero emissions by midcentury. As the largest cumulative emitter, the United States has the opportunity to lead the global fight against climate change. It has set an interim emissions target of 50–52 percent below 2005 levels by 2030 toward a net-zero goal. The recent trio of federal legislative actions—the Infrastructure Investment and Jobs Act of 2021 (IIJA), the CHIPS and Science Act of 2022 (CHIPS), and the Inflation Reduction Act of 2022 (IRA)—in addition to federal regulations and executive orders, state and local government policies, and private sector activities— put the United States in a position to claim such international leadership. Modeling analyses suggest that the federal policies could provide 70–80 percent of the emissions reductions toward the 2030 target, putting the country close to a 30-year trajectory to net zero. Concurrent to reducing emissions, the policies also aim to meet societal needs such as creating domestic jobs, eliminating energy and environmental injustices, increasing U.S. economic competitiveness, revitalizing the energy and industrial sectors, and improving human health. Achieving all of these intended outcomes will require overcoming formidable innovation and implementation challenges and ensuring that the policy portfolio produces as designed.

Through an assessment of current federal, state, and local climate and energy policies, this report from the National Academies of Sciences, Engineering, and Medicine's Committee on Accelerating Decarbonization in the United States: Technology, Policy, and Societal Dimensions identifies gaps and barriers to implementation that would prevent the nation from attaining its climate, economic, and societal goals. It follows from the committee's first report, released in February 2021,[1] which laid out federal actions needed during the 2020s to put the nation on a fair and equitable path to decarbonization by midcentury. Both reports were tasked with examining "societal, institutional, behavioral, and equity drivers and implications of deep decarbonization" and emphasize the need for a strong social contract to maintain support for the decades-long transition to a decarbonized energy system that is fair, equitable, and just.

---

[1] See National Academies of Sciences, Engineering, and Medicine, 2021, *Accelerating Decarbonization of the U.S. Energy System*, Washington, DC: The National Academies Press, https://doi.org/10.17226/25932.

To that end, the committee organized this report around five major objectives of decarbonization policy—GHG emission reductions, equity and fairness, health, employment, and public engagement—that cut across eight sectors: electricity, buildings, land use, transportation, industry, finance, fossil fuels, and non-federal actors. Chapters on objectives are tailored to readers most interested in the impacts of the transition on equity, justice, health, and employment, and the need for public engagement, but also discuss relevant practical, technical, institutional, and legal constraints to achieving the societal objective. Similarly, the sectoral chapters are tailored to experts on technologies and policies to reduce emissions in that sector but, where appropriate, also assess how these technologies and policies will impact equity, employment, health, and public engagement. While causing some redundancy, this organization provides a fuller picture to specialists who will read only portions of the report.

The committee's analysis resulted in approximately 80 recommendations directed toward a variety of government, nonprofit, and private-sector actors. These recommendations can be grouped into the following 10 broad categories, which represent main themes of the report:

- A Broadened Policy Portfolio
- Rigorous and Transparent Analysis and Reporting for Adaptive Management
- Ensuring Procedural Equity in Planning and Siting New Infrastructure and Programs
- Ensuring Equity, Justice, Health, and Fairness of Impacts
- Siting and Permitting Reforms for Interstate Transmission
- Tightened Targets for the Buildings and Industrial Sectors and a Backstop for the Transport Sector
- Managing the Future of the Fossil Fuel Sector
- Building the Needed Workforce and Capacity
- Reforming Financial Markets
- Research, Development, and Demonstration Needs

In developing its findings and recommendations, the committee recognized the inherent risks and uncertainties associated with such an unprecedented, long-term, whole-of-society transition. These include *execution risk*—that the nation will be unable to execute current climate and energy policy at the necessary pace and scale, or that the policies will not work as intended; *technological risk*—that non-emitting technologies might not be ready in time at the right price; *political, judicial, and societal polarization risk*—that political and judicial actions or societal pressures will change the policy landscape; and *risk from events outside the energy system*—that war, disease, and other disruptions will inevitably arise and impact national and global energy systems. Mitigating these risks will require adaptive management and governance to coordinate and evaluate policy implementation and to communicate progress on outcomes. A comprehensive, system-wide evaluation of decarbonization policies and programs will also be critical for monitoring cross-sector impacts, sustaining a social license to operate, and keeping the nation on track to achieve its goal of an equitable net-zero transition.

While the destination is clear and a solid foundation has been set, the road ahead will not be easy. Individuals, businesses, and organizations across all sectors of the economy will have to work with government to implement, adapt, and expand on existing local, state, and federal climate and energy policies. But the potential benefits are great: energy services that are clean, affordable, and equitable; reduced impacts from climate change; better health and employment opportunities; and cleaner air. This report's recommendations provide advice on filling policy gaps, overcoming implementation barriers, and establishing adaptive management strategies so that the United States can realize its net-zero emissions goal and all Americans can benefit from an equitable energy system.

Summary

When it passed in 2022, the Inflation Reduction Act (IRA) revolutionized U.S. climate and energy policy as the largest legislative action in the nation's history to mitigate climate change, with anticipated public support for clean energy investments ranging from almost $400 billion to more than $1 trillion over the coming decade. The IRA is complemented by the bipartisan Infrastructure Investment and Jobs Act (IIJA, November 2021; more than $62 billion in appropriations for Department of Energy [DOE] climate and energy programs); CHIPS and Science Act (CHIPS, August 2022; $54.2 billion in appropriations for domestic semiconductor production and $170 billion in 5-year authorizations for research and development); and a number of executive orders and regulations from the Biden administration, including Executive Order (EO) 14008, which, among other things, established the Justice40 Initiative. As discussed in Chapter 1, these amounts are not necessarily additive considering that IRA, IIJA, and CHIPS do not use equivalent funding mechanisms. States, localities, and other entities have also enacted policies to advance deep decarbonization, such as California's zero-emissions vehicles mandate and actions of state policy makers in Texas supporting the rapid growth of wind farms there. Most modeling analyses indicate that this policy portfolio will cause a dramatic shift in the trajectory of U.S. greenhouse gas (GHG) emissions and place the nation close to a 30-year path to net-zero emissions, *but only if formidable challenges of innovation and implementation can be overcome and the policy portfolio produces as designed.*

The stakes could not be higher. Most nations of the world have announced a goal of zero net GHG emissions by midcentury because of overwhelming scientific evidence that climate change is dangerous, and human caused. Annual global net emissions must decline to zero in approximately 30 years to keep Earth's mean surface temperature from climbing above the Paris Agreement's preferred target of 1.5°C (reaffirmed at COP 27 in November 2022), and in so doing prevent the most serious effects of accelerating climate change. The United States has the largest cumulative, and twelfth largest per capita, GHG emissions of any country, and the second largest annual emissions (after China), but until passage of the IRA, the United States was not at the vanguard of national actions to combat climate change.

Successful implementation of current U.S. policy would establish the nation as the international leader in the fight against climate change and would be vital to achieve global emissions reductions. The policy portfolio also has additional objectives that reach far beyond climate mitigation. It is intended to improve the lives of ordinary people by increasing the number of high-paying domestic jobs, increasing U.S. economic competitiveness, revitalizing our energy and industrial sectors, eliminating the environmental injustices in our current energy system, putting a fair and equitable system in its place, and improving people's health.

This report offers an assessment of what current federal, state, and local climate and energy policies could accomplish, together with actions and implementation by the private sector and civil society. It focuses specifically on gaps in the current policy portfolio and barriers to implementation that would prevent the nation from attaining its climate, economic, and humanistic goals. The report offers additional policies that could fill gaps and overcome barriers, most of which could be implemented under existing federal legislation through actions by the executive branch and/or state and local governments, although some would require Congress. Significant gaps and barriers are to be expected because nothing of this scale and with this diversity of goals has ever been attempted. Watching for and, over time, filling those gaps will be essential to overall success.

The economic opportunities of a transition to net zero stem largely from recent revolutionary changes in the cost of technologies and equipment that are not fueled by fossil energy. From 2010–2021, the levelized cost of energy for onshore wind and utility-scale solar dropped by nearly 70 and 90 percent respectively, to become cost-competitive with or cheaper than new fossil power projects over most of the globe. As a result, by the early to mid-2030s, the United States could rely on wind and solar electricity generation, together with existing hydro and nuclear assets, to supply at least 80 percent of the country's electricity demand at inflation-adjusted retail costs similar to today. Technological options also are in advanced development to eliminate emissions from the last 20 percent of power supply, using technologies such as advanced nuclear power, batteries and other energy storage technologies to manage long-term fluctuations in demand, natural gas with carbon capture and storage (CCS), and green hydrogen or biogas combusted in turbines. However, realizing the potential of these dispatchable power options will require relentless research, development, demonstration, and deployment (RDD&D); public engagement; and learnings at scale after deployment.

Over the same 11 years, electric vehicle (EV) lithium-ion battery packs dropped in cost by approximately 80 percent, and the price of lithium-ion batteries for all applications fell by 98 percent between 1991–2018 as they migrated from consumer electronics to packs storing more than 100 kWh of electricity in battery electric vehicles (BEVs) (enough to drive more than 300 miles in a passenger vehicle). Dramatic cost reductions in batteries have made BEVs cost-competitive with new light- and medium-duty vehicles powered by internal combustion engines (ICEs). The average cost of owning and operating a light-duty BEV is likely lower than comparable ICE vehicles for some models and will be for most over the next 5 years (2023–2028) depending on vehicle class and other factors.

Thus, within the same decade, the cheapest options for new equipment in two sectors responsible for approximately 70 percent of U.S. CO_2 emissions—power generation and motor vehicles—are switching from being fossil-fuel powered to non-emitting alternatives. Decarbonized electric power, together with ongoing improvements of heat pumps and generally improving energy efficiency, also unlock the potential for emissions-free buildings. Using data even a few years old could lead to claims that the energy transition would be prohibitively expensive, as those data would not reflect current costs given their rapid recent descent.

This technological revolution converged with two other trends to convince elected officials of the need for immediate action. First, climate change–induced increases in the frequency and severity of extreme weather became obvious even to casual observers, and climate activism flourished against the backdrop of immediate and personal dangers that could be mitigated by cost-effective measures. Second, while these technological, scientific, and political changes were taking place, the United States was confronting with renewed vigor the consequences of its history of discrimination and the ongoing systemic problems that persist because of it. The U.S. energy system today contains considerable environmental injustice, such as the disproportionate exposure to fossil-produced air pollution that afflicts communities of color, with some of this owing to overtly racist policies like redlining during the 20th century and discriminatory and predatory lending and investment practices that continue today.

These issues have sharpened the goal, from a 30-year transition to net-zero emissions, to a *fair, equitable, and just* 30-year transition. In January 2021, the Biden administration established the Justice40 Initiative as official U.S. policy, which states that people and "disadvantaged communities that are marginalized, underserved, and overburdened by pollution" receive 40 percent of benefits from some federal investments in climate change, clean energy, energy efficiency, transit, affordable housing, workforce development, and remediation and reduction of legacy pollution.[1] The congressional framers of the IRA followed suit, by directing up to $60 billion of IRA funding to environmental justice priorities.

---

[1] For more information on the Justice40 Initiative, see https://www.whitehouse.gov/environmentaljustice/justice40.

Thus, both the stakes of success and the costs of failure are high. The United States is attempting the first fair, equitable, and just technological transition in its history with a narrow portfolio of policies that relies extensively on subsidies. If successful, the transition will affect almost every part of the U.S. economy and leave the country with an affordable and accessible energy system that produces zero net GHG emissions. It will also afford important co-benefits, such as reduced emissions of ambient air pollutants that cause illness and death; revitalized energy, building, and industrial sectors; increased resilience to environmental and social stressors; net increases in employment; and fair, equitable, and just treatment of both displaced fossil fuel workers and low-income and historically marginalized populations.

This is the second report from a National Academies of Sciences, Engineering, and Medicine committee that was constituted in 2020 to address "societal, institutional, behavioral, and equity drivers and implications of deep decarbonization." (See the statement of task in Box 1-1 of Chapter 1.) The committees that wrote the first and second reports shared a majority of members and had an approximately equal number of experts in energy technologies as in policy and the social science of energy policies, including policies affecting equity and energy justice. The committee was tasked to examine how the nation might achieve an equitable transition to net zero, not whether it should do so. The committee also was not tasked to assess policies designed to address impacts of climate change. The motivation for the focus on equity, fairness, and justice is at least partly pragmatic, given the need to maintain public and political support during a transition that will affect every part of society. Inclusive and equitable approaches, moreover, are key to preempting or minimizing implementation challenges that would delay or derail decarbonization projects.

SUMMARY OF FIRST REPORT

The first report, *Accelerating Decarbonization of the U.S. Energy System*,[2] was released in February 2021 and focused on federal actions needed during the 2020s to put the nation on a fair and equitable path to decarbonization by midcentury. It identified "no-regrets" actions during this first decade of a transition to net-zero emissions that would be robust to uncertainty about the system's final technological mix. It outlined a set of technological and socioeconomic goals to support a just and equitable transition that are robust to alternative future technology options. On the technology side, the committee set goals for carbon-free electricity; electrifying transportation, buildings, and industry; investing in energy efficiency and productivity; deploying critical

---

[2] See National Academies of Sciences, Engineering, and Medicine, 2021, *Accelerating Decarbonization of the U.S. Energy System*, Washington, DC: The National Academies Press, https://doi.org/10.17226/25932.

infrastructure; and expanding the innovation toolkit through clean energy research, demonstration, and deployment. The committee's socioeconomic goals were to strengthen the U.S. economy; promote equity and inclusion; support communities, businesses, and workers; and maximize cost-effectiveness.

The committee recognized that a strong social contract would be essential to maintain support for an energy transition covering 3 decades. Thus, it proposed policies "to build a more competitive U.S. economy, to increase the availability of high-quality jobs, to build an energy system without the social injustices that permeate the current system, and to allow those individuals and businesses that are marginalized today to share equitably in future benefits." The diverse portfolio of policy recommendations called for both system-wide and sector-specific policies that establish the U.S. commitment to a rapid, just, and equitable transition; set rules and standards for technology planning and deployment; invest in research, technology, people, and infrastructure; and assist and build capacities for families, businesses, communities, cities, and states to ensure that disadvantaged communities do not suffer disproportionate burdens. The committee continues to endorse the goals and policies recommended in the first report, while acknowledging that the list of new policies needed today is fundamentally shaped by the radical changes to the policy landscape since its publication. Table 1-1 in Chapter 1 compares the policies recently adopted with the recommendations in the committee's first report.

SUMMARY OF CURRENT FEDERAL CLIMATE AND ENERGY POLICY

The extensive decarbonization policy portfolio that the United States has today did not come together until the summer of 2022 with the passage of the IRA and CHIPS, which complement the IIJA passed in November 2021. The potential impacts of these three pieces of legislation are nothing short of transformative for the energy sector and technology innovation in general. They lay out an expansive domestic industrial policy that puts climate, innovation, manufacturing, and wealth creation across all parts of the nation and economy as a central mission. Combined with the aspirations of related executive orders, this package seeks to develop a more equitable, fair, and just framework for the energy system transition.

Of the three pieces of legislation, the IRA contains by far the most significant and wide-reaching policies to decarbonize the U.S. economy. It provides incentives for purchasing, producing, developing, and deploying clean energy technologies and makes investments in environmental justice and low-income and historically marginalized communities. Modeling studies estimate that successful implementation of the act would put the United States on track to achieving 70–80 percent of the emissions reductions necessary to reach its 2030 emissions target of ~3.3 Gt CO_2e (50–52 percent below 2005 emissions levels) along a trajectory to its midcentury net-zero goal.

CHIPS incentivizes domestic research, development, and manufacturing of semiconductors used in clean energy and a broad range of other modern technologies. It further boosts the country's leadership in science and technology by authorizing investments in research and development, workforce training, and commercialization of a wide range of technologies. These include not only advanced energy technologies but also artificial intelligence, quantum computing, and biotechnology engineering that will impact decarbonization in unexpected ways. CHIPS recognizes the need to diversify the innovation ecosystem by authorizing the creation of regional technology hubs and increasing opportunities for disadvantaged students and communities.

The IIJA is designed to improve roads, bridges, and other components of the nation's aging infrastructure. It is wide-ranging and often future-leaning in its scope by expanding broadband; providing grants for battery manufacturing and recycling facilities; investing in carbon capture, transport, utilization, and storage infrastructure; and deploying EV charging stations, in addition to repairing roads and bridges. While some of these investments may run counter to decarbonization goals in the near term, the legislation establishes regional clean hydrogen and direct air capture hubs, which will provide critical learning as the country takes on the challenges of harder to decarbonize energy uses within industry and transportation in the 2030s. The IIJA further solidified the role of the Department of Energy (DOE) in moving beyond its traditional focus on research, development, and early-stage demonstration to latter-stage demonstration, deployment, and commercialization.

In addition to legislation, other federal actions support the nation's clean energy, equity, and climate priorities. Within days of entering office, the Biden administration issued EO 14008 focusing on steps that the federal government can take both domestically and internationally to address the climate crisis. Importantly, it made Justice40 official U.S. policy and established the Interagency Working Group (IWG) on Coal and Power Plant Communities and Economic Revitalization, recognizing the need to support coal mining and power plant workers and communities in the energy transition. Justice40 and the IWG, together with the equity, justice, and fairness provisions in the IRA and other executive orders, represent a step change in equitable energy and climate policy. In response, federal agencies have created new offices and hired staff focused on energy justice and equity. Other executive orders and regulatory actions have targeted federal procurement power as a catalyst for developing a domestic clean energy economy, fuel economy and GHG standards for light- and medium-/heavy-duty vehicles, and emissions standards for existing and new fossil-fueled power plants and industrial facilities.

SECOND REPORT

In early 2021, it became clear that the 117th Congress was likely to enact significant new climate and energy legislation before the release of the committee's second report. For the second report to be relevant and useful under this scenario, the assessment of "a wider spectrum of technological, policy, social, and behavioral dimensions of deep decarbonization" called for in the statement of task would need to address whatever path to "deep decarbonization" had been chosen by Congress and the administration. The committee spent the next 18 months preparing the foundation for an analysis of a comprehensive federal policy portfolio, and then the past 8 months completing a draft report for the portfolio that is now federal law. The committee subsequently submitted the report for a comprehensive external review and has modified the analysis as needed to reflect reviewer comments and updates to policy, regulations, and other climate-related actions. Also, given the first report's focus on federal action, the committee examined the contributions of states, localities, the private sector, and civil society to mitigating climate change.

The committee identified five objectives of decarbonization policy—GHG emission reductions,[3] equity and fairness, health, employment, and public engagement—and eight sectors—electricity, buildings, land use, transportation, industry, finance, fossil fuels, and non-federal actors. This report has chapters focused on objectives that cut across sectors, and chapters on the sectors that cut across objectives (think of a matrix, with objectives as rows and sectors as columns). Although this structure entails some redundancy, it facilitates understanding by specialists who may read only part of the report. It also helps emphasize the crosscutting and systems-level characteristics of deep decarbonization. Sectoral chapters support the interests of sectoral specialists, while also sharing with them that many daunting sectoral barriers are likely to be social and cultural, such as the development of public resistance to a process that seems unfair, unequitable, or unjust. Chapters on objectives are tailored to the interests of social scientists, environmental justice activists, and others who are interested in the fairness of the transition; its impacts on equity, justice, health, and employment; and the need for effective public engagement, while confronting them with myriad practical, technical, institutional, and legal constraints that cannot be ignored. A prime example of the latter is the fact that a minimum amount of electricity production must be dispatchable, which means that one can turn it on whenever needed in order to meet the demands of a complex economy and just society, even during conditions when the winds are quiet, the skies are hazy, and the temperatures are extreme for several consecutive days.

---

[3] This report primarily covers CO_2 emissions, with some discussion of non-CO_2 GHGs where relevant.

After passage of the IRA, IIJA, and CHIPS, it was clear from their budgets, comprehensive sectoral coverage, and narrow policy portfolio that the dominant risk to achieving the maximum possible emissions reduction is inadequate or failed implementation. Congressional passage significantly reduces the political risk of repeal, while the focus on low-carbon electricity and electrification during the 2020s, which already costs less than new emitting fossil alternatives, largely puts off the risk that essential non-emitting technologies might not be ready in time at the right price until after 2030. Hence, the committee's second and final report is not just about progress but also about gaps and barriers that would prevent successful implementation, where success is measured against the five separate objectives: reduced emissions of GHGs; fairness, equity, and justice; health; the number and quality of jobs; and transparent public engagement in planning and decision-making. For each gap and barrier, the report offers a recommended remedy.

Recommendations Summary[4]

A summary of approximately 80 recommendations is provided in Table S-1, with a list of the actors responsible for implementation and the sectors and objectives that each is designed to address. Recommendations are also sorted into 10 broad categories that are described below.

A Broadened Policy Portfolio. The committee's first report recommended a broad set of policies, including taxes, standards, and incentives, with some redundancy to make the portfolio more robust to the failure or repeal of any one component. For example, manufacturing standards for home heating appliances would ensure a transition to heat pumps even if the carbon tax proved unable to overcome consumer inertia. The narrow policy portfolio in the IRA, IIJA, and CHIPS—exclusively tax incentives and other subsidies, with the exception of a fee for fugitive methane emissions—lacks the backstops of a diverse portfolio and thus makes achieving its emissions reduction goals more uncertain. Also, a 30-year transition will require that some critical elements possess the political durability that only congressional action can provide. Recommendation 1-1 repeats two recommendations for Congress from the first report: a national GHG emissions budget and an economy-wide carbon tax with provisions to protect people with low incomes and energy-intensive businesses exposed to import competition. The committee fully appreciates the political headwinds currently facing these actions. Within the body of the current report, the committee also reiterates recommendations from its first report for clean energy standards for electricity, zero-emissions vehicle sales mandates, zero-emissions appliance standards, and the

---

[4] The text in this section was changed during editorial review to improve clarity and alignment with information in other sections of the report.

creation of a National Transition Corporation. Recommendations in Table S-1 that would broaden the federal policy portfolio include 1-1, 5-3, 5-8, 6-1, 7-5, 7-7, 8-2, 8-6, 8-8, 9-1, 9-3, 10-2, 10-3, 10-6, 10-7, 10-9, and 12-2.

Rigorous and Transparent Analysis and Reporting for Adaptive Management. Few of the policies in the IRA have ever been implemented at the scale and pace required by the law and necessary to achieve climate mitigation goals, while also ensuring energy justice and equity. Federal agencies and, especially, state governments currently lack the capacity needed to administer the funds and implement programs. State and local motivations behind implementation are highly heterogenous, with some areas energetically supportive and others opposed. Overt or passive public resistance to the deployment of critical infrastructure could materialize in some locations or sectors. The climate and energy programs in the IRA and IIJA are scattered across the federal government, with no durable entity to gather data, monitor, and analyze them and periodically report on progress against GHG emissions, equity, justice, employment, health, and public engagement goals. This will limit the nation's ability to learn what works and what does not, to course-correct, and to design effective policies for the subsequent 2 decades of the transition. Recommendations 1-2 and 1-3 are for Congress to designate an enduring entity to oversee and execute rigorous and transparent data analysis, monitoring, and reporting about investments and progress, in much the same way that the U.S. Global Change Research Program (USGCRP) was empowered by Congress to report periodically on climate change and its impacts. Other recommendations that support Recommendations 1-2 and 1-3 include 2-1, 2-2, 2-3, 2-6, 3-2, 7-1, 7-3, 7-5, 8-1, 8-3, 8-4, 9-5, 10-1, 10-5, 10-6, 10-7, 11-4, and 11-5.

Ensuring Procedural Equity in Planning and Siting New Infrastructure and Programs. The social sciences and technical literature on planning and developing new infrastructure during technological transitions shows that public consensus and support require a careful collaborative process managed by specially trained people, and with active participation by the diversity of people in the local community. It is also important to involve affected publics in planning for energy development early, rather than coming to them with fully baked project proposals. Without robust process, policy implementers may lose or fail to gain the trust of a local community before learning what its members want and would support. Communities are then left opposing infrastructure like community solar, which could pay them revenue, reduce energy bills, or even be owned by them, because they (understandably) do not trust the people or process that promotes it, rather than deciding based on the advantages and disadvantages for their community. Although robust public process takes time, it increases the probability of success; there is some evidence that good process yields trust and awareness that can facilitate subsequent siting in the same location.

Recommendations 5-1 through 5-8 would implement this process, with support from 2-3, 2-4, 2-5, 6-6, 7-1–7-3, 9-5, 11-1–11-3, and 12-1.

Ensuring Equity, Justice, Health, and Fairness of Impacts. The current federal policy portfolio contains many provisions designed to ensure a fair, just, and equitable distribution of costs and benefits from the transition, and to eliminate current injustice in our energy system. Recommendations 2-1 through 2-6 are designed to strengthen these provisions and reduce barriers to their successful implementation. The report also recommends that Justice40 or an equivalent target be made durable (institutionalized) through an act of Congress (Recommendation 2-1). Public health is a critical component of justice and fairness in many parts of the energy transition. Perhaps the single gravest environmental injustice in the U.S. energy system is that up to 355,000 deaths per year are caused by air pollution from fossil fuels combustion, which disproportionately occur in communities of color and low-income households. A large fraction of this pollution will be eliminated by actions stimulated by the IRA during the 2020s because of coal plant closures. This will make a significant down payment on Justice40, above and beyond the funding in the law that is directed explicitly to environmental justice. Recommendations 3-1 through 3-3 would make sure that health impacts are assessed when technological and new infrastructure decisions are made, and Recommendation 10-4 would promote development of technologies that reduce both CO_2 and co-pollutant emissions. Last, Recommendations 4-2, 4-3, 5-2, 5-4, 5-8, 6-4, 6-5, 9-5, 9-6, 10-4, 12-1, 12-3, 12-6, and 12-7 would mitigate the harms to workers and communities from the loss of fossil-dependent jobs.

Siting and Permitting Reforms for Interstate Transmission. Perhaps the single greatest risk to a successful energy transition during the 2020s is the risk that the nation fails to site, modernize, and build out the electrical grid. Except where new transmission has been shown as needed to keep the lights on, adding transmission is complicated by the need to secure cooperation from numerous individual landowners and affected publics—many of whom may perceive greater cost than benefit from high-voltage transmission lines. The need for adding new transmission capacity and pathways during the 2020s is unprecedented, given the committee's goal of at least 75 percent clean power by 2030, laid out in the first report. Studies show that without significant new transmission capacity, renewables deployment would be delayed, just as electrification of transport and heating are starting to increase demands for power. The net result could be *increased* generation by fossil electricity plants and increased national fossil emissions during the 2020s, which would make the entire effort appear to be a failure, even assuming that investments in energy efficiencies occur in conjunction with electrification. This would also prolong and increase the environmental injustice of exposure to dangerous particulate emissions from fossil power plants.

Expansion of the high-voltage interstate transmission grid is needed in addition to, rather than instead of, modernization of local electricity distribution systems, deployment of energy resources (such as solar and storage) close to customers, and much more aggressive adoption of energy efficiency. The committee recommends siting and permitting reforms through the collection of executive, state, and private-sector actions in Recommendation 6-2 with support from Recommendations 5-5–5-7, 6-3–6-6, and 7-6.

Tightened Targets for the Buildings and Industrial Sectors and a Backstop for the Transport Sector. The 2030 sectoral emissions goals set by the Biden administration can be achieved or mostly achieved by the current policy portfolio, in part because the goals for "harder-to-decarbonize" sectors, such as industry and buildings are not particularly stringent. Recommendations 7-1 through 7-5 would facilitate effective implementation of components of the IRA and the IIJA directed at buildings and the built environment, especially those that would further equity and fairness objectives. Chapter 7 also includes 10 technically achievable actions in buildings and the built environment that could eliminate up to 1 billion metric tons of today's CO_2 emissions per yer, mostly by increasing energy efficiency and decreasing demand for energy services. Although several of the actions would face formidable political headwinds if proposed today as new federal policies, most could be implemented by states, municipalities, and property owners. By increasing energy efficiency, these actions would decrease current and future demand for electricity, which would ease the pressures to site new transmission, distribution, and generation capacity so quickly. In the transport sector, the Biden administration has set a goal for zero-emission vehicles (ZEVs) to comprise 50 percent of 2030 sales. Because of uncertainties about the pace of deployment of charging infrastructure and consumer adoption of ZEVs, some published analyses predict that the 2030 goal will be achieved, while others predict a significant shortfall. Recommendation 9-1 calls for federal executive action to establish a ZEV standard to backstop the tax credits in the IRA, while Recommendations 9-2 through 9-5 call for state, local, and private actions to promote the growth of ZEV sales. Recommendations 10-1–10-4, 10-6, 10-7, and 10-9 call for specific congressional actions to accelerate decarbonization of the industrial sector. These might attract bipartisan interest as part of a package to revitalize U.S. industry. The less ambitious emissions reduction goals for the buildings and industrial sectors in the 2020s mean that large reductions will be required after 2030. In addition, large atmospheric CO_2 removals in the 2040s will likely be needed from technologies, like direct air capture, that are unproven at a commercial scale.

Managing the Future of the Fossil Fuel Sector. The committee's first report concluded that, in the 2020s, approximately the same actions are required in scenarios that assume exclusively renewable sources in 2050 as in those that assume a mix of renewables, nuclear power, and fossil energy with carbon capture and storage. For this reason, the fact that the IRA, IIJA, and CHIPS include incentives for non-renewable options does not significantly take any decarbonized technology option off the table. Most estimates indicate a continued role for fossil fuels—particularly oil and gas—in meeting energy demand throughout the next decade but significant uncertainty in the 2030s and beyond. Recommendations 12-1, 12-3, and 12-5–12-7 would help manage some consequences of these demand reductions, including safe operation of municipal gas distribution networks despite a declining base of rate payers, reforms to taxes on petroleum products, remediation of abandoned fossil facilities, and transition planning and assistance for communities and states now heavily dependent on fossil extraction and production. Recommendations 12-2 and 12-4 would avoid investments that are not essential to meet current demand and might end up excluded from the final net-zero mix.

Building the Needed Workforce and Capacity. The federal government is rapidly adding the capacity it needs to implement current climate and energy policy. However, the nation as a whole lacks the trained workers needed to implement fairness, equity, justice, and public engagement provisions. Implementation of Recommendations 2-4, 5-1, 5-6, 5-9, 5-10, 7-4, and 7-5 would supply the needed training. Recommendations 4-1–4-4, 10-8, and 12-1 would provide training for workers needed by decarbonized industries and retraining for current fossil fuel workers. Last, Recommendations 13-1–13-5 are needed to remedy the severe capacity shortage in most subnational governments, which will be responsible for administering most of the programs in current policy.

Reforming Financial Markets. The financial sector directs the flow of capital and financial services to businesses and households throughout the United States and has increasingly focused on the risks and opportunities associated with the net-zero transition. Historically, some communities have not had equal access to these services, an inequality that the energy transition must address. Targeted programs can address these inequities, and Recommendations 11-1 through 11-3 focus on this outcome. Additionally, better data and information can allow investors and regulators to fully understand climate-related risks and opportunities in the financial sector, and Recommendations 11-4 and 11-5 aim to improve and standardize data collection and disclosure. Last, financial regulators need to improve their monitoring and supervision of climate risks, and Recommendation 11-6 addresses needed scenario analysis and stress testing to understand the vulnerability of key financial institutions and the sector as a whole.

Research, Development, and Demonstration Needs. DOE is implementing many of the new or expanded RD&D programs required in the first decade of a transition to net zero, so that the nation will be ready for the second 2 decades. However, additional investments will be required to address medium- to long-term challenges, such as developing new methods to make low-carbon products using green chemistry or engineering. Furthermore, the breadth of the energy transition requires an RD&D portfolio broader than DOE's domain, including for example land-use practices that store carbon while improving agricultural productivity, research on artificial meat and dairy food products, and ways to reduce food waste and shift toward more plant-based diets. In general, the most formidable barriers to successful implementation are not just technical, but rather within the domain of the social sciences, where DOE and the federal government investments have historically been small or absent. Recommendation 5-9 would greatly enhance investments in energy-related social sciences, while 3-3, 6-7, 7-6, 8-4, 8-5, 8-7, 9-6, 10-1, 10-2, 10-4, and 10-6 would fill specific technological gaps in RD&D, which in some cases (e.g., for land-use-related approaches) will require identifying which technologies to pursue and how.

TABLE S-1 Summary of Recommendations for Policies Designed to Meet Net-Zero Carbon Emissions Goal and How the Policies Support Specific Sectors, Objectives, and Overarching Categories

Short-Form Recommendation	Actor(s) Responsible for Implementing Recommendation	Sector(s) Addressed by Recommendation	Objective(s) Addressed by Recommendation	Overarching Categories Addressed by Recommendation
1-1: Enact Two Federal Policies Recommended in the First Report: National Greenhouse Gas Emissions Budget and Economy-Wide Carbon Tax	Congress, Department of the Treasury, Environmental Protection Agency (EPA)	• Electricity • Buildings • Land use • Transportation • Industry • Finance • Fossil fuels	• Greenhouse gas (GHG) reductions	A Broadened Policy Portfolio
1-2: Leverage the Evidence Act to Execute Data Collection and Evaluation on Decarbonization Investments and Programs	Congress and Office of Management and Budget (OMB)	• Electricity • Buildings • Land use • Transportation • Industry • Finance • Fossil fuels	• GHG reductions • Equity • Health • Employment • Public engagement	Rigorous and Transparent Analysis and Reporting for Adaptive Management
1-3: Identify and Provide Resources for a Central Entity to Provide Timely, Public-Facing Information on the Nation's Progress Toward Decarbonization	Congress and single other agency (e.g., Energy Information Administration [EIA], Global Change Research Program, OMB)	• Electricity • Buildings • Land use • Transportation • Industry • Finance • Fossil fuels	• GHG reductions • Equity • Health • Employment • Public engagement	Rigorous and Transparent Analysis and Reporting for Adaptive Management

2-1: Codify the Justice40 Initiative	Congress	• Buildings • Transportation • Land use • Transportation • Industry • Finance • Fossil fuels	• Equity • Health • Employment	Rigorous and Transparent Analysis and Reporting for Adaptive Management Ensuring Equity, Justice, Health, and Fairness of Impacts
2-2: Develop a Federal Baseline Set of Metrics for Disadvantaged Communities for Program Design and Evaluation	Council on Environmental Quality (CEQ)	• Electricity • Buildings • Land use • Transportation • Industry • Finance • Fossil fuels • Non-federal actors	• Equity	Rigorous and Transparent Analysis and Reporting for Adaptive Management Ensuring Equity, Justice, Health, and Fairness of Impacts
2-3: Implement Federal Legislation for Equitable Outcomes	Federal policy makers	• Electricity • Buildings • Land use • Transportation • Industry • Finance • Fossil fuels • Non-federal actors	• Public engagement • GHG reductions • Equity	Rigorous and Transparent Analysis and Reporting for Adaptive Management Ensuring Equity, Justice, Health, and Fairness of Impacts Ensuring Procedural Equity in Planning and Siting New Infrastructure and Programs
2-4: Build Multi-Level Capacity to Support Community-Led Transitions	Congress, National Transition Corporation, EPA and Department of Energy (DOE), state legislatures	• Non-federal actors	• GHG reductions • Equity • Health • Employment • Public engagement	Ensuring Procedural Equity in Planning and Siting New Infrastructure and Programs Ensuring Equity, Justice, Health, and Fairness of Impacts Building the Needed Workforce and Capacity

continued

TABLE S-1 Continued

Short-Form Recommendation	Actor(s) Responsible for Implementing Recommendation	Sector(s) Addressed by Recommendation	Objective(s) Addressed by Recommendation	Overarching Categories Addressed by Recommendation
2-5: Develop Equitable Technical Assistance Guidelines	Federal Interagency Thriving Communities Network, White House Environmental Justice Advisory Committee (WHEJAC)	• Electricity • Buildings • Land use • Transportation • Industry • Finance • Fossil fuels • Non-federal actors	• Equity • Public engagement	Ensuring Procedural Equity in Planning and Siting New Infrastructure and Programs Ensuring Equity, Justice, Health, and Fairness of Impacts
2-6: Evaluate the Equity Impacts of the Just Energy Transition	Omnibus entity, WHEJAC	• Electricity • Buildings • Land use • Transportation • Industry • Finance • Fossil fuels • Non-federal actors	• Equity • Transparency • Health • Employment	Rigorous and Transparent Analysis and Reporting for Adaptive Management Ensuring Equity, Justice, Health, and Fairness of Impacts
3-1: Phase Out Incentives for the Highest Greenhouse Gas Emitting Animal Protein Sources	Congress and U.S. Department of Agriculture (USDA)	• Land use	• GHG reduction • Health	Ensuring Equity, Justice, Health, and Fairness of Impacts

3-2: Increase Use of Health Impact Assessment Tools in Energy Project Decision-Making	Congress, Centers for Disease Control and Prevention (CDC), National Center for Environmental Health/Agency for Toxic Substances and Disease Registry (NCEH/ATSDR), Department of Health and Human Services (HHS) Office of Climate Change and Health Equity	• Electricity • Buildings • Transportation • Industry • Fossil fuels	• Equity • Health	Rigorous and Transparent Analysis and Reporting for Adaptive Management Ensuring Equity, Justice, Health, and Fairness of Impacts
3-3: Assess Occupational Health Risks Associated with Clean Energy Technologies	CDC, NCEH/ATSDR, Occupational Safety and Health Administration (OSHA)	• Electricity • Buildings • Transportation • Industry • Fossil fuels	• Equity • Health • Employment	Ensuring Equity, Justice, Health, and Fairness of Impacts Research, Development, and Demonstration Needs
4-1: Support the Development of Net-Zero Curriculum and Skill Development Programs for K–12 Students	Department of Education, local governments, and school districts	• Electricity • Buildings • Transportation • Industry • Non-federal actors	• Equity • Employment • Public engagement	Building the Needed Workforce and Capacity
4-2: Invest in Linking People from Disadvantaged Communities to Quality Jobs	Congress	• Electricity • Buildings • Transportation • Industry • Non-federal actors	• Equity • Employment	Ensuring Equity, Justice, Health, and Fairness of Impacts Building the Needed Workforce and Capacity

continued

TABLE S-1 Continued

Short-Form Recommendation	Actor(s) Responsible for Implementing Recommendation	Sector(s) Addressed by Recommendation	Objective(s) Addressed by Recommendation	Overarching Categories Addressed by Recommendation
4-3: Extend Unemployment Insurance Duration for Fossil Fuel–Related Layoffs and Develop Decarbonization Workforce Adjustment Assistance Program	Congress	• Transportation • Fossil fuels	• Equity • Employment • Public engagement	Ensuring Equity, Justice, Health, and Fairness of Impacts Building the Needed Workforce and Capacity
4-4: Collect and Report Data on Net-Zero-Relevant Professions	DOE	• Electricity • Buildings • Transportation • Industry • Non-federal actors	• Equity • Employment	Building the Needed Workforce and Capacity
5-1: Encourage Prospective, Inclusive Dialogue at National and Regional Levels	National Climate Task Force (NCTF), DOE, and EPA	• Non-federal actors	• Equity • Employment • Public engagement	Ensuring Procedural Equity in Planning and Siting New Infrastructure and Programs Building the Needed Workforce and Capacity
5-2: Accelerate the Development, Implementation, Assessment, and Sharing of Energy System Policy and Approaches That Deliver Local Benefits	Subnational government, elected officials and their representative coalitions, federal partners	• Non-federal actors	• Equity • Public engagement	Ensuring Equity, Justice, Health, and Fairness of Impacts Ensuring Procedural Equity in Planning and Siting New Infrastructure and Programs

				A Broadened Policy Portfolio
5-3: Fix Policy Gaps That Limit Role of Public Land in Decarbonization	Congress and state legislatures	• Electricity • Non-federal actors • Land use	• Equity • Public engagement	Ensuring Procedural Equity in Planning and Siting New Infrastructure and Programs
5-4: Address Barriers to Local Benefits from Renewable Energy Facilities	State legislatures	• Non-federal actors	• Equity • Public engagement	Ensuring Equity, Justice, Health, and Fairness of Impacts Ensuring Procedural Equity in Planning and Siting New Infrastructure and Programs
5-5: Convene a National Working Group on Siting Process Innovation with Input from State Energy Officials	DOE, CEQ, Federal Energy Regulatory Commission (FERC), National Association of Regulatory Utility Commissioners (NARUC), and National Association of State Energy Officials (NASEO)	• Non-federal actors • Electricity	• Equity • Public engagement	Ensuring Procedural Equity in Planning and Siting New Infrastructure and Programs Siting and Permitting Reforms for Interstate Transmission
5-6: Mandate and Allocate Resources for a National Assessment on the Public Engagement Workforce and Gaps	Congress, DOE, NCTF	• Electricity • Non-federal actors	• Equity • Employment • Public engagement	Ensuring Procedural Equity in Planning and Siting New Infrastructure and Programs Siting and Permitting Reforms for Interstate Transmission Building the Needed Workforce and Capacity

continued

TABLE S-1 Continued

Short-Form Recommendation	Actor(s) Responsible for Implementing Recommendation	Sector(s) Addressed by Recommendation	Objective(s) Addressed by Recommendation	Overarching Categories Addressed by Recommendation
5-7: Develop Collaborative Regional Renewable Energy Deployment Plans	Civil society leaders and philanthropic organizations	• Non-federal actors	• Equity • Public engagement	Ensuring Procedural Equity in Planning and Siting New Infrastructure and Programs Siting and Permitting Reforms for Interstate Transmission
5-8: Address the Priorities of Native American and Environmental Justice Communities	Congress and federal program designers	• Electricity	• Equity • Public engagement	A Broadened Policy Portfolio Ensuring Procedural Equity in Planning and Siting New Infrastructure and Programs Ensuring Equity, Justice, Health, and Fairness of Impacts
5-9: Invest in and Integrate Social Science Research into Transition Decision-Making	DOE, Department of Transportation (DOT), Department of Defense (DoD), EPA, and National Science Foundation (NSF)	• Non-federal actors	• Equity • Employment • Public engagement	Building the Needed Workforce and Capacity Research, Development, and Demonstration Needs
5-10: Establish an Energy Systems Education Network	DOE and Department of Education	• Electricity • Buildings • Transportation • Industry • Non-federal actors	• Public engagement	Building the Needed Workforce and Capacity

				A Broadened Policy Portfolio
6-1: Adopt National Policy to Limit Power-Sector Greenhouse Gas Emissions	Congress	• Electricity	• GHG reductions • Health	
6-2: Support the Expansion of the Transmission Grid	FERC, DOE, states, transmission companies, public stakeholders, and Department of the Interior (DOI)	• Electricity • Non-federal actors	• GHG reductions • Health • Public engagement	Siting and Permitting Reforms for Interstate Transmission
6-3: Expand Regional Power Markets Consistent with Decarbonization Objectives	Congress, FERC, regional transmission organizations (RTOs)	• Electricity • Non-federal actors	• GHG reductions • Equity • Health	Siting and Permitting Reforms for Interstate Transmission
6-4: Provide Rate Options to Encourage Flexible Demand While Ensuring Affordable Electricity	Decision makers on utility rates (i.e., state utility regulators for jurisdictional investor-owned utilities and boards of cooperatives, municipal electric utilities, and other publicly owned utilities)	• Electricity • Non-federal actors	• Equity	Ensuring Equity, Justice, Health, and Fairness of Impacts Siting and Permitting Reforms for Interstate Transmission
6-5: Support Equitable Deployment of Distributed Energy Resources	States, localities, and tribal governments	• Electricity • Non-federal actors	• GHG reductions • Equity • Health • Public engagement	Ensuring Equity, Justice, Health, and Fairness of Impacts Siting and Permitting Reforms for Interstate Transmission

continued

TABLE S-1 Continued

Short-Form Recommendation	Actor(s) Responsible for Implementing Recommendation	Sector(s) Addressed by Recommendation	Objective(s) Addressed by Recommendation	Overarching Categories Addressed by Recommendation
6-6: Support Planning, Public Participation, and Investment in Modernizing Local Grids	Decision makers on utility service provision (i.e., state utility regulators for jurisdictional investor-owned utilities and boards of cooperatives, municipal electric utilities, and other publicly owned utilities)	• Electricity • Non-federal actors	• GHG reductions • Equity • Health • Public engagement	Ensuring Procedural Equity in Planning and Siting New Infrastructure and Programs Siting and Permitting Reforms for Interstate Transmission
6-7: Invest in Research, Development, and Demonstration of On-Demand Electric Generating Technologies and Long-Duration Storage Technologies	Congress	• Electricity	• GHG reductions • Equity	Research, Development, and Demonstration Needs
7-1: Ensure Clarity and Consistency for the Implementation of Building Decarbonization Policies	DOE	• Buildings	• GHG reductions • Equity • Public engagement	Rigorous and Transparent Analysis and Reporting for Adaptive Management Ensuring Procedural Equity in Planning and Siting New Infrastructure and Programs Tightened Targets for the Buildings and Industrial Sectors and a Backstop for the Transport Sector

7-2: Promote an Equitable Focus Across Building Decarbonization Policies	DOE	• Buildings • Non-federal actors	Ensuring Procedural Equity in Planning and Siting New Infrastructure and Programs Tightened Targets for the Buildings and Industrial Sectors and a Backstop for the Transport Sector
		• GHG reductions • Equity • Public engagement	
7-3: Expand and Evaluate the Weatherization Assistance Program	DOE	• Buildings	Rigorous and Transparent Analysis and Reporting for Adaptive Management Ensuring Procedural Equity in Planning and Siting New Infrastructure and Programs Tightened Targets for the Buildings and Industrial Sectors and a Backstop for the Transport Sector
		• GHG reductions • Equity • Health	
7-4: Coordinate Subnational Government Agencies to Align Decarbonization Policies and Implementation	State and municipal government offices	• Buildings • Non-federal actors	Tightened Targets for the Buildings and Industrial Sectors and a Backstop for the Transport Sector Building the Needed Workforce and Capacity
		• Equity • Employment	

continued

TABLE S-1 Continued

Short-Form Recommendation	Actor(s) Responsible for Implementing Recommendation	Sector(s) Addressed by Recommendation	Objective(s) Addressed by Recommendation	Overarching Categories Addressed by Recommendation
7-5: Build Capacity for States and Municipalities to Adopt and Enforce Increased Regulatory Rigor for Buildings and Equipment	Congress	• Buildings • Non-federal actors	• GHG reductions • Equity • Health • Employment	A Broadened Policy Portfolio Rigorous and Transparent Analysis and Reporting for Adaptive Management Tightened Targets for the Buildings and Industrial Sectors and a Backstop for the Transport Sector Building the Needed Workforce and Capacity
7-6: Increase Research, Development, Demonstration, and Deployment for Built Environment Decarbonization Interventions	Congress	• Buildings • Non-federal actors	• Equity • Health • Employment • Public engagement	Siting and Permitting Reforms for Interstate Transmission Research, Development, and Demonstration Needs
7-7: Extend Current Decarbonization Incentives Beyond the Next Decade While Scaling Up Mandates	Congress	• Buildings	• GHG reductions • Equity • Health • Employment	A Broadened Policy Portfolio Tightened Targets for the Buildings and Industrial Sectors and a Backstop for the Transport Sector
8-1: Convene an Expert Group to Recommend Ways to Measure Additional Forest Sinks	Secretary of Agriculture	• Land use	• GHG reductions	Rigorous and Transparent Analysis and Reporting for Adaptive Management

8-2: Prioritize Ecosystem-Level Carbon Storage	Secretary of Agriculture	• Land use	• GHG reductions	A Broadened Policy Portfolio
8-3: Establish a Permanent, National-Scale, High-Quality Soil Monitoring Network	USDA	• Land use	• GHG reductions	Rigorous and Transparent Analysis and Reporting for Adaptive Management
8-4: Build Out Long-Term Agricultural Field Experiments	USDA	• Land use • Non-federal actors	• GHG reductions	Rigorous and Transparent Analysis and Reporting for Adaptive Management Research, Development, and Demonstration Needs
8-5: Fund Research to Quantify Indicators That Influence Adoption of Regenerative Agriculture Practices	USDA	• Land use	• GHG reductions • Equity • Public engagement	Research, Development, and Demonstration Needs
8-6: Incentivize the Abatement of CH_4 and N_2O Emissions and Improve Soil Carbon Sequestration	USDA	• Land use	• GHG reductions • Equity	A Broadened Policy Portfolio
8-7: Release a Comprehensive Research, Development, Demonstration, and Deployment Program for Biomass Energy with Carbon Capture and Storage	DOE	• Land use	• GHG reductions	Research, Development, and Demonstration Needs

continued

TABLE S-1 Continued

Short-Form Recommendation	Actor(s) Responsible for Implementing Recommendation	Sector(s) Addressed by Recommendation	Objective(s) Addressed by Recommendation	Overarching Categories Addressed by Recommendation
8-8: Convene an Expert Group to Recommend Policies That Could Encourage Sustainable Diets	Secretary of Agriculture	• Land use	• GHG reductions • Health	A Broadened Policy Portfolio
9-1: Accelerate the Adoption of Battery Electric Vehicles	Federal, state, and local governments	• Transportation • Finance • Non-federal actors	• GHG reductions • Equity • Health • Public engagement	A Broadened Policy Portfolio Tightened Targets for the Buildings and Industrial Sectors and a Backstop for the Transport Sector
9-2: Promote Vehicle Electrification at Ports and Airports	Ports and airports and their state and local government owners	• Transportation • Non-federal actors	• GHG reductions • Health	Tightened Targets for the Buildings and Industrial Sectors and a Backstop for the Transport Sector
9-3: Pursue Cost-Effective Efficiency Improvements to Reduce Greenhouse Gas Emissions	Private companies and state and local governments	• Buildings • Transportation • Fossil fuels • Non-federal actors	• GHG reductions • Equity • Health	A Broadened Policy Portfolio Tightened Targets for the Buildings and Industrial Sectors and a Backstop for the Transport Sector
9-4: Pursue Infrastructure Design, Standards, Specifications, and Procedures That Effectively Reduce Transportation Carbon Emissions	State DOTs, American Association of State Highway and Transportation Officials, American Road and Transportation Builders Association, and other specialized transportation infrastructure materials and construction associations	• Transportation • Industry • Non-federal actors	• GHG reductions	Tightened Targets for the Buildings and Industrial Sectors and a Backstop for the Transport Sector

9-5: Enhance Transportation Equity and Environmental Justice Through Programs, Planning, and Services	States and local governments	• Buildings • Transportation • Finance • Non-federal actors	• GHG reductions • Equity • Health • Public engagement	Rigorous and Transparent Analysis and Reporting for Adaptive Management Ensuring Procedural Equity in Planning and Siting New Infrastructure and Programs Ensuring Equity, Justice, Health, and Fairness of Impacts Tightened Targets for the Buildings and Industrial Sectors and a Backstop for the Transport Sector
9-6: Support Advances in Battery Design and Recycling, Fuel Cell Electric Vehicles, and Net-Zero Liquid Fuels	DOE and NSF	• Land use • Transportation • Industry	• GHG reductions	Ensuring Equity, Justice, Health, and Fairness of Impacts Research, Development, and Demonstration Needs
10-1: Develop and Enable Cost-Competitive Process and Waste Heat Solutions	DOE and industrial companies	• Buildings • Industry	• GHG reductions	Rigorous and Transparent Analysis and Reporting for Adaptive Management Tightened Targets for the Buildings and Industrial Sectors and a Backstop for the Transport Sector Research, Development, and Demonstration Needs

continued

TABLE S-1 Continued

Short-Form Recommendation	Actor(s) Responsible for Implementing Recommendation	Sector(s) Addressed by Recommendation	Objective(s) Addressed by Recommendation	Overarching Categories Addressed by Recommendation
10-2: Invest in Energy and Materials Efficiency and Industrial Electrification	Congress and DOE	• Buildings • Industry • Finance • Non-federal actors • Transportation	• GHG reductions	A Broadened Policy Portfolio Tightened Targets for the Buildings and Industrial Sectors and a Backstop for the Transport Sector Research, Development, and Demonstration Needs
10-3: Spur Innovation to Achieve Price-Performance Parity for Low-Carbon Solutions	Congress, DOE, non-governmental organizations (NGOs), industry associations (e.g., American Chemistry Council [ACC], American Iron and Steel Institute [AISI], Portland Cement Association [PCA], National Association of Manufacturers [NAM], and others), and industry	• Industry • Finance • Non-federal actors	• GHG reductions	A Broadened Policy Portfolio Tightened Targets for the Buildings and Industrial Sectors and a Backstop for the Transport Sector
10-4: Pursue Technologies That Reduce Both Greenhouse Gas and Air Pollution Emissions	DOE, NGOs, industry, industry associations (e.g., ACC, AISI, PCA, NAM, and others), and engineering companies	• Industry • Non-federal actors	• GHG reductions • Health	Ensuring Equity, Justice, Health, and Fairness of Impacts Tightened Targets for the Buildings and Industrial Sectors and a Backstop for the Transport Sector Research, Development, and Demonstration Needs

10-5: Use Mass-Based Rather Than Concentration-Based NO$_x$ Standards	Regulatory and permitting organizations	• Industry • Electricity • Transportation	• GHG reductions • Health	Rigorous and Transparent Analysis and Reporting for Adaptive Management
10-6: Develop and Standardize Life-Cycle Assessment Approaches for Carbon Intensity of Industrial Products	DOE, EPA, National Institute of Standards and Technology (NIST), and other relevant agencies	• Industry • Buildings • Transportation • Non-federal actors	• GHG Reductions	A Broadened Policy Portfolio Rigorous and Transparent Analysis and Reporting for Adaptive Management Tightened Targets for the Buildings and Industrial Sectors and a Backstop for the Transport Sector Research, Development, and Demonstration Needs
10-7: Establish a Program Connecting Market-Pull Approaches to the Deployment of Low-Carbon Technologies	Congress, DOE, Department of Commerce (DOC), General Services Administration (GSA), DoD, and DOT	• Buildings • Transportation • Industry • Non-federal actors • Finance	• GHG reductions	A Broadened Policy Portfllio Rigorous and Transparent Analysis and Reporting for Adaptive Management Tightened Targets for the Buildings and Industrial Sectors and a Backstop for the Transport Sector

continued

TABLE S-1 Continued

Short-Form Recommendation	Actor(s) Responsible for Implementing Recommendation	Sector(s) Addressed by Recommendation	Objective(s) Addressed by Recommendation	Overarching Categories Addressed by Recommendation
10-8: Develop Effective Workforce Development Programs for Industry	Congress, DOE, labor associations, NGOs, industry leaders, and academia	• Industry • Non-federal actors	• Employment	Building the Needed Workforce and Capacity
10-9: Implement a Product-Based Tradeable Performance Standard for Domestic Manufacturing and Foreign Trade	Congress, DOE, DOC, and EPA	• Industry • Finance	• GHG reductions	A Broadened Policy Portfolio Tightened Targets for the Buildings and Industrial Sectors and a Backstop for the Transport Sector
11-1: Expand and Extend Funding and Financing Assistance for Actions Benefiting Low-Income and Disadvantaged Households and Communities	Congress and EPA	• Buildings • Transportation • Finance • Non-federal actors	• Equity	Ensuring Procedural Equity in Planning and Siting New Infrastructure and Programs Reforming Financial Markets
11-2: Disclose Equity Indicators for Federal Funding of Clean Energy	OMB	• Electricity • Buildings • Transportation • Finance • Non-federal actors	• Equity	Ensuring Procedural Equity in Planning and Siting New Infrastructure and Programs Reforming Financial Markets
11-3: Address Limited Access Faced by Low-Income and Marginalized Households	Treasury Advisory Group on Racial Equity	• Finance	• Equity	Ensuring Procedural Equity in Planning and Siting New Infrastructure and Programs Reforming Financial Markets

Recommendation	Actor			Related Themes
11-4: Fill Gaps in Federal Financial Risk Data and Information Collection Rules	Federal agency decision makers that are members of the Financial Stability Oversight Council (FSOC)	• Finance		Rigorous and Transparent Analysis and Reporting for Adaptive Management Reforming Financial Markets
11-5: Strengthen Climate Disclosure Rules and Standardize Data and Methods	Securities and Exchange Commission (SEC) and Commodity Futures Trading Commission	• Finance • Non-federal actors	• GHG reductions	Rigorous and Transparent Analysis and Reporting for Adaptive Management Reforming Financial Markets
11-6: Implement Financial Stability Oversight Council Recommendations to Ensure the Stability of U.S. Financial Markets	FSOC members and the Federal Reserve	• Finance		Reforming Financial Markets
12-1: Authorize and Provide Appropriations for State Transition Offices to Address Coal, Oil, and Natural Gas Community Transitions	Congress and state transition offices	• Fossil fuels • Non-federal actors	• Equity • Employment	Ensuring Procedural Equity in Planning and Siting New Infrastructure and Programs Ensuring Equity, Justice, Health, and Fairness of Impacts Building the Needed Workforce and Capacity Managing the Future of the Fossil Fuel Sector

continued

TABLE S-1 Continued

Short-Form Recommendation	Actor(s) Responsible for Implementing Recommendation	Sector(s) Addressed by Recommendation	Objective(s) Addressed by Recommendation	Overarching Categories Addressed by Recommendation
12-2: Consider Whether Proposed Natural Gas Pipeline Projects Are Needed, Incorporate Greenhouse Gas Emissions Impacts into National Environmental Policy Act, and Require the Use of Depreciation Periods for Pipeline Application Reviews	Congress and FERC	• Fossil fuels • Transportation	• GHG reductions	A Broadened Policy Portfolio Managing the Future of the Fossil Fuel Sector
12-3: Require Utilities and Service Providers to Plan for the Transition	State regulators of natural gas distribution utilities and fossil fuel supplier/ service providers	• Electricity • Buildings • Transportation • Industry • Fossil fuels • Non-federal actors	• GHG reductions • Equity • Health • Public engagement	Ensuring Equity, Justice, Health, and Fairness of Impacts Managing the Future of the Fossil Fuel Sector
12-4: Consider Adoption of Moratoria on New Gas Lines in Previously Unserved Areas	States and communities	• Fossil fuels • Non-federal actors	• GHG reductions	Managing the Future of the Fossil Fuel Sector
12-5: Modify the Design of Taxes on Gasoline, Diesel, and Petroleum Products	Congress and states	• Transportation • Fossil fuels • Non-federal actors	• GHG reductions	Managing the Future of the Fossil Fuel Sector
12-6: Require Recipients of Federal Funding to Provide Advance Notice of Facility Closures	Congress and recipients of federal agency funding	• Fossil fuels • Non-federal actors	• GHG reductions • Equity • Employment • Public engagement	Ensuring Equity, Justice, Health, and Fairness of Impacts Managing the Future of the Fossil Fuel Sector

Recommendation	Actors	Sectors	Co-Benefits	Objectives
12-7: Fund the Decommissioning, Cleanup, and Just Transition for Communities Historically Dependent on Fossil Fuels	Congress, state legislatures, state agencies, and state regulators	• Finance • Fossil fuels • Non-federal actors	• GHG reductions • Equity • Health • Public engagement	Ensuring Equity, Justice, Health, and Fairness of Impacts Managing the Future of the Fossil Fuel Sector
13-1: Establish an Ongoing Process to Integrate Feedback into Federal Application and Technical Assistance Processes	Executive Office of the President	• Non-federal actors	• Equity • Public engagement	Building the Needed Workforce and Capacity
13-2: Disburse Capacity-Building Funds for State, Local, and Community Recipients Flexibly and Speedily	DOE, EPA, DOT, USDA, and other federal agencies	• Non-federal actors	• Equity • Employment • Public engagement	Building the Needed Workforce and Capacity
13-3: Designate an Official or Entity to Track Decarbonization Program Opportunities and Deadlines	Governors, mayors, and county officials; states, counties, and cities	• Electricity • Buildings • Transportation • Industry • Non-federal actors	• Equity • Public engagement	Building the Needed Workforce and Capacity
13-4: Structure Competitive Opportunities as Non-Competitive Planning Grants Followed by Competitive Grants	Federal agencies	• Electricity • Buildings • Transportation • Industry • Non-federal actors	• Equity	Building the Needed Workforce and Capacity
13-5: Continue to Expand Reliable and Flexible Funding to Subnational Governments	Congress and federal contracting officials	• Electricity • Buildings • Transportation • Industry • Non-federal actors	• Equity • Employment • Public engagement	Building the Needed Workforce and Capacity

1

Introduction

This is the second of two reports from the National Academies of Sciences, Engineering, and Medicine committee constituted in 2020 to analyze possible ways for the United States to decarbonize its energy system. The committee interpreted "deep decarbonization" in the statement of task (see Box 1-1) to mean a decline to net-zero U.S. greenhouse gas (GHG) emissions by 2050, consistent with the target announced by most developed nations.[1] The committee that wrote the first report included approximately the same number of policy experts—particularly those focused on how policy affects equity, fairness, and justice—as it did scientists and engineers. Importantly, the committee was convened to study how the nation might achieve an equitable transition to net zero, not whether it should do so. Policies to reduce impacts of climate change and to promote climate adaptation are outside the committee's task, even though their human dimensions overlap with mitigation policy in multiple ways, including normative commitments to equity, justice, economic development, and place-based issues.

Like previous large-scale technological revolutions, the transition to a net-zero emissions energy system will create new industries and jobs throughout the U.S. economy, while leaving older technologies behind. It is difficult to imagine how such a transition could maintain public and political support for 3 decades without an equitable and fair sharing of benefits and costs and a focus on supporting workers and communities. The study's focus on equity and fairness can thus be motivated on purely pragmatic grounds, in addition to ethical or moral grounds. Furthermore, our current energy system has significant injustice built into it, such as the disproportionate exposure to air pollution from combustion of fossil fuels suffered by communities of color (Liu et al. 2021), in part because of redlining and other discrimination (Fears 2022; Lane et al. 2022). These issues of environmental and energy justice must be redressed to gain the trust and support of the large number of people who have been harmed.

---

[1] Net-zero emissions mean that any ongoing atmospheric release of greenhouse gases (GHGs) covered by the United Nations Framework Convention on Climate Change (UNFCCC): carbon dioxide (CO_2), methane (CH_4), nitrous oxide (N_2O), and several fluorinated gases, must be balanced by removals of CO_2 from the atmosphere. Under the UNFCCC, positive emissions balance negative emissions if the two have equal 100-year global warming potentials (GWPs). This report primarily covers CO_2 emissions, with some discussion of non-CO_2 GHGs where relevant (Chapters 3, 7, 8, 10, 12).

BOX 1-1
STATEMENT OF TASK

Building off the needs identified at the *Deployment of Deep Decarbonization Technologies* workshop in July 2019, the National Academies of Sciences, Engineering, and Medicine will appoint an ad hoc consensus committee to assess the technological, policy, social, and behavioral dimensions to accelerate the decarbonization of the U.S. economy. The focus is on emission reduction and removal of CO_2, which is the largest driver of climate change and the greenhouse gas most intimately integrated into the U.S. economy and way of life. The scope of the study is necessarily broad and takes a systemic, cross-sector approach. The committee will summarize the status of technologies, policies, and societal factors needed for decarbonization and recommend research and policy needs. It will focus its findings and recommendations on near- and mid-term (5–20 years) high-value policy improvements and research investments and approaches required to put the United States on a path to achieve long-term net-zero emissions. This consensus study will also provide the foundation for a larger National Academies' initiative on Deep Decarbonization. The committee will produce an interim report and a final report. The interim report will provide an assessment of no-regrets policies, strategies, and research directions that provide benefits across a spectrum of low-carbon futures. The final report will assess a wider spectrum of technological, policy, social, and behavioral dimensions of deep decarbonization and their interactions. Specific questions that will be addressed in the final report include the following:

- *Sectoral interactions and systems impacts*—How do changes in one sector (e.g., transportation) impact other sectors (e.g., electric power) and what positive and negative systems-level impacts arise through these interactions; how should the understanding of sectoral interactions impact choices related to technologies and policies?
- *Technology research, development, and deployment at scale*—What are the technological challenges and opportunities for achieving deep decarbonization, including in challenging activities like air travel and heavy processing; what research, development, and demonstration efforts can accelerate the technologies; how can financing and capital effectively support decarbonization; what are key metrics for tracking progress in deployment and scale-up of technologies and key measurements for tracking emissions?
- *Social, institutional, and behavioral dimensions*—What are the societal, institutional, behavioral, and equity drivers and implications of deep decarbonization; how do the impacts of deep decarbonization differ across states, regions, and urban versus rural areas and how can equity issues be identified and the uneven distribution of impacts be addressed; and what is the role of the private sector in achieving emissions reductions, including companies' influence on their external supply chains; what are the economic opportunities associated with deep decarbonization; and what are the workforce and human capital needs?
- *Policy coordination and sequencing at local, state, and federal levels*—What near-term policy developments at local, state, and federal levels are driving decarbonization; how can policies be sequenced to best achieve near-, medium-, and long-term goals; and what synergies exist between mitigation, adaptation, resilience, and economic development?

This second committee study was conducted during a period of unprecedented and revolutionary change in U.S. climate and energy policy—caused primarily by the passage of the Inflation Reduction Act (IRA) in August 2022, in combination with the Infrastructure Investment and Jobs Act (IIJA) in November 2021, the CHIPS and Science Act (CHIPS) in August 2022, and federal executive and regulatory actions. Although the IRA is the primary mechanism that will directly impact GHG emissions, it is the combination of these policies that remakes the federal science and technology landscape in the United States. The committee's first report was released before this revolution, and the second (this report) was written after.

SUMMARY OF THE FIRST REPORT

The first report, released in February 2021, almost 9 months before the passage of IIJA and 18 months before the passage of the IRA and CHIPS, was written by a committee of experts in technology, policy, and social sciences, and included many of the committee members that also wrote the second report.[2] The committee held its first meeting in March 2020, which was an in-person meeting, and subsequently held three additional virtual committee meetings and many subgroup calls. The committee's report focused on federal and executive branch actions needed during the 2020s to put the nation on a fair and equitable path to decarbonization by midcentury.

Contents of First Report

The first report offers a technical blueprint and federal policy manual for the first 10 years of a 30-year effort to replace the current U.S. energy system with one that has net-zero anthropogenic GHG emissions (NASEM 2021). It begins with discussions of the current U.S. GHG emissions inventory, the historical and potential future changes in emissions within different sectors, and the committee's choice of a 30-year timeframe to net zero as its emissions reduction goal. It provides an illustrative pathway to this overall emissions goal and notes the role of four key ingredients to meet that goal: deep reductions in CO_2 emissions, declines in non-CO_2 GHGs, maintenance or expansion of land carbon sinks, and expansion of negative emissions technologies. The 30-year timeframe is justified

---

[2] Some members from the original committee resigned and other new members were added to the committee after the first report was published.

on two fronts: first, analysis shows that reaching net-zero global anthropogenic emissions eliminates the most severe impacts of climate change (IPCC 2018), and second, a 30-year horizon for the energy transition leverages the normal pace of much asset replacement and avoids significant premature retirement of existing assets. The report also discusses the economics and capital requirements of a transition to net zero; sectoral targets, technology options, and uncertainties; and the societal dimensions of deep decarbonization. It shows how decarbonization can provide a net economic benefit, take advantage of the country's unique assets, and be accomplished in a manner that improves equity and opportunity. Within the report, the committee identifies "no-regrets" actions that would be robust to addressing uncertainties about the energy system's final technological mix and hedging actions designed to keep open as many viable paths to net zero as possible. It also identifies sector-specific research priorities and technological goals for expanding the innovation toolkit, particularly for sectors where low GHG emissions alternatives are in pilot stages or nascent industries.

First Report Goals and Policies

The first report laid out five technology goals and four socioeconomic goals critical to achieving midcentury decarbonization. On the technology side, the committee set goals for producing carbon-free electricity; electrifying energy services in transportation, buildings, and industry; investing in energy efficiency and productivity; planning, permitting, and building critical infrastructure; and expanding the decarbonization toolkit through investments in clean energy research, development, demonstration, and deployment (RDD&D). The committee's four socioeconomic goals were to strengthen the U.S. economy; promote equity and inclusion; support communities, businesses, and workers; and maximize cost-effectiveness.

The committee further recognized that a strong social contract would be essential to maintain support for an energy transition covering 3 decades. It proposed policies "to build a more competitive U.S. economy, to increase the availability of high-quality jobs, to build an energy system without the social injustices that permeate the current system, and to allow those individuals and businesses that are marginalized today to share equitably in future benefits" (NASEM 2021, p. 1). The diverse portfolio of policy recommendations (shown in Appendix C) called for both system-wide and sector-specific policies that would establish the U.S. commitment to a rapid, just, and equitable transition; set rules and standards for technology planning and deployment; invest in research, technology, people, and infrastructure; and assist and build capacities for families, businesses, communities, cities, and states to ensure that disadvantaged communities[3] do not suffer disproportionate burdens. The committee continues to endorse the goals and policies recommended in the first report, while acknowledging that the list of new policies needed today is fundamentally shaped by the radical changes to the policy landscape since its publication.

Overarching policies in the first report included an emissions budget, an economy-wide price on carbon, a national Green Bank, education and training programs to develop a clean-energy workforce, and increased federal investments in clean energy RDD&D. The committee also called for the establishment of groups both within the federal government and an independent corporation to help ensure a just and equitable transition through analysis, evaluation, capacity building, and investment. The recommended sector-specific policies included regulations for clean electricity generation; requirements for labor employed with government funding; standards for zero-emissions vehicles and efficient electric appliances; actions to improve the regulation, design, and functioning of clean electricity markets; and investments to increase energy efficiency in low-income households, expand rural broadband access, and electrify tribal lands. Some—but not all—of the committee's first report recommendations have been incorporated into recent legislation (as discussed in Table 1-1 below).

Dissemination of First Report

Throughout the spring and summer of 2021, committee members held approximately 50 briefings on the first report with philanthropic organizations sponsoring the study, congressional staff and committees, federal agencies, non-governmental organizations, and other stakeholders. Notably, the committee convened several listening sessions with climate and environmental justice experts and groups focused on fairness, equity, and justice in the energy transition, which led to an expansion of the committee to include more such expertise. These discussions and others led to the list of topics covered in the second report—including state and local decarbonization efforts, terrestrial carbon sinks, health impacts of the energy system, workforce and employment economics, and sector-specific technologies and policies—and to the addition of new committee members with expertise in these areas.

---

[3] This report typically uses the term "disadvantaged communities" to maintain consistency with federal agency guidance issued by the Council on Environmental Quality (CEQ), Office of Management and Budget (OMB), and the White House Office of Domestic Climate Policy (Climate Policy Office [CPO]). Disadvantaged communities are those that are marginalized, underserved, and overburdened by pollution and have other socioeconomic burdens (e.g., low income, high unemployment) (CEQ n.d.). To identify a disadvantaged community for federal programs and funding, CEQ recommends using the Climate and Economic Justice Screening Tool (CEJST), an interactive mapping tool that qualifies a census tract as a "disadvantaged community" if it is above the threshold for one or more environmental or climate indicators and above the threshold for the socioeconomic indicators (OMB et al. 2023).

First Report's Role in the Second Report

This first report provides the guideposts for the second—the fundamental set of goals, conclusions, and priority recommendations. It also defines the focus of the study to be on the near- to mid-term timeframe—specifically, what needs to be done in the 2020s to meet the interim goal of ~50 percent emissions reduction from 2005 levels by 2030 and put the nation on a trajectory to meet the net-zero goal by mid-century. In particular, the first report concluded that the technology pathway with the lowest uncertainty, a "no-regrets" pathway, is one that over the next decade focuses on decarbonizing electricity, electrifying end uses, increasing energy efficiency, and undertaking a robust RDD&D agenda to prepare for the additional technological and societal solutions needed for the 2030s and beyond. The contents, context, and framing of the first report serve as the foundation and basis for the discussion and analysis in the second report. Consistent with the first report and the tasking for the committee, the second report focuses on deep decarbonization just within the United States.

COMMITTEE'S APPROACH TO SECOND REPORT

Development of Second Report

The statement of task mandates that the second report "will assess a wider spectrum of technological, policy, social, and behavioral dimensions of deep decarbonization" than the first report. Given the evidence in early 2021 that a majority in Congress intended to pass comprehensive climate legislation, the committee realized that the second report would be most useful if its wider assessment were to include a detailed analysis of any new comprehensive climate and energy policy portfolio. While waiting for the new legislation, the committee prepared to rapidly finish a draft for review of its second report once new legislation was passed by Congress or when it became clear that there would be none.

During 2021 and the first half of 2022, the committee reviewed injustices embedded in our energy system and the history and preferred policies of the environmental justice movement. The committee developed detailed sectoral analyses of options to decarbonize electricity, industry, transport, buildings, agriculture, forestry, the financial sector, and the fossil fuel industries. It analyzed possible roles for non-federal actors—including state and local governments, non-governmental organizations (NGOs), and private companies—and examined technologies and policies to reduce emissions of methane, nitrous oxide, and fluorinated gases. Last, the committee studied policy options to accomplish five objectives that cut across sectors: emissions reductions, equity and fairness, health, employment in good-paying jobs, and public engagement. The last of these is critical to build and maintain public support but should also be thought of as a separate objective, as studies show that people value the ability to participate in decision-making, no matter what the outcome. Public consultation is often more important than providing monetary benefits through revenue sharing in successfully siting new infrastructure (Chapter 5).

During this interim period, the committee held information-gathering webinars and workshops to hear from additional experts. Webinar topics were wide-ranging and included leveraging financial systems for decarbonization; soil carbon offsets; government, nonprofit, and philanthropic perspectives on implementing a just and equitable energy transition; manufacturing and industrial decarbonization; public engagement strategies; and priority research and development for building technologies. (Appendix D provides a list of these webinars.) The committee also hosted a one-day workshop, *Pathways to an Equitable and Just Transition: Principles, Best Practices, and Inclusive Stakeholder Engagement*, that convened researchers and stakeholders focused on public health and safety, jobs and workforce, equitable access, and energy affordability to discuss "actionable recommendations to operationalize equity and justice in the energy transition with inclusive stakeholder engagement" (NASEM 2022).

Because of this preparatory work, the committee was able within a short timeframe to analyze the radically altered landscape created by the IRA, IIJA, and CHIPS, and develop a draft of its second report.

Structure and Key Issues for Second Report

Report Structure

Each of the 12 chapters that follow includes an analysis of gaps and barriers, which requires an assessment of the efficacy of the relevant provisions in the IRA, IIJA, and CHIPS against their goals. Each chapter also offers recommendations about how to close gaps and overcome barriers. Gaps, barriers, and recommendations cover actions required under the IRA, IIJA, and CHIPS during the 2020s as well as actions needed to prepare for the subsequent 2 decades.

Chapters 2–5 correspond to the final four of the five crosscutting objectives. Chapter 2 focuses on equity and energy justice. It offers a scholarly review of the environmental justice movement and the inequities that have been built into our current energy system, some because of past or ongoing discrimination. Chapter 3 focuses on health, Chapter 4 on employment, and Chapter 5 on public engagement. While the emissions reduction objective does not have its own chapter, it is the singular issue that cuts across every chapter in the report. Chapters 6–13 focus on sectors: electricity, buildings and the built environment, land use such as agriculture and forestry, transportation, industry, the financial sector, the fossil fuel industry, and non-federal actors. Each sectoral chapter assesses how well current policies will achieve the carbon emissions objectives for the sector, identifies gaps and barriers, and provides recommendations to close or overcome them. Owing to the complexity of emissions from this sector, the land use chapter includes discussion of the spectrum of GHG emissions and mitigation options from agriculture, the uptake of carbon from terrestrial sinks, and land requirements for renewable energy. This includes discussion of the GHG impacts of dietary choices. Where appropriate, each sectoral chapter also assesses likely progress toward equity, employment, health, and public engagement objectives—creating some intended redundancy with the material in Chapters 2–5. The report can best help sectoral experts who do not read Chapters 2–5 by emphasizing that among the greatest risks to sectoral progress are loss of public support and institutional constraints, both of which can affect the viability and durability of climate policies. Sustained public support requires public perception that the transition is fair and equitable, that it brings material benefits to health and employment that compensate for inevitable losses, and that people have sufficient say in decisions affecting their lives and communities.

Approach to Key Overarching Issues

Key elements of the committee's approach to its charge included how it addresses system interactions and cross-sector impacts; how it deals with key vulnerabilities and uncertainties, including those related to politics and polarization; and how it incorporates the role of non-federal actors. The approach to these issues and others is guided by the study charge, information gathered from the public during webinars and listening sessions, committee expertise, and the literature. The report covers what the committee concluded to be the most critical technology, policy, and societal dimensions of accelerating decarbonization within the energy and related sectors. Given the inevitable limitations of time and resources, there are many issues only briefly covered in the text, each of which could be expanded into a full treatise.

The committee's report structure recognizes the most important system interactions and cross-sector impacts. The recognition that decarbonization would have vast impacts on justice and equity, public health, and jobs motivated chapters devoted to these system-level impacts. The requirement for a strong social contract for the systemic transition to clean energy motivated a chapter devoted to public engagement. And the recognition that decarbonization itself provides an existential challenge to fossil fuels motivated a single chapter on that sector. Each of these chapters considers a few critical system interactions. For example, the health chapter discusses the need for expanded analysis of emerging occupational health risks associated with clean energy technologies, and the transportation chapter highlights connections between transportation and energy justice, public heath, and the built environment. However, discussing the full litany of potential ways that policies or technologies adopted in one sector in the future might impact another sector, the full energy system, or society at large was beyond the scope and time constraints of the study. Moreover, such interactions can be very difficult to predict or evaluate in advance. This underscores the report's inclusion of policies that would comprehensively evaluate progress, in part to facilitate adaptive management.

The key vulnerabilities and uncertainties emphasized in this report are the implementation of existing decarbonization policies and the closing of gaps between policy commitments and the level of effort required to meet 2030 and 2050 emissions reductions and climate goals. These issues are discussed throughout the report and are the focus of findings and recommendations; however, they are not the only vulnerabilities and uncertainties affecting the transition to a decarbonized energy system. For example, recognizing the role of the politics of climate change in executing current policies and adopting new ones, especially in this time of polarized political and public discourse, the committee discussed at length the need for public engagement and a strong social contract and value proposition. Although such steps will clearly not guarantee success, to not undertake them would doom this effort to failure. Geopolitical concerns related energy and national security also influence decisions about decarbonization; it is outside the scope of this study to address these concerns fully, but key components are highlighted in relevant chapters. Furthermore, as noted in the Preface, the study took place amid the COVID-19 pandemic and the Ukraine war, both of which profoundly impacted the energy system. The committee recognizes that changes outside the energy systems (e.g., wars, famines, pandemics, and natural disasters) will fundamentally alter the trajectory of the energy system. While these types of events will occur over the decades-long transition, it is beyond the committee's expertise and resources to unravel the potential magnitude and duration of impacts from these external forces.

The report discusses climate mitigation efforts by a diverse set of actors, including federal and subnational governments, private sector, and philanthropy. Much of the general discussion of non-federal actors is contained the final chapter, but roles for the private sector in specific technology areas are described throughout the report. The chapter on public engagement recognizes the importance of mobilizing the participation and support of the public; however, the report does not address the role of individual voluntary efforts to mitigate emissions. It also does not consider the complex behavioral issues that motivate individuals and the private sector to act in the absence of mandates or incentives. The committee leaves the behavioral elements, including the equity implications, to others who could do these complex topics justice.

Changes in the Federal Policy Relevant for the Second Report

The current comprehensive decarbonization policy portfolio of the United States did not come together until August 2022, with the passage of the Inflation Reduction Act (P.L. 117-169) and CHIPS and Science Act (P.L. 117-167), to complement the Infrastructure Investment and Jobs Act (P.L. 117-58), which passed in November 2021.[4] Of these three, the IRA contains by far the most significant and wide-reaching policies to decarbonize the U.S. economy, with climate and energy related investments totaling approximately $271 billion to $1.2 trillion in tax credits and $121 billion in direct spending, loans, and other investments (Bistline et al. 2023; CBO 2022; Goldman Sachs 2023; Jiang et al. 2022). It provides incentives for purchasing, producing, and developing clean energy technologies, and makes investments in disadvantaged communities. CHIPS appropriates $54.2 billion to incentivize domestic semiconductor manufacturing and authorizes $170 billion over 5 years for investments in science, technology, engineering, and medicine (STEM) programs, workforce development, and technology R&D (Badlam et al. 2022; Senate Commerce Committee 2022). The IIJA provides funding for a range of infrastructure projects including repairing roads and bridges; upgrading power and public transit infrastructure; expanding broadband; deploying electric vehicle charging stations; and cleaning up Superfund sites (White House 2021a). It includes appropriations for approximately $62 billion for Department of Energy (DOE) climate and energy programs (DOE 2022a). It also increases the scope and authorization for DOE's Loan Programs Office—for example, by establishing the CO_2 Infrastructure Finance and Innovation Act (CIFIA) program—and authorizes regional clean hydrogen and direct air capture hubs funded via public–private partnerships and managed by DOE's new Office of Clean Energy Demonstrations.

Each of these pieces of legislation will ultimately have unique and substantial impacts on U.S. climate, energy, and technology policies. For example, the IIJA, although not

---

[4] It should be noted that the IRA, IIJA, and CHIPS are not equivalent in funding mechanisms. The IRA primarily consists of spending programs (appropriations) and tax expenditures. Spending programs can allocate federal resources to projects and activities up to the amount of their appropriation. By contrast, tax expenditures, such as the production tax credits in IRA, typically have no limit on the amount that could be claimed by taxpayers. The IIJA consists of a mix of authorizations and appropriations, while CHIPS contains primarily authorizations. Authorizations are laws that establish or continue a federal program or agency and are typically passed by Congress for a set period of time, but authorizations require appropriations before funds can be spent. Appropriations are laws that actually provide the money for government programs and must be passed by Congress every year in order for the government to continue to operate.

expected to provide significant emissions reductions (Larsen et al. 2022; Larson et al. 2021; Mahajan et al. 2022), has caused a fundamental shift within DOE, with the addition of a new undersecretary position focused on clean energy infrastructure, and with the adoption of a new strategy prioritizing demonstration and deployment (DOE 2022a). The new DOE Under Secretary for Infrastructure oversees several new offices—the Office of Clean Energy Demonstration, Grid Deployment Office, Office of Manufacturing and Energy Supply Chains, and Office of State and Community Energy Programs—as well as several existing offices. CHIPS, with $280 billion in funding authorized over 10 years (Badlam et al. 2022), is an investment in the future of U.S. innovation and manufacturing, emphasizing R&D in cutting-edge technologies, especially semiconductors, and establishing programs to support a robust, diverse STEM workforce. It also establishes regional technology hubs to spread technology innovation across a wider geographical area in the country. Last, the IRA is widely considered the most significant piece of climate legislation in U.S. history. Its more than $390 billion in energy and climate investments (Bistline et al. 2023; CBO 2022) encompasses a wide variety of technologies, prioritizes low-income and disadvantaged communities, and—as discussed further below—makes significant progress toward national 2030 emissions goals and moves the nation much or most of the way to a trajectory that reaches net zero at midcentury (Larsen et al. 2022; Larson et al. 2021; Mahajan et al. 2022; Chapters 6–12).

In addition to legislation, several executive orders support the nation's equity and climate priorities. In January 2021, with Executive Order (EO) 14008, the White House made "Justice40" official U.S. policy when it announced that 40 percent of benefits of federal investments in "climate change, clean energy and energy efficiency, clean transit, affordable and sustainable housing, training and workforce development, remediation and reduction of legacy pollution, and the development of critical clean water and wastewater infrastructure" will be directed to people and "disadvantaged communities that are marginalized, underserved, and overburdened by pollution" (White House n.d.). The Biden administration further signaled its commitment to diversity, equity, inclusion, and accessibility (DEIA) through EO 13985, which provides direction to federal agencies to advance DEI practices in hiring and training of employees (White House 2021b). Justice40, together with the equity, justice, and fairness provisions in the IRA and executive orders, represent a step change in energy and climate policy. In response, agencies have created new offices (e.g., DOE's Office of Energy Justice Policy and Analysis and the Federal Energy Regulatory Commission's Office of Public Participation) and are actively hiring staff focused on energy justice and equity.

EO 14057, issued in December 2021, enlists the federal government's procurement power as a catalyst for developing a domestic clean energy economy and sets the following national targets for the operations of the federal government: "100 percent carbon pollution-free electricity by 2030, 100 percent zero-emission vehicle acquisitions by 2035, net-zero emissions from procurement by 2050, net-zero emissions building portfolio by 2045, and net-zero emissions from overall operations by 2050" (White House 2021c). It further outlines overarching objectives for federal procurement and operations, which include "building a climate- and sustainability-focused workforce" and "advancing environmental justice and equity" (White House 2021c).

Continuing regulatory efforts also play an important role in meeting decarbonization targets. For example, NHTSA and EPA have released new vehicle fuel economy and GHG emissions standards, respectively, under their existing legislative authorities. In April 2023, EPA proposed more stringent, performance-based GHG and criteria pollutant standards under the Clean Air Act for model year 2027–2032 light-, medium-, and heavy-duty vehicles. EPA projected in model year (MY) 2032 that the standards could result in nearly 70 percent BEV sales in the light-duty fleet, 40 percent in the medium-duty van and pickup fleet, 50 percent ZEV sales in vocational vehicles, 34 percent ZEV sales in day cab tractors, and 25 percent ZEV sales for sleeper cab tractors in MY 2032 (EPA 2023a–d). Following a review mandated in EO 13990, DOT revised the fuel economy standards for MY 2024–2026, which would result in a fleet-wide average fuel economy of 49 miles per gallon for MY 2026, and, according to DOT projections, yield an 8 percent reduction in CO_2 emissions from passenger cars and light trucks between 2021 and 2100 compared to the alternative of leaving the less stringent Safer Affordable Fuel Efficient Vehicles Rule in place (EO 13990 2021; NHTSA 2022). Under the same regulatory review required by EO 13990, in 2022 EPA restored its waiver of preemption of California's GHG and ZEV standards, allowing their Advanced Clean Cars (ACC) program to continue as well as allowing other states to adopt the California standards pursuant to Clean Air Act Section 177 (EPA 2022). In July 2023, NHTSA continued to update its regulations under its existing authority from the Energy Policy and Conservation and Energy Independence and Security Acts, proposing an 18 percent increase in fuel economy from MY 2027–2032, with trucks requiring greater yearly fuel economy increases than cars (NHTSA 2023). As another example, in May 2023, EPA proposed new standards for regulating CO_2 emissions from new and existing fossil-fueled power plants, which are projected to yield 617 million metric tons of CO_2 emissions reductions and $85 billion in climate and health benefits through 2042 (EPA 2023e).

Table 1-1 compares the policies recently adopted by Congress, the executive branch, and federal agencies with the recommendations in the committee's first report. The color-coding in Table 1-1 indicates where implemented policies fully align with the committee's recommendation (green), where implemented policies are related but different from the committee's recommendation (yellow), and where no relevant policies have yet been implemented (red). Individual chapters of this report offer detailed analysis of the information in Table 1-1.

TABLE 1-1 Comparison of Policies from Committee's First Report with Those Implemented in Legislation and/or Executive Action

Committee's First Report Recommendation	Relevant Policy in IIJA, CHIPS, IRA, or Executive Action
U.S. CO$_2$ and other GHG emissions budget reaching net zero by 2050.	None
Economy-wide price on carbon.	None
Establish 2-year federal National Transition Task Force to assess vulnerability of labor sectors and communities to the transition to carbon neutrality.	• National Climate Task Force (EO 14008) • Interagency Working Group on Coal and Power Plant Communities and Economic Revitalization (EO 14008)

Establish White House Office of Equitable Energy Transitions.
  • Establish criteria to ensure equitable and effective energy transition funding.
  • Sponsor external research to support development and evaluation of equity indicators and public engagement.
  • Report annually on energy equity indicators and triennially on transition impacts and opportunities.
	• Justice40 Initiative (EO 14008)
	• White House Office of Domestic Climate Policy (EO 14008)
	• White House Environmental Justice Advisory Council (EO 14008)
	• White House Environmental Justice Interagency Council (EO 14008)
	• Office of Environmental Justice with the Department of Justice Initiative (EO 14008)

Establish an independent National Transition Corporation to ensure coordination and funding in the areas of job losses, critical location infrastructure, and equitable access to economic opportunities and wealth, and to create public energy equity indicators.	None
Set clean energy standard for electricity generation, designed to reach 75 percent zero-emissions electricity by 2030 and decline in emissions intensity to net-zero emissions by 2050.	• Civil Nuclear Credit Program (IIJA, §40323) • Hydroelectric Production Incentives (IIJA, §40331) • Zero-Emission Nuclear Power Production Credit (IRA, §13105) • Clean Electricity Production Credit (IRA, §13701) • Clean Electricity Investment Credit (IRA, §13702)

continued

TABLE 1-1 Continued

Committee's First Report Recommendation	Relevant Policy in IIJA, CHIPS, IRA, or Executive Action
Set national standards for LD, MD, HD zero-emissions vehicles, and extend and strengthen stringency of CAFE standards. LD ZEV standard ramps to 50 percent of sales in 2030; medium- and heavy-duty to 30 percent of sales in 2030.	• Corporate Average Fuel Economy (CAFE) Standards for MY 2024–2026 (87 Fed. Reg. 25710) • Proposed Rule: Multi-Pollutant Emissions Standards for Model Years 2027 and Later Light-Duty and Medium-Duty Vehicles (88 Fed. Reg. 29184) • Notice of Proposed Rulemaking for Greenhouse Gas Emissions Standards for Heavy-Duty Vehicles—Phase 3 (88 Fed. Reg. 25926) • Notice of Proposed Rulemaking for CAFE Standards for Passenger Cars and Light Trucks for MY 2027–2032 and Fuel Efficiency Standards for Heavy-Duty Pickup Trucks and Vans for MY 2030–2035 (88 Fed. Reg. 56128) • Federal government target of 100 percent zero-emission vehicle acquisitions by 2035 (EO 14057) • Clean Vehicle Credit (IRA, §13401) • Credit for Previously Owned Clean Vehicles (IRA, §13402) • Qualified Commercial Clean Vehicles (IRA, §13403) • U.S. Postal Service Clean Fleets (IRA, §70002)
Set manufacturing standards for zero-emissions appliances, including hot water, cooking, and space heating. DOE establishes appliance minimum efficiency standards, ramping down to achieve close to 100 percent all-electric in 2050.	None
Enact three near-term actions on new and existing building energy efficiency, two by DOE/EPA and one by GSA. • Direct DOE/EPA to expand outreach of and support for adoption of benchmarking and transparency standards by state and local govts through the expansion of Portfolio Manager.	• Energy Efficiency and Conservation Block Grant Program (IIJA, §40552) • Grants for energy efficiency improvements and renewable energy improvements at public school facilities (IIJA, §40541) • Energy efficiency materials pilot program (IIJA, §40542) • Assisting Federal Facilities with Energy Conservation Technologies Grant Program (IIJA, §40554)

TABLE 1-1 Continued

Committee's First Report Recommendation	Relevant Policy in IIJA, CHIPS, IRA, or Executive Action
• Direct DOE/EPA to further investigate the development of model carbon-neutral standards for new and existing buildings that, in turn, could be adopted by states and local authorities. Policies targeting retrofits of existing buildings will be in the final report.	• Extension, increase, and modification of nonbusiness energy property credit (IRA, §13301) • Energy efficient commercial buildings deduction (IRA, §13303) • Extension, increase, and modifications of new energy efficient home credit (IRA, §13304) • Home energy performance-based whole house rebates (IRA, §50121) • High-efficiency electric home rebate program (IRA, §50122) • State-based home energy efficiency contractor training grants (IRA, §50123) • Assistance for latest and zero building energy code adoption (IRA, §50131)
Enact five congressional actions to advance clean electricity markets, and to improve their regulation, design, and functioning.	Office of Public Participation at FERC (IIJA, §40432)
Deploy advanced electricity meters for the retail market and support the ability of state regulators to review proposals for time/location-varying retail electricity prices.	Utility Demand Response (IIJA, §40104)
Recipients of federal funds and their contractors must meet labor standards, including Davis-Bacon Act prevailing wage requirements; sign Project Labor Agreements where relevant; and negotiate Community Benefits (Workforce) Agreements where relevant.	• IRA has prevailing wage and apprenticeship requirements to access "bonus rate" credits for renewable electricity production tax credit, PTC (45), carbon oxide sequestration credit (45Q), zero emission nuclear power production credit (45U), and energy investment tax credit, ITC (48) • EO 14008, §206: "Agencies shall, consistent with applicable law, applyw and enforce the Davis-Bacon Act and prevailing wage and benefit requirements. The Secretary of Labor shall take steps to update prevailing wage requirements."

continued

TABLE 1-1 Continued

Committee's First Report Recommendation	Relevant Policy in IIJA, CHIPS, IRA, or Executive Action
Report and assess financial and other risks associated with the net-zero transition and climate change by private companies, government agencies, and the Federal Reserve. Private companies receiving federal funds must also report their clean energy R&D by tech category (wind, solar).	• SEC Proposed Rule: The Enhancement and Standardization of Climate-Related Disclosures for Investors • EO 14030: Climate-Related Financial Risk • Financial Stability Oversight Council (FSOC) 2021 Report on Climate-Related Financial Risk • Federal Reserve Pilot Climate Scenario Analysis • Federal Reserve System Draft Principles for Climate-Related Financial Risk Management for Large Financial Institutions • Federal Deposit Insurance Corporation (FDIC) Statement of Principles for Climate-Related Financial Risk Management for Large Financial Institutions
Ensure that Buy America and Buy American provisions are applied and enforced for key materials and products in federally funded projects.	Made in America Office (IIJA, §70923)
Establish an environmental product declaration library to create the accounting and reporting infrastructure to support the development of a comprehensive Buy Clean policy.	Buy Clean Task Force, charged with recommending policies and procedures for considering embodied emissions and pollutants of construction materials in Federal procurement (EO 14057)
Establish a federal Green Bank to finance low- or zero-carbon technology, business creation, and infrastructure.	Greenhouse Gas Reduction Fund (IRA, §60103)
Amend the Federal Power Act and Energy Policy Act by making changes to facilitate needed new transmission infrastructure.	Siting of interstate electric transmission facilities (IIJA, §40105)
Plan, fund, permit, and build additional electrical transmission, including long-distance high-voltage direct current (HVDC). Require fair public participation measures to ensure meaningful community input.	• Transmission facilitation program (IIJA, §40106) • Power marketing administration transmission borrowing authority (IIJA, §40110) • Transmission facility financing (IRA, §50151) • Grants to facilitate the siting of interstate electricity transmission lines (IRA, §50152)

TABLE 1-1 Continued

Committee's First Report Recommendation	Relevant Policy in IIJA, CHIPS, IRA, or Executive Action
	• Interregional and offshore wind electricity transmission planning, modeling, and analysis (IRA, §50153) • Federal permitting improvement steering council environmental review improvement fund mandatory funding (IRA, §70007) • Environmental review implementation funds (IRA, §60505) • Environmental Protection Agency efficient, accurate, and timely reviews (IRA, §60115)
Expand EV charging network for interstate highway system.	• National Electric Vehicle Formula Program (IIJA, Title VIII) • Grants for charging and fueling infrastructure (IIJA, §11401) • Establishment of Joint Office of Energy and Transportation (IIJA, Title VIII) • Alternative Fuel Refueling Property Credit (IRA, §13404)
Expand broadband for rural and low-income customers to support advanced metering.	• Grants for broadband deployment (IIJA, §60102) • Private activity bonds for qualified broadband projects (IIJA, §80401) • Enabling middle mile broadband infrastructure (IIJA, §60401) • Tribal Broadband Connectivity Program (IIJA, §60201) • Digital Equity Act of 2021 (IIJA, Title III) • Broadband affordability (IIJA, §60502)
Plan and assess the requirements for national CO_2 transport network, characterize geologic storage reservoirs, and establish permitting rules. Require fair public participation measures to ensure meaningful community input.	• Carbon capture technology program (IIJA, §40303) • Carbon dioxide transportation infrastructure finance and innovation (IIJA, §40304) • Carbon storage validation and testing (IIJA, §40305) • Secure geologic storage permitting (IIJA, §40306) • Geologic sequestration on the outer Continental Shelf (IIJA, §40307) • Federal permitting improvement (IIJA, §70801)
Establish educational and training programs to train the net-zero workforce, with reporting on diversity of participants and job placement success.	• Energy auditor training grant program (IIJA, §40503) • Building, training, and assessment centers (IIJA, §40512) • Career skills training (IIJA, §40513) • Directorate for Technology, Innovation, and Partnerships (CHIPS, §10381–10399A)

continued

TABLE 1-1 Continued

Committee's First Report Recommendation	Relevant Policy in IIJA, CHIPS, IRA, or Executive Action
	• Clean Energy Technology Transfer Coordination (CHIPS, §10715) • Rural STEM education research (CHIPS, §10511–10517) • Clean Energy Technology Transfer Coordination (CHIPS, §10715) • Establishment of expansion awards pilot program as a part of the Hollings Manufacturing Extension Partnership (CHIPS, §10251) • Broadening Participation in Science (CHIPS, §10501–10510)
Revitalize clean energy manufacturing.	• State manufacturing leadership (IIJA, §40534) • Battery processing and manufacturing (IIJA, §40207) • Advanced energy manufacturing and recycling grant program (IIJA, §40209) • Low-emissions steel manufacturing research program (CHIPS, §10751) • Creating helpful incentives to produce semiconductors for America fund (CHIPS, §102) • Advanced manufacturing investment credit (CHIPS, §107) • Extension of advanced energy project credit (IRA, §13501) • Advanced manufacturing production credit (IRA, §13502) • Domestic manufacturing conversion grants (IRA, §50143) • Advanced industrial facilities deployment program (IRA, §50161) • Funding for implementation of the American Innovation and Manufacturing Act (IRA, §60109)
Increase clean energy and net-zero transition RD&D that integrates equity indicators.	• Office of Clean Energy Demonstrations (IIJA, §41201) • Regional clean hydrogen hubs (IIJA, §40314) • Regional direct air capture hubs (IIJA, §40308) • Energy storage demonstration projects (IIJA, §41001) • Advanced reactor demonstration program (IIJA, §41002) • Direct air capture technology prize competition (IIJA, §41005) • Carbon capture demonstration and pilot programs (IIJA, §41004) • Renewable energy projects (IIJA, §41007)

TABLE 1-1 Continued

Committee's First Report Recommendation	Relevant Policy in IIJA, CHIPS, IRA, or Executive Action
	• Industrial emissions demonstration programs (IIJA, §41008) • Regional clean energy innovation program (CHIPS, §10622) • National clean energy incubator program (CHIPS, §10713)
Increase funds for low-income households for energy expenses, home electrification, and weatherization.	• Weatherization assistance program (IIJA, §40551) • Improving Energy Efficiency or Water Efficiency or Climate Resilience of Affordable Housing (IRA, §30002) • Home Energy Performance-Based, Whole-House Rebates (IRA, §50121) • High-Efficiency Electric Home Rebate Program (IRA, §50122)
Increase electrification of tribal lands.	• Tribal Electrification Program (IRA, §80003) • High-Efficiency Electric Home Rebate Program (IRA, §50122)
Establish National Laboratory support to subnational entities for planning and implementation of net-zero transition.	Clean Energy to Communities program (DOE-EERE and NREL)
Establish 10 regional centers to manage socioeconomic dimensions of the net-zero transition.	None
Establish net-zero transition office in each state capital.	None
Establish local community block grants for planning and to help identify especially at-risk communities. Greatly improve environmental justice (EJ) mapping and screening tool and reporting to guide investments.	• Justice40 Initiative, Climate and Economic Justice Screening Tool (EO 14008) • State energy program (IIJA, §40109) • Environmental and climate justice block grants (IRA, §60201) • Neighborhood access and equity grant program (IRA, §60501) • Rural Energy for America program (IRA, §22002) • Grants to reduce air pollution at ports (IRA, §60102) • Clean heavy-duty vehicles (IRA, §60101) • Low emissions electricity program (IRA, §60107)

NOTE: Red shading indicates no related policy implemented, yellow shading indicates policy implemented related to but different from the one recommended, and green shading indicates implemented policy is the same or very similar to the one recommended.

While the IRA is strikingly comprehensive, with nearly the same emissions reduction impacts and technology objectives as the committee's first report (as discussed further below), its policy portfolio (like those in IIJA and CHIPS) was strongly shaped by political constraints, including the budget "reconciliation" rules under which it was passed and the types of policies that attracted a voting majority. Consequently, the largest difference between the IRA's policy portfolio and the portfolio in the committee's first report is that the IRA relies almost exclusively on tax credits and other incentives, whereas the committee recommended a variety of standards and incentives, as well as a carbon tax[5] and statutes to create new institutions. Nonetheless, the portfolios in the first report and IRA are broadly similar in that both contain policies with the same goals and intent (Table 1-1). For example, both contain policies designed to decarbonize electricity and transport, and to electrify buildings and industry at about the same rates. Both would establish a Green Bank, expand electric vehicle charging networks, and increase funding for tribal electrification and for energy expenses, home electrification, and weatherization of low-income households.

> Finding 1-1: The climate policy portfolio in the Inflation Reduction Act depends entirely on incentives, with the exception of a fee on fugitive methane emissions, whereas the committee's first report recommended a broad portfolio of incentives, taxes, regulatory standards, and statutes creating new institutions/entities.

The committee's broad policy portfolio in the first report was specifically formulated to include some redundancy and complementarity in case some components do not work as intended. For example, the clean power standard would only have an impact if the recommended carbon tax proved insufficient to decarbonize power generation at the accelerated rate necessary. Moreover, each kind of policy brings a different set of advantages and risks. Taxes are politically difficult to enact and sustain but are economically efficient and reach every part of an economy affected by a tax. Standards and incentives are easier to enact and sustain, but are less cost-efficient, and risk disrupting technological progress by diverting resources from a path that subsequently proves to be better. In addition, incentives carry the risk that they might fail to elicit the intended response by consumers or businesses. Under existing statutory

---

[5] In its first report, the committee recommended a carbon tax of $40 per ton CO_2 starting in 2021 and rising at 5 percent per year, a level that could ameliorate equity and competitiveness concerns and generate about $200 billion per year in revenue. For more details, see NASEM (2021, pp. 184–186).

authorities, regulators may not have the ability to promulgate ambitious climate-change-related standards.

Recommendation 1-1: *Enact Two Federal Policies Recommended in the First Report: National Greenhouse Gas Emissions Budget and Economy-Wide Carbon Tax.* **Congress should enact the following two federal policies that are described in detail in the committee's first report:**

  a. **A greenhouse gas budget for the U.S. economy.**
  b. **An economy-wide tax on carbon emissions, starting at $40/tCO$_2$ and rising 5 percent per year. Additional policies would also be needed if the tax were enacted to protect low-income families (such as a pre-determined per-capita rebate of all or part of it) and import/export exposed businesses.**

These additional provisions would backstop the incentives in current policy and help fill gaps between incentives. In particular, the recommended carbon tax would backstop all of the incentives for emissions reductions in the current bills and extend beyond the IRA sunset date of 2032, thus accelerating coal-fired power plant retirements relative to the current policy landscape of subsidizing renewables and existing nuclear. Other recommendations included in the committee's first report and also repeated later within this report include an independent National Transition Corporation (Chapter 2), regional planning initiatives that would lay the groundwork for just and inclusive energy transitions (Chapter 5), standards to limit emissions from the electric sector and vehicles (Chapters 6 and 9), and manufacturing standards for appliances (Chapter 7).

The committee fully recognizes how difficult it would be at present for Congress to pass the policies in Recommendation 1-1 and those cited above. However, the committee views these policies as key enablers of an efficient, equitable, and affordable path to decarbonization.

Impact of Legislation on Emissions Trajectory

Prior to the passage of the IRA, U.S. GHG emissions had been slowly decreasing since 2005, despite approximately flat primary energy consumption. A linear extrapolation of the historical trajectory between 2005 and 2022 hits zero sometime after 2100. This pace of decarbonization is well short of three widely discussed targets requiring nearly identical emissions reductions: (1) the U.S. target under the Paris Agreement; (2) the Biden administration's announced goal to reduce GHG emissions by just over

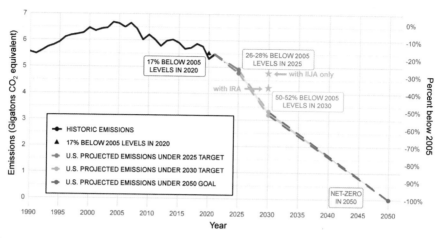

FIGURE 1-1 Historic and projected emissions levels.
NOTES: The black line shows historic U.S. emissions. The blue, green, and red dashed lines represent goals for 2025, 2030, and 2050, respectively. The yellow stars show projected 2030 emissions with IIJA policies only (upper, 28 percent below 2005 levels) and with both IIJA and IRA policies (lower, 37 percent below 2005 levels) (Bistline et al. 2023). Projected emissions in 2030 without either IIJA or IRA policies are about 0.1 gigatons of CO_2 equivalents per year[6] higher than the upper star. SOURCE: Courtesy of Department of State (2021).

50 percent from 2005 levels by 2030; and (3) a trajectory that falls linearly to net-zero emissions in 2050. Meeting the near-term (2030) target would put the United States on a trajectory generally consistent with the long-term (2050) target of net-zero emissions. Figure 1-1 shows this linear trajectory to net zero in 2050 and projections of the emissions impacts of IRA and IIJA policies from a multimodel analysis (Bistline et al. 2023).

Six modeling studies—by Energy Innovation, the REPEAT Project at Princeton University, Rhodium Group, Brookings, the U.S. Energy Information Administration (EIA), and a multi-model analysis using nine independent models—conclude that the IRA would yield approximately 70–80 percent of the emissions reductions necessary to achieve the first two of the above targets, and the first 10 years of the third (Bistline et al. 2023; DeCarolis and LaRose 2023; Jenkins et al. 2023; Larsen et al.

---

[6] The use of carbon dioxide equivalent (CO_2e) is a metric for describing the global warming potential of different GHGs in a common measure by defining the number of mass units of CO_2 that would have the equivalent global warming impact of one unit of another GHG. While simple to describe, GWPs depend on timeframe. The committee has adopted a 100-year timeframe for reporting CO_2e, the standard used in the Paris Accord and other climate agreements. The committee recognizes Ocko et al. (2017) and others recommend reporting estimates using the CO_2e metric for multiple timeframes.

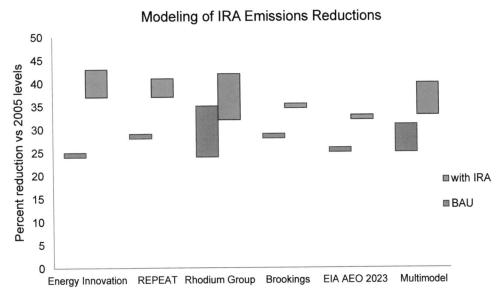

FIGURE 1-2 Modeling projections of U.S. GHG emissions reductions by 2030 relative to 2005 emissions from IRA provisions compared to a business-as-usual scenario. SOURCES: Data from Bistline et al. (2023a,b), DeCarolis and LaRose (2023), Mahajan et al. (2022), Jenkins et al. (2023), Larsen et al. (2022).

2022; Mahajan et al. 2022). As summarized in Figure 1-2, these groups project between 32–43 percent reductions in GHG emissions below 2005 levels by 2030, compared to approximately 25–35 percent reductions in a business-as-usual (BAU) case. In addition, all six groups predict that most reductions as of 2030 will come from the electricity sector. The committee believes that with effective implementation of the provisions of the IRA, the United States is likely to be close to the trajectory required to achieve the 2050 emission reduction targets set by the White House. However, there are significant risks.

RISKS AND OBJECTIVES FOR MEETING POLICY OBJECTIVES

The IRA's climate and energy provisions were formulated to put the nation on, or nearly on, the first 10 years of a 3-decade path to net-zero emissions. However, four categories of risks—technological, political/public resistance, execution, and external events—could prevent the nation from achieving this goal. A primary objective of this report is to identify those risks and propose solutions for overcoming them.

Risks

Technological Risk—the risk that essential non-emitting technologies might not be ready in time at the right price. A technological revolution over the past 2 decades has already brought humanity renewable electricity and electric transport that is cost-competitive or cheaper than fossil alternatives, which explains the IRA's focus on deployment of clean electricity, electrification of heating, and electric vehicles during the 2020s (Chapters 6 and 9). However, after the IRA's 10 years of tax incentives wind down and the investments that they have stimulated have put much new capacity into operation, the nation will still need a host of zero-emissions technologies, which are in various stages of readiness, in order to reach its 2050 net-zero emissions commitment. These include non-emitting options for dispatchable electricity generation, such as advanced nuclear reactors; methods and machinery for net-zero manufacturing, such as the use of hydrogen to produce high-temperature industrial heat; ways to remove CO_2 from the atmosphere, such as direct air capture (DAC); and options for zero-emitting heavy trucks, marine shipping, and aviation, such as batteries or hydrogen fuel cells for heavy trucks, and biofuels or net-zero synthetic fuels for shipping and aviation.

Four factors significantly reduce technological risk (Chapters 6, 7, 9, and 10).

- Most technological gaps (i.e., need for on-demand electricity, long-term energy storage, fuels for aviation/shipping/freight) have multiple options in advanced stages of research and development.
- There is a massive international RDD&D effort on zero-carbon technologies, including large new expenditures in the IRA, IIJA, and CHIPS.
- Learning by doing will accelerate progress because the IRA, IIJA, and CHIPS include funds for rapid deployment of many of the needed technologies at considerable scale during the 2020s.
- Addressing and developing solutions for solving these technological challenges are magnets for the innovative and entrepreneurial.

Political, Judicial, and Societal Polarization Risks—the risk of a change in policy landscape. This could include, for example, repeal of climate and energy provisions in the IRA, federal executive branch or state agency action that limits IRA implementation, or new legislation that inhibits climate mitigation efforts. Such changes could come about owing to electoral changes in government and/or judicial review. The risk of political reversal is lower for congressional statutes than for executive actions. Legislation to repeal the IRA would require united support of the president, Senate, and House of Representatives, or two-thirds support in the Senate and House to override a presidential veto. However, even if the IRA remains intact, its incentives may not be extended beyond the law's 2032 sunset date. This could ultimately deter investment in technologies eligible for tax credits, as their lifetimes typically exceed the current tax credit duration. Judicial challenges to legislation or regulation could also slow or roll back decarbonization policies. Notably, Supreme Court decisions in *West Virginia v. EPA* and *Biden v. Nebraska*, which use the "major questions" doctrine to argue for clear, specific congressional authorization for agency action, may limit agencies' ability to implement regulations that are not explicitly called for in legislative text.

More fundamentally, public sentiment against climate mitigation policy could inspire federal, state, and local politicians to create roadblocks or outright opposition to such policies. The committee is cognizant that the public and political discourse around topics like climate change is often polarized. The energy transition must be seen as just, equitable, and fair, or public support for it will ultimately be lost, followed inevitably by lost political support. The committee is also cognizant that mitigation policies must be durable, reflecting the simple fact that activities required to greatly reduce emissions will continue long after the coalition that enabled the policies to be adopted are no longer around (Carlson and Burtraw 2019; Patashnik 2008). This brings us to the third category of risk.

Execution Risk—the risk that the nation will be unable to execute the energy and climate policies in the IRA, IIJA, and CHIPS and the related regulatory initiatives at the intended pace and scale, or that the policies will not work as intended because of a wide variety of behavioral, organizational, and political factors. The White House and federal agencies clearly view these *execution risks* as the most important and daunting vulnerabilities facing the current policy portfolio, given their public statements and the focused energy of their implementation effort. The committee concurs with this view, and so has focused its second report on *barriers and gaps*. A barrier is anything that stands in the way of successful implementation and that might prevent the nation from accomplishing the first 10 years of a fair and equitable 30-year path toward a net-zero energy system. A gap is a missing component in the legislation. Because the IRA is such a comprehensive bill for the first decade of the transition, most gaps are not as simple as, for example, an omitted sector or GHG. Instead, most reflect the absence of policy that could overcome an anticipated barrier or an effort that must be undertaken during the 2020s to continue decarbonization during the 2030s and beyond. Some examples of gaps and barriers are as follows:

- *Gap:* To meet a net-zero trajectory, national emissions will need to drop by nearly 2 $GtCO_2e/y$ by 2030, and modeling analyses to date conclude that current federal legislation will leave approximately 20–30 percent of this job undone (Bistline et al. 2023; DeCarolis and LaRose 2023; Jenkins et al. 2023; Larsen et al. 2022; Mahajan et al. 2022). Could state, local, and voluntary private actions fill this gap, or is additional federal legislation required?

- *Barrier:* Commercial-scale renewables projects in the United States take an extended time to plan, permit, construct, and connect to the electricity grid. One study estimates that projects currently require 5–8 years for completion and 89 percent are abandoned before completion, many because of the difficulties of permitting and siting (Jenkins et al. 2021; Chapter 6). The record annual deployment of new wind and solar capacity is approximately 25 GW in both 2020 and 2021, but the average pace must accelerate by 100–300 percent during the 2020s to put the nation on a path to net zero, with the larger number corresponding to a 100 percent renewable energy system, and the smaller corresponding to a cheapest system that includes some nuclear electricity and fossil assets with carbon capture and storage (CCS) (Jenkins et al. 2021). Organized opposition to commercial-scale renewable infrastructure and new transmission lines has already emerged and can be expected to accelerate as landscapes in different parts of the country are visually transformed (Chapters 2 and 5). Identifying policy reforms that can effectively transform permitting processes to facilitate *both* meaningful public engagement and infrastructure deployment at pace is a major challenge. Meaningful public engagement takes time and is likely to slow deployment, at least initially (Chapter 5). On the other hand, inadequate public participation risks escalating conflicts that can frustrate total progress over the longer term and is a major existing factor in environmental injustice (Chapter 2). This dilemma represents more than simply tension between technocratic and societal objectives. For example, slowing deployment of clean infrastructure in the interest of fairness also prolongs fossil air pollution deaths, which disproportionately afflict disadvantaged communities (Chapter 3). How can inclusive and consultative processes be implemented to speed rather than slow deployment, and in so doing, simultaneously address technocratic and societal goals?
- *Barrier:* During the energy transition, most locations will become primary energy producers for the first time, because wind and solar are present everywhere to a greater or lesser degree, whereas commercial fossil resources are concentrated in a smaller number of locations. For most locations, the shift to renewables will increase economic activity and employment, as well as bring other changes to landscapes and communities (Chapters 4 and 12). However, as legacy industries shut down, some communities will see net job losses, like those that have experienced the closure of a coal mine or coal-fired power plant in recent decades. They may also face rapid and abrupt changes in tax revenue that threaten the

viability of critical public services (Chapter 12). Political and financial interests that oppose the policies in the IRA will use these losses to organize and build opposition. Does the current policy portfolio do enough to minimize this risk by providing opportunities for workers and communities to make a fair and equitable transition?

- *Gap:* The IRA earmarks direct investments into disadvantaged communities that range from $40 billion–$42.5 billion (Chi 2022; EELP 2022), with other estimates as high as $60 billion going to environmental justice priorities (out of total expenditures estimated to range from $400 billion to more than $1 trillion). However, the ambiguity of the tax credit provisions makes calculations of direct benefits difficult, and expenditures that do not directly target disadvantaged communities may benefit them (e.g., air quality improvements from accelerated vehicle electrification and renewable power deployment as a result of EV and zero-carbon electricity tax credits). Will this achieve the administration's goal that 40 percent of benefits flow to low-income and historically marginalized people and communities? Are these funds sufficient to gain the trust of communities that have suffered unjust and discriminatory harm? Will these funds be effectively deployed to advance decarbonization in a manner that addresses historical injustices, poverty, and the need to support local economies?

- *Barrier:* Federal, state, and local governments currently lack the staff and expertise to effectively spend and administer IRA, IIJA, and CHIPS funds (Chapters 2 and 13). Moreover, because the country is politically and economically heterogenous, some state and local governments will rapidly build capacity and use it to pursue and spend these funds, while others will not build sufficient capacity and offer support ranging from indifference to active resistance (Chapter 13). Are the provisions in the IRA, IIJA, and CHIPS sufficient to overcome this barrier? What additional actions could federal, state, and local governments take to increase capacity at all levels of government and to garner state and local support?

- *Gap:* Note that Recommendation 1-1 is designed to fill an execution gap. Specifically, will tax credits and other incentives in the IRA, IIJA, and CHIPS deliver the required pace of change, given behavioral inertia by consumers, organizational inertia in the electricity sector, and inevitable targeted messaging by entrenched financial interests? This gap would be filled by the more diverse policy portfolio in Recommendation 1-1, as well as the more detailed recommendations in Chapter 5 to strengthen public engagement in the transition.

Risk from Events Outside of the Energy System—as noted in the Preface to this second report, the committee's work occurred during a time of disruptions within national and global energy systems caused by outside events such as the invasion of Ukraine and the global COVID-19 pandemic. While it appears that activities within the U.S. energy system are returning to normal with the receding impacts of COVID-19, there will undoubtedly be some mixture of wars, disasters, and other disruptions that will punctuate the decades-long transition to net zero. Furthermore, geopolitical considerations of energy and national security will often underlie decisions about decarbonization policy. These risks are unpredictable and will have to be managed during the clean energy transition as they arise.

Objectives

To clarify the analysis of gaps and barriers, it is important to define as explicitly as possible the objectives to be addressed by the committee's recommended actions. The committee defines five objectives: (1) carbon and GHG emissions by sector, (2) equity and fairness, (3) health, (4) employment, and (5) public engagement and acceptance.

Carbon Emissions Objectives

There is an infinite number of net-zero emissions trajectories that the country could follow between now and 2050. Examples include Larsen et al. (2021), Larson et al. (2021), Lempert et al. (2019), and Williams et al. (2021). Most assume a roughly linear decline in emissions until 2050. They also tend to focus on meeting anticipated BAU demand for energy services,[7] rather than asking that Americans do with less. Most of these scenarios are similar because prices of available technology and current emissions strongly constrain the cheapest pattern of deployment. Unless otherwise stated, this report uses the sector-specific scenarios in "The Long-Term Strategy of the United States: Pathways to Net-Zero Greenhouse Gas Emissions by 2050" issued by the White House in late 2021 because these encompass most published scenarios and reflect current U.S. policy goals (Figures 1-3 and 1-4). It is important to note that while this report references the updated emissions pathways published by the White House, the overall emissions trajectories laid out in the committee's first report and the key

---

[7] The projected energy demand is derived from the U.S. Energy Information Administration's National Energy Modeling System, which incorporates continued improvements in energy efficiency (e.g., for vehicles, appliances, and other equipment) over time.

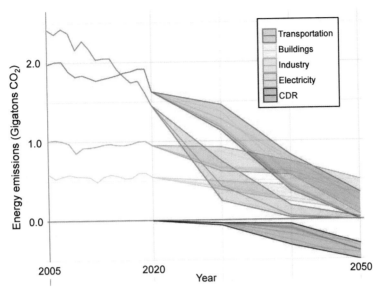

FIGURE 1-3 Projected ranges of CO_2 emissions over time by sector. NOTE: CDR stands for carbon dioxide removal and includes industrial practices like direct air capture and biomass energy with carbon capture and storage. SOURCE: Courtesy of Department of State (2021).

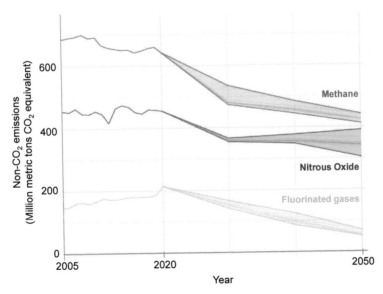

FIGURE 1-4 Projected ranges of methane, nitrous oxide, and fluorinated gas emissions over time. These are the three categories of non-CO_2 greenhouse gases regulated under the United Nations Framework Convention on Climate Change. SOURCE: Courtesy of Department of State (2021).

ingredients of the net-zero transition for the U.S economy described in that report are fundamentally the same (see Figure 2.2 in NASEM 2021).

The policies recommended in this report are consistent with the projected emissions objectives illustrated in Figures 1-3 and 1-4. The anticipated rapid emissions reductions during the 2020s in the electricity sector and during the 2020s and 2030s in the transport sector reflect the current relatively low costs for renewable infrastructure and lithium-ion batteries (Chapters 6 and 9). The levelized cost of energy (LCOE) for onshore wind has decreased between 62–63 percent globally from 1983–2021 and 2009–2023, respectively (IRENA 2022; Lazard 2023), and 70 percent in the United States between 1998–2021 (LBNL 2022a). The LCOE of utility-scale solar decreased by 83–88 percent globally in 2009–2023 and 2010–2021 (IRENA 2022; Lazard 2023), and 85 percent in the United States from 2010–2021 (LBNL 2022b), with the global price of solar photovoltaic (PV) modules dropping 99.8 percent between 1975 and 2021 (*Our World in Data* n.d.). DOE (2023) estimates an 89 percent decline in cost of EV lithium-ion battery packs between 2008–2022, while other sources estimate a 79 percent decrease in lithium-ion battery pack and cell price between 2013–2022 (BloombergNEF 2022) and 98 percent decline in lithium-ion battery prices between 1991–2018 (Ziegler and Trancik 2021a,b).

The slower emissions reductions during the 2020s and 2030s for industry reflect both the high costs for decarbonization options and slow turnover of industrial infrastructure. The slow turnover of this infrastructure is important because retrofit or replacement is most economic when equipment becomes obsolete (Chapter 10). The relatively slow decline of emissions from buildings in Figure 1-3 reflects slow stock turnover and limited retrofits because of anticipated property-owner inertia (Chapter 7).[8] In contrast, the rapid decline in emissions from the electricity sector in Figure 1-3 reflects the robust deployment of renewable electricity owing to the reductions in costs for those technologies and the tax credits provided within the IRA.[9] It should be noted, however, that non-cost barriers to the deployment of these technologies—including supply chain development, need for skilled labor and enhanced public engagement, regulatory approvals, and engineering, procurement, and construction of both generation and transmission—will have to be addressed to achieve the projected emissions reductions. The relatively slow reductions for nitrous

---

[8] Chapter 7 makes the case that much more could be done in the buildings sector over the next 10 years, and that this would reduce the need for a rapid deployment of carbon dioxide removal technology during the 2040s (Figure 1-2), which is risky to count on given current prices and technology.

[9] While Chapter 6 reflects this optimism toward the decarbonization of electricity, it does recommend a national emissions limit for the electricity sector to ensure the outcome shown in Figure 1-3 as well as the need for an omnibus solution to support expansion of the transmission grid.

oxide and methane reflect agricultural emissions from ongoing agricultural demand (Chapter 8). The White House Long-Term Strategy report (DOS and EOP 2021) also has a scenario for the net sink from land use, land use change, and forestry (LULUCF), which grows from 759 $MtCO_2e/y$ today to 940 $MtCO_2e/y$ in 2050 (see Chapter 8). The midpoints (darkest-colored lines) within the ranges for 2050 sectoral emissions in Figure 1-3 and for GHG-specific emissions in Figure 1-4, together with an LULUCF net emission of -940 $MtCO_2e/y$, sum to zero net emissions.

While the overarching emissions reduction goal of 50 percent by 2030 and net zero by midcentury is a singular objective, the discussion above makes clear that it is really the summation of sector-specific emissions reductions. This overarching goal will necessarily be met by sector-specific policies and technologies, and cost minimizations within sectors and across the whole system. In this way, the objective to reduce emissions to net zero by midcentury (or 50 percent by 2030) can be thought of as a constraint with the goal to minimize cost while maximizing desirable societal objectives of equity, employment, health, and public engagement. Indeed, technology analyses are typically formulated as constrained optimizations that solve for the mix of technologies needed to meet various emissions goals, with social welfare as the objective function in the constrained optimization (often specified as per capita consumption in economic models). While this formulation is useful to develop effective policies, once policies are in place, they must be evaluated in part by how well they achieve emissions reductions. Thus, the constraint in the mathematical analysis supporting a policy becomes, in practical terms, an objective of the policy once it is implemented.

Equity and Fairness Objectives

What actions in the pathways to decarbonization would adequately address historical, eliminate current, and prevent future injustices? These objectives are much more difficult to quantify than emissions targets. The committee came up with four specific objectives, however inadequate. First, new infrastructure built under the IRA, IIJA, and CHIPS should not replicate the disproportionate exposure to fossil-related health and safety hazards suffered by disadvantaged communities (Chapters 2 and 3). Second, an adequate fraction of the benefits of spending under the IRA, IIJA, and CHIPS should go to low-income and historically marginalized groups and communities, with the current federal goal being 40 percent. Where benefits are difficult to quantify, the fraction of IRA, IIJA, and CHIPS spending that goes directly to these groups and communities provides an alternative objective to assessing benefits in the near term, but historically unprecedented efforts to assess actual benefits are still necessary (Chapter 2). As discussed further below, rigorous evaluation will be needed to understand and quantify the benefits of widespread clean energy deployment to disadvantaged communities. Third, historically marginalized people and communities should have equitable access to new jobs and not suffer disproportionately from job losses (Chapter 4). Fourth, policies should attempt to minimize the harms that displaced fossil fuel workers suffer during the transition (also a component of the Employment Objectives below).

Health Objectives

The objective is to maximize health and minimize harm, including illness, disability, and death caused by fossil fuel-related pollution (Chapter 3). Other health risks include the mental and physical impacts related to the losses of employment and livelihoods in fossil fuel communities. Major health benefits are possible in the transition, particularly in reduced exposure to air pollution from fossil fuel combustion. Other health benefits, both physical and mental, can also accrue through active transportation, changes in nutrition policies, and improvements in the characteristics of the built environment, like urban tree cover and improved walkability. As the development and deployment of low-carbon technologies increases, health harms across the full life cycle of these technologies should be minimized, including risks from new clean energy industries (e.g., in mining and manufacturing operations) and from the introduction of these new technologies to the public (e.g., safety concerns).

Employment Objectives

These objectives are to maximize employment in high-quality jobs that are created by the transition and to minimize the disruption caused by losses of fossil-dependent jobs (Chapter 4). Modeling analysis suggests that although gains will exceed losses in most locations and in the nation, losses will be concentrated in three areas of the country that produce most of the gas, oil, and coal—Appalachia, the Gulf Coast and adjacent areas, and the inter-mountain west (Chapter 4; Mayfield et al. 2021, 2023). However, most of these losses will occur later than 2030, except in coal producing areas in Appalachia and the inter-mountain west that have seen contraction for decades (Chapter 4; Larson et al. 2021; Mayfield et al. 2021, 2023). At smaller scale, some communities will suffer net losses throughout the country—for example, in small towns that lose a dominant employer such as a coal-fired power plant. Also, losses will likely occur in some occupations sprinkled at low density throughout the country, such as in automotive repair shops that do not transition successfully to electric transport or experience lower demand owing to lower main-tenance needs of EVs, but most of these losses will occur after 2030 because of the average lifetime of fossil vehicles.

Public Engagement Objectives

These objectives are to develop robust public engagement practices that involve people and groups across the country in the goals, design, and implementation of the energy system transition. Such practices are fundamental to develop and maintain a robust social contract for deep decarbonization with the people of the United States (Chapter 5). The committee chose to emphasize public engagement rather than public acceptance or support for several reasons. First, people, communities, and regions have heterogeneous preferences. Some communities may remain against components of a net-zero energy system, such as nuclear electricity because of perceived danger, or large-scale wind and solar because of the visual transformation of the landscape. Public support for the transition as a whole may thus hinge on respecting public rejection of parts of it. Second, as stated above, people value having a say in decisions that affect their lives, *independent of the final outcome* (Chapter 5). Third, the committee heard from officials in cities such as Fresno, California, that inclusive participation in siting de-cisions initially slows deployment and may result in some kinds of projects being can-celled at the outset (Bedsworth et al. 2023). Nonetheless, inclusive consultation speeds subsequent infrastructure deployment, both because officials are aware of what the community will support, and because they have gained the public's trust.

NEED FOR COMPREHENSIVE EVALUATION AND ADAPTIVE MANAGEMENT

The scope, scale, and pace of the transition that will be required to decarbonize the U.S. energy system—and the associated uncertainties—are unprecedented. Thus, adap-tive management and governance—that is, an iterative learning process producing improved understanding and management over time—is crucial to coordinate and monitor implementation and feedbacks in the face of the complexity and uncertainties associated with climate change and societal transition throughout the country. In order to stay on the trajectory to an equitable net-zero emissions goal, there is a need to respond where policies and technologies do not work as intended, where current poli-cies fall short of achieving the full set of objectives, and where emergent issues create unanticipated problems and opportunities (e.g., larger than anticipated cost reductions for renewables). There is also a need for not just ex post evaluation and monitoring, but also for ex ante estimates or "scoring" of new proposals. As the nation implements the policies laid out in the IRA, IIJA, CHIPS, and other actions, the challenge of collecting and reporting on the use of federal and other dollars and the outcomes accomplished with those investments and expenditures should be front and center. Comprehensive and system-wide evaluation of decarbonization policies and programs is essential to cross-sector and systems level impacts, as well as to sustain a social license to operate.

Given the untrammeled ground that transitioning to net-zero emissions must traverse, a high degree of humility and commitment to learning and adjustment is in order. The committee acknowledges its limited ability to anticipate the many possible ways in which decarbonization could be at least partially derailed. As a result, adaptive management guided by continuous monitoring and evaluation offers the most likely path to success. Such efforts will be possible only if the nation makes a significant investment to gather the information necessary across all programs and activities fundamental to decarbonization, has the opportunity to modify programs based on these evaluations, and accommodates changing technological and socioeconomic conditions. To maintain public support and engagement, it is critical that monitoring and evaluation cover socioeconomic aspects of the transition in addition to emissions and technology deployment. This includes the need to develop data collection and evaluation into program design and expedite and expand data collection activities.

The committee's first report recommended that Congress establish an executive-level Office of Equitable Energy Transitions to serve as the designated centralized authority to establish criteria to ensure equitable and effective allocation of energy transition funding and monitor progress. No such authority was included within the IRA, IIJA, or CHIPS, although, as shown in Table 1-1, elements of EO 14008 establish some aspects of the committee's recommended office. An executive-level, designated authority would ensure that the nation's approach to monitoring, evaluation, and communication aligns with the scope of the challenge and investment of decarbonization.

Nonetheless, the current administration is clearly focused on implementation of the recent legislation. The federal government now features many new staff, working groups, task forces, and committees to facilitate implementation of recent bills, provide oversight both within and across departments and agencies, and produce information to the public on descriptions and application processes of funding opportunities. For example, upon passage of the IRA, the Biden administration's White House appointed a Senior Advisor for Clean Energy Innovation and Implementation specifically to oversee implementation of the act's clean energy and climate provisions, including developing the regulations required to distribute funding (White House 2022b). The administration released a guidebook for the public and local, state, and tribal governments to take advantage of the funding available through the legislation and maintains an updated list of available funding opportunities (White House 2022c, 2023a). It also established the "Bipartisan Infrastructure Law Maps Dashboard," which depicts locations of announced and awarded funding and provides information about each funded project (White House 2023b).

The Biden administration has also taken significant steps to address key elements necessary for tracking and evaluating impacts of decarbonization policy, as shown in Table 1-2. Notably, the White House Environmental Justice Interagency Council (IAC), established by EO 14008, is tasked with developing performance metrics to ensure accountability and publishing an annual public performance scorecard on the implementation of Justice40. As discussed in Chapter 2, an evaluation of clean energy spending would illustrate the distribution of the benefits of all programs, including those covered by Justice40. Many IRA tax expenditure policies do not target disadvantaged communities, but as they transform the energy system, they will deliver benefits to these communities.

In addition, the Foundations for Evidence-Based Policymaking Act (Evidence Act) (January 2019, P.L. 115-435) is currently being implemented by the Office of Management and Budget (OMB), as directed by the law. The Evidence Act requires each agency to develop an evaluation policy, evidence-building plan (i.e., learning agenda), evaluation plan, and capacity assessment. The purpose of learning agendas in particular is to develop evidence to answer questions about how the agency meets its mission and how the agency's programs, policies, and regulations function (OMB 2019, p. 14). Thus, the law "institutionalizes program evaluation as a critical element of learning agendas throughout the federal government" (Aldy 2022). The emerging portfolio of products assessing the performance of the American Rescue Plan (White House 2022a,b) and their influence on recent policy implementation suggest that the agency learning agenda offers an effective model for policy evaluation. See Table 1-2 for additional information on the Evidence Act and OMB's important role in implementation of the law.

And still, vesting the aggregation, synthesis, translation, and communication of metrics in a single, congressionally mandated entity and process is necessary to ensure that a decarbonization progress report is available to the U.S. public and the world on a consistent and sustained basis, regardless of which party controls the executive branch. Federal policy makers have previously seen fit to establish data-collection and/or forecasting entities (e.g., EIA, Bureau of Labor Statistics) when information has been viewed as important for public and private actors in the economy and societal goals. However, as noted in Table 1-2, these statistical offices do not span the whole of government. Given the broad changes likely to occur from the new federal

TABLE 1-2 Scorecard Detailing the Key Evaluation Elements of U.S. Decarbonization Policy That Are Covered by the Current Federal Policy Portfolio and Elements That Are Still Missing from the Federal Policy Portfolio

Key Evaluation Elements	Status	Comments
Accountability and oversight of spending	☑	Inflation Reduction Act (IRA)—$25 million for GAO to support oversight of distribution and use of funds and evaluate whether impacts of funds are equitable. Infrastructure Investment and Jobs Act (IIJA)—The Infrastructure Implementation Task Force (Executive Order [EO] 14052) priorities include efficient and equitable investment of public dollars, including through the Justice40 Initiative and effective coordination with state, local, tribal, and territorial governments in implementing investments.
Creation of indicators and establishment of targets	☑	Each federal agency is responsible for creating indicators and establishing targets. EO 14008—To identify geographically defined disadvantaged communities for any covered programs under the Justice40 Initiative and for programs where a statute directs resources to disadvantaged communities, federal agencies are expected to use the Climate and Economic Justice Screening Tool (CEJST). The White House Environmental Justice Interagency Council (IAC) is responsible for developing clear performance metrics to ensure accountability in the implementation of Justice40.
Data collection and tracking	◆	Each federal agency is responsible for data collection and tracking. However, no entity ensures that data collection and tracking is consistent for the purposes of presenting an aggregate picture. The Evidence Act requires • agency Open Data Plans to make federal data publicly available by default, and their data inventories searchable; • designation of a Chief Data Officer in each agency; and • establishment of an Advisory Committee on Data for Evidence Building and Chief Data Officer Council by OMB. Additional federal data collection efforts include the following: • IIJA §40553 established the Energy Jobs Council under DOE to survey, analyze, and report on employment and demographics in the U.S. energy, energy efficiency, and motor vehicle sectors.

TABLE 1-2 Continued

Key Evaluation Elements	Status	Comments
		• IRA §60401 appropriates $32.5 million for CEQ Environmental and Climate Data Collection to (1) support data collection efforts relating to disproportionate negative environmental harms and climate impacts and cumulative impacts of pollution and temperature rise; (2) establish, expand, and maintain efforts to track disproportionate burdens and cumulative impacts and provide academic and workforce support for analytics and informatics infrastructure and data collection systems; and (3) support efforts to ensure that any mapping or screening tool is accessible to community-based organizations and community members. • IRA appropriates $25 million for OMB to oversee IRA implementation and tracking of labor, equity, and environmental standards and performance. • U.S. Global Change Research Program (USGCRP) Global Change Information System provides data collected from individual government agencies and programs. • Federal institutions are implementing recommendations from the interagency Equitable Data Working Group (EO 13985).
Evaluation and learning • mitigation progress and outcomes • trends in technology and infrastructure deployment • societal outcomes (justice, fairness, equity) • vulnerability of labor sectors and communities	◆	Each federal agency is responsible for evaluating programs and learning what works and what does not. Additional pilot research programs may be needed to facilitate further learning. Moreover, opportunities for learning across agencies may be missed absent a comprehensive effort. Consistency in reporting and evaluation is also necessary for creating a comprehensive picture. The Evidence Act requires agencies to undertake program evaluation, with coordination and standards established by the law's lead implementer, OMB. It includes a required biennial OMB report to Congress and establishes a Chief Data Officer Council within OMB to enable a whole-of-government approach to data generally, and could be applied specifically to climate and clean energy. In addition to having statutory authority to implement the Evidence Act, OMB spans the whole of government, in contrast to an individual agency's statistical office (e.g., EIA, Bureau of Labor Statistics [BLS]) or the USGCRP. OMB can also focus efforts on ex post empirical performance evaluation, which is less common among some statistical agencies such as EIA.

continued

TABLE 1-2 Continued

Key Evaluation Elements	Status	Comments
• transition impacts and opportunities • pilot programs and research to support development and evaluation of equity indicators and public engagement		EO 14008—The Interagency Working Group on Coal and Power Plant Communities and Economic Revitalization (EO 14008) is responsible for assessing opportunities to ensure benefits and protections for coal and power plant workers.
Communication/ reporting short- and long-term outlooks containing an assessment of GHG mitigation and societal/equity outcomes using individual agency data and evaluations	◆	Each federal agency is responsible for reporting on progress and/ or outcomes associated with relevant agency decarbonization programs. As of the writing of this report, no single entity has been tasked with reporting on (1) the GHG mitigation and societal outcomes associated with recent decarbonization policy and (2) short- and long-term outlooks of the U.S. decarbonization trajectory. Moreover, no entity is tasked with ensuring consistency that allows a comprehensive picture to emerge. The USGCRP, with a 2022 budget of $3.7 billion, is congressionally mandated to support expanded coordination across federal agencies to design and implement research and dissemination programs that advance knowledge of climate and global change impacts, risks, and responses, including (1) emissions mitigation and interventions to reduce atmospheric greenhouse gas concentrations and warming and (2) the social context, consequences, and efficacy of various adaptation, mitigation, and intervention measures, including their impacts on equity. The USGCRP coordinates with 14 federal agencies to produce a single report, the National Climate Assessment (NCA), every 4 years. It should be noted that the NCAs involve a large number of non-government experts, and the Department of Treasury, which is charged with implementing many of the IRA's energy tax credits, does not participate in the USGCRP.

TABLE 1-2 Continued

Key Evaluation Elements	Status	Comments
		Additional related reporting instruments include • OMB biennial report to Congress • USGCRP National Climate Assessment and National Nature Assessment • EIA Energy Outlooks • DOE Quadrennial Energy Review • EPA Greenhouse Gas Reporting Program • EPA Community Notification Program for Frontline and Fenceline Communities (EO 14008) • White House Environmental Justice Interagency Council (IAC) annual public performance scorecard on the implementation of Justice40 (EO 14008) • CEQ and OMB annual Environmental Justice Scorecard (EO 14008)

NOTES: The column on the left lists key evaluation elements of U.S. decarbonization policy; the column on the right lists federal entities and actions that are already carrying out specific evaluation elements; and the middle "Status" column indicates whether the listed federal actions align with the committee's recommendations on evaluation and reporting. (The green check indicates that the evaluation component is being addressed via the current policy portfolio, and the orange diamond indicates a gap between the current policy portfolio and the committee's recommendations.)

statutes, such information collection would also be valuable in terms of public and private investment and programmatic activities (e.g., tax credits and the consumer investments related to them, and Greenhouse Gas Reduction Fund investments and lending activities) in infrastructure and clean energy. Moreover, only a consistent, comprehensive, and coordinated compilation provides the information necessary to inform future policy choices.

A full assessment of the adequacy of funding for robust evaluation as well as public-facing data reporting and communications lies outside the scope of this report. However, as an example, the separate $25 million allocated to both OMB and GAO (IRA §70004–70005) to track progress on the IRA amounts to roughly 0.013 percent of the projected $391 billion in IRA energy and climate-related funding, falling short of the 1 percent of program administration resources recommended by the Commission on Evidence-Based Policymaking in its 2017 final report, and the 3.7 percent norm for foundation spending on evaluation (Commission on Evidence-Based Policymaking

2017; Twersky and Arbreton 2014).[10] The Data Foundation, a nonprofit think-tank seeking to "improve government and society by using data to inform public policy-making," recommends federal agencies determine and articulate evaluation funding needs to OMB because the scope and scale of data collection and evaluation will inevitably vary across agencies and programs (Fatherree and Hart 2019). Additionally, a so-called Evidence Incentive Fund[11] could be established as a potential funding mechanism for agency program evaluations and learning agendas, as was recommended by the Commission on Evidence-Based Policymaking in its 2017 final report. Nevertheless, it is important to remember that federal agencies using IIJA and IRA funds to administer programs, provide technical assistance, and perform monitoring and evaluation reduces the funds available to state and local governments, community-based organizations, and private sector actors who are implementing many of the programs mandated under these laws.

Furthermore, absent a clearly (and publicly) identified lead authority for streamlined, externally facing communication of multi-agency/multi-program funding outcomes, it is unclear how public stakeholders will access trusted, non-partisan, salient information about progress on and outcomes of decarbonization via the current model of tasking multiple agency-level as well as interagency task forces with related but separate missions. Ensuring the salience of the federal assistance portfolio to both groups and places in search of not only environmental but also climate, energy, and transition justice depends on integrating—or at least aligning—ongoing definitional and evaluation exercises in order to facilitate program implementation and evaluation and—vitally important—to communicate equity and mitigation outcomes.

While it is too early to assess their durability, credibility, and efficacy, the nascent communication efforts by the current administration and key agencies, notably CEQ—such as the Climate and Economic Justice Screening Tool (CEQ 2022) and the proposed Environmental Justice Scorecard (*Federal Register* 2022)—are important experiments in streamlining information desired by public stakeholders. As much as possible, the information delivered should acknowledge the reality that people encounter federal investments through their lived experience as workers and residents

---

[10] Some of the costs of program evaluation are fixed costs, and thus the percentages of program budgets dedicated to evaluation may reflect, in part, the scale of the programs.

[11] "Evidence Incentive Funds in each department are conceptualized by the Commission to operate similarly to Working Capital Funds or Salary and Expense accounts. The funds could be created by taking up to 10 percent of unobligated balances at the end of a fiscal year to be allocated for future evidence-generating activities" (Commission on Evidence-Based Policymaking 2017, p. 104).

of places, not in terms of agency silos. This recognition should translate into easy mechanisms to enable public stakeholders to share and translate project information via trusted, existing communication tools.

Finding 1-2: Monitoring, Evaluating, and Communicating the Nation's Progress on Decarbonization and Policy Outcomes. The urgency and scope of the nation's deep decarbonization transition create an imperative to pursue comprehensive and innovative approaches to tracking and communicating progress on decarbonization policies. Failure on the part of national and state governments to pursue adaptive management through rigorous evaluation and monitoring of policy outcomes, or to give the public and key stakeholders meaningful information to demonstrate tangible progress fundamentally threatens the prospect of a successful shift to a net-zero trajectory over the next 10 years.

Recommendation 1-2: *Leverage the Evidence Act to Execute Data Collection and Evaluation on Decarbonization Investments and Programs.* **Congress should authorize and fund the Office of Management and Budget, the lead agency implementing the Evidence Act, to develop guidance for all federal agencies on evaluating decarbonization policy spending and impacts, including investments and program funding, greenhouse gas emissions (including life-cycle emissions as relevant), costs, equity, and other societal outcomes. Data collection protocols should be developed in tandem with program design, and all data collection and analysis protocols should be made with a clear evaluation strategy in mind. An important component of the evaluation is the assessment of emerging systems-level impacts and cross-sectoral issues that may impede progress on decarbonization. Agencies could solicit input on evaluation strategies during comment periods and requests for information associated with proposed program rules.**

   a. **Department of the Treasury information from its administration of clean energy tax credits could be one source of data, given their centrality in implementation of the Inflation Reduction Act (IRA).**
   b. **Other recommendations later in the report provide guidance on specific elements of evaluation, such as the priorities for evaluation of equity indicators (2-2), investments in clean energy technologies within disadvantaged communities (2-6), effectiveness of the Weatherization Assistance Program (7-3), implementation of incentive programs for residential and commercial buildings initiated in the IRA (7-7) effectiveness of forest and land carbon sinks (8-3), and integration of state**

and local government feedback into federal application and technical assistance processes (13-1).

c. The primary purpose of this effort is to inform adaptive management of these programs and policies to ensure that investments result in effective, efficient, and equitable decarbonization. The data collected for evaluation can also serve as an input for public-facing reporting discussed in Recommendation 1-3.

Recommendation 1-3: *Identify and Provide Resources for a Central Entity to Provide Timely, Public-Facing Information on the Nation's Progress Toward Decarbonization.* Congress should authorize and fund a single enduring entity to collect, aggregate, interpret, and communicate publicly accessible descriptive statistics about the pace and scale of decarbonization of the U.S. economy.

a. In addition to collecting and analyzing data on its own, this entity should make use of the data, analyses, and evaluations produced under Recommendation 1-2.

b. Given the importance of accurate and comprehensive data to attaining verifiable net-zero emissions targets, the collection and availability of data (on the distribution of economic and environmental benefits, health effects including ex-post health impacts, life-cycle inventories and assessments, among others), and continued development of relevant analysis methods must also be a priority.

c. The entity's public communications should include short- and long-term outlooks and reviews of agencies' progress toward equitable decarbonization in the United States, and explicitly characterize trends in greenhouse gas emissions, infrastructure deployment, employment, and equity metrics.

Potential candidates for the lead entity include the U.S. Energy Information Administration, the U.S. Global Change Research Program, and the Office of Management and Budget, as well as new institutions, such as an Office of Equitable Energy Transitions in the executive branch.

CONCLUDING REMARKS

When this National Academies' committee was first convened, a U.S. transition to net zero seemed far away. But then, Congress passed historic legislation that sets the nation on a fundamentally new course and establishes the United States as an international leader in the fight against climate change. The rest of this report shows how current federal legislation and executive actions, along with actions at the subnational levels, will touch nearly all facets of the energy economy, while providing the energy services the nation needs at a price it can afford. If it all works as planned, most Americans will still receive the energy services they expect but will live in a nation—and hopefully a world—with reduced impacts from climate change, with cleaner air, and with better health and employment. Many historically marginalized and low-income Americans will, for the first time, experience a fair, equitable, and just energy system. This path is risky because the task is complicated, vast, fast-paced, and never traveled before. It will take a national effort, involving all Americans and the commitment of our nation's business, industrial, and energy sectors, as well as an adaptive policy approach, to successfully execute. This report documents gaps between current policies and likely barriers to implementation. Its recommendations are designed to fill gaps, overcome barriers, and prepare the nation for the adaptive management it will need over the next 30 years to achieve a net-zero energy system.

This report includes a particular focus on execution risks because there are so many ways in which execution could prove inadequate. The climate and energy provisions in the IRA, IIJA, and CHIPS are intended to create unprecedented changes that will affect all parts of the nation's economy and many aspects of daily life. The portfolio is designed to shepherd what would arguably be the first deliberately fair and equitable technological transition in the nation's history. It must rely on a mix of policy instruments and institutions that have never been tested at this scale and executed within a polity designed to limit centralized control.

SUMMARY OF RECOMMENDATIONS

TABLE 1-3 Summary of Recommendations from Chapter 1

Short-Form Recommendation	Actor(s) Responsible for Implementing Recommendation	Sector(s) Addressed by Recommendation	Objective(s) Addressed by Recommendation	Overarching Categories Addressed by Recommendation
1-1: Enact Two Federal Policies Recommended in the First Report: National Greenhouse Gas Emissions Budget and Economy-Wide Carbon Tax	Congress, Department of the Treasury, Environmental Protection Agency	• Electricity • Buildings • Land use • Transportation • Industry • Finance • Fossil fuels	• Greenhouse gas (GHG) reductions	A Broadened Policy Portfolio
1-2: Leverage the Evidence Act to Execute Data Collection and Evaluation on Decarbonization Investments and Programs	Congress and Office of Management and Budget (OMB)	• Electricity • Buildings • Land use • Transportation • Industry • Finance • Fossil fuels	• GHG reductions • Equity • Health • Employment • Public engagement	Rigorous and Transparent Analysis and Reporting for Adaptive Management
1-3: Identify and Provide Resources for a Central Entity to Provide Timely, Public-Facing Information on the Nation's Progress Toward Decarbonization	Congress and single other agency (e.g., Energy Information Administration, Global Change Research Program, OMB)	• Electricity • Buildings • Land use • Transportation • Industry • Finance • Fossil fuels	• GHG reductions • Equity • Health • Employment • Public engagement	Rigorous and Transparent Analysis and Reporting for Adaptive Management

REFERENCES

Aldy, J.E. 2022. "Learning How to Build Back Better Through Clean Energy Policy Evaluation." Working Paper 22-15. Resources for the Future. https://media.rff.org/documents/WP_22-15.pdf.

Americorps. 2014. "Budgeting for Evaluation." https://americorps.gov/sites/default/files/document/2014_11_12_BudgetingforEvaluationAudioDescription_ORE.pdf.

Badlam, J., S. Clark, S. Gajendragadkar, A. Kumar, S. O'Rourke, and D. Swartz. 2022. "The CHIPS and Science Act: What Is It and What Is in It?" *McKinsey Insights*. https://www.mckinsey.com/industries/public-and-social-sector/our-insights/the-chips-and-science-act-heres-whats-in-it.

Bedsworth, L., M. Arias, J. R. DeShazo, R. Winston, and I. Saunders. 2023. "Transformative Climate Communities—Lessons Learned and Best Practices." Webinar. January 6. Washington, DC: National Academies of Sciences, Engineering, and Medicine. https://www.nationalacademies.org/event/01-06-2023/accelerating-decarbonization-in-the-united-states-technology-policy-and-societal-dimensions-transformative-climate-communities-lessons-learned-and-best-practices.

Bistline, J., G. Blanford, M. Brown, D. Burtraw, M. Domeshek, J. Farbes, A. Fawcett, et al. 2023a. "Emissions and Energy Impacts of the Inflation Reduction Act." *Science* 380(6652):1324–1327. https://doi.org/10.1126/science.adg3781.

Bistline, J., N. Mehrotra, and C. Wolfram. 2023b. "Economic Implications of the Climate Provisions of the Inflation Reduction Act." *Brookings Papers on Economic Activity*, spring ed. Brookings Institution. https://www.brookings.edu/bpea-articles/economic-implications-of-the-climate-provisions-of-the-inflation-reduction-act.

Bloomberg NEF. 2022. "Lithium-Ion Battery Pack Prices Rise for First Time to an Average of $151/KWh." *BloombergNEF* (blog), December 6. https://about.bnef.com/blog/lithium-ion-battery-pack-prices-rise-for-first-time-to-an-average-of-151-kwh.

Carlson, A., and D. Burtaw, eds. 2019. *Lessons from the Clean Air Act Building Durability and Adaptability into US Climate and Energy Policy*. Cambridge, UK: Cambridge University Press.

CBO (Congressional Budget Office). 2022. "Summary Estimated Budgetary Effects of Public Law 117-169." https://www.cbo.gov/system/files/2022-09/PL117-169_9-7-22.pdf.

CEQ (Council on Environmental Quality). 2022. "Explore the Map." Climate and Economic Justice Screening Tool. https://screeningtool.geoplatform.gov.

Chi, S. 2022. "IRA: Our Analysis of the Inflation Reduction Act." Just Solutions Collective. https://assets-global.website-files.com/5fd7d64c5a8c62dc083d7a25/63232854dd4d104128f01b8c_JSC%20-%20Analysis%20of%20the%20Inflation%20Reduction%20Act%20-r3.pdf.

Child Care Technical Assistance Network. 2016. "How Much Will an Evaluation Cost?" https://childcareta.acf.hhs.gov/systemsbuilding/systems-guides/evaluation-and-improvement/what-program-evaluation/how-much-will.

Commission on Evidence-Based Policymaking. 2017. "The Promise of Evidence-Based Policymaking: Report of the Commission on Evidence-Based Policymaking." https://www2.census.gov/adrm/fesac/2017-12-15/Abraham-CEP-final-report.pdf.

DeCarolis, J., and A. LaRose. 2023. "Annual Energy Outlook 2023 with Projections to 2050." Presented at the Annual Energy Outlook 2023 Release at Resources for the Future, U.S. Energy Information Administration. March 16. https://www.eia.gov/outlooks/aeo/pdf/AEO2023_Release_Presentation.pdf.

DOE (Department of Energy). 2022a. "DOE Optimizes Structure to Implement $62 Billion in Clean Energy Investments from Bipartisan Infrastructure Law." https://www.energy.gov/articles/doe-optimizes-structure-implement-62-billion-clean-energy-investments-bipartisan.

DOE. 2022b. "Inflation Reduction Act Keeps Momentum Building for Nuclear Power." https://www.energy.gov/ne/articles/inflation-reduction-act-keeps-momentum-building-nuclear-power.

DOE. 2023. "FOTW #1272, January 9, 2023: Electric Vehicle Battery Pack Costs in 2022 Are Nearly 90% Lower Than in 2008, According to DOE Estimates." https://www.energy.gov/eere/vehicles/articles/fotw-1272-january-9-2023-electric-vehicle-battery-pack-costs-2022-are-nearly.

DOS and EOP (Department of State and Executive Office of the President). 2021. "The Long-Term Strategy of the United States, Pathways to Net-Zero Greenhouse Gas Emissions by 2050." https://www.whitehouse.gov/wp-content/uploads/2021/10/US-Long-Term-Strategy.pdf.

EELP (Environmental and Energy Law Program). 2022. "Environmental Justice (EJ) Provisions of the 2022 Inflation Reduction Act." http://eelp.law.harvard.edu/wp-content/uploads/EELP-IRA-EJ-Provisions-Table.pdf.

EO (Executive Order) 13990. 86 Federal Register 14 (January 25, 2021). "Protecting Public Health and the Environment and Restoring Science to Tackle the Climate Crisis." *Federal Register* 86(14):7037–7043.

EPA (Environmental Protection Agency). 2022. "California State Motor Vehicle Pollution Control Standards: Advanced Clean Car Program; Reconsideration of a Previous Withdrawal of a Waiver of Preemption; Notice of Decision." *Federal Register* 87:14332. https://www.govinfo.gov/content/pkg/FR-2022-03-14/pdf/2022-05227.pdf.

EPA. 2023a. "EPA Proposes New Carbon Pollution Standards for Fossil Fuel-Fired Power Plants to Tackle the Climate Crisis and Protect Public Health." https://www.epa.gov/newsreleases/epa-proposes-new-carbon-pollution-standards-fossil-fuel-fired-power-plants-tackle.

EPA. 2023b. "Greenhouse Gas Emissions Standards for Heavy-Duty Vehicles—Phase 3." *Federal Register* 88(81):25926–26161. https://www.govinfo.gov/content/pkg/FR-2023-04-27/pdf/2023-07955.pdf.

EPA. 2023c. "Multi-Pollutant Emissions Standards for Model Years 2027 and Later Light-Duty and Medium-Duty Vehicles." EPA-420-F-23-009. https://nepis.epa.gov/Exe/ZyPDF.cgi?Dockey=P1017626.pdf.

EPA. 2023d. "Multi-Pollutant Emissions Standards for Model Years 2027 and Later Light-Duty and Medium-Duty Vehicles." *Federal Register* 88(87):29184–29446. https://www.govinfo.gov/content/pkg/FR-2023-05-05/pdf/2023-07974.pdf.

EPA. 2023e. "Proposed Standards to Reduce Greenhouse Gas Emissions from Heavy-Duty Vehicles for Model Year 2027 and Beyond." EPA-420-F-23-011. https://nepis.epa.gov/Exe/ZyPDF.cgi?Dockey=P101762L.pdf.

Evaluation.gov. 2023. "Learning Agendas." https://www.evaluation.gov/evidence-plans/learning-agenda.

Fatherree, K., and N.R. Hart. 2019. "Funding the Evidence Act." Data Foundation. https://www.datafoundation.org/funding-the-evidence-act-paper-2019.

Fears, D. 2022. "Redlining Means 45 Million Americans Are Breathing Dirtier Air, 50 Years After It Ended." *The Washington Post*, March 9. https://www.washingtonpost.com/climate-environment/2022/03/09/redlining-pollution-environmental-justice.

Federal Register. 2022. "Environmental Justice Scorecard Feedback." CEQ-2022-0004. *Federal Register* 87:47397. https://www.federalregister.gov/documents/2022/08/03/2022-16635/environmental-justice-scorecard-feedback.

Goldman Sachs. 2023. "The US Is Poised for an Energy Revolution." *Goldman Sachs*. https://www.goldmansachs.com/intelligence/pages/the-us-is-poised-for-an-energy-revolution.html.

IRENA (International Renewable Energy Agency). 2022. *Renewable Power Generation Costs in 2021*. Abu Dhabi.

Jenkins, J., E.D. Larson, C. Greig, E. Mayfield, and S.W. Pacala. 2021. "Mission Net-Zero America: The Nation-Building Path to a Prosperous, Net-Zero Emissions Economy." *Joule* 5(11):2755–2761. https://doi.org/10.1016/j.joule.2021.10.016.

Jenkins, J.D., E.N. Mayfield, J. Farbes, R. Jones, N. Patankar, Q. Xu, and G. Schivley. 2022. "Preliminary Report: The Climate and Energy Impacts of the Inflation Reduction Act of 2022." Princeton, NJ: REPEAT Project, Zero-Carbon Energy Systems Research and Optimization Laboratory. August. https://repeatproject.org/docs/REPEAT_IRA_Preliminary_Report_2022-08-04.pdf.

Jenkins, J.D., E.N. Mayfield, J. Farbes, G. Schivley, N. Patankar, and R. Jones. 2023. "Climate Progress and the 117th Congress: The Impacts of the Inflation Reduction Act and Infrastructure Investment and Jobs Act." REPEAT Project. Princeton, NJ: Princeton University. July. https://doi.org/10.5281/zenodo.8087805.

Jiang, B., M. Mandloi, R. Carlson, M. Hope, M. Ziffer, N. Campanella, A. Rosa, et al. 2022. "US Inflation Reduction Act—A Tipping Point in Climate Action." Credit Suisse. https://www.credit-suisse.com/about-us-news/en/articles/news-and-expertise/us-inflation-reduction-act-a-catalyst-for-climate-action-202211.html.

Lane, H.M., R. Morello-Frosch, J.D. Marshall, and J.S. Apte. 2022. "Historical Redlining Is Associated with Present-Day Air Pollution Disparities in U.S. Cities." *Environmental Science and Technology Letters* 9(4):345–350. https://doi.org/10.1021/acs.estlett.1c01012.

Larsen, J., B. King, E. Wimberger, H. Pitt, H. Kolus, A. Rivera, N. Dasari, C. Jahns, K. Larsen, and W. Herndon. 2021. "Pathways to Paris: A Policy Assessment of the 2030 US Climate Target." Rhodium Group. https://rhg.com/research/us-climate-policy-2030.

Larsen, J., B. King, H. Kolus, N. Dasari, G. Hiltbrand, and W. Herndon. 2022. "A Turning Point for US Climate Progress: Assessing the Climate and Clean Energy Provisions in the Inflation Reduction Act." Rhodium Group. https://rhg.com/research/climate-clean-energy-inflation-reduction-act.

Larson, E., C. Greig, J. Jenkins, E. Mayfield, A. Pascale, C. Zhang, J. Drossman, et al. 2021. "Net-Zero America: Potential Pathways, Infrastructure, and Impacts, Final Report." Princeton, NJ: Princeton University. https://netzeroamerica. princeton.edu/the-report.

Lazard. 2023. "2023 Levelized Cost of Energy+." https://www.lazard.com/research-insights/2023-levelized-cost-of-energyplus.

LBNL (Lawrence Berkeley National Laboratory). 2022a. "Land-Based Wind Market Report." https://emp.lbl.gov/wind-technologies-market-report.

LBNL. 2022b. "Utility-Scale Solar 2022." https://emp.lbl.gov/utility-scale-solar.

Lempert, R., B.L. Preston, J. Edmonds, L. Clarke, T. Wild, M. Binsted, E. Diringer, and B. Townsend. 2019. "Pathways to 2050: Scenarios for Decarbonizing the U.S. Economy." Center for Climate and Energy Solutions. https://www.c2es.org/document/pathways-to-2050-scenarios-for-decarbonizing-the-u-s-economy.

Liu, J., L.P. Clark, M.J. Bechle, A. Hajat, S.Y. Kim, A.L. Robinson, L. Sheppard, A.A. Szpiro, and J.D. Marshall. 2021. "Disparities in Air Pollution Exposure in the United States by Race/Ethnicity and Income, 1990–2010." *Environmental Health Perspectives* 129(12):127005. https://doi.org/10.1289/EHP8584.

Mahajan, M., O. Ashmoore, J. Rissman, R. Orvis, and A. Gopal. 2022. "Updated Inflation Reduction Act Modeling Using the Energy Policy Simulator." San Francisco: Energy Innovation. https://energyinnovation.org/wp-content/uploads/2022/08/Updated-Inflation-Reduction-Act-Modeling-Using-the-Energy-Policy-Simulator.pdf.

Mayfield, E., J. Jenkins, E. Larson, and C. Greig. 2023. "Labor Pathways to Achieve Net-Zero Emissions in the United States by Mid-Century." *Energy Policy* 177(June 1):113516. https://doi.org/10.1016/j.enpol.2023.113516.

NASEM (National Academies of Sciences, Engineering, and Medicine). 2021. *Accelerating Decarbonization of the U.S. Energy System*. Washington, DC: The National Academies Press.

NASEM. 2022. "Pathways to an Equitable and Just Transition Workshop." July 26. https://www.nationalacademies.org/event/07-26-2022/accelerating-decarbonization-in-the-united-states-technology-policy-and-societal-dimensions-pathways-to-an-equitable-and-just-transition-workshop.

NHTSA (National Highway Traffic Safety Administration). 2022. "Corporate Average Fuel Economy Standards Model Years 2024–2026." *Federal Register* 87:25710. https://www.federalregister.gov/documents/2022/05/02/2022-07200/corporate-average-fuel-economy-standards-for-model-years-2024-2026-passenger-cars-and-light-trucks.

NHTSA. 2023. "Corporate Average Fuel Economy Standards for Passenger Cars and Light Trucks for Model Years 2027–2032 and Fuel Efficiency Standards for Heavy-Duty Pickup Trucks and Vans for Model Years 2030–2035." Notice of Proposed Rulemaking. Department of Transportation. https://www.nhtsa.gov/sites/nhtsa.gov/files/2023-07/CAFE-2027-2032-HDPUV-2030-2035-NPRM-web-version.pdf.

Ocko, I., S. Hamburg, D.J. Jacob, D.W. Keith, N.O. Keohane, M. Oppenheimer, J.D. Roy-Mayhew, et al. 2017. "Unmask Temporal Trade-Offs in Climate Policy Debates." *Science* 356(6337):492–493. https://doi.org/10.1126/science.aaj2350.

OMB (Office of Management and Budget). 2019. "Phase 1 Implementation of the Foundations for Evidence-Based Policymaking Act of 2018: Learning Agendas, Personnel, and Planning Guidance." OMB Memorandum M-19-23. https://www.whitehouse.gov/wp-content/uploads/2019/07/m-19-23.pdf.

OMB, CEQ, and CPO (Office of Management and Budget, Council on Environmental Quality, and Climate Policy Office). 2023. "Addendum to the Interim Implementation Guidance for the Justice40 Initiative, M-21-28, on Using the Climate and Economic Justice Screening Tool (CEJST)." Memorandum for the Heads of Departments and Agencies. https://www.whitehouse.gov/wp-content/uploads/2023/01/M-23-09_Signed_CEQ_CPO.pdf.

Our World in Data. n.d. "Solar (Photovoltaic) Panel Prices." https://ourworldindata.org/grapher/solar-pv-prices. Accessed June 5, 2023.

Pasanen, T. n.d. "Monitoring and Evaluation: Five Reality Checks for Adaptive Management." ODI. https://odi.org/en/insights/monitoring-and-evaluation-five-reality-checks-for-adaptive-management. Accessed January 19, 2023.

Patashnik, E. 2008. *Reforms at Risk: What Happens After Major Policy Changes Are Enacted*. Princeton, NJ: Princeton University Press.

Senate Commerce Committee. 2022. "CHIPS and Science Act of 2022 Summary." https://www.commerce.senate.gov/services/files/2699CE4B-51A5-4082-9CED-4B6CD912BBC8.

Twersky, F., and A. Arbreton. 2014. "Benchmarks for Spending on Evaluation." William and Flora Hewlett Foundation. https://hewlett.org/wp-content/uploads/2016/08/Benchmarks%20for%20Spending%20on%20Evaluation_2014.pdf.

White House. 2021a. "Fact Sheet: The Bipartisan Infrastructure Deal." https://www.whitehouse.gov/briefing-room/statements-releases/2021/11/06/fact-sheet-the-bipartisan-infrastructure-deal.

White House. 2021b. "Fact Sheet: President Biden Signs Executive Order Advancing Diversity, Equity, Inclusion, and Accessibility in the Federal Government." https://www.whitehouse.gov/briefing-room/statements-releases/2021/06/25/fact-sheet-president-biden-signs-executive-order-advancing-diversity-equity-inclusion-and-accessibility-in-the-federal-government.

White House. 2021c. "Fact Sheet: President Biden Signs Executive Order Catalyzing America's Clean Energy Economy Through Federal Sustainability." https://www.whitehouse.gov/briefing-room/statements-releases/2021/12/08/fact-sheet-president-biden-signs-executive-order-catalyzing-americas-clean-energy-economy-through-federal-sustainability.

White House. 2022a. "Advancing Equity Through the American Rescue Plan." https://www.whitehouse.gov/wp-content/uploads/2022/05/Advancing-Equity-Through-The-American-Rescue-Plan.pdf.

White House. 2022b. "American Rescue Plan Equity Learning Agenda." https://www.whitehouse.gov/wp-content/uploads/2022/05/American-Rescue-Plan-Equity-Learning-Agenda.pdf.

White House. 2022c. "Biden-Harris Administration Releases Inflation Reduction Act Guidebook for Clean Energy and Climate Programs." https://www.whitehouse.gov/briefing-room/statements-releases/2022/12/15/biden-harris-administration-releases-inflation-reduction-act-guidebook-for-clean-energy-and-climate-programs.

White House. 2022d. "President Biden Announces Senior Clean Energy and Climate Team." https://www.whitehouse.gov/briefing-room/statements-releases/2022/09/02/president-biden-announces-senior-clean-energy-and-climate-team.

White House. 2023a. "Inflation Reduction Act Guidebook." January. https://www.whitehouse.gov/cleanenergy/inflation-reduction-act-guidebook.

White House. 2023b. "Maps of Progress." https://www.whitehouse.gov/build/maps-of-progress.

White House. n.d. "Justice40: A Whole-of-Government Initiative." https://www.whitehouse.gov/environmentaljustice/justice40. Accessed January 20, 2023.

Williams, J.H., R.A. Jones, B. Haley, G. Kwok, J. Hargreaves, J. Farbes, and M.S. Torn. 2021. "Carbon-Neutral Pathways for the United States." *AGU Advances* 2(1):21. https://doi.org/10.1029/2020AV000284.

Ziegler, M.S., and J.E. Trancik. 2021a. "Data Series for Lithium-Ion Battery Technologies." Harvard Dataverse. https://doi.org/10.7910/DVN/9FEJ7C.

Ziegler, M.S., and J.E. Trancik. 2021b. "Re-Examining Rates of Lithium-Ion Battery Technology Improvement and Cost Decline." *Energy and Environmental Science* 14(4):1635–1651. https://doi.org/10.1039/D0EE02681F.

2

Energy Justice and Equity

ABSTRACT

The U.S. federal government has put forth a whole-of-government equity agenda through a series of executive orders and legislation. This national focus on equitable outcomes of the transition to net-zero energy systems will directly and indirectly allow for the incorporation of equity into future federal, state, and local decarbonization action. However, without the intentional attention to energy justice or its principles, a just energy transition cannot be achieved in the United States.

The transition to a net-zero-carbon future offers multiple socioeconomic benefits, including improved public health and energy affordability, but action is needed to ensure energy justice can be advanced. This requires disadvantaged communities, local governments, and community-based organizations to be engaged in defining where, when, and how to prioritize federal and state resources during the energy transition. Inclusive and equitable approaches are key to preventing potential implementation challenges or project derailment.

Place-based decarbonization approaches that address the priorities and concerns of affected communities will help support an equitable transition to a net-zero future that avoids worsening existing inequities or creating new ones. For the transition to be a success, there needs to be a bottom-up approach beginning with community-led programs sharing lessons learned coupled with a top-down process beginning with federal adoption and implementation of energy justice principles. This chapter outlines principles and best practices of energy justice and reviews the opportunities and barriers associated with different energy justice approaches. Table 2-3, at the end of the chapter, summarizes all the recommendations that appear in this chapter to support a just energy transition.

INTRODUCTION

While the transition to a net-zero energy system will offer multiple socioeconomic advantages for society, such as access to affordable energy options, economic and employment progress, and improvement of human health (Õunmaa 2021), it is not a foregone conclusion that these opportunities will extend to all. Without an intentional and concerted movement away from inequitable energy structures and policies, disadvantaged populations—such as racial, ethnic, and low-income communities—may experience even more burdens from the new energy system than they do today. Recent federal action provides a critical down payment on a just and equitable transition. However, the critical opportunity to enhance societal and economic outcomes comes with equity challenges that require careful attention and intentional action.

For example, exposure to air pollution from fossil fuel combustion is one of the most significant disparities of the current energy system (see Chapter 3). A large-scale change in the production, distribution, and use of energy will likely require the elimination of most fossil fuel use. This will significantly reduce nearly all associated air pollution, providing positive impacts nationally. However, disproportionate and negative health and socioeconomic outcomes of societal and technological change have been well-documented in low-income populations and communities of color (Lerner 2010; Méndez 2020; Romero-Lankao et al. 2022). Negative outcomes of the U.S. energy transition, especially increased air pollution from the construction of new infrastructure or from the continued combustion of fossil fuels, can undermine the success of an equitable carbon-neutral future.

This transition is not merely a technological transformation of the energy sector; it is a fundamental and wholesale transformation that will affect numerous sectors and nearly every household. This chapter will address the challenges associated with two of the core societal goals identified in the first report: ensuring a just and equitable transition to carbon neutrality and ensuring that workers, communities, and businesses impacted by the transition are fully supported during the transition (NASEM 2021). It begins by outlining the evolution of the just transition movement and introducing key components and principles of energy justice. It reviews existing energy inequities in several areas, including energy affordability and accessibility to low-carbon technologies, and discusses how energy policies and programs can be developed to redress inequities and avoid creating new ones. The chapter then assesses recent federal actions that can support a just energy transition and equitable implementation of decarbonization actions. The chapter concludes with recommendations for how to build capacity to support a community-level decarbonization action and develop evaluation standards to detail the nation's progress toward a just energy transition.

EVOLUTION OF THE JUST TRANSITION MOVEMENT: FROM ENVIRONMENTAL JUSTICE TO ENERGY JUSTICE

The transition of U.S. energy systems to a net-zero-carbon future offers access to affordable energy options, economic and employment progress, and the improvement of human health (Õunmaa 2021). However, challenges exist and cannot be ignored as the transition proceeds. Initially used by coal communities and the labor movement, the modern just transition concept focuses on inclusive processes to achieve climate and energy goals and envelops three overlapping concepts: environmental justice, climate justice, and energy justice. Decarbonizing the energy system is an opportunity to move forward on energy justice—the provision of safe, affordable, and sustainable energy—for the nation and internationally. Box 2-1 provides key terms that will be used throughout the chapter. This section briefly traces the scholarship of the just transition movement and the key concepts of the environmental and climate justice. It then introduces the energy justice movement and key principles that need to be applied to decarbonization policy to achieve a just energy transition.

History of the Just Transition and Environmental and Climate Justice Movements

In the United States, the original use of the term "just transition" began with the labor movement in the late 1970s, which advocated for the protection, support, and compensation of displaced workers and communities when a society makes significant policy decisions resulting in job loss in energy-related businesses (Carley and Konisky 2020). It

---

BOX 2-1
KEY TERMS

Energy justice—the provision of safe, affordable, and sustainable energy to all individuals (Jenkins 2018) through the incorporation of recognitional, procedural, and distributional equity into energy design, ownership, governance, and implementation.

Energy transition—efforts by jurisdictions to transform or develop their energy sector away from fossil fuels (Bozeman et al. 2022) with a large-scale technological and societal change in the production, distribution, and use of energy.

Equity—being fair and unbiased regarding access, opportunities, risks, and burdens for an individual or group, especially as a function of an organization or system (Romero-Lankao and Nobler 2021).

Just energy transition—a process of transforming the energy system by ensuring that all communities, workers, and social groups are fairly included in the processes toward and outcomes of the net-zero future through the incorporation of the principles of energy justice.

Justice—ensuring that all individuals and groups have the necessary and sufficient capability to achieve the lives they value (Sen 2009).

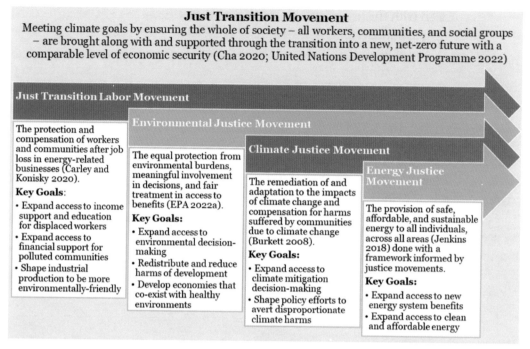

FIGURE 2-1 Temporal illustration of when just transition movements were introduced, and the definition for and the key goals of each movement.

has come to be used more broadly with scholars arguing that a just transition ensures that "workers and communities supported by a declining industry are able to transition into a new economy with a comparable level of economic security or retire with dignity" (Cha 2020, p. 149) and with advocacy groups emphasizing that a just transition means "meeting climate goals by ensuring the whole of society—all communities, all workers, all social groups—are brought along with the pivot to a net-zero future" (UNDP 2022). Environmental, climate, and energy justice have emerged as guiding principles for just transition action and conceptually exist in parallel (Jenkins et al. 2018). Figure 2-1 shows how the just transition movement has evolved beyond its origins in concerns with impacts on workers to encompass a broader array of principles for just and equitable change.[1]

The contemporary environmental justice movement can be traced to the 1980s. Its earliest claims dealt with "environmental racism," which emphasizes how racial minority communities are targeted, intentionally or unintentionally, for disproportionate exposure to pollutants or degraded environments (Bullard 2002). Environmental

---

[1] For a literature review about the just transition movement in North America, see Wilgosh et al. (2022). For an overview of just transition frameworks, see Henry et al. (2020).

racism is coupled with the systemic exclusion of Black, Indigenous, Latino, and other communities of color in decisions on environmental policymaking, enforcement, and remediation (Méndez 2020). These communities have historically been subject to and have mobilized against institutional processes that have resulted in residential segregation, unsafe housing stock, inadequate transportation options, displacement, disinvestment, and neglect (Covington 2009; Ong et al. 2023; Romero-Lankao et al. 2022; Walker 2009). The movement stresses the need for vulnerable communities to have access to environmental decision-making and for the harms and benefits of environmental development to be fairly distributed.

Environmental justice groups have long maintained that climate change mitigation measures might inadvertently increase localized air pollution unless both hazards are understood through the lens of human health. However, such connections have only recently been foregrounded in climate change policymaking in the United States (Méndez 2020; Yoder 2022). EPA's finding that carbon dioxide is a pollutant under the Clean Air Act marked a shift toward recognizing this entwined nature by identifying greenhouse gas (GHG) emissions as a pollution associated with fossil fuel combustion (EPA 2009). More recently, the Climate Pollution Reduction Grant program codified this language and shift in perspective by defining greenhouse gases as "the air pollutants carbon dioxide, hydrofluorocarbons, methane, nitrous oxide, perfluorocarbons, and sulfur hexafluoride" (P.L. 117-58 §60114). The concept of "climate pollution" bridges conversations focused on environmental justice issues and those focused on climate mitigation by relying on the broadly held concerns about the effects of multiple sources of "pollution" on individuals' health (Méndez 2020; Yoder 2022).

The climate justice movement, which emerged in the late 1990s, recognizes the global and disproportionate responsibility for and impacts of climate change (Baker et al. 2019; Schlosberg and Collins 2014). The movement acknowledges that countries and communities historically contributing least to climate change are more likely to be most impacted by climate change (Birkmann et al. 2022). Climate justice discussions have revolved around two issues: responsibility for climate mitigation and climate adaptation. For mitigation responsibility, in 2019, the combined GHG emissions from the least-developed countries[2] contributed 3.3 percent to global

---

[2] The Intergovernmental Panel on Climate Change (IPCC) defines least-developed countries as those meeting the following criteria: "(1) a low income criterion below a certain threshold of gross national income per capita of 750 to 900 (USD), (2) a human resource weakness based on indicators of health, education, and adult literacy, and (3) an economic vulnerability weakness based on indicators on instability of agricultural production, instability of export of goods and services, economic importance of non-traditional activities, merchandise export concentration, and the handicap of economic smallness." These criteria and the list of least-developed countries are designated by the Economic and Social Council of the United Nations (IPCC 2022).

GHG emissions (IPCC 2022), whereas the United States alone emitted 12.5 percent of global emissions (Ge et al. 2022). Regarding adaptation, the poorest and most vulnerable communities are the most at risk to the impacts of climate change, including in high-income countries (Birkmann et al. 2022; Carley and Konisky 2020; Romero-Lankao and Norton 2018). This is especially true for female, Latino, Black, and LGBTQ+ individuals within U.S. communities (Goldsmith et al. 2022; Méndez et al. 2020). Inadequate infrastructure and supportive aid—such as safe housing, emergency response systems, and health care—increase the vulnerability of these groups to climate change (Birkmann et al. 2022). Like the environmental justice movement, this movement advocates for affected communities to have access to climate change mitigation decision-making to ensure that policy efforts address disproportionate climate harms.

The Energy Justice Movement

In recent years, the just transition movement has increasingly focused on energy, in what some have referred to as "a new front-line in environmental justice research and activism" (Sze and London 2008). The concept of energy justice focuses separately on energy concerns among the broader issues addressed in the environmental justice movement (Bickerstaff et al. 2013; Jenkins et al. 2018) by integrating social equity principles into energy systems. An overlap of environmental justice and energy justice is the siting of energy infrastructure: an energy justice approach considers whether the location of energy infrastructure makes energy more affordable or accessible for historically disadvantaged households, whereas an environmental justice approach broadly considers whether the location of energy infrastructure unequally burdens a nearby community. Key to the energy justice movement is access to new energy system benefits and access to clean and affordable energy for everyone. With energy justice, energy systems can support economic growth in addition to energy security for individuals and communities. See Chapter 5 for more information about community energy projects, community benefits agreements, and energy sovereignty for tribal nations.

An energy transition can provide enormous opportunities for cleaner energy sources, new employment, and technological innovation (Cha 2020; Miller 2022, 2023). However, it also can exacerbate existing disparities afflicting communities of color and low-income neighborhoods or reduce access to opportunities that accompany energy transitions (Carley and Konisky 2020). Despite the shared manifestations of racial–ethnic and income-based disparities, research has shown that racial–ethnic factors have a larger effect on disparities than income-based factors. For example, a review of national exposure to air pollution from 1990 to 2010 found absolute exposure disparities were larger for racial and ethnic groups than for income categories (Liu et al. 2021; see Chapter 3 for more information). Due to the absence of racial–ethnic indicators in the federal government's definition of disadvantaged communities, the discussions in this chapter focus on income-based disparities of the U.S. energy system. However, it is important to acknowledge and understand the distinction between racial–ethnic and income-based factors to develop appropriate solutions.

The committee defines a "just energy transition" as the process of transforming the energy system by ensuring that all communities, workers, and social groups are included in the processes toward and outcomes of the net-zero future through the incorporation of the principles of energy justice. Incorporating energy justice principles in the energy transition will provide the nation an opportunity to prioritize human-centered approaches in energy system design and policymaking so that the costs and benefits of energy services are distributed fairly (Tarekegne et al. 2021), thus making it just. Furthermore, Table 2-1 illustrates four principles of energy justice, their focus, and guiding questions.

The energy justice principles provide an analytical and decision-making framework for researchers, advocates, policy makers, and communities to understand the human and social dimensions of energy systems and their inequities (Sovacool and Dworkin 2014).[3] This chapter largely focuses on recognitional, procedural, and distributional equity in its discussion of barriers, examples, and recommended solutions. However, restorative equity provides important context-setting for recognitional, procedural, and distributional equity and is therefore the foundation of all equity frameworks (Spurlock et al. 2022). The integration of energy justice principles needs to be both a bottom-up approach beginning with community-led programs sharing lessons learned and best practices and a top-down process beginning with federal adoption and implementation of these principles.

The term "intersectionality" describes how structures and systems of oppression— such as racism, sexism, homophobia, xenophobia, and redlining—heighten the effects of discrimination, exclusion, and social inequality on communities marginalized by multiple systems (Cooper 2016; Crenshaw 1989; Dhamoon 2011; Goldsmith et al. 2022; Roman 2017). An intersectional approach to energy justice emphasizes how multiple systems of marginalization and human identities interact to increase

---

[3] For more information, see Carley (2022); Heffron and McCauley (2017); Romero-Lankao and Nobler (2021); and Schlosberg (2007).

TABLE 2-1 Principles of Energy Justice, Their Focus, and Related Guiding Questions

Principle	Focus	Guiding Questions
Recognitional (or Structural) Equity	Understand structural determinants of exclusion and vulnerability and specific needs associated with energy services among social groups (Energy Equity Project 2022) and institutionalize accountability (Park 2014).	• Who is vulnerable and excluded and how? • Who is privileged and how?
Procedural Equity	Promote diverse representation and a meaningful voice for impacted communities among decision makers and energy service providers (Energy Equity Project 2022).	• Who is at the table? • What power do they have in influencing planning, decision-making, implementation, and evaluation?
Distributional Equity	How the benefits and harms of the energy system are distributed (Energy Equity Project 2022).	• Who bears the brunt of the burdens and how? • Who receives the most benefits and how?
Restorative (or Transgenerational) Equity	Decision makers ensure that all potential harms and injustices are addressed in prevention and mitigation plans (Energy Equity Project 2022) and generational impacts are considered (Park 2014).	• Who will remedy the foregoing injustices, and how? • How can we rectify past injustices caused by the energy system?

exposure to environmental harms and reduce access to energy and environmental benefits (Crenshaw 2017; Goldsmith et al. 2022; Kaijser and Kronsell 2014). Efforts have analyzed how intersectionality affects distributional inequalities to create energy inequities and have made recommendations to target the social and political practices of exclusion through which these inequalities are generated (Schlosberg and Collins 2014; Walker 2009). Relatedly, "[e]nergy democracy" recognizes such intersectional factors as it focuses attention on strengthening inclusive decision-making processes and democratic institutions, often through decentralized energy projects (Berthod et al. 2022; Nadesan et al. 2023). (See Chapter 5 for more on energy democracy and engaging the public in the energy transition.)

ENERGY INEQUITIES: HISTORICAL AND RECENT TRENDS

Some aspects of the energy transition could aggravate, rather than address, energy inequities if decarbonization actions are not intentionally focused on equity and justice (Carley et al. 2018b; Romero-Lankao et al. 2022; Sovacool et al. 2022). For example, the energy transition can reinforce inequities in access to affordable, accessible, and safe energy; it may create new health risks; and may limit opportunities for workforce development in both communities dependent on fossil fuels and new, renewable energy systems (Carley and Konisky 2020; Carley et al. 2021; Cha 2022). It is important to recognize the historical factors and patterns that led to present-day disparities to avoid creating new or worsening existing inequities during and following the energy transition. This section analyzes four historical and recent trends in energy-relevant inequities: energy affordability; accessibility, acceptability, and adoption; public health and community resilience; and jobs and workforce development.

Energy Affordability

The current energy system has led to disparities in energy affordability, the ability to afford one's energy bills, with disadvantaged communities experiencing most of the negative costs. Intersecting systemic inequities result in households' experiencing unequal access to basic energy, unequal ability to meet basic energy needs, and unequal availability of the income needed to obtain energy. Energy burden, energy insecurity, and energy poverty are increasingly severe instances of social inequities that in turn relate to unequal vulnerability to other stressors. Analyzing energy system impacts across intersecting socioeconomic and demographic metrics[4] allows for the identification of individuals who are most vulnerable, underserved, or marginalized (Hernández et al. 2014; Jenkins 2018). Energy poverty, the lack of affordable, reliable, and environmentally sound energy services (Reddy 2000), will be directly addressed through the incorporation of energy justice into the U.S. energy transition.

Energy insecurity is the inability to adequately meet basic household energy needs over time (Hernández et al. 2016). It is attributed to several factors, including inefficient housing and appliances leading to more inefficient energy use; lack of financial resources to afford air conditioning and heat pumps; and unequal access to cooling or heating that may lead some residents to dangerously under-heat or under-cool their homes (Hernández et al. 2014). These energy inequities amplify other existing health, educational, and socioeconomic disparities and further reinforce obstacles to civic participation in society (Bouzarovski 2018). For example, households with children are more likely to engage in dangerous financial

---

[4] A literature review of 10 reports on energy criteria noted that the terms metric, indicator, and index were frequently understood to have the same definition across the reports (Tarekegne et al. 2021). This report primarily uses the term "metric." The committee notes that CEQ utilizes "indicator" in the development and publication of the Climate and Economic Justice Screening Tool.

and behavioral coping strategies, to be disconnected from energy services, and to be energy insecure (Carley et al. 2022; Konisky et al. 2022; Memmott et al. 2021). Furthermore, children in moderately and severely energy insecure households are more likely to experience food insecurity, hospitalizations, and developmental concerns than children in energy secure homes (Cook et al. 2008; Hernández et al. 2016; Smith et al. 2007).

Energy burden, the percentage of gross household energy costs spent on energy, is a metric that operationalizes energy affordability and identifies groups in need of targeting policies and investments to reduce high energy burdens (Cong et al. 2022). According to a survey by Indiana University's Energy Justice Lab, nearly 40 percent of Latino households and more than 26 percent of Black households said that they were unable to pay their electricity bill (Carley et al. 2022) and thus experience high energy burden. Additionally, compared to White respondents, Latino and Black respondents were 80 percent and 30 percent, respectively, more likely to have their service disconnected by their utility provider, which often comes with additional fees to restore electricity services (Carley et al. 2022). Poor housing conditions, including lack of insulation and old rooftops, and a lack of transportation options, such as accessible public transit and safe biking, tend to perpetuate high energy burdens (Drehobl and Ross 2016).

Households within disadvantaged communities in the United States often spend a larger fraction of their household income on utilities for heating, cooling, and other home energy services than the general population (Drehobl and Ross 2016; Drehobl et al. 2020). Data from the Department of Energy (DOE) Low-Income Energy Affordability Data (LEAD) Tool,[5] designed to improve the understanding of states, communities, and stakeholders about energy characteristics, shows that the average energy burden for low-income households is 8.6 percent (DOE 2020), which is more than double the national median of 3.1 percent. Figure 2-2 illustrates the comparison of the national median energy burden with the median energy burden of certain groups. However, average energy burden does not accurately reflect existing discrepancies in utility rates, especially between rural and coastal urban areas. During the development and implementation of programs addressing energy burden, regional differences must be considered and adjusted for.

Increases in the cost of energy often force families to decide whether to spend more of their household income on energy or on something else such as rent,

---

[5] See the Department of Energy's "Low-Income Energy Affordability Data Tool" at https://www.energy. gov/scep/slsc/lead-tool.

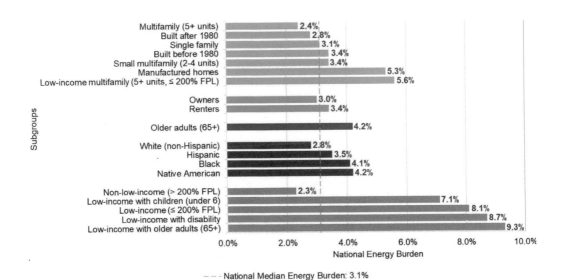

FIGURE 2-2 National energy burdens across subgroups compared with the national median energy burden. Orange bars show energy burden for low-income populations. Red bars show energy burden by race and ethnicity. The purple bar shows energy burden for older adults. Blue bars show energy burden for renters and owners. Green bars show energy burden by housing type. NOTE: FPL, Federal Poverty Level. SOURCE: Data from Drehobl et al. (2020).

education, food, and transportation (Brown et al. 2020). Disadvantaged communities tend to have older or less energy-efficient homes, which increases household energy expenses. These households often either cannot afford to upgrade to energy efficient products or are renters and do not have the ability to do so. Subsidies and programs such as the Low Income Home Energy Assistance Program (LIHEAP) and the Weatherization Assistance Program (WAP) are designed to mitigate these burdens (Brown et al. 2020; Hernández and Bird 2010). However, these programs have historically been underfunded and tend to be intentionally short-term solutions that ensure that utilities—not the households—are protected against potential debts and disconnection of services. Programs designed to mitigate energy burden also suffer from significant implementation failures (Carley et al. 2022; Farley et al. 2021) and tend to be hard for consumers to navigate.

The benefits of the energy transition to date have not always been equally distributed. The energy burdens of inefficient appliances, homes, and vehicles persist in low-income populations and for households of color. Conversely, wealthier consumers can afford the relatively high up-front costs of energy-saving and emission-reducing technologies (e.g., high-efficiency air conditioners, smart

meters, electric vehicles, heat pumps, and rooftop solar), which have lower operating costs and so decrease their total energy costs (Carley and Konisky 2020;
Drehobl and Ross 2016; Ross et al. 2018). Such energy-saving devices are often
cost prohibitive for and not prioritized by low-income households, especially
when a working fossil fuel–based device (e.g., gas-powered furnace) is already in
place (Agyeman et al. 2016; Lukanov and Krieger 2019; Morrissey et al. 2020). To
address the challenge of energy affordability and associated burdens and achieve
energy justice, it is important to recognize these existing burdens and the intersecting factors that influence them.

Accessibility, Acceptability, and Adoption

Disadvantaged communities are often economically excluded from, reluctant to
adopt, or unaware of opportunities to install low-carbon technologies. This might
be owing to fear of hidden costs, program limitations, lack of trust in government,
inadequate outreach and information, insufficient capacity, and inequitable
and predatory financing (Madrid 2017; Méndez et al. 2020; Vogelsong 2022). For
example, split incentives between owner and tenant create barriers to the energy
transition, as building owners do not have any incentive to pay for retrofits, energy efficiency, or safety improvements if only tenants receive the benefits from
decreased energy bills (Besley 2010; Boudet 2019; Segreto et al. 2020). Solutions
for the tensions caused by split incentives need to be designed to create tangible
benefits for both parties. (For more information about split incentives, see
Chapter 7.)

Inadequate or lacking community engagement might result in rejection of and
even opposition to new technologies for reasons that include, but are not limited
to, the high costs associated with smart home devices and the physical look of
renewable infrastructure (Boudet 2019; Devine-Wright 2005; Devine-Wright and
Devine-Wright 2009). Examples include public opposition to wind energy and
the siting of bioenergy infrastructures, owing to concerns about security, privacy,
noise, and potential health and socioeconomic impacts (Boudet 2019; Selfa et al.
2011; Wüste and Schmuck 2012). Similar dynamics exist in other energy industries
that will be heavily impacted by the energy transition, including the automobile
and oil and gas industries. For the committee's recommendations on opportunities and practices to overcome obstacles to meaningful public engagement, see
Chapter 5. For ensuring access to subsidies for improving building efficiency, see
Chapter 7, and for equity in access to electric vehicles and public transportation,
see Chapter 9.

Public Health and Community Resilience

The effect of air pollution, extreme heat, and other environmental stressors on communities is often determined by socio-spatial inequalities[6] in exposure experienced and the capacity to mitigate health risks (Harvey 2008; Logan and Molotch 2005). These effects and disparities are further associated with intersectional factors such as race, gender, and income. For instance, the average exposures to diesel particles are higher than average for non-White, lower-income households living along transportation corridors (Romero-Lankao et al. 2022). Furthermore, the legacies of past discriminatory practices often prevent disadvantaged communities from reaping the rewards of tree shade, open space, good-quality housing, energy-efficient building envelopes, and cleaner air (Church et al. 2000; Lucas 2012; Morello-Frosch et al. 2011). Recent analysis by Romitti et al. (2022) shows similar inequalities in the access to residential air conditioning in metropolitan areas where heat extremes and urban heat island effects are increasing. Such systemic inequities contribute to higher adverse health impacts and lower community resilience, where community resilience is defined as capacity to draw on income, education, and other socioeconomic resources to adapt to pollution, extreme heat, energy outages, and other disruptions (Harlan et al. 2013; Hayden et al. 2011; Qin et al. 2015; Romero-Lankao et al. 2016). See Chapter 3 for more information about the disparities in public health impacts from pollution and extreme heat.

In the 1930s, the Home Owners' Loan Corporation, a now-defunct government-sponsored entity, graded neighborhoods according to levels of mortgage risk (Hillier 2003; Jackson 1985; Michney and Winling 2020). "Areas with African Americans, as well as those with older housing and poorer households, were consistently given a fourth grade, or 'hazardous,' rating and colored red" in a practice known as redlining (Hillier 2003, p. 395). Meanwhile, those same neighborhoods were often targeted as sites for undesirable land uses such as major freeway construction (Bullard 2004). Such projects resulted in the displacement of and exposure to construction pollution for redlined communities (Jackson 1980, 1985; Katznelson 2005; Massey and Denton 1998; Michney and Winling 2020; Rothstein 2017). The effects of these practices are still evident today. Figure 2-3 shows the interplay of the Home Owners' Loan Corporation Risk Rating and disproportionate nitrogen oxides (NO_x) exposure for communities of color in Berkeley and Oakland, California.

Another outcome of construction and land-use projects in redlined areas is the development of urban heat islands, areas with higher-than-average temperatures (Guhathakurta and Gober 2007; Hoffman et al. 2020; Hsu et al. 2021; Romero-Lankao et al. 2012).

---

[6] The manifestation of social inequalities into spatial patterns (Han 2022).

FIGURE 2-3 Maps of air pollution exposure and population density of people of color show that areas that the Home Owners' Loan Corporation deemed "hazardous" in 1930, outlined in red, are now some of the areas with high NO_x air pollution and high concentrations of people of color. SOURCE: Zhong and Popovich (2022), © 2022 The New York Times Company. All rights reserved. Used under license.

These high temperatures have fatal outcomes; in the United States, exposure to extreme heat led to about 17,000 premature deaths in 2020 (Shindell et al. 2021). On the other hand, adaptation to heat with air conditioning requires more energy and if that energy continues to be generated with fossil fuels, this runs the risk of perpetuating adverse health impacts from air pollution from fossil fuel combustion. An increase in surfaces covered with vegetation and permeable, reflective materials can decrease the temperature of heat islands and the need to overuse air conditioning indoors. However, such features are less common in disadvantaged neighborhoods which leads to higher temperature that can lead to higher morbidity and mortality risks. See Chapters 3 and 7 for more on urban planning and building retrofits to address heat islands.

Jobs and Workforce Development

National energy production and carbon-intensive industries provide significant economic output and jobs, but national-level trends in the green economy play out unevenly across geographies, leading to opportunities for some while leaving others behind (Cameron and van der Zwann 2015; E2 2019; NASEM 2023). Communities dependent on the fossil fuel industry have expressed concerns about the disruption that will be faced during the transition, including the potential lack of access to high-quality job opportunities that offer similar economic stability as the jobs lost (Cha 2022).[7] For example, owing to the reduced demand for local services by laid-off coal employees, communities in Appalachia experienced loss of retail and commercial occupations (Carley et al. 2018a; Lobao et al. 2016). In addition, the loss of fossil fuel jobs and production will impact local tax revenues (Pollin and Callaci 2019). If not implemented equitably, the energy transition will have a detrimental impact on the culture, identity, and sense of place of these communities.

OVERCOMING BARRIERS FOR A JUST ENERGY TRANSITION

Decarbonization measures can reduce GHG emissions while simultaneously providing new career opportunities, improving public health, increasing energy accessibility and affordability, and reducing energy justice disparities (NASEM 2021, 2023). For the transition to be a success, intentional learning from past projects, engagement with all stakeholders, and the inclusion of input into the development and implementation of transition programs and policies are needed (Krieger 2022).

Operationalizing Energy Justice in the U.S. Transition

Operationalizing the energy justice principles at all levels of policymaking and program development is critical for the energy transition to have equitable outcomes. The multistep process starts with eliciting community values and aspirations as well as defining equitable goals and results in the creation of measurable progress (Aguayo 2022). The energy justice principles need to be systemically and effectively incorporated into the planning, implementation, and evaluation of decarbonization policies. Occasionally, this will result in incremental changes to existing practices and programs; at other times, it will require across-the-board restructuring of plans. Policies will need to be adjustable and responsive to the goals and needs of individual communities that have different histories, experiences, and priorities.

Across all four dimensions examined in the previous section—energy affordability; accessibility, acceptability, and adoption; public health and community resilience; and jobs and workforce development—disadvantaged communities would benefit most from actions

---

[7] See Chapter 4 for more about the impacts the transition will have on workforce and Chapter 12 for the impacts the transition will have on the on the fossil fuel industry.

that thoroughly and consistently communicate the available decarbonization programs and technologies, the purpose and goal of each option, and the benefits and costs of each option. For instance, programs directed toward disadvantaged community needs can include (1) making home visits to support maintenance and energy-efficient upgrades or retrofits aimed at reducing indoor pollution or other environmental health issues; or (2) preventive safety aid, information, and training to support the safe installation of clean energy technologies. Table 2-2 gives examples of how the energy justice principles can be put into practice to produce equitable program outcomes.

Impacted communities also benefit from programs that address emissions mitigation and additionally provide other direct, long-term benefits. For example, the Department of Housing and Urban Development (HUD) and DOE are piloting demonstrations for housing interventions that combine the Lead Hazard Reduction Healthy Homes programs and the WAP (HUD 2023). Through the exclusive engagement of low-income households and a coordinated assessment of outcomes, HUD and DOE hope to determine if the streamlined delivery of home services achieve cost effectiveness and meet public health and energy efficiency objectives. Future decarbonization programs that prioritize engagement from communities most impacted by the current energy system and provide multifaceted, long-term solutions and benefits help address all three tenets of energy justice.

Stakeholder's Environmental Justice Concerns

Several environmental justice and energy justice organizations have raised concerns about aspects of recent decarbonization policies and their impacts on the environment and on disadvantaged communities. Although advocates acknowledge the positive investments in air pollution monitoring, urban tree planting programs, and measures that address legacy harms and climate change, there are concerns that some investments "are not aligned with centering overburdened communities in decision-making or transitioning away from fossil fuels" (We Act for Environmental Justice 2022). For example, a study of communities with biofuel development showed that newly funded energy facilities did not create the levels of employment promised, but instead led to issues of water availability and quality, road damage, and livestock feed cost increases within these communities (Kulcsar et al. 2016).

In a letter to DOE Secretary Granholm, the Environmental Justice Leadership Forum (EJLF), a national coalition of environmental justice leadership, states "[e]nergy sources should not be classified as 'clean' if they increase pollution burden, expand fossil fuel reliance or infrastructure, or exacerbate health risks" (EJLF 2022, p. 2). The EJLF

TABLE 2-2 Operationalizing Energy Justice in Energy-Relevant Programs

Principle	Operationalization	Program Suggestions
Recognitional Equity	Incorporate baseline assessments of existing environmental burdens and harms on local communities and ensuring that clean energy projects and investments do not further add to those burdens, and if they do, the project location and scale should be reconsidered.	• Clearly define the vulnerable and disadvantaged groups a program intends to benefit and engage with. • Design programs, community-adapted strategies, and capacity-building methods tailored to specific disadvantaged groups. • Prioritize emissions reductions in vulnerable areas.
Procedural Equity	Embrace all four themes of procedural justice—participation, information, fair decision-making, and local context (Elmallah and Rand 2022)—while centering local knowledge and concerns in project development.	• Provide real-time information about peak energy use rates to change behavior and save money on energy bills. • Create pilots and examples of how investment in technology works at the neighborhood-level and communicate outcomes. • Develop simplified application forms and increase communication between programs to allow for one qualification to authorize another. • Build trust with homeowners and community members.
Distributional Equity	Design policies and programs that (1) compensate communities negatively impacted by the energy transition; (2) reduce energy costs and burdens of low-income households; (3) ensure equitable distribution of the benefits of clean energy technologies while avoiding harms to communities; and (4) ensure participation of people of color and individuals from under-served communities in jobs in the growing clean energy economy.	• Develop programs that offer home visits to support maintenance or retrofits. • Provide training on transition technologies and preventive safety information. • Develop tailored funding assistance for new technology purchases and reducing energy burdens for disadvantaged communities. • Develop training, financing, and educational programs for transition and clean energy jobs that are tailored to specific communities.

proposes that certain actions be taken to address community impacts and concerns if such technologies continue to be supported. They endorse developing robust community engagement prior to the creation of implementation strategies; conducting a comprehensive analysis prior to funding to understand and minimize adverse impacts of the project or program; and listening sessions to learn of the actual community impacts with the goal of actionable remediation of concerns (EJLF 2022).

Social life-cycle assessment (S-LCA) is an emerging analysis technique that leverages life-cycle assessment's full accounting of a technology or system's inflows, outflows, and quantification of impacts throughout its full life cycle. While LCA has typically been used to quantify environmental impact categories, social and equity impacts are present in the system and outcomes being assessed, even if not noted explicitly (Bozeman et al. 2022). S-LCA creates a connection between the established methodological approaches for life-cycle accounting with impact categories for equity, societal, and justice-related outcomes. Significant concerns have been raised regarding decarbonization transition technologies and practices that may have a higher risk of harms to disadvantaged communities and the communities that host the infrastructure required for these technologies.[8] These include technologies that: manage carbon such as carbon capture and storage (CCS) and carbon utilization; produce non-fossil energy carriers such as hydrogen and biofuels; generate electricity like nuclear power generation; and enable other decarbonization technologies, like pipelines and mining (EJLF 2022). The underlying concerns associated with these technologies range include the following:

1. Direct environmental impacts of siting industrial infrastructure in or near disadvantaged communities;
2. Direct societal impacts from community participation in siting decisions;
3. Indirect impacts of enabling continued fossil fuel development and combustion; and
4. Reliance on future implementation of negative emissions technologies that may not come to fruition.

Box 2-2 presents the equity impacts of CCS that may have direct and indirect environmental and social impacts on communities.

In this light, recent federal decarbonization policies within the Infrastructure Investment and Jobs Act (IIJA) (P.L. 117-58) and the Inflation Reduction Act (IRA) (P.L. 117-169) fail to fully address the fact that disadvantaged communities face disproportionate impacts from the fossil fuel industry. For example, IIJA includes provisions for nuclear energy

---

[8] For more about the health considerations for communities that host extraction operations, see Chapter 3. For more about community energy and collective benefits for communities, see Chapter 5.

BOX 2-2
EQUITY IMPACTS OF CARBON CAPTURE AND STORAGE

Although decarbonization technologies will reduce GHG emissions, they will have different impacts on communities owing to varying environmental, health, and quality of life benefits and harms. CCS will not likely be a first choice for emissions reduction from emitting processes because other low-emission generating resources are lower cost and avoid both GHG emissions and local air pollution. However, there are instances where the net benefits mean that CCS is the best option to reduce facility emissions. For example, CCS can be used to mitigate or offset the emissions of industrial processes with limited non-emitting options, such as the production of cement or plastics. The examination of CCS illustrates both the direct and indirect environmental and energy justice impacts that have to be addressed for decarbonization technologies to contribute to a just energy transition. CCS is discussed in detail in Chapter 10.

CCS is a technological approach to carbon management that collects and often concentrates CO_2 from waste gas streams of combustion or other industrial processes or from the atmosphere or oceans. Captured CO_2 is then stored so that it cannot accumulate in the atmosphere and lead to climate change. CCS technologies may be well suited to mitigating GHGs and other pollutants. A recent study by the Energy Futures Initiative, citing work by the Clean Air Task Force, estimates a 10–96 percent reduction in local air pollutants associated with the gas pretreatment and capture processes required for operation of a CCS project applied to a cement plant (Brown et al. 2023). A study of the technical requirements for, and the costs and benefits of capturing health-harming co-pollutants along with CO_2 capture showed that for all industries examined, there are positive health benefits (Bennett et al. 2023). CCS technologies that combust fuels in pure oxygen, such as oxy-fuel combustion facilities, also reduce NO_x and SO_x emissions (NETL 2023). The reduction of local air pollutants from fossil fuel combustion is discussed in detail in Chapter 3.

Key benefits of CCS facilities include systemic benefits such as GHG mitigation, and the ability of power plants to operate on a dispatchable basis and support around-the-clock generation, as well as community benefits such as workforce development, tax and other community revenues, and for many capture technologies implemented on flue gas streams, significant reductions in local health-harming air pollutant emissions. However, CCS processes can have direct and indirect environmental risks, including potentially polluting local air and water sources from CO_2 capture, transportation, or storage processes. Additionally, CCS technologies also pose the risk of displacing an alternative technology that produces no emissions, such as power generation from solar or wind.

Environmental justice groups have expressed concerns about the potential of CCS technologies and infrastructure, including direct air capture, to perpetuate harms to disadvantaged communities (e.g., see Amsalem and Bogdan Tejeda 2022; Chemnick 2023; Climate Justice Alliance 2023; Natter 2023). As investments continue to be made, CCS technologies will need to enhance benefits and reduce or mitigate harms to equitably serve affected communities, including potentially producing value for the community through community ownership of facilities and workforce development opportunities. Additionally, transparency is needed about the benefits, harms, and trade-offs during project planning, development, siting, and permitting, as well as operation and closure, especially when impacting disadvantaged communities (CEQ 2022). Both community engagement and transparency adhere to the procedural equity principle. Furthermore, the incorporation of energy justice principles into CCS implementation will increase the opportunity for distributional equity by ensuring that benefits will be experienced by host communities. Chapter 5 discusses the role of public engagement for host communities to determine expected benefits and risks.

and logging on public lands and IRA facilitates the potential continued investment in oil, methane-derived hydrogen, and biofuel as energy sources. There are concerns that these investments do not prioritize GHG emissions mitigation, which could lead to increases in pollutants and hazardous waste in communities already suffering the greatest impacts of fossil fuel combustion. For more information about the IIJA and IRA, see the section "Assessment of Recent Federal Actions."

Furthermore, environmental justice groups are wary that federal agencies may be slow to identify disadvantaged communities' priorities and needs, which will slow the delivery of equitable benefits (Walls 2022). Organizations have also focused on equitable implementation, including the need for local, state, and community capacity-building to support disadvantaged communities and community-based organizations as they apply for funding and access technical assistance (Walker et al. 2022). This could be done through the incorporation of procedural justice principles into the monitoring of funding allocation and recognitional justice provisions to include affected communities in program design. Effective implementation can be further achieved through the increase in interagency coordination. See the section "Building Community Capacity to Develop Community-Driven Programs" for more information about coordination.

> Finding 2-1: It is critical that the energy transition to a net-zero future be just, minimizing harm and fostering equity across all populations, regions, and economies of the country. To achieve this, stakeholders, especially disadvantaged community members, need to be engaged when defining where, when, and how to prioritize federal and state resources and investments during the energy transition. Inclusive and equitable approaches, moreover, are key to preempting or minimizing the potential for implementation challenges or the derailment of decarbonization projects altogether.

FEDERAL ACTIONS, GAPS, AND RECOMMENDED SOLUTIONS

The committee's first report included findings and recommendations on the need to advance decarbonization in 2021–2030 (NASEM 2021). Recommendations focused on an equitable and just energy transition include the following:

- The establishment of a 2-year National Transition Taskforce to assess vulnerability of labor sectors and communities to the transition;
- The establishment of a White House–level Office of Equitable Energy Transitions;
- The establishment of an independent National Transition Corporation to ensure coordination and funding in the areas of job losses, critical infrastructure, and equitable access to economic opportunities and wealth;

- The creation of public energy equity metrics;
- The establishment of educational and training programs to train the net-zero workforce, with reporting on diversity of participants and job placement success; and
- The increase of research, development, and deployment in clean energy and net-zero transitions that integrates equity metrics.

The committee continues to find these recommendations relevant, even considering the new legislative and executive actions. This section will review and assess what has been done on energy justice and equity at the federal level since the release of the first report and will propose solutions for identified gaps.

Assessment of Recent Federal Actions

Significant steps have been taken by the federal government to support the nation's decarbonization agenda. By going beyond a narrow focus on GHG emissions mitigation to include quality jobs, public health, and environmental justice, these efforts open unique opportunities to leverage synergies and intersections between the Sustainable Development Goals of the United Nations[9] and the advancement of U.S. climate ambitions. The recent federal actions to reduce GHGs present multiple opportunities for equity and justice co-benefits with goals to reduce both fossil fuel use and air pollution from combustion. See Chapter 3 for a detailed discussion of the co-benefits, positive additional health impacts of decarbonization policies, of climate mitigation policies. However, continued conscious and targeted efforts are needed to move away from past and current inequitable social structures and constraints.

Executive Orders

Executive Order 13985

Executive Order (EO) 13985—Advancing Racial Equity and Support for Underserved Communities Through the Federal Government—was signed to develop a whole-of-government equity agenda that requires federal agencies to assess whether and to what extent their programs and policies target barriers to opportunities and

---

[9] The 2030 Agenda for Sustainable Development was adopted by members of the United Nations in 2015. The document details 17 Sustainable Development Goals (SDGs) that recognize that climate change strategies that improve health, reduce inequalities, and foster economic growth must also end poverty and other deprivations (UN n.d.). For implementation progress reports for the SDGs, visit https://sdgs.un.org/goals.

benefits for underserved communities[10] (EO 13985 2021). EO 13985 defines the role of the White House Domestic Policy Council as coordinating the formulation and implementation of domestic policy objectives and coordinating efforts to embed equity principles, policies, and approaches across the federal government. Reflective of the recognitional and procedural equity principles, EO 13985 requires the heads of federal agencies to evaluate opportunities to increase coordination, communication, and engagement with community-based organizations, including through a participatory process with members of historically underrepresented and underserved communities (EO 13985 2021). It also establishes the Interagency Working Group on Equitable Data (Equitable Data Working Group) to consult with agencies and provide recommendations on inadequacies in existing federal data collection programs, policies, and infrastructure. The work done by the Equitable Data Working Group will be critical for a national assessment of the progress made toward a just energy transition. See the section "Evaluation of the Just Energy Transition" below for more detail.

Executive Order 14008

Adding to the whole-of-government equity agenda, EO 14008—Tackling the Climate Crisis at Home and Abroad—created a comprehensive approach to addressing environmental justice concerns by establishing the following groups (EO 14008 2021):

- *National Climate Task Force* to "facilitate planning and implementation of key Federal actions to reduce climate pollution; increase resilience to the impacts of climate change; protect public health; conserve our lands, water, oceans, and biodiversity; deliver environmental justice; and spur well-paying union jobs and economic growth."
- *Interagency Working Group on Coal and Power Plant Communities and Economic Revitalization* to "coordinate the identification and delivery of Federal resources to revitalize the economies of coal, oil and gas, and power plant communities."
- *White House Environmental Justice Interagency Council* to "develop clear performance metrics to ensure accountability and publish an annual public performance scorecard on its implementation."
- *White House Environmental Justice Advisory Council (WHEJAC)* to provide recommendations to the White House Environmental Justice Interagency Council "on how to increase the Federal Government's efforts to address current and historic environmental injustice."

---

[10] Within the language of the EO, the term "underserved communities" refers to populations sharing a particular characteristic, as well as geographic communities, that have been systematically denied a full opportunity to participate in aspects of economic, social, and civic life.

The establishment of each of the above groups supports the application of the recognitional and distributional equity principles to federal legislation. For instance, each group provides the opportunity for the benefits and burdens of climate policy to be identified during the development of legislative actions and after implementation. Furthermore, all groups recognize the disparity of climate change impacts, and their goals focus on certain groups of communities impacted most by climate change.[11]

As mandated by EO 14008, the Chair of CEQ, the Director of the Office of Management and Budget (OMB), and the National Climate Advisor published recommendations in consultation with the WHEJAC and affected disadvantaged communities "on how certain federal investments might be made with a goal that 40 percent of benefits flow to disadvantaged communities" (EO 14008 2021). The published recommendations included implementation guidance for the Justice40 Initiative and its related covered programs that are federal programs investing in one or more of seven areas: climate change, clean energy and energy efficiency, clean transit, affordable and sustainable housing, training and workforce development, remediation and reduction of legacy pollution, and the development of critical clean water and wastewater infrastructure (Young et al. 2021). The procedural equity principle is evident in the Justice40 Initiative through the requirement for guidelines to be designed through the consultation of affected communities and requirement for covered programs to engage in stakeholder engagement.

Executive Order 14091

EO 14091—Further Advancing Racial Equity and Support for Underserved Communities Through the Federal Government—details the progress made to advance equity and what remains to be done (EO 14091 2023). It recognizes the strides made to incorporate equity into the federal actions and acknowledges that some communities, especially underserved and rural communities, are still facing barriers accessing and benefiting from federal programs and policies. To fill the gap, EO 14091 requires federal agencies to submit annual Equity Action Plans[12] enumerating

---

[11] As part of EO 14008, OMB was directed to publish an annual Environmental Justice Scorecard (EJ Scorecard) (EO 14008 2021). Phase One of the EJ Scorecard, released in early 2023, provides a baseline assessment of federal action in 2021 and 2022 that support the nation's environmental justice goals, including progress toward the Justice40 Initiative. To view the EJ Scorecard and its initial baseline report on 24 federal agencies, see https://ejscorecard.geoplatform.gov/scorecard.

[12] The Urban Institute has reviewed the 2022 equity action plans of 24 federal agencies published in response to EO 13985 and compiled a collection of analyses and recommendation to support the 2023 equity plans mandated by EO 14091. For each agency reviewed, there is a two-page summary of the principles, pillars, and metrics for equity that are included in the equity action plan. To view these summaries, see Urban Institute (n.d.).

implementation barriers to equitable policy outcomes and providing strategies to address barriers that advance equity through evidence-based approaches and reduce administrative burdens (EO 14091 2023). The order also focuses on opportunities to strengthen partnerships with underserved communities and to help rural communities identify resources that build community wealth. Last, EO 14091 requires the Director of the Office of Science and Technology Policy (OSTP) to report on the progress of federal equitable data practices. The continued focus on including equity frameworks into federal action will support the equitable implementation of recent and future net-zero policies and programs. Additionally, the availability of data regarding federal policies will support an evaluation of the energy transition (see the section "Evaluation of the Just Energy Transition" below).

Executive Order 14096

EO 14096—Revitalizing Our Nation's Commitment to Environmental Justice for All—identifies environmental justice as something the federal government is responsible for ensuring, stressing the right for every person to have "clean air to breathe ... and an environment that is healthy and sustainable, climate-resilient" (EO 14096 2023). The order attempts to establish a government-wide approach to environmental justice by requiring federal agencies to:

- Identify and address disproportionate, adverse health and environmental effects of federal activities, including the cumulative impacts of burdens on communities;
- Identify and address barriers that impair the ability of communities to receive equitable access to benefits, including those related to climate mitigation and resilience;
- Consider adopting or requiring measures to avoid or mitigate adverse environmental and health effects of federal activities on communities; and
- Provide opportunities for community engagement, including by fully considering input provided during decision-making processes and providing technical assistance.

The act also requires agencies to carry out such reviews as part of the National Environmental Policy Act (NEPA). See Chapter 5 for more information about NEPA and its impact on engaging the public in the energy transition. Federal agencies are mandated to submit to CEQ and make public an Environmental Justice Strategic Plan detailing priority actions and metrics to address environmental justice every 4 years (EO 14096 2023). This approach toward environmental justice and the requirement for agencies to make public their approach and metrics can be applied to the energy justice as the nation moves forward with the transition to a net-zero energy system.

Limitations of Programs Initiated Through Executive Orders

Recent executive actions present an innovative approach to addressing equity and justice concerns, especially through the Justice40 Initiative. This approach has the potential to be supported by the equity assessment focus of EO 13985 and future whole-of-government actions focused on energy justice can be modeled after the requirements for federal agencies in EO 14082. However, unlike legislation, executive orders only govern the conduct of the federal executive branch, including the federal agencies, and can be repealed by future administrations. Specifically, the Justice40 Initiative and other equity-focused initiatives run the risk of being overturned or ignored following a change in presidential administration. The codification of a durable program allows for the outcome of the program to be evaluated and modified to better meet the target.

> Finding 2-2: EO 14008 presents an innovative approach to addressing environmental justice concerns by requiring that 40 percent of the benefits from covered programs go to disadvantaged communities. Because actions put in place by executive orders are not enacted through statute, there is a risk that a change in presidential administrations will result in these requirements being ignored. Specifically, without a more durable legislative mandate for the Justice40 Initiative or an alternative quantitative target, federal agencies may not honor the policy that programs be implemented in ways that directly benefit disadvantaged communities.

> **Recommendation 2-1:** *Codify the Justice40 Initiative.* **Congress should enact legislation that codifies either the Justice40 Initiative or an alternative, equally stringent quantitative target to provide a clear standard that the entire federal government will use to measure progress against fairness, equity, and justice goals. Federal legislation should also require the collection of and reporting on standardized metrics for measuring and evaluating direct benefits and negative impacts on jobs, public health, energy affordability, and access to technologies for disadvantaged and frontline communities.**

Legislation

Infrastructure Investment and Jobs Act

The Infrastructure Investment and Jobs Act (IIJA),[13] also commonly known as the Bipartisan Infrastructure Deal, is a $1.2 trillion investment in the nation's roads, bridges and rails, and targeted investments to advance environmental justice, tackle the climate crisis, and support community resilience (Tomer et al. 2021; White House 2021). The IIJA's environmental and climate justice appropriations include $39 billion to modernize public transit (§11130, §11133, §11206, §11403); $21 billion to the environmental remediation of brownfield sites through the Superfund program (§80201); and a total of $64.41 billion for broadband infrastructure (§60201, §60401), access (§60102, §60304, §60305, §60105), and affordability (§60502). Additionally, the IIJA creates the Reconnecting Communities Pilot Program to fund the design and planning of transportation infrastructure, and the demolition and reconstruction of infrastructure that had divided communities (§11509) and the Clean School Bus Program to transition existing school buses to clean and zero-emission school buses (§71101). It is estimated that $240 billion of the total appropriations will address environmental justice priorities (White House 2021).

While the appropriations and authorizations of the IIJA do not specifically advance energy justice, the intentional focus on investing in communities and ensuring effective implementation indirectly advance energy justice by aligning with the distributional equity principle. For example, EO 14052—Implementation of the Infrastructure Investment and Jobs Act—requires agencies to equitably invest IIJA appropriations, including through the Justice40 Initiative (EO 14052 2021). To adhere to this, federal agencies released their estimates for Justice40 compliance; for example, EPA stated that more than 40 percent of its IIJA appropriations supported underserved communities by 2022 (EPA 2022b). If implemented appropriately, several IIJA provisions will lay the foundation for a just energy transition.

---

[13] It should be noted that the IIJA and IRA are not equivalent in funding mechanisms. The IIJA consists of a mix of authorizations and appropriations while the IRA primarily consists of spending programs (appropriations) and tax expenditures. Appropriations are laws that provide money for government programs and must be passed by Congress every year in order for the government to continue to operate. Spending programs can allocate federal resources to projects and activities up to the amount of their appropriation. By contrast, tax expenditures, such as the production tax credits in the IRA, typically have no limit on the amount that could be claimed by taxpayers.

Inflation Reduction Act

The IRA includes large investments in clean energy technologies that will reduce the use of fossil fuels, lower energy costs for families, create good-paying jobs, and tackle the climate crisis (White House 2022a). The IRA directs nearly $400 billion in appropriations and authorizations to clean energy, including to reduce carbon emissions and support environmental justice objectives (Elliot et al. 2022). Chi (2022) estimates $40 billion in appropriated funding will have direct and indirect impacts on disadvantaged communities while supporters of the IRA claim that $60 billion in appropriations will go to environmental justice priorities (Walls 2022), but there is acknowledged ambiguity in the IRA's tax credit provisions and a lack of clarity about what a direct benefit is which make calculations difficult to agree upon. See Appendix F for the committee's evaluation of the impacts, direct and indirect, that the IRA's appropriated funds and tax expenditures will have on underserved, low-income, and disadvantaged communities.

IRA spending programs that support fossil fuel reduction present multiple opportunities for equity and justice co-benefits. For example, the act includes appropriations to improve CEQ's stakeholder and community engagement (§60402) and climate resilience investments in Indigenous communities (§80001). The IRA also creates a Greenhouse Gas Reduction Fund to issue grants to state, local, regional, and tribal governments, and to non-governmental organizations that provide financial or technical support enabling under-resourced and disadvantaged communities to benefit from or deploy zero-emissions technologies (§60103). To support the implementation of specific provisions, EO 14082—Implementation of the Energy and Infrastructure Provisions of the Inflation Reduction Act of 2022—was signed to prioritize the implementation of IRA provisions that make progress toward reducing national GHG emissions and achieve a carbon-free electricity sector by 2035; advance environmental and climate justice; increase and improve equitable access to high-quality job opportunities; reduce energy costs while increasing energy security; and coordinate with non-federal and private-sector stakeholders to build sustainable and resilient communities (EO 14082 2022).

Because the IRA statutory language was constrained by the budget reconciliation process, federal agencies are left responsible for the identification of key languages, processes, and requirements for awarding grants that would aid communities during the transition. For example, the IRA delegates to federal implementers the responsibility of defining "disadvantaged community" and "energy community." As an example, since enactment of the IRA, the term "energy community" has been defined in at least two ways: one included in the statutory language in the IRA (U.S. Congress 2022) and one proposed by Resources for the Future that takes a scaled approach for some factors to be more inclusive of different communities (Raimi and Pesek 2022). Additionally, the IRA leaves wide discretion to states to decide how some programs are designed, implemented, and assessed. See Appendix G for federal definitions of disadvantaged communities. See Chapter 13 for more discussion of the role of non-federal entities in the implementation of the IRA provisions.

As mentioned earlier in the section "Stakeholders' Environmental Justice Concerns," there are concerns that federal agencies may be slow to identify the priorities of communities in need of decarbonization action. This concern is heightened by the multiplicity of ways to identify these communities. For example, EO 14008 tasked CEQ with the creation of the Climate and Economic Justice Screening Tool (CEJST) to help federal agencies identify disadvantaged communities[14] (EO 14008 2021). However, many federal agencies with Justice40 covered programs had already developed their own screening tools by the time CEJST exited its beta phase. Appendix G summarizes the federal agencies with covered programs and compares their selected indicators with the ones released by CEQ in November 2022. Other federal agencies determined to have Justice40 covered programs without public or comprehensive definitions of disadvantaged communities include the Department of Homeland Security, the Department of Health and Human Services, HUD, and the Department of Agriculture. Although there exists some overlap between the communities identified by various screening tools, it would be beneficial to the implementation and evaluation of outcomes from federal transition actions if there were core identifiers, used by all federal agencies as a foundation on which they can build for a program-specific definition of disadvantaged community.

> Finding 2-3: The ability to define and identify disadvantaged communities is essential to measure the direct impacts of federal policies and programs on disadvantaged communities. However, the evaluation of the federal decarbonization action, especially the Justice40 Initiative, is constrained by the lack of a robust definition of disadvantaged communities and centralized screening tools to map these communities. The committee recognizes that there cannot be a single definition for disadvantaged communities that applies to all programs because federal and state programs may have different target populations and related burden indicators for target population identification. However, the multiplicity of non-compatible definitions, including the one used for the Climate and Economic Justice Screening Tool, makes it difficult to accurately measure

---

[14] As of March 2023, CEQ's Climate and Economic Justice Screening Tool Version 1.0 is one of more than 30 environmental justice screening tools across federal, state, and local agencies (Dean and Esling 2023). For more information about how these screening tools intersect, see the Environmental Policy Innovation Center's EJ Tools Map at https://epic-tech.shinyapps.io/ej-tools-beta.

the impacts of federal actions, specifically on disadvantaged communities, across programs and agencies. It is important to advance a set of core metrics to identify disadvantaged communities to use in program design and evaluation, and decision analysis.

Recommendation 2-2: *Develop a Federal Baseline Set of Metrics for Disadvantaged Communities for Program Design and Evaluation.* **To enable consistent program design and evaluation, the White House Council on Environmental Quality should develop a standardized set of core metrics for programs serving disadvantaged communities to be used in all federal activities, to the extent feasible under statutes governing each agency. The use of the core metrics would be required, and agencies would be encouraged to select additional context-specific metrics to match program needs. For federal programs that engage with states and localities with existing disadvantaged community metrics, the program design should include a rationale for why the state or local designations could be used in place of those recommended by the Council on Environmental Quality.**

Implementation of Federal Decarbonization Policies

To avoid creating new or worsening existing burdens faced by disadvantaged communities, future federal actions need to make a concerted effort to equitably design and implement climate-related and decarbonization policy. Specifically, that it is critical policy makers include the equity principles when designing processes for the implementation of these polices. Recent executive-level actions—that is, EO 14052, EO 14082, and EO 14091—attempt to address the concerns of equitable design and implementation in federal policymaking in general. Additionally, although funded programs and services within the 2022 American Rescue Plan (ARP) Act (P.L. 117-2 2021) were designed to counteract the effects of COVID-19, the White House ARP Implementation Team developed the below equitable implementation actions that can be applied to all federal legislation (White House 2022a):

- Establish program goals and measurable targets, and track program progress against these goals and targets.
- Foster awareness and capacity to access programs and services, particularly among underserved individuals and communities.
- Allocate and leverage resources and funding and design tools to spend resources equitably.
- Collect and analyze sufficient data to determine whether and how disparities change across key outcomes and impact measures.

- Create feedback mechanisms for regular internal review, including soliciting feedback from underserved individuals and communities for continuous improvements.
- Build on evidence to advance equity in program design and implementation.
- Ensure that data are collected and strategies are evaluated to adapt and improve programs.

Given that all decarbonization approaches have benefits and harms, policies need to be designed and implemented through inclusive and ongoing engagement such that they deliver those benefits and harms equitably and justly. Additionally, owing to the impermanence of executive orders, it will be critical for actions to support equitable implementation of federal actions to be made consistent and permanent.

Finding 2-4: The decarbonization provisions in the Infrastructure Investment and Jobs Act and Inflation Reduction Act are likely to slow the U.S. emissions that contribute to climate change, primarily through reduced fossil fuel combustion, with local air quality benefits that are likely to have positive equity and justice impacts. Many provisions are also specifically directed at equity and justice, including block grants, subsidies to improve technology uptake, and top-up funding for projects in disadvantaged communities. However, some provisions, such as those for offshore drilling and carbon capture and storage, do not align with environmental justice goals, including undermining climate mitigation goals and creating or continuing pollution that threatens public health and quality of life for disadvantaged communities.

Recommendation 2-3: *Implement Federal Legislation for Equitable Outcomes.* Federal policy makers should include equity principles in the design and implementation of decarbonization policies. Specifically, policy makers should review the implementation actions as developed by the White House American Rescue Plan Implementation Team and apply the actions to existing and future policies. Federal agencies should engage communities most impacted by energy inequities as key stakeholders to ensure that the voices of affected communities are meaningfully heard and develop policies and programs that are informed by and responsive to concerns raised.

ENSURING EQUITABLE ACCESS TO AND OUTCOMES OF THE JUST ENERGY TRANSITION

At the early stages of the implementation of decarbonization policy, it is critical that federal, state, and local authorities develop monitoring and evaluation mechanisms aimed to identify equity gaps in the development and implementation of acts, orders, and other relevant action; outline areas for improvement; and hold federal, state, and local authorities accountable for implementing these activities. These requirements challenge federal, state, and local authorities to create sufficient institutional and financial capacity to implement programs and to develop a comprehensive assessment of the overall equity potential of these bills. This section highlights the barriers and solutions for community-led transition actions. It includes the bottom-up incorporation of energy justice through transformation projects and the top-down coordination of transition resources. The section concludes with the critical role evaluation and adaptive management plays in ensuring the energy transition is just.

Building Community Capacity to Develop Community-Led Transitions

Strategies that prioritize place-based interventions can reduce disparities faster than sector-based decarbonization strategies (Wang et al. 2022). However, attempting federal-led and community-led decarbonization actions simultaneously will increase the success of the energy transition. To support all decarbonization action, human and fiscal capital must be available. In its first report, the committee recommended that Congress support actions to overcome barriers created by a lack of capacity-building including funding research to support the regional coordination of the transition; establishing equitable energy transition offices in each state; and funding community block grants for local decarbonization planning (NASEM 2021). Few congressional actions have directly focused on supporting community capacity in climate change mitigation except for the IRA (e.g., the Environmental and Climate Justice Block Grants [§60201]). However, many state and federal initiatives have been launched to advance capacity building and provide support to communities in need. There is also a role for nongovernment entities, to support holistic community transition programs.

State Initiatives

Despite the challenges produced by the complex political landscape of the United States, states and cities that have taken climate-mitigation actions represent two-thirds of the nation's population and economy (Zhao et al. 2022). In fact, many states have adopted holistic, community-driven approaches to develop and implement transition solutions. For example, Louisiana's Strategic Adaptations for Future Environments (LA SAFE) program integrates risk-mitigation planning for stormwater management, housing, transportation, economic development, education, recreation, and culture for holistic community resilience solutions (LA SAFE 2019). New York State passed the Climate Leadership and Community Protection Act supporting an equitable and inclusive transition focused on distributing no less than 35 percent of clean energy benefits of spending to disadvantaged communities (S.B. No. S6599 2019). From this legislation, the Cap-and-Invest Program was established to "apply a price to the amount of pollution" with proceeds supporting critical investments in "climate mitigation, energy efficiency, clean transportation, and other projects" that ensure that the program is affordable for all state citizens and delivers benefits to disadvantaged communities (Cap-and-Invest n.d.). See Chapter 13 for more about incentivizing state action to support climate-mitigation at local and regional levels.

Drawing lessons and best practices from existing programs and adapting them to different state and regional contexts could prove beneficial for the nation's energy transition. For example, California's Transformative Climate Communities Program (TCC) invests in community-led transformation by funding development and infrastructure projects that have multiple environmental, health, and socioeconomic benefits. Box 2-3 describes the key features of TCC, which include targeting disadvantaged communities for the funding of transformation projects and requiring the evaluation of funded project through their completion. Although not all states have similar financial structures and may not have the same climate priorities as California, TCC is a possible approach other states can adopt to their own energy transformation.

Even with the development of innovative community programs and competitive funding, some communities may still struggle to access available opportunities owing to unclear or overburdensome application requirements, lack of time to apply for funding, or unawareness of funding opportunities available. For example, the administrative burden of applying for funding—the time and cognitive load required to complete forms and acquire, collect, and submit supporting documents—often prevents participation from under-resourced communities in decarbonization programs or increases the occurrence of temporary solutions that may not sufficiently address the community's priorities (NASEM 2023b). Barriers to the access of funding and programs need to be sufficiently considered during the development and implementation of new transition programs for equitable access and outcomes.

A human-centered, bottom-up approach that considers and appropriately incorporates community concerns and priorities would be reflective of the procedural equity principle (NASEM 2023b). However, as mentioned in Box 2-3, two key challenges of developing community-driven programs and multi-stakeholder partnerships are overcoming feelings of mistrust between historically underserved communities and different levels of government and securing sufficient and continued program funding. Political polarization and feelings of disrespect drive distrust and disengagement, but these can be overcome with pragmatic understanding of how people define their

BOX 2-3
CALIFORNIA'S TRANSFORMATIVE CLIMATE COMMUNITIES PROGRAM

The Transformative Climate Communities Program (TCC) empowers communities most impacted by climate change to choose their own goals, strategies, and projects to reduce greenhouse gas emissions and local air pollution. Funded by the state's Greenhouse Gas Reduction Fund, TCC is directed by California Assembly Bill 2722 to make at least 35 percent of climate change investments in the state's disadvantaged communities, low-income communities, and low-income households (A.B. No. 2722; California Climate Investments n.d.). All project areas, which are no more than 5 square miles, must include census tracts that are within the top 25 percent of disadvantaged communities (SGC 2023).

The process is largely community-led with continued support from the California Strategic Growth Council (SGC) which guides applicants in their selection of recommended strategies and development of a proposal for at least three projects that address the SGC objectives: (1) reduce greenhouse gas emissions; (2) improve public health and environmental benefits; and (3) expand economic opportunity and shared prosperity (SGC 2021). Once a project is approved, implementing TCC facilitates the development of relationships between project areas, implementers, and the SGC that incorporates trust. In such a multi-faceted process, there "has to be continuing transparency and accountability" between all partners involved so that the history of a community cannot be ignored in the development and implementation of community-level improvement projects (Saunders 2023). Table 2-3-1 highlights Transform Fresno and the anticipated benefits of the selected projects.

TABLE 2-3-1 Transform Fresno Funded Projects and Anticipated Outputs, Outcomes, and Impacts

Transformative Climate Communities Program (TCC) Funded Projects	Anticipated Outputs	Anticipated Outcomes and Impacts
• Active Transportation • Affordable Housing and Sustainable Communities • Food Waste Prevention and Rescue • Low Carbon Transportation • Rooftop Solar and Energy Efficiency • Urban and Community Forestry • Urban Greening	• 57 new housing units • 42 new battery-electric vehicles for a car-sharing network • 1,458 new street trees • 784 kW of solar power on affordable multi-family and single-family homes • 200 TCC area individuals trained for residential solar installation projects	• 20,816 metric tons of avoided GHG emissions[a] • 14,832,662 miles of averted travel in passenger miles • $4,826,413 in energy cost savings for solar PV and street tree beneficiaries • 337 direct jobs, 112 indirect jobs, and 190 induced jobs supported by TCC funding

[a] Measured in CO_2e.

SOURCE: Committee generated from Luskin Center for Innovation (2022).

continued

Each grant recipient is required to designate a third-party Evaluation Partner to conduct an analysis of the process, outcome, and impact of selected strategies by gathering both quantitative data and qualitative feedback (SCG 2021, 2023) which is then communicated to stakeholders and policy makers. Evaluation reports have shown that once trust is established, the speed of progress on implementing projects, including those not funded through TCC, significantly increases. Additionally, implementation challenges identified by evaluation include how to secure continued funding for the projects and how to overcome a community's mistrust of the local government (NASEM 2023). The availability of evaluation results supports efforts to improve the design and implementation of current and future TCC supported projects (Luskin Center for Innovation 2020). TCC offers a blueprint for climate investments that help redress historic injustices through stakeholder and community decision-making[a] in all aspects of the program design and implementation "to ensure grant funds provide direct, meaningful, and assured benefits to disadvantaged communities" (SGC 2023, p. 49). Additionally, the publicly available evaluation of project implementation allows the best practices of TCC to be reviewed and outcomes makes TCC good example for other states to review and potentially apply lessons learned to their own holistic community transformation programs.

---

[a] The Strategic Growth Council provides a list of proven engagement methods that facilitate direct engagement and participation from community residents (see SCG 2023, Appendix C).

---

problems and priorities while creating regional solutions that are designed to be relevant to local communities (Beckfield 2022). Creating institutions and programs with the human capacity to provide a space to give the community a voice and staff members to be able to listen and address concerns will be critical during the transition as community-focused mitigation programs are implemented. Furthermore, implementing programs with the appropriate amount of available financial and human resources will support the development of trust among stakeholders and increase opportunities for equitable outcomes. See Chapter 5 for more details on how to create and expand human capacity to listen to community concerns.

Federal Initiatives

The Federal Interagency Thriving Communities Network was developed to coordinate the planning, implementation, and technical support of initiatives funded and created by the IIJA, IRA, and ARP (DOT n.d.). It offers the opportunity for disadvantaged communities to access place-based technical assistance and capacity-building resources from a variety of federal agencies including DOE, the Department of Transportation

(DOT), and HUD. As part of the Federal Interagency Thriving Communities Network, the joint EPA-DOE Environmental Justice Thriving Communities Technical Assistance Center (TCTAC) Program works in coordination with DOT's Thriving Communities Program to provide technical assistance for transformative projects and capacity-building to under-resourced and disadvantaged communities (DOT 2022; EPA 2023). The development of a single center offering both the EPA and DOT technical assistance programs will streamline access to training, assistance, and capacity building for underserved and disadvantaged community members.[15]

Federal support of community capacity-building and engagement is further evidenced by DOE's Inclusive Energy Innovation Prize, which incentivizes and rewards community-based pathways within disadvantaged communities with funding to implement and evaluate proposed plans (DOE 2023a). Additionally, DOE's Clean Energy to Communities (C2C) Program provides communities with tailored assistance through three levels of engagement: (1) in-depth technical partnerships that provide "cross-sector modeling, analysis, and validation" and direct funding to help "teams of local governments, electric utilities, and community-based organization reach their goals and/or overcome specific challenges" through multi-year collaborations; (2) peer-learning cohorts that organize regular meetings for "small groups of local governments, electric utilities, or community-based organizations" to develop a collaborative environment "to develop program proposals, action plans, strategies, and/or best practices on a predetermined clean energy topic"; and (3) short-term assistance through which technical experts are matched with communities "to help address near-term clean energy questions or challenges" (DOE-EERE n.d.).

Non-Governmental Actors

Non-governmental actors can influence transition policy and mitigation directly by funding community and state transformation initiatives that produce scalable and replicable solutions (Hale 2016) and indirectly by supporting research and dissemination of best practices that build capacity and catalyze supportive political coalitions (Chan et al. 2015). The committee's first report acknowledged the critical role for nongovernment organizations, stating that they were key to mobilizing public support and instrumental to closing the funding gap for organizations supporting

---

[15] In April 2023, EPA announced that 17 new TCTAC hosts that would receive at least $10 million each to remove barriers to accessibility for underserved and under-resourced communities during the energy transition. The new partners include national organizations that have the capacity to assist tribes during the transition. For the complete list of the new regional and national TCTACs, see https://www.epa.gov/newsreleases/biden-harris-administration-announces-177-million-17-new-technical-assistance-centers.

communities in addressing climate change (NASEM 2021).[16] These organizations include colleges and universities, philanthropic foundations, and nonprofit organizations, as well as state, local, and tribal governments. They are essential to the energy transition because they can address areas where there is no market solution and focus on promoting equity and energy justice (Lewis 2022).

The philanthropic sector can help ensure that decarbonization policies are developed and implemented justly in addition to providing financial support for projects and programs. Through the funding of projects, activities, and initiatives, foundations can shape transition action and the development of best practices that focus on the theory of change, a description of how or why a desired change will happen in a specific context that is used as a framework for project planning, implementation, and evaluation. Additionally, philanthropic organizations can be partners to society, government, and the private sector to accelerate the transition (DeBacker 2022) by taking on a critical role of influencing who is invested in and how much funding is dedicated during the transition (Beckman 2022). However, since project funding mostly goes to regions supportive of climate-mitigation strategies, the effectiveness of philanthropic funding to motivate holistic change across the nation is lessened.

An additional barrier is the transparency of climate funding allocation by non-governmental actors. Increased transparency for funding trends may increase the success of the just energy transition by making foundations appear more trustworthy to community organizations looking for funding. To better align with the principles of energy justice, funding needs to be more equally distributed to ensure communities most impacted by energy injustices to be involved in the development of local solutions. To address this challenge, the Donors of Color Network announced a Climate Funders Justice Pledge in 2021, which has participating foundations commit to give at least 30 percent of their funding to groups that are centered on racial and economic climate justice, similar to the Justice40 Initiative, with an optional commitment to transparency about where their funding goes (Donors of Color Network 2021). At the time of this writing, some of the nation's biggest funders have not committed to being transparent about the percent of their dollars going

---

[16] Following the release of the committee's first report, the House Committee on Natural Resources held an oversight hearing to discuss how to provide communities targeted by the environmental justice movement with a voice to speak out against and support federal policies (Committee on Natural Resources 2022). During this hearing, the first report's finding that decarbonization cannot be achieved without inclusive policy was quoted to support the argument that justice, equity, diversity, and inclusion should be central to federal efforts.

to environmental justice organizations.[17] Furthermore, Taylor and Blondell (2023) found that most of the funding from foundations goes to large environmental organizations with a small fraction going to environmental justice organizations and organizations with less than $1 million in annual revenues.

In addition to providing support for community capacity building, a critical role for all non-federal actors will be to identify and communicate the areas of need during the energy transition. For example, these actors can contribute to strategizing about where federal-level interagency coordination is needed the most with the acknowledgement that some communities have already begun incorporating procedural justice into their energy transition or have already developed initiatives that have seen equitable outcomes for the transition to clean energy. Any additional support needed by these transitioning and transforming communities must not come at the expense of those who have not yet started energy transition activities. See Chapter 5 for more information about how these actors can support meaningful public engagement in decarbonization action and processes.

Multi-Level Coordination to Support the Energy Transition

As the committee emphasized in its first report, the lack of meaningful federal coordination of transition processes with local- and regional-scale institutions will impede efforts to address the needs and concerns of disadvantaged communities as decarbonization programs expand in pace and scope (NASEM 2021). The current federal policy encourages regional and local planning that integrates energy into community-based and holistic approaches to address climate resilience, environmental justice, and economic opportunity in disadvantaged communities. However, increased multi-level coordination of existing programs would better support access to these programs and better ensure communities are not experiencing high administrative burdens to access funding. Additionally, stronger collaboration efforts between nongovernment and government actors are needed to ensure best practices and lessons learned are effectively communicated to increase the success of the energy transition. For example, as mentioned at the committee's *Pathways to a Just and Equitable Transition* workshop, the benefit of the sovereign structure of native communities is that they can demonstrate successful just transition laws and policies with sufficient and continued financial support from the federal government and philanthropic organizations (NASEM 2023b). See Chapter 5 for more information about the progress that tribal nations have made in the

---

[17] For more information about the Climate Funders Justice Pledge and a living list of foundations who have committed to the pledge, see https://www.climate.donorsofcolor.org/whos-pledged.

development of energy sovereignty and energy security and related recommendations about how tribal knowledge should be used in decarbonization policy.

To address the barrier of community capacity, the first report recommended the establishment of an independent National Transition Corporation (NTC) "to ensure coordination and funding in the areas of the areas of job losses, critical location infrastructure, and equitable access to economic opportunities and wealth creation" (NASEM 2021, p. 190). The proposed NTC features the capacity to build lasting, meaningful partnerships and trust with disadvantaged and transitioning communities through its mandate to provide technical assistance, relational and capacity-building, and financing functions. When paired with the joint EPA-DOE TCTAC Program or programs modeled after TCC, the NTC has the potential to fund programs that provide place-specific technical assistance and programs that encourage community-led transformation projects. For these transformation programs to be successful, it will be important for regional actors to coordinate with the NTC to identify and prioritize community needs that are both unique and overarching.

> Finding 2-5: The development of meaningful federal coordination of transition processes across local- and regional-scale institutions will support efforts to engage with and include disadvantaged communities as decarbonization programs expand in pace and scope. To this end, a National Transition Corporation model would be a federal complement to state and community initiatives during the transition. By necessity, state- and local-level efforts need access to a range of perspectives and resources; this includes needing support from many different parts of the federal government and coordination with private and civil society actors. Furthermore, it is critical for implementers to understand the priorities of under-resourced communities, or risk friction that could prevent community participation in decarbonization programs. This is best achieved through a coordinated effort that continuously communicates lessons learned and best practices.

Recommendation 2-4: *Build Multi-Level Capacity to Support Community-Led Transitions.* To enable a lasting and effective commitment to community-led solutions to energy transitions, Congress should:

   a. **Authorize a National Transition Corporation (NTC) to consolidate resources, finances, technical assistance, and strategy in an entity with experienced, multi-sectoral leadership. NTC would have the scope to allocate funds in modes (duration, amount, program design) better aligned with the aspirations, needs, and constraints of communities based on where they are in the energy transition.**

b. Adequately fund, continue, and expand the Thriving Communities Technical Assistance Center (TCTAC) Program model so that every community, especially under-resourced communities, can access robust, evidence-based, and culturally competent technical assistance necessary to develop effective mitigation investment plans. TCTACs will act as the connector between NTC and regional or local organizations that need support during the transition.

c. Create a program for states to facilitate holistic, community-driven mitigation projects in disadvantaged communities, called Climate Opportunity Zones (COZs). Modeled after California's Transformative Climate Communities program, COZs would foster multi-stakeholder partnerships, planning, and investments that reduce greenhouse emissions and demonstrate co-benefits for the economy, workforce, and health.

Finding 2-6: The current federal policy encourages regional and local planning that integrates energy into community-based and holistic approaches to address climate resilience, environmental justice, and economic opportunity in disadvantaged communities. However, increased inclusive and equitable approaches to technical assistance can better support under-resourced communities in transition and transformation programs. Inclusive and equitable approaches to technical assistance, capacity-building, and program development are key to preempting or minimizing the potential for implementation challenges or the derailment of projects altogether.

Recommendation 2-5: *Develop Equitable Technical Assistance Guidelines.* **The Federal Interagency Thriving Communities Network should work with state and local agencies to develop guidelines that make it easy to access and obtain technical assistance resources. These guidelines should be developed through an inclusive process that engages disadvantaged communities, stakeholders, and staff from local, state, and federal agencies. Furthermore, the White House Environmental Justice Advisory Committee should review and advise these guidelines to ensure that they adhere to an equity framework.**

Evaluation of the Just Energy Transition

The data collection of relevant metrics, evaluation of outcomes and progress, and communication of results are critical components of any effective policy action, especially when a new policy is first implemented. Evaluation of programmatic data is critical for adaptive management and for planning action during the transition's second 2 decades. Affected communities must be consulted during the design of the evaluative process for it to produce equitable measures that reflect and support their priorities.

There is a need to assess if energy transition actions are resulting in equitable and just outcomes for the nation, especially for disadvantaged communities. All principles of energy justice need to be operationalized in evaluation with a twofold goal of (1) determining if policies are equitable in their design, development, impacts, and outcomes; and (2) establishing the process to monitor and revise program design and implementation. Adaptive management, an iterative learning process producing improved understanding and management over time, can help the nation stay on the trajectory to an equitable net-zero emissions goal while also being able to revise policies and rethink technologies that do not work as intended. Both the diversity of policies intended to promote equity and justice during the energy transition and their distribution across many agencies and locations increase the need for a single entity that monitors, aggregates, synthesizes, and translates equity metrics to evaluate these policies. See Chapter 1 for more about the need for comprehensive evaluation and adaptive management.

Metrics serve multiple objectives such as holding decision-making publicly accountable, locating target populations, assessing policy design and development, or evaluating how disadvantaged communities are faring in the short term (outputs) and the long term (outcomes). The evaluation of metrics can also serve as justification for continued and increased financial support from federal agencies and philanthropic organizations. In support of efforts to develop an energy equity framework, the Pacific Northwest National Laboratory conducted a literature review and identified three equity metric types (Tarekegne et al. 2021):

- *Target population identification* metrics locate or describe a target population. The identification of relevant populations must happen at the beginning of the process.
- *Investment decision-making* metrics measure the potential impact of investments and assess the distributional effects of investments across groups. These metrics need to be considered during program and project design and can also be used to measure the short-term impacts on the target population.
- *Program impact assessment* metrics measure the benefits that directly reach people. These metrics are analyzed after the implementation of a program or project and should continue to be collected to determine the performance and success of the program or project and the long-term impacts on the target population.

Furthermore, a review of 57 distinct equity metrics found 24 target population iden-tification metrics; 25 investment decision-making metrics; and 8 program impact assessment metrics (Tarekegne et al. 2021). The authors suggest that baseline equity measurements can be developed by collecting and analyzing demographic and energy-related metrics such as income, race, geographic location, energy access, energy affordability, access to renewable energy, and community engagement. Thus, equity metrics are fundamental to operationalize notions of justice into concrete at-tributes, determinants, or outcomes.

To support an equity evaluation of the U.S. transition, the collection of data on the equity impacts of investments and the equity outcomes of programs is still needed because there are multiple screening tools to locate disadvantaged com-munities for targeted energy equity programs. Box 2-4 describes CEQ's Climate and Economic Justice Screening, which identifies disadvantaged communities and will be used for the implementation of Justice40 covered programs, as an example of the iterative process of identifying target population identification metrics. Eq-uity data, especially sociodemographic data, will need to be standardized in addi-tion to being collected (Bozeman et al. 2022). Data collection for program impact assessment will support analyses of federal agency compliance with the Justice40 Initiative.

The CEJST provides an opportunity to have one set of data for federal target popula-tion identification and investment decision-making. However, environmental justice advocates note that CEJST Version 1.0 does not explicitly consider racial demo-graphics as a factor for disadvantaged communities despite "evidence that race is the strongest and most consistent predictor of environmental burdens" (Sadasivam 2023). The absence of explicit mention of race in screening tools and program de-sign is common, even within programs designed to address environmental racism. At the federal level, race-neutral criteria are often selected to develop tools that will survive "legal challenges that would stymie their efforts" (Friedman 2022) (e.g., Su-preme Court ruling on the use of affirmative action in college admissions [Supreme Court Docket Number 20-1199 2023]). Although legal experts agree with the prag-matic approach to federal and state[18] programs, advocates stress that discussions of

---

[18] At the state level, the California Communities Environmental Health Screening Tool (CalEnviroScreen) does not include indicators of race, ethnicity, or age. However, the California Environmental Protection Agency analyzes and publishes supplemental reports on the relationship between screening tool scores, race, and ethnicity to show how accurately the screening tool identifies communities of color that are impacted by environmental injustices (CalEPA 2018). For more information, see R. Liévanos, 2018, "Retool-ing CalEnviroScreen: Cumulative Pollution Burden and Race-Based Environmental Health Vulnerabilities in California," *International Journal of Environmental Research and Public Health* 15(4), https://doi.org/10.3390/ijerph15040762.

BOX 2-4
CEQ'S CLIMATE AND ECONOMIC JUSTICE SCREENING TOOL

To help federal agencies identify disadvantaged communities, EO 14008 directed the CEQ to create a Climate and Economic Justice Screening Tool (CEJST). CEQ released CEJST Version 1.0[a] in November 2022, which included updated datasets, categorizations, and features responding directly to feedback received during the public comment period on the beta version and listening sessions (White House 2022b). Recommendations from the WHEJAC to add historic redlining data, identify Tribal Nations, display demographic information, and enhance data on climate vulnerability were also included in Version 1.0. However, this tool specifically excludes race and ethnicity as a consideration for vulnerability status; it is understood that these factors are excluded owing to potential legal challenges that would hinder the use of CEJST by federal agencies.

CEJST Version 1.0 uses datasets for 30 burden indicators, which are organized into eight categories. See Figure 2-4-1 for a depiction of the eight burden categories and related indicators. A community is designated as disadvantaged if it is in a census tract scoring at or above the threshold for one or more burden indicators and is at or above the threshold for an associated socioeconomic indicator. Low income was the socioeconomic indicator for all categories, except for workforce, which used high school education. The current version of the tool identifies 27,251 census tracts as disadvantaged or partially disadvantaged (White House 2022b), meaning that 33 percent of the nation's population is within a disadvantaged community. The CEJST will continue to be updated annually based on public feedback and information collected by the National Academies' Committee on Utilizing Advanced Environmental Health and Geospatial Data and Technologies to Inform Community Investment[b] (NASEM 2023a) with full transparency on the methodology and datasets supported by the U.S. Digital Service (White House 2022c).

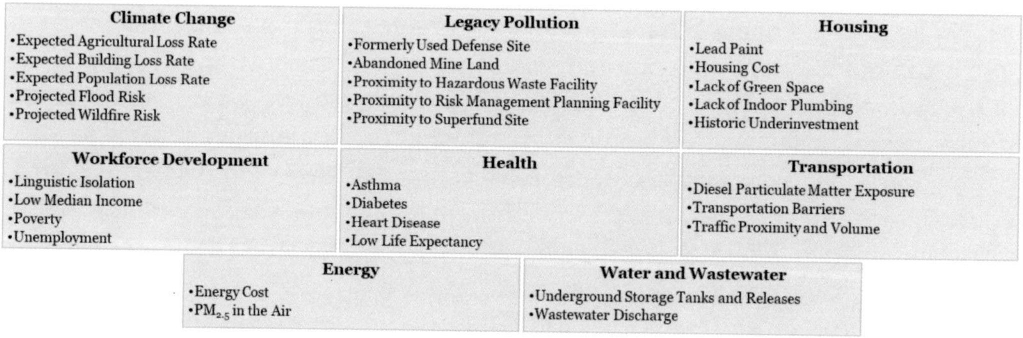

Climate Change	Legacy Pollution	Housing
•Expected Agricultural Loss Rate •Expected Building Loss Rate •Expected Population Loss Rate •Projected Flood Risk •Projected Wildfire Risk	•Formerly Used Defense Site •Abandoned Mine Land •Proximity to Hazardous Waste Facility •Proximity to Risk Management Planning Facility •Proximity to Superfund Site	•Lead Paint •Housing Cost •Lack of Green Space •Lack of Indoor Plumbing •Historic Underinvestment

Workforce Development	Health	Transportation
•Linguistic Isolation •Low Median Income •Poverty •Unemployment	•Asthma •Diabetes •Heart Disease •Low Life Expectancy	•Diesel Particulate Matter Exposure •Transportation Barriers •Traffic Proximity and Volume

Energy	Water and Wastewater
•Energy Cost •$PM_{2.5}$ in the Air	•Underground Storage Tanks and Releases •Wastewater Discharge

FIGURE 2-4-1 Climate and Economic Justice Screening Tool burden categories and indicators.
SOURCE: Data from Climate and Economic Justice Screening Tool.

During the beta phase of CEJST, OMB released a memo in July 2021 for the Heads of Departments and Agencies urging federal agencies with Justice40 covered programs to develop and publicize their own methodology for calculating the benefits of the programs accruing in disadvantaged communities (Young et al. 2021). This resulted in a majority of federal agencies with Justice40 covered programs developing their own screening tools (see Appendix G). However, an addendum to the 2021 memo encouraged federal agencies to use CEJST Version 1.0 to identify disadvantaged communities for Justice40 covered programs and other federal programs where resources are directed to disadvantaged communities (Young et al. 2023). Additionally, CEQ encourages federal agencies to "use the entire list of disadvantaged communities identified by the CEJST as a starting point" while noting that agencies may use their own data to prioritize certain communities from the list (CEQ 2023b, p. 5). Thus, CEQ provides a ceiling list of disadvantaged communities for federal agencies with covered programs to consider in the evaluation of program impacts.

---

[a] View the Climate and Economic Justice Screening Tool at https://screeningtool.geoplatform.gov/en/#3/33.47/-97.5.

[b] For more about the National Academies' Committee on Utilizing Advanced Environmental Health and Geospatial Data and Technologies to Inform Community Investment, visit the study's webpage at https://www.nationalacademies.org/our-work/utilizing-advanced-environmental-health-and-geospatial-data-and-technologies-to-inform-community-investment.

---

justice must recognize race as a factor of inequities and that efforts without explicit focus on race will not ultimately prioritize disadvantaged communities of color for Justice40 programs (Friedman 2022; Sadasivam 2023). Navigating the political environment to create mapping tools and programs that identify the on-the-ground experiences of disadvantaged communities without explicitly using racial and ethnic demographic data will impact what program evaluations will report as outcomes and impacts.

There is still a need for program impact assessment data to be collected and analyzed. EO 13985 tasks OMB to conduct a federal equity assessment of programs and policies through the consultation of the heads of agencies (EO 13985 2021). OMB found that the most promising evaluations of equity (1) consider historical legacies of disparities, prospective assessment of new interventions, and inclusive data initiatives—for example, developing and utilizing methodological innovations to impute missing data values, and methods that address equity in program eligibility; and (2) assess whether eligible groups receive benefits (OMB 2021). The report concludes with recommendations to federal, state, and local authorities that support the continued exploration of equity evaluation practices, including the following:

- Continually identify methods to assess equity for program improvement;
- Prioritize the expertise, capacity, and capabilities to improve data collection and analysis for equity considerations;
- Prioritize the expertise, capacity, and capabilities needed to engage stakeholders meaningfully; and
- Sustain and institutionalize equity in planning and workforce initiatives.

These recommendations provide a good foundation for the evaluation of programs and policies designed to support the just transition to a net-zero energy system.[19]

The Equitable Data Working Group is critical to determine what data federal agencies are already collecting, and what data needs to be collected for an overall review of equity and justice in the nation's energy transition. EO 14091 requires OSTP to coordinate with the Equitable Data Working Groups to implement strategies that address federal capacity-building needs for the collection and assessment of programmatic data (EO 14091 2023). Researchers have already noted the need for the consolidation of concepts relevant to the evaluation of equity, particularly for the use in life-cycle analyses (Bozeman et al. 2022). Both standardization of data practices and chosen equity concepts will be important for the development of a just energy transition evaluation.[20] This standardization can start at the federal level and support state and local development and use. Recent community-level initiatives also offer insights that can inform future programmatic and policy efforts to create a just energy transition through a bottom-up approach. However, as noted in the committee's Pathways to a Just and Equitable Transition workshop, the communication of these lessons learned will benefit from publicly accessible and standardized data (NASEM 2023a). For instance, metrics on energy costs, usage, and needs across different households and communities can provide critical guidance and best practices for the design of solutions that are most appropriate for the context of people's lives.

---

[19] In July 2023, DOE introduced its new Office of Energy Justice Policy and Analysis (OEJPA), which will collaborate with members of minority and disadvantaged communities to "achieve equity-centered Federal energy policy, research and development, and demonstration and deployment activities" (DOE n.d.). OEJPA will analyze the "socio-economic and environmental effects of energy programs, policies, and regulations" on communities and will additionally ensure Justice40 benefits flow to disadvantaged communities (DOE 2023b).

[20] For a review of the consolidated knowledge about equity applications and a 10-step process for developing standard sociodemographic data practices, see Bozeman et al. (2022).

Equity Evaluation of the Just Energy Transition

A foundation exists for a cross-agency evaluation on decarbonization investments and outcomes for an equitable and just energy transition. To support adaptive management of transition actions, a progressive assessment tool for the justice implication of policies and programs and the future impacts of the energy transition may be required (Heath 2022). Examples of energy transition analyses from local, state, and global organizations include

- Initiative for Energy Justice's Energy Justice Scorecard, which assesses existing or proposed energy policy based on the tenets of energy justice (see Baker et al. 2019);
- Maryland's Just Transition Analysis, which models the impact of the transition on fossil-fuel-reliant industries and workforce (see Irani et al. 2021);
- International Energy Agency's Tracking Clean Energy Progress assessment, which categorizes components of the energy system based on whether they are on target for the 2050 net-zero scenario (see IEA 2023); and
- World Benchmarking Alliances' Just Transition Assessment, which focuses on the transition of companies to a low-carbon future (see WBA 2021).

Empowering disadvantaged communities in decision-making will be critical to any equity evaluation to motivate actionable remediation in federal programs or implementation processes that are not achieving the desired or promised equitable outcomes. To build on this foundation and make an evaluation digestible by affected communities and groups, streamlined information about the transition needs to be developed by a single entity and communicated to stakeholders through trusted, existing communication tools.

> Finding 2-7: A critical component of understanding and improving the energy transition is to evaluate policy design, process, outcome, and impact regarding energy justice and equity. This understanding of equitable outcomes is an important aspect of a periodic evaluation of energy system decarbonization. For evaluations of equity, analysis of only quantitative data is insufficient. The analysis of both qualitative and quantitative data through participatory workshops, focus groups, and other elicitation techniques including the engagement of community stakeholders in their development, is critical to creating a comprehensive understanding of the implementation and outcomes of climate mitigation programs, and to redressing any failings of federal programs.

> **Recommendation 2-6:** *Evaluate the Equity Impacts of the Just Energy Transition.* **A single entity should collect, analyze, and communicate the equity data of**

federal program implementation and outcome. This evaluation should include quantitative and qualitative data and analysis, and selected metrics should be advised by the White House Environmental Justice Advisory Council and community-level stakeholders. The communication of the evaluation results should include regular reviews of progress toward equitable decarbonization in the United States that explicitly address the trends in energy burden reduction, workforce development and employment, community health and resilience. Additional impact and outcome metrics should be reported by federal agencies as relevant to specific decarbonization program and policy goals.

THE FUTURE OF ENERGY JUSTICE BEYOND THE 2020S

The nation's transition to a decarbonized energy system will require a fundamental shift in the way the burdens, concerns, priorities, and benefits of affected groups are considered. A just energy transition will require planning, implementation, and evaluation processes to be collaborative with the public, especially with its disadvantaged members, to support a bottom-up approach. Additionally, energy justice principles need to be incorporated into policymaking to achieve top-down integration and implementation. Laying these foundations now through baseline definitions, equitable implementation, and capacity building is critical to ensuring a just energy transition. Implementing energy justice principles will require a significant shift in the timescales involved in and approaches to policymaking. Meaningful participation, moreover, from all stakeholders takes time and new governance structures (see Chapter 5). Such collaborative processes determine how goals, barriers, burdens, and benefits are defined and evaluated. The integration of knowledge and expertise throughout policymaking and regulatory processes will be key to assuring procedural, recognitional, and distributional justice are incorporated into the nation's just energy transition. Table 2-3 summarizes all the recommendations that appear in this chapter to support a just energy transition.

SUMMARY OF RECOMMENDATIONS ON ENERGY JUSTICE AND EQUITY

TABLE 2-3[a] Summary of Recommendations on Energy Justice and Equity

Short-Form Recommendation	Actor(s) Responsible for Implementing Recommendation	Sector(s) Addressed by Recommendation	Objective(s) Addressed by Recommendation	Overarching Categories Addressed by Recommendation
2-1: Codify the Justice40 Initiative	Congress	• Electricity • Buildings • Land use • Transportation • Industry • Finance • Fossil fuels	• Equity • Health • Employment	Rigorous and Transparent Analysis and Reporting Ensuring Equity, Justice, Health, and Fairness of Impacts
2-2: Develop a Federal Baseline Set of Metrics for Disadvantaged Communities for Program Design and Evaluation	Council on Environmental Quality	• Electricity • Buildings • Land use • Transportation • Industry • Finance • Fossil fuels • Non-federal actors	• Equity	Rigorous and Transparent Analysis and Reporting Ensuring Equity, Justice, Health, and Fairness of Impacts
2-3: Implement Federal Legislation for Equitable Outcomes	Federal policy makers	• Electricity • Buildings • Land use • Transportation • Industry • Finance • Fossil fuels • Non-federal actors	• Public engagement • GHG reductions • Equity	Rigorous and Transparent Analysis and Reporting Ensuring Equity, Justice, Health, and Fairness of Impacts Ensuring Procedural Equity in Planning and Siting New Infrastructure and Programs

[a] The text in this table was changed during editorial review to improve clarity and alignment with information in other sections of the report.

continued

TABLE 2-3 Continued

Short-Form Recommendation	Actor(s) Responsible for Implementing Recommendation	Sector(s) Addressed by Recommendation	Objective(s) Addressed by Recommendation	Overarching Categories Addressed by Recommendation
2-4: Build Multi-Level Capacity to Support Community-Led Transitions	Congress, National Transition Corporation, Environmental Protection Agency, Department of Energy, state legislatures	• Non-federal actors	• GHG reductions • Equity • Health • Employment • Public engagement	Ensuring Procedural Equity in Planning and Siting New Infrastructure and Programs Ensuring Equity, Justice, Health, and Fairness of Impacts Building the Needed Workforce and Capacity
2-5: Develop Equitable Technical Assistance Guidelines	Federal Interagency Thriving Communities Network, White House Environmental Justice Advisory Committee (WHEJAC)	• Electricity • Buildings • Land use • Transportation • Industry • Finance • Fossil fuels • Non-federal actors	• Equity • Public engagement	Ensuring Procedural Equity in Planning and Siting New Infrastructure and Programs Ensuring Equity, Justice, Health, and Fairness of Impacts
2-6: Evaluate the Equity Impacts of the Just Energy Transition	Omnibus entity, WHEJAC	• Electricity • Buildings • Land use • Transportation • Industry • Finance • Fossil fuels • Non-federal actors	• Equity • Health • Employment	Rigorous and Transparent Analysis and Reporting Ensuring Equity, Justice, Health, and Fairness of Impacts

REFERENCES

A.B. No. (California Assembly Bill Number) 2722. 2016. "Transformative Climate Communities Program." 2015–2016 Regular Session. https://openstates.org/ca/bills/20152016/AB2722.

Aguayo, L. 2022. "Operationalizing Equity." Presented at Accelerating Decarbonization in the United States: Technology, Policy, and Societal Dimensions | Pathways to an Equitable and Just Transition Workshop. July 26. https://www.nationalacademies.org/event/07-26-2022/accelerating-decarbonization-in-the-united-states-technology-policy-and-societal-dimensions-pathways-to-an-equitable-and-just-transition-workshop.

Agyeman, J., D. Schlosberg, L. Craven, and C. Matthews. 2016. "Trends and Directions in Environmental Justice: From Inequity to Everyday Life, Community, and Just Sustainabilities." *Annual Review of Environment and Resources* 41(1):321–340. https://doi.org/10.1146/annurev-environ-110615-090052.

Amsalem, G., and V. Bogdan Tejeda. 2022. "EPA Urged to Reject Carbon Capture Projects in Central California." *Center for Biological Diversity.* https://biologicaldiversity.org/w/news/press-releases/epa-urged-to-reject-carbon-capture-projects-in-central-california-2022-06-29.

Baker, S., S. DeVar, and S. Prakash. 2019. *The Energy Justice Workbook.* Initiative for Energy Justice. https://iejusa.org/workbook.

Bauer, G.R. 2014. "Incorporating Intersectionality Theory into Population Health Research Methodology: Challenges and the Potential to Advance Health Equity." *Social Science Medicine* 110:10–17. https://doi.org/10.1016/j.socscimed.2014.03.022.

Beckfield, D. 2022. "Addressing Social Barriers to Decarbonization." Presented at Accelerating Decarbonization in the United States: Technology, Policy, and Societal Dimensions | Pathways to an Equitable and Just Transition Workshop. July 26. https://www.nationalacademies.org/event/07-26-2022/accelerating-decarbonization-in-the-united-states-technology-policy-and-societal-dimensions-pathways-to-an-equitable-and-just-transition-workshop.

Beckman, D. 2022. "Opening Remarks." Presented at Accelerating Decarbonization in the United States: Technology, Policy, and Societal Dimensions: Implementing a Just and Equitable Energy Transition—Philanthropic Perspectives. June 13. https://www.nationalacademies.org/event/06-13-2022/accelerating-decarbonization-in-the-united-states-technology-policy-and-societal-dimensions-implementing-a-just-and-equitable-energy-transition-philanthropic-perspectives.

Bennett, J., R. Kammer, J. Eidbo, M. Ford, S. Henao, N. Holwerda, E. Middleton, et al. 2023. *Carbon Capture Co-Benefits.* Great Plains Institute. https://carboncaptureready.betterenergy.org/wp-content/uploads/2023/08/Carbon-Capture-Co-Benefits.pdf.

Berthod, O., T. Blanchet, H. Busch, C. Kunze, C. Nolden, and M. Wenderlich. 2022. "The Rise and Fall of Energy Democracy: 5 Cases of Collaborative Governance in Energy Systems." *Environmental Management* 71:551–564. https://doi.org/10.1007/s00267-022-01687-8.

Besley, J.C. 2010. "Public Engagement and the Impact of Fairness Perceptions on Decision Favorability and Acceptance." *Science Communication* 32(2):256–280. https://doi.org/10.1177/1075547009358624.

Bickerstaff, K., G. Walker, and H. Bulkeley. 2013. "Introduction: Making Sense of Energy Justice." In *Energy Justice in a Changing Climate: Social Equity and Low-Carbon Energy*, K. Bickerstaff, G. Walker, H. Bulkeley, eds. London, UK: Bloomsbury Publishing.

Birkmann, J., E. Liwenga, R. Pandey, E. Boyd, R. Djalante, F. Gemenne, W. Leal Filho, P.F. Pinho, L. Stringer, and D. Wrathall. 2022. "Poverty, Livelihoods and Sustainable Development." Pp. 1171–1274 in *Climate Change 2022: Impacts, Adaptation and Vulnerability.* Contribution of Working Group II to the Sixth Assessment Report of the Intergovernmental Panel on Climate Change. Cambridge, UK, and New York: Cambridge University Press. https://doi.org/10.1017/9781009325844.010.

Boudet, H.S. 2019. "Public Perceptions of and Responses to New Energy Technologies." *Nature Energy* 4(6):446–455. https://doi.org/10.1038/s41560-019-0399-x.

Bouillass, G., I. Blanc, and P. Perez-Lopez. 2021. "Step-by-Step Social Life Cycle Assessment Framework: A Participatory Approach for the Identification and Prioritization of Impact Subcategories Applied to Mobility Scenarios." *International Journal of Life Cycle Assessment* 26:2408–2435. https://link.springer.com/article/10.1007/s11367-021-01988-w.

Bouzarovski, S. 2018. *Energy Poverty: (Dis)Assembling Europe's Infrastructural Divide.* Cham, Switzerland: Palgrave Macmillan. https://doi.org/10.1007/978-3-319-69299-9.

Bozeman III, J.F., E. Nobler, and D. Nock. 2022. "A Path Toward Systemic Equity in Life Cycle Assessment and Decision-Making: Standardizing Sociodemographic Data Practices." *Environmental Engineering Science* 39(9):759–769. https://doi.org/10.1089/ees.2021.0375.

Brown, J.D., S. Carpenter, S.D. Comello, S. Hovorka, S. Mackler, and M.H. Schwartz. 2023. *Turning CCS Projects in Heavy Industry and Power into Blue Chip Financial Investments.* Energy Futures Initiative. https://energyfuturesinitiative.org/wp-content/uploads/sites/2/2023/02/20230212-CCS-Final_Full-copy.pdf.

Brown, M.A., A. Soni, A.D. Doshi, and C. King. 2020. "The Persistence of High Energy Burdens: A Bibliometric Analysis of Vulnerability, Poverty, and Exclusion in the United States." *Energy Research & Social Science* 70:101756. https://doi.org/10.1016/j.erss.2020.101756.

Bullard, R.D. 2002. "Confronting Environmental Racism in the Twenty-First Century." *Global Dialogue* 4(1):34–48.

Bullard, R.D. 2004. "Making Environmental Justice a Reality in the 21st Century." *Sustain: A Journal of Environmental and Sustainability Issues* 10:5–12.

Burkett, M. 2008. "Just Solutions to Climate Change: A Climate Justice Proposal for a Domestic Clean Development Mechanism." *Buffalo Law Review* 56:169–243.

CalEPA (California Environmental Protection Agency). 2018. "Analysis of Race/Ethnicity, Age, and CalEnviroScreen 3.0 Scores." California Environmental Protection Agency: Office of Environmental Health Hazard Assessment. https://oehha.ca.gov/media/downloads/calenviroscreen/document-calenviroscreen/raceageces3analysis.pdf.

California Climate Investments. n.d. "Cap-and-Trade Dollars at Work." https://www.caclimateinvestments.ca.gov. Accessed March 15, 2023.

Cameron, L., and B. van der Zwaan. 2015. "Employment Factors for Wind and Solar Energy Technologies: A Literature Review." *Renewable and Sustainable Energy Reviews* 45:160–172. https://doi.org/10.1016/j.rser.2015.01.001.

Cap-and-Invest. n.d. "Reducing Pollution, Investing in Communities, Creating Jobs, and Preserving Competitiveness." New York State. https://capandinvest.ny.gov. Accessed March 15, 2023.

Carley, S. 2022. "Specialty Grand Challenge: Energy Transitions, Human Dimensions, and Society." *Frontiers in Sustainable Energy Policy* 1. https://doi.org/10.3389/fsuep.2022.1063207.

Carley, S., and D.M. Konisky. 2020. "The Justice and Equity Implications of the Clean Energy Transition." *Nature Energy* 5(8):569–577. https://doi.org/10.1038/s41560-020-0641-6.

Carley, S., T.P. Evans, M. Graff, and D.M. Konisky. 2018a. "A Framework for Evaluating Geographic Disparities in Energy Transition Vulnerability." *Nature Energy* 3(8):621–627. https://doi.org/10.1038/s41560-018-0142-z.

Carley, S., T.P. Evans, and D.M. Konisky. 2018b. "Adaptation, Culture, and the Energy Transition in American Coal Country." *Energy Research & Social Science* 37:133–139. https://doi.org/10.1016/j.erss.2017.10.007.

Carley, S., C. Engle, and D.M. Konisky. 2021. "An Analysis of Energy Justice Programs Across the United States." *Energy Policy* 152:112219. https://doi.org/10.1016/j.enpol.2021.112219.

Carley, S., M. Graff, D.M. Konisky, and T. Memmott. 2022a. "Behavioral and Financial Coping Strategies Among Energy-Insecure Households." *Proceedings of the National Academy of Sciences* 119(36):e2205356119. https://www.pnas.org/doi/abs/10.1073/pnas.2205356119.

Carley, S., D.M. Konisky, and T. Memmott. 2022b. *Household Energy Insecurity Survey.* Indiana University Energy Justice Lab. https://energyjustice.indiana.edu/doc/ejl-energy-insecurity-report-winter-2022.pdf.

CEQ (Council on Environmental Quality). 2022. "Carbon Capture, Utilization, and Sequestration Guidance." CEQ-2022-0001. *Federal Register* 87(32):8808–8811.

CEQ. 2023. *Instructions to Federal Agencies on Using the Climate and Economic Justice Screening Tool ("CEJST Instructions").* Council on Environmental Quality. https://static-data-screeningtool.geoplatform.gov/data-versions/1.0/data/score/downloadable/CEQ-CEJST-Instructions.pdf.

CEQ. n.d. "Methodology." Council on Environmental Quality. https://screeningtool.geoplatform.gov/en/methodology. Accessed March 15, 2023.

Cha, J.M. 2020. Chapter 4, "Just Transition: Tools for Protecting Workers and Their Communities at Risk of Displacement Due to Climate Policy." Pp. 147–175 in *Putting California on the High Road: A Jobs and Climate Action Plan for 2030*, C. Zabin, ed. University of California, Berkeley, Labor Center. https://laborcenter.berkeley.edu/putting-california-on-the-high-road-a-jobs-and-climate-action-plan-for-2030.

Cha, M. 2022. "Challenges and Opportunities for a Just Transition." Presented at Accelerating Decarbonization in the United States: Technology, Policy, and Societal Dimensions | Pathways to an Equitable and Just Transition Workshop. July 26. https://www.nationalacademies.org/event/07-26-2022/accelerating-decarbonization-in-the-united-states-technology-policy-and-societal-dimensions-pathways-to-an-equitable-and-just-transition-workshop.

Chan, S., H. van Asselt, T. Hale, K.W. Abbott, M. Beisheim, M. Hoffmann, B. Guy, et al. 2015. "Reinvigorating International Climate Policy: A Comprehensive Framework for Effective Nonstate Action." *Global Policy* 6(4):466–473. https://doi.org/10.1111/1758-5899.12294.

Chemnick, J. " 'Down Your Throat': Biden Pushes CCS on Polluted Places." *E&E News: Climate Wire.* https://www.eenews.net/articles/down-your-throat-biden-pushes-ccs-on-polluted-places.

Chi, S. 2022. "IRA: Our Analysis of the Inflation Reduction Act." Just Solutions Collective. https://justsolutionscollective.org/solution/ira-our-analysis-of-the-inflation-reduction-act.

Church, A., M. Frost, and K. Sullivan. 2000. "Transport and Social Exclusion in London." *Transport Policy* 7(3):195–205. https://doi.org/10.1016/S0967-070X(00)00024-X.

Climate Justice Alliance. 2023. "Climate Justice Leaders Denounce Carbon Capture and Storage Plans as Scams, Warn Reliance on Unproven Technology Exacerbates Climate Change and Environmental Injustices." https://climate-justicealliance.org/cj-leaders-denounce-ccs-at-whejac.

Committee on Natural Resources. 2022. "Justice, Equity, Diversity, and Inclusion in Environmental Policy-Making: The Role of Environmental Organizations and Grantmaking Foundations." Oversight Hearing before the U.S. House of Representatives Committee on Natural Resources. https://www.congress.gov/117/chrg/CHRG-117hhrg46811/CHRG-117hhrg46811.pdf.

Cong, S., D. Nock, Y.L. Qiu, and B. Xing. 2022. "Unveiling Hidden Energy Poverty Using the Energy Equity Gap." *Nature Communications* 13(1):2456. https://doi.org/10.1038/s41467-022-30146-5.

Cook, J.T., D.A. Frank, P.H. Casey, R. Rose-Jacobs, M.M. Black, M. Chilton, S. Ettinger de Cuba, et al. 2008. "A Brief Indicator of Household Energy Security: Associations with Food Security, Child Health, and Child Development in U.S. Infants and Toddlers." *Pediatrics* 122(4):e867–e875. https://doi.org/10.1542/peds.2008-0286.

Cooper, B. 2016. Chapter 19, "Intersectionality." Pp. 385–406 in *The Oxford Handbook of Feminist Theory*, L. Disch and M. Hawkesworth, eds. Oxford: Oxford University Press.

Covington, K.L. 2009. "Spatial Mismatch of the Poor: An Explanation of Recent Declines in Job Isolation." *Journal of Urban Affairs* 31(5):559–587. https://doi.org/10.1111/j.1467-9906.2009.00455.x.

Crenshaw, K. 1989. "Demarginalizing the Intersection of Race and Sex: A Black Feminist Critique of Antidiscrimination Doctrine, Feminist Theory, and Antiracist Politics." *University of Chicago Legal Forum* 1989(1):31.

Crenshaw, K.W. 2017. *On Intersectionality: Essential Writings.* New York: New Press.

Dean, B., and P. Esling. 2023. "CEJST Is a Simple Map, with Big Implications—and Attention to Cumulative Burdens Matters." *Intersections.* https://cnt.org/blog/cejst-is-a-simple-map-with-big-implications-and-attention-to-cumulative-burdens-matters.

DeBacker, L. 2022. "Opening Remarks." Presented at Accelerating Decarbonization in the United States: Technology, Policy, and Societal Dimensions: Implementing a Just and Equitable Energy Transition—Philanthropic Perspectives. June 13. https://www.nationalacademies.org/event/06-13-2022/accelerating-decarbonization-in-the-united-states-technology-policy-and-societal-dimensions-implementing-a-just-and-equitable-energy-transition-philanthropic-perspectives.

Devine-Wright, P. 2005. "Beyond NIMBYism: Towards an Integrated Framework for Understanding Public Perceptions of Wind Energy." *Wind Energy* 8(2):125–139. https://doi.org/10.1002/we.124.

Devine-Wright, P., and H. Devine-Wright. 2009. "Public Engagement with Community-Based Energy Service Provision: An Exploratory Case Study." *Energy & Environment* 20(3):303–317. https://doi.org/10.1260/095830509788066402.

Dhamoon, R.K. 2011. "Considerations on Mainstreaming Intersectionality." *Political Research Quarterly* 64(1):230–243. https://doi.org/10.1177/1065912910379227.

DOE (Department of Energy). 2023a. "Inclusive Energy Innovation Prize | American-Made Challenges." https://american-madechallenges.org/challenges/inclusiveenergyinnovation/index.html.

DOE. 2023b. "Introducing DOE's Office of Energy Justice Policy and Analysis." DOE Office of Economic Impact and Diversity. https://content.govdelivery.com/accounts/USDOEOEID/bulletins/35f0f6b.

DOE. n.d. "Office of Energy Justice Policy and Analysis." https://www.energy.gov/diversity/office-energy-justice-policy-and-analysis. Accessed March 15, 2023.

DOE-EERE (Department of Energy Office of Energy Efficiency and Renewable Energy). 2022. "Fact Sheet: Low-Income Energy Affordability Data (LEAD) Tool." https://lead.openei.org/docs/LEAD-Factsheet.pdf.

DOE-EERE. n.d. "Clean Energy to Communities (C2C) Program." https://www.energy.gov/eere/clean-energy-communities-program. Accessed March 15, 2023.

Donors of Color Network. 2021. "Donors of Color Network Launches Climate Funders Justice Pledge to Shift Millions to Environmental and Justice Groups Led by People of Color." https://www.climate.donorsofcolor.org/donors-of-color-network-launches-climate-funders-justice-pledge-to-shift-millions-to-environmental-justice-groups-led-by-people-of-color.

DOT (Department of Transportation). 2022. "Thriving Communities Program." https://www.transportation.gov/grants/thriving-communities.

DOT. n.d. "Federal Interagency Thriving Communities Network." https://www.transportation.gov/federal-interagency-thriving-communities-network. Accessed March 15, 2023.

Drehobl, A., and L. Ross. 2016. *Lifting the High Energy Burden in America's Largest Cities: How Energy Efficiency Can Improve Low Income and Underserved Communities.* Washington, DC: American Council for an Energy Efficient Economy.

Drehobl, A., L. Ross, and R. Ayala. 2020. *How High Are Household Energy Burdens? An Assessment of National and Metropolitan Energy Burden Across the United States.* Washington, DC: American Council for an Energy Efficient Economy.

E2 (Environmental Entrepreneurs). 2019. "Clean Jobs America 2019." https://www.e2.org/wp-content/uploads/2019/04/E2-2019-Clean-Jobs-America.pdf.

EJLF (Environmental Justice Leadership Forum). 2022. "False Solutions in Justice40 Letter." https://www.weact.org/wp-content/uploads/2022/09/EJLF-False-Solutions-in-Justice40-Letter-091922.pdf.

Elliot, N.M., B.A. Viola, H. Coulter, I. Lane, J.D. Odintz, M. Franco, C. Armstrong, et al. 2022. "The Inflation Reduction Act: Summary of Budget Reconciliation Legislation." Holland and Knight LLP. https://www.hklaw.com/-/media/files/insights/publications/2022/08/080822inflationreductionactsummary.pdf.

Elmallah, S., and J. Rand. 2022. "'After the Leases Are Signed, It's a Done Deal': Exploring Procedural Injustices for Utility-Scale Wind Energy Planning in the United States." *Energy Research and Social Science* 89(July):102549. https://doi.org/10.1016/j.erss.2022.102549.

Energy Equity Project. 2022. "Energy Equity Framework: Combining Data and Qualitative Approaches to Ensure Equity in the Energy Transition." University of Michigan—School for Environment and Sustainability. https://energyequityproject.com/wp-content/uploads/2022/08/220174_EEP_Report_8302022.pdf.

EO (Executive Order) 13985. 2021. "Advancing Racial Equity and Support for Underserved Communities Through the Federal Government." *Federal Register* 86:7009–7013.

EO 14008. 2021. "Tackling the Climate Crisis at Home and Abroad." *Federal Register* 86:7619–7633. https://www.federalregister.gov/documents/2021/02/01/2021-02177/tackling-the-climate-crisis-at-home-and-abroad.

EO 14052. 2021. "Implementation of the Infrastructure Investment and Jobs Act." *Federal Register* 86:64335–64336. https://www.federalregister.gov/documents/2021/11/18/2021-25286/implementation-of-the-infrastructure-investment-and-jobs-act.

EO 14082. 2022. "Implementation of the Energy and Infrastructure Provisions of the Inflation Reduction Act of 2022." *Federal Register* 87:56861–56864. https://www.federalregister.gov/documents/2022/09/16/2022-20210/implementation-of-the-energy-and-infrastructure-provisions-of-the-inflation-reduction-act-of-2022.

EO 14091. 2023. "Further Advancing Racial Equity and Support for Underserved Communities Through the Federal Government." *Federal Register* 88:10825–10833. https://www.federalregister.gov/documents/2023/02/22/2023-03779/further-advancing-racial-equity-and-support-for-underserved-communities-through-the-federal.

EO 14096. 88 Federal Register 25251 (April 21, 2023). "Revitalizing Our Nation's Commitment to Environmental Justice for All." https://www.federalregister.gov/documents/2023/04/26/2023-08955/revitalizing-our-nations-commitment-to-environmental-justice-for-all.

EPA (Environmental Protection Agency). 2009. "Endangerment and Cause or Contribute Findings for Greenhouse Gases Under Section 202(a) of the Clean Air Act: Final Rule." *Federal Register* 74(239):66496–66546.

EPA. 2022a. "Learn About Environmental Justice." https://www.epa.gov/environmentaljustice/learn-about-environmental-justice.

EPA. 2022b. "Year One Anniversary Report: Bipartisan Infrastructure Law." https://www.epa.gov/system/files/documents/2022-11/V-4_BIL_FirstAnniversaryReport_Nov142022.pdf.

EPA. 2023. "The Environmental Justice Thriving Communities Technical Assistance Centers Program." https://www.epa.gov/environmentaljustice/environmental-justice-thriving-communities-technical-assistance-centers.

Farley, C., J. Howat, J. Bosco, N. Thakar, J. Wise, and J. Su. 2021. *Advancing Equity in Utility Regulation*. GRID Modernization Laboratory Consortium. https://doi.org/10.2172/1828753.

Friedman, L. 2022. "White House Takes Aim at Environmental Racism, But Won't Mention Race." *The New York Times*. February 15. https://www.nytimes.com/2022/02/15/climate/biden-environment-race-pollution.html.

Ge, M., J. Friedrich, and L. Vigna. 2022. "4 Charts Explain Greenhouse Gas Emissions by Countries and Sectors." https://www.wri.org/insights/4-charts-explain-greenhouse-gas-emissions-countries-and-sectors.

Goforth, T., and D. Nock. 2022. "Air Pollution Disparities and Equality Assessments of U.S. National Decarbonization Strategies." *Nature Communications* 13(1):7488. https://doi.org/10.1038/s41467-022-35098-4.

Goldsmith, L., V. Raditz, and M. Méndez. 2022. "Queer and Present Danger: Understanding the Disparate Impacts of Disasters on LGBTQ+ Communities." *Disasters* 46(4):946–973. https://doi.org/10.1111/disa.12509.

Guhathakurta, S., and P. Gober. 2007. "The Impact of the Phoenix Urban Heat Island on Residential Water Use." *Journal of the American Planning Association* 73(3):317–329. https://doi.org/10.1080/01944360708977980.

Hale, T. " 'All Hands on Deck': The Paris Agreement and Nonstate Climate Action." *Global Environmental Politics* 16(3):12–22.

Han, S. 2022. "Spatial Stratification and Socio-Spatial Inequalities: The Case of Seoul and Busan in South Korea." *Humanities and Social Sciences Communications* 9(1):23. https://doi.org/10.1057/s41599-022-01035-5.

Harlan, S.L., J.H. Declet-Barreto, W.L. Stefanov, and D.B. Petitti. 2013. "Neighborhood Effects on Heat Deaths: Social and Environmental Predictors of Vulnerability in Maricopa County, Arizona." *Environmental Health Perspectives* 121(2):197–204. https://doi.org/10.1289/ehp.1104625.

Harvey, D. 2008. "The Right to the City." *New Left Review* (53):23–40.

Hayden, M.H., H. Brenkert-Smith, and O.V. Wilhelmi. 2011. "Differential Adaptive Capacity to Extreme Heat: A Phoenix, Arizona, Case Study." *Weather, Climate, and Society* 3(4):269–280. https://doi.org/10.1175/WCAS-D-11-00010.1.

Heath, G. 2022. "Lessons from LA100." Presented at Accelerating Decarbonization in the United States: Technology, Policy, and Societal Dimensions | Pathways to an Equitable and Just Transition Workshop. July 26. https://www.nationalacademies.org/event/07-26-2022/accelerating-decarbonization-in-the-united-states-technology-policy-and-societal-dimensions-pathways-to-an-equitable-and-just-transition-workshop.

Heffron, R.J., and D. McCauley. 2017. "The Concept of Energy Justice Across the Disciplines." *Energy Policy* 105:658–667. https://doi.org/10.1016/j.enpol.2017.03.018.

Henry, M.S., M.D. Bazilian, and C. Markuson. 2020. "Just Transitions: Histories and Futures in a Post-COVID World." *Energy Research and Social Science* 68:101668. https://doi.org/10.1016/j.erss.2020.101668.

Hernández, D., and S. Bird. 2010. "Energy Burden and the Need for Integrated Low-Income Housing and Energy Policy." *Poverty and Public Policy* 2(4):668–688. https://doi.org/10.2202/1944-2858.1095.

Hernández, D., Y. Aratani, and Y. Jiang. 2014. *Energy Insecurity Among Families with Children*. National Center for Children in Poverty. https://www.nccp.org/wp-content/uploads/2020/05/text_1086.pdf.

Hernández, D., Y. Jiang, D. Carrión, D. Phillips, and Y. Aratani. 2016. "Housing Hardship and Energy Insecurity Among Native-Born and Immigrant Low-Income Families with Children in the United States." *Journal of Children and Poverty* 22(2):77–92. https://doi.org/10.1080/10796126.2016.1148672.

Hillier, A.E. 2003. "Redlining and the Homeowners' Loan Corporation." *Journal of Urban History* 29(4):394–420.

Hoffman, J.S., V. Shandas, and N. Pendleton. 2020. "The Effects of Historical Housing Policies on Resident Exposure to Intra-Urban Heat: A Study of 108 U.S. Urban Areas." *Climate* 8(1):12. https://doi.org/10.3390/cli8010012.

Hsu, A., G. Sheriff, T. Chakraborty, and D. Manya. 2021. "Disproportionate Exposure to Urban Heat Island Intensity Across Major US Cities." *Nature Communications* 12(1):2721. https://doi.org/10.1038/s41467-021-22799-5.

HUD (Department of Housing and Urban Development). 2023. "FY 2021 Healthy Homes and Weatherization Co-operation Demonstration." https://www.hud.gov/program_offices/spm/gmomgmt/grantsinfo/fundingopps/fy21_hhweatherization.

IEA (International Energy Agency). 2023. "Tracking Clean Energy Progress—Topics." https://www.iea.org/topics/tracking-clean-energy-progress.

IPCC (Intergovernmental Panel on Climate Change). 2022. "Summary for Policymakers." Pp. 3–33 in *Climate Change 2022: Mitigation of Climate Change.* Contribution of Working Group III to the Sixth Assessment Report of the Intergovernmental Panel on Climate Change. Cambridge, UK, and New York: Cambridge University Press. https://doi.org/10.1017/9781009157926.001.

Irani, D., J. Leh, C. Menking, and N. Shokry. 2021. "Just Transition Study Update." In *The Greenhouse Gas Emissions Reduction Act: 2030 GGRA Plan.* Regional Economic Studies Institute: Townson University. https://mde.maryland.gov/programs/air/ClimateChange/Documents/2030%20GGRA%20Plan/Just%20Transition%20Study%20Update.pdf.

Jackson, K. 1985. *Crabgrass Frontier: The Suburbanization of the United States.* New York: Oxford University Press.

Jackson, K.T. 1980. "Federal Subsidy and the Suburban Dream: The First Quarter-Century of Government Intervention in the Housing Market." *Records of the Columbia Historical Society, Washington, DC* 50:421–451.

Jenkins, K. 2018. "Setting Energy Justice Apart from the Crowd: Lessons from Environmental and Climate Justice." *Energy Research and Social Science* 39:117–121. https://doi.org/10.1016/j.erss.2017.11.015.

Jenkins, K., B.K. Sovacool, and D. McCauley. 2018. "Humanizing Sociotechnical Transitions Through Energy Justice: An Ethical Framework for Global Transformative Change." *Energy Policy* 117:66–74. https://doi.org/10.1016/j.enpol.2018.02.036.

Kaijser, A., and A. Kronsell. 2014. "Climate Change Through the Lens of Intersectionality." *Environmental Politics* 23(3):417–433. https://doi.org/10.1080/09644016.2013.835203.

Katznelson, I. 2005. *When Affirmative Action Was White: An Untold History of Racial Inequality in Twentieth-Century America.* New York: W.W. Norton & Company.

Konisky, D.M., S. Carley, M. Graff, and T. Memmott. 2022. "The Persistence of Household Energy Insecurity During the COVID-19 Pandemic." *Environmental Research Letters* 17(10):104017. https://doi.org/10.1088/1748-9326/ac90d7.

Krieger, N. 2022. "Panel Discussion." Presented at Accelerating Decarbonization in the United States: Technology, Policy, and Societal Dimensions | Pathways to an Equitable and Just Transition Workshop. July 26. https://www.nationalacademies.org/event/07-26-2022/accelerating-decarbonization-in-the-united-states-technology-policy-and-societal-dimensions-pathways-to-an-equitable-and-just-transition-workshop.

Kulcsar, L.J., T. Selfa, and C.M. Bain. 2016. "Privileged Access and Rural Vulnerabilities: Examining Social and Environmental Exploitation in Bioenergy Development in the American Midwest." *Journal of Rural Studies* 47:291–299. https://doi.org/10.1016/j.jrurstud.2016.01.008.

LA SAFE (Louisiana's Strategic Adaptations for Future Environments). 2019. "Regional and Parish Adaptation Strategies." https://lasafe.la.gov.

Lerner, S. 2010. *Sacrifice Zones: The Front Lines of Toxic Chemical Exposure in the United States.* Cambridge, MA: MIT Press.

Lewis, J. 2022. "Panel Discussion." Presented at Accelerating Decarbonization in the United States: Technology, Policy, and Societal Dimensions: Implementing a Just and Equitable Energy Transition—Nonprofit Perspectives. March 29. https://www.nationalacademies.org/event/03-29-2022/accelerating-decarbonization-in-the-united-states-technology-policy-and-societal-dimensions-just-transition-webinar-series-non-profit-perspectives.

Liévanos, R.S. 2018. "Retooling CalEnviroScreen: Cumulative Pollution Burden and Race-Based Environmental Health Vulnerabilities in California." *International Journal of Environmental Research and Public Health* 15(4):762. https://doi.org/10.3390/ijerph15040762.

Liu, J., L.P. Clark, M.J. Bechle, A. Hajat, S.-Y. Kim, A.L. Robinson, L. Sheppard, A.A. Szpiro, and J.D. Marshall. 2021. "Disparities in Air Pollution Exposure in the United States by Race/Ethnicity and Income, 1990–2010." *Environmental Health Perspectives* 129(12):127005. https://ehp.niehs.nih.gov/doi/abs/10.1289/EHP8584.

Lobao, L., M. Zhou, M. Partridge, and M. Betz. 2016. "Poverty, Place, and Coal Employment Across Appalachia and the United States in a New Economic Era: Poverty, Place, and Coal Employment." *Rural Sociology* 81(3):343–386. https://doi.org/10.1111/ruso.12098.

Logan, J., and H. Molotch. 2005. "The City as a Growth Machine: From Urban Fortunes: The Political Economy of Place (1987)." Chapter 11 in *The Urban Sociology Reader*, J. Lin and C. Mele, eds. London, UK: Routledge.

Lucas, K. 2012. "Transport and Social Exclusion: Where Are We Now?" *Transport Policy* 20:105–113. https://doi.org/10.1016/j.tranpol.2012.01.013.

Lukanov, B.R., and E.M. Krieger. 2019. "Distributed Solar and Environmental Justice: Exploring the Demographic and Socio-Economic Trends of Residential PV Adoption in California." *Energy Policy* 134:110935. https://doi.org/10.1016/j.enpol.2019.110935.

Luskin Center for Innovation (University of California, Los Angeles, Luskin Center for Innovation). 2020. "Transform Fresno: A Baseline and Progress Report on Early Implementation of the Transformative Climate Communities Program Grant." Los Angeles, CA: University of California, Los Angeles, Luskin Center for Innovation.

Luskin Center for Innovation. 2022. "Transform Fresno: 2022 Progress Report on Implementation of the Transformative Climate Communities Program Grant." Los Angeles, CA: University of California, Los Angeles, Luskin Center for Innovation.

Madrid, J. 2017. "People of the Sun: Ensuring Latino Access and Participation in the Solar Energy Revolution." *Harvard Journal of Hispanic Policy* 29:37–46.

Massey, D.S., and N.A. Denton. 1998. *American Apartheid: Segregation and the Making of the Underclass*. Cambridge, MA: Harvard University Press.

Memmott, T., S. Carley, M. Graff, and D.M. Konisky. 2021. "Sociodemographic Disparities in Energy Insecurity Among Low-Income Households Before and During the COVID-19 Pandemic." *Nature Energy* 6(2):186–193. https://doi.org/10.1038/s41560-020-00763-9.

Méndez, M. 2020. *Climate Change from the Streets: How Conflict and Collaboration Strengthen the Environmental Justice Movement*. New Haven, CT: Yale University Press.

Méndez, M., G. Flores-Haro, and L. Zucker. 2020. "The (In)Visible Victims of Disaster: Understanding the Vulnerability of Undocumented Latino/A and Indigenous Immigrants." *Geoforum* 116:50–62. https://doi.org/10.1016/j.geoforum.2020.07.007.

Michney, T.M., and L. Winling. 2020. "New Perspectives on New Deal Housing Policy: Explicating and Mapping HOLC Loans to African Americans." *Journal of Urban History* 46(1):150–180. https://doi.org/10.1177/0096144218819429.

Miller, C.A. 2022. "Redesigning Political Economy: The Promise and Peril of a Green New Deal for Energy." Chapter 13 in *The Green New Deal and the Future of Work*, C. Calhoun and B.Y. Fong, eds. New York: Columbia University Press.

Miller, C.A. 2023. Chapter 12, "The Future of Energy Ownership." Pp. 107–114 in *Energy Democracies for Sustainable Futures*, M. Nadesan, M.J. Pasqualetti, and J. Keahey, eds. Elsevier, Inc.

Morello-Frosch, R., M. Zuk, M. Jerrett, B. Shamasunder, and A.D. Kyle. 2011. "Understanding the Cumulative Impacts of Inequalities in Environmental Health: Implications for Policy." *Health Affairs* 30(5):879–887. https://doi.org/10.1377/hlthaff.2011.0153.

Morrissey, J., E. Schwaller, D. Dickson, and S. Axon. 2020. "Affordability, Security, Sustainability? Grassroots Community Energy Visions from Liverpool, United Kingdom." *Energy Research and Social Science* 70:101698. https://doi.org/10.1016/j.erss.2020.101698.

Nadesan, M., M.J. Pasqualetti, and J. Keahey, eds. 2023. *Energy Democracies for Sustainable Futures*. Elsevier, Inc.

NASEM (National Academies of Sciences, Engineering, and Medicine). 2021. *Accelerating Decarbonization of the U.S. Energy System*. Washington, DC: The National Academies Press.

NASEM. 2022. "Pathways to an Equitable and Just Transition Workshop." Hosted by Accelerating Decarbonization in the United States: Technology, Policy, and Societal Dimensions. https://www.nationalacademies.org/event/01-06-2023/accelerating-decarbonization-in-the-united-states-technology-policy-and-societal-dimensions-transformative-climate-communities-lessons-learned-and-best-practices.

NASEM. 2023a. "Panel Discussion." Presented at Accelerating Decarbonization in the United States: Technology, Policy, and Societal Dimensions: Transformative Climate Communities—Lessons Learned and Best Practices Webinar. January 6.

NASEM. 2023b. *Pathways to an Equitable and Just Energy Transition: Principles, Best Practices, and Inclusive Stakeholder Engagement: Proceedings of a Workshop*. Washington, DC: The National Academies Press.

NASEM. n.d. "Utilizing Advanced Environmental Health and Geospatial Data and Technologies to Inform Community Investment." https://www.nationalacademies.org/our-work/utilizing-advanced-environmental-health-and-geospatial-data-and-technologies-to-inform-community-investment. Accessed March 15, 2023.

Natter, A. 2023. "Neighbors Don't Want to Be 'Test Dummies' for Biden's Carbon Removal Hubs." *Bloomberg News*. https://www.bloomberg.com/news/articles/2023-08-23/biden-s-co2-capture-hubs-to-fight-climate-change-meet-local-resistance.

NETL (National Energy Technology Laboratory). 2023. "Oxy-Combustion." https://netl.doe.gov/node/7477.

OMB (Office of Management and Budget). 2021. *Study to Identify Methods to Assess Equity: Report to the President*. Washington, DC: Office of Management and Budget.

Ong, P., A. Cheng, A. Comandon, and S. Gonzalez. 2023. Chapter 6, "South Los Angeles Since the Sixties: Half a Century of Change?" Pp. 111–136 in *California Policy Options 2019*, D.J.B. Mitchel, ed. Los Angeles, CA: University of California, Los Angeles, Luskin School of Public Affairs.

Õunmaa, L. 2021. "What Are the Socio-Economic Impacts of An Energy Transition?" *UNDP Europe and Central Asia*. https://www.undp.org/eurasia/blog/what-are-socio-economic-impacts-energy-transition.

Park, A. 2014. "Equity in Sustainability: An Equity Scan of Local Government Sustainability Programs." Urban Sustainability Directors Network. https://www.usdn.org/uploads/cms/documents/usdn_equity_scan_sept_2014_final.pdf.

P.L. (Public Law) 117-2. 2021. "American Rescue Plan Act of 2021." https://www.congress.gov/bill/117th-congress/house-bill/1319/text.

P.L. 117-169. 2022. "Inflation Reduction Act of 2022." https://www.congress.gov/bill/117th-congress/house-bill/5376/text.

P.L. 117-58. 2021. "Infrastructure Investment and Jobs Act." https://www.congress.gov/bill/117th-congress/house-bill/3684/text.

Pollin, R., and B. Callaci. 2019. "The Economics of Just Transition: A Framework for Supporting Fossil Fuel–Dependent Workers and Communities in the United States." *Labor Studies Journal* 44(2):93–138. https://doi.org/10.1177/0160449X18787051.

Qin, H., P. Romero-Lankao, J. Hardoy, and A. Rosas-Huerta. 2015. "Household Responses to Climate-Related Hazards in Four Latin American Cities: A Conceptual Framework and Exploratory Analysis." *Urban Climate* 14:94–110. https://doi.org/10.1016/j.uclim.2015.05.003.

Raimi, D., and S. Pesek. 2022. *What Is an "Energy Community"? Alternative Approaches for Geographically Targets Energy Policy*. Washington, DC: Resources for the Future.

Reddy, A.K.N. 2000. Chapter 2, "Energy and Social Issues." Pp. 39–61 in *World Energy Assessment: Energy and the Challenge of Sustainability*. New York: United Nations Development Programme.

Román, E.M. 2017. *Race and Upward Mobility: Seeking, Gatekeeping, and Other Class Strategies in Postwar America*. Stanford, CA: Stanford University Press.

Romero-Lankao, P., and E. Nobler. 2021. *Energy Justice: Key Concepts and Metrics Relevant to EERE Transportation Projects*. Golden, CO: National Renewable Energy Laboratory.

Romero-Lankao, P., and R. Norton. 2018. "Interdependencies and Risk to People and Critical Food, Energy, and Water Systems: 2013 Flood, Boulder, Colorado, USA." *Earth's Future* 6(11):1616–1629. https://doi.org/10.1029/2018ef000984.

Romero-Lankao, P., H. Qin, and K. Dickinson. 2012. "Urban Vulnerability to Temperature-Related Hazards: A Meta-Analysis and Meta-Knowledge Approach." *Global Environmental Change* 22(3):670–683. https://doi.org/10.1016/j.gloenvcha.2012.04.002.

Romero-Lankao, P., D.M. Gnatz, O. Wilhelmi, and M. Hayden. 2016. "Urban Sustainability and Resilience: From Theory to Practice." *Sustainability* 8(12):1224.

Romero-Lankao, P., A. Wilson, and D. Zimny-Schmitt. 2022. "Inequality and the Future of Electric Mobility in 36 U.S. Cities: An Innovative Methodology and Comparative Assessment." *Energy Research and Social Science* 91:102760. https://doi.org/10.1016/j.erss.2022.102760.

Romitti, Y., I. Wing, K. Spangler, and G. Wellenius. 2022. "Inequality in the Availability of Residential Air Conditioning Across 115 US Metropolitan Areas." *PNAS Nexus* 1. https://doi.org/10.1093/pnasnexus/pgac210.

Ross, L., A. Drehobl, and B. Stickles. 2018. *The High Cost of Energy in Rural America: Household Energy Burdens and Opportunities for Energy Efficiency*. Washington, DC: American Council for and Energy-Efficient Economy.

Rothstein, R. 2017. *The Color of Law: A Forgotten History of How Our Government Segregated America*. New York and London: Liveright Publishing Corporation, W.W. Norton.

Sadasivam, N. 2023. "Why the White House's Environmental Justice Tool Is Still Disappointing Advocates." *Grist*. https://grist.org/equity/white-house-environmental-justice-tool-cejst-update-race.

Sala, S., A. Vasta, L.A. Mancini, J.L. Dewulf, and E.J. Rosenbaum. 2015. "Social Life Cycle Assessment: State of the Art and Challenges for Supporting Product Policies." Joint Research Centre, European Commission. EUR 27624. Luxembourg: Publications Office of the European Union. JRC99101.

Saunders, I. 2023. "Panel Discussion." Presented at Accelerating Decarbonization in the United States: Technology, Policy, and Societal Dimensions | Transformative Climate Communities—Lessons Learned and Best Practices. January 6. https://www.nationalacademies.org/event/01-06-2023/accelerating-decarbonization-in-the-united-states-technology-policy-and-societal-dimensions-transformative-climate-communities-lessons-learned-and-best-practices.

S.B. (New York State Senate Bill) No. S6599. 2019. New York State Senate. 2019–2020 Legislative Session. https://legislation.nysenate.gov/pdf/bills/2019/S6599.

Schlosberg, D. 2007. *Defining Environmental Justice: Theories, Movements, and Nature.* Oxford, UK: Oxford University Press.

Schlosberg, D., and L.B. Collins. 2014. "From Environmental to Climate Justice: Climate Change and the Discourse of Environmental Justice." *WIREs Climate Change* 5(3):359–374. https://doi.org/10.1002/wcc.275.

Segreto, M., L. Principe, A. Desormeaux, M. Torre, L. Tomassetti, P. Tratzi, V. Paolini, and F. Petracchini. 2020. "Trends in Social Acceptance of Renewable Energy Across Europe—A Literature Review." *International Journal of Environmental Research and Public Health* 17(24):9161. https://doi.org/10.3390/ijerph17249161.

Selfa, T., L. Kulcsar, C. Bain, R. Goe, and G. Middendorf. 2011. "Biofuels Bonanza?: Exploring Community Perceptions of the Promises and Perils of Biofuels Production." *Biomass and Bioenergy* 35(4):1379–1389. https://doi.org/10.1016/j.biombioe.2010.09.008.

Sen, A. 2009. *The Idea of Justice.* Cambridge, MA: Harvard University Press.

SGC (California Strategic Growth Council). 2021. *Transformative Climate Communities Program: Round 4 DRAFT Program Guidelines—FY 2021–2022.* https://sgc.ca.gov/programs/tcc/docs/20211116-TCC_Round_4_Draft_Guidelines.pdf.

SGC. 2023. *Transformative Climate Communities Program: Round 5 Implementation Guidelines.* https://sgc.ca.gov/programs/tcc/docs/20230308-TCC_R5_Guidelines.pdf.

Shindell, D., M. Ru, Y. Zhang, K. Seltzer, G. Faluvegi, L. Nazarenko, G.A. Schmidt, et al. 2021. "Temporal and Spatial Distribution of Health, Labor, and Crop Benefits of Climate Change Mitigation in the United States." *Proceedings of the National Academy of Sciences* 118(46):e2104061118. https://doi.org/10.1073/pnas.2104061118.

Smith, L.A., E.W. Brown, J.T. Cook, L. Rosenfeld, E. Feinberg, E. Goodman, M. Kotelchuck, et al. 2007. *Unhealthy Consequences: Energy Costs and Child Health, a Child Health Impact Assessment of Energy Costs and the Low-Income Home Energy Assistance Program.* Boston, MA: Children's Health Watch.

Sovacool, B.K., and M.H. Dworkin. 2014. *Global Energy Justice: Problems, Principles, and Practices.* Cambridge, UK: Cambridge University Press. https://doi.org/10.1017/CBO9781107323605.

Sovacool, B.K., P. Newell, S. Carley, and J. Fanzo. 2022. "Equity, Technological Innovation and Sustainable Behaviour in a Low-Carbon Future." *Nature Human Behaviour* 6(3):326–337. https://doi.org/10.1038/s41562-021-01257-8.

Spurlock, C.A., S. Elmallah, and T.G. Reames. 2022. "Equitable Deep Decarbonization: A Framework to Facilitate Energy Justice-Based Multidisciplinary Modeling." *Energy Research and Social Science* 92:102808. https://doi.org/10.1016/j.erss.2022.102808.

Supreme Court Docket Number 20-1199. 2023. *Students for Fair Admissions, Inc. v. President and Fellows of Harvard College.* https://www.supremecourt.gov/DocketPDF/20/20-1199/274255/20230731100842602_EFiling%2020-1199%20Rev%20COSTS%20CA1%207.31.pdf.

Sze, J., and J.K. London. 2008 "Environmental Justice at the Crossroads." *Sociology Compass* 2/4:1331–1354.

Tarekegne, B.W., G.R. Pennell, D.C. Preziuso, and R.S. O'Neil. 2021. *Review of Energy Equity Metrics.* Pacific Northwest National Laboratory.

Taylor, D.E., and M. Blondell. 2023. *Examining Disparities in Environmental Grantmaking: Where the Money Goes.* New Haven, CT: Yale School of the Environment.

Tomer, A., C. George, J.W. Kane, and A. Bourne. 2021. "America Has an Infrastructure Bill. What Happens Next?" https://www.brookings.edu/articles/america-has-an-infrastructure-bill-what-happens-next.

Urban Institute. n.d. *Federal Agencies' Equity Action Plans: Analysis, Summaries, Recommendations.* https://www.urban.org/projects/federal-agencies-equity-action-plans-analysis-summaries-recommendations. Accessed March 15, 2023.

UN (United Nations). n.d. *The 17 Goals—History.* United Nations Department of Economic and Social Affairs. https://sdgs.un.org/goals. Accessed March 15, 2023.

UNDP (United Nations Development Programme). 2022. "What Is a Just Transition? And Why Is It Important?" *UNDP Climate.* https://climatepromise.undp.org/news-and-stories/what-just-transition-and-why-it-important.

Vogelsong, S. 2022. "Predatory Residential Solar Installers Could Sow Mistrust, Advocates Fear: More Guardrails Sought for Industry." *Virginia Mercury*. https://www.virginiamercury.com/2022/05/25/as-residential-solar-grows-more-popular-advocates-worry-about-consumer-protections.

Walker, C., L. Shaver, and C. Macomber. 2022. "How Prepared Are US Cities to Implement the Justice40 Initiative?" *World Resources Institute Technical Perspective*. https://www.wri.org/technical-perspectives/how-prepared-are-us-cities-implement-justice40-initiative.

Walker, G. 2009. "Beyond Distribution and Proximity: Exploring the Multiple Spatialities of Environmental Justice." *Antipode* 41(4):614–636. https://doi.org/10.1111/j.1467-8330.2009.00691.x.

Walls, M.A. 2022. "Climate Policy, Environmental Justice, and the Inflation Reduction Act." *Resources*. August 9. https://www.resources.org/common-resources/climate-policy-environmental-justice-and-the-inflation-reduction-act.

Wang, Y., J.S. Apte, J.D. Hill, C.E. Ivey, R.F. Patterson, A.L. Robinson, C.W. Tessum, and J.D. Marshall. 2022. "Location-Specific Strategies for Eliminating US National Racial-Ethnic PM2.5 Exposure Inequality." *Proceedings of the National Academy of Sciences* 119(44):e2205548119. https://www.pnas.org/doi/abs/10.1073/pnas.2205548119.

WBA (World Benchmarking Alliance). 2021. *Just Transition Assessment 2021: Are High-Emitting Companies Putting People at the Heart of Decarbonization?* World Benchmarking Alliance. https://assets.worldbenchmarkingalliance.org/app/uploads/2021/11/2021_JustTransitionAssessment.pdf.

We Act for Environmental Justice. 2022. "We Act for Environmental Justice Responds to Inflation Reduction Act of 2022." https://www.weact.org/2022/07/we-act-for-environmental-justice-responds-to-inflation-reduction-act-of-2022.

White House. 2021a. "The Bipartisan Infrastructure Law Advances Environmental Justice." https://www.whitehouse.gov/briefing-room/statements-releases/2021/11/16/the-bipartisan-infrastructure-law-advances-environmental-justice.

White House. 2021b. "Fact Sheet: The Bipartisan Infrastructure Deal." https://www.whitehouse.gov/briefing-room/statements-releases/2021/11/06/fact-sheet-the-bipartisan-infrastructure-deal.

White House. 2022a. *Advancing Equity Through the American Rescue Plan*. Washington, DC. https://www.whitehouse.gov/wp-content/uploads/2022/05/advancing-equity-through-the-american-rescue-plan.pdf.

White House. 2022b. "Biden-Harris Administration Launches Version 1.0 of Climate and Economic Justice Screening Tool, Key Step in Implementing President Biden's Justice40 Initiative." https://www.whitehouse.gov/ceq/news-updates/2022/11/22/biden-harris-administration-launches-version-1-0-of-climate-and-economic-justice-screening-tool-key-step-in-implementing-president-bidens-justice40-initiative.

White House. 2022c. "Fact Sheet: The Inflation Reduction Act Supports Workers and Families." https://www.whitehouse.gov/briefing-room/statements-releases/2022/08/19/fact-sheet-the-inflation-reduction-act-supports-workers-and-families.

Wilgosh, B., A.H. Sorman, and I. Barcena. 2022. "When Two Movements Collide: Learning from Labour and Environmental Struggles for Future Just Transitions." *Futures* 137:102903. https://doi.org/10.1016/j.futures.2022.102903.

Wüste, A., and P. Schmuck. 2012. "Bioenergy Villages and Regions in Germany: An Interview Study with Initiators of Communal Bioenergy Projects on the Success Factors for Restructuring the Energy Supply of the Community." *Sustainability* 4(2):244–256. https://doi.org/10.3390/su4020244.

Yoder, K. 2022. "'It Makes Climate Change Real': How Carbon Emissions Got Rebranded as 'Pollution.'" *Grist*. https://grist.org/health/how-carbon-emissions-got-rebranded-climate-pollution-ira.

Young, S.D., B. Mallory, and G. McCarthy. 2021. "Interim Implementation Guidance for the Justice40 Initiative." M-21-28 Memorandum for the Heads of Departments and Agencies.

Young, S.D., B. Mallory, and A. Zaidi. 2023. "Addendum to the Interim Implementation Guidance for the Justice40 Initiative, M-21-28, on Using the Climate and Economic Justice Screening Tool (CEJST)." M-23-06. Memorandum for the Heads of Departments and Agencies.

Zhao, A., S. Kennedy, K. O'Keefe, M. Borreo, K. Clark-Sutton, R. Cui, C. Dahl, et al. 2022. *An "All-In" Pathway to 2030: Beyond 50 Scenario*. University of Maryland, Center for Global Sustainability. https://cgs.umd.edu/sites/default/files/2022-11/CGS_UMD_Beyond50report_Nov2022.pdf.

Zhong, R., and N. Popovich. 2022. "How Air Pollution Across America Reflects Racist Policy from the 1930s: A New Study Shows How Redlining, A Depression-Era Housing Policy, Contributed to Inequalities That Persist Decades Later in U.S. Cities." *The New York Times*. March 9. https://www.nytimes.com/2022/03/09/climate/redlining-racism-air-pollution.html.

3

Public Health Co-Benefits and Impacts of Decarbonization

ABSTRACT

The energy system—which incorporates transportation, industry, buildings, and agriculture—supports daily activities that have both beneficial and adverse impacts on public health. The energy transition to a net-zero future provides an opportunity to address multiple health and energy challenges simultaneously. Numerous health benefits, also referred to as co-benefits, are possible with the energy transition, including improvements in air quality, water quality, physical fitness, and green space and living conditions. One of the primary benefits of decarbonizing the U.S. economy is preventing premature deaths related to fossil fuel production and combustion.

It is crucial to minimize human health risks, including health inequities during the energy transition. The energy transition is an opportunity not only to avoid repeating past injustices and disparities but also to create a more equitable and health-promoting energy system overall. New energy policies and technologies come with potential trade-offs for climate mitigation and health that must also be considered. To address and overcome barriers of the energy transition, the committee recommends health impact assessments be conducted during the development of transition programs and deployment of technologies to monitor the health outcomes of decarbonization actions. Table 3-2, at the end of the chapter, summarizes all the recommendations that appear in this chapter to support the inclusion of health considerations in decarbonization efforts.

INTRODUCTION

How we generate and use energy impacts our health in a multitude of adverse ways: from the environmental and health hazards associated with the extraction and processing of resources (Epstein et al. 2011; Healy et al. 2019), to the pollutants produced during power generation, and ultimately to the use of power and fuels in support of our daily lives such as through heating and cooling homes and buildings, public and private transportation, and operating medical technology. Air pollution, mainly particulate matter, contributes to an estimated 53,200–355,000 annual premature deaths in the United States (Mailloux et al. 2022; Vohra et al. 2021).

The hazards associated with our existing energy system tend to disproportionately impact disadvantaged communities,[1] including ethnic and racial minorities and low-income households in the United States and abroad (Agyeman et al. 2002; Healy et al. 2019; Lane et al. 2022; Mohai et al. 2009). Discriminatory policies can contribute to increased health risks for vulnerable communities that live near these hazards, even long after the policies have ended (Huang and Sehgal 2022; Lane et al. 2022; Wilson et al. 2008). To prevent further injustice, procedures for siting new energy technology and remediation of past damage must consider how risks and benefits are distributed with income and race and ethnicity (McCauley and Heffron 2018). Without active correction and the intentional inclusion of and consideration for affected communities, existing disparities will persist and continue the nation's legacy of desperate and unjust health and economic damages—or even worsen inequities and create new disparities.

In addition to the impacts of the current energy system, climate change has adverse impacts on health. These impacts include increased risk of premature death and exposure to extreme climate events and environmental hazards. This section provides background on current impacts of energy access and air quality on health. This section will also provide a brief background about the health impacts of climate change itself, but the focus of this chapter will be the health impacts from air pollution of fossil fuel combustion.

Health Impacts of Energy Access

Energy access is vital for health and well-being. Basic needs for adequate heating, cooling, and some life-sustaining medical equipment require reliable and affordable energy. Without energy, additional health hazards can arise such as lack of clean water for hygiene; increased exposure to heat during heatwaves, which can exacerbate chronic health conditions (Jessel et al. 2019); and coping strategies during cold weather that increase risk of house fires (Carley et al. 2022). Climate change associated with fossil fuel energy sources may challenge reliable energy access, increase heat waves, and reduce access to clean water in many communities. As further discussed in Chapter 2, vulnerable populations are particularly at risk for limited energy access.

---

[1] Communities that are marginalized, underserved, and overburdened by pollution and experience other socioeconomic burdens, such as low income or high unemployment.

Even households not identified as energy insecure based on income metrics may still limit their energy use, potentially risking more heat- and cold-related illnesses (Cong et al. 2022). The uncertainty and challenges of controlling energy costs can have mental health impacts—including anxiety, chronic stress, and depression—as well as physical impacts from the effects of heat and cold, and when households are forced to choose to spend their income either on food or on energy (Hernández et al. 2016).

> Finding 3-1: Energy access and affordability persist as barriers to low-income communities in achieving health and economic stability. Health risks include heat and cold stress, anxiety, increase of fire risk, and lack of reliable access to energy for medical devices.

Health Impacts of Air Pollution

While climate change mitigation is primarily focused on methods for reducing greenhouse gas (GHG) emissions, sometimes referred to as climate pollutants, the same measures to reduce GHGs often reduce many co-emitted "traditional" air pollutants as well. "Traditional" air pollutants include the six explicitly regulated by the National Ambient Air Quality Standards (also known as criteria pollutants): particulate matter (PM), sulfur dioxide (SO_2), ground-level ozone (O_3), nitrogen dioxide (NO_2), carbon monoxide (CO), and lead (Pb) (EPA 2022a). These pollutants have distinct direct health impacts, typically via acute or chronic inhalation. Criteria air pollutants tend to be short lived (e.g., hours, days, months) in the atmosphere and exert much of their impact regionally. See Box 3-1 below for information about criteria air pollutants. In contrast, GHGs last from 12 years to thousands of years in the atmosphere (EPA 2022b) and exert global effects. GHGs include carbon dioxide (CO_2) and methane (CH_4). Some pollutants could be considered both traditional air pollutants and climate pollutants, including ozone, precursors to ozone (e.g., CH_4), and black carbon, a component of fine particulate matter emitted from sources that burn fossil fuel. Clarifying the differences between traditional air pollutants and climate pollutants is also important for public perception and support of health-based decarbonization policies (Dryden et al. 2018). Furthermore, the transience of traditional air pollutants can be beneficial for decarbonization policies because immediate health co-benefits can be achieved from reduction of fossil fuel emissions and are highly relevant on a local scale.

In the United States, ambient air pollution, especially from fine particulate matter ($PM_{2.5}$), is among the top environmental risk factors for premature death. Estimates may differ based on how pollution concentration is calculated, the number of health outcomes included, and the exposure response function used in the study (Pozzer et al. 2023). See Table 3-1 for a compilation of estimates from various studies.

147

BOX 3-1
FOSSIL FUEL COMBUSTION AND CRITERIA AIR POLLUTANTS

Particulate matter (PM) is a complex mixture of particles derived from a variety of sources—including fossil fuel combustion, wildfires, windblown dust, agriculture, and chemical reactions of other pollutants like ammonia and sulfur dioxide. PM is one of the top environmental health concerns as it is estimated to contribute to as many as 8.9 million premature deaths per year globally (Burnett et al. 2018). Fine particulate matter ($PM_{2.5}$), comprising particles 2.5 microns or less, presents the greatest health concern, because it can infiltrate the lungs deeper than larger particles. Well-established causes of death associated with $PM_{2.5}$ include ischemic heart disease, stroke, chronic obstructive pulmonary disease (COPD), lung cancer, and lower respiratory infections (McDuffie et al. 2021). Some fossil fuel combustion PM sources may be more dangerous than others, even after correcting for the mass of $PM_{2.5}$ they produce, although more research is needed (Thurston and Bell 2020; Wang et al. 2022; West et al. 2016).

Sulfur dioxide (SO_2) is a gas released upon burning of fossil fuels, particularly from coal-fired power plants, which emit 66 percent of U.S. SO_2 emissions (EPA 2023a). Diesel combustion and industrial processes such as metal extraction, pulp and paper mills, and gasoline extraction also emit SO_2 (WHO 2000). SO_2 has been identified as a potential contributor to developing and exacerbating asthma (Andersson 2006; Casey et al. 2020; Gorai et al. 2014). Short-term exposure to SO_2 is linked to an increase in asthma-associated emergency room visits and hospital admissions (Zheng et al. 2021) and is positively associated with all-cause and respiratory mortality (Orellano 2021).

Ground-level ozone (O_3) forms from precursors emitted from fossil fuel sources, particularly tailpipe emissions containing nitrogen oxides (NO_x). While ozone in the stratosphere forms the UV-protective ozone layer, ground-level ozone (or tropospheric ozone) is a health hazard. Under favorable conditions of heat and sunlight, ozone is formed from the combination of NO_x and either volatile organic compounds (VOCs), CO, or methane. O_3 causes respiratory harm through worsening asthma and COPD and causing inflammation. It has also been linked to causing premature death from short and long-term exposure (EPA 2013).

Nitrogen dioxide (NO_2) is formed during fossil fuel combustion from oxidation of nitrogen contained in the fuel and/or from the reaction of N_2 and O_2 in air at high temperatures. NO_2 and other nitrogen oxides (NO_x) are often associated with traffic-related air pollution. NO_x can interact with VOCs to form acid rain. NO_x can irritate the respiratory tract, aggravating asthma and potentially causing the development of asthma (EPA n.d.(b)).

Carbon monoxide (CO) is released from the incomplete combustion of carbon-containing fuels, often associated with traffic-related air pollution. Sources of indoor CO emissions include furnaces, gas water heater, and gas stoves. In high doses, especially in enclosed environments, CO can cause fatigue, headaches, confusion, and death (EPA n.d.(c)). Although elevated levels of CO outdoors are uncommon, this can be an issue, particularly for people with cardiovascular disease, who may have a harder time getting oxygen to their heart.

Lead (Pb) is a metal that can be suspended in the air and absorbed and accumulated in the body. Lead can cause irreversible brain damage, as well as damage to liver and kidneys, immune system, and reproductive system (EPA n.d.(a)). Since the EPA began phasing out leaded gasoline in 1973, lead levels in the air dropped 98 percent between 1980 and 2014 (EPA n.d.(a)). Lead can still also be found in soil and resuspended in the air, leaded fuels are still used in piston-engine aircraft, and lead is a pollutant from certain types of ore and metal processing.

TABLE 3-1 Estimated Premature Deaths from $PM_{2.5}$ Pollution in the United States: Total and Attributable to Fossil Fuels

Study	Total Estimated Premature Deaths from $PM_{2.5}$ Emissions (thousands)	Estimated $PM_{2.5}$ Deaths Attributable to Fossil Fuels (thousands)
McDuffie et al. (2021)	47 in 2017	N/A
Thakrar et al. (2020)	100[a,b] in 2015	N/A
Goodkind et al. (2019)	107[c] in 2011	N/A
Fann et al. (2018)	121 in 2014	N/A
Tessum et al. (2019)	131 in 2015	N/A
Shindell et al. (2021)	191 in 2020	N/A
Mailloux et al. (2022)	205 in 2016	53 in 2016
Burnett et al. (2018)	213[d] in 2015	N/A
Lelieveld et al. (2019)	283[e,f] in 2015	194[e] in 2015
Vohra et al. (2021)	N/A	355[g,h] in 2012

[a] Estimate includes primary $PM_{2.5}$ and secondary $PM_{2.5}$ precursors (NO_x, NH_3, NMVOC, and SO_x).

[b] Estimate attributes 99,900 deaths to anthropogenic $PM_{2.5}$ from the transportation, electricity, food and agriculture, residential, and industrial and commercial sectors.

[c] Estimate attributes 60,990 deaths to pollution from energy consumption.

[d] Estimate includes mortality data from the United States and Canada.

[e] Estimate includes ozone (O_3) pollution.

[f] Estimate attributes 230,000 deaths to anthropogenic $PM_{2.5}$, which includes agriculture, residential energy use, and non-fossil emissions.

[g] Estimate includes mortality data for long-term exposure to $PM_{2.5}$ from fossil fuel combustion.

[h] Estimate includes mortality data for populations older than 14 years old.

Despite the varying estimates of attributable premature deaths attributable to $PM_{2.5}$, decreasing fossil fuel combustion is a key target for reducing PM emissions because it can be more easily controlled than natural sources.

A few studies evaluate the impact specific energy sectors and process have on the premature deaths and costs in the United States. For example, Penn et al. (2017) estimates that $PM_{2.5}$ from electricity production, mainly driven by SO_2 emissions forming secondary $PM_{2.5}$, cause 21,000 premature deaths per year in the United States. Another estimate finds that human-caused $PM_{2.5}$ emissions contributed to $886 billion in costs with 57 percent of the impacts attributable to electricity generation and transportation (Goodkind et al. 2019). Goodkind et al. (2019) also point out that air pollution from electricity generation and industry may be easier to control than $PM_{2.5}$ emissions from road dust or residential wood burning. These health impacts do not include additional damages from other co-emitted criteria air pollutants.

While there is a range of estimates owing to differing methodologies, the evidence indicates that a reduction in GHG emissions could have significant positive health outcomes from reduction in co-emitted air pollutants (Gallagher and Holloway 2022). A retrospective analysis, for example, found that between 2007 and 2015 the improvements in air quality from increasing replacement of coal-generated power with wind and solar in the United States have prevented between 3,000 and 12,700 deaths and saved $29.1 billion–$112.8 billion (2015$) of health costs (Millstein et al. 2017). The benefits of clean air, especially reductions in particulate matter are large: the Environmental Protection Agency's (EPA's) cost-benefit analysis of the Clean Air Act estimates that 85 percent of the economic benefits of the Clean Air Act can be attributed to reductions in premature mortality from particulate matter (DeMocker and Neumann 2011). For more information about the air quality impacts of certain energy sectors, see the section "Health Co-Benefits of Decarbonization" below.

Current Disparities in Exposure to Air Pollution

Decarbonization policies can reduce disparities in exposure to air pollution and create health and equity-related co-benefits. Effective policies will target households and communities experiencing the greatest harm. Low-income, racial, and ethnic minority households often live in older, less energy efficient homes, where energy efficiency upgrades would improve both the health and financial stability of the household (Lewis et al. 2020; Tonn et al. 2014). Furthermore, Black Americans and Hispanic Americans face excess exposure to $PM_{2.5}$ relative to the pollution caused by their consumption of goods and services (56 percent and 63 percent, respectively) (Tessum et al. 2019). Increasingly, decarbonization strategies are location-specific, and at least one analysis reports that these are more effective than broad regional or sector-specific strategies, especially when trying to achieve multiple goals (Wang et al. 2022).

Despite the regional and state variation, racial and ethnic minority groups historically have the highest national average exposure to all six criteria pollutants (Liu et al. 2021). The health risks are also disproportionate: people of color have higher rates of emergency department visits for asthma and other diseases (Nardone et al. 2020) and are more likely to be living with at least one chronic condition that enhances their susceptibility to air pollution, including asthma, diabetes, and heart disease (Erqou et al. 2018). The evidence of socioeconomic disparities in respiratory health may be, in part, explained by disparities in exposure to air pollution (Bravo et al. 2016; Liu et al. 2021; Ringquist 2005; Woodruff et al. 2003). The following sections examine the existing disparities in exposure to indoor and outdoor air pollution.

Indoor Air Quality

While much research on air quality centers on effects of ambient air quality, these same pollutants can be found in indoor environments. Residents of the United States are estimated to spend 87 percent of their time indoors (Klepeis et al. 2001), and while outdoor air quality influences indoor air quality, there can be greater variation indoors than outdoors (O'Dell et al. 2023). This means indoor air pollution can reach very high levels in some rooms or dwellings, with corresponding health impacts (Ilacqua et al. 2022; NASEM and NAE 2022). Indoor combustion (e.g., unvented gas fireplaces) can release CO and NO_2 at levels higher than health-based standards, even when appliances are correctly operated (Francisco 2010; Lebel et al. 2022). A recent meta-analysis on gas stove use and asthma found that 12.7 percent of childhood asthma could be attributed to gas stove use (Gruenwald et al. 2023). A National Academies' report (2022a) recommends that researchers and practitioners engage disadvantaged communities in studies on indoor environments and in developing research priorities for indoor air quality standards. A better understanding of the factors impacting air quality indoors—such as the type of heating, cooling, ventilation, and filtration systems; building materials and maintenance practices; occupant density and housing type; and the source, proximity, and scale of outdoor contaminants—would be useful. Furthermore, the electrification of home appliances can improve indoor air quality and reduce the health risks associated with indoor air pollution.

Ambient Air Quality and Facility Siting

Since 1982, the environmental justice movement in the United States has identified disproportionate siting of hazardous facilities, particularly sites for energy production and petrochemical facilities, near historically disadvantaged populations (Agyeman et al. 2002; GAO 1983; James et al. 2012; Linder et al. 2008; Mohai et al. 2009), with communities of color up to 75 percent more likely to live near pollution from the fossil fuel industry (Fleischman and Franklin 2017). Socioeconomic and racial disparities to outdoor air pollution are related to inequities in the proximity of communities to these environmental hazards (Brender et al. 2011). However, as noted in Chapter 2, disparities in absolute exposure to air pollution have been found to be larger for racial and ethnic groups than for income categories. For example, racial–ethnic exposure disparities to air pollution from emissions sources are found to be consistent across incomes and within urban and rural areas (Liu et al. 2021; Tessum et al. 2021). Furthermore, a study quantifying $PM_{2.5}$ exposure disparity by emission type found that the industry, light-duty gasoline vehicles, construction, and heavy-duty diesel vehicles sectors are responsible for the largest emission disparities for Black, Hispanic, and Asian populations when compared to White populations (Tessum et al. 2021). Changes in passenger vehicle activity during COVID-19 lockdowns also revealed further disparities in impact of NO_x emissions from heavy-duty vehicles (Kerr et al. 2021).

> Finding 3-2: Siting of electricity generating facilities and other large industries, as well highways and roadways with light- and heavy-duty vehicles in the United States has a legacy of disproportionately harming Black, Indigenous, and low-income communities through higher exposure to criteria air pollutants, especially $PM_{2.5}$. Existing disparities in exposure to air pollution need to be recognized to ensure that the siting of decarbonization infrastructure does not worsen them during the transition.

Redlining, Air Pollution, and Heat Islands

The legacy effects of redlining—a 1930s process through which areas with high populations of people of color, older housing, and/or poorer neighborhoods were deemed hazardous for home loans—include increased exposure to urban heat islands (Hoffman et al. 2020), increased exposure to air pollution (Lane et al. 2022; Rothstein 2017), and higher rates of asthma-related emergency room visits (Nardone et al. 2020). The impact and occurrence of heat islands can be reduced through the strategic placement of reflective surfaces and green space. However, redlined areas often do not have these features and are therefore more impacted by the adverse outcomes of heat islands, especially heat-related mortality. Figure 3-1 shows a comparison of greenspace and the occurrence of heat islands within Richmond, Virginia.

Adaptation to these adverse effects of redlining commonly lead to energy intensive actions, such as increased air conditioning to combat high outdoor temperatures (Abel et al. 2018). Furthermore, as shown in Chapter 2 (see Figure 2-3), redlining has caused air pollution exposure to persist in racial-ethnic minority communities despite the national decrease of $PM_{2.5}$ exposure overall (Tessum et al. 2021). See the section

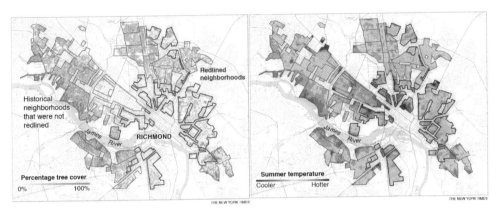

FIGURE 3-1 Legacy impacts of redlining in Richmond, Virginia: (a) tree cover, and (b) summer temperatures compared to the city average. SOURCE: Plumer and Popovich (2020). © 2022 The New York Times Company. All rights reserved. Used under license.

"Built Environment Co-Benefits" below for more information about how the effects of redlining and the current energy system can be mitigated with the transition to a net-zero energy system.

Decarbonization policies can reduce current disparities, with the most effective policies acknowledging these disparities and targeting emission sources causing the most harm (Tessum et al. 2021). Additionally, subnational analysis of national decarbonization strategies and their impacts show the importance of considering co-pollutant emissions when decarbonizing the electricity sector to reduce the inequities of $PM_{2.5}$ exposures (Goforth and Nock 2022). That is, although $PM_{2.5}$ has been identified as a risk factor for premature death, decarbonization actions that simultaneously target specific emission sources and reduce other criteria pollutions associated with fossil fuel combustion will support the reduction of air pollution disparities. Furthermore, siting of new energy facilities need to consider the health impacts neighboring communities will face and engage with communities to inform them of the risks and benefits of new energy facilities will be the most successful during the transition. See Chapter 5 for more information about engaging affected communities in energy transition decision-making.

Climate Change and Air Pollution Intertwined

Adverse health impacts from climate change are extensive and provide some of the most compelling motivation for climate mitigation. Although this chapter primarily focuses on the health impacts of the transition to net-zero energy, Box 3-2 briefly reviews some of the major health impacts linked to climate change.

BOX 3-2
HEALTH RISKS FROM CLIMATE EXTREMES

Warming since 1850–1900 has increased the frequency and intensity of extreme climatic events globally, including extreme heat and cold, heavy precipitation, floods, droughts, desertification, dust storms, and wildfires (Diffenbaugh et al. 2017; Ebi et al. 2021), and climate change is projected to exacerbate this trend (Cissé et al. 2022). These climatic events cause significant human mortality and adversely affect human health (Alderman et al. 2012; Bressler 2021; Ebi et al. 2021; GCRP 2016). Health costs from premature mortality, health care, and lost wages during these events can reach billions of dollars (Knowlton et al. 2011; Limaye et al. 2019). Long-term negative effects include increased respiratory illnesses; vector-borne, water-borne, and food-borne diseases; food insecurity, and detrimental mental health impacts (Cissé et al. 2022; Limaye and De Alwis 2021).

As the intensity and frequency of heatwaves increases, death tolls and hospitalizations from heat are also increasing globally (Hayashida et al. 2019; Kollanus et al. 2021; Li et al. 2012; Limaye et al. 2019; Vogel et al. 2019). For example, in 2020, about 17,000 premature deaths were attributable to heat exposure (Shindell et al. 2021) and in June 2021, the number of heat-related emergency department visits was 69 times higher than during the same period in 2019 (Schramm et al. 2021). Adaptation to heat with air conditioning use requires more energy, which could in turn increase air pollutants from that energy use, leading to a 5–9 percent increase in air-pollution-related mortality from building electricity demand (Abel et al. 2018), if fossil fuels continue to be used to generate electricity.

Predicting the impact of policies on prevention of direct climate change health impacts, like excess heat-related deaths or improved air quality, can be challenging. One study finds that, while air pollution impacts are more immediate, U.S. premature deaths from heat exposure increases from about 20,000 premature deaths annually in this decade to 100,000–150,000 premature deaths per year by 2070 even with U.S. and global decarbonization action (Shindell et al. 2021). Comparatively, the Climate Impact Lab's Lives Saved Calculator,[a] which uses a damage function to calculate deaths and health costs of future climate change, predicts that globally 7.4 million annual premature deaths and $3.7 trillion in annual adaptation costs (e.g., building cooling centers, installing air conditioning) could be avoided if the United States achieves net-zero emissions by 2050 (Climate Impact Lab 2022).

Increasing aridity is likely to lead to increased drought, dust, and wildfires in some regions, with associated health issues. Wildfires have increased dramatically in the past decades, driven in part by climate change, and continue to increase (Burke et al. 2021; Ford et al. 2018; Romanello et al. 2022). Wildfire activity is associated with premature mortality and increased hospital admissions for respiratory and cardiovascular incidents from smoke (Neumann et al. 2021). Rising aridity in the U.S. Southwest (Overpeck and Udall 2020) is expected to increase fine and coarse dust levels, triggering additional mortality and hospitalizations for cardiovascular conditions and asthma (Achakulwisut et al. 2019). Furthermore, drought, fires, and excess heat are likely to stress agricultural production, impacting food security, human health (including outdoor workers), and livestock health (Bezner Kerr et al. 2022; Gowda et al. 2018).

BOX 3-2 Continued

Over half of all human pathogenic disease can be aggravated by climate change (Mora et al. 2022). Rising temperatures will increase water-related illnesses (Limaye and De Alwis 2021; Trtanj et al. 2016) and food-borne diseases (Cissé et al. 2022). Additionally, allergies and respiratory diseases, such as asthma, are also predicted to be enhanced by rising temperatures, which induce longer pollen seasons and higher pollen concentrations (Anderegg et al. 2021; Ziska et al. 2011) and increased CO_2 levels, which increase the potency of aeroallergens (Bielory et al. 2012). Furthermore, climate change is expected to alter the seasonal and geographical activities of vectors, including mosquitoes and ticks, affecting the transmission of the infections that they carry, and may increase human exposure to vector-borne illnesses (Kraemer et al. 2019).

In addition to impacting physical health, climate change–related extreme events affect mental health through multiple pathways: extreme events, heat, and climate anxiety. Extreme events can lead to depression and posttraumatic stress disorder (Lowe et al. 2019). Other consequences of climate change and extreme events, such as displacements and malnutrition, were also linked to several mental health problems (Cissé et al. 2022). Hotter days were correlated with an increase in self-reported mental health issues in the United States and globally (Li et al. 2020; Obradovich et al. 2018). Last, the view of climate change as an existential threat was suggested to increase levels of stress, anxiety, and hopelessness, of particular concern for young people (Hickman et al. 2021; Ojala et al. 2021; Palinkas and Wong 2020).

---

[a] To view the Lives Saved calculator, see https://lifesaved.impactlab.org.

Despite the extensive negative health effects of climate change, research shows the health effects from fossil fuel combustion alone are much larger than those associated with climate change. For example, Shindell et al. (2021) found health benefits from improved air quality outweighed those related to avoided climate change. Additionally, the economic benefits of avoided climate change, which includes the monetized impacts of avoided heat exposure and extreme weather events, have been found to be smaller values than the economic benefits of reduced fossil fuel combustion (EPA 2011; Markandya et al. 2018; Vandyck et al. 2018). However, there are decarbonization pathways that will directly address the impacts of fossil fuel combustion while indirectly lessening the adverse health impacts of climate change. For example, decarbonization pathways that increase green space may also reduce allergy and respiratory impacts, and mental health impacts associated with climate change. The rest of this chapter will focus on the health and monetized benefits predicted to follow from decarbonizing the U.S. energy system.

HEALTH CO-BENEFITS FROM DECARBONIZATION

Large health benefits come from the associated reduction in air pollution when fossil fuel combustion is reduced. For example, renewable energy sources such as wind and solar reduce GHGs and have the co-benefit of reduced ambient air pollution emissions relative to fossil fuel combustion. While reduced air pollution will likely provide the largest amount of health co-benefits, others, such as green space and infrastructure for active travel, can yield additional mental and physical well-being co-benefits (Grabow et al. 2012; Nieuwenhuijsen 2021; Raifman et al. 2021; Younkin et al. 2021). This section describes co-benefits of decarbonization for air quality, built environment, transportation, water quality, nutrition, and occupational health.

Air Quality Co-Benefits

As the country transitions to net-zero emissions and most fossil fuel combustion is phased out, positive air quality health co-benefits are universally expected, although studies relying on different methodologies or modeling different policies show considerable variation. Some studies use models to assess how future policies and actions—such as phasing out fossil fuel combustion (Goodkind et al. 2019; Mailloux et al. 2022; Penn et al. 2017; Shindell et al. 2021), replacing some energy generation with emissions-free renewables (Abel et al. 2018; Driscoll et al. 2015; Prehoda and Pearce 2017), or increasing energy efficiency (Abel et al. 2019)—affect air pollutant emissions and subsequent human exposures.

While estimates of the number of deaths avoided and health costs vary, simulation modeling indicates that future decarbonization of electricity generation would prevent thousands of deaths per year in the United States. Shindell et al. (2021) find that decarbonizing in the United States to maintain a 2°C pathway could prevent 4.5 million premature deaths and 1.4 million hospitalizations and emergency room visits. Comparatively, Mailloux et al. (2022) predicts nationwide efforts to eliminate energy-related emissions across the electric, transportation, building, and industrial sectors could result in 53,000 avoided premature deaths per year and approximately $610 billion annual savings. Figure 3-2 depicts the projected decrease in ambient $PM_{2.5}$ from the simultaneous removal of $PM_{2.5}$, SO_2, and NO_x emissions from energy-related sectors: electricity fuel use, industrial fuel use, residential fuel use, on-road vehicles, non-road vehicles, and oil and gas production and refining. Other benefits of decarbonizing are also being quantified. In addition to $56 trillion to $163 trillion in public health benefits from 2020–2100, decarbonizing in the United States could prevent

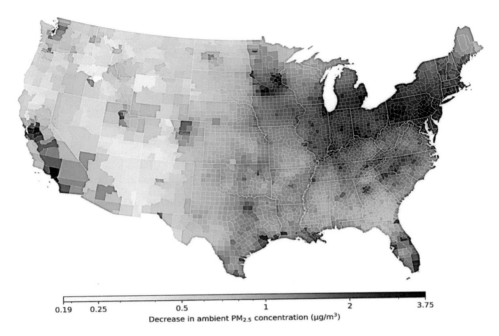

FIGURE 3-2 Projected decrease in ambient PM$_{2.5}$ concentration from the simultaneous removal of PM$_{2.5}$, SO$_2$, and NO$_x$ emissions from energy-related sectors.
NOTE: The energy-related sectors included in this estimate are electricity fuel use, industrial fuel use, residential/commercial fuel use, on-road vehicles, non-road vehicles, and oil and gas production and refining. SOURCE: Courtesy of Mailloux et al. (2022), https://doi.org/10.1029/2022GH000603. CC BY-NC 4.0.

300 million lost workdays, and 440 million tons of crop losses (Shindell et al. 2021) over the next 20 years.

Numerous studies have quantified health benefits for specific decarbonization policies, such as replacing fossil fuel combustion for energy generation with renewable options, and have found that benefits vary based on region, technology, and methodology (e.g., model assumptions, extent of life-cycle assessment). For example, Wiser et al. (2016a) finds that $77 billion–$298 billion in air quality and public health benefits will result from future solar energy use but acknowledges the uncertainty from existing estimates of GHG impacts and air pollution. Similarly, McCubbin and Sovacool (2013) estimate avoided health and non-health externalities for wind power plants in Idaho are between $18 million and $104 million and in California are between $560 million and $4.38 billion, noting that the ambiguity is owing to emission rate and location, and estimate of effect.

Impacts from the Infrastructure and Jobs Act and Inflation Reduction Act

The recent major U.S. climate-change-related laws, the Infrastructure Investment and Jobs Act (IIJA) (P.L. 117-58) and the Inflation Reduction Act (IRA) (P.L. 117-169), have health co-benefits associated with many of the provisions in the bills.[2] Together, the IIJA and IRA invest in the nation's energy and transportation sectors through appropriations and authorization which will have direct and indirect impacts on GHG emission reduction goals. The health co-benefits from these investments include avoided premature deaths from $PM_{2.5}$ and increased opportunity for active travel. However, a majority of the provisions have yet to be implemented and the true impacts of the bill cannot yet be reported. In lieu of a peer-reviewed analysis of the bills, policy analysis and previous estimates of potential co-benefits can help with estimating the health impacts. This section highlights modeling studies that estimate the health impacts of the IIJA and IRA.

The IRA and IIJA include appropriations that are likely to help both reduce GHG emissions and improve air quality (see Appendix H). The provisions include appropriations for fleets of zero-emission medium- and heavy-duty vehicles, appropriations for light-duty EVs, and spending programs to reduce air pollution at ports; tax credits for electricity produced from renewables or new solar and wind facilities in low-income communities; appropriations for construction of renewable energy facilities and energy efficient buildings; spending programs for improving air quality monitoring for underserved populations and at schools and block grants for environmental and climate justice projects; and appropriations for clean energy projects, including the EPA Greenhouse Gas Reduction Fund, which was modeled on successful green banks. Multiple groups have attempted to quantify the air quality co-benefits from the IRA. Rhodium Group's analysis suggests that SO_2 and NO_2 emissions will be reduced from 2021–2030 from the IRA, as compared to baseline modeling without the IRA (Larsen et al. 2022).

The improper implementation of IIJA and IRA has been identified as a key barrier to accessing the health and decarbonization benefits of the provisions. For previous climate-related policies, it has been observed that the monetized health impacts from air quality improvements have been found to either partially offset (Sergi et al. 2020; Shindell et al. 2021; Thompson et al. 2016) or exceed the up-front cost of implementing policies and funding incentives (Abel et al. 2018; Buonocore et al. 2019; Wang

---

[2] It should be noted that the IIJA and IRA are not equivalent in funding mechanisms. The IIJA consists of a mix of authorizations and appropriations while the IRA primarily consists of spending programs (appropriations) and tax expenditures. Appropriations are laws that provide money for government programs and must be passed by Congress every year in order for the government to continue to operate. Spending programs can allocate federal resources to projects and activities up to the amount of their appropriation. By contrast, tax expenditures typically have no limit on the amount that could be claimed by taxpayers.

et al. 2020). According to a presentation from the REPEAT Project, the IRA could have health benefits of reducing 5,800 premature deaths from particulate matter annually by 2030 (Jenkins et al. 2022). Comparatively, Resources for the Future (RFF) estimate that IRA policies focused on electricity generation will lead to at most 1,300 avoided deaths with an associated $12 billion–$22 billion in health benefits in the year 2030 alone (Roy et al. 2022). An NREL modeling report estimates that the cumulative impacts of both the IIJA and the IRA may result in 4,200–18,000 avoided premature deaths and $45 billion–$190 billion in avoided health damages, and estimated reductions in SO_2 and NO_x from the 2023 to 2030 (Steinberg et al. 2023). Additionally, Energy Innovation indicated that the IRA could prevent at most 3,900 premature deaths, 100,000 asthma attacks, and 417,00 lost workdays in the year 2030 alone owing to corresponding reductions in air pollution, specifically particulate matter (Mahajan et al. 2022). The benefits of recent U.S. decarbonization policies are expected to continue beyond the modeled 2030 outcomes, assuming no changes occur to remove them. It is also expected that additional legislation will be passed to support a national net-zero energy system by 2050. Calculating the effects of policy implemented beyond 2030 increases the cumulative benefits.

Other components of the IIJA and IRA specifically target low-income, disadvantaged, and environmental justice communities, which could help with health equity goals. However, because most of the bills' programs offer only incentives, such as tax credits, there will not be equitable outcomes unless the incentives are appropriately implemented and work as intended. For example, if incentives to electrify transportation and home heating are highly successful while simultaneously incentives to site and deploy renewables are not, air quality could decrease and fossil fuel use could increase through 2030. (See Chapter 5 for more about siting processes.) If incentives work as planned, improvements in air quality, active travel, and mitigation of climate change will provide substantial positive health benefits.

Some provisions in both pieces of legislation may have additional health impacts from infrastructure that supports active travel, such as biking and walking. Provisions in the IIJA that can be used to improve active travel include: the Surface Transportation Block Grant program (§11109); Congestion Mitigation and Air Quality Improvement program (§11115); Safe Routes to Schools Program (§11119); bicycle transportation and pedestrian walkways (§11133); and funding to increase streets for safe and accessible transportation (§11206). See the section "Transportation Co-Benefits" for more about active travel and related health benefits.

> Finding 3-3: Serious and widespread negative health consequences would continue from fossil fuel use under the "business as usual" greenhouse gas emissions projections. Recent climate legislation, especially the IRA, will have direct and indirect impacts on U.S. emissions, mainly through air quality improvements.

The improvements in air quality anticipated from implementation of the IRA will prevent thousands of premature deaths and thus provide significant monetary benefits. Even larger health and monetary co-benefits could occur with further reductions in fossil fuel combustion and deployment of technologies to manage combustion emissions, both of which are supported by various IRA and IIJA provisions to meet the net-zero emissions goal.

Regional Variation

Health benefits from emissions reductions will depend on regional factors, including weather, population density, and the type of emissions sources. For example, atmospheric chemistry with natural precursor emissions (like biogenic VOCs) can affect the formation of ozone. Wind speed and direction also affect where air pollution ends up. When densely populated regions are affected, the value of total co-benefits is larger because more people are affected; likewise, the underlying characteristics of the population that affect vulnerability to air pollution (e.g., the elderly, or children) can also affect the calculated health benefits. Lastly, regional variation in emissions sources can affect the co-benefit estimate—using wind or solar to replace a highly polluting coal-fired power plant versus a somewhat less polluting natural gas-fired plant.

Health benefits from replacing existing electricity sources with wind, rooftop solar, and utility solar energy in 2017 have been estimated across the 10 U.S. electrical grid regions (Buonocore et al. 2019). Figure 3-3 illustrates the results of the study, highlighting how different renewable energy technologies relate to CO_2 reductions and monetized health benefits. As shown below, renewable energy deployment offers the highest health benefits in the Great Lakes Mid-Atlantic region, followed by the Upper Midwest, and then the Northeast. The main differences in health benefits are owing to the fuel type and corresponding emissions displaced, and size of the population affected. For example, the Great Lakes Mid-Atlantic electrical grid region (including the Ohio River Valley region) benefits from decarbonization as a higher concentration of coal plants would be replaced with cleaner energy, and substantial populations downwind are affected. Likewise, in the Northeast, gas and oil would be reduced, affecting the high population density region, resulting in high health benefits per ton of CO_2.

Localized health benefits from future decarbonization efforts will vary by location. For example, between 32 percent and 95 percent of the health benefits from eliminating emissions in a region will remain in that region, with marked state-to-state variability

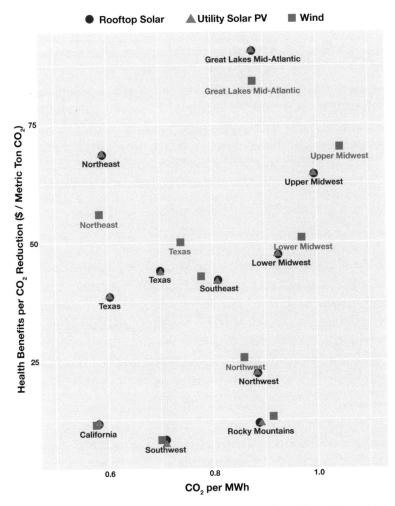

FIGURE 3-3 CO_2 reductions and health benefits per ton of CO_2 reduced, by region and renewable energy type. NOTES: Red circles represent rooftop solar, green triangles represent utility solar PV, and blue squares represent wind. Points for rooftop solar and utility solar PV overlap. Not all points are labeled to prevent overplotting. SOURCE: Courtesy of Buonocore et al. (2019), https://doi.org/10.1088/1748-9326/ab49bc. CC BY 3.0.

(Mailloux et al. 2022), as pollutants easily cross county (Sergi et al. 2020) and state boundaries (Dedoussi et al. 2020). The regional nature of air pollution impacts has equity implications, as different decarbonization strategies may not bring emissions reduction benefits to all demographic groups at the same scale and/or pace (Goforth and Nock 2022). It will be important to consider both health and equity impacts of decarbonization technologies during this energy transition.

Built Environment Co-Benefits

The built environment encompasses homes, workplaces, neighborhoods, and metropolitan and regional geographies. Globally, 30 percent of final energy consumption and 26 percent of energy-related emissions come from the planning, design, maintenance, and disposition of the built environment (EIA n.d.). This section identifies the health co-benefits that stem from improving the two key challenges within the built environment: heat islands and energy use. See the section "Identifying Potential Health Risks" for more about retrofitting existing buildings. See Chapter 7 for more information about additional challenges that will be faced during the energy transition.

Urban Green Space

Local decarbonization and health action can reduce urban heat islands along with the GHG emissions associated with the energy demand for cooling. Urban heat islands can be mitigated with cool surfaces (e.g., roofs and pavement designed to help reflect rather than absorb sunlight) as well as green space. Cool surfaces can reduce urban air temperatures and up to 6 percent of GHG emissions, according to one study (Azarijafari et al. 2021; EPA 2008). However, the true reduction of GHG emissions and heat islands will be context-specific, with some neighborhoods seeing less reduction than others.

Creating green space in built environments can provide many potential benefits related to health and decarbonization, including reducing urban heat, reducing air pollution, and benefiting mental health. Green spaces reduce urban heat by increasing evaporative cooling, creating shade, and altering wind around buildings, and this heat reduction has both energy and health impacts. Urban forests[3] in the United States reduce electricity use by 38.8 million MWh annually, with the average reduction in residential energy use from trees estimated at 7.2 percent (Nowak et al. 2017). For health impacts, several studies indicate that green space, along with other sociodemographic factors, may decrease heat-related mortality risk (Choi et al. 2022; Gronlund et al. 2015). Additionally, greater tree canopy cover is associated with reduced ambulance calls during extreme heat events (Graham et al. 2016). However, a case study in Chicago demonstrates that a green roof on a new building did not create the same heat mitigation effect as the field the building was constructed on (McConnell et al. 2022). It will be important to understand the limit of green spaces to mitigate GHG emissions or reduce heat islands in built-up landscapes.

---

[3] Trees in cities and suburban areas.

Other potential co-benefits from green space include improved air quality, mental health, and stormwater management. Urban vegetation can also help remove air pollutants from the atmosphere. The EPA estimates that urban forests in the U.S. net annual sequestration is 37,580,224 metric tons of carbon (EPA 2023b). Green space also has been found to be associated with lower levels of depression, anxiety, and stress, after controlling for many confounding factors (Beyer et al. 2014). Green infrastructure can additionally help stormwater management, improving water quality and reducing runoff (Kuehler et al. 2017). Considering the health impacts in total, a study in Portland, Oregon, finds that one premature death can be avoided for every 100 trees planted, with older trees providing more value (Donovan et al. 2022).

While creating green space offers multiple benefits, there are limitations and potential trade-offs involving green space depending on the specific choice of plants (Wolf et al. 2020). Many factors in how well vegetation can provide heat, air quality, and water quality benefits, as well as the overall maintenance and water usage required, may depend on the type, diversity, and density of vegetation (Fineschi and Loreto 2020; Rambhia et al. 2023). Two potential disbenefits include increase in pollen, which can exacerbate asthma and allergies (Sousa-Silva 2021; van Dorn 2017), and emissions of VOCs, which may contribute to ozone (Drewniak et al. 2014; Sousa-Silva 2021; van Dorn 2017). Selection of trees that are known to be less allergenic or diversifying tree planting overall may improve urban health.

Energy Efficiency

Substantial health co-benefits can result from decarbonizing the built environment, specifically from energy efficient buildings. These co-benefits include improved mental and physical health, and reduced risk of dehydration and excess winter mortality in hot climates (IEA 2019). However, challenges exist in the implementation energy efficiency policies buildings. For example, the primary use of voluntary financial incentives requires extensive coordination to be successful (see Chapter 7).

This report will examine the impacts of energy efficiency in buildings by exploring the sector most pertinent to health—the health care system. Within the buildings sector, the health care system contributes 8.5 percent of U.S. GHG emissions when estimates include buildings, purchased electricity, and supply chain emissions, which is nearly double the global average of health care emissions of about 4.5 percent (Eckelman et al. 2020). According to the Commercial Buildings Energy Consumption Survey, inpatient health care was the third highest energy intensity per square foot for commercial buildings in 2018, after food service and food sales, and was also the sector with the largest decrease since 2012 (EIA 2022). Note that other chapters will discuss the emissions associated with electricity (Chapter 6) and the built environment (Chapter 7) as separate sectors in of themselves.

The health care sector can act as a leader in promoting innovation for health, equity, and climate change. Many organizations have already taken steps to reduce GHG emissions and support health equity in this sector. For example, the Department of Health and Human Services announced the Health Sector Climate Pledge in April 2022 to encourage organizations to commit to lowering their GHG emissions and to building more climate resilient infrastructure (HHS n.d.). Within a year, more than 100 stakeholders—hospitals, health centers, insurance companies, pharmaceutical companies, and so on—have signed the pledge. In addition to the commitment of federal health systems, about 15 percent of U.S. hospitals have committed to reducing GHG emissions (HHS n.d.). Addressing the full life-cycle energy demand from the health sector—from facilities and products (e.g., anesthetic gases), to electricity and steam sources, and ultimately to supply chains—is the motivation of the National Academy of Medicine's Grand Challenge on Climate Change, Human Health, and Equity[4] as well as other initiatives within the health care sector.

Transportation Co-Benefits

Like decarbonization of electricity generation, decarbonization of transportation can also have health co-benefits from fewer emissions. Transport decarbonization can also yield benefits through physical fitness. In addition to air pollution, transportation health considerations include physical fitness and avoided emissions from road dust. One study that explored reducing short car trips in 11 cities in the upper midwestern United States found that the combined benefits from improved air quality plus the added physical fitness benefits from making 50 percent of short trips via biking could result in a total of nearly 1,300 lives and $8 billion saved annually from avoided morbidity and mortality (Grabow et al. 2012). However, owing to the relative lack of research in transportation health co-benefits compared to air quality co-benefits, there are limited tools and literature to assess the immediate impacts of the initiatives. Furthermore, the health impacts of transportation modes, such as mass transit and active transport, can be challenging to model across large scales.

---

[4] The National Academy of Medicine's Grand Challenge on Climate Change, Human Health, and Equity seeks to improve and protect human health and equity through a multi-year global initiative. For more details, see https://nam.edu/programs/climate-change-and-human-health.

Vehicle Transportation

Several pollutants are associated with the transportation sector, particularly tailpipe emissions from fossil fuel combustion, including $PM_{2.5}$, NO_x, and ozone. One study estimates that 22,000 out of 115,000 deaths attributable to $PM_{2.5}$ and ozone came from the transportation sector, with health damages costing approximately $210 billion (2015$) in the United States in 2015 (Anenberg et al. 2019). At a finer spatial resolution, many reviews have noted the near-highway health effects from motor vehicles emissions, especially in urban areas (e.g., Brugge et al. 2007; HEI Panel on the Health Effects of Traffic-Related Air Pollution 2010; Khreis et al. 2020). Studies have shown how decarbonization of vehicles, especially the electrification of the heavy-duty vehicle fleet, can provide significant health and equity co-benefits (Ramirez-Ibarra and Saphores 2023; Zhu et al. 2022) and will play a large role in decarbonization, given the large amount of GHG emissions from the transportation sector and the large percentage of transportation emissions resulting from passenger, freight, and other vehicle trips. However, even with a fully electric fleet, vehicles would still produce transportation-related pollution, especially coarse particulates, from brake dust, tire abrasion, and road dust (Liu et al. 2022; Timmers and Achten 2016).

Active Transportation

Active travel, including walking and cycling, can reduce emissions associated with gasoline combustion and improve health through increased physical activity (Castillo et al. 2021; Celis-Morales et al. 2017; Dinu et al. 2019; Hamer and Chida 2008; Kelly et al. 2014; Mueller et al. 2017). A systematic review and meta-analysis found that people who participated in active transportation had an 8 percent reduction in all-cause mortality, a 9 percent reduction in risk of cardiovascular disease, and a 30 percent reduction in risk of diabetes (Dinu et al. 2019). For cyclists, a 24 percent reduction in all-cause mortality and a 25 percent reduction in cancer mortality was also identified (Dinu et al. 2019). Additionally, several studies find the health benefits of active transport are much higher than the health risks of traffic collisions while walking or cycling (Maizlish et al. 2022; Mizdrak et al. 2019; Mueller et al. 2015). Accounting for both avoided deaths from increased physical activity and additional traffic deaths, an analysis of multiple scenarios of active transportation infrastructure investments found that the benefits of avoided deaths greatly exceed the costs of building the infrastructure (Raifman et al. 2021).

In the United States, 52 percent of trips are less than 3 miles, with only 2 percent of trips greater than 50 miles (DOE-EERE 2022). Given that most emissions are from longer trips unsuitable for biking and walking, active transportation is likely to have a small impact on GHG emissions from transportation. Based on Bureau of Transportation Statistics data, doubling active transportation and public transit is likely to replace less than 5 percent of miles traveled in personal vehicles and a similar percentage of GHG emissions (BTS 2017). Nonetheless, the health benefits from active travel are substantial, and health promotion priority for walkable and bikeable communities within transportation decarbonization planning is elevated because small incremental increases in routine exercise at the population level offer substantial health benefits (Mueller et al. 2017; Pedersen and Febbraio 2012).

To what extent active transportation replaces personal vehicle travel depends on the regional planning and infrastructure. Examples of short-term changes of increases in bike lanes in some areas have demonstrated that large, rapid increases in cycling are possible when the infrastructure exists to support people's ability to bike safely (Kraus and Koch 2021). Safety for cycling could be improved with physical separation for bike lanes; one study estimates that benefits would be 10–25 percent greater than the costs (Macmillan et al. 2014). Increases in mass transit ridership may also increase physical activity if replacing automobile trips. While riders of mass transit may not be active during a bus ride or train ride, 90 percent of transit riders walk to or from their transit stops (NASEM 2021). In the United States, this is a median time of 19 minutes of daily walking to and from transit (Hirsch et al. 2018; Xiao et al. 2019).

Few studies have explored the equity implications of active transportation. People of low income often rely on active travel owing to lack of vehicle access, and while they may benefit from physical activity, there may be other equity concerns, such as lack of nearby access to healthy food or health care (Hansmann et al. 2022). There are also many barriers to active transportation, including lack of safe street infrastructure (e.g., sidewalks, bike lanes, etc.), pollution exposure, exclusionary zoning, crime and policing, harassment, and racism (Agyeman and Doran 2021; Barajas 2021; Brown 2016). Some of these barriers can be overcome with equitable urban planning that considers the health impacts of and access to active transportation. However, the needs and constraints for each community will vary depending on their transportation needs, climate, and health goals, and thus active travel and transportation infrastructure decisions need to be made at the local level.

Water Quality Co-Benefits

Water and energy are interrelated concepts: water is used within all phases of energy production and energy is required to pump and deliver water (DOE 2014). If the transition to net-zero is not completed with an intentional focus on impacts on water use and availability, the United States may risk increasing existing water stress[5] (IEA 2020). Although most studies of impacts of climate change on water quality focus on the hydrologic cycle (e.g., increased contamination from flooding), climate mitigation strategies may also impact water quality, with corresponding health effects. Decarbonization may impact water quality directly via pollution or more indirectly through changes in water withdrawal and consumption, which affects the availability of quality water sources.

Some current energy sources directly produce water pollutants. Steam electric power plants (powered by fossil or nuclear fuels) generate an estimated 30 percent of all toxic pollutants that are discharged into surface waters from industry (EPA 2015; Massetti et al. 2017). Toxic metals (including lead and arsenic) are of special concern in water, as they are consumed by fish and wildlife and accumulate to dangerous levels, and may be eaten by people in the community, with deadly effects on health (CDC n.d.). Mining and other extractive activities can also negatively impact water quality as a result of acid mine drainage, contamination from tailing ponds, and pollution from oil and gas well runoffs (de Oliviera Bredariol 2022). Relatedly, some processes to mitigate and reduce water pollution, such as desalination, are energy intensive, expensive, and often come with environmental impacts related to waste management (Molinos-Senanta and SalaGarrido 2017; Shahzad et al. 2017).

The reductions in fossil fuel use expected in a net-zero energy system could lead to improvements in water quality from decreased discharges of toxic pollutants. The electric sector in the United States is one of the largest withdrawers of water (removing and returning to a source, often at higher temperatures), although it consumes (removal without return) only 6 percent of the nation's water (Cameron et al. 2014). Several analyses find potential for decreasing water withdrawals and consumption with increasing renewable use (especially wind and, to some extent, solar), although the extent varies by region based on the portfolio of technology (Barbose et al. 2016; Ou et al. 2018; Wiser et al. 2016a). Increased mining for the critical minerals needed for clean energy technologies could negate some of these potential water quality benefits, although there are recommended technologies and policies to mitigate environmental impacts of mining (IEA 2021). Likewise, agriculture for biofuels, mining, and fracking processes can have adverse impacts on water quality. For example, runoff can contribute to excess nutrient contamination and algal blooms. (See the section "Identifying Potential Health Risks" below.)

---

[5] For more information about the water–energy nexus and the predicted impacts energy transitions to net-zero will have on water availability, see https://www.iea.org/articles/introduction-to-water-and-energy.

Improved Nutrition Co-Benefits

Shifting from diets high in animal products, particularly red meat and processed meat, to more plant-based diets can reduce GHG emissions and improve health. This shift can also provide other environmental benefits, including improved water quality and decreased nutrient runoff. The EPA estimates that 11.2 percent of national GHG emissions from 2020 are attributable to agriculture with this sector's electricity-related emissions accounting for 0.6 percent (USDA 2022). See Chapter 8 for more on the GHG and land-use impacts of agriculture. A realistic healthy diet, as defined by the *Dietary Guidelines for Americans*, could reduce U.S. food-system energy use by 3 percent, and, if further maximized for energy efficiency, a 74 percent reduction in food system energy use could be achieved (Canning et al. 2017).

There are numerous significant health benefits of sustainable and plant-based diets (Gao et al. 2018; Tilman and Clark 2014) and for diets when meat consumption is reduced (Biesbroek et al. 2014; Scarbough et al. 2012). Most significant is a decreased risk of all-cause mortality. For example, adhering to the EAT-Lancet Commission's sustainable diet guidelines could prevent 11.6 million deaths per year among adults worldwide (Willet and Rockström 2019). Similarly, according to a UK-based study, following the World Health Organization's dietary recommendations could also reduce deaths, improve life expectancy by 8 months, and reduce GHG emissions (Milner et al. 2015). Additionally, three prospective cohort studies in the United States found that a healthy plant-based diet reduced the risk of developing type 2 diabetes by about 20 percent (Satija et al. 2016).

There are GHG trade-offs associated with all diets, including harvest emissions, transport emissions, and issues of food waste. Global transport of food accounts for 19 percent of total food-system emissions with high-income countries accounting for 46 percent of international food-miles and food-miles emissions (Li et al. 2022). Domestically, Li et al. (2022) found that the emissions from food-miles for fruits and vegetables were 0.61 Gt CO_2e, whereas the food-mile emissions for meat were 0.007 Gt CO_2e. In comparison, the food production emissions for meat were 2.00 Gt CO_2e versus 0.03 Gt CO_2e for fruits and vegetables (Li et al. 2022). In addition to transportation-related emissions, roughly one-third of food in the United States is never eaten, representing a significant waste of resources and embodied emissions. See Chapter 8 for more information about reducing food waste to support climate mitigation.

Current policies indirectly subsidize the costs of animal products and encourage the use of concentrated animal feeding operations (CAFOs) (Horrigan et al. 2002; Sealing 2008; Story et al. 2008). CAFOs enable a large and cheap supply of meat in the U.S. food system and contribute a variety of air and water quality issues, as well as increase the risk of emerging infectious diseases (Hribar and Schultz 2010). Because animal products produced by this system also cause environmental health impacts, federal spending on agriculture has not been optimized to promote human health (Mozaffarian et al. 2019). For example, current spending on corn and soy to support CAFOs outweighs spending on other fruits and vegetables. Policies must be modified to accurately include the health and environmental impacts in food prices (i.e., healthy and sustainable will also be cheap and affordable). The current system does not reflect "the true cost of food," which would incorporate the cost of negative health and environmental externalities (Rockefeller Foundation 2021). Changes to the system need to consider the benefits of incorporating health and environmental impacts and the risks of heightening food security disparities with increased food prices.

> Finding 3-4: Shifting from diets high in animal products to more plant-based diets can reduce GHG emissions, especially methane, from food production and improve health. Plant-based diets are associated with a lower risk of type 2 diabetes, coronary heart disease, stroke, and cancer.

> **Recommendation 3-1:** *Phase Out Incentives for the Highest Greenhouse Gas (GHG)-Emitting Animal Protein Sources.* **Congress and the U.S. Department of Agriculture should phase out incentives for the highest GHG-emitting animal protein sources, such as beef, and increase incentives toward sustainable, low-emission production of fruits and vegetables. This could reduce the risk for farmers, create greater access to fruit and vegetables, and reduce production and consumption of high-emission animal foods.**

Occupational Health Co-Benefits

Moving to a decarbonized energy system will decrease jobs in industries with substantial health risks. Analysis of the comparative risk of severe injuries has found that fossil fuels jobs entail greater health risk (higher overall fatalities and higher fatalities per unit of energy produced) than jobs in wind or solar electricity, hydropower, and nuclear energy (Burgherr and Hirschberg 2014). For coal, accidents and fatalities occur primarily in mining, while oil and natural gas have the largest proportion of accidents (e.g., spills, leaks) during transportation and storage (Burgherr and Hirschberg 2014). Although accidents comprise a small portion of overall health impacts compared to ambient air pollution, they can have major impacts on local environmental health (e.g., the *Deepwater Horizon* spill). In addition to accidents, coal miners are at risk of lung diseases including pneumoconiosis and COPD from their exposure to coal mine dust (NIOSH n.d.(b)). The risk of lung diseases may persist for miners of the critical minerals needed to create some net-zero energy technologies. See the section "Manufacturing Net-Zero Energy Technologies" below.

Maximizing Health Co-Benefits of the Energy Transition

Any approach to decarbonizing the electricity sector will decrease co-pollutant emissions compared to a scenario in which no additional carbon policies are implemented; however, some of these approaches may be more effective and faster than others at reducing regional, racial/ethnic, and socioeconomic disparities in exposure to air pollution (Goforth and Nock 2022). Goforth and Nock (2022) note that, given the equity implications of these different approaches, decision makers need to consider both national (e.g., total air pollutant reductions) and regional impacts of electricity sector decarbonization scenarios.

Health impact assessments (HIAs) are intended to analyze how decisions affect population health, including health impacts specific to disadvantaged communities (CDC 2016). These assessments can identify key health outcomes that need to be considered during program design and evaluated after program implementation. HIAs also help build public trust and acceptance and support data analysis to understand the scale of health benefits and harms relevant for decision-making on facility siting and the stringency of standards (Nkykyer and Dannenberg 2019). The use of HIAs that support adaptive management will be critical for the adjustment of projects that are not on target or are creating unforeseen adverse impacts. (See Chapter 1 about the critical role of adaptive management.) However, health impact assessment tools would benefit from technological advancement to improve ease of use and speed of results.[6]

> Finding 3-5: Decarbonizing the U.S. energy system has the potential to provide substantial health co-benefits including access to safe active transportation options, reduced heat islands, and reduced air and water pollution. However, there are risks to human health that need to be avoided during the energy transition. To maximize health co-benefits and minimize health risks, public health experts need to be engaged early in the decision-making process and often to broaden the consideration of health and equity impacts into planning decisions. Furthermore, affected communities need to be considered priority decision makers and should be consulted for relevant decarbonization actions that may pose adverse health risks. The coordinated engagement between experts and affected stakeholders will advance both public health and equity approaches of the energy transition. Therefore, successful decarbonization policies must include selection criteria and formal evaluation with health benefits in mind, in addition to equity and efficacy for carbon emissions reduction.

---

[6] EPA compiled a report enumerating the existing HIA tools and resources with the goal to generate a comprehensive list for HIA practitioners to use throughout the process. For more information, see Pepe et al. (2016).

Recommendation 3-2: *Increase Use of Health Impact Assessment Tools in Energy Project Decision-Making.* **Health impact assessment tools should be incorporated into the program design and evaluation processes of decarbonization policy with consideration for the full life-cycle impacts. The inclusion of health impacts into existing life-cycle assessments for decarbonization technologies will ensure that benefits and costs are considered. This will support adaptive management efforts by providing insight into which programs are not having the intended effects on public health. To support the increased inclusion of advanced health assessments in energy decisions**

a. **Congress should allocate new funds to the Centers for Disease Control and Prevention's National Center for Environmental Health/ Agency for Toxic Substances and Disease Registry to add a Climate Mitigation Health Co-Benefits component to the existing state health department-level Building Resilience Against Climate Effects program.**
b. **The Department of Health and Human Services Office of Climate Change and Health Equity should establish and convene meetings for a new interagency working group with the goal of developing a rapid health impact assessment tool to assess the health and equity risks and benefits arising from deep decarbonization and to mitigate risk to communities.**

IDENTIFYING POTENTIAL HEALTH RISKS OF DECARBONIZATION

While decarbonization is generally expected to produce large positive health co-benefits, some decarbonization strategies could contribute to health harms. To understand the multifaceted benefits of decarbonization, it is important to consider the full life-cycle costs and benefits of emerging energy technologies,[7] including related to public and personal health impacts, in comparison to fossil-dependent energy processes. Many energy technologies that do not burn fossil fuels are associated with reduced exposure to pollution (Chapman et al. 2018; Hawkins et al. 2013; Lee et al. 2012; Romero-Lankao et al. 2022). However, some of these technologies can also create risks to human health and quality of life. One example is biofuels, which unlike wind or solar, still do emit some air pollutants and would require additional land

---

[7] For a comparison of the health benefits and disbenefits of multiple energy generation technologies and processes, see Smith et al. (2013).

devoted to agriculture for energy crops, which could have negative impacts on water quality, water availability, and food security (Hill et al. 2009; Luderer et al. 2019). This section discusses the adverse health impacts that need to be considered during the energy transition.

Continued Combustion of Fuels

A net-zero future will likely still require combustion of fuels for certain applications to maintain reliable and affordable energy services. For example, biofuels or low-carbon synthetic fuels may be required to decarbonize aviation and some heavy-duty transportation, and natural gas burning power plants with carbon capture and sequestration (CCS) may be needed as a firm source of power to back up a mostly renewable grid. Use of low-carbon synthetic fuels or biofuels and combustion of fossil fuels with CCS could reduce or eliminate net CO_2 emissions, yet may still lead to release of air pollutants, including PM, NO_x, and ammonia (NH_3), that have the potential to degrade air or water quality (Driscoll et al. 2015; Tzanidakis et al. 2013; Veltman et al. 2010). In some cases, chemical- and site-related emissions from hydraulic fracturing, also known as fracking, have been linked to drinking water contamination and negative infant health outcomes, including preterm births, low birth weight, and lymphoma (Apergis et al. 2019; Currie et al. 2017; Hill and Ma 2022; Li et al. 2017; Schuele et al. 2022; University of Pittsburgh 2023).

Net-zero-carbon fuels, either derived from biomass or synthesized from CO_2 and H_2, can emit NO_x, CO, SO_x, and PM when combusted. Implementing CCS on existing fossil fuel powered plants will require using more energy than an equivalent energy output of a non-CCS equipped plant, and could increase emissions of NO_x, NH_3, and PM throughout the fuel life cycle, depending on the source of energy to run the capture unit (Tzanidakis et al. 2013). At the capture location itself, NO_x and PM emissions are likely to decrease because of pretreatments used to purify flue gas (EEA 2020) or system designs that integrate CO_2 capture with NO_x and SO_x removal (Shaw 2009). One case study found that, with just the pollution control equipment required for the carbon capture technology to function, NO_x emissions decrease by 10 percent, SO_2 and filterable PM emissions by 96 percent, and condensable PM emissions by 46 percent (Brown et al. 2023). Depending on the carbon capture technology used, emissions of other co-pollutants, such as NH_3 and volatile organic compounds, may increase owing to solvent degradation or reactions of the solvent and flue gas (Benquet et al. 2021; Gibbins and Lucquiaud 2022; Gorset et al. 2014; NETL 2020; Spietz et al. 2017). However, there are strategies for mitigating these emissions. See Chapter 10 for more about mitigating the emissions of criteria pollutants.

In some cases, use of low-carbon fuels and biofuels and deployment of CCS may fail to maximize the health co-benefits and to some extent, impede decarbonization.[8] For example, large subsidies on biofuels, particularly first-generation biofuels, could miss health benefits that would be obtained with other renewable energy options and contribute to worse harms via air and water pollution and higher food prices—while also failing to meet GHG emission targets (Lark et al. 2022). See Chapters 8 and 9 for more about biofuels and land use strategies.

Increased production of biomass for fuels would require additional U.S. land devoted to agriculture for energy crops, which could have negative impacts on water quality, water availability, and food security. A systematic review found that 56 percent of 224 publications reported negative impacts of biofuels on food security (Ahmed et al. 2021). This finding was not significantly different based on which fuels derived from feedstocks directly compete with food production, like corn ethanol, versus inedible biomass. Additionally, the agriculture practices can lead to the contamination of wells by runoff containing excess nitrogen from fertilizer and animal waste. This can lead to methemoglobinemia (blue baby syndrome) and harmful algal blooms that often produce their own adverse health impacts and are associated with some cancers in adults (Carmichael and Boyer 2016; Temkin et al. 2019; Ward et al. 2018).

Similarly, outfitting combusting facilities with CCS has fewer health co-benefits than decarbonization strategies that rely on retirement of combusting facilities enabled by increasing energy efficiency and increasing renewable energy use (Driscoll et al. 2015). While CCS could be targeted for industries where decarbonization is particularly technically challenging, its deployment could potentially delay retirement of polluting facilities where cleaner alternatives exist. For example, in 2020 Wyoming legislated that electric utilities generate power from coal plants with CCS, despite increased costs for consumers and increased pollution, rather than enabling a transition to renewables (Kusnetz 2022). Furthermore, data from a coal plant with carbon capture capabilities indicates that a net of only 10.5 percent of CO_2 emissions are captured,[9]

---

[8] See Chapter 2 for more information about the environmental and equity concerns surrounding CCS and Appendix E for more information about challenges associated with other decarbonization technologies.

[9] Jacobson (2019) determined that the low net capture rates are due to uncaptured combustion emissions from the natural gas used to power the carbon capture equipment, uncaptured upstream emissions, and uncaptured coal combustion emissions.

and that for the same energy costs, wind and solar can reduce more CO_2 without the air pollution (Jacobson 2019). Furthermore, CCS comes with potential environmental risks (Warner et al. 2020):

- Contaminating underground sources of drinking water;
- Moving radon closer to the surface where it could affect businesses and homes;
- Releasing CO_2 to the soil or the atmosphere; and
- Physical damage, such as landslides, sinkholes, and earthquakes.

Regulation can reduce risk, through site characterization, monitoring, and safe operational practices (Warner et al. 2020). (See Box 2-2 for additional discussion on CCS, environmental risk, and equity.)

> Finding 3-6: Decarbonization technologies that involve continued combustion, such as power generation with biofuels and carbon capture and storage, if focused solely on mitigating CO_2, can potentially harm human health through continued or increased emission of harmful non-CO_2 air pollutants, water contamination, and food insecurity. Similarly, the recent incentives for biomass fuels could have negative impacts on water quality, food security, and human health by encouraging the growth of feedstocks that compete with food production.

Manufacturing Net-Zero Energy Technologies

Increased demand for minerals, especially critical minerals, needed in the production of solar panels and batteries—for example, lithium, nickel, cobalt, and gallium—could increase mining and the subsequent impacts from mining, including significant environmental damage and health risks from contaminated water[10] (Luckeneder et al. 2021; Martinez-Alier 2001). See Chapter 9 for more information about the critical minerals and supply chain associated with electric vehicles. The mining of certain minerals categorized as critical for net-zero technologies is concentrated in a few places: 60 percent of the worlds' cobalt comes from the Democratic Republic of Congo (Brinn 2023), while lithium mining is highest in Chile, Australia, China, and Argentina (Dall-Orsoletta et al. 2022). Cobalt and lithium mining has "effects on human health and local biodiversity, water consumption,

---

[10] Analogous effects are associated with fossil fuel mining and extraction, and such activities are likely to decrease in a decarbonized future.

energy intensity, and conflicts with local and indigenous people" (Dall-Orsoletta et al. 2022, p. 5). Without equity-centered transition programs and attention to local labor and environmental standards and impacts, these mining activities could affect the health and quality of life of communities where they are extracted (Mayyas et al. 2019; Sharma and Manthiram 2020).

Globally, the demand for minerals is increasing interest in mining in areas where mines have been closed as well as in areas with no prior mining activity. This has led to residents expressing concern about potential damage to their communities. See Chapter 12 for more information about recovering critical minerals from mine tailings and other sources. Many of these regions may lack adequate or updated mining regulations to protect public health and the environment (Healy and Baker 2021; NASEM 2022a). In February 2022, the Department of the Interior announced an Interagency Working Group on Mining Reform to inform potential updates to regulations and permitting (DOI 2022). Communities need to be informed about and involved in the process of siting new mines and related infrastructure. See Chapter 5 for more information about inclusive siting and development practices. CDC's National Institute for Occupational Safety and Health (NIOSH) is a research agency whose work is supported by five advisory committees that provide advice and guidance on topical areas, including occupational and mine safety and health (NIOSH n.d.(a)). NIOSH, in coordination with other relevant research groups, can play a critical role in the assessment of the risks associated with decarbonization technologies, especially occupational health risks.

Although this report focuses on the United States, the full life-cycle assessment of health impacts needs to take into account international impacts. A transition to greater use of rooftop solar can reduce GHGs and water consumption; however, when the full life cycle of solar is considered, the water demands and environmental hazards would likely be transferred to countries where the solar panels are being manufactured, creating global health and equity concerns (Frisvold and Marquez 2013; Vengosh and Weinthal 2023). Greater consideration is also required for the end of the life cycle of these in-demand materials, especially for metals, which can be recycled but often end up in landfills (Reck and Graedel 2012; Seeberger et al. 2016). Specifically, 95 percent of critical minerals in lithium-ion batteries can be recycled at the commercial scale (Brin 2023). One way to mitigate mineral extraction and waste from solar panels and batteries would be to encourage manufacturers to develop standardized modules to enable easier recycling. (See Chapter 9 for more information about mitigating mineral extraction.) Likewise, the planning for the decommissioning and restoration of the sites of mining, natural resource extraction, oil and gas production, and fossil power plants is important for the renewal of communities and avoiding continued health harm from insufficient environmental remediation. Additionally, more mass transit and active travel could reduce the demand for vehicle manufacture.

Finding 3-7: Some components for low-carbon energy technology show potential health harms for workers or the general population that must be mitigated and weighed against other expected benefits, especially the benefit of reduced life-cycle harms from fossil fuels, which include contamination during resource extraction or oil spills.

Recommendation 3-3: *Assess Occupational Health Risks Associated with Clean Energy Technologies.* The Centers for Disease Control and Prevention's National Institute for Occupational Safety and Health should assess health risks associated with the manufacturing and deployment of clean energy technologies. This assessment should characterize potential occupational health risks relating to the extraction of raw materials, the manufacturing and installation of technologies, and the final disposal of waste products in an approach consistent with life-cycle assessment practices. Furthermore, the analysis should also identify preventative interventions for addressing such occupational risks.

Building Retrofits

Unintended health risks may occur with indoor air quality when buildings have improved sealing against the air but can be managed with precaution. While energy efficiency retrofits generally provide health benefits, increased air tightness in homes could also increase concentrations of unwanted indoor air pollutants, including radon and VOCs like formaldehyde (Fisk et al. 2020; Symonds et al. 2019). Building retrofitters need to check for this risk in radon-prone areas, or in buildings with older or damaged foundations, and add mechanical ventilation (Ferguson et al. 2020; IOM 2011; Rapp et al. 2012). Weatherization that avoids exacerbating indoor air pollution, combined with decarbonization, produces the greatest health benefits while minimizing the potential hazards. See Chapter 7 for more information about building retrofits and weatherization.

Transportation Electrification

For transportation electrification, the benefits depend on how energy is generated. The provisions in the IRA and IIJA intend for electric vehicles (EVs) to be powered by a grid with mostly renewables, which would provide both GHG and health benefits. Even today, an average EV is lower emitting than an average internal combustion engine vehicle (ICEV) in every region of the United States; furthermore, more than 90 percent of the U.S. population lives in a region where the average EV is lower emitting that the most efficient ICEV when traveling at 59 mph (Reichmuth et al. 2022). However, increasing EVs without decarbonizing electric power would limit the health and GHG benefits of vehicle electrification, primarily shifting emissions from where the vehicles are driven to where the fossil power is generated (Brinkman et al. 2010; Nopmongcol et al. 2017; Razeghi et al. 2016; Thompson et al. 2009; Weis et al. 2015).

Studies differ on the likely impact of vehicle electrification when paired with fossil-fuel intensive electric power, including at the extreme case of EVs fueled entirely by coal-powered electricity. Some studies estimating health impacts of EVs charged by coal power indicate 80 percent higher environmental health costs relative to an ICEV fleet (Tessum et al. 2014), while others indicate that coal-powered EVs still lead to health improvements relative to the average ICEV, given that transportation emissions tend to occur much closer to where people breathe than power plant emissions (Shindell 2015). Studies examining more realistic cases of EVs powered by today's grid mix and likely future fuel mixes of the power sector generally show that EVs reduce both GHG and health-harming emissions relative to ICEVs (e.g., Funke et al. 2023; Peters et al. 2020). To fully reap both health and decarbonization benefits, transportation electrification must be accompanied by decarbonization of the electricity sector, as this report calls for. See Chapter 9 for more information about decarbonizing the transportation sector and Chapter 11 for more information about decarbonizing the U.S. electricity grid.

Equity Considerations

Decarbonization, improved public health, and reducing inequity cannot be accomplished by individual choices alone, requiring consideration of system-wide processes. While people can be asked to reduce their carbon footprint and better their health, this places the burden on individuals to overcome obstacles to these activities—especially challenging for those with limited time, money, and in areas where accessing energy efficient appliances, fresh vegetables, and safe biking routes may not be easy. This is also a challenge in wealthier, mostly White neighborhoods; even with newer, energy efficient houses, the larger size of these houses means that they still have high total emissions per resident, relative to Black neighborhoods, creating an "emissions paradox" (Goldstein et al. 2022). Systematic approaches to alleviating barriers are needed and health and equity impacts need to be prioritized over technological solutions through requirements for health- and equity-focused analyses during decision-making (Rudolph 2022). Such analyses must emphasize benefits to communities whose health and economic priorities have been met with resistance or misinformation from the fossil fuel industry (NASEM 2021). To improve health during the energy transition, the nation must both be aware of the costs of decarbonization technology and reduce inequity.

SUMMARY OF RECOMMENDATIONS ON PUBLIC HEALTH CO-BENEFITS AND IMPACTS OF DECARBONIZATION

TABLE 3-2[a] Summary of Recommendations on Public Health Co-Benefits and Impacts of Decarbonization

Short-Form Recommendation	Actor(s) Responsible for Implementing Recommendation	Sector(s) Addressed by Recommendation	Objective(s) Addressed by Recommendation	Overarching Categories Addressed by Recommendation
3-1: Phase Out Incentives for the Highest Greenhouse Gas–Emitting Animal Protein Sources	Congress and U.S. Department of Agriculture (USDA)	• Land use	• GHG reduction • Health	Ensuring Equity, Justice, Health, and Fairness of Impacts
3-2: Increase Use of Health Impact Assessment Tools in Energy Project Decision-Making	Congress, Centers for Disease Control and Prevention (CDC), National Center for Environmental Health/Agency for Toxic Substances and Disease Registry (NCEH/ATSDR), Department of Health and Human Services Office of Climate Change and Health Equity	• Electricity • Buildings • Transportation • Industry • Fossil fuels	• Equity • Health	Rigorous and Transparent Analysis and Reporting for Adaptive Management Ensuring Equity, Justice, Health, and Fairness of Impacts
3-3: Assess Occupational Health Risks Associated with Clean Energy Technologies	CDC, NCEH/ATSDR, Occupational Safety and Health Administration	• Electricity • Buildings • Transportation • Industry • Fossil fuels	• Equity • Health • Employment	Ensuring Equity, Justice, Health, and Fairness of Impacts Research, Development, and Demonstration Needs

[a] The text in this table was changed during editorial review to improve clarity and alignment with information in other sections of the report

REFERENCES

Abel, D., T. Holloway, M. Harkey, A. Rrushaj, G. Brinkman, P. Duran, M. Janssen, and P. Denholm. 2018. "Potential Air Quality Benefits from Increased Solar Photovoltaic Electricity Generation in the Eastern United States." *Atmospheric Environment* 175:65–74. https://doi.org/10.1016/j.atmosenv.2017.11.049.

Abel, D.W., T. Holloway, J. Martínez-Santos, M. Harkey, M. Tao, C. Kubes, and S. Hayes. 2019. "Air Quality-Related Health Benefits of Energy Efficiency in the United States." *Environmental Science and Technology* 53(7):3987–3998. https://doi.org/10.1021/acs.est.8b06417.

Achakulwisut, P., S.C. Anenberg, J.E. Neumann, S.L. Penn, N. Weiss, A. Crimmins, et al. 2019. "Effects of Increasing Aridity on Ambient Dust and Public Health in the U.S. Southwest Under Climate Change." *GeoHealth* 3(5):127–144. https://doi.org/10.1029/2019GH000187.

Agyeman, J., R.D. Bullard, and B. Evans. 2002. "Exploring the Nexus: Bringing Together Sustainability, Environmental Justice, and Equity." *Space and Polity* 6(1):77–90. https://doi.org/10.1080/13562570220137907.

Agyeman, J., and A. Doran. 2021. " 'You Want Protected Bike Lanes, I Want Protected Black Children. Let's Link': Equity, Justice, and the Barriers to Active Transportation in North America." *Local Environment* 26(12):1480–1497. https://doi.org/10.1080/13549839.2021.1978412.

Ahmed, S., T. Warne, E. Smith, H. Goemann, G. Linse, M. Greenwood, J. Kedziora, et al. 2021. "Systematic Review on Effects of Bioenergy from Edible Versus Inedible Feedstocks on Food Security." *npj Science of Food* 5(1):9. https://doi.org/10.1038/s41538-021-00091-6.

Alderman, K., L.R. Turner, and S. Tong. 2012. "Floods and Human Health: A Systematic Review." *Environment International* 47:37–47. https://doi.org/10.1016/j.envint.2012.06.003.

Anderegg, W.R.L., J.T. Abatzoglou, L.D.L. Anderegg, L. Bielory, P.L. Kinney, and L. Ziska. 2021. "Anthropogenic Climate Change Is Worsening North American Pollen Seasons." *Proceedings of the National Academy of Sciences* 118(7):e2013284118. https://doi.org/10.1073/pnas.2013284118.

Andersson, E. 2006. "Incidence of Asthma Among Workers Exposed to Sulphur Dioxide and Other Irritant Gases." *European Respiratory Journal* 27(4):720–725. https://doi.org/10.1183/09031936.06.00034305.

Anenberg, S.C., P. Achakulwisut, M. Brauer, D. Moran, J.S. Apte, and D.K. Henze. 2019. "Particulate Matter-Attributable Mortality and Relationships with Carbon Dioxide in 250 Urban Areas Worldwide." *Scientific Reports* 9(1):11552. https://doi.org/10.1038/s41598-019-48057-9.

Apergis, N., T. Hayat, and T. Saeed. 2019. "Fracking and Infant Mortality: Fresh Evidence from Oklahoma." *Environmental Science and Pollution Research* 26(31):32360–32367. https://doi.org/10.1007/s11356-019-06478-z.

Azarijafari, H., X. Xu, J. Gregory, and R. Kirchain. 2021. "Urban-Scale Evaluation of Cool Pavement Impacts on the Urban Heat Island Effect and Climate Change." *Environmental Science and Technology* 55(17):11501–11510. https://doi.org/10.1021/acs.est.1c00664.

Barajas, J.M. 2021. "Biking Where Black: Connecting Transportation Planning and Infrastructure to Disproportionate Policing." *Transportation Research Part D: Transport and Environment* 99:103027. https://doi.org/10.1016/j.trd.2021.103027.

Barbose, G., R. Wiser, J. Heeter, T. Mai, L. Bird, M. Bolinger, A. Carpenter, et al. 2016. "A Retrospective Analysis of Benefits and Impacts of U.S. Renewable Portfolio Standards." *Energy Policy* 96:645–660. https://doi.org/10.1016/j.enpol.2016.06.035.

Benquet, C., A.B.N. Knarvik, E. Gjernes, O.A. Hvidsten, E.R. Kleppe, and S. Akhter. 2021. "First Process Results and Operational Experience with CESAR1 Solvent at TCM with High Capture Rates (ALIGN-CCUS Project)." *Proceedings of the 15th Greenhouse Gas Control Technologies Conference 15–18 March 2021*. https://dx.doi.org/10.2139/ssrn.3814712.

Benzer Kerr, R., T. Hasegawa, R. Lasco, I. Bhatt, D. Deryng, A. Ferrell, H. Gurney-Smith, et al. 2022. "Food, Fibre, and Other Ecosystem Products." Pp. 713–906 in *Climate Change 2022: Impacts, Adaptation and Vulnerability*, D.C. Roberts, H.-O. Pörtner, M., E.S. Poloczanska Tignor, K. Mintenbeck, A. Alegría, M. Craig, S. Langsdorf, et al., eds. Cambridge, UK, and New York: Cambridge University Press.

Beyer, K., A. Kaltenbach, A. Szabo, S. Bogar, F. Nieto, and K. Malecki. 2014. "Exposure to Neighborhood Green Space and Mental Health: Evidence from the Survey of the Health of Wisconsin." *International Journal of Environmental Research and Public Health* 11(3):3453–3472. https://doi.org/10.3390/ijerph110303453.

Bielory, L., K. Lyons, and R. Goldberg. 2012. "Climate Change and Allergic Disease." *Current Allergy and Asthma Reports* 12(6):485–494. https://doi.org/10.1007/s11882-012-0314-z.

Biesbroek, S., H.B. Bueno-de-Mesquita, P.H.M. Peeters, W.M.M. Verschuren, Y.T. van der Schouw, G.F.H. Kramer, M. Tyszler, and E.H.M. Temme. 2014. "Reducing Our Environmental Footprint and Improving Our Health: Greenhouse Gas Emission and Land Use of Usual Diet and Mortality in EPIC-NL: A Prospective Cohort Study." *Environmental Health* 13(1):27. https://doi.org/10.1186/1476-069X-13-27.

Bravo, M.A., R. Anthopolos, M.L. Bell, and M.L. Miranda. 2016. "Racial Isolation and Exposure to Airborne Particulate Matter and Ozone in Understudied US Populations: Environmental Justice Applications of Downscaled Numerical Model Output." *Environment International* 92–93:247–255. https://doi.org/10.1016/j.envint.2016.04.008.

Brender, J.D., J.A. Maantay, and J. Chakraborty. 2011. "Residential Proximity to Environmental Hazards and Adverse Health Outcomes." *American Journal of Public Health* 101(Suppl. 1):S37–S52. https://doi.org/10.2105/ajph.2011.300183.

Bressler, R.D. 2021. "The Mortality Cost of Carbon." *Nature Communications* 12(1). https://doi.org/10.1038/s41467-021-24487-w.

Brinkman, G.L., P. Denholm, M.P. Hannigan, and J.B. Milford. 2010. "Effects of Plug-In Hybrid Electric Vehicles on Ozone Concentrations in Colorado." *Environmental Science and Technology* 44(16):6256–6262. https://doi.org/10.1021/es101076c.

Brinn, J. 2023. "Building Batteries Better: Doing the Best with Less." Natural Resources Defense Council. https://www.nrdc.org/sites/default/files/2023-07/ev-battery-supply-chains-report.pdf.

Brown, J.D., S. Carpenter, S.D. Comello, S. Hovorka, S. Mackler, and M.H. Schwartz. 2023. *Turning CCS Projects in Heavy Industry and Power into Blue Chip Financial Investments*. Energy Futures Initiative.

Brown, V., B.Z. Diomedi, M. Moodie, J.L. Veerman, and R. Carter. 2016. "A Systematic Review of Economic Analyses of Active Transport Interventions That Include Physical Activity Benefits." *Transport Policy* 45:190–208. https://doi.org/10.1016/j.tranpol.2015.10.003.

Brugge, D., J.L. Durant, and C. Rioux. 2007. "Near-Highway Pollutants in Motor Vehicle Exhaust: A Review of Epidemiologic Evidence of Cardiac and Pulmonary Health Risks." *Environmental Health* 6(1):23. https://doi.org/10.1186/1476-069X-6-23.

BTS (Bureau of Transportation Statistics). 2017. "Table 1-40: U.S. Passenger-Miles (Millions)." https://www.bts.gov/archive/publications/national_transportation_statistics/table_01_40.

Buonocore, J.J., E.J. Hughes, D.R. Michanowicz, J. Heo, J.G. Allen, and A. Williams. 2019. "Climate and Health Benefits of Increasing Renewable Energy Deployment in the United States." *Environmental Research Letters* 14(11):114010. https://doi.org/10.1088/1748-9326/ab49bc.

Burgherr, P., and S. Hirschberg. 2014. "Comparative Risk Assessment of Severe Accidents in the Energy Sector." *Energy Policy* 74:S45–S56. https://doi.org/10.1016/j.enpol.2014.01.035.

Burke, M., A. Driscoll, S. Heft-Neal, J. Xue, J. Burney, and M. Wara. 2021. "The Changing Risk and Burden of Wildfire in the United States." *Proceedings of the National Academy of Sciences* 118(2):e2011048118. https://doi.org/10.1073/pnas.2011048118.

Burnett, R., H. Chen, M. Szyszkowicz, N. Fann, B. Hubbell, C.A. Pope, J.S. Apte, et al. 2018. "Global Estimates of Mortality Associated with Long-Term Exposure to Outdoor Fine Particulate Matter." *Proceedings of the National Academy of Sciences* 115(38):9592–9597. https://doi.org/10.1073/pnas.1803222115.

Cameron, C., W. Yelverton, R. Dodder, and J.J. West. 2014. "Strategic Responses to CO_2 Emission Reduction Targets Drive Shift in U.S. Electric Sector Water Use." *Energy Strategy Reviews* 4:16–27. https://doi.org/10.1016/j.esr.2014.07.003.

Canning, P., S. Rehkamp, A. Waters, and H. Etemadnia. 2017. "The Role of Fossil Fuels in the U.S. Food System and the American Diet." *Economic Research Report*: U.S. Department of Agriculture Economic Research Service. https://www.ers.usda.gov/webdocs/publications/82194/err-224.pdf.

Carley, S., M. Graff, D.M. Konisky, and T. Memmott. 2022. "Behavioral and Financial Coping Strategies Among Energy-Insecure Households." *Proceedings of the National Academy of Sciences* 119(36):e2205356119. https://doi.org/10.1073/pnas.2205356119.

Carmichael, W.W., and G.L. Boyer. 2016. "Health Impacts from Cyanobacteria Harmful Algae Blooms: Implications for the North American Great Lakes. *Harmful Algae* 54:194–212. https://doi.org/10.1016/j.hal.2016.02.002.

Casey, J.A., J.G. Su, L.R.F. Henneman, C. Zigler, A.M. Neophytou, R. Catalano, R. Gondalia, et al. 2020. "Improved Asthma Outcomes Observed in the Vicinity of Coal Power Plant Retirement, Retrofit and Conversion to Natural Gas." *Nature Energy* 5(5):398–408. https://doi.org/10.1038/s41560-020-0600-2.

Castillo, M.D., S.C. Anenberg, Z.A. Chafe, R. Huxley, L.S. Johnson, I. Kheirbek, M. Malik, et al. 2021. "Quantifying the Health Benefits of Urban Climate Mitigation Actions: Current State of the Epidemiological Evidence and Application in Health Impact Assessments." *Frontiers in Sustainable Cities* 3:768227. https://doi.org/10.3389/frsc.2021.768227.

CDC (Centers for Disease Control and Prevention). 2016. "Health Impact Assessment." https://www.cdc.gov/healthyplaces/hia.htm.

CDC. n.d. "Water Contamination and Diseases." https://www.cdc.gov/healthywater/drinking/contamination.html.

Celis-Morales, C.A., D.M. Lyall, P. Welsh, J. Anderson, L. Steell, Y. Guo, R. Maldonado, et al. 2017. "Association Between Active Commuting and Incident Cardiovascular Disease, Cancer, and Mortality: Prospective Cohort Study." *BMJ* 357:j1456. https://doi.org/10.1136/bmj.j1456.

Chapman, A.J., B.C. McLellan, and T. Tezuka. 2018. "Prioritizing Mitigation Efforts Considering Co-Benefits, Equity and Energy Justice: Fossil Fuel to Renewable Energy Transition Pathways." *Applied Energy* 219:187–198. https://doi.org/10.1016/j.apenergy.2018.03.054.

Choi, H.M., W. Lee, D. Roye, S. Heo, A. Urban, A. Entezari, A.M. Vicedo-Cabrera, et al. 2022. "Effect Modification of Greenness on the Association Between Heat and Mortality: A Multi-City Multi-Country Study." *eBioMedicine* 84:104251. https://doi.org/10.1016/j.ebiom.2022.104251.

Cissé, G., R. McLeman, H. Adams, P. Aldunce, K. Bowen, D. Campbell-Lendrum, S. Clayton, et al. 2022. Chapter 7. "Health, Wellbeing and the Changing Structure of Communities." In *Climate Change 2022: Impacts, Adaptation and Vulnerability*, Intergovernmental Panel on Climate Change Working Group II. Cambridge, UK, and New York: Cambridge University Press.

Climate Impact Lab. 2022. "Measuring the Life-Saving Effects of Reducing Greenhouse Gas Emissions in the U.S. Climate Impact Lab." https://impactlab.org/news-insights/lives-saved-calculator.

Cong, S., D. Nock, Y.L. Qiu, and B. Xing. 2022. "Unveiling Hidden Energy Poverty Using the Energy Equity Gap." *Nature Communications* 13(1):2456. https://doi.org/10.1038/s41467-022-30146-5.

Currie, J., M. Greenstone, and K. Meckel. 2017. "Hydraulic Fracturing and Infant Health: New Evidence from Pennsylvania." *Science Advances* 3(12):e1603021. https://doi.org/10.1126/sciadv.1603021.

Dall-Orsoletta, A., P. Ferreira, and G. Gilson Dranka. 2022. "Low-Carbon Technologies and Just Energy Transition: Prospects for Electric Vehicles." *Energy Conversion and Management: X* 16:100271. https://doi.org/10.1016/j.ecmx.2022.100271.

de Oliviera Bredariol, T. 2022. "Reducing the Impact of Extractive Industries on Groundwater Resources." Paris, France: International Energy Agency. https://www.iea.org/commentaries/reducing-the-impact-of-extractive-industries-on-groundwater-resources.

Dedoussi, I.C., S.D. Eastham, E. Monier, and S.R.H. Barrett. 2020. "Premature Mortality Related to United States Cross-State Air Pollution." *Nature* 578(7794):261–265. https://doi.org/10.1038/s41586-020-1983-8.

DeMocker, J., and J. Neumann. 2011. *The Benefits and Costs of the Clean Air Act from 1990 to 2020, Final Report—Rev. A.* Environmental Protection Agency Office of Air and Radiation. https://www.epa.gov/sites/default/files/2015-07/documents/fullreport_rev_a.pdf.

Diffenbaugh, N.S., D. Singh, J.S. Mankin, D.E. Horton, D.L. Swain, D. Touma, A. Charland, et al. 2017. "Quantifying the Influence of Global Warming on Unprecedented Extreme Climate Events." *Proceedings of the National Academy of Sciences* 114(19):4881–4886. https://doi.org/10.1073/pnas.1618082114.

Dinu, M., G. Pagliai, C. Macchi, and F. Sofi. 2019. "Active Commuting and Multiple Health Outcomes: A Systematic Review and Meta-Analysis." *Sports Medicine* 49(3):437–452. https://doi.org/10.1007/s40279-018-1023-0.

DOE (Department of Energy). 2014. "Ensuring the Resiliency of Our Future Water and Energy Systems." https://www.energy.gov/policy/articles/ensuring-resiliency-our-future-water-and-energy-systems.

DOE-EERE (Department of Energy Office of Energy Efficiency and Renewable Energy). 2022. "More Than Half of all Daily Trips Were Less Than Three Miles in 2021." https://www.energy.gov/eere/vehicles/articles/fotw-1230-march-21-2022-more-half-all-daily-trips-were-less-three-miles-2021.

DOI (Department of the Interior). 2022. "Interior Department Launches Interagency Working Group on Mining Reform." https://www.doi.gov/pressreleases/interior-department-launches-interagency-working-group-on-mining-reform.

Donovan, G.H., J.P. Prestemon, and A.R. Kaminski. 2022. "The Natural Environment and Social Cohesion: Tree Planting Is Associated with Increased Voter Turnout in Portland, Oregon." *Trees, Forests and People* 7:100215. https://doi.org/10.1016/j.tfp.2022.100215.

Drewniak, B.A., P.K. Snyder, A.L. Steiner, T.E. Twine, and D.J. Wuebbles. 2014. "Simulated Changes in Biogenic VOC Emissions and Ozone Formation from Habitat Expansion of Acer Rubrum (Red Maple)." *Environmental Research Letters* 9(1):014006. https://doi.org/10.1088/1748-9326/9/1/014006.

Driscoll, C.T., J.J. Buonocore, J.I. Levy, K.F. Lambert, D. Burtraw, S.B. Reid, H. Fakhraei, and J. Schwartz. 2015. "US Power Plant Carbon Standards and Clean Air and Health Co-Benefits." *Nature Climate Change* 5(6):535–540. https://doi.org/10.1038/nclimate2598.

Dryden, R., M.G. Morgan, A. Bostrom, and W. Bruine De Bruin. 2018. "Public Perceptions of How Long Air Pollution and Carbon Dioxide Remain in the Atmosphere." *Risk Analysis* 38(3):525–534. https://doi.org/10.1111/risa.12856.

Ebi, K.L., J. Vanos, J.W. Baldwin, J.E. Bell, D.M. Hondula, N.A. Errett, K. Hayes, et al. 2021. "Extreme Weather and Climate Change: Population Health and Health System Implications." *Annual Review of Public Health* 42(1):293–315. https://doi.org/10.1146/annurev-publhealth-012420-105026.

Eckelman, M.J., K. Huang, R. Lagasse, E. Senay, R. Bubrow, and J.D. Sherman. 2020. "Health Care Pollution and Public Health Damage in the United States: An Update." *Health Affairs* 39(12):2071–2079. https://doi.org/10.1377/hlthaff.2020.01247.

EEA (European Environment Agency). 2022. "Carbon Capture and Storage Could Also Impact Air Pollution." https://www.eea.europa.eu/highlights/carbon-capture-and-storage-could.

EIA (U.S. Energy Information Administration). 2022. *2018 Commercial Buildings Energy Consumption Survey: Consumption and Expenditures Highlights.* https://www.eia.gov/consumption/commercial/data/2018/pdf/CBECS%202018%20CE%20Release%202%20Flipbook.pdf.

EPA (Environmental Protection Agency). 2008. *Reducing Urban Heat Islands: Compendium of Strategies.* Draft. https://www.epa.gov/heat-islands/heat-island-compendium.

EPA. 2011. *The Benefits and Costs of the Clean Air Act from 1990–2020: Summary Report.* https://www.epa.gov/sites/default/files/2015-07/documents/summaryreport.pdf.

EPA. 2013. *Integrated Science Assessment for Ozone and Related Photochemical Oxidants.* Washington, DC.

EPA. 2015. "Final Effluent Limitations Guidelines and Standards for the Steam Electric Power Generating Industry." https://www.epa.gov/sites/default/files/2015-10/documents/steam-electric-final-rule-factsheet_10-01-2015.pdf.

EPA. 2022a. "Climate Change Indicators: Greenhouse Gases." https://www.epa.gov/climate-indicators/greenhouse-gases.

EPA. 2022b. "Initial List of Hazardous Air Pollutants with Modifications." https://www.epa.gov/haps/initial-list-hazardous-air-pollutants-modifications.

EPA. 2023a. Inventory Greenhouse Gas Emissions and Sinks: 1990–2021. https://www.epa.gov/ghgemissions/inventory-us-greenhouse-gas-emissions-and-sinks-1990-2021.

EPA. 2023b. "U.S. Electric Power Industry Estimated Emissions by State (EIA-767, EIA-906, EIA-920, and EIA-923)." https://www.eia.gov/electricity/data/state.

EPA. n.d.(a). "Basic Information About Lead Air Pollution." https://www.epa.gov/lead-air-pollution/basic-information-about-lead-air-pollution. Accessed March 15, 2023.

EPA. n.d.(b). "Basic Information about NO_2." https://www.epa.gov/no2-pollution/basic-information-about-no2. Accessed March 15, 2023.

EPA. n.d.(c). "Carbon Monoxide's Impact on Indoor Air Quality." https://www.epa.gov/indoor-air-quality-iaq/carbon-monoxides-impact-indoor-air-quality. Accessed March 15, 2023.

Epstein, P.R., J.J. Buonocore, K. Eckerle, M. Hendryx, B.M. Stout III, R. Heinberg, R.W. Clapp, et al. 2011. "Full Cost Accounting for the Life Cycle of Coal." *Annals of the New York Academy of Sciences* 1219(1):73–98. https://doi.org/10.1111/j.1749-6632.2010.05890.x.

Erqou, S., J.E. Clougherty, O. Olafiranye, J.W. Magnani, A. Aiyer, S. Tripathy, E. Kinnee, K.E. Kip, and S.E. Reis. 2018. "Particulate Matter Air Pollution and Racial Differences in Cardiovascular Disease Risk." *Arteriosclerosis, Thrombosis, and Vascular Biology* 38(4):935–942. https://doi.org/10.1161/atvbaha.117.310305.

Fann, N., E. Coffman, B. Timin, and J.T. Kelly. 2018. "The Estimated Change in the Level and Distribution of PM(2.5)-Attributable Health Impacts in the United States: 2005–2014." *Environmental Research* 167:506–514. https://doi.org/10.1016/j.envres.2018.08.018.

Ferguson, L., J. Taylor, M. Davies, C. Shrubsole, P. Symonds, and S. Dimitroulopoulou. 2020. "Exposure to Indoor Air Pollution Across Socio-Economic Groups in High-Income Countries: A Scoping Review of the Literature and a Modelling Methodology." *Environment International* 143:105748. https://doi.org/10.1016/j.envint.2020.105748.

Fineschi, S., and F. Loreto. 2020. "A Survey of Multiple Interactions Between Plants and the Urban Environment." *Frontiers in Forests and Global Change* 3. https://doi.org/10.3389/ffgc.2020.00030.

Fisk, W.J., B.C. Singer, and W.R. Chan. 2020. "Association of Residential Energy Efficiency Retrofits with Indoor Environmental Quality, Comfort, and Health: A Review of Empirical Data." *Building and Environment* 180:107067. https://doi.org/10.1016/j.buildenv.2020.107067.

Fleischman, L., and M. Franklin. 2017. *Fumes Across the Fence-Line: The Health Impacts of Air Pollution from Oil and Gas Facilities on African American Communities*. National Association for the Advancement of Colored People and Clean Air Task Force.

Ford, B., M. Val Martin, S.E. Zelasky, E.V. Fischer, S.C. Anenberg, C.L. Heald, and J.R. Pierce. 2018. "Future Fire Impacts on Smoke Concentrations, Visibility, and Health in the Contiguous United States." *GeoHealth* 2(8):229–247. https://doi.org/10.1029/2018gh000144.

Francisco, P.W., J.R. Gordon, and B. Rose. 2010. "Measured Concentrations of Combustion Gases from the Use of Unvented Gas Fireplaces." *Indoor Air* 20(5):370–379. https://doi.org/10.1111/j.1600-0668.2010.00659.x.

Frisvold, G.B., and T. Marquez. 2013. "Water Requirements for Large-Scale Solar Energy Projects in the West." *Journal of Contemporary Water Research and Education* 151(1):106–116. https://doi.org/10.1111/j.1936-704X.2013.03156.x.

Funke, C., J. Linn, S. Robson, E. Russell, D. Shawhan, and S. Witkin. 2023. "What Are the Climate, Air Pollution, and Health Benefits of Electric Vehicles?" https://media.rff.org/documents/WP_23-01_5fpM1Y5.pdf.

Gallagher, C.L., and T. Holloway. 2022. "U.S. Decarbonization Impacts on Air Quality and Environmental Justice." *Environmental Research Letters* 17(11):114018. https://doi.org/10.1088/1748-9326/ac99ef.

GAO (General Accounting Office). 1983. *Siting of Hazardous Waste Landfills and Their Correlation with Racial and Economic Status of Surrounding Communities*. Gaithersburg, MD: General Accounting Office.

Gao, J., S. Kovats, S. Vardoulakis, P. Wilkinson, A. Woodward, J. Li, S. Gu, et al. 2018. "Public Health Co-Benefits of Greenhouse Gas Emissions Reduction: A Systematic Review." *Science of the Total Environment* 627:388–402. https://doi.org/10.1016/j.scitotenv.2018.01.193.

GCRP (U.S. Global Change Research Program). 2016. "The Impacts of Climate Change on Human Health in the United States: A Scientific Assessment." Washington, DC: U.S. Global Change Research Program.

Gibbins, J., and M. Lucquiaud. 2022. *BAT Review for New-Build and Retrofit Post-Combustion Carbon Dioxide Capture Using Amine-Based Technologies for Power and CHP Plants Fueled by Gas and Biomass and for Post-Combustion Capture Using Amine-Based and Hot Potassium Carbonate Technologies on EfW Plants as Emerging Technologies Under IED for the UK*. Version 2.0. https://ukccsrc.ac.uk/best-available-technology-bat-information-for-ccs.

Goforth, T., and D. Nock. 2022. "Air Pollution Disparities and Equality Assessments of US National Decarbonization Strategies." *Nature Communications* 13(1):7488. https://doi.org/10.1038/s41467-022-35098-4.

Goldstein, B., T.G. Reames, and J.P. Newell. 2022. "Racial Inequity in Household Energy Efficiency and Carbon Emissions in the United States: An Emissions Paradox." *Energy Research and Social Science* 84:102365. https://doi.org/10.1016/j.erss.2021.102365.

Goodkind, A.L., C.W. Tessum, J.S. Coggins, J.D. Hill, and J.D. Marshall. 2019. "Fine-Scale Damage Estimates of Particulate Matter Air Pollution Reveal Opportunities for Location-Specific Mitigation of Emissions." *Proceedings of the National Academy of Sciences* 116(18):8775–8780. https://doi.org/10.1073/pnas.1816102116.

Gorai, A., F. Tuluri, and P. Tchounwou. 2014. "A GIS Based Approach for Assessing the Association Between Air Pollution and Asthma in New York State, USA." *International Journal of Environmental Research and Public Health* 11(5):4845–4869. https://doi.org/10.3390/ijerph110504845.

Gorset, O., J.N. Knudsen, O.M. Bade, and I. Askestad. 2014. "Results from Testing of Aker Solutions Advanced Amine Solvents at CO_2 Technology Centre Mongstad." *Energy Procedia* 63(2014):6267–6280. https://doi.org/10.1016/j.egypro.2014.11.658.

Gowda, P., J. Steiner, M. Grusak, M. Boggess, and T. Farrigan. 2018. Chapter 10, "Agriculture and Rural Communities." Pp. 391–437 in *Impacts, Risks, and Adaptation in the United States: Fourth National Climate Assessment*, D.R. Reidmiller, C.W. Avery, D.R. Easterling, K.E. Kunkel, K.L.M. Lewis, T.K. Maycock, and B.C. Stewart, eds. Washington, DC: U.S. Global Change Research Program.

Grabow, M.L., S.N. Spak, T. Holloway, B. Stone, A.C. Mednick, and J.A. Patz. 2012. "Air Quality and Exercise-Related Health Benefits from Reduced Car Travel in the Midwestern United States." *Environmental Health Perspectives* 120(1):68–76. https://doi.org/10.1289/ehp.1103440.

Graham, D.A., J.K. Vanos, N.A. Kenny, and R.D. Brown. 2016. "The Relationship Between Neighbourhood Tree Canopy Cover and Heat-Related Ambulance Calls During Extreme Heat Events in Toronto, Canada." *Urban Forestry and Urban Greening* 20:180–186. https://doi.org/10.1016/j.ufug.2016.08.005.

Gronlund, C.J., V.J. Berrocal, J.L. White-Newsome, K.C. Conlon, and M.S. O'Neill. 2015. "Vulnerability to Extreme Heat by Socio-Demographic Characteristics and Area Green Space Among the Elderly in Michigan, 1990–2007." *Environmental Research* 136:449–461. https://doi.org/10.1016/j.envres.2014.08.042.

Gruenwald, T., B.A. Seals, L.D. Knibbs, and H.D. Hosgood III. 2023. "Population Attributable Fraction of Gas Stoves and Childhood Asthma in the United States." *International Journal of Environmental Research and Public Health* 20(1):75. https://doi.org/10.3390/ijerph20010075.

Hamer, M., and Y. Chida. 2008. "Active Commuting and Cardiovascular Risk: A Meta-Analytic Review." *Preventive Medicine* 46(1):9–13. https://doi.org/10.1016/j.ypmed.2007.03.006.

Hansmann, K.J., M. Grabow, and C. McAndrews. 2022. "Health Equity and Active Transportation: A Scoping Review of Active Transportation Interventions and Their Impacts on Health Equity." *Journal of Transport and Health* 25:101346. https://doi.org/10.1016/j.jth.2022.101346.

Hawkins, T.R., B. Singh, G. Majeau-Bettez, and A.H. Strømman. 2013. "Comparative Environmental Life Cycle Assessment of Conventional and Electric Vehicles." *Journal of Industrial Ecology* 17(1):53–64. https://doi.org/10.1111/j.1530-9290.2012.00532.x.

Hayashida, K., K. Shimizu, and H. Yokota. 2019. "Severe Heatwave in Japan." *Acute Medicine and Surgery* 6(2):206–207. https://doi.org/10.1002/ams2.387.

Healy, J., and M. Baker. 2021. "As Miners Chase Clean-Energy Minerals, Tribes Fear a Repeat of the Past." *The New York Times*, December 27. https://www.nytimes.com/2021/12/27/us/mining-clean-energy-antimony-tribes.html.

Healy, N., J.C. Stephens, and S.A. Malin. 2019. "Embodied Energy Injustices: Unveiling and Politicizing the Transboundary Harms of Fossil Fuel Extractivism and Fossil Fuel Supply Chains." *Energy Research and Social Science* 48:219–234. https://doi.org/10.1016/j.erss.2018.09.016.

HEI (Health Effects Institute) Panel on the Health Effects of Traffic-Related Air Pollution. 2010. *Traffic-Related Air Pollution: A Critical Review of the Literature on Emissions, Exposure, and Health Effects*. HEI Special Report 17. Boston, MA: Health Effects Institute.

Hernández, D., D. Phillips, and E.L. Siegel. 2016. "Exploring the Housing and Household Energy Pathways to Stress: A Mixed Methods Study." *International Journal of Environmental Research and Public Health* 13(9):916.

HHS (Department of Health and Human Services). n.d. "Health Sector Commitments to Emissions Reduction and Resilience." https://www.hhs.gov/climate-change-health-equity-environmental-justice/climate-change-health-equity/actions/health-sector-pledge/index.html. Accessed March 15, 2023.

Hickman, C., E. Marks, P. Pihkala, S. Clayton, R.E. Lewandowski, E.E. Mayall, B. Wray, C. Mellor, and L. van Susteren. 2021. "Climate Anxiety in Children and Young People and Their Beliefs About Government Responses to Climate Change: A Global Survey." *The Lancet: Planetary Health* 5(12):E863–E873. https://doi.org/10.1016/S2542-5196(21)00278-3.

Hill, E.L., and L. Ma. 2022. "Drinking Water, Fracking, and Infant Health." *Journal of Health Economics* 82:102595. https://doi.org/10.1016/j.jhealeco.2022.102595.

Hill, J., S. Polasky, E. Nelson, D. Tilman, H. Huo, L. Ludwig, J. Neumann, H. Zheng, and D. Bonta. 2009. "Climate Change and Health Costs of Air Emissions from Biofuels and Gasoline." *Proceedings of the National Academy of Sciences* 106(6):2077–2082. https://doi.org/10.1073/pnas.0812835106.

Hirsch, J.A., D.N. DeVries, M. Brauer, L.D. Frank, and M. Winters. 2018. "Impact of New Rapid Transit on Physical Activity: A Meta-Analysis." *Preventive Medicine Reports* 10:184–190. https://doi.org/10.1016/j.pmedr.2018.03.008.

Hoffman, J.S., V. Shandas, and N. Pendleton. 2020. "The Effects of Historical Housing Policies on Resident Exposure to Intra-Urban Heat: A Study of 108 US Urban Areas." *Climate* 8(1):12. https://doi.org/10.3390/cli8010012.

Horrigan, L., R.S. Lawrence, and P. Walker. 2002. "How Sustainable Agriculture Can Address the Environmental and Human Health Harms of Industrial Agriculture." *Environmental Health Perspectives* 110(5):445–456. https://doi.org/10.1289/ehp.02110445.

Hribar, C., and M. Schultz. 2010. *Understanding Concentrated Animal Feeding Operations and Their Impact on Communities.* Bowling Green, OH: National Association of Local Boards of Health.

Huang, S.J., and N.J. Sehgal. 2022. "Association of Historic Redlining and Present-Day Health in Baltimore." *PLOS ONE* 17(1):e0261028. https://doi.org/10.1371/journal.pone.0261028.

IEA (International Energy Agency). 2019. "Multiple Benefits of Energy Efficiency: Health and Wellbeing." https://www.iea.org/reports/multiple-benefits-of-energy-efficiency/health-and-wellbeing

IEA. 2020. "Introduction to the Water-Energy Nexus." https://www.iea.org/articles/introduction-to-the-water-energy-nexus.

IEA. 2021. "The Role of Critical Minerals in Clean Energy Transitions." World Energy Outlook Special Report. Paris, France: International Energy Agency. https://www.iea.org/reports/the-role-of-critical-minerals-in-clean-energy-transitions.

IEA. n.d. "Buildings." https://www.iea.org/energy-system/buildings. Accessed March 15, 2023.

Ilacqua, V., N. Scharko, J. Zambrana, and D. Malashock. 2022. "Survey of Residential Indoor Particulate Matter Measurements 1990–2019." *Indoor Air* 32(7):e13057. https://doi.org/10.1111/ina.13057.

IOM (Institute of Medicine). 2011. *Climate Change, the Indoor Environment, and Health.* Washington, DC: The National Academies Press.

Jacobson, M.Z. 2019. "The Health and Climate Impacts of Carbon Capture and Direct Air Capture." *Energy and Environmental Science* 12(12):3567–3574. https://doi.org/10.1039/C9EE02709B.

James, W., C. Jia, and S. Kedia. 2012. "Uneven Magnitude of Disparities in Cancer Risks from Air Toxics." *International Journal of Environmental Research and Public Health* 9(12):4365–4385.

Jenkins, J., E.N. Mayfield, J. Farbes, R. Jones, N. Patankar, Q. Xu, and G. Schivley. 2022. *Preliminary Report: The Climate and Energy Impacts of the Inflation Reduction Act of 2022.* Princeton, NJ: REPEAT Project.

Jessel, S., S. Sawyer, and D. Hernández. 2019. "Energy, Poverty, and Health in Climate Change: A Comprehensive Review of an Emerging Literature." *Frontiers in Public Health* 7:357. https://doi.org/10.3389/fpubh.2019.00357.

Kelly, P., S. Kahlmeier, T. Götschi, N. Orsini, J. Richards, N. Roberts, P. Scarborough, and C. Foster. 2014. "Systematic Review and Meta-Analysis of Reduction in All-Cause Mortality from Walking and Cycling and Shape of Dose Response Relationship." *International Journal of Behavioral Nutrition and Physical Activity* 11(1):132. https://doi.org/10.1186/s12966-014-0132-x.

Kerr, G.H., D.L. Goldberg, and S.C. Anenberg. 2021. "COVID-19 Pandemic Reveals Persistent Disparities in Nitrogen Dioxide Pollution." *Proceedings of the National Academy of Sciences* 118(30):e2022409118. https://doi.org/10.1073/pnas.2022409118.

Khreis, H., M.J. Nieuwenhuijsen, J. Zietsman, and T. Ramani. 2020. Chapter 1, "Traffic-Related Air Pollution: Emissions, Human Exposures, and Health: An Introduction." Pp. 1–21 in *Traffic-Related Air Pollution*, H. Khreis, M. Nieuwenhuijsen, J. Zietsman, and T. Ramani, eds. Elsevier.

Klepeis, N.E., W.C. Nelson, W.R. Ott, J.P. Robinson, A.M. Tsang, P. Switzer, J.V. Behar, S.C. Hern, and W.H. Engelmann. 2001. "The National Human Activity Pattern Survey (NHAPS): A Resource for Assessing Exposure to Environmental Pollutants." *Journal of Exposure Science and Environmental Epidemiology* 11(3):231–252. https://doi.org/10.1038/sj.jea.7500165.

Knowlton, K., M. Rotkin-Ellman, L. Geballe, W. Max, and G.M. Solomon. 2011. "Six Climate Change–Related Events in the United States Accounted for About $14 Billion in Lost Lives and Health Costs." *Health Affairs* 30(11):2167–2176. https://doi.org/10.1377/hlthaff.2011.0229.

Kollanus, V., P. Tiittanen, and T. Lanki. 2021. "Mortality Risk Related to Heatwaves in Finland—Factors Affecting Vulnerability." *Environmental Research* 201:111503. https://doi.org/10.1016/j.envres.2021.111503.

Kraemer, M.U.G., R.C. Reiner, O.J. Brady, J.P. Messina, M. Gilbert, D.M. Pigott, D. Yi, et al. 2019. "Past and Future Spread of the Arbovirus Vectors *Aedes aegypti* and *Aedes albopictus*." *Nature Microbiology* 4(5):854–863. https://doi.org/10.1038/s41564-019-0376-y.

Kraus, S., and N. Koch. 2021. "Provisional COVID-19 Infrastructure Induces Large, Rapid Increases in Cycling." *Proceedings of the National Academy of Sciences* 118(15):e2024399118. https://doi.org/10.1073/pnas.2024399118.

Kuehler, E., J. Hathaway, and A. Tirpak. 2017. "Quantifying the Benefits of Urban Forest Systems as a Component of the Green Infrastructure Stormwater Treatment Network." *Ecohydrology* 10(3):e1813. https://doi.org/10.1002/eco.1813.

Kusnetz, N. 2022. "In a Bid to Save Its Coal Industry, Wyoming Has Become a Test Case for Carbon Capture, But Utilities Are Balking at the Price Tag." *Inside Climate News*, May 29. https://insideclimatenews.org/news/29052022/coal-carbon-capture-wyoming.

Lane, H.M., R. Morello-Frosch, J.D. Marshall, and J.S. Apte. 2022. "Historical Redlining Is Associated with Present-Day Air Pollution Disparities in U.S. Cities." *Environmental Science and Technology Letters* 9(4):345–350. https://doi.org/10.1021/acs.estlett.1c01012.

Lark, T.J., N.P. Hendricks, A. Smith, N. Pates, S.A. Spawn-Lee, M. Bougie, E.G. Booth, C.J. Kucharik, and H.K. Gibbs. 2022. "Environmental Outcomes of the US Renewable Fuel Standard." *Proceedings of the National Academy of Sciences* 119(9):e2101084119. https://doi.org/10.1073/pnas.2101084119.

Larsen, J., B. King, H. Kolus, N. Dasari, G. Hiltbrand, and W. Herndon. 2022. "A Turning Point for US Climate Progress: Assessing the Climate and Clean Energy Provisions in the Inflation Reduction Act." Rhodium Group. https://rhg.com/research/climate-clean-energy-inflation-reduction-act.

Lebel, E.D., C.J. Finnegan, Z. Ouyang, and R.B. Jackson. 2022. "Methane and NO_x Emissions from Natural Gas Stoves, Cooktops, and Ovens in Residential Homes." *Environmental Science and Technology* 56(4):2529–2539. https://doi.org/10.1021/acs.est.1c04707.

Lee, G., S. You, S.G. Ritchie, J.-D. Saphores, R. Jayakrishnan, and O. Ogunseitan. 2012. "Assessing Air Quality and Health Benefits of the Clean Truck Program in the Alameda Corridor, CA." *Transportation Research Part A: Policy and Practice* 46(8):1177–1193. https://doi.org/10.1016/j.tra.2012.05.005.

Lelieveld, J., K. Klingmüller, A. Pozzer, R.T. Burnett, A. Haines, and V. Ramanathan. 2019. "Effects of Fossil Fuel and Total Anthropogenic Emission Removal on Public Health and Climate." *Proceedings of the National Academy of Sciences* 116(15):7192–7197. https://doi.org/10.1073/pnas.1819989116.

Lewis, J., D. Hernández, and A.T. Geronimus. 2020. "Energy Efficiency as Energy Justice: Addressing Racial Inequities Through Investments in People and Places." *Energy Efficiency* 13(3):419–432. https://doi.org/10.1007/s12053-019-09820-z.

Li, B., S. Sain, L.O. Mearns, H.A. Anderson, S. Kovats, K.L. Ebi, M.Y.V. Bekkedal, M.S. Kanarek, and J.A. Patz. 2012. "The Impact of Extreme Heat on Morbidity in Milwaukee, Wisconsin." *Climatic Change* 110(3):959–976. https://doi.org/10.1007/s10584-011-0120-y.

Li, M., S. Ferreira, and T.A. Smith. 2020. "Temperature and Self-Reported Mental Health in the United States." *PLOS ONE* 15(3):e0230316. https://doi.org/10.1371/journal.pone.0230316.

Li, M., N. Jia, M. Lenzen, A. Malik, L. Wei, Y. Jin, and D. Raubenheimer. 2022. "Global Food-Miles Account for Nearly 20% of Total Food-Systems Emissions." *Nature Food* 3(6):445–453. https://doi.org/10.1038/s43016-022-00531-w.

Li, X., S. Huang, A. Jiao, X. Yang, J. Yun, Y. Wang, X. Xue, et al. 2017. "Association Between Ambient Fine Particulate Matter and Preterm Birth or Term Low Birth Weight: An Updated Systematic Review and Meta-Analysis." *Environmental Pollution* 227:596–605. https://doi.org/10.1016/j.envpol.2017.03.055.

Limaye, V.S., W. Max, J. Constible, and K. Knowlton. 2019. "Estimating the Health-Related Costs of 10 Climate-Sensitive U.S. Events During 2012." *GeoHealth* 3(9):245–265. https://doi.org/10.1029/2019GH000202.

Limaye, V., and D. De Alwis. 2021. *The Costs of Inaction: The Economic Burden of Fossil Fuels and Climate Change on Health in the United States*. Natural Resources Defense Council.

Linder, S.H., D. Marko, and K. Sexton. 2008. "Cumulative Cancer Risk from Air Pollution in Houston: Disparities in Risk Burden and Social Disadvantage." *Environmental Science and Technology* 42(12):4312–4322. https://doi.org/10.1021/es072042u.

Liu, J., L.P. Clark, M.J. Bechle, A. Hajat, S.-Y. Kim, A.L. Robinson, L. Sheppard, A.A. Szpiro, and J.D. Marshall. 2021. "Disparities in Air Pollution Exposure in the United States by Race/Ethnicity and Income, 1990–2010." *Environmental Health Perspectives* 129(12):127005. https://ehp.niehs.nih.gov/doi/abs/10.1289/EHP8584.

Liu, Y., H. Chen, Y. Li, J. Gao, K. Dave, J. Chen, T. Li, and R. Tu. 2022. "Exhaust and Non-Exhaust Emissions from Conventional and Electric Vehicles: A Comparison of Monetary Impact Values." *Journal of Cleaner Production* 331:129965. https://doi.org/10.1016/j.jclepro.2021.129965.

Lowe, S.R., J.L. Bonumwezi, Z. Valdespino-Hayden, and S. Galea. 2019. "Posttraumatic Stress and Depression in the Aftermath of Environmental Disasters: A Review of Quantitative Studies Published in 2018." *Current Environmental Health Reports* 6(4):344–360. https://doi.org/10.1007/s40572-019-00245-5.

Luckeneder, S., S. Giljum, A. Schaffartzik, V. Maus, and M. Tost. 2021. "Surge in Global Metal Mining Threatens Vulnerable Ecosystems." *Global Environmental Change* 69:102303. https://doi.org/10.1016/j.gloenvcha.2021.102303.

Luderer, G., M. Pehl, A. Arvesen, T. Gibon, B.L. Bodirsky, H.S. de Boer, O. Fricko, et al. 2019. "Environmental Co-Benefits and Adverse Side-Effects of Alternative Power Sector Decarbonization Strategies." *Nature Communications* 10(1):5229. https://doi.org/10.1038/s41467-019-13067-8.

Macmillan, A., J. Connor, K. Witten, R. Kearns, D. Rees, and A. Woodward. 2014. "The Societal Costs and Benefits of Commuter Bicycling: Simulating the Effects of Specific Policies Using System Dynamics Modeling." *Environmental Health Perspectives* 122(4):335–344. https://doi.org/10.1289/ehp.1307250.

Mahajan, M., O. Ashmoore, J. Rissman, R. Orvis, and A. Gopal. 2022. *Modeling the Inflation Reduction Act Using the Energy Policy Simulator.* Energy Innovation Policy and Technology.

Mailloux, N.A., D.W. Abel, T. Holloway, and J.A. Patz. 2022. "Nationwide and Regional $PM_{2.5}$-Related Air Quality Health Benefits from the Removal of Energy-Related Emissions in the United States." *GeoHealth* 6(5). https://doi.org/10.1029/2022gh000603.

Maizlish, N., L. Rudolph, and C. Jiang. 2022. "Health Benefits of Strategies for Carbon Mitigation in US Transportation, 2017–2050." *American Journal of Public Health* 112(3):426–433. https://doi.org/10.2105/ajph.2021.306600.

Markandya, A., J. Sampedro, S. J. Smith, R. Van Dingenen, C. Pizarro-Irizar, I. Arto, and M. González-Eguino. 2018. "Health Co-Benefits from Air Pollution and Mitigation Costs of the Paris Agreement: A Modelling Study." *Lancet: Planetary Health* 2(3):E126–E133. https://doi.org/10.1016/S2542-5196(18)30029-9.

Martinez-Alier, J. 2001. "Mining Conflicts, Environmental Justice, and Valuation." *Journal of Hazardous Materials* 86(1):153–170. https://doi.org/10.1016/S0304-3894(01)00252-7.

Massetti, E., M.A. Brown, M. Lapsa, I. Sharma, J. Bradbury, C. Cunliff, and Y. Li. 2017. *Environmental Quality and the U.S. Power Sector: Air Quality, Water Quality, Land Use and Environmental Justice.* Oak Ridge, TN: Oak Ridge National Laboratory.

Mayyas, A., D. Steward, and M. Mann. 2019. "The Case for Recycling: Overview and Challenges in the Material Supply Chain for Automotive Li-Ion Batteries." *Sustainable Materials and Technologies* 19:e00087. https://doi.org/10.1016/j.susmat.2018.e00087.

McCauley, D., and R. Heffron. 2018. "Just Transition: Integrating Climate, Energy and Environmental Justice." *Energy Policy* 119:1–7. https://doi.org/10.1016/j.enpol.2018.04.014.

McConnell, K., C.V. Braneon, E. Glenn, N. Stamler, E. Mallen, D.P. Johnson, R. Pandya, J. Abramowitz, G. Fernandez, and C. Rosenzweig. 2022. "A Quasi-Experimental Approach for Evaluating the Heat Mitigation Effects of Green Roofs in Chicago, Illinois." *Sustainable Cities and Society* 76:103376. https://doi.org/10.1016/j.scs.2021.103376.

McCubbin, D., and B.K. Sovacool. 2013. "Quantifying the Health and Environmental Benefits of Wind Power to Natural Gas." *Energy Policy* 53:429–441. https://doi.org/10.1016/j.enpol.2012.11.004.

McDuffie, E.E., R.V. Martin, J.V. Spadaro, R. Burnett, S.J. Smith, P. O'Rourke, M.S. Hammer, et al. 2021. "Source Sector and Fuel Contributions to Ambient PM2.5 and Attributable Mortality Across Multiple Spatial Scales." *Nature Communications* 12(1):3594. https://doi.org/10.1038/s41467-021-23853-y.

Millstein, D., R. Wiser, M. Bolinger, and G. Barbose. 2017. "The Climate and Air-Quality Benefits of Wind and Solar Power in the United States." *Nature Energy* 2(9):17134. https://doi.org/10.1038/nenergy.2017.134.

Milner, J., R. Green, A.D. Dangour, A. Haines, Z. Chalabi, J. Spadaro, A. Markandya, and P. Wilkinson. 2015. "Health Effects of Adopting Low Greenhouse Gas Emission Diets in the UK." *BMJ Open* 5(4):e007364. https://doi.org/10.1136/bmjopen-2014-007364.

Mizdrak, A., T. Blakely, C.L. Cleghorn, and L.J. Cobiac. 2019. "Potential of Active Transport to Improve Health, Reduce Healthcare Costs, and Reduce Greenhouse Gas Emissions: A Modelling Study." *PLOS ONE* 14(7):e0219316. https://doi.org/10.1371/journal.pone.0219316.

Mohai, P., D. Pellow, and J.T. Roberts. 2009. "Environmental Justice." *Annual Review of Environment and Resources* 34(1):405–430. https://doi.org/10.1146/annurev-environ-082508-094348.

Molinos-Senante, M., and R. Sala-Garrido. 2017. "Energy Intensity of Treating Drinking Water: Understanding the Influence of Factors." *Applied Energy* 202:275–281. https://doi.org/10.1016/j.apenergy.2017.05.100.

Mora, C., T. McKenzie, I.M. Gaw, J.M. Dean, H. von Hammerstein, T.A. Knudson, R.O. Setter, et al. 2022. "Over Half of Known Human Pathogenic Diseases Can Be Aggravated by Climate Change." *Nature Climate Change* 12(9):869–875. https://doi.org/10.1038/s41558-022-01426-1.

Mozaffarian, D., T. Griffin, and J. Mande. 2019. "The 2018 Farm Bill—Implications and Opportunities for Public Health." *JAMA* 321(9):835. https://doi.org/10.1001/jama.2019.0317.

Mueller, N., D. Rojas-Rueda, T. Cole-Hunter, A. de Nazelle, E. Dons, R. Gerike, T. Götschi, L. Int Panis, S. Kahlmeier, and M. Nieuwenhuijsen. 2015. "Health Impact Assessment of Active Transportation: A Systematic Review." *Preventive Medicine* 76:103–114. https://doi.org/10.1016/j.ypmed.2015.04.010.

Mueller, N., D. Rojas-Rueda, X. Basagaña, M. Cirach, T. Cole-Hunter, P. Dadvand, D. Donaire-Gonzalez, et al. 2017. "Urban and Transport Planning Related Exposures and Mortality: A Health Impact Assessment for Cities." *Environmental Health Perspectives* 125(1):89–96. https://doi.org/10.1289/ehp220.

Nardone, A., J.A. Casey, R. Morello-Frosch, M. Mujahid, J.R. Balmes, and N. Thakur. 2020. "Associations Between Historical Residential Redlining and Current Age-Adjusted Rates of Emergency Department Visits Due to Asthma Across Eight Cities in California: An Ecological Study." *Lancet Planetary Health* 4(1):e24–e31. https://doi.org/10.1016/S2542-5196(19)30241-4.

NASEM (National Academies of Sciences, Engineering, and Medicine). 2021. *Transit and Micromobility*. Washington, DC: The National Academies Press.

NASEM. 2022a. *The Potential Impacts of Gold Mining in Virginia*. Washington, DC: The National Academies Press.

NASEM. 2022b. *Why Indoor Chemistry Matters*. Washington, DC: The National Academies Press.

NASEM and NAE (National Academy of Engineering). 2022. *Indoor Exposure to Fine Particulate Matter and Practical Mitigation Approaches: Proceedings of a Workshop*. Washington, DC: The National Academies Press.

National Energy Technology Laboratory. 2020. *2020 Carbon Capture Program R&D Compendium of Carbon Capture Technology*. Department of Energy.

Neumann, J.E., M. Amend, S. Anenberg, P.L. Kinney, M. Sarofim, J. Martinich, J. Lukens, J.-W. Xu, and H. Roman. 2021. "Estimating PM2.5-Related Premature Mortality and Morbidity Associated with Future Wildfire Emissions in the Western US." *Environmental Research Letters* 16(3):035019. https://doi.org/10.1088/1748-9326/abe82b.

Nieuwenhuijsen, M.J. 2021. "Green Infrastructure and Health." *Annual Review of Public Health* 42(1):317–328. https://doi.org/10.1146/annurev-publhealth-090419-102511.

NIOSH (National Institute for Occupational Safety and Health). n.d.(a). "About NIOSH." https://www.cdc.gov/niosh/about/default.html. Accessed March 15, 2023.

NIOSH. n.d.(b). "Mining Research Program." Centers for Disease Control and Prevention National Institute for Occupational Safety and Health. https://www.cdc.gov/niosh/mining/researchprogram. Accessed March 15, 2023.

Nkyekyer, E.W., and A.L. Dannenberg. 2019. "Use and Effectiveness of Health Impact Assessment in the Energy and Natural Resources Sector in the United States, 2007–2016." *Impact Assessment and Project Appraisal* 37(1):17–32. https://doi.org/10.1080/14615517.2018.1519221.

Nopmongcol, U., J. Grant, E. Knipping, M. Alexander, R. Schurhoff, D. Young, J. Jung, T. Shah, and G. Yarwood. 2017. "Air Quality Impacts of Electrifying Vehicles and Equipment Across the United States." *Environmental Science and Technology* 51(5):2830–2837. https://doi.org/10.1021/acs.est.6b04868.

Nowak, D.J., N. Appleton, A. Ellis, and E. Greenfield. 2017. "Residential Building Energy Conservation and Avoided Power Plant Emissions by Urban and Community Trees in the United States." *Urban Forestry and Urban Greening* 21:158–165. https://doi.org/10.1016/j.ufug.2016.12.004.

Obradovich, N., R. Migliorini, M.P. Paulus, and I. Rahwan. 2018. "Empirical Evidence of Mental Health Risks Posed by Climate Change." *Proceedings of the National Academy of Sciences* 115(43):10953–10958. https://doi.org/10.1073/pnas.1801528115.

O'Dell, K., B. Ford, J. Burkhardt, S. Magzamen, S.C. Anenberg, J. Bayham, E.V. Fischer, and J.R. Pierce. 2023. "Outside In: The Relationship Between Indoor and Outdoor Particulate Air Quality During Wildfire Smoke Events in Western US Cities." *Environmental Research: Health* 1(1):015003. https://doi.org/10.1088/2752-5309/ac7d69.

Ojala, M., A. Cunsolo, C.A. Ogunbode, and J. Middleton. 2021. "Anxiety, Worry, and Grief in a Time of Environmental and Climate Crisis: A Narrative Review." *Annual Review of Environment and Resources* 46(1):35–58. https://doi.org/10.1146/annurev-environ-012220-022716.

Orellano, P., J. Reynoso, and N. Quaranta. 2021. "Short-Term Exposure to Sulphur Dioxide (SO_2) and All-Cause and Respiratory Mortality: A Systematic Review and Meta-Analysis." *Environment International* 150:106434. https://doi.org/10.1016/j.envint.2021.106434.

Ou, Y., W. Shi, S.J. Smith, C.M. Ledna, J.J. West, C.G. Nolte, and D.H. Loughlin. 2018. "Estimating Environmental Co-Benefits of U.S. Low-Carbon Pathways Using an Integrated Assessment Model with State-Level Resolution." *Applied Energy* 216:482–493. https://doi.org/10.1016/j.apenergy.2018.02.122.

Overpeck, J.T., and B. Udall. 2020. "Climate Change and the Aridification of North America." *Proceedings of the National Academy of Sciences* 117(22):11856–11858. https://doi.org/10.1073/pnas.2006323117.

Palinkas, L.A., and M. Wong. 2020. "Global Climate Change and Mental Health." *Current Opinion in Psychology* 32:12–16. https://doi.org/10.1016/j.copsyc.2019.06.023.

Pedersen, B.K., and M.A. Febbraio. 2012. "Muscles, Exercise and Obesity: Skeletal Muscle as a Secretory Organ." *Nature Reviews Endocrinology* 8(8):457–465. https://doi.org/10.1038/nrendo.2012.49.

Penn, S.L., S. Arunachalam, M. Woody, W. Heiger-Bernays, Y. Tripodis, and J.I. Levy. 2017. "Estimating State-Specific Contributions to $PM_{2.5}$- and O_3-Related Health Burden from Residential Combustion and Electricity Generating Unit Emissions in the United States." *Environmental Health Perspectives* 125(3):324–332. https://doi.org/10.1289/EHP550.

Peters, D.R., J.L. Schnell, P.L. Kinney, V. Naik, and D.E. Horton. 2020. "Public Health and Climate Benefits and Trade-Offs of U.S. Vehicle Electrification." *GeoHealth* 4:e2020GH000275. https://doi.org/10.1029/2020GH000275.

Plumer, B., and N. Popovich. 2020. "How Decades of Racist Housing Policy Left Neighborhoods Sweltering." *The New York Times*. August 24. https://www.nytimes.com/interactive/2020/08/24/climate/racism-redlining-cities-global-warming.html.

Pope, S., J. Rhodus, F. Fulk, B. Mintz, and S. O'Shea. 2016. *The Health Impact Assessment (HIA) Resource and Tool Compilation: A Comprehensive Toolkit for New and Experienced HIA Practitioners in the U.S.* Environmental Protection Agency. https://www.epa.gov/sites/default/files/2017-07/documents/hia_resource_and_tool_compilation.pdf.

Pozzer, A., S.C. Anenberg, S. Dey, A. Haines, J. Lelieveld, and S. Chowdhury. 2023. "Mortality Attributable to Ambient Air Pollution: A Review of Global Estimates." *GeoHealth* 7(1):e2022GH000711. https://doi.org/10.1029/2022GH000711.

Prehoda, E.W., and J.M. Pearce. 2017. "Potential Lives Saved by Replacing Coal with Solar Photovoltaic Electricity Production in the U.S." *Renewable and Sustainable Energy Reviews* 80:710–715. https://doi.org/10.1016/j.rser.2017.05.119.

Raifman, M., K.F. Lambert, J.I. Levy, and P.L. Kinney. 2021. "Mortality Implications of Increased Active Mobility for a Proposed Regional Transportation Emission Cap-and-Invest Program." *Journal of Urban Health* 98(3):315–327. https://doi.org/10.1007/s11524-020-00510-1.

Rambhia, M., R. Volk, B. Rismanchi, S. Winter, and F. Schultmann. 2023. "Supporting Decision-Makers in Estimating Irrigation Demand for Urban Street Trees." *Urban Forestry and Urban Greening* 82:127868. https://doi.org/10.1016/j.ufug.2023.127868.

Ramirez-Ibarra, M., and J.-D.M. Saphores. 2023. "Health and Equity Impacts from Electrifying Drayage Trucks." *Transportation Research Part D: Transport and Environment* 116:103616. https://doi.org/10.1016/j.trd.2023.103616.

Rapp, V.H., B.C. Singer, C. Stratton, and C.P. Wray. 2012. *Assessment of Literature Related to Combustion Appliance Venting Systems*. Berkeley, CA: Lawrence Berkeley National Laboratory.

Razeghi, G., M. Carreras-Sospedra, T. Brown, J. Brouwer, D. Dabdub, and S. Samuelsen. 2016. "Episodic Air Quality Impacts of Plug-In Electric Vehicles." *Atmospheric Environment* 137:90–100. https://doi.org/10.1016/j.atmosenv.2016.04.031.

Reck, B.K., and T.E. Graedel. 2012. "Challenges in Metal Recycling." *Science* 337(6095):690–695. https://doi.org/10.1126/science.1217501.

Reichmuth, D., J. Dunn, and D. Anair. 2022. *Driving Cleaner: Electric Cars and Pickups Beat Gasoline on Lifetime Global Warming Emissions*. Union of Concerned Scientists.

Ringquist, E.J. 2005. "Assessing Evidence of Environmental Inequities: A Meta-Analysis." *Journal of Policy Analysis and Management* 24(2):223–247. https://doi.org/10.1002/pam.20088.

Rockefeller Foundation. 2021. "Executive Summary." P. 6 in *True Cost of Food: Measuring What Matters to Transform the U.S. Food System*. Rockefeller Foundation.

Romanello, M., C. Di Napoli, P. Drummond, C. Green, H. Kennard, P. Lampard, D. Scamman, et al. 2022. "The 2022 Report of the Lancet Countdown on Health and Climate Change: Health at the Mercy of Fossil Fuels." *Lancet* 400(10363):1619–1654. https://doi.org/10.1016/S0140-6736(22)01540-9.

Romero-Lankao, P., A. Wilson, and D. Zimny-Schmitt. 2022. "Inequality and the Future of Electric Mobility in 36 U.S. Cities: An Innovative Methodology and Comparative Assessment." *Energy Research and Social Science* 91:102760. https://doi.org/10.1016/j.erss.2022.102760.

Rothstein, R. 2017. *The Color of Law: A Forgotten History of How Our Government Segregated America*. First ed. New York and London: Liveright Publishing Corporation, W.W. Norton.

Roy, N., M. Domeshek, D. Burtraw, K. Palmer, K. Rennert, J.-S. Shih, and S. Villanueve. 2022. *Beyond Clean Energy: The Financial Incidence and Health Effects of the IRA*. Resources for the Future.

Rudolph, L. 2022. "Panel Discussion." Presented at Accelerating Decarbonization in the United States: Technology, Policy, and Societal Dimensions | Pathways to an Equitable and Just Transition Workshop. July 26. https://www.nationalacademies.org/event/07-26-2022/accelerating-decarbonization-in-the-united-states-technology-policy-and-societal-dimensions-pathways-to-an-equitable-and-just-transition-workshop.

Satija, A., S.N. Bhupathiraju, E.B. Rimm, D. Spiegelman, S.E. Chiuve, L. Borgi, W.C. Willett, J.E. Manson, Q. Sun, and F.B. Hu. 2016. "Plant-Based Dietary Patterns and Incidence of Type 2 Diabetes in US Men and Women: Results from Three Prospective Cohort Studies." *PLOS Medicine* 13(6):e1002039. https://doi.org/10.1371/journal.pmed.1002039.

Scarborough, P., S. Allender, D. Clarke, K. Wickramasinghe, and M. Rayner. 2012. "Modelling the Health Impact of Environmentally Sustainable Dietary Scenarios in the UK." *European Journal of Clinical Nutrition* 66(6):710–715. https://doi.org/10.1038/ejcn.2012.34.

Schramm, P., A. Vaidyanathan, L. Radhakrishnan, A. Gates, K. Hartnett, and P. Breysse. 2021. "Heat-Related Emergency Department Visits During the Northwestern Heat Wave—United States, June 2021." *Morbidity and Mortality Weekly Report*. http://dx.doi.org/10.15585/mmwr.mm7029e1.

Schuele, H., C.F. Baum, P.J. Landrigan, and S.S. Hawkins. 2022. "Associations Between Proximity to Gas Production Activity in Counties and Birth Outcomes Across the US." *Preventive Medicine Reports* 30:102007. https://doi.org/10.1016/j.pmedr.2022.102007.

Sealing, K. 2008. "Attack of the Balloon People: How America's Food Culture and Agricultural Policy Threaten the Food Security of the Poor, Farmers and the Indigenous Peoples of the World." *Vanderbilt Journal of Transnational Law* 40(4):1015–1037.

Seeberger, J., R. Grandhi, S.S. Kim, W.A. Mase, T. Reponen, S.-M. Ho, and A. Chen. 2016. "Special Report: E-Waste Management in the United States and Public Health Implications." *Journal of Environmental Health* 79(3):8–17.

Sergi, B., I. Azevedo, S.J. Davis, and N.Z. Muller. 2020. "Regional and County Flows of Particulate Matter Damage in the US." *Environmental Research Letters* 15(10):104073. https://doi.org/10.1088/1748-9326/abb429.

Shahzad, M.W., M. Burhan, L. Ang, and K.C. Ng. 2017. "Energy-Water-Environment Nexus Underpinning Future Desalination Sustainability." *Desalination* 413:52–64. https://doi.org/10.1016/j.desal.2017.03.009.

Sharma, S.S., and A. Manthiram. 2020. "Towards More Environmentally and Socially Responsible Batteries." *Energy and Environmental Science* 13(11):4087–4097. https://doi.org/10.1039/D0EE02511A.

Shaw, D. 2009. "Cansolv CO_2 Capture: The Value of Integration." *Energy Procedia* 1(1):237–246. https://doi.org/10.1016/j.egypro.2009.01.034.

Shindell, D.T., 2015. "The Social Cost of Atmospheric Release." *Climatic Change* 130:313–326.

Shindell, D., M. Ru, Y. Zhang, K. Seltzer, G. Faluvegi, L. Nazarenko, G.A. Schmidt et al. 2021. "Temporal and Spatial Distribution of Health, Labor, and Crop Benefits of Climate Change Mitigation in the United States." *Proceedings of the National Academy of Sciences* 118(46):e2104061118. https://doi.org/10.1073/pnas.2104061118.

Smith, K.R., H. Frumkin, K. Balakrishnan, C.D. Butler, Z.A. Chafe, I. Fairlie, P. Kinney, et al. 2013. "Energy and Human Health." *Annual Review of Public Health* 34:159–188. https://doi.org/10.1146/annurev-publhealth-031912-114404.

Sousa-Silva, R., A. Smargiassi, D. Kneeshaw, J. Dupras, K. Zinszer, and A. Paquette. 2021. "Strong Variations in Urban Allergenicity Riskscapes Due to Poor Knowledge of Tree Pollen Allergenic Potential." *Scientific Reports* 11(1):10196. https://doi.org/10.1038/s41598-021-89353-7.

Spietz, T., S. Dobras, L. Więcław-Solny, and A. Krótki. 2017. "Nitrosamines and Nitramines in Carbon Capture Plants." *Environmental Protection and Natural Resources* 28(4):43–50. https://doi.org/10.1515/oszn-2017-0027.

Steinberg, D.C., M. Brown, R. Wiser, P. Danohoo-Vallett, P. Gagnon, A. Hamilton, M. Mowers, C. Murphy, and A. Prasana. 2023. *Evaluating Impacts of the Inflation Reduction Act and Bipartisan Infrastructure Law on the U.S. Power System*. Golden, CO: National Renewable Energy Laboratory.

Story, M., K.M. Kaphingst, R. Robinson-O'Brien, and K. Glanz. 2008. "Creating Healthy Food and Eating Environments: Policy and Environmental Approaches." *Annual Review of Public Health* 29:253–272. https://doi.org/10.1146/annurev. publhealth.29.020907.090926.

Symonds, P., D. Rees, Z. Daraktchieva, N. McColl, J. Bradley, I. Hamilton, and M. Davies. 2019. "Home Energy Efficiency and Radon: An Observational Study." *Indoor Air* 29(5):854–864. https://doi.org/10.1111/ina.12575.

Temkin, A., S. Evans, T. Manidis, C. Campbell, and O.V. Naidenko. 2019. "Exposure-Based Assessment and Economic Valuation of Adverse Birth Outcomes and Cancer Risk Due to Nitrate in United States Drinking Water." *Environmental Research* 176:108442. https://doi.org/10.1016/j.envres.2019.04.009.

Tessum, C.W., J.D. Hill, and J.D. Marshall. 2014. "Life Cycle Air Quality Impacts of Conventional and Alternative Light-Duty Transportation in the United States." *Proceedings of the National Academy of Sciences* 111(52):18490–18495. https://doi.org/10.1073/pnas.1406853111.

Tessum, C.W., J.S. Apte, A.L. Goodkind, N.Z. Muller, K.A. Mullins, D.A. Paolella, S. Polasky, et al. 2019. "Inequity in Consumption of Goods and Services Adds to Racial–Ethnic Disparities in Air Pollution Exposure." *Proceedings of the National Academy of Sciences* 116(13):6001–6006. https://doi.org/10.1073/pnas.1818859116.

Tessum, C.W., D.A. Paolella, S.E. Chambliss, J.S. Apte, J.D. Hill, and J.D. Marshall. 2021. $PM_{2.5}$ Polluters Disproportionately and Systemically Affect People of Color in the United States. *Science Advances* 7(18):eabf4491. https://doi.org/10.1126/sciadv.abf4491.

Thakrar, S.K., S. Balasubramanian, P.J. Adams, I.M. Azevedo, N.Z. Muller, S.N. Pandis, S. Polasky, C.A. Pope III, A.L. Robinson, J.S. Apte, and C.W. Tessum. 2020. "Reducing Mortality from Air Pollution in the United States by Targeting Specific Emission Sources." *Environmental Science and Technology Letters* 7(9):639–645. https://pubs.acs.org/doi/10.1021/acs.estlett.0c00424.

Thompson, T., M. Webber, and D.T. Allen. 2009. "Air Quality Impacts of Using Overnight Electricity Generation to Charge Plug-In Hybrid Electric Vehicles for Daytime Use." *Environmental Research Letters* 4(1):014002. https://doi.org/10.1088/1748-9326/4/1/014002.

Thompson, T.M., S. Rausch, R.K. Saari, and N.E. Selin. 2016. "Air Quality Co-Benefits of Subnational Carbon Policies." *Journal of the Air and Waste Management Association* 66(10):988–1002. https://doi.org/10.1080/10962247.2016.1192071.

Thurston, G.D., and M.L. Bell. 2021. "The Human Health Co-Benefits of Air Quality Improvements Associated with Climate Change Mitigation." Pp. 181–202 in *Climate Change and Global Public Health*, K.E. Pinkerton and W.N. Rom, eds. Cham, Switzerland: Springer International Publishing.

Tilman, D., and M. Clark. 2014. "Global Diets Link Environmental Sustainability and Human Health." *Nature* 515(7528):518–522. https://doi.org/10.1038/nature13959.

Timmers, V.R.J.H., and P.A.J. Achten. 2016. "Non-Exhaust PM Emissions from Electric Vehicles." *Atmospheric Environment* 134:10–17. https://doi.org/10.1016/j.atmosenv.2016.03.017.

Tonn, B., E. Rose, B. Hawkins, and B. Conlon. 2014. *Health and Household-Related Benefits Attributable to the Weatherization Assistance Program*. Oak Ridge, TN: Oak Ridge National Laboratory.

Trtanj, J., L. Jantarasami, J. Brunkard, T. Collier, J. Jacobs, E. Lipp, S. McLellan, et al. 2016. Chapter 6, "Climate Impacts on Water-Related Illness." Pp. 157–188 in *The Impacts of Climate Change on Human Health in the United States: A Scientific Assessment*, A. Cummins, J. Balbus, J.L. Gamble, C.B. Beard, J.E. Bell, D. Dodgen, et al., eds. Washington, DC: U.S. Global Change Research Program.

Tubiello, F.N., C. Rosenzweig, G. Conchedda, K. Karl, J. Gütschow, P. Xueyao, G. Obli-Laryea, et al. 2021. "Greenhouse Gas Emissions from Food Systems: Building the Evidence Base." *Environmental Research Letters* 16(6):065007. https://doi.org/10.1088/1748-9326/ac018e.

Tzanidakis, K., T. Oxley, T. Cockerill, and H. ApSimon. 2013. "Illustrative National Scale Scenarios of Environmental and Human Health Impacts of Carbon Capture and Storage." *Environment International* 56:48–64. https://doi.org/10.1016/j.envint.2013.03.007.

University of Pittsburgh. 2023. "Hydraulic Fracturing Epidemiology Research Studies: Childhood Cancer Case-Control Study." University of Pittsburgh School of Public Health. https://paenv.pitt.edu/assets/Report_Cancer_outcomes_2023_August.pdf.

USDA (U.S. Department of Agriculture). 2022. "Climate Change." U.S. Department of Agriculture Economic Research Service. https://www.ers.usda.gov/topics/natural-resources-environment/climate-change.

van Dorn, A. 2017. "Urban Planning and Respiratory Health." *Lancet Respiratory Medicine* 5(10):781–782. https://doi.org/10.1016/S2213-2600(17)30340-5.

Vandyck, T., K. Keramidas, A. Kitous, J.V. Spadaro, R. Van Dingenen, M. Holland, and B. Saveyn. 2018. "Air Quality Co-Benefits for Human Health and Agriculture Counterbalance Costs to Meet Paris Agreement Pledges." *Nature Communications* 9(1):4939. https://doi.org/10.1038/s41467-018-06885-9.

Veltman, K., B. Singh, and E.G. Hertwich. 2010. "Human and Environmental Impact Assessment of Postcombustion CO_2 Capture Focusing on Emissions from Amine-Based Scrubbing Solvents to Air." *Environmental Science and Technology* 44(4):1496–1502. https://doi.org/10.1021/es902116r.

Vengosh, A., and E. Weinthal. 2023. "The Water Consumption Reductions from Home Solar Installation in the United States." *Science of the Total Environment* 854:158738. https://doi.org/10.1016/j.scitotenv.2022.158738.

Vogel, M.M., J. Zscheischler, R. Wartenburger, D. Dee, and S.I. Seneviratne. 2019. "Concurrent 2018 Hot Extremes Across Northern Hemisphere Due to Human-Induced Climate Change." *Earth's Future* 7(7):692–703. https://doi.org/10.1029/2019EF001189.

Vohra, K., A. Vodonos, J. Schwartz, E.A. Marais, M.P. Sulprizio, and L.J. Mickley. 2021. "Global Mortality from Outdoor Fine Particle Pollution Generated by Fossil Fuel Combustion: Results from GEOS-Chem." *Environmental Research* 195:110754. https://doi.org/10.1016/j.envres.2021.110754.

Wang, T., Z. Jiang, B. Zhao, Y. Gu, K.-N. Liou, N. Kalandiyur, D. Zhang, and Y. Zhu. 2020. "Health Co-Benefits of Achieving Sustainable Net-Zero Greenhouse Gas Emissions in California." *Nature Sustainability* 3(8):597–605. https://doi.org/10.1038/s41893-020-0520-y.

Wang, Y., J.S. Apte, J.D. Hill, C.E. Ivey, R.F. Patterson, A.L. Robinson, C.W. Tessum, and J.D. Marshall. 2022. "Location-Specific Strategies for Eliminating US National Racial-Ethnic $PM_{2.5}$ Exposure Inequality." *Proceedings of the National Academy of Sciences* 119(44):e2205548119. https://doi.org/10.1073/pnas.2205548119.

Ward, M.H., R.R. Jones, J.D. Brender, T.M. De Kok, P.J. Weyer, B.T. Nolan, C.M. Villanueva, and S.G. Van Breda. 2018. "Drinking Water Nitrate and Human Health: An Updated Review." *International Journal of Environmental Research and Public Health* 15(7):1557. https://www.mdpi.com/1660-4601/15/7/1557.

Warner, T., D. Vikara, A. Gionan, R. Dillmore, R. Walter, T. Stribley, and M. McMillen. 2020. "Overview of Potential Failure Modes and Effects Associated with CO_2 Injection and Storage Operations in Saline Formations." Pittsburgh, PA: National Energy Technology Laboratory. https://www.energy.gov/lpo/articles/overview-potential-failure-modes-and-effects-associated-co2-injection-and-storage.

Weis, A., J.J. Michalek, P. Jaramillo, and R. Lueken. 2015. "Emissions and Cost Implications of Controlled Electric Vehicle Charging in the U.S. PJM Interconnection." *Environmental Science and Technology* 49(9):5813–5819. https://doi.org/10.1021/es505822f.

West, J.J., A. Cohen, F. Dentener, B. Brunekreef, T. Zhu, B. Armstrong, M.L. Bell, et al. 2016. "What We Breathe Impacts Our Health: Improving Understanding of the Link Between Air Pollution and Health." *Environmental Science and Technology* 50(10):4895–4904. https://doi.org/10.1021/acs.est.5b03827.

WHO (World Health Organization). 2000. *Air Quality Guidelines for Europe.* 2nd ed. World Health Organization.

Willett, W., and J. Rockström. 2019. "Healthy Diets from Sustainable Food Systems: Summary Report of the EAT-Lancet Commission." EAT-Lancet Commission on Healthy Diets from Sustainable Food Systems. https://eatforum.org/content/uploads/2019/07/EAT-Lancet_Commission_Summary_Report.pdf.

Wilson, S., M. Hutson, and M. Mujahid. 2008. "How Planning and Zoning Contribute to Inequitable Development, Neighborhood Health, and Environmental Injustice." *Environmental Justice* 1(4):211–216. https://doi.org/10.1089/env.2008.0506.

Wiser, R., G. Barbose, J. Heeter, T. Mai, L. Bird, M. Bolinger, A. Carpenter, G. Heath, D. Keyser, J. Macknick, A. Mills, and D. Millstein. 2016a. "A Retrospective Analysis of Benefits and Impacts of U.S. Renewable Portfolio Standards." NREL/TP-6A20-65005. Lawrence Berkeley National Laboratory and National Renewable Energy Laboratory. https://www.nrel.gov/docs/fy16osti/65005.pdf.

Wiser, R., D. Millstein, T. Mai, J. Macknick, A. Carpenter, S. Cohen, W. Cole, B. Frew, and G. Heath. 2016b. "The Environmental and Public Health Benefits of Achieving High Penetrations of Solar Energy in the United States." *Energy* 113:472–486. https://doi.org/10.1016/j.energy.2016.07.068.

Wolf, K.L., S.T. Lam, J.K. McKeen, G.R.A. Richardson, M. van den Bosch, and A.C. Bardekjian. 2020. "Urban Trees and Human Health: A Scoping Review." *International Journal of Environmental Research and Public Health* 17(12):4371. https://www.mdpi.com/1660-4601/17/12/4371.

Woodruff, T.J., J.D. Parker, A.D. Kyle, and K.C. Schoendorf. 2003. "Disparities in Exposure to Air Pollution During Pregnancy." *Environmental Health Perspectives* 111(7):942–946. https://doi.org/10.1289/ehp.5317.

Xiao, C., Y. Goryakin, and M. Cecchini. 2019. "Physical Activity Levels and New Public Transit: A Systematic Review and Meta-Analysis." *American Journal of Preventive Medicine* 56(3):464–473. https://doi.org/10.1016/j.amepre.2018.10.022.

Younkin, S.G., H.C. Fremont, and J.A. Patz. 2021. "The Health-Oriented Transportation Model: Estimating the Health Benefits of Active Transportation." *Journal of Transport and Health* 22:101103. https://doi.org/10.1016/j.jth.2021.101103.

Zheng, X.-Y., P. Orellano, H.-L. Lin, M. Jiang, and W.-J. Guan. 2021. "Short-Term Exposure to Ozone, Nitrogen Dioxide, and Sulphur Dioxide and Emergency Department Visits and Hospital Admissions Due to Asthma: A Systematic Review and Meta-Analysis." *Environment International* 150:106435. https://doi.org/10.1016/j.envint.2021.106435.

Zhu, S., M. Mac Kinnon, A. Carlos-Carlos, S.J. Davis, and S. Samuelsen. 2022. "Decarbonization Will Lead to More Equitable Air Quality in California." *Nature Communications* 13(1):5738. https://doi.org/10.1038/s41467-022-33295-9.

Ziska, L., K. Knowlton, C. Rogers, D. Dalan, N. Tierney, M.A. Elder, W. Filley, et al. 2011. "Recent Warming by Latitude Associated with Increased Length of Ragweed Pollen Season in Central North America." *Proceedings of the National Academy of Sciences* 108(10):4248–4251. https://doi.org/10.1073/pnas.1014107108.

4

Workforce Needs, Opportunities, and Support

ABSTRACT

Investments in clean technologies and infrastructure have the potential not only to address the climate crisis by reducing greenhouse gas emissions but also to create and support millions of jobs in the rapidly growing clean energy economy and stimulate economic growth. The transition to net zero will have uneven impacts across sectors, demographics, and geographies of the U.S. workforce over varying timeframes. Employment impacts will depend on the pathways for decarbonization in specific sectors of the economy and will be influenced by many factors, including technologies utilized, timing/pace of transition, and siting decisions. Job growth is expected to be geographically heterogeneous, and not geographically congruent with fossil losses. New jobs may differ from lost jobs in terms of skills and wages, and new jobs may not be created at the same time as at-risk jobs are being lost. Size, location, and timing of employment growth will be impacted by many factors such as domestic content, worker productivity increases, and siting decisions.

To attract and retain the workforce necessary to accomplish the ambitious goals associated with the transition to a net-zero economy, it is vital to ensure that the clean energy economy generates high-quality, accessible jobs and career paths. Last, action is needed to ensure that the transition enhances the inclusiveness and diversity of the clean energy workforce. This is only possible with the active engagement of all stakeholders in the "workforce development" ecosystem, including governments, employers, and workers. Providing employment options that meet workforce needs is important in maintaining the social contract necessary for accomplishing the coming decades of transition.

The clean energy transition will not be without challenges. There is widespread agreement that historical trends of net fossil fuel job losses will continue and may accelerate, although there is still uncertainty about the timing, geographical distribution, and magnitude of these losses. Policies and programs need to address the losses that will inevitably occur in employment, economic base, and public revenue for workers and communities with close ties to the fossil fuel industry. Historical analyses of large shocks to local economies display that the existing safety net is ill-equipped to address the scale and scope of these impacts.

INTRODUCTION

The transition to net zero will have uneven impacts across sectors, demographics, and geographies of the U.S. workforce over varying timeframes. Employment impacts will depend on the pathways for decarbonization in specific sectors of the economy and will be influenced by many factors, including technologies utilized, timing/pace of transition, and siting decisions. There is widespread agreement that historical trends of net fossil fuel job losses will continue and may accelerate, although there is still uncertainty about the timing, geographical distribution, and magnitude of these losses (see Chapter 12 for technical details). Job growth is expected to be geographically heterogeneous, and not geographically congruent with fossil losses. New jobs may differ from lost jobs in terms of skills and wages, and new jobs may not be created at the same time as at-risk jobs are being lost. Size, location, and timing of employment growth will be impacted by many factors such as domestic content, worker productivity increases, and siting decisions (Mayfield et al. 2023).

Workforce impacts of the transition to net zero go beyond what we may think of as energy jobs. Upstream materials and supply chain jobs to produce equipment for a net-zero economy will be impacted, as will the way we produce carbon intensive industrial materials such as steel and cement (see Chapter 10 for details). Other jobs directly impacted include internal combustion engine (ICE) vehicle manufacturing and related automotive jobs (such as auto mechanics) as more electric vehicles enter the fleet (see Chapter 9 for details). Additionally, direct energy jobs and indirect manufacturing jobs support ancillary jobs throughout the economy, and these jobs will be impacted, along with communities that rely on tax revenue from affected businesses. Communities whose economies rely heavily on the fossil fuel industry are at risk, and impacts are already unfolding. (See Chapter 13 for discussions of policies directed toward coal communities, and Chapter 5 for details on working effectively with communities throughout the transition.)

The net-zero transition presents opportunities to achieve social objectives in addition to providing significant environmental and economic benefits. Not only does it have the potential to be an engine of economic growth, but it also has the potential to provide high-quality jobs[1] that contribute significantly to social inclusion. However,

---

[1] As defined in the first report of this committee, "a high-quality job entails, at a minimum, a safe and secure working environment, family-sustaining wages and comprehensive benefits, regular schedules and hours, and skills-development opportunities that enable wage advancement and career development" (NASEM 2021, p. 45).

this is only possible with the active engagement of all stakeholders in the "workforce development" ecosystem, including governments, educational institutions, employers, and workers. All workers are balancing a series of factors in their employment such as location; wages; working time; safety and working conditions; insurance coverage; the degree of job security; the type of contract; social security coverage; business culture and job design; skill development and career advancement opportunities; and access to paid leave, parental leave, and sick leave (AFL-CIO 2017; Aspen Institute 2022; Congdon et al. 2020; Gammarano 2020; ILO 2020; United Way Worldwide 2012). Providing employment options that meet workforce needs is important in maintaining the social contract necessary for accomplishing the coming decades of transition.

Policy choices can greatly affect the employment impacts of transition. Employment opportunities in clean energy and other net-zero-relevant fields are projected to grow significantly in the next decades as the decarbonization push intensifies, but employers already face hiring, recruitment, and retention challenges. The Inflation Reduction Act (IRA) provisions are expected to ramp up demand for apprenticeships, unions, and other high-road employment pathways, but many state and local workforce and apprenticeship systems lack the capacity and resources to produce sufficient volume of skilled, trained, and well-compensated workers. This chapter discusses the factors that affect employment impacts of a transition and what might be done to maximize benefits and minimize losses for workers and communities. Table 4-1, at the end of the chapter, summarizes all the recommendations in this chapter regarding building the workforce needed to accomplish decarbonization objectives, as well as supporting workers negatively impacted by the transition.

FIRST REPORT RECOMMENDATIONS

The committee's first report was released in February 2021 and included findings and recommendations on how to advance decarbonization in 2021–2030 while achieving four societal goals: strengthening the U.S. economy, promoting equity and inclusion, supporting communities, businesses, and workers directly impacted by the transition, and maximizing cost effectiveness (NASEM 2021). Each goal is intertwined with the workers who will build and maintain a net-zero economy.

The goal to support communities, businesses, and workers had four sub-goals that inform the recommendations relevant to workforce:

- To provide workers and communities accurate information about how clean energy transitions could impact them and access to viable economic transition strategies;

- To directly address during transition planning risks to "highly vulnerable"[2] locations where the economic transition to carbon neutrality will exacerbate existing economic disadvantages and health disparities;
- To hold companies accountable for ensuring that fossil fuel energy infrastructures are properly decommissioned and that their long-term environmental impacts are remediated to prevent the creation of persistent environmental contamination and associated health impacts for local populations; and
- To develop strategies to ensure that local, tribal, and state governments are able to replace lost revenue from plant, mine, and other industrial facility closures.

The first report included three overarching recommendations to Congress to understand and address impacts of transition on labor and workforce: (1) establish a 2-year federal National Transition Task Force to assess the vulnerability of labor sectors and communities to the transition; (2) establish a White House–level Office of Equitable Energy Transitions that would establish criteria for funding, sponsor research, and report on equity and transition impacts indicators; and (3) establish an independent National Transition Corporation to ensure coordination and funding to mitigate job losses, deploy and decommission infrastructure, and provide equitable access to economic opportunities and wealth, and to create public energy equity indicators. (See NASEM [2021] for additional details on these recommendations.)

One labor-specific recommendation is aimed to ensure that jobs created in transition will be high-quality jobs and that workers capture maximum benefits from federal funds: Federal grants, loans, tax incentives, and other support for projects should (1) be conditioned on recipients and their contractors meeting strong labor standards (including Davis-Bacon Act prevailing wage requirements and compliance with all labor, safety, environmental, and civil rights statutes); (2) require that federally funded construction and infrastructure project developers sign Project Labor Agreements (PLAs) where relevant; and (3) require recipients of federal incentives to negotiate Community Benefits (or Workforce) Agreements, where relevant. Similarly, a recommendation for domestic content requirements was included to bolster U.S. supply chains and maintain and create upstream jobs in materials, component, and equipment manufacturing.

---

[2] See Table 3.3.1, "Vulnerable Groups in the Context of an Energy Transition," in NASEM (2021, pp. 126–128) for details; these locations include those disproportionately impacted by transition such as communities that rely on fossil fuel jobs and/or tax revenue (especially rural or disconnected communities), Native American nations, communities with high energy costs, and so on.

On workforce development and training, the first report recommended that Congress establish education and training programs through appropriations to the Department of Energy (DOE) and the National Science Foundation (NSF) to supply the net-zero workforce, with reporting on diversity of participants and job placement success. This included establishing a 10-year GI Bill–type program for anyone seeking a degree in a net-zero area; establishing new and innovative community college, college, and university programs focused on knowledge and skills for a net-zero economy; providing funds for interdisciplinary doctoral and postdoctoral training programs to support decarbonization and energy justice; and supporting doctoral and postdoctoral fellowships in science and engineering, policy, and related areas. The report also recommended that the Department of Homeland Security should eliminate or ease visa restrictions for international students who want to study climate change and clean energy at the undergraduate and graduate levels, where appropriate. Last, to help communities and policy makers understand the current workforce, plan for transition, and evaluate effectiveness of interventions, the report recommended that Congress should pass the Promoting American Energy Jobs Act of 2019 to reestablish the Energy Jobs Strategy Council under DOE, require energy and employment data collection and analysis, and provide a public report on energy and employment in the United States.

Many other recommendations of the first report would have implications for the workforce even if the recommendations themselves do not contain specific workforce components. Examples include actions to expand clean energy, weatherize homes, and revitalize U.S. manufacturing that would create jobs overall, but would reduce demand for fossil fuels and therefore reduce demand for fossil jobs. Also, all workforce impacts will ripple through the economy. These impacts are discussed later in this chapter as well as in the sector-specific chapters.

POLICY PROGRESS SINCE 2021 REPORT

Federal, state, and local policy makers all play important roles in creating a pipeline of workers who can build the clean energy economy, enabling the transition of erstwhile fossil fuel workers to jobs in the clean energy economy, and ensuring that high-quality job creation and retention are an integral part of their climate and clean energy agendas. Many states and cities have forged partnerships with the private sector, unions, utilities, nonprofits, colleges, and other stakeholders to support their workforce development programs. These efforts vary greatly, and many do not go far enough. Chapter 13 provides greater detail regarding subnational actions supporting decarbonization. The following section discusses relevant federal actions since the committee's first report was published.

Federal Actions

Infrastructure Investment and Jobs Act

The Infrastructure Investment and Jobs Act (IIJA), passed in November 2021, authorizes funding to establish and extend workforce training and development efforts across several new and existing programs. The IIJA could create more than 800,000 jobs at its peak impact in the middle of the decade, reducing unemployment by a few tenths of a percentage point (Zandi and Yaros 2021).

Under DOE, the IIJA established a 21st Century Energy Workforce Advisory Board to provide recommendations to the Secretary of Energy on DOE's strategy in support of current and future energy sector workforce needs (DOE 2022). The IIJA supports workforce training in many existing and new DOE programs (American-Made Challenges n.d.; ARC 2021; DOE n.d.(c), n.d.(d); DOT Maritime Administration 2023; FTA 2021, 2023, n.d.; NGA 2023):

- The Energy Auditor Training Grant Program was authorized $40 million to establish a competitive grant program for eligible states to train workers to conduct energy audits or surveys of commercial and residential buildings.
- The Building Training and Assessment Centers Program was authorized $10 million to provide grants to institutions of higher education to establish building training and assessment centers that would educate and train building technicians and engineers on how to implement modern building technologies.
- The Career Skills Training Program was authorized $10 million to award grants to pay the federal share of associated career skills training programs under which students concurrently receive classroom instruction and on-the-job training for the purpose of obtaining an industry-related certification to install energy efficient buildings technologies.
- The Industrial Research and Assessment Centers Program was authorized $150 million in cooperative agreements for institutions of higher education, community colleges, trade schools, and union training programs to identify opportunities for optimizing energy efficiency and environmental performance at manufacturing and other industrial facilities, and an additional $400 million in implementation grants to small and medium manufacturers assessed by these institutions to implement suggestions.
- Several other DOE programs had funding authorized through the IIJA for various decarbonization efforts that can be used for workforce development efforts and/or are available to higher education institutions, including Battery Materials

Processing Grants, Battery Manufacturing and Recycling Grants, and the Electric Drive Vehicle Battery Recycling and 2nd Life Apps Program.

- Some DOE programs require a workforce development plan as part of the application for the program, including Regional Clean Hydrogen Hubs.
- The Energy Champions Leading the Advancement of Sustainable Schools Prize (Energy CLASS Prize) Program offers a prize that can be used to hire and train select school administration and facilities personnel as energy managers.
- Under the Federal Transit Administration, the Low or No Emissions Program was authorized $5.25 billion and the Buses and Bus Facilities Program was authorized $2 billion (5339(b) and (c)). These competitive grants provide funding to state and local governments for the purchase or lease of zero-emission and low-emission transit buses, including acquisition, construction, and leasing of required supporting facilities. Five percent of Low–No competitive grants are required to support workforce development and training, to ensure that diesel mechanics and other transit workers are not left behind in the transition to new technology. More than $3.3 billion of this funding has already been made available as of July 2023 and nearly 25 percent of projects funded for fiscal year 2023 explicitly include workforce elements such as Project Labor Agreements, use of registered apprenticeships, and/or expansion or establishment of workforce training programs.
- Under the Department of Transportation, the IIJA authorized $2.25 billion for the Port Infrastructure Development Program, which includes worker training to support electrification technology.
- The Appalachian Regional Commission, an economic development partnership agency of the federal government and 13 state governments that invests with local, regional, and state partners to transform Appalachian communities, create jobs, and strengthen the regional economy, was authorized $1 billion through the IIJA.
- While not specific to decarbonization, states can use funds from several IIJA formula and grant programs to invest in their infrastructure workforce. Additionally, the IIJA encourages, but does not require, 5-year Human Capital Plans to outline transportation and infrastructure workforce needs.

Creating Helpful Incentives to Produce Semiconductors and Science Act

The CHIPS and Science Act, passed in August 2022, incentivizes domestic semiconductor manufacturing by authorizing investments totaling $280 billion over the next 10 years in science, technology, engineering, and medicine (STEM) programs, workforce development, and technology research and development (R&D). This includes authorization for $174 billion in investments to support STEM, R&D, and workforce and economic programs primarily at NSF and DOE and smaller amounts to the Economic Development Administration and the National Institute of Standards and Technology (Badlam et al. 2022a). Another $53 billion is focused on semiconductor manufacturing, R&D, and workforce development. Specific funding relevant to the decarbonization workforce includes $200 million to the CHIPS for America Workforce and Education Fund to develop the domestic semi-conductor workforce; $2 billion to the Department of Defense for microelectron-ics research, fabrication, and workforce training; and, under the Department of Commerce (DOC), $2 billion for the National Semiconductor Technology Center to expand workforce training and development opportunities, $2.5 billion for the National Advanced Packaging Manufacturing Program, and funding for the Manufac-turing USA Semiconductor Institute for development and deployment of training (U.S. Senate Committee on Commerce, Science, and Transportation 2022). Training under these programs will cover a wide variety of fields necessary for domestic semiconductor production, including materials science, electrical engineering, software development, and factory machine operation (Shivakumar et al. 2022). While workforce development and training are mentioned in regard to several of the funding streams above, it is yet unclear how much of this funding will be directed toward workforce development versus other priorities.

Inflation Reduction Act

The IRA, passed in August 2022, appropriates almost $400 billion for clean energy through more than a dozen federal agencies (Badlam et al. 2022b). This funding flows either through the states, which then distribute funds to local governments, communities, or companies, or directly to private entities and/or individuals (Harvey et al. 2022). Appropriations and incentives from the IRA could create millions of jobs across the United States over the next decade (Mahajan et al. 2022; Pollin et al. 2022a; Shrestha et al. 2022).

For a number of IRA programs, labor standards such as prevailing wage and appren-ticeship utilization and domestic content standards (which require certain materials or products be made in the United States) must be met for a project/entity to be eligible for the "bonus" rate. The domestic content standards are intended to shore up domestic supply chains and create upstream jobs in materials and manufacturing. Some programs also have additional bonuses available for projects/facilities located in certain types of disadvantaged communities, including Energy Communities.[3] Additionally, the Justice40 Initiative aims to ensure that 40 percent of IRA benefits go to disadvantaged communities. Other programs require that details or analysis of the proposed project's potential impact on affected communities be included in the proposal, and some require engagement with those communities.

Harvey et al. (2022) describes several tax credits and deductions under the Department of the Treasury that incorporate labor standards (e.g., Clean Energy PTCs and ITCs, Carbon Oxide Sequestration Credit [45Q], Zero-Emission Nuclear Power PTC [45U], and Clean Hydrogen Production Credit [45V]). To receive the bonus rate on tax credits for eligible clean energy deployment, developers must pay a prevailing wage and employ a percentage of registered apprentices on the project. Additionally, two "bonus" tax credits offer additional incentives: the Low-Income Communities Bonus Credit for projects located in communities with significant poverty (Department of the Treasury 2023), and the Energy Community Tax Credit Bonus for projects located in communities that have lost fossil jobs, where a coal plant has closed, or that are host to a brownfield site (Interagency Working Group on Coal and Power Plant Communities and Economic Revitalization 2023) (see Chapter 6 for more details). The Domestic Content Bonus offers an additional 10 percent if a project uses domestically sourced iron and steel, and an increasing percentage of domestically manufactured goods (20–40 percent in 2023 up to 55 percent for projects beginning construction after 2026) (IRS 2023). A "Direct Pay" option makes the tax credits more accessible for projects that meet these domestic content standards (DOE Solar Energy Technologies Office 2023). The Clean Vehicle Tax Credit (30D) has final assembly and component conditions: the Battery Components Credit requires that by 2024, 50 percent of battery components come from North America, rising to 100 percent by 2029; and the Critical Minerals Credit requires that an increasing percentage of critical minerals within the battery are mined in countries with which the United States has a free trade agreement or are recycled in North America—40 percent by 2024 and 80 percent by 2027 (BGA 2022). The Alternative Fuel Infrastructure Tax Credit (30C) offers additional credit for projects guaranteeing prevailing wage and apprentice labor hours (AFDC n.d.). The Commercial Buildings Energy-Efficient Tax Deduction (179D) requires prevailing

---

[3] The IRA provides three criteria to define an Energy Community. Meeting any one of these criteria qualifies: (1) An industrial brownfield site, as defined in a prior statute, 42 U.S.C. 9601(39). (2) A county that meets two criteria: i. At any time from 2010 on, it had 0.17 percent or greater direct employment OR 25 percent or greater local tax revenues related to the extraction, processing, transport, or storage of coal, oil, or natural gas AND ii. now has unemployment rate that exceeds U.S. average. (3) A census tract (plus all adjoining census tracts) where i. a coal mine closed in 2000 or later, OR ii. a coal plant closed in 2010 or later.

wage and apprenticeship utilization, and the New Energy Efficient Home Tax Credit (45L) requires prevailing wage for multi-family buildings (DOE n.d.(a), n.d.(b)).

Under DOE, the extension and expansion of the Advanced Energy Project Credit (48c) offers a bonus credit of 30 percent (compared to base credit of 6 percent) if prevailing wage and apprenticeship utilization standards are met. Additionally, $4 billion is reserved for manufacturing investments to boost job growth and economic opportunities in Energy Communities. The Advanced Industrial Facilities Deployment Program application requires measuring benefits to the local community, and the Energy Infrastructure Reinvestment Program application requires analysis of how the project will engage with and affect associated communities. The Advanced Technology Vehicles Manufacturing Program and the Domestic Manufacturing Conversion Grants require prevailing wage for construction work. The Transmission Line and Intertie Incentives requires prevailing wage. Under the U.S. Department of Agriculture, the Assistance for Rural Electric Cooperatives and Electric Loans for Rural Electric Energy require prevailing wage.

IRA appropriations for workforce development include the following: under DOE, $200 million for State-Based Home Energy Efficiency Contractor Training Grants to provide state energy offices with grants for the training of contractors to carry out energy efficiency upgrades in residential and commercial buildings (Sec. 50123, DOE n.d.(e)); and under the Environmental Protection Agency, $1 billion for the Clean Heavy-Duty Vehicle Program, with $400 million set aside for communities located in nonattainment areas, for grants and rebates for up to 100 percent of costs for clean heavy-duty vehicles (e.g., school buses and garbage trucks) as well as associated maintenance, workforce training, and planning (Sec. 60101, EPA 2023). The IRA also appropriates funding for multiple programs including youth, national service, pre-apprenticeship, and apprenticeship programs related to climate resilience and mitigation; however, these are not decarbonization activities and are out of scope for this report.

Executive Orders

- *Executive Order (EO) 14005: Ensuring the Future Is Made in All of America:* Increases domestic content requirements on federal procurement (EO 14005 2021).
- *EO 14008: Tackling the Climate Crisis at Home and Abroad:* Calls for an all-of-government approach to "create well-paying union jobs to build a modern and sustainable infrastructure, deliver an equitable, clean energy future, and put the United States on a path to achieve net-zero emissions, economy-wide, by no later than 2050" (EO 14008 2021).

- *EO 14025: Worker Organizing and Empowerment:* Established the Task Force on Worker Organizing and Empowerment to identify ways the federal government could fully utilize its authority to encourage worker organizing and collective bargaining (EO 14025 2021). On February 7, 2022, the task force released its report, detailing nearly 70 recommendations for revising labor laws and regulations (White House 2022).
- *EO 14052: Implementation of the Infrastructure Investment and Jobs Act:* Emphasizes the importance of high labor standards, including prevailing wages and the free and fair chance to join a union in the implementation of the Infrastructure Investment and Jobs Act (EO 14052 2021).
- *EO 14063: Use of Project Labor Agreements for Federal Construction Projects:* Requires PLAs on large federally contracted construction projects (EO 14063 2022).

Other Legislation

In December 2022, the congressional appropriations omnibus passed with a provision for a 1-year extension on funding for the Trade Adjustment Assistance (TAA), the 60-plus-year federal program to support workers whose employment is impacted by trade (P.L. 117-328). However, the omnibus language did not include the necessary authorizing language to extend administering TAA, and thus the program is still under termination provisions at time of writing (DOL 2022, 2023). If reauthorized, TAA could be strengthened in several ways to better serve workers through this transition; see Recommendation 4-3 below.

> Finding 4-1: There have been significant efforts recently to incentivize businesses to utilize domestic content, pay prevailing wages, and incorporate apprenticeships into projects. However, the majority of the policy action does not involve durable, mandatory labor standards set by regulatory action, but rather limited-duration tax incentives with labor elements contained to bonus credits earned through voluntary action.

CURRENT ENERGY EMPLOYMENT AND TRENDS

In 2022, more than 8.1 million U.S. workers were employed in the energy workforce, including professional, construction, utility, operations, and production occupations associated with energy infrastructure, production, and use and the manufacturing of motor vehicles. More than 40 percent of total energy jobs in 2022 were in net-zero emissions–aligned areas (DOE 2023). Net-zero emissions–aligned jobs are those related to renewable energy, grid technologies and storage; traditional electricity transmission and non-fossil distribution; nuclear energy; a subset of energy efficiency; biofuels; and plug-in hybrid, fully electric, and hydrogen fuel cell vehicles and components.

Some concerns exist in the energy workforce today that will need to be addressed for a successful transition to a net-zero energy system. As detailed in the following sections, these concerns include

- Hiring challenges across energy jobs, particularly in manufacturing and construction;
- Job quality issues of new clean energy jobs;
- Diversity and inclusion;
- Deindustrialization and decline in U.S. manufacturing and domestic supply chain capacity; and
- Skills gaps.

A primary concern is whether the United States has a skilled workforce willing and capable to produce, install, and operate the materials, products, equipment, and infrastructure needed for a net-zero economy. A Deloitte study found that 80 percent of the needed skills for the short- and medium-term transition already exist in the workforce (Deloitte 2022). This means that most workers already on the job today wouldn't need complete retraining to remain in their jobs or find new work, but rather upskilling through micro-credentials or on-the-job training. A 2020 report from the University of California, Berkeley, Center for Labor Research and Education, found that "[t]he vast majority of the jobs that will be involved in work to lower greenhouse gas emissions across the economy are in traditional occupations where specific 'low carbon' knowledge and skills are only one component of a broader occupational skill set" (Zabin et al. 2020). In other words, it is more about greening existing jobs than creating a whole set of new jobs.

However, the 2023 U.S. Energy and Employment Report documents hiring difficulty in every industry across types of energy jobs (although it does not provide averages across all industries) and documents an increase in employers reporting hiring difficulties from 72 percent in 2016 to 88 percent in 2022 (BW Research 2016; DOE 2023). The problem appears to be most pervasive for motor vehicles employers, where 94 percent of respondents report hiring difficulties, followed by the energy efficiency industry (~92 percent), electric power generation (~87 percent), fuels (85 percent), and electricity transmission, distribution, and storage (~83 percent) (DOE 2023). As demand increases for workers in the non-fossil sector, wages are expected to increase to reflect these needs. Wages can help incentivize the supply of workers, but these adjustments can be sluggish. Moreover, higher wages are often passed through to consumers in the form of higher product prices. Investing in training and skill development today can minimize these skill gaps and wage pressure.

Fossil Trends

As discussed in Chapter 12, there are clear trends across the three fossil fuels: Coal production continues to decline, while natural gas and oil production have not only avoided declines but in fact have increased in the past decade to meet both domestic needs and provide strategic energy security internationally. Employment in fossil fuel extraction has been declining, impacting geographic areas with concentrated fossil jobs.

Over the past 40 years, coal employment has experienced major contractions: the first occurred in the 1980s, when oil prices fell from their 1970s highs and caused a major reduction in coal demand (Black et al. 2005); the second came in the 2010s, as natural gas and renewable energy increasingly supplanted coal in generating electricity (Fell and Kaffine 2018). These previous contractions have been shocks to the system, leaving workers and communities in the lurch (see Chapter 12 for details). Oil refineries in California are starting to close as the state aims to be carbon neutral by 2045, spurring conversation about "unplanned transition" versus a "just transition" (Gerdes 2020; Martin 2022). Another big question is the extent to which new and growing jobs will require similar or transferable skills that fossil fuel workers have. Resources for the Future found that, exempting technical skills, the skills necessary for disappearing fossil fuel jobs do not align with the skills needed for fast-growing occupations at similar pay rates, many of which require skills in customer service or management (Greenspon and Raimi 2022). See the section "Supporting Workers Impacted by Labor Disruptions Associated with the Net-Zero Transition" for a more in-depth discussion of the policy options for transition-related job loss.

Renewable Trends

As noted above, net-zero emissions–aligned jobs made up more than 40 percent of total energy jobs in 2022 and are the top growth areas in energy jobs (DOE 2023). Generally, workers in clean energy earn higher than the national average: while the mean hourly wage of all workers nationally was $23.86 in 2016, workers in clean energy production, energy efficiency, and environmental management earned $28.41, $25.90, and $27.45, respectively (Muro et al. 2019). There are examples of clean energy investments creating high-quality jobs, spurring economic recovery, and growing the clean economy (Boom 2021).

However, there are concerns about the quality of jobs in clean energy, especially when compared to fossil jobs being lost (BGA 2021; Shrestha et al. 2022). Clean energy jobs look good compared to the national average because the economy overall has many low-wage, no-benefits service jobs (BGA 2021). Many incumbent energy jobs that will be lost tend to be high wage, with benefits, and high rates of unionization, and the clean energy jobs being created do not always provide the same array of benefits. Job quality may be of particular concern for solar photovoltaic installation jobs, as solar projects generally rely on short-term workforces retained through temp agencies or subcontractors. These firms, which are known for unfair hiring practices and minimal transparency, offer jobs with low wages, minimal benefits, limited training, and no job security (Gurley 2022; Harris 2022). These short-term jobs do not typically lead to long-term work or cultivate transferable job skills for a career (Gurley 2022). Without effective policies, the transition to a low-carbon economy could reduce the quality of jobs in the energy sector. Box 4-1 describes the role of apprenticeships and pre-apprenticeships in supporting high-quality jobs.

The IRA has attempted to address this issue by tying better labor standards to various tax credits, which is one approach. A stronger approach would require new labor laws. The Protecting the Right to Organize (PRO) Act, which passed the House in the 117th Congress and has since been reintroduced in the 118th, includes several provisions to make it easy for workers to join unions.[4] One of them is a policy that would override right-to-work laws that currently exist in 27 states (NCSL 2023). While state right-to-work laws do not prohibit workers from joining a union, they provide workers the option not to pay union dues that collectively help pay for the costs of bargaining and negotiating contracts. With fewer workers paying dues, unions have found it difficult to sustain themselves. The Economic Policy Institute (Gould and Kimball 2015) has estimated that right-to-work laws have led to a 3.1 percent decline in wages for both union and non-union workers after accounting for differences in labor market characteristics, cost of living, and demographics. The PRO Act would also allow independent contractors and gig workers the right to collectively bargain.

In addition to job quality concerns, there are also diversity concerns in the clean energy workforce; a 2021 analysis found that about 61 percent of clean energy workers were white (non-Hispanic), whereas Black, Hispanic/Latinx, and women workers were all less represented in clean energy than across the rest of the economy (E2 2021). In 2022, Black workers held only 9 percent of jobs in wind technology and energy efficiency and

---

[4] H.R.842—117th Congress (2021–2022). Protecting the Right to Organize Act of 2021. https://www.congress.gov/bill/117th-congress/house-bill/842. H.R.20—118th Congress (2023–2024). Richard L. Trumka Protecting the Right to Organize Act of 2023. https://www.congress.gov/bill/118th-congress/house-bill/20.

BOX 4-1
APPRENTICESHIPS, PRE-APPRENTICESHIPS, AND EQUITY

Apprenticeship programs are widely considered a promising pathway to high-quality jobs, as they pay workers to learn to become highly skilled and enter a career path (Inclusive Economics 2021). These programs have the potential to help alleviate labor shortage concerns in manufacturing and skilled trades; fill skills gaps; direct young people into careers with high retirement rates; and advance diversity, equity, and inclusion in the growing clean energy workforce. Registered apprenticeships are approved by the Department of Labor or a state agency; can involve businesses, industry experts, unions, education institutions, and other local partners; and are industry vetted and validated. They are paid and lead to a credential such as a nationally recognized certificate (DOL n.d.(b)). While registered apprenticeship programs are becoming increasingly diverse (Jones et al. 2021), particularly union apprenticeships (BGA 2021; Bilginsoy et al. 2022), continued action is needed to overcome historical inequities and ensure that apprentices and the future trades workforce are more representative of the general population (Seleznow and McCane 2021). One tool to help achieve this is pre-apprenticeship programs, which can prepare job seekers, particularly those from disadvantaged communities, and set them up for success in completing apprenticeship programs (Foster et al. 2020; Inclusive Economics 2021). These programs can be associated with specific apprenticeship programs and include wrap-around services such as childcare and transportation (DOL n.d.(a)).

A study of energy efficiency programs in California found approximately two-thirds of the jobs generated directly by energy efficiency investments to be in traditional building and construction trades (e.g., electricians, sheet metal workers, plumbers, carpenters, stationary engineers, and others), with only around one-sixth in professional occupations, and only 2 percent in specialized energy efficiency occupations like energy auditor. This result shows that very niche training programs that cater to a specific clean energy technology, such as training for solar panel installation, may be less effective than programs like apprenticeships and pre-apprenticeships that provide broader training and equip workers with skills that can move between clean technologies (Zabin et al. 2014).

---

only 8 percent in solar technology jobs; women held 27 percent of energy efficiency jobs, 31 percent of solar jobs, and 33 percent of wind technology jobs (DOE 2023). Despite improvements in recent years—more than half of clean energy jobs added in 2022 went to women—the participation rate of women and Black workers remains well below the national workforce average, highlighting the persistence of these disparities (DOE 2023). There is also evidence of discrepancies in roles and career progression among employees across race and gender. In the solar industry, for example, Black workers are less likely to hold management-, director-, and president-level positions than White workers, and women are less likely to hold these positions than men (IREC 2021).

Critical Minerals and Mining Trends

Critical minerals are necessary in producing wind turbines, solar panels, electric vehicles, smart home devices, sensors and digital controls, batteries, and myriad other technologies that will support decarbonization. While U.S. mining and geological engineering employment is expected to grow more slowly than all U.S. occupations over the next decade, about 500 openings are projected each year on average, mostly stemming from the need to replace workers who transfer to different occupations or exit the labor force, such as to retire (BLS 2022). A 2013 National Research Council study on emerging workforce trends in the domestic energy and mining industries named a wide array of challenges, including aging and retiring workforce and faculty; a decrease in mining, mineral engineering, and economic geology programs; negative perceptions with respect to the nature of the work; and foreign competition for U.S. talent (NRC 2013). Workforce challenges persist owing not only to these factors but also to the physically demanding nature and often remote locations of these jobs (Sicurella 2021). More than 200,000 of today's domestic mining workforce will be retiring and need replacement by 2029; however, in 2021, just over 300 degrees in mining and mineral engineering were awarded in the United States, far from the rate needed to maintain the workforce, much less meet expected growth in demand for domestically sourced or processed critical minerals (Data USA n.d.; Hale 2023; Society for Mining, Metallurgy, and Exploration 2014). DOC (2019) made recommendations aimed at several key goals for growing the critical minerals workforce: (1) bolster education in mining engineering, geology, and other fields related to critical minerals mining and manufacturing; (2) promote interdisciplinary collaboration among material science, computer science, and related disciplines to modernize the minerals supply sector industry and make the field more attractive to new talent; (3) implement personnel and management reform to ensure appropriate human capital to support exploration and development of critical minerals on federal lands; and (4) facilitate sustained interaction with critical mineral stakeholders and the general public.

Manufacturing Jobs

As discussed in Chapter 10, manufacturing the equipment and building out the infrastructure needed to create a net-zero economy will be a key part of the transition, and there are opportunities to create and maintain high-quality jobs as well as global competitiveness if done well. Several current trends in manufacturing could be barriers to implementing a transition to net zero and need to be addressed, including the disappearance of the manufacturing wage premium, deindustrialization and offshoring of plants, and persistent hiring difficulties.

Manufacturing jobs once provided a key path for middle-class growth and prosperity (Barrett and Bivens 2021); however, the manufacturing wage premium has disappeared in recent years, contributing to the increase in overall wage inequality and potentially adding to the decline of U.S. manufacturing (Bayard et al. 2022). Over the past 20 years, more than 5 million U.S. manufacturing jobs have disappeared and nearly 70,000 factories have closed (Scott et al. 2022). The U.S. economy shifted toward lower-wage, service-sector jobs with fewer benefits and lower rates of unionization than manufacturing jobs, resulting in lower average wages for all workers without a 4-year degree (Scott et al. 2022). This decline has disproportionally impacted workers and families of color (Scott et al. 2022; Taylor 2016).

For many manufacturing companies that remain in the United States, workforce challenges are negatively impacting operations and growth. In the National Association of Manufacturers 2021 Manufacturers' Outlook Survey, nearly 45 percent of respondents reported having to turn down business opportunities owing to insufficient staff (NAM 2021). In the same survey, about 71 percent of respondents noted that staffing shortages also have negative impacts on the timeliness of product deliveries and on production processes (NAM 2021). The figure is a bit higher among respondents of the 2022 Workforce Institute survey, which found that labor shortages impacted production demands for 84 percent of manufacturers; for 76 percent of these respondents, the impact on their bottom line was considered "moderate" or "severe" (Workforce Institute 2022).

The millions of middle-income American workers currently employed in the automotive industry still earn average wages that are higher than the national averages (Walter et al. 2020), but these averages obscure a more nuanced reality. While unionized, full-time workers continue to earn high wages with benefits and enjoy decent working conditions, not all workers have access to these benefits. Weakened labor standards in U.S. manufacturing have eroded job opportunities and job quality in the sector (and across the entire economy) (Cutcher-Gershenfeld et al. 2015). As a result, real earnings have been declining for all autoworkers (Ruckelshaus and Leberstein 2014). Most automotive manufacturing jobs created since 2009 have been non-union or temporary (Ruckelshaus and Leberstein 2014; Walter et al. 2020). While previously the majority of autoworkers were employed in assembly, auto parts workers now account for 72 percent of jobs in the sector, where they are more likely to be temporary and earn significantly less than assembly workers (Ruckelshaus and Leberstein 2014).

> Finding 4-2: There is a pay gap between some fossil fuel jobs and clean energy jobs. Low wages are often accompanied by lack of benefits. Prioritizing high-quality jobs, including union jobs, in the clean energy sector can lead to better outcomes for both workers and the environment while ensuring a just transition for fossil fuel workers.

EMPLOYMENT IMPLICATIONS IN A NET-ZERO TRANSITION

Analyses by Rhodium Group, REPEAT, Energy Innovation, and Chapters 6–12 of this report conclude that recent legislation is likely to move the nation much or a majority of the way to a net-zero trajectory. While there are no comprehensive peer-reviewed jobs analyses of recent legislation, major job gains are projected: Energy Innovation found that the provisions in the IRA could create 1.4 to 1.5 million new jobs in 2030 concentrated in the manufacturing, construction, and service industries (Mahajan et al. 2022). Energy Futures Initiative found that the IRA could create 1.46 million more jobs than a BAU scenario, and construction, manufacturing, and the electric utility sector would be key sectors for growth (Foster et al. 2023). Political Economy Research Institute (PERI) and the BlueGreen Alliance (BGA) found that robust application of the IRA's strong labor standards could create more than 9 million jobs throughout the economy over the next decade—an average of nearly 1 million jobs each year (Pollin et al. 2022a). A World Resources Institute (WRI) study found that federal policies relying on a combination of tax credits for low-carbon technologies (as included in the IRA) and infrastructure investments (as included in the IIJA) could generate an additional 900,000 net jobs by 2035, compared to a reference scenario without these laws (Shrestha et al. 2022). This does not account for the additional economic benefits generated by incentivizing domestic manufacturing of clean energy technologies and their supply chains. When domestic manufacturing is factored in, then an additional 3.1 million net jobs are created in 2035 compared to a 2035 reference scenario (Shrestha et al. 2022).

Studies of employment along potential paths to net zero offer a useful upper bound even if they do not specifically include recent legislation:

- Mayfield et al. (2023) modeled employment impacts of the Princeton Net-Zero America scenarios, finding that a transition to net zero in 2050 supports an annual average job creation of approximately 3 million direct jobs during the first decade, and approximately 4–8 million direct jobs during the 2040s. This study does not include energy efficiency or vehicles jobs. The study provides separate results for different technologies, regions, and individual states; estimates training, education, and experience requirements for jobs that are created; and estimates policies that would maximize employment benefits (Mayfield et al. 2023).
- The WRI study mentioned above also concluded that a net-zero emissions scenario would result in 2.3 million more net jobs (direct, indirect, and induced) than the reference case between 2020 and 2035; new clean energy jobs would be concentrated in construction of buildings and electricity (Shrestha et al. 2022).

- The Decarb America Research Initiative found that decarbonizing the U.S. economy would create a net increase of more than 2 million jobs economy-wide (includes direct, indirect, and induced) by midcentury. This study focuses on electricity generation and includes vehicles but excludes oil/petroleum (Chan et al. 2022).
- The Sustainable Development Solutions Network (SDSN) developed the America's Zero Carbon Action Plan, which found that a transition to net zero by 2050 would generate about 2.5 million direct and indirect jobs per year compared to a reference case, and over 4 million jobs per year if induced jobs are included. The authors concluded that many industrial jobs would be created in the Appalachian region and Midwest (SDSN 2020).

Studies looking at global net-zero transition reported similar findings to U.S.-focused studies regarding large net gains in employment. However, gross job losses would also be substantial. A 2022 study from the McKinsey Global Institute found that the transition would have uneven effects on sectors, geographies, and communities, especially at the beginning (Krishnan et al. 2022). It also found that shifts in employment would likely be substantially higher in a disorderly transition. The report notes that economic impacts should be considered in perspective with job dislocations from other trends—including automation, remote work, and e-commerce—which could lead to considerably more losses than the global transition to net zero. A 2022 study from Deloitte found that 13 million jobs in the United States are vulnerable to climate change impacts or transition effects. While the U.S. share is lower than other regions of the world, unmanaged transition increases the risk to these jobs, while proactive policy decisions through the 2020s can create a job dividend more than 30 years earlier than a passive transition (Deloitte 2022). A 2021 IEA report highlights the benefits of managing transition impacts, enabling companies to find qualified workers, using existing practices to center people in transition, supporting long-term engagement and strong social dialogue, and using detailed energy employment data (Cozzi and Motherway 2021).

Thus, both domestic and international analyses broadly agree that net increases in jobs are likely, together with contractions of fossil jobs. Fossil job contractions are concentrated in a few regions, while the job increases are likely more dispersed. Moreover, fossil contractions generally occur after 2030, except for coal jobs, which have been declining for decades (see Chapter 12 for details). An illustrative example is Mayfield et al. (2023), originally produced as part of Princeton University's Net-Zero America Project. Mayfield et al. find that from 2020 to 2030, employment stays constant or grows in most states except coal-producing states in the Appalachian basin, where employment slightly declines but rebounds in the 2030s. By 2050, energy jobs grow both as a fraction of total jobs in the economy and in most states, but this growth happens in boom-bust cycles. The largest interim job losses occur in rural states with large upstream fossil fuel industries, like West Virginia. Employment and wage losses in fossil fuels are offset in aggregate by growth in low carbon sectors. These results are broadly consistent with the conclusions of the literature in this area and offer granular insights about how decarbonization impacts employment by resource sector, geography, and timing. A detailed look at this study illustrates the level of geographic, temporal, and sectoral heterogeneity expected in employment during the net-zero transition.

Figures 4-1 and 4-2 depict results from Mayfield et al. (2023) on total annual employment in energy jobs and employment by resource sector, respectively, for each U.S. state. The following discussion further explains the results shown in these two figures.

Fossil Fuels (see also Chapter 12). Coal is the smallest fossil fuel sector in terms of current employment. Coal employment has been declining for the past 3 decades and is projected to continue to decline by half in 2030 and by more than 80 percent by 2050, even in the reference scenario of Mayfield et al. (2023). Coal employment is clustered geographically in the Appalachian and Powder River basins, and coal mining is a dominant industry in some communities. Employment in coal-fired power generation is also spread across 45 states.

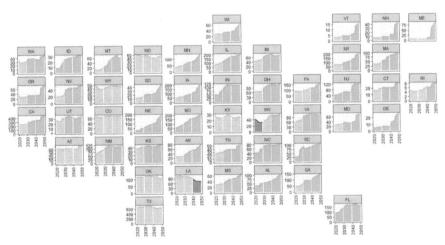

FIGURE 4-1 Spatial distribution of employment: Annual employment in energy jobs for the least-constrained net-zero scenario modeled by Mayfield et al. (2023), which permits nuclear and carbon capture and storage in addition to renewables.
NOTE: Green means employment >15 percent above 2021, yellow means within ±15 percent of 2021, and red means >15 percent below 2021.
SOURCE: Reprinted from *Energy Policy*, Vol. 177, Mayfield et al., "Labor Pathways to Achieve Net-Zero Emissions in the United States by Mid-Century," p. 12, Copyright 2023, with permission from Elsevier.

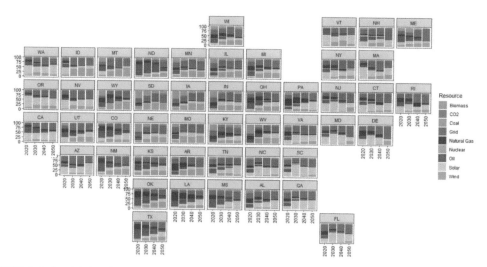

FIGURE 4-2 Spatial distribution of employment: Distribution of employment by resource sector in the least-constrained net-zero scenario modeled by Mayfield et al. (2023), which permits nuclear and carbon capture and storage in addition to renewables.
SOURCE: Reprinted from *Energy Policy*, Vol. 177, Mayfield et al., "Labor Pathways to Achieve Net-Zero Emissions in the United States by Mid-Century," p. 12, Copyright 2023, with permission from Elsevier.

Oil is the largest fossil fuel sector in terms of employment and is spread across all states, with some clustering in areas that produce oil and in areas with major industry concentration. The reference scenario of Mayfield et al. (2023) shows a 40 percent decline in oil sector employment by 2050, and the net-zero scenarios show 60–90 percent declines, largely influenced by the rate of transportation electrification and the level of future exports. In areas like North Dakota, the oil sector employs a large share of the labor force through midcentury.

Natural gas employment is influenced by rate of heating electrification, renewable siting constraints, and natural gas exports, leading to vastly different potential pathways. Natural gas employment declines over both the short and long term in both the reference and net-zero scenarios in Mayfield et al. (2023): in the reference scenario, employment declines steadily by 15 percent between 2020 and 2050. However, in the net-zero scenarios, natural gas employment declines 15–30 percent by 2030 and 50–80 percent by 2050. Most of the decline is upstream; employment in the gas-fueled electric power industry only slightly declines as carbon capture and storage (CCS) technology expands.

Low-Carbon Sectors. The solar sector rapidly scales up in the net-zero scenarios studied by Mayfield et al. (2023), with utility-scale solar capacity increasing 10-fold by 2030 and 20- to 120-fold by 2050, as domestic manufacturing share of related components multiplies 2 to 4 times by 2030 and 3 to 18 times by 2050. By 2030, the solar sector has the highest employment of all resource sectors. Solar jobs are spread across the United States with a concentration in the southern half of the United States, and solar employment increases from 250,000–500,000 jobs currently to 700,000–2,500,000 jobs in 2050.

The wind sector experiences a rapid expansion, increasing capacity 2- to 4-fold by 2030 and 6- to 27-fold by 2050, as domestic manufacturing share of related components multiplies 4 to 10 times by 2030 and 4 to 46 times by 2050. Wind resources increase employment mid-continent, and manufacturing increases employment across the East, Midwest, and Great Lakes. Wind has the potential to mitigate job losses from fossil fuels in some areas, such as Wyoming. Nationwide, wind employment increases from 100,000–150,000 currently to 350,000–2,200,000 jobs in 2050. Depending on the proportion of domestic content (materials and goods made in the United States) used in offshore wind energy development and construction, Stefek et al. (2022) estimate that from 2024 to 2030, the offshore wind energy industry will need an annual average of between 15,000 (25 percent domestic content) and 58,000 full-time workers (100 percent domestic content). As most of these new offshore wind jobs are expected to be added in the manufacturing and supply chain sectors (as well as project development, installation, ports and vessels, operations, and maintenance), realizing this job growth depends on the construction of new manufacturing and supply chain facilities within the United States.

In these scenarios, the electric grid undergoes a 2×–4× infrastructure expansion, and employment related to transmission increases in all states and nearly doubles in the largest states like Texas and California. Electric grid employment increases from 450,000–600,000 currently to 1,100,000–3,500,000 jobs in 2050.

Nuclear is a relatively small sector in terms of employment today, and its fate depends on constraints that the nation places on it and other technologies. For example, the 100 percent renewables scenario in Mayfield et al. (2023) assumes that the public will prohibit nuclear electricity. As a result, nuclear employment declines by 20–40 percent by 2030 and approximately 95 percent by 2050. In the scenario in which renewable deployment is constrained beneath the maximum historic rate—for example, because of public pressure on deployment—both nuclear deployment and employment would expand by 10×. In the least-constrained scenario, shown in Figure 4-2, operations and slow decommissioning provide steady employment from 2020 to 2040. Thus, across the full range of scenarios, today's nuclear employment of 50,000 could either decline almost entirely or increase to 500,000 by 2050.

Biomass is another small sector today that could see growth and transformation from corn ethanol production to advanced biofuels from woody, non-woody, and other feedstocks. Biomass has the potential to provide concentrated benefits to rural communities owing to colocation of farming activities. In the scenario depicted in Figure 4-2, biomass employment increases from an average of 80,000–90,000 per year in the 2020s to 160,000–220,000 per year in the 2050s.

Carbon dioxide capture, transport, and sequestration or use currently employs few people, but is expected to increase to substantial levels in all scenarios in Mayfield et al. (2023) that permit geologic sequestration. However, in the Princeton America studies, much of this is CO_2 captured from biomass late in the transition. In scenarios where carbon capture infrastructure gets built, employment increases in areas with existing natural gas pipeline infrastructure, extending from the mid-continent to the East coast. While skills for carbon dioxide jobs are similar to natural gas jobs, employment in this sector is predicted to remain much smaller than declines in the natural gas workforce. Carbon dioxide employment could increase to 60,000–110,000 in 2050. Note that direct air capture, geothermal energy, and hydropower were not included in this analysis, as the models did not show significant new capacity. As previously mentioned, many factors affect potential capacity and jobs impacts and these technologies may warrant additional study.

The Mayfield study focuses on energy supply and thus excludes vehicles and energy efficiency, which are the largest current sectors of energy employment (DOE 2023), and both likely to see changes owing to the transition. As the United States shifts from ICE vehicles to battery electric vehicles (BEVs), employment in the U.S. automotive sector could increase by more than 150,000 jobs in 2030—if U.S. BEV purchases rise to 50 percent by 2030, the United States produces the same share of the battery supply chain as it currently does ICE components, and the U.S.-assembled market share increases 10 percentage points (to 60 percent) (Barrett and Bivens 2021). If these conditions are not met, the U.S. auto sector could lose roughly 75,000 jobs by 2030. Modeling a transition to net zero, a WRI study found that by 2035 there is a loss of 2 million net jobs (direct, indirect, and induced) associated with ICE vehicle manufacturing, maintenance, and sales (Shrestha et al. 2022). However, increasing the share of domestic battery manufacturing from 25 percent (base model assumption) to 50 percent leads to an additional 850,000 jobs, and increasing to 75 percent leads to 1.7 million jobs gained, countering 87 percent of losses. The domestic content and manufacturing requirements to harness bonus rates for tax credits included in the IRA will support realizing the job gains projected in WRI's scenario. The WRI study also reports an increase of 3.6 million jobs from EV charging infrastructure deployment (Shrestha et al. 2022). Similarly, the Decarb America report finds that EVs could generate hundreds of thousands of jobs with a domestic supply chain (Chan et al. 2022). Employment in energy efficiency activities such as weatherization, building retrofits, electrification, and industrial energy efficiency will be critical in transition to net zero. The WRI net-zero scenario found that the largest job gains are in construction of buildings and electricity infrastructure and that the majority of construction jobs are well-paying (Shrestha et al. 2022). The Decarb America report finds that energy efficiency jobs drive gains in the first decade of transition as infrastructure is built out (Chan et al. 2022). The SDSN study found that investments in energy efficiency would generate approximately 800,000 new jobs per year (SDSN 2020).

While most net-zero studies include manufacturing jobs to produce equipment and infrastructure needed for the net-zero economy, most do not address specifically job impacts of transition on the industrial sector. The SDSN report finds that effective industrial policies could increase total job creation by up to about 10 percent, and reports that in their modeling, many industrial jobs are created in the Appalachian and Midwest regions (SDSN 2020).

> Finding 4-3: The employment impacts of the transition to a net-zero economy will be uneven geographically, temporally, and sectorally. There will not be a 1:1 replacement of lost fossil energy jobs with new clean energy jobs. Some workers and communities will be disproportionately at risk. Proactive policy and transition management is critical to mitigating negative impacts and improving equitable outcomes.

> Finding 4-4: Decarbonization will impact not only direct energy jobs but also jobs in supporting communities and peripheral industries. This means that the energy transition will implicate a wide swath of the U.S. economy.

Findings and Recommendations Regarding Attracting and Retaining the Workforce Needed to Accomplish the Transition

Recommendation 4-1: *Support the Development of Net-Zero Curriculum and Skill Development Programs for K–12 Students.* **The transition to a net-zero economy will take decades; children born in the 2020s will participate in the workforce of the 2040s, and their preparation to be a part of the net-zero economy needs to begin today. The Department of Education should provide support for state and local governments and school districts to develop curricula and skill development programs that prepare K–12 students for careers in the net-zero economy. Local governments and school districts should engage local employers and workforces when crafting decarbonization-relevant workforce development programs to ensure that they meet community needs and that program participants will have career paths in their communities.**

Finding 4-5: There is significant under-representation of women and people of color in the growing clean energy workforce. Accomplishing the transition to a net-zero economy will require trained and qualified workers across the country, in all communities. Developing a diverse and representative clean energy workforce will support the social contract necessary to maintain support for the transition long-term.

Recommendation 4-2: *Invest in Linking People from Disadvantaged Communities to Quality Jobs.* **Congress should invest in linking people from historically disadvantaged communities to quality jobs through Registered Apprenticeship Programs and pre-apprenticeships. This could include incentivizing or requiring the use of Project Labor, Community Benefits, and Community Workforce Agreements with equity-focused stipulations; the use of Registered Apprenticeships; the adjustment of wage reimbursement rates for professions with historically low wages; and the expansion of fair chance hiring policies to Registered Apprenticeships.**

SUPPORTING WORKERS IMPACTED BY LABOR DISRUPTIONS ASSOCIATED WITH THE NET-ZERO TRANSITION

All published analyses of the impact of current climate and energy policy on employment broadly predict continued fossil fuel job losses but differ on details. Some baseline scenarios also predict declines in fossil employment even without the new policies in the IRA, IIJA, and CHIPS (e.g., Mayfield et al. 2023). For example, IEA's Stated Policies Scenario (STEPS) predicts that demand for fossil fuels would have declined or remained constant through 2040 (Raimi 2021). Fossil jobs at risk include coal, oil, and natural gas jobs in upstream mining and drilling, refining and transporting, energy generation, and use in the transportation sector (including ICE vehicle manufacturing) as well as downstream supply chain and related jobs (such as auto mechanics). Mayfield et al. (2023) found that mining sector (i.e., oil, gas, coal upstream activities) jobs comprise a declining portion of jobs over time. Raimi (2021, p. 2) uses IEA's Sustainable Development Scenario as the basis for its analysis and found that "because coal has the highest carbon content and can be substituted easily in the power sector (where most coal is used), these communities will be the first to face a transition due to climate policy," and oil and natural gas would follow behind.

Geography of Job Losses

Nationally, fossil fuel jobs account for less than 1 percent of total employment. However, this is highly regionally variable, and in some communities, 10–20 percent of employment is directly involved in fossil fuels (Raimi et al. 2022). Just 10 states contain 73 percent of oil, coal, and natural gas production (Foster et al. 2020). Raimi et al. (2022) associates these geographic hot spots with four regions: Intermountain West, Texas, Appalachia, and the Gulf Coast. As shown in Chapter 12, certain types of fossil jobs are geographically concentrated (fossil fuel mining and extraction), some are dispersed but can be clustered (fossil fuel power plant generation), and others are widely dispersed across the country (gas station workers, auto mechanics).

Job losses in industries like ICE vehicle manufacturing will be more concentrated in areas that have vehicle manufacturing, like the Great Lakes and industrial Northeast. Other job losses will be more widely dispersed, such as in communities with coal-fired power plants, ICE parts manufacturers, auto mechanics, and so on. Other attributes of geography such as low economic diversity, isolation, and lack of training opportunities create risk for communities and workers.

The Costs of Job Loss for Workers

Job loss, especially in fossil sectors, is likely to be extremely costly to an individual/household. Evidence from researchers examining mass layoffs in a variety of contexts show that earnings losses for affected workers are large and persistent—about 20 percent below their forecasted earnings trajectory up to 20 years later, the sum of which amounts to around $110,000–$140,000 in earnings losses for someone making $50,000 annually (von Wachter et al. 2009). Job loss may also come with loss of health insurance and retirement benefits, and may even produce direct health effects (Sullivan and von Wachter 2009). Even if there is an overall increase in jobs nationally through the transition, aggregate outcomes do not reflect individual experiences, and those individuals/households who will experience a major disruption need to be directly supported.

Outside of the fossil sectors directly affected, there are likely to be effects in other industries and communities that directly and indirectly support and/or benefit from fossil fuels and services. These include, but are not limited to, non-tradeable goods and services in a local community, such as restaurants, construction, nursing homes, and so on. Loss of a major local employer may also cause a contraction in the local tax base, which is often a primary source of public-school funding. The effects of job loss can also impact families and children in myriad ways. For example, Stevens and Schaller (2009) and Oreopoulos et al. (2008) provide evidence that parental layoffs have a causal effect on their children's test scores and their subsequent adult earnings.

These combined effects have the potential to significantly erode the social fabric that connects communities, and policy makers need to think carefully about potential solutions to minimize these impacts. Economic gains from getting people back to work are partly the present and future gains to the income of workers. However, the broader social gains can include stronger families, a better network of informal job connections, a decline in state-level spending on Medicaid and welfare payments, reduced drug use and crime, and other benefits. Recent evidence suggests that large disruptions in U.S. manufacturing, primarily owing to expansion of international trade with China, causally led to increases in "deaths of despair" from fatal drug overdoses and enrollment in Social Security Disability Insurance (SSDI) (Pierce and Schott 2020).

At the same time, the existing federal and state policies designed to mitigate these costs, such as unemployment insurance (UI), are insufficient to compensate for these foregone earnings and other job-related benefits. The federal-state UI system helps many people who have lost their jobs by temporarily replacing a part of their wages. Workers in most states are eligible for up to 26 weeks of benefits from the regular state-funded unemployment compensation program, although 10 states provide fewer weeks, and 2 provide more. On average, UI benefits replace about 40 percent of a worker's pre-layoff wages, but they vary substantially by state. When accounting for the time-limited nature of UI income and the non-wage benefits that are part of the compensation package for many workers (benefits like employer-provided health insurance and retirement contributions), the true pay replacement rate is much lower (EPI n.d.). In the process of searching for jobs, many workers are likely to exhaust UI benefits. Research suggests that upon exhaustion of UI, families' consumption falls, and the incidence of poverty rises (CBO 2004; Ganong and Noel 2019; Gruber 1997). There are also risks of dropping out of the labor force altogether, which have even larger fiscal costs described below. These effects are particularly large for single-earner families with children.

Lessons from Past Experience

The scale of the labor force transition away from fossil production is likely to be unprecedented in nature, but it may be helpful to look at past experiences to learn about likely impacts as well as potential solutions. There are at least two major upheavals in the past 40 years that may be helpful in this regard: the early 1980s and 2010s decline in coal production in the Eastern United States and the more recent and widespread erosion of manufacturing employment associated with trade exposure to Chinese import competition.

Hanson (2023) considers the consequences of the post-1980 decline of coal in order to see how the safety net might address job loss from the energy transition. During the past 40 years, coal mining has had two major contractions. The first occurred in the 1980s, when oil prices fell from their 1970s highs and caused a major reduction in coal demand (Black et al. 2005); the second came in the 2010s, as natural gas and renewable energy increasingly supplanted coal in generating electricity (Fell and Kaffine 2018). Following the first shock, employment and earnings fell precipitously in coal counties, which then saw sharp increases in uptake of government transfers across a wide set of programs, such as Social Security Disability Insurance (Black et al. 2002, 2003; Jacobsen and Parker 2016; Pierce and Schott 2020). At the time, some analysts worried that monetary support for coal communities was insufficient, while others raised concerns that government assistance would create a culture of welfare dependence. Hanson (2023) shows how regions exposed to the four-decade coal bust have seen long-run reductions in earnings and employment rates, increases in government income assistance, expanded Medicare and Medicaid usage, and substantial decreases in population, especially among younger workers.

More recently, evidence has emerged as to the devastating impact that China's accession to the World Trade Organization and associated import competition has had on U.S. manufacturing (Acemoglu et al. 2016; Alden 2016; Autor et al. 2013; Pierce and Schott 2016). Rising imports from China caused higher unemployment, lower labor force participation, and reduced wages in local labor markets that contained import-competing manufacturing industries—ultimately responsible for nearly a quarter of the decline in U.S. manufacturing employment (Autor et al. 2013). Between 600,000 and 1 million U.S. manufacturing jobs disappeared between 1990 and 2007 (Autor et al. 2013).

Conventional views suggested that labor markets would adjust to these forces—workers who lost jobs from trade competition or coal decline would move to industries or labor markets less exposed to trade or coal. The initial decrease in labor demand and a corresponding increase in labor supply from a newly available set of workers may depress wages, but the effect of these shocks should ultimately be diffuse. However, increasing evidence demonstrates that these adjustments are highly heterogeneous and incomplete (Autor et al. 2021), with only modest migration from affected areas, mostly by foreign-born workers and younger native-born adults (ages 25–39). Expansions in import competition led to many localized recessions: displaced workers spent less on restaurants, entertainment, home renovations, childcare, and other services, pushing the economy into a downward spiral of further job losses and spending cuts.

Currently, the United States has targeted transition assistance programs to help with the disruptive costs of job displacement for some affected workers. The largest and most well-known of these programs is TAA, a federal transfer program established under the 1962 Trade Expansion Act that provides assistance to workers "who lose their jobs or whose hours of work and wages are reduced as a result of increased imports or shifts in production out of the United States." In fiscal year 2010, nearly $1 billion in annual cash transfers were appropriated to subsidize an estimated 230,000 qualified workers to enroll in retraining programs after trade-related layoffs.

Hyman (2022) provides a comprehensive evaluation of the TAA program using detailed longitudinal microdata on workers' employment and earnings histories. He finds that TAA-approved workers have $50,000 greater cumulative earnings 10 years after the fact—driven by both higher incomes and greater labor force participation. Overall, this program includes many of the interventions described in the following section "Policy Tools and Mitigation of Impacts": extended UI benefits and active labor market programs, such as intensive job training. More recently, part of TAA has been further experimenting with wage insurance programs that have also shown promise (Hyman et al. 2021, 2023).

The picture that emerges from these historical experiences is bleak: large shocks to local economies have led to a collapse in local labor markets where gradual outmigration ultimately left behind a population that is disproportionately old, sick, and poor (Hanson 2023). The existing safety net was and is ill-equipped to address the scale and scope of these impacts.

Policy Tools and Mitigation of Impacts

What can be done to mitigate some of the transitional costs to the workforce and communities in light of the concerns listed above? Myriad policy tools are available that research and practice have shown to be effective at minimizing costs to affected workers and communities while also being cost effective from the standpoint of government expenditures. Key to these suggestions is minimizing labor force exit and/or preventing an increase in the number of long-term unemployed.[5] Policies that incorporate transparency and certainty in the timing of impacts, described in more detail below, are also likely to aid the workforce transition.

Extending and Potentially Reforming Unemployment Insurance Benefits in Fossil Communities

There has been a long-standing concern that extending the duration and increasing the benefits of UI will induce workers to delay seeking new jobs and thereby elevate unemployment rates and prolong economic recovery. Yet, extensive literature now

---

[5] The following discussion is based on congressional testimony from von Wachter (2011) and recent work by Hanson (2023).

suggests that extending UI prevents large declines in individual and/or local consumption for the substantial number of workers at risk of exhausting their benefits (Ganong and Noel 2019; Kovalski and Sheiner 2022). If this is the case, not all of the disemployment effects of UI generosity represent a distortion but may be a sign that UI helps to alleviate credit constraints that prevent individuals from self-insuring against unemployment shocks (i.e., UI benefits provide a form of social insurance). Relatedly, UI extensions can also provide a degree of demand stabilization for local economies.

Extensions in UI duration can also prevent individuals who are at risk of dropping out of the labor force entirely from entering more costly (and permanent) government programs such as SSDI or claiming Social Security benefits early. Therefore, these extensions could imply cost savings for the Social Security trust fund that need to be incorporated into calculations of the budgetary effect of UI extensions (von Wachter 2011).[6]

Recessions and local economic shocks also tend to lead to early retirement from the labor force, especially for less educated men (Autor et al. 2012; Yagan 2019). For example, von Wachter (2007) shows that in past U.S. recessions, a 5-point rise in state unemployment rates is causally related to a 5 percentage point reduction in the employment-population ratio of 60- to 64-year-old high-school graduates. The majority of these workers do not return to the labor force and are likely to claim SSDI (if eligible) or Social Security benefits early. Extensions in UI durations may prevent some of these workers from dropping out of the labor force completely.

Most extensions to UI occur at the state level. There appears to be scope to further tune UI benefits to local economic conditions. Because job loss caused by the energy transition is likely to be highly concentrated in specific local labor markets, state-level triggers may be too crude to help the regions that will suffer high levels of worker displacement, entailing larger than necessary fiscal expenditures. Criteria like those provided to define Energy Communities in the IRA could similarly be used to target social assistance and workforce support programs and policies described below.[7]

---

[6] With a monthly job finding rate of 10 percent (i.e., the job finding rate suggested by Hall [2005] at the trough of the 1982 recession), an extension of benefits by 6 months would imply that about half of the individuals looking for a job upon benefit expiration would find a job.

[7] The IRA provides three criteria to define an Energy Community. Meeting any one of these criteria qualifies: (1) An industrial brownfield site, as defined in a prior statute, 42 U.S.C. 9601(39). (2) A county that meets two criteria: i. at any time from 2010 on, it had 0.17 percent or greater direct employment OR 25 percent or greater local tax revenues related to the extraction, processing, transport, or storage of coal, oil, or natural gas AND ii. now has unemployment rate that exceeds U.S. average. (3) A census tract (plus all adjoining census tracts) where i. a coal mine closed in 2000 or later, OR ii. a coal plant closed in 2010 or later.

Transition Assistance Programs

As noted above, the largest and most well-known program to alleviate the disruptive costs of job displacement is TAA. While TAA contains several program components, its primary benefit is coverage of training costs for every year a qualified worker is retraining, up to a statutory maximum of 3 years. Median annual coverage from 2001 to 2016 was $7,500/recipient-year, including up to 2 years for "basic" retraining, and an additional year for "remedial" training (if deemed necessary) or "completion" training for workers who are close to completing a credentialed curriculum but have exhausted basic benefits.

To receive TAA benefits, workers (or their surrogates) must file petitions at the Department of Labor (DOL) within 1 year of their trade-related separation from a given employer, at which point a DOL investigator is tasked with determining whether applicants were laid off by companies whose decline in sales was owing to increased imports or outsourcing. Once workers are approved for TAA, state career centers (e.g., American Job Centers, One-Stop Career Centers) guide workers to potential training program matches based on prior experience, with workers having the final say over where to train. Once enrolled, training subsidies and regular Trade Readjustment Allowance (TRA) payments are administered through local state career centers, where workers recoup paychecks. Recipients are also entitled to expanded UI benefits while training (TRAs), conditional on providing regular proof of training enrollment (Hyman 2022, p. 7). Extended UI is available for up to 3 years, including the standard initial 26-week UI duration.

Wage Insurance Program

Because UI reimburses only a modest fraction of the long-run earnings losses associated with job displacement, some have suggested wage insurance systems to counteract these effects (LaLonde 2007). Wage insurance provides a temporary subsidy covering a portion of the wage decline to workers whose reemployment wages are lower than their predisplacement wages. Proponents argue that wage insurance not only financially compensates workers facing wage reductions after job displacement but also incentivizes job search, shortens unemployment durations, and supports workers for whom job training may be less effective (Kletzer and Litan 2001). By reducing what employers need to initially pay to attract and hire a displaced worker, the wage insurance benefit can help employers and employees find a match and provide a financial bridge to the worker until he or she is able to command a higher salary without the subsidy, thanks to on-the-job training and experience.

Since 2002, the TAA program has included a wage insurance program available to workers aged 50 and over who were laid off in a trade-related displacement (known as Alternative or Reemployment TAA). This national program is the largest and longest-running wage insurance program in the world. Hyman et al. (2021, 2023) uses the age-eligibility cutoffs in this program to explore the causal effect of these policies on employment and earnings trajectories of trade-affected workers in the United States. While all TAA-certified workers had access to training and extended UI payments (see above for a more detailed discussion of TAA), only those over age 50 had the additional option of receiving wage insurance.

The authors find that workers eligible for wage insurance are 25 percent more likely to be employed in the years just after displacement, and their earnings are a sustained 20 percent higher, relative to similar workers who are not eligible for the program. Most of these differences in earnings are accounted for by the higher probability of employment, rather than job quality improvements, suggesting that these programs can be effective in getting displaced workers back into the labor force and out from the long-term unemployed. The magnitudes of the earnings effects are large enough for this program to "pay for itself," through both increased government revenue associated with higher income tax revenue per participant as well as lower social expenditures in the form of UI or Supplemental Security Income/SSDI.

Active Labor Market Programs to Prevent Long-Term Unemployment

An increasing amount of evidence suggests that active labor market programs can be successful in helping displaced and disadvantaged workers to find employment in new occupations and at wage rates that are higher than they otherwise would have commanded. Many workers displaced by the energy transition will need to retool their skills for new occupations. A better equipped local workforce may help a region to rebound more quickly from the loss of key export industries. Research and past experience have demonstrated at least three types of programs that are able to achieve lasting increases in employment while potentially saving money for the UI system: (1) retraining programs, (2) job search assistance, and (3) reemployment bonuses.

Evidence from recent randomized control trials shows that specific types of job training programs yield high returns: raising wages for low-wage workers and sometimes paying for themselves within 5 years (Katz et al. 2022). These programs provide training in sector-specific skills demanded by local employers, who sometimes help to define the training, and offer wrap-around services regarding career readiness, career counseling, job placement, and post-placement job advancement. There is also evidence from other countries that well-designed, sectoral training programs have improved individual employment outcomes (Card et al. 2018).

Even though there are clear success stories regarding training, there are also concerns about scaling and implementation (Kanengiser and Schaberg 2022). Employers' wariness to participate in these programs (i.e., by guaranteeing to hire certain numbers of qualified graduates) is a common problem. Another problem may stem from the dizzying number of agencies (i.e., Workforce Development Boards) involved in administering these programs, with jurisdictional boundaries that bear little relation to the geographic structure of local labor markets and often do not align with the regional structure of local economic development agencies (Hanson 2023). There are also relatively low enrollment rates for training programs from eligible employees. Participants may fear that the opportunity costs from program participation may exceed any future gains in productivity or workforce advancement. People are often not paid during the time it takes to train, and when they do receive stipends, they are unlikely to cover the cost of living. Drop-out rates can also be high (Heckman et al. 2000). Gazmararian (2022) found that in places like southwest Pennsylvania where there were layoffs in the coal industry, economic decline persisted despite the presence of long-running development programs like the POWER Initiative through the Appalachian Regional Commission or the Assistance for Coal Communities program by the Economic Development Administration.

Other forms of training include community and 4-year colleges. "In community colleges, job-specific training often takes the form of certificate programs. These are practical courses of study of less than 2 years in length, which target specific occupations such as construction, manufacturing, repair, transportation, and other vocational trades" (Hanson 2023, p. 20). Prior research finds that when local economic conditions deteriorate, enrollment in certificate programs tends to rise. These effects are largest for programs that provide certification in industries where employment is expanding, suggesting that workers are using the programs to move between occupations in response to changing economic conditions (Acton 2021; Foote and Grosz 2020). However, most community colleges are geared to prepare students for entry into 4-year colleges and universities and allocate substantially fewer resources to certificate programs (Schanzenbach and Turner 2022). Goolsbee et al. (2019) develop a detailed proposal for expanding the training capacity of community colleges, as part of a broader agenda to increase and strengthen education of the U.S. workforce.

Federal financial aid can also play a vital role in assisting displaced workers in updating or modifying their skills via higher education. However, in many states the UI system does not continue to pay benefits when individuals enroll in school. Policies enacted in the American Recovery and Reinvestment Act of 2009 have led to reforms in several states that continue payment of UI benefits for workers obtaining certain types of training for up to 26 weeks. It is worthwhile to consider further initiatives to encourage efforts by UI recipients to obtain retraining (von Wachter 2011).

For some workers, a long period of time may elapse before they find a new job. These workers may have lost motivation, hope, or a realistic view of what wages to expect in the labor market. If targeted to workers most likely to exhaust UI benefits, bonuses that pay workers for finding a new job can reconnect long-term unemployed workers to the labor force, raising employment and reducing the cost for the UI system (von Wachter 2011). Evidence on "reemployment bonus experiments" suggests that short-term subsidies raise employment (e.g., Meyer 1995) but may only be cost effective if targeted to workers most likely to exhaust their benefits (DOL 1995; O'Leary et al. 2005). This sort of exit strategy built into the UI system may be particularly useful for older laid-off workers who face strong wage penalties and low employment rates (von Wachter 2011). It may also help to address concerns regarding the effect of extending UI benefits on the employment rate itself.

Prevention of Layoffs Through Work-Sharing Programs

One way to reduce the costs of the decarbonization of the workforce is to slow the pace of job destruction by using work-sharing programs (also known as "short-time compensation"). For example, the cost of UI benefits for a typical worker is a small fraction of the total earnings lost owing to a layoff over the remainder of the individual's working life (Kovalski and Sheiner 2022). If the same benefits were paid during employment to avoid job loss, this would substantially reduce the cost to workers. These programs would prevent the decline in spending power associated with layoffs, avoid dislocation and long-lasting earnings losses of laid-off workers, and may be cost-effective from society's point of view (von Wachter 2011).

Such a system of work-sharing has already been instituted in 17 states.[8] However, the current system may have to be extended and publicized to have a visible impact on forecasted job destruction in fossil communities and to have a substantial impact on employment (von Wachter 2021). At the same time, these benefits to the workforce

---

[8] See DOL (1997) for an overview of short-time compensation programs in different states. The German experience is the most cited example of a successful implementation of a work sharing program (see Möller [2010] for a discussion).

have to be considered against the ongoing social costs of fossil-intensive production that these compensation programs would ostensibly support.

A Safety Net for Workers Not Yet in the Workforce

The impact of decarbonization on the families and young individuals who remain in affected communities may also be worth considering. First, the current system of financial aid for college could be used to help prevent children of low-income background or of families who experienced a job loss from dropping out of college (von Wachter 2011). Research has documented a robust correlation of parental income with college attendance, especially of lower-income individuals, and this relationship appears to have strengthened over time (Deming and Dynarski 2009). Financial aid can be an important buffer against labor market shocks affecting parental income or students' own ability to work while in school. Another concern is that many resources available especially for lower-income students are currently provided at the state level, such as subsidized community colleges or merit scholarships. If decarbonization and the resulting fiscal implications impact state budgets, these resources may be at risk.

Migration Subsidies

Conventional views of labor markets suggest myriad ways that communities could adjust to a major change in economic landscape. One potential adjustment is through worker migration, moving to higher opportunity areas in search of gainful employment. In reality, however, migration has been relatively sluggish to respond in communities affected by trade or coal shocks, and this sluggishness has increased over time (Raimi 2022). When workers without a college degree lose their jobs, few choose to move elsewhere, even when local market conditions are poor. As a result, the proportion of the working-age population that have jobs (i.e., the employment to population ratio) has fallen significantly in affected communities. One potential explanation for this pattern is that families prefer to stay in their communities for other reasons, such as affordability or proximity to family and jobs. An alternative explanation is that they do not move to high-opportunity areas because of barriers that prevent them from making such moves.

Economists and policy makers have recently proposed migration subsidies or vouchers as one possible solution to overcome some of these barriers. There is some precedent, as the U.S. government spends approximately $20 billion each year on the Housing Choice Voucher Program, which provides rental assistance to low-income families with a goal of expanding residential choice and giving low-income families access to higher-opportunity areas. Evidence as to the effectiveness of these vouchers is promising, although it mostly comes from experimental evidence tied to families currently living in public housing, rather than economically distressed communities. For example, the Department of Housing and Urban Development designed the Moving to Opportunity (MTO) experiment to determine whether providing low-income families assistance in moving to better neighborhoods could improve their economic and health outcomes (HUD n.d.). The MTO experiment was conducted between 1994 and 1998 in five large U.S. cities. Approximately 4,600 families living in high-poverty public housing projects were randomly assigned to one of three groups: an experimental voucher group that was offered a subsidized housing voucher that came with a requirement to move to a census tract with a poverty rate below 10 percent, a Section 8 voucher group that was offered a standard housing voucher with no additional contingencies, and a control group that was not offered a voucher (but retained access to public housing).

Researchers found that the experimental voucher group in the MTO experiment experienced improved mental health, physical health, and subjective well-being of adults as well as family safety (Katz et al. 2001). More recently, researchers have shown how these vouchers had long-run effects on children's outcomes; moving a child to a low-poverty area when young (at age 8 on average) using a subsidized voucher like the MTO experimental voucher increases the child's total lifetime earnings by about $302,000 (Chetty et al. 2015). The additional tax revenue generated from these earnings increases would itself offset the incremental cost of the subsidized voucher relative to providing public housing.

Place-Based Policies

An alternative to addressing job loss by targeting individuals is to target exposed regions through "place-based" policies, which condition assistance on the state of the local economy. This is a general term that is meant to include policies such as tax incentives to recruit or to retain companies, subsidized lending for real estate development, and technical assistance to local business (Bartik 2020; place-based policies are also discussed in Chapters 5 and 13). Because decarbonization is likely to reduce the economic vibrancy of regions currently specialized in fossil fuels, the role of place-based policies would be to help communities develop a new economic base and replace the well-paying jobs that have been lost (Hanson 2023). Research has shown that large, place-based policies (e.g., Tennessee Valley Authority) can have long-lived effects on regional specialization (Bianchi and Giorcelli 2022; Kline and Moretti 2014), which suggests that they have the potential to catalyze local investments in individuals (labor) and businesses (capital).

Examples from past policy successes and failures reveal several challenges associated with design and implementation. If the approach is not careful, these programs may intensify zero-sum tax competition among regions to attract firms (Kim 2021) or be manipulated by elected officials for political gain (Slattery 2020). Relatedly, policy implementation tends to be badly fragmented across state and federal agencies, which often fail to coordinate their efforts and instead frequently design incentive structures that cause them to work at cross purposes (Hanson 2023).[9]

Place-Based Policies: Business Tax Incentives

A common approach to incentivizing new business and capital is to provide tax incentives to a large company in return for promised investments in new productive capacity, the expansion of existing operations, or the creation of R&D facilities (Slattery 2020). The hope is that if the company breaks ground, it will attract upstream industry suppliers and downstream industry buyers, potentially generating industry agglomeration that could raise regional employment, productivity, and wages. There is a range of evidence suggesting that policies targeting specific industries in various regions succeeded in expanding regional output and/or productivity in the target area well beyond the duration of the policies (Bianchi and Giorcelli 2022; Freedman 2017; Garin and Rothbaum 2020; Greenstone et al. 2010). It is unclear whether business tax incentives simply move targets from one location to another (at substantial taxpayer expense) or truly expand aggregate output nationally. That being said, place-based tax incentives may also be justified from an equity perspective by transferring resources to communities in which needy households are clustered (Gaubert et al. 2021).

While some place-based policies such as business tax incentives can be useful in stimulating new capital deployment within a region (e.g., new or expanding businesses), there are also policies that may be helpful in raising the productivity of incumbent businesses. For example, "in economically distressed regions, local entrepreneurs may have difficulty securing loans to launch a new business while owners of existing firms may face challenges in financing business improvements or expansions" (Hanson 2023, p. 30).

The decline of a region's existing industrial base may reduce housing values and associated equity, stifling business formation (Davis and Haltiwanger 2019). In the

---

[9] "In the United States, the practice of local economic development tends to be organized around five major areas: business retention and recruitment, workforce development, financial and technical assistance to small business, infrastructure development, and financial incentives to invest in low-income areas (Bartik 2020). . . . These areas tend to be managed by different bureaucracies, funded from different sources, and guided by different and often conflicting incentives" (Hanson 2023, pp. 27–28).

aftermath of localized economic downturns, there may be cause to subsidize services to businesses that have a demonstrated interest in expanding local employment.

Hanson (2023) highlights several existing government programs intended to provide a wide range of support to small- and medium-size businesses. The Small Business Administration guarantees loans to qualifying small businesses and runs more than 900 Small Business Development Centers, often housed in community colleges or universities. These centers provide technical assistance and consulting services to local firms. The Economic Development Administration funds similar business services through its grants to colleges and universities. The Manufacturing Extension Program run by the National Institute of Standards and Technology is specific to industrial production and helps companies upgrade their technology through a national network of centers. In theory, expanded versions of these programs could help regions adversely affected by the energy transition, although evidence as to the efficacy and cost-effectiveness of these programs remains limited. Learning more about how well these programs work, ideally through gold-standard, randomized controlled trials should be a high priority.[10]

Place-Based Policies: Fossil Retirement Subsidies

Another potential place-based policy to help mitigate some of the devastating impacts to fossil communities are place-based transfers designed to incentivize the closure of fossil facilities such as coal-fired power plants. While closing a facility would lead to job loss for affected workers, there are at least two reasons why these policies can be helpful. First, some of these facility retirement subsidies can be earmarked for worker transition assistance.[11] Second, subsidies can be designed in a way that provides important information as to the timing of the shutdown that allows not only affected workers but also affected communities to begin planning.[12] For example, Germany has been experimenting with reverse auctions, as a type of subsidy, to compensate early retirement of hard coal and small-scale lignite power plants. A reverse auction is a type of auction in

---

[10] There is evidence from developing countries about the impact of these wraparound consulting services (Bloom et al. 2013, 2020; Iacovone et al. 2022). For example, Hanson (2023) describes a variety of randomized control trials that show how supplying consulting services to medium-size businesses in developing countries leads to long-lasting improvements in economic performance.

[11] Colorado has recently required that regulated utilities submit a workforce transition plan and authorizes "rate recovery" for the expenses. As a consequence, the utility has a financial incentive to attend to impacted communities and encourage workforce development, in order to recoup the costs from closing down a coal plant early (Righetti et al. 2021).

[12] The U.S. Worker Adjustment and Retraining Notification Act (WARN) has a similar mandate, requiring most employers with 100 or more employees to provide 60-day advance notification of planned closings and mass layoffs of employees. However, 60 days is not a sufficient length of time for workers or communities to plan for workforce transitions.

which sellers (i.e., coal facilities) bid for the prices at which they are willing to retire their plants. At the end of the auction, the seller with the lowest amount wins the auction and receives the payment, and a closure date for the facility is announced publicly. After the fifth round of auctions, the German government can force compulsory power plant closures without financial compensation. By providing both "carrots" in the form of higher maximum bids early in the auction rounds and "sticks" in the form of compulsory closures without compensation at the end of the auction rounds, Germany has provided strong incentives for retirement. Box 4-2 expands on the lessons policy makers can take away from Germany's experience with coal phaseout.

Summary: Policy Tools and Mitigation of Impacts

The transitional costs associated with reallocating the workforce away from the fossil sector will likely impose substantial and lasting costs on affected workers and communities in terms of earnings, health, and strain on their families. Displaced workers earn significantly less than similar workers who have not been displaced, even years after the separation occurred (Jacobson et al. 1993; von Wachter et al. 2011). Over the past several decades, import competition from China (Autor et al. 2013), the automation of manufacturing production (Autor and Dorn 2013), and the shift in electricity generation away from coal (Black et al. 2005) have caused locally concentrated job loss in the United States (Richardson and Anderson 2021), which has led to lasting declines in employment rates, earnings, and social conditions in the local labor markets that were exposed to these shocks. Younger and more educated workers were most likely to migrate away from these regions, but for the older and less-educated workers left behind, localized distress has persisted for decades beyond the actual displacement events (Hanson 2023). Without significant changes to the existing safety net and transition assistance, the transition from fossil fuels seems likely to repeat the now familiar story of industry decline and regional hardship.

Manufacturing support and training programs can help catalyze job creation and job alternatives for displaced workers. However, these programs do not address the concerns raised by existing research on the fate of displaced workers and communities in the recent past. Research and prior policy experiments suggest cost-effective ways to alleviate the burden for these workers. For example, a well-designed and well-funded transition assistance program (see Recommendation 4-3) may offer the best hope for reducing the harms to displaced workers who find the transition to new jobs difficult for one reason or another. Each of the policy options described will require an act of Congress. In many cases, changes in policy to address these challenges need to be paired with making necessary administrative data and program information available to allow researchers to give better assessments of the full costs and benefits of these future programs.

BOX 4-2
POLICY LESSONS FROM GERMANY'S COAL PHASEOUT

Germany's Commission on Growth, Structural Change and Employment ("Coal Commission") was established by the federal government in 2018 with representation from a wide range of stakeholders to build a consensus on the phase-out of coal and promote a just transition. The commission developed a plan to end coal-fired power generation by 2038 and provide targeted support for coal-dependent regions and some 32,800 coal industry workers (German Coal Commission 2019).

There are three clear takeaways:

1. *Clear expectations.* The phase-out will occur through auctions and direct compensation to coal companies to reduce capacity over time. For many of the coal plants, an exact retirement date has been set, giving communities and the companies time to prepare.

2. *Compensation.*
   a. $5 billion to coal companies for retirement.
      i. Compensation is distributed through auctions, where the government awards funds to the "bidder" (operator) that proposes to retire the most GW of capacity at the lowest cost.
      ii. Maximum compensation amount is specified in the law, and it declines with every year of the auction process, incentivizing operators to seek an early shut-down. As of 2027, hard coal power plants are to be shut down by regulatory order without compensation.
   b. $47 billion to diversify the regions' economies and create new jobs over the coming 2 decades as coal is phased out.
      i. Around $30 billion of the fund goes to infrastructure and other projects determined by the national government, and $16.5 billion is set aside for regional investment.
      ii. Regions can apply for investment in projects across nine categories from tourism to research, allowing each area to decide how to grow its economy according to its own strengths rather than a top-down vision.
   c. Coal workers will receive 5 billion euros ($6 billion) in compensation for losing their jobs and/or retiring early.

3. *Strong existing safety net for the labor force.*
   a. More generous unemployment insurance program and robust network of job training.
   b. Health insurance and pension program not tied to employment.

Finding 4-6: Reallocating the workforce away from the fossil sector will likely impose substantial and lasting costs on affected workers and communities in terms of earnings, health, and strain on their families. Without significant changes to the existing safety net and transition assistance, the transition from fossil fuels seems likely to repeat the now familiar story of industry decline and regional hardship.

Recommendation 4-3: *Extend Unemployment Insurance Duration for Fossil Fuel–Related Layoffs and Develop Decarbonization Workforce Adjustment Assistance Program.* Congress should authorize and appropriate a comprehensive transition assistance program for workers whose employment is negatively impacted by the transition to net-zero emissions. This transition program should include extending unemployment insurance duration for workers affected by fossil fuel–related layoffs and continuing payments for those who choose to enroll in skill development courses or higher education programs following job loss. The program should include wage insurance to support laid-off workers who find new employment where pay is not commensurate to their previous employment. It should also include resources to scale up active labor market programs that have demonstrated recent success in improving worker outcomes.

CONCLUDING FINDINGS AND RECOMMENDATIONS

Finding 4-7: Employment impacts of decarbonization will depend greatly on pathways to decarbonization, timing, and the extent to which the transition is managed and coordinated across entities in the workforce pipeline, including governments, the private sector, research institutions, and training program entities such as community colleges and labor unions. High-quality information and data about workforce supply and demand is critical to ensuring positive outcomes for workers, communities, and companies as well as evaluating success of policies, programs, and funding initiatives.

Recommendation 4-4: *Collect and Report Data on Net-Zero-Relevant Professions.* The Department of Energy should expand on its existing energy workforce data collection and analysis efforts through the U.S. Energy and Employment Report by

a. Collecting up-to-date and actionable data on net-zero-relevant professions, including data on employment by industry and occupation for businesses that produce low- or no-emissions goods and services;

data on the occupations and wages of jobs related to net-zero-relevant technologies and practices; and career information publications related to emissions reductions, decarbonization, and climate change;

b. **Conducting analyses to inform where and when job gains and losses may occur related to decarbonization (e.g., $ value of different types of jobs in fossil and non-fossil industries, job losses);**

c. **Creating estimates to inform state and local workforce development programs and the private sector of workforce preferences and capabilities (e.g., demand for different types of jobs); and**

d. **Evaluating the efficacy of workforce interventions throughout the transition.**

Table 4-1 summarizes all the recommendations in this chapter regarding building the workforce needed to accomplish decarbonization objectives, as well as supporting workers negatively impacted by the transition.

SUMMARY OF RECOMMENDATIONS ON WORKFORCE NEEDS, OPPORTUNITIES, AND SUPPORT

TABLE 4-1 Summary of Recommendations on Workforce Needs, Opportunities, and Support

Short-Form Recommendation	Actor(s) Responsible for Implementing Recommendation	Sector(s) Addressed by Recommendation	Objective(s) Addressed by Recommendation	Overarching Categories Addressed by Recommendation
4-1: Support the Development of Net-Zero Curriculum and Skill Development Programs for K–12 Students	Department of Education, local governments, and school districts	• Electricity • Buildings • Transportation • Industry • Non-federal actors	• Equity • Employment • Public engagement	Building the Needed Workforce and Capacity
4-2: Invest in Linking People from Disadvantaged Communities to Quality Jobs	Congress	• Electricity • Buildings • Transportation • Industry • Non-federal actors	• Equity • Employment	Ensuring Equity, Justice, Health, and Fairness of Impacts Building the Needed Workforce and Capacity
4-3: Extend Unemployment Insurance Duration for Fossil Fuel–Related Layoffs and Develop Decarbonization Workforce Adjustment Assistance Program	Congress	• Transportation • Fossil fuels	• Equity • Employment • Public engagement	Ensuring Equity, Justice, Health, and Fairness of Impacts Building the Needed Workforce and Capacity
4-4: Collect and Report Data on Net-Zero-Relevant Professions	Department of Energy	• Electricity • Buildings • Transportation • Industry • Non-federal actors	• Equity • Employment	Building the Needed Workforce and Capacity

REFERENCES

Acemoglu, D., D. Autor, D. Dorn, G.H. Hanson, and B. Price. 2016. "Import Competition and the Great US Employment Sag of the 2000s." *Journal of Labor Economics* 34(S1):S141–S198. https://doi.org/10.1086/682384.

Acton, R.K. 2021. "Community College Program Choices in the Wake of Local Job Losses." *Journal of Labor Economics* 39(4):1129–1154. https://doi.org/10.1086/712555.

AFDC (Alternative Fuels Data Center). n.d. "Alternative Fuel Infrastructure Tax Credit." Department of Energy. https://afdc.energy.gov/laws/10513. Accessed March 28, 2023.

AFL-CIO. 2017. "Resolution 1: Workers' Bill of Rights | AFL-CIO." AFL-CIO. October 25. https://aflcio.org/resolutions/resolution-1-workers-bill-rights. Accessed July 18, 2023.

Alden, E.H. 2016. *Failure to Adjust: How Americans Got Left Behind in the Global Economy*. Lanham, MD: Rowman & Littlefield.

American-Made Challenges. n.d. "Energy CLASS Prize." Herox.Com. https://www.herox.com/energy-class. Accessed March 28, 2023.

ARC (Appalachian Regional Commission). 2021. "Legislative Update: President Biden Signs Infrastructure Investment and Jobs Act, Which Includes $1 Billion for ARC Over Five Years." Appalachian Regional Commission. https://www.arc.gov/news/legislative-update-president-biden-signs-infrastructure-investment-and-jobs-act-which-includes-1-billion-for-arc-over-five-years.

Autor, D.H., and D. Dorn. 2013. "The Growth of Low-Skill Service Jobs and the Polarization of the US Labor Market." *American Economic Review* 103(5):1553–1597.

Autor, D.H., D. Dorn, and G.H. Hanson. 2013. "The China Syndrome: Local Labor Market Effects of Import Competition in the United States." *American Economic Review* 103(6):2121–2168. https://doi.org/10.1257/aer.103.6.2121.

Autor, D.H., D. Dorn, and G. Hanson. 2021. "On the Persistence of the China Shock." Cambridge, MA: National Bureau of Economic Research. https://doi.org/10.3386/w29401.

Badlam, J., S. Clark, S. Gajendragadkar, A. Kumar, S. O'Rourke, and D. Swartz. 2022a. "The CHIPS and Science Act: What Is It and What Is in It?" *McKinsey Insights*. https://www.mckinsey.com/industries/public-and-social-sector/our-insights/the-chips-and-science-act-heres-whats-in-it.

Badlam, J., J. Cox, A. Kumar, N. Mehta, S. O'Rourke, and J. Silvis. 2022b. "What's in the Inflation Reduction Act (IRA) of 2022." *McKinsey Insights*. https://www.mckinsey.com/industries/public-and-social-sector/our-insights/the-inflation-reduction-act-heres-whats-in-it.

Barrett, J., and J. Bivens. 2021. "The Stakes for Workers in How Policymakers Manage the Coming Shift to All-Electric Vehicles." Economic Policy Institute. https://www.epi.org/publication/ev-policy-workers.

Bartik, T.J. 2020. "Using Place-Based Jobs Policies to Help Distressed Communities." *Journal of Economic Perspectives* 34(3):99–127. https://doi.org/10.1257/jep.34.3.99.

Bayard, K., T. Cajner, V. Gregorich, and M.D. Tito. 2022. "Are Manufacturing Jobs Still Good Jobs? An Exploration of the Manufacturing Wage Premium." Federal Reserve Board of Governors. https://www.federalreserve.gov/econres/feds/are-manufacturing-jobs-still-good-jobs-an-exploration-of-the-manufacturing-wage-premium.htm.

BGA (BlueGreen Alliance). 2021. "Making Clean Deliver: Improving Clean Energy Job Quality and Growing the Clean Energy Manufacturing Supply Chain in the United States." https://www.bluegreenalliance.org/site/making-clean-deliver.

BGA. 2022. "Fact Sheet: Clean Vehicle Provisions in the Inflation Reduction Act." https://www.bluegreenalliance.org/wp-content/uploads/2022/08/BGA-Clean-Vehicle-IRA-Factsheet-10722-FINAL.pdf.

Bianchi, N., and M. Giorcelli. 2022. "The Dynamics and Spillovers of Management Interventions: Evidence from the Training Within Industry Program." *Journal of Political Economy* 130(6):1630–1675. https://doi.org/10.1086/719277.

Bilginsoy, C., D. Bullock, A.T. Wells, and R. Zullo. 2022. "Diversity, Equity, and Inclusion Initiatives in the Construction Trades." NABTU. https://www.dropbox.com/s/no051eil9jtf1f5/ICERES%20Study%20%22Diversity%2C%20Equity%2C%20and%20Inclusion%20Initiatives%20in%20the%20Construction%20Trades%22.pdf?dl=0.

Black, D., K. Daniel, and S. Sanders. 2002. "The Impact of Economic Conditions on Participation in Disability Programs: Evidence from the Coal Boom and Bust." *American Economic Review* 92(1):27–50. https://doi.org/10.1257/000282802760015595.

Black, D.A., J.A. Smith, M.C. Berger, and B.J. Noel. 2003. "Is the Threat of Reemployment Services More Effective Than the Services Themselves? Evidence from Random Assignment in the UI System." *American Economic Review* 93(4):1313–1327.

Black, D., T. McKinnish, and S. Sanders. 2005. "The Economic Impact of the Coal Boom and Bust." *Economic Journal* 115(503):449–476. https://doi.org/10.1111/j.1468-0297.2005.00996.x.

Bloom, N., B. Eifert, A. Mahajan, D. McKenzie, and J. Roberts. 2013. "Does Management Matter? Evidence from India." *Quarterly Journal of Economics* 128(1):1–51. https://doi.org/10.1093/qje/qjs044.

Bloom, N., A. Mahajan, D. McKenzie, and J. Roberts. 2020. "Do Management Interventions Last? Evidence from India." *American Economic Journal: Applied Economics* 12(2):198–219. https://doi.org/10.1257/app.20180369.

BLS (Bureau of Labor Statistics). 2022. "Mining and Geological Engineers: Occupational Outlook Handbook." https://www.bls.gov/ooh/architecture-and-engineering/mining-and-geological-engineers.htm#tab-1.

BLS. 2023. "Automotive Industry: Employment, Earnings, and Hours: U.S. Bureau of Labor Statistics." https://www.bls.gov/iag/tgs/iagauto.htm.

Boom, M. 2021. "United: Union Jobs Improve the Clean Energy Economy." R: 21-09-A. Natural Resources Defense Council. https://www.nrdc.org/sites/default/files/united-union-jobs-clean-energy-report.pdf.

BW Research. 2016. "U.S. Energy and Employment Report 2016." https://www.energy.gov/policy/us-energy-and-employment-report-2016.

Card, D., J. Kluve, and A. Weber. 2018. "What Works? A Meta Analysis of Recent Active Labor Market Program Evaluations." *Journal of the European Economic Association* 16(3):894–931. https://doi.org/10.1093/jeea/jvx028.

Chan, I., M. Sagatelova, A. Laska, L. Walter, L. Jantarasami, M. Tesfaye, C. Normile, and S. Davis. 2022. "Employment Impacts in a Decarbonized Economy." Decarb America. https://decarbamerica.org/wp-content/uploads/2022/06/Employment-Impacts-in-a-Decarbonized-Economy.pdf.

Chetty, R., N. Hendren, and L. Katz. 2015. "The Effects of Exposure to Better Neighborhoods on Children: New Evidence from the Moving to Opportunity Experiment." Cambridge, MA: National Bureau of Economic Research. https://doi.org/10.3386/w21156.

Congdon, W.J, D. Nightingale, M.M. Scott, J. Shakesprere, B. Katz, and P. Loprest. 2020. "Understanding Good Jobs: A Review of Definitions and Evidence." The Urban Institute. https://www.urban.org/sites/default/files/publication/102603/understanding-good-jobs-a-review-of-definitions-and-evidence_2.pdf.

Cozzi, L., and B. Motherway. 2021. "The Importance of Focusing on Jobs and Fairness in Clean Energy Transitions." IEA. https://www.iea.org/commentaries/the-importance-of-focusing-on-jobs-and-fairness-in-clean-energy-transitions.

Cutcher-Gershenfeld, J., D. Brooks, and M. Molloy. 2015. "The Decline and Resurgence of the U.S. Auto Industry." Briefing Paper 399. Washington, DC: Economic Policy Institute. https://www.epi.org/publication/the-decline-and-resurgence-of-the-u-s-auto-industry.

Data USA. n.d. "Mining and Mineral Engineering." https://datausa.io/profile/cip/mining-mineral-engineering. Accessed August 2, 2023.

Davis, S., and J. Haltiwanger. 2019. "Dynamism Diminished: The Role of Housing Markets and Credit Conditions." Cambridge, MA: National Bureau of Economic Research. https://doi.org/10.3386/w25466.

Deloitte. 2022. "Work Toward Net Zero: The Rise of the Green Collar Workforce in a Just Transition." https://www.deloitte.com/global/en/issues/climate/work-toward-net-zero.html.

Deming, D., and S. Dynarski. 2009. "Into College, Out of Poverty? Policies to Increase the Postsecondary Attainment of the Poor." Cambridge, MA: National Bureau of Economic Research. https://doi.org/10.3386/w15387.

Department of the Treasury. 2023. "Treasury, Energy Release Guidance on Inflation Reduction Act Programs to Incentivize Investments in Underserved Communities, Hard-Hit Coal Communities." https://home.treasury.gov/news/press-releases/jy1269.

DOC (Department of Commerce). 2019. *A Federal Strategy to Ensure Secure and Reliable Supplies of Critical Minerals.* Washington, DC: Department of Commerce. https://www.commerce.gov/data-and-reports/reports/2019/06/federal-strategy-ensure-secure-and-reliable-supplies-critical-minerals.

DOE (Department of Energy). 2022. "U.S. Department of Energy Begins Formation of the 21st Century Energy Workforce Advisory Board." https://www.energy.gov/policy/articles/us-department-energy-begins-formation-21st-century-energy-workforce-advisory-board.

DOE. 2023. "2023 U.S. Energy and Employment Report." https://www.energy.gov/sites/default/files/2023-06/2023%20USEER%20REPORT-v2.pdf.

DOE. n.d.(a). "45L Tax Credits for Zero Energy Ready Homes." https://www.energy.gov/eere/buildings/45l-tax-credits-zero-energy-ready-homes. Accessed July 18, 2023.

DOE. n.d.(b). "179D Commercial Buildings Energy-Efficiency Tax Deduction." https://www.energy.gov/eere/buildings/179d-commercial-buildings-energy-efficiency-tax-deduction. Accessed July 18, 2023.

DOE. n.d.(c). "Clean Energy Infrastructure Programs at Department of Energy." https://www.energy.gov/clean-energy-infrastructure/clean-energy-infrastructure-programs-department-energy. Accessed March 28, 2023.

DOE. n.d.(d). "Grants for Energy Improvements at Public School Facilities." https://www.energy.gov/scep/grants-energy-improvements-public-school-facilities. Accessed March 28, 2023.

DOE. n.d.(e). "State-Based Home Energy Efficiency Contractor Training Grants." https://www.energy.gov/scep/state-based-home-energy-efficiency-contractor-training-grants. Accessed March 28, 2023.

DOE Solar Energy Technologies Office. 2023. "Federal Solar Tax Credits for Businesses." https://www.energy.gov/sites/default/files/2023-04/Federal-Solar-Tax-Credits-for-Businesses-4-23.pdf.

DOL (Department of Labor). 2022. "Training and Employment Guidance Letter No. 13-21." http://www.dol.gov/agencies/eta/advisories/training-and-employment-guidance-letter-no-13-21.

DOL. 2023. "Trade Adjustment Assistance for Workers (TAA) Program: Impacts of Program Termination." https://www.dol.gov/sites/dolgov/files/ETA/tradeact/pdfs/TAA_Termination_Fact_Sheet.pdf.

DOL. n.d.(a). "Explore Pre-Apprenticeship." https://www.apprenticeship.gov/employers/explore-pre-apprenticeship. Accessed March 29, 2023.

DOL. n.d.(b). "Registered Apprenticeship Program." https://www.apprenticeship.gov/employers/registered-apprenticeship-program. Accessed March 29, 2023.

DOT (Department of Transportation) Maritime Administration. 2023. "Frequently Asked Questions—Port Infrastructure Development Grants | MARAD." https://www.maritime.dot.gov/PIDP%20Grants/FAQs.

E2. 2021. "Help Wanted: Diversity in Clean Energy." E2, Alliance to Save Energy, American Association of Blacks in Energy, Energy Efficiency for All, Black Owners of Solar Services, and BW Research Partnership. https://e2.org/wp-content/uploads/2021/09/E2-ASE-AABE-EEFA-BOSS-Diversity-Report-2021.pdf.

EPA (Environmental Protection Agency). 2023. "Clean Heavy-Duty Vehicle Program." Overviews and Factsheets. https://www.epa.gov/inflation-reduction-act/clean-heavy-duty-vehicle-program.

EPI (Economic Policy Institute). n.d. "Section 5. Benefit Levels: Increase UI Benefits to Levels Working Families Can Survive On." https://www.epi.org/publication/section-5-benefit-levels-increase-ui-benefits-to-levels-working-families-can-survive-on. Accessed March 29, 2023.

Fell, H., and D.T. Kaffine. 2018. "The Fall of Coal: Joint Impacts of Fuel Prices and Renewables on Generation and Emissions." *American Economic Journal: Economic Policy* 10(2):90–116. https://doi.org/10.1257/pol.20150321.

Foote, A., and M. Grosz. 2020. "The Effect of Local Labor Market Downturns on Postsecondary Enrollment and Program Choice." *Education Finance and Policy* 15(4):593–622. https://doi.org/10.1162/edfp_a_00288.

Foster, D., S. Nabahe, and B.S.H. Ng. 2020. "Energy Workforce Development in the 21st Century." MIT CEEPR Working Paper Series. Roosevelt Project Special Series. Massachusetts Institute of Technology and Harvard University. https://ceepr.mit.edu/workingpaper/energy-workforce-development-in-the-21st-century.

Foster, D., A. Maranville, and S.F. Savitz. 2023. "Jobs, Emissions, and Economic Growth—What the Inflation Reduction Act Means for Working Families." Policy Paper. Washington, DC: Energy Futures Initiative. https://energyfuturesinitiative.org/reports/jobs-emissions-and-economic-growth-what-the-inflation-reduction-act-means-for-working-families-jobs-emissions-and-economic-growth.

Freedman, M. 2017. "Persistence in Industrial Policy Impacts: Evidence from Depression-Era Mississippi." *Journal of Urban Economics* 102(November):34–51. https://doi.org/10.1016/j.jue.2017.08.001.

FTA (Federal Transit Administration). 2021. "Fact Sheet: Buses and Bus Facilities Program." December 9. https://www.transit.dot.gov/funding/grants/fact-sheet-buses-and-bus-facilities-program.

FTA. n.d. "Low or No Emission Vehicle Program—5339(c)." https://www.transit.dot.gov/lowno. Accessed March 28, 2023.

Gammarano, R. 2020. "Measuring Job Quality: Difficult But Necessary." ILOSTAT. January 27. https://ilostat.ilo.org/measuring-job-quality-difficult-but-necessary.

Ganong, P., and P. Noel. 2019. "Consumer Spending During Unemployment: Positive and Normative Implications." *American Economic Review* 109(7):2383–2424. https://doi.org/10.1257/aer.20170537.

Garin, A., and J. Rothbaum. 2020. "Was the Arsenal of Democracy an Engine of Mobility? Public Investment and the Roots of Mid-Century Manufacturing Opportunity." October. https://conference.nber.org/conf_papers/f134699.pdf.

Gaubert, C., P. Kline, and D. Yagan. 2021. "Place-Based Redistribution." Cambridge, MA: National Bureau of Economic Research. https://doi.org/10.3386/w28337.

Gazmararian, A.F. 2022. "Sources of Partisan Change: Evidence from Unmanaged Energy Transitions in American Coal Country." SSRN Scholarly Paper. https://doi.org/10.2139/ssrn.4164734.

Gerdes, J. 2020. "California's Oil Industry Could Become a Just Transition Model." *Energy Monitor* (blog), September 21. https://www.energymonitor.ai/policy/just-transition/californias-oil-industry-could-become-a-just-transition-model.

Glasmeier, A.K., and MIT (Massachusetts Institute of Technology). 2020a. "Living Wage Calculation for Washington-Arlington-Alexandria, DC." https://livingwage.mit.edu/metros/47900.

Goolsbee, A., G. Hubbard, and A. Ganz. 2019. "A Policy Agenda to Develop Human Capital for the Modern Economy." The Aspen Institute. https://www.aspeninstitute.org/longform/expanding-economic-opportunity-for-more-americans/a-policy-agenda-to-develop-human-capital-for-the-modern-economy.

Gould, E., and W. Kimball. 2015. "'Right-to-Work' States Still Have Lower Wages." Economic Policy Institute. https://www.epi.org/publication/right-to-work-states-have-lower-wages.

Greenspon, J., and D. Raimi. 2022. "Matching Geographies and Job Skills in the Energy Transition." WP 22-25. Resources for the Future. https://media.rff.org/documents/WP_22-25_PnkcURf.pdf.

Greenstone, M., R. Hornbeck, and E. Moretti. 2010. "Identifying Agglomeration Spillovers: Evidence from Winners and Losers of Large Plant Openings." *Journal of Political Economy* 118(3):536–598. https://doi.org/10.1086/653714.

Gruber, J. 1997. "The Consumption Smoothing Benefits of Unemployment Insurance." *American Economic Review* 87(1):192–205.

Gurley, L.K. 2022. "Shifting America to Solar Power Is a Grueling, Low-Paid Job." *Vice* (blog), June 27. https://www.vice.com/en/article/z34eyx/shifting-america-to-solar-power-is-a-grueling-low-paid-job.

Hale, T. 2023. "The United States Needs More Than Mining Engineers to Solve Its Critical Mineral Challenges." Center for Strategic and International Studies. https://www.csis.org/analysis/united-states-needs-more-mining-engineers-solve-its-critical-mineral-challenges.

Hall, R. 2005. "Job Loss, Job Finding, and Unemployment in the U.S. Economy Over the Past Fifty Years." Cambridge, MA: National Bureau of Economic Research. https://doi.org/10.3386/w11678.

Hanson, G. 2023. "Local Labor Market Impacts of the Energy Transition: Prospects and Policies." Cambridge, MA: National Bureau of Economic Research. https://doi.org/10.3386/w30871.

Harris, L. 2022. "Workers on Solar's Front Lines." American Prospect. https://prospect.org/labor/workers-on-solars-front-lines.

Harvey, A., J. Walsh, J. Eckdish, K. Connolly, B. Beachy, E. Steen, R. Johnson, et al. 2022. "A User Guide to the Inflation Reduction Act: How New Investments Will Deliver Good Jobs, Climate Action, and Health Benefits." BGA. https://www.bluegreenalliance.org/wp-content/uploads/2023/02/BGA-IRA-User-Guide-Print-FINAL-Web.pdf.

Heckman, J., N. Hohmann, J. Smith, and M. Khoo. 2000. "Substitution and Dropout Bias in Social Experiments: A Study of an Influential Social Experiment." *Quarterly Journal of Economics* 115(2):651–694. https://doi.org/10.1162/003355300554764.

H.R.842—117th Congress. 2021–2022. "Protecting the Right to Organize Act of 2021." https://www.congress.gov/bill/117th-congress/house-bill/842.

H.R.20—118th Congress. 2023–2024. "Richard L. Trumka Protecting the Right to Organize Act of 2023." https://www.congress.gov/bill/118th-congress/house-bill/20.

HUD (Department of Housing and Urban Development). n.d. "Moving to Opportunity (MTO)." https://www.huduser.gov/portal/datasets/mto.html. Accessed July 18, 2023.

Hyman, B.G. 2018. "Can Displaced Labor Be Retrained? Evidence from Quasi-Random Assignment to Trade Adjustment Assistance." *Proceedings, Annual Conference on Taxation and Minutes of the Annual Meeting of the National Tax Association* 111:1–70.

Hyman, B.G. 2022. "Can Displaced Labor Be Retrained? Evidence from Quasi-Random Assignment to Trade Adjustment Assistance." Center for Economic Studies. CES 22-05. https://www2.census.gov/ces/wp/2022/CES-WP-22-05.pdf.

Hyman, B.G., B.K. Kovak, A. Leive, and T. Naff. 2021. "Wage Insurance and Labor Market Trajectories." *AEA Papers and Proceedings* 111(May):491–495. https://doi.org/10.1257/pandp.20211093.

Iacovone, L., W. Maloney, and D. McKenzie. 2022. "Improving Management with Individual and Group-Based Consulting: Results from a Randomized Experiment in Colombia." *Review of Economic Studies* 89(1):346–371. https://doi.org/10.1093/restud/rdab005.

ILO (International Labour Organization). 2020. "Decent Work." International Labour Organization. 2020. https://www.ilo.org/global/topics/decent-work/lang—en/index.htm.

Inclusive Economics. 2021. "High-Road Workforce Guide for City Climate Action." https://www.usdn.org/uploads/cms/documents/workforce-guide_4.12.21_form.pdf.

Interagency Working Group on Coal and Power Plant Communities and Economic Revitalization. 2023. "Energy Community Tax Credit Bonus." Energy Communities. https://energycommunities.gov/energy-community-tax-credit-bonus.

IREC (Interstate Renewable Energy Council). 2022. "12th Annual National Solar Jobs Census 2021." https://irecusa.org/wp-content/uploads/2022/National-Solar-Jobs-Census-2021.pdf.

IRS (Internal Revenue Service). 2023. IRS Notice 2023-38, 2023-22 IRB 1. https://www.irs.gov/pub/irs-drop/n-23-38.pdf.

Jacobsen, G.D., and D.P. Parker. 2016. "The Economic Aftermath of Resource Booms: Evidence from Boomtowns in the American West." *Economic Journal* 126(593):1092–1128. https://doi.org/10.1111/ecoj.12173.

Jacobson, L.S., R.J. LaLonde, and D.G. Sullivan. 1993. "Earnings Losses of Displaced Workers." *American Economic Review* 83(4):685–709.

Jones, J., A. Hertel-Fernandez, and C. DeCarlo. 2021. "Equity Snapshot: Apprenticeships in America." *DOL* (blog), November 4. http://blog.dol.gov/2021/11/03/equity-snapshot-apprenticeships-in-america.

Kanengiser, H., and K. Schaberg. 2022. *Employment and Earnings Effects of the WorkAdvance Demonstration After Seven Years.* Washington, DC: MDRC.

Katz, L.F., J.R. Kling, and J.B. Liebman. 2001. "Moving to Opportunity in Boston: Early Results of a Randomized Mobility Experiment." *Quarterly Journal of Economics* 116(2):607–654.

Katz, L.F., J. Roth, R. Hendra, and K. Schaberg. 2022. "Why Do Sectoral Employment Programs Work? Lessons from WorkAdvance." *Journal of Labor Economics* 40(S1):S249–S291. https://doi.org/10.1086/717932.

Kim, D. 2021. "Economic Spillovers and Political Values in Government Competition for Firms: Evidence from the Kansas City Border War." Working Paper. Iowa State University. https://conference.nber.org/conf_papers/f160953.pdf.

Kletzer, L.G., and R.E. Litan. 2001. "A Prescription to Relieve Worker Anxiety." Peterson Institute for International Economics. March. https://www.piie.com/publications/policy-briefs/prescription-relieve-worker-anxiety.

Krishnan, M., H. Samandari, J. Woetzel, S. Smit, D. Pacthod, D. Pinner, T. Nauclér, et al. 2022. "The Net-Zero Transition: What It Would Cost, What It Could Bring." McKinsey Global Institute. https://www.mckinsey.com/capabilities/sustainability/our-insights/the-net-zero-transition-what-it-would-cost-what-it-could-bring.

LaLonde, R.J. 2007. *The Case for Wage Insurance.* CSR, no. 30. New York: Council on Foreign Relations.

Mahajan, M., O. Ashmoore, J. Rissman, R. Orvis, and A. Gopal. 2022. "Modeling the Inflation Reduction Act Using the Energy Policy Simulator." Energy Innovation.

Martin, J. 2022. "California Needs a Petroleum Phaseout Plan." Union of Concerned Scientists. August 24. https://blog.ucsusa.org/jeremy-martin/california-needs-a-petroleum-phaseout-plan.

Mayfield, E., J. Jenkins, E. Larson, and C. Greig. 2023. "Labor Pathways to Achieve Net-Zero Emissions in the United States by Mid-Century." *Energy Policy* 177(June):113516. https://doi.org/10.1016/j.enpol.2023.113516.

Meyer, B.D. 1995. "Lessons from the U.S. Unemployment Insurance Experiments." *Journal of Economic Literature* 33(1):91–131.

Möller, J. 2010. "The German Labor Market Response in the World Recession—De-Mystifying a Miracle." *Zeitschrift Für ArbeitsmarktForschung* 42(4):325–336. https://doi.org/10.1007/s12651-009-0026-6.

Muro, M., A. Tomer, R. Shivaram, and J.W. Kane. 2019. "Advancing Inclusion Through Clean Energy Jobs." Washington, DC: The Brookings Institution. https://www.brookings.edu/research/advancing-inclusion-through-clean-energy-jobs.

NAM (National Association of Manufacturers). 2021. "NAM Manufacturers' Outlook Survey Fourth Quarter 2021." https://www.nam.org/wp-content/uploads/2021/12/Manufacturers_Outlook_Survey_December_2021.pdf.

NASEM (National Academies of Sciences, Engineering, and Medicine). 2021. *Accelerating Decarbonization of the U.S. Energy System*. Washington, DC: The National Academies Press. https://doi.org/10.17226/25932.

NCSL (National Conference of State Legislatures). 2023. "Right-to-Work Resources." National Conference of State Legislatures. January 9. https://www.ncsl.org/labor-and-employment/right-to-work-resources.

NGA (National Governors Association). 2023. "Workforce Development in the IIJA, CHIPS and IRA." February 8. https://www.nga.org/publications/workforce-development-in-the-iija-chips-and-ira.

NRC (National Research Council). 2013. *Emerging Workforce Trends in the U.S. Energy and Mining Industries: A Call to Action*. Washington, DC: The National Academies Press. https://doi.org/10.17226/18250.

O'Leary, C.J., P.T. Decker, and S.A. Wandner. 2005. "Cost-Effectiveness of Targeted Reemployment Bonuses." *Journal of Human Resources* 40(1):270–279.

Oreopoulos, P., M. Page, and A.H. Stevens. 2008. "The Intergenerational Effects of Worker Displacement." *Journal of Labor Economics* 26(3):455–483. https://doi.org/10.1086/588493.

Pierce, J.R., and P.K. Schott. 2016. "The Surprisingly Swift Decline of US Manufacturing Employment." *American Economic Review* 106(7):1632–1662. https://doi.org/10.1257/aer.20131578.

Pierce, J.R., and P.K. Schott. 2020. "Trade Liberalization and Mortality: Evidence from US Counties." *American Economic Review: Insights* 2(1):47–64. https://doi.org/10.1257/aeri.20180396.

Pollin, R., C. Lala, and S. Chakraborty. 2022a. *Job Creation Estimates Through Proposed Inflation Reduction Act*. Political Economy Research Institute (PERI), University of Massachusetts Amherst. https://peri.umass.edu/publication/item/1633-job-creation-estimates-through-proposed-inflation-reduction-act.

Raimi, D. 2021. "Mapping the US Energy Economy to Inform Transition Planning." Resources for the Future. https://www.rff.org/publications/reports/mapping-the-us-energy-economy-to-inform-transition-planning.

Raimi, D., S. Carley, and D. Konisky. 2022. "Mapping County-Level Vulnerability to the Energy Transition in US Fossil Fuel Communities." *Scientific Reports* 12(1):15748. https://doi.org/10.1038/s41598-022-19927-6.

Richardson, J., and L. Anderson. 2021. *Supporting the Nation's Coal Workers and Communities in a Changing Energy Landscape*. Washington, DC: Union of Concerned Scientists. www.ucsusa.org/resources/support-coal-workers.

Righetti, T., T. Stoellinger, and R. Godby. 2021. "Adapting to Coal Plant Closures: A Framework for Understanding State Resistance to the Energy Transition." *Environmental Law* 51(4):957–990.

Ruckelshaus, C., and S. Leberstein. 2014. "Manufacturing Low Pay: Declining Wages in the Jobs That Built America's Middle Class." National Employment Law Project. https://www.nelp.org/wp-content/uploads/2015/03/Manufacturing-Low-Pay-Declining-Wages-Jobs-Built-Middle-Class.pdf.

Schanzenbach, D.W., and S. Turner. 2022. "Limited Supply and Lagging Enrollment: Production Technologies and Enrollment Changes at Community Colleges during the Pandemic." Cambridge, MA: National Bureau of Economic Research. https://doi.org/10.3386/w29639.

Scott, R.E., V. Wilson, J. Kandra, and D. Perez. 2022. "Botched Policy Responses to Globalization Have Decimated Manufacturing Employment with Often Overlooked Costs for Black, Brown, and Other Workers of Color." Washington, DC: Economic Policy Institute. https://www.epi.org/publication/botched-policy-responses-to-globalization.

SDSN (Sustainable Development Solutions Network). 2020. "America's Zero Carbon Action Plan." https://www.unsdsn.org/Zero-Carbon-Action-Plan.

Seleznow, E., and L. McCane. 2021. "It's Time to Close Apprenticeship's Long-Overlooked Equity Gaps." *The Hill*. August 30. https://thehill.com/blogs/congress-blog/labor/569939-its-time-to-close-apprenticeships-long-overlooked-equity-gaps.

Shivankumar, S., C. Wessner, and T. Howell. 2022. "Reshoring Semiconductor Manufacturing: Addressing the Workforce Challenge." Washington, DC: Center for Strategic and International Studies. https://csis-website-prod.s3.amazonaws.com/s3fs-public/publication/221006_Shivakumar_Reshoring_SemiconductorManufacturing.pdf?VersionId=_8r_xAqZZpAxyMFMMtgzuKjlWt.B18Od.

Shrestha, R., J. Neuberger, and D. Saha. 2022. *Federal Policy Building Blocks to Support a Just and Prosperous New Climate Economy in the United States*. World Resources Institute. https://www.wri.org/research/federal-policy-building-blocks-support-just-prosperous-new-climate-economy-united-states.

Sicurella, S. 2021. "America's Security May Depend on Critical Minerals. But Mine Workers Are Scarce." *NPR*. August 18. https://www.npr.org/2021/08/18/1022352668/americas-security-may-depend-on-critical-minerals-but-mine-workers-are-scarce.

Slattery, C. 2020. "Bidding for Firms: Subsidy Competition in the U.S." Working Paper, University of California, Berkeley. https://cailinslattery.com/research.

Society for Mining, Metallurgy, and Exploration. 2014. "Workforce Trends in the U.S. Mining Industry." https://www.smenet.org/What-We-Do/Technical-Briefings/Workforce-Trends-in-the-US-Mining-Industry.

Stefek, J., C. Constant, C. Clark, H. Tinnesand, C. Christol, and R. Baranowski. 2022. "U.S. Offshore Wind Workforce Assessment." https://doi.org/10.2172/1893828.

Stevens, A.H., and J. Schaller. 2009. "Short-Run Effects of Parental Job Loss on Children's Academic Achievement." Cambridge, MA: National Bureau of Economic Research. https://doi.org/10.3386/w15480.

Sullivan, D., and T. von Wachter. 2009. "Job Displacement and Mortality: An Analysis Using Administrative Data." *Quarterly Journal of Economics* 124(3):1265–1306. https://doi.org/10.1162/qjec.2009.124.3.1265.

Taylor, G.D. 2016. "Unmade in America: Industrial Flight and the Decline of Black Communities." Alliance for American Manufacturing. http://s3-us-west-2.amazonaws.com/aamweb/uploads/research-pdf/UnmadeInAmerica.pdf.

United Way Worldwide. 2012. *Financial Stability Focus Area: Family-Sustaining Employment*. Alexandria, VA: United Way Worldwide. https://unway.3cdn.net/077876d3896dda796a_hjm6btvnb.pdf.

U.S. Senate Committee on Commerce, Science, and Transportation. 2022. "The CHIPS Act of 2022: Section-by-Section Summary." https://www.commerce.senate.gov/services/files/592E23A5-B56F-48AE-B4C1-493822686BCB.

von Wachter, T. 2007. "The Effect of Economic Conditions on the Employment of Workers Nearing Retirement Age." *SSRN Electronic Journal*. https://doi.org/10.2139/ssrn.1294717.

von Wachter, T. 2011. "Testimony Before the Budget Committee 'Challenges for the U.S. Economic Recovery.'" http://www.econ.ucla.edu/tvwachter/testimony/Von_Wachter_Testimony_Before_Senate_Budget_Committee_2011.pdf.

von Wachter, T. 2021. "Long-Term Employment Effects from Job Losses During the COVID-19 Crisis? A Comparison to the Great Recession and Its Slow Recovery." *AEA Papers and Proceedings* 111(May):481–485. https://doi.org/10.1257/pandp.20211091.

von Wachter, T., J. Song, and J. Manchester. 2009. "Long-Term Earnings Losses Due to Mass Layoffs During the 1982 Recession: An Analysis Using U.S. Administrative Data from 1974 to 2004." Department of Economics, Columbia University. April. http://www.econ.ucla.edu/tvwachter/papers/mass_layoffs_1982.pdf.

von Wachter, T., J. Song, and J. Manchester. 2011. "Trends in Employment and Earnings of Allowed and Rejected Applicants to the Social Security Disability Insurance Program." *American Economic Review* 101(7):3308–3329. https://doi.org/10.1257/aer.101.7.3308.

Walter, K., T. Higgins, B. Bhattacharyya, M. Wall, and R. Cliffton. 2020. *Electric Vehicles Should Be a Win for American Workers*. Washington, DC: Center for American Progress. https://www.americanprogress.org/article/electric-vehicles-win-american-workers.

White House. 2022. "White House Task Force on Worker Organizing and Empowerment Report." https://www.whitehouse.gov/wp-content/uploads/2022/02/White-House-Task-Force-on-Worker-Organizing-and-Empowerment-Report.pdf.

Workforce Institute. 2022. "Is Stability in Sight? Surveying the Future State of Manufacturing." https://workforceinstitute.org/wp-content/uploads/2022-Manufacturing-Survey_final_rev3.pdf.

Yagan, D. 2019. "Employment Hysteresis from the Great Recession." *Journal of Political Economy* 127(5):2505–2558. https://doi.org/10.1086/701809.

Zabin, C., J.H.F. Hammerling, M.E. Scott, and B. Jones. 2014. "Workforce Issues and Energy Efficiency Programs: A Plan for California's Utilities." University of California, Berkeley, Center for Labor Research and Education. https://laborcenter.berkeley.edu/workforce-issues-and-energy-efficiency-programs-a-plan-for-californias-utilities.

Zabin, C., R. Auer, J.M. Cha, R. Collier, R. France, J. MacGillvary, H. Myers, J. Strecker, and S. Viscelli. 2020. "Putting California on the High Road: A Jobs and Climate Action Plan for 2030." University of California, Berkeley, Center for Labor Research and Education. https://laborcenter.berkeley.edu/putting-california-on-the-high-road-a-jobs-and-climate-action-plan-for-2030.

Zandi, M., and B. Yaros. 2021. "Macroeconomic Consequences of the Infrastructure Investment and Jobs Act and Build Back Better Framework." Moody's Analytics. https://www.moodysanalytics.com/-/media/article/2021/macroeconomic-consequences-of-the-infrastructure-investment-and-jobs-act-and-build-back-better-framework.pdf.

5

Public Engagement to Build a Strong Social Contract for Deep Decarbonization

ABSTRACT

One of the greatest threats to a successful transition to a net-zero economy is failing to mobilize the participation and support of the people who call the United States home. In every corner of the nation, decarbonization efforts will ask households to buy and use new technologies; businesses and workers to transform energy systems; and institutions in the public and private sector to collaboratively imagine, plan for, and invest in clean energy futures. Furthermore, diverse communities will be asked to assent to and support new policies, programs, and infrastructure construction, and to adapt to the resulting changes to society, the economy, and the environment. Without full participation in these intertwined and interdependent activities, the United States may fall short of implementing decarbonization at the pace, scale, depth, and universality necessary to achieve carbon neutrality by 2050.

Public engagement is a crucial element of the social contract necessary to sustain the political will for decarbonization. It is needed to prepare and marshal individuals and communities to act; deliver tangible and meaningful benefits to all; and acknowledge, mitigate, and compensate for the disruptions, risks, losses, and added burdens many will experience. To participate in decision-making, people will need new knowledge, capabilities, opportunities, and resources. Industries and governments will need new methods to meaningfully engage publics, new skilled professionals to do this work, and robust research and educational programs to guide their efforts. Getting this work done in the coming decades is a daunting human challenge, but it is just as crucial as developing and implementing the technologies needed for deep decarbonization. Meeting this challenge will entail continuous and robust public engagement opportunities offered by governments, the energy and electricity industries, and civil society.[1]

---

[1] Civil society is the composition of communities and organization not associated with government. Civil society organizations include schools, advocacy groups, churches, and cultural institutions (Ingram 2020), as well as labor unions and indigenous groups (Longley 2022).

Although many of the elements of recent policy initiatives create opportunities to engage and invest various publics in clean energy futures, the human challenge of decarbonization has received only a tiny fraction of the investment in federal and subnational policy and private action. Inadequate public engagement curtails opportunities to advance creative, place-based energy systems and their potential advantages for equitable decarbonization. Furthermore, public engagement literature shows that policies and reforms that reduce public engagement risk slowing transition processes. Without additional resources and determined strategies, current public engagement efforts risk exacerbating public resistance to the pace and scale of systemic change necessary for deep decarbonization. Table 5-2, at the end of the chapter, summarizes the committee's recommendations that appear in this chapter to support innovative public engagement in decarbonization.

INTRODUCTION

The people of the United States are essential contributors to and participants in the decarbonization of the U.S. energy system. Without their active involvement and support, the nation will not achieve the policy, technology, and societal changes necessary to fashion a carbon neutral economy by 2050. To engage the public well is to build a strong social contract for the whole-of-society commitment necessary for deep and rapid decarbonization of the U.S. energy system. Unfortunately, in many parts of the country, the opposite is occurring: a growing number of people are feeling left out of decisions that are affecting the communities, places, and landscapes where they live and work—decisions that they see as having little to no local benefit. The number of communities placing significant new restrictions on actions required to achieve deep decarbonization is growing rapidly, especially in the deployment of renewable energy (Aidun et al. 2022; Lopez and Levine 2022a,b; Zullo 2023). These restrictions reflect the fact that the United States is not just an abstract territory or population: it is a land of urban and rural places—with associated histories, communities, resources, and industries to which many people have considerable attachment and concern for risks from new kinds of energy projects.

Effective response to public engagement concerns requires that public- and private-sector institutions and civil society establish new ways and capabilities to draw people into processes, including for deliberating the pathways and specific actions needed to achieve carbon neutrality. Only through such innovations will the diverse members of the public at large feel able to *meaningfully* contribute to and see themselves as a part of the decarbonization project and the decisions that shape future U.S. energy systems and the associated societal and economic futures (Devine-Wright 2011).

(See Chapter 2 for the equity dimensions of strengthening meaningful public engagement in deep decarbonization.) Furthermore, the nation needs to simultaneously move forward with distributed-, community-, and utility-scale decarbonization projects that incorporate public engagement early and often. Without synergistic, innovative public engagement opportunities, the nation's ability to achieve deep decarbonization may be put at risk.

This chapter maps out the public engagement innovations required to facilitate a social contract for deep decarbonization, which go well beyond "social acceptance" of technology. Robust public engagement practices are necessary to involve people in the setting of transition goals for and the design and implementation of the energy system transition. The committee's first report introduced the joint goals of accelerating decarbonization and facilitating a just and equitable transition, goals that are often considered to be in conflict. The tension between pace and process poses many real challenges for those implementing energy transition policy. However, this chapter emphasizes that failure to prioritize justice, equity, and a multi-faceted and multi-scalar[2] approach to engagement will in fact slow decarbonization and highlights the ways to make meaningful engagement processes more effective.

The chapter begins with a summary of lessons and priority actions for public engagement innovation, followed by a brief assessment of progress toward the first report's goals and recommendations. The bulk of the chapter consists of four sections that review public engagement theory and practice that could substantially enhance the ability of policy and energy institutions to involve U.S. publics in deep decarbonization: (1) Strengthening Energy Democracy Through Inclusive Policy Dialogue; (2) Community Energy, Energy Sovereignty, and Collective Benefits; (3) Meaningful Engagement in Siting and Permitting; and (4) Building the Nation's Expertise in the Human Dimensions of Decarbonization. Box 5-1 summarizes key lessons from practice and scholarship that provide the framework for this chapter.

Status of Prior Committee Recommendations Related to Public Engagement

New decarbonization technologies and infrastructure programs will involve much of the U.S. public in extensive changes to the energy system. Furthermore, fairness and justice are essential to a net-zero energy future. The committee's first report addressed the scale, fairness, and justice aspects of decarbonization and identified an important

---

[2] Related to multiple scales, including individual, local, regional, and national.

BOX 5-1
CORE LESSONS ABOUT PUBLIC ENGAGEMENT

- *A coordinated and comprehensive transition to net zero will serve the public well.* The significance and complexity of decarbonization requires all levels of government, the energy sector, and the media to provide the public with accurate depictions of decarbonization progress, risks being addressed, and benefits being provided.
- *People value being consulted no matter what the outcome.* When processes are understood to be accessible, transparent, fair, and inclusive, actions to decarbonize the U.S. energy system will be more widely viewed as acceptable.
- *Trust can ultimately reduce time to achieve consensus about key decisions.* While it takes time and effort to build such trust, the perceived legitimacy of the public process depends on trust and the character of the relationships among stakeholders.
- *The development of new energy infrastructure is fundamentally a social process.* When planning and engagement occur early and often, attachment to places can be leveraged as a catalyst for technological processes.
- *Decarbonization processes will be slow in pace without appropriate public engagement opportunities.* While meaningful engagement does not guarantee consensus, its absence can increase opposition to project development.
- *Projects must deliver tangible or visible public benefits aligned with publicly identified priorities.* Projects that provide tangible or visible public benefits have a greater likelihood of securing support from communities.
- *It is essential to be better equipped to learn as the transition progresses.* This requires greater investment in knowledge of how transitions are affecting the public and investment in methods to estimate future impacts.

---

set of broad policy goals to support an equitable transition (see Appendix C). Many aspects of the first report's recommendations explicitly and implicitly called for inclusive public engagement:

- A White House–level Office of Equitable Energy Transitions;
- A National Transition Corporation to ensure coordination and funding for assistance to communities and regions;
- National laboratory support to subnational entities for planning and implementation of net-zero transition;
- Educational and training programs to train the net-zero workforce;
- Ten regional centers to manage socioeconomic dimensions of the net-zero transition;
- High-profile regional public dialogues and listening sessions to discuss decarbonization pathways and goals;

- Net-zero transition offices in each state capitol;
- Local community block grants for planning; and
- Opportunities to grow community-owned and planned energy systems.

The report also recommended that the energy industry follow best practices in stakeholder engagement and suggested ways to overcome barriers to participation facing disadvantaged populations.

Despite substantial changes in the federal policy landscape that heighten the importance of effective public engagement, growth in federal support of public engagement is not commensurate with the major energy investments made in recent legislation. Appendix I summarizes the committee's evaluation of the implementation of the first report's public engagement objectives in the Infrastructure Investment and Jobs Act (IIJA) (P.L. 117-58), the Inflation Reduction Act (IRA) (P.L. 117-169), and other relevant federal actions. In summary, the committee finds recent legislation falls short of what is needed to empower the public to effectively participate in deep decarbonization. The outlier is the objective to "invest in community block grants that support local transition planning, community-based action, and community-benefiting economic and technological change," which was codified in several sections in the IRA (e.g., Greenhouse Gas Reduction Fund [§60103], Climate Pollution Reduction Grants [§60114], Environmental and Climate Justice Block Grants [§60201], and Neighborhood Access and Equity Grants [§60501]).

The 2023 Policy Landscape: Innovations, Barriers, Opportunities, and Requirements

The current policy landscape of budgetary statutes and executive orders is limiting the reach of public engagement. The IIJA and the IRA authorize and appropriate essential funding for infrastructure deployment but only feature modest opportunities for engaging the public, primarily through funding requirements that distribute benefits via access to technologies and economic opportunities. Public engagement is a significant barrier to IIJA and IRA implementation in both areas of low and high readiness to capitalize on available funding. In parts of the country primed to capitalize on funding for projects, the increased activity is likely to generate new siting and permitting conflicts. In other areas of the country, shortfalls in human and organizational capacity and readiness to act will limit the impact of many of these laws' provisions. Gaps in engagement also create barriers at the local and community level and exacerbate equity concerns. For example, many of the provisions for technology adoption will primarily engage wealthy households and businesses, given the laws' reliance on subsidies. Executive Orders (EOs) 13985 and 14008 created parameters and strategies to advance equity and established supporting task forces, initiatives, and working groups to support equitable outcomes. However, as Chapter 2 points out, the administration's executive-level approach to equity and justice is not codified in law and faces significant implementation challenges. Furthermore, the scope of publics that must be effectively engaged in supporting decarbonization extends far beyond disadvantaged communities; public engagement needs to provide opportunities for every potential stakeholder, regardless of income status or region, to play a role in decision-making processes throughout the transition to a net-zero energy future.

Much of the implementation of the IRA and IIJA will ultimately be carried out by state and local governments and other subnational actors,[3] accentuating the gap between leaders and laggards. The continued politicization of climate action poses a major obstacle to the transfer of knowledge from states that are further along the transition to others that have made less progress (Gustafson et al. 2019). Deploying decarbonization in a fractured political landscape is an opportunity for innovative public engagement. However, this engagement will come in the form of support and opposition to aspects of the transitions, specifically the deployment and adoption of new technologies.[4] This is not an insignificant concern; decarbonization is a whole-of-the-nation challenge. To be successful, areas of the country with a Republican majority will play a substantial role in many facets of the transition to net zero. This is especially true for the siting of extensive infrastructure within communities and the adoption of electric vehicles (EVs) and heat pumps by households. Box 5-2 summarizes the opportunities made available by recent U.S. policy and the chapter recommendations associated with them.

It is critical for the White House and federal agencies to develop a comprehensive approach to public engagement that makes the U.S. public full partners in deep decarbonization. What is needed goes beyond simply adding public engagement as a requirement to federal grants and providing modest funding for community empowerment, as important as those efforts are. To address the challenges that threaten progress on the social contract needed for rapid deep decarbonization, the committee

---

[3] Subnational actors, also known as non-federal actors, are states, cities, corporations, philanthropic and religious organizations, and academic institutes (Cyrs and Elliot 2018), as well as regions, tribal nations, and civil society (Kok and Ludwig 2022). The role of subnational actors is further discussed in Chapter 13.

[4] Case in point: A group of Wyoming state legislators recently introduced a bill proposing the phase *out* of electric vehicles (Wyoming State Legislature 2023). Mounting, and increasingly coordinated, resistance to renewable energy deployment in response to aggressive state mandates is visible in numerous media reports (e.g., see Catenacci [2022]; Clifford [2022]; French [2023]; Gearino [2022]; Gelles [2022]; Rittman [2023]) and is the subject of increased scholarly attention (e.g., see Crawford et al. [2022]; Nilson [2022]).

BOX 5-2
SUMMARY OF RECOMMENDATIONS TO IMPROVE PUBLIC ENGAGEMENT

Executive-Level and Congressional Actions
- Convene a federally sponsored national public dialogue to engage all residents in a robust vision for decarbonization (Recommendation 5-1a).
- Enact legislation to facilitate development of geothermal, solar, or wind energy on public lands (Recommendation 5-3).
- Mandate a national public engagement workforce assessment (Recommendation 5-6a).
- Require best practices in meaningful engagement in federal environmental review practices (Recommendation 5-8).

Department of Energy (DOE) Initiatives
- Develop regional planning networks to convene inclusive multi-stakeholder dialogues around place-based decarbonization (Recommendation 5-1b).
- Develop an assessment-informed national workforce development program focused on public engagement professions and professional competencies (Recommendation 5-6b).
- Fund legal clinics at public institutions to provide technical assistance in collective and community benefit programs (Recommendation 5-6c).
- Develop place-based internships to deliver immediate capacity for local dialogue and planning efforts (Recommendation 5-6b).
- Support public engagement in DOE deployment strategies (Recommendation 5-6b).
- Convene a national working group on innovation in generation facility siting processes with input from state energy officials (Recommendation 5-5).
- Encourage rapid analysis and action plan to address public access and engagement challenges in decarbonization decision-making (Recommendation 5-1c).

Federal Research and Development and Capacity-Building Investments
- Integrate human dimensions research and graduate training into all clean energy technology research, innovation, and deployment programs (Recommendation 5-9a).
- Deploy an independent research program focused on the basic human and social sciences of energy (Recommendation 5-9b).
- Develop a network of regional, university-led research centers for energy transitions (Recommendation 5-9c).
- Deploy a national energy science, technology, engineering, and mathematics education network and program (Recommendation 5-10).

Civil Society Initiatives
- Pilot regional planning networks to test models and lay the groundwork for subsequent federal action with philanthropic support (Recommendation 5-7a).
- Develop collaborative regional land and resource use plans focused on renewable energy deployment opportunities (Recommendation 5-7b).

BOX 5-2 Continued

Priorities for Subnational Actors
- Encourage the development, implementation, assessment, and sharing of policy and practice that deliver local benefits (Recommendation 5-2).
- Ensure that renewable energy facilities contribute to public services and provide funding for economic diversification (Recommendation 5-3).
- Reform fiscal policy to increase direct local benefits for hosting renewable energy facilities (Recommendation 5-4).

---

offers a series of detailed recommendations to address gaps in the current approach to public engagement. These recommendations are organized around four areas of innovation that are essential to successfully decarbonize the U.S. energy system:

1. *Strengthening energy democracy through inclusive public dialogues*: What are the opportunities and challenges for expanding energy democracy?
2. *Community energy, energy sovereignty, and collective benefits*: How can the transition to carbon neutrality promote projects that meaningfully advance local stakeholders' goals that go beyond the rapid deployment of clean energy technologies?
3. *Meaningful engagement in siting and permitting*: How can design, siting, approval, and construction of new decarbonization infrastructures better engage publics in ways that manage conflict productively, meaningfully incorporate public input, and enhance trust and fairness?
4. *Building the nation's expertise in the human dimensions of decarbonization*: What is needed to ensure (a) that efforts to transition the U.S. energy system are robustly informed about how the transition impacts people and the roles people need to play in getting it done, and (b) that the public has the competency and literacy to be effective partners in deep decarbonization?

This chapter describes the need for ambitious, broad, and differentiated forms of public engagement linked to transition planning and implementation. Recent scholarship emphases the need to frame public engagement in energy system transformation in terms of "wider ecologies of multiple interrelating practices of . . . participation that are constitutive of, shape, and are shaped by energy systems" (Chilvers et al. 2018, p. 208). Public engagement policies and practices encompass a great deal of government and private-sector activity. Unfortunately, in far too many cases and places, publics desiring to engage in clean energy debates and decision-making still need to actively advocate for and sometimes push themselves into processes, rather than being invited in.

The following sections focus on the opportunities for engagement present, nascent, or absent in the nation's current climate and transition policy portfolio. The Creating Helpful Incentives to Produce Semiconductors and Science Act (CHIPS and Science Act) (P.L. 117-167, 2022) is also discussed as an exemplar of the kind of comprehensive research and development (R&D) policy initiative necessary to build the knowledge base for a national engagement strategy. While the policy recommendations focus primarily on federal actors, the committee also notes the important role of civil society and subnational entities.

Finding 5-1: Public engagement that considers the complexity of human dimensions of energy systems and their intersection with lives and livelihoods of people is critical to the success of the transition. Yet, the current national decarbonization policy portfolio lacks a comprehensive strategy and adequate workforce and resources for engaging the public to advance and maintain a social contract for deep decarbonization. There is potential for innovative public engagement to be developed and incorporated into a social contract to support the pace and scale of infrastructural investment and construction needed for the transition of the national energy system to net zero.

STRENGTHENING ENERGY DEMOCRACY THROUGH INCLUSIVE POLICY DIALOGUES

The Challenge

Energy democracy is the ability of democratic publics to meaningfully participate in governing U.S. energy systems. Efforts to enhance and expand energy democracy start from the recognition that "energy is inescapably political" (Nadesan et al. 2023, p. xxxvii) and, therefore, that decarbonization should be governed in a manner that is consistent with societies' broad commitments to democratic norms and principles. The idea that new ways of organizing energy systems could support the growth of democratic societies and be carried out in ways that would enhance democracy has a long history (e.g., see Lilienthal 1944). Recent analyses of energy democracy have highlighted the substantial power, scope, scale, and influence of energy systems in contemporary economies and societies (Miller 2022) and the growing efforts of activists and citizens to open and/or decentralize energy governance, decision-making, systems, and operations (Burke and Stephens 2018; Szulecki 2018).

Central to energy democracy is inclusive policy dialogue, supporting avenues for the public to inform, deliberate, and contribute to choices about future trajectories of energy systems. Inclusive policy dialogue encourages all members of the public, particularly those left out of policy discussions, to deliberate and help shape policy proposals and implementation (Forester 1999; McCoy and Scully 2002). Researchers have identified four elements of inclusive dialogue: participation, information, fair decision-making, and local context (Elmallah and Rand 2022). This is an admittedly tricky issue to operationalize. Democracy in the United States is subject to intense and divisive polarization, so it is risky to presume a set of shared norms and principles or the capacity to act on shared norms in constructive ways (Sides et al. 2022). By that logic, however, it is even more important to protect the integrity of energy deliberation processes through a deliberate commitment to the mechanisms described here.

Engagement mechanisms that catalyze equitable deep decarbonization address each of the elements of inclusive dialogue. Such procedures have been used in small deliberative groups of the general public and open sessions of e-governance. See Box 5-3 for an example of the challenges and opportunities associated with developing new settings for engagement in policy design. Mechanisms of public deliberation must focus on improving deliberative processes, as well as outcomes. "This means more inclusion and procedural integrity, increasing participants' knowledge and their commitment to democratic norms, and providing symbolic value as a means of legitimizing institutions forced to make difficult decisions" (Gastil 2018, p. 273). In the context of decarbonization, inclusive policy dialogue includes two-way, multi-sited,[5] and continuous engagements that connect policy to affected publics from the local to the regional to the national scale.

Change is under way in the energy and electricity sectors that aims to open governance and decision-making to broader and more inclusive public participation, especially regarding decarbonization. As the world reimagines and redesigns how it produces and consumes energy, many communities and organizations have seen the desirability of expanding efforts to engage different facets of the public. Globally, governments are also increasingly looking for new ways of public involvement in developing and deliberating the future of energy, using methods such as citizens' assemblies (Lacelle-Webster and Warren 2021) and community visioning (Trutnevyte et al. 2011), on scales from cities (Sandover et al. 2021) to countries (Devaney et al. 2020; Duvic-Paoli 2022; Shehabi and Al-Masri 2022).

Innovative inclusion can help deepen the impact of already-established best practices in industrial development, including strategic environmental assessment (SEA), a procedure to assess the environmental impacts of a program, policy, or plan. For example, the SEA process, conducted at national or regional scales, "acts in anticipation of

---

[5] Offered in diverse venues to accommodate the different capacities and constraints of participants.

BOX 5-3
CREATIVE TECHNOLOGY USE TO FACILITATE INCLUSIVE POLICY DEVELOPMENT

The U.S. House of Representatives Natural Resources Committee has recently experimented with holding inclusive public dialogues. In 2014, Representative Raúl Grijalva, chair of the House Natural Resources Committee, introduced what he thought was the perfect bill to address environmental justice concerns (Meeker 2021). However, the bill failed. In 2019, he and the late Representative Donald McEachin tried again with a new process that engaged environmental justice groups from around the nation before the new bill was drafted. The working group of environmental justice organizations was invited to join in-person and online convenings with congressional staffers to exchange expertise, experiences, and perspectives. Together, they identified guiding principles that were incorporated into the text of the committee's draft bill. This draft bill text was then shared through an online platform through which members of the working group could comment directly on the proposed bill text. The platform received more than 350 comments, which the committee incorporated into the text of H.R. 2021—Environmental Justice for All Act.[a]

This process has been generally well-received: the online platform was created by POPVOX, Inc., a private technology company, and was viewed as a non-partisan forum that offered transparency to the process and the data (Sobczyk 2020). Additionally, non-experts were able to access the platform to participate in the same forums as experts. Through the working group, the public learned about the decisions and trade-offs that policy makers must make, gaining understanding of democratic practice. Yet, the process is not without concerns, perhaps the most visible of which is that it was not initiated in a bipartisan manner—only Democratic Representatives and their staff participated in the forum.

Moving toward a just energy future will require bipartisan involvement in deliberative processes. While the online nature of the POPVOX forum allowed for participation from people across the country, technology is not without social dimensions that can act as obstacles to participation. For example, broadband is not evenly distributed across the country nor is recreational time evenly distributed across economic status. Last, scaling online forums would require balancing how to identify participants for working groups and the role of anonymity in certain processes. These factors, in addition to concerns about fraud and administrative burdens associated with high volumes of comments are under consideration in the context of regulatory rulemaking (ACUS 2021).

Other experiments are also being explored, more directly related to energy and decarbonization. For example, the Department of Energy is currently building a novel consent-based siting process for examining future potential nuclear waste repository sites in the United States (DOE 2022). This process has the foundation of earlier innovation in public consultation and participatory technology assessment (Richter et al. 2022). Continued innovation will support the diversity of public engagement opportunities that are available during the transition to a net-zero energy future.

---

[a] The bill was reintroduced as H.R. 1705—A. Donald McEachin Environmental Justice for All Act.

future problems, needs, or challenges and creates and examines alternatives leading to the preferred option" (Noble 2000, p. 210). This prospective, integrated approach is associated with "greater efficiency in resource use, shortened the duration of the project level assessment process and proactively contributed to achieving improved environmental practices" when compared to conventional project impact assessments (Fischer et al. 2020, p. 35). Additionally, SEAs can provide the clarification of the necessary policy reforms for industry deployment. For example, in Saskatchewan, Canada's SEA process produced a blueprint for coordinated institutional reforms necessary to enable a successful regional transition to renewable energy (Nwanekezie et al. 2022).

While SEAs do not eliminate controversy, they can mitigate against time lost to the contentious politics created by after-the-fact rulemaking—something already evident in tensions between state and local governments about renewable energy laws (Dawson 2023; Paullin 2022). For example, the relegation of key decisions about natural resources to state and local politics has created a highly uneven and uncertain regulatory landscape for shale gas developers to navigate (Rabe 2014). The risks associated with shale gas development—increased consumption of water, induced earthquakes, air quality impacts, and increased truck traffic, noise, and dust (DOE n.d.(c))—have created public controversy. In fact, the state of Texas eventually conducted a SEA, recognizing a need to "improve the broad understanding and awareness of the impacts of shale production" (TAMEST 2017, p. 15).

Generative dialogue, conversations that create and expand understanding through meaningful inquiry, is a key aspect of inclusive engagement. Examples in Arizona, Canada, the United Kingdom, and Australia illustrate applications of generative dialogue in different dimensions and settings of the energy transition. The 2011 Arizona Town Hall brought together more than 100 policy, business, civil society, and energy leaders to discuss strategies for advancing the state's energy future (Miller and Moore 2011). Other future-oriented initiatives in the state have solicited diverse stakeholder participation in creating scenarios of the future of solar energy (Miller et al. 2015), identifying potential economic pathways for decarbonization (Miller et al. 2022a), and imagining the impacts of future renewable energy development on urban and rural life (Eschrich and Miller 2019, 2021). In Canada, a national dialogue about the energy future in 2017 was attended by more than 380,000 people who identified public values and principles to guide efforts to design and build Canada's national energy future (Government of Canada 2017). This public engagement mechanism included multiple venues: in-person sessions, online comment submission forums, and polls and quizzes. Through this set of events, a 14-member "Generation Energy Council" collected input from citizens to be used to inform its recommendations to Canada's decision makers (Government of Canada 2018).

Generative dialogue can also take the form of citizens' climate assemblies, which incentivize representative, small groups of the public to participate in the policy-shaping process at national and local levels. Citizens' climate assemblies have been established in the United Kingdom[6] and have the potential to (Devine-Wright 2022):

- Provide upstream engagement to develop an understanding of concerns and values before projects are proposed;
- Enable net-zero policy to have broader legitimacy and better inclusion;
- Make information, including about national or state energy policy, more accessible;
- Bridge the gap between the national and local level;
- Give participants a chance to form their own informed views about a given technology through interactions with expert witnesses; and
- Identify which technologies are suitable for the county location.

Through all the above strategies, climate assemblies intend to generate socially acceptable plans for infrastructure development.

Elsewhere, there is support for deliberative dialogue about energy futures in regions experiencing widespread abandonment of fossil fuel facilities. In Australia, practitioners and academics—and in some cases, industry—support consideration of social impacts of and public perspectives on mine closures in addition to the policy focus on environmental rehabilitation in coal-dependent areas (Cameron and Gibson 2005; Measham et al. 2021). For example, AGL Energy Ltd., an Australian publicly traded utility, recently commissioned a study of community perspectives on reclamation options for three Latrobe Valley coal mines and surrounding lands. Community perspectives were "obtained through a series of focus groups with key stakeholders, including community organizations, environmental groups, government authorities, business groups, primary producers and Traditional Owners; and a web-based survey, completed by over 560 participants" (Reeves et al. 2022, p. 173). The resulting study generated a community-driven plan for further consultation about options for remediation to include "an iterative consultative or co-design process to capture values, share opportunities and address concerns" with "[t]he voices of youth and Traditional Owners . . . at the forefront" (Reeves et al. 2022, p. 184).

---

[6] See for example the UK Climate Assembly, consisting of 108 participants representative of the UK population, and the Devon Climate Assembly, made up of 70 participants representative of Devon County (Devine-Wright 2022; Devine-Wright and Moseley 2019).

Opportunities and Barriers in Current Policy

Federal Opportunities and Barriers

Recent legislative action does not provide formal direction and support for generative public dialogue activities, as shown in Appendix I. Instead, the most explicit federal commitments to inclusive dialogue on the energy transition can be found in executive directives and interagency initiatives. For example, the White House Environmental Justice Advisory Council (WHEJAC), established by EO 14008, brings together a council of experts who "have knowledge about or experience in environmental justice, climate change, disaster preparedness, racial inequity, among other areas of expertise" (EPA 2021). It encourages the experts to provide advice and input on policy development and implementation, including providing in-depth recommendations on key policy initiatives such as the Justice40 Initiative. Public comment at WHEJAC meetings demonstrates that the council is attracting and facilitating input from representatives of communities who have previously lacked meaningful access to federal policy conversations.[7] However, WHEJAC and other advisory groups or initiatives created through executive orders lack administrative support and adequate resources, undermining their efficacy (CEQ 2022; WHEJAC 2022a). In addition, the emphasis on environmental and climate justice, while critical, does not always encompass all energy transition questions and issues.

Several new offices have been established to facilitate federal support of public engagement in the energy transition. The U.S. Federal Energy Regulatory Commission (FERC) recently established a new Office of Public Participation to build capacity to facilitate public involvement in FERC processes. The goal of the office is to help the public better understand the institution and to reform agency rules and practices to ensure that the agency hears from the publics that it needs to hear from in order to make good decisions (FERC 2022). The Department of Energy (DOE) has also begun to include commitments to community engagement as an important criterion in reviewing federal energy R&D investments, although the ultimate efficacy of the resulting engagement practices remains to be determined (DOE-OCED n.d.). Recently, DOE introduced its new Office of Energy Justice Policy and Analysis, which will collaborate with members of minority and disadvantaged communities to evaluate policy impacts and administer programs that advance energy justice and equity (DOE 2023).

---

[7] For example, the May 11, 2022, meeting minutes summarize the updates and the public comment period, during which the public expressed concerns related to environmental justice in their communities. For more information, see WHEJAC (2022b).

The Interagency Working Group (IWG) on Coal and Power Plant Communities and Economic Revitalization is an exception to the gap in federal prioritization of accessible and continuous public dialogue on the energy transition. As further discussed in Chapter 12, the IWG is an explicit acknowledgment that maintaining a social contract in support of decarbonization demands engaging impacted communities by "[r]ecognizing the importance of meeting these communities where they are" in the energy transition (IWG 2023, p. 5).[8] The IWG focuses on economic and technical assistance to communities with high numbers of "workers directly employed in coal mining and power generation, and also the workers in related jobs in logistics and services, residents who are dependent on coal-related tax revenue" (IWG 2021, p. 1). However, the IWG was not designed to facilitate prospective policy dialogue and is currently limited to locations that host coal mines and coal-fired power plants. Regardless, this is a significant strategy that merits continued investment, financial and otherwise, from the federal government and is a good model for general and targeted engagement during the transition.

State and Regional Opportunities and Barriers

As introduced in Chapter 2, some states have enacted legislation that promotes community engagement to facilitate energy transition planning. These initiatives often focus on including historically excluded populations and centering their priorities in program and policy development. Following the passage of the Climate Leadership and Community Protection Act of 2019 (S. 6599, 2019–2020 Sen., Reg. Sess. §1), New York created a Climate Justice Working Group that includes representatives from environmental justice communities across the state to provide strategic advice to state policy makers regarding the economic, social, and environmental impacts of the transition (New York State 2022). As part of Washington state's Climate Commitment Act (2021), which incorporates just transition principles into utility and energy sector regulation, the Environmental Justice Council was established to provide formal

---

[8] The IWG recently released its year 2 report outlining the activities the working group and the Biden administration have undertaken since the signing of EO 14008. The report includes the progress made in terms of keeping the promises made in its first report to the President. Of note, the IWG oversaw a set of roundtables that discussed the funding opportunities made available by the new U.S. Economic Development Administration office, established to provide a foundation for durable regional economies. These actions allowed the working group to keep its promise to "launch a series of town hall meetings . . . to both listen to the concerns of key constituencies and identify federal resources communities could immediately access" (IWG 2023, p. 3). View the IWG report at https://energycommunities.gov/wp-content/uploads/2023/04/IWG-Two-Year-Report-to-the-President.pdf.

advice on the implementation of climate policies (Washington State Department of Ecology n.d.). The Climate Commitment Act also includes dedicated support for tribal participation in climate project planning.

In addition to legislation, some states have committed to engaging the public through collaborative, multi-scalar regional planning for the energy transition. California's Transformative Climate Communities program is discussed in detail in Chapter 2. Re-Imagine Appalachia, a coalition of civil society, elected officials, and activists, is focused on creating regional dialogue about how the energy transition can "boost economic opportunity and benefit working people" through deliberate policy choices (Brown 2021). Notably, the coalition regularly uses digital convenings to solicit input from diverse stakeholders about specific policies and update their "Blueprint" for the region's economic transition, which connects local priorities to broader policy opportunities and priorities (ReImagine Appalachia 2021). States and communities are more likely to access support from new federal programs when the programs have prioritized the network-building, visioning, and capacity necessary to ensure equitable and effective investments. Box 5-4 highlights an engagement practice from Canada's Participant Funding and Policy Dialogue program, which provides compensation for participation.

There are substantial opportunities for states, localities, and tribes to leverage the appropriated funding of the IIJA and the IRA for participatory and innovative planning and visioning for local and regional energy transitions. These opportunities include the $150 million Reconnecting Communities Pilot program (IIJA §11509), $16 billion to address legacy pollution (IIJA §40601, §40701), and $11 billion total funding available for community block grants (IRA §60114,[9] §60201, and §60501). Areas with high levels of existing capacity and bridging social capital[10] are expected to access and deploy these funds with ease. Conversely, lower-resourced and less-networked areas will struggle to access, let alone implement, the funds to support inclusive dialogues about the energy transition. It is critical to augment capacity gaps for the transition to include opportunities for generative dialogues (see Recommendation 2-4 in Chapter 2). Until these opportunities to take advantage of federal funding are translated into effective state, local, and community action, however, many critical aspects of state clean energy and climate development will remain inaccessible to many individuals and communities.

---

[9] At the time of writing, the Environmental Protection Agency (EPA) indicates this money will include non-competitive funds for planning, followed by competitive implementation grants (EPA 2023).

[10] A social network of individuals with different demographic characteristics.

BOX 5-4
CANADIAN MODEL FOR COMPENSATION FOR PUBLIC PARTICIPATION IN LOCAL CLIMATE ACTION

Stakeholders who would otherwise be excluded from participation owing to social and economic circumstances can be compensated for their time and expertise. Canada's Participant Funding and Policy Dialogue Programs provide travel support, stipends, and other resources for individuals and groups. This is done under the principle that enabling public participation means "assessments can be more open, balanced, credible and of higher quality" (Impact Assessment Agency of Canada 2022). An expert report on best practices in citizen engagement in local climate action planning provides this valuable insight on the role of compensation:

> The majority of the Citizens' Assemblies cited in this report have provided a small honorarium or "gift" to compensate participants for their time, usually as a monetary reward or sometimes as vouchers. One of the reasons behind this practice is a simple acknowledgement of the significant time and commitment involved, and because payment can help sustain participant involvement. Importantly, it helps to deliver inclusivity, by ensuring that people on low incomes can participate and are not deterred by the prospect of foregone earnings…. A further reason for providing payment is that without this, only those who are intrinsically motivated by the topic may volunteer, resulting in a sample is biased toward those with more pro-social or communitarian views or with stronger views on the topic at hand…. The flip side, however, is that some people may take part purely for the financial incentive and may therefore not be committed to the process. (Devine-Wright and Moseley 2019, p. 21)

The Participant Funding and Policy Dialogue programs and other proactive policies to advance public inclusion in climate planning provide an adaptable template for nascent U.S. efforts to meaningfully engage publics in decarbonization action (see Recommendation 5-5).

Utility Opportunities and Barriers

Electric utilities across the country are also developing new strategies for engaging with communities who are impacted by their decisions. Target communities include those where coal-fired power plants are closing (e.g., the Salt River Project's Coal Community Transition initiative [SRP n.d.]), low-income and minority communities that experience high energy burdens (e.g., Sacramento Municipal Utility District's Building Sustainable Communities program [SMUD n.d.]), and frontline communities grappling with the long-term challenges of pollution and other environmental risks (e.g., New York Power Authority's Environmental Justice program [NYPA n.d.]). However, the development of investor-owned utility regulation and ratemaking is overseen by state public utility commissions (PUCs) or public service commissions whose processes, authorities, and functions resemble those of courts (EPA 2010). As a result, participating in utility regulatory decision-making processes in most states is complex, expensive, technical, and may require representation by an attorney. These costs and barriers are prohibitive for community-based organizations and individuals. Some mechanisms exist to make the process more accessible, including through intervenor compensation rules, the establishment by some state legislatures of non-governmental state Citizens Utility Boards (CUBs) to advocate consumer interests and priorities in PUC settings (e.g., Minnesota's CUB [CUB Minnesota n.d.]), and grant programs to improve public participation (e.g., the California PUC's Equity Initiatives and Clean Energy Access Grant Program, currently under development [CPUC 2023]). In general, however, the effectiveness of these mechanisms is modest.

Findings and Recommendations

The examples and conceptual underpinnings of inclusive policy dialogue described above underscore the importance of the public contribution to energy transition dialogue and visioning. These approaches help to align public values and policy goals; build public understanding and awareness; incorporate community perspectives into policies and infrastructures; uncover potential roadblocks or policy gaps; allow communities to shape and design meaningful co-benefits; and coordinate across scale, region, and sector with multi-scalar planning activities. However, neither the IIJA nor IRA provide formal direction or support for generative dialogue.

Finding 5-2: The United States is failing to engage in sufficient public dialogue to facilitate the pace, scale, and equity ambitions of deep decarbonization by 2050. More determined and consistent prioritization of and support for regional planning is needed to compensate for the uneven levels of preexisting technical and social capacity and political will across the nation. Successful regional dialogues currently under way in metro, remote and tribal, and rural regions provide models and templates upon which to build.

Recommendation 5-1: *Encourage Prospective, Inclusive Dialogue at National and Regional Levels.* **The National Climate Task Force (NCTF), Department of Energy (DOE), and Environmental Protection Agency (EPA) should pursue multiple avenues to encourage prospective, inclusive dialogue at the national and regional levels.**

a. **The NCTF should convene a formal national public conversation on the energy transition using state-of-the-art public relations, communications, and engagement strategies to appeal to diverse sectors and social groups and to meaningfully draw them into the public discussion. It should prioritize involvement of groups often left out of energy planning activities such as rural populations, fenceline communities, workers in carbon-intensive industries, and the nation's youth.**

b. **DOE and EPA should establish regional systems planning networks and convene multi-stakeholder dialogues around place-based decarbonization strategies so that subnational actors and Indigenous nations can build the necessary capacity to take full advantage of the Infrastructure Investment and Jobs Act and the Inflation Reduction Act. Participatory planning efforts should aim to identify positive intersections among mitigation and energy service priorities, including economic development, public health, accessibility, and climate resilience.**

c. **DOE's Office of Energy Justice and Equity should direct a rapid analysis and detailed action plan to address public access and engagement challenges in state public utility and public service commission proceedings and other key sites of decision-making for decarbonization. Engagement of a multi-sectoral steering committee with representatives from public interest groups, civil society, the utility industry, and regulatory agencies will ensure the effort's credibility and impact.**

COMMUNITY ENERGY, ENERGY SOVEREIGNTY, AND COLLECTIVE BENEFITS

The Challenge

Extending energy democracy to the participation of individuals in small-scale energy systems and large-scale transition projects is key to the clean energy future. This includes increased opportunities for small groups or communities to own and operate energy processes and to directly benefit from decarbonization actions, including through community and tribal energy systems that provide collective benefits. For low-income communities, the benefits of clean energy development have the potential to permanently lower energy burdens, with proper policy support (Biswas et al. 2022). Similarly, localized renewable energy infrastructures are increasingly understood as opportunities for Indigenous nations to pursue self-determination and sovereignty, as well as economic development and resilience.

Community Energy Systems

There is a global explosion of interest in decentralized energy production, such as distributed solar energy, as an important element of democratizing involvement in and control of energy systems (Lotfi et al. 2020). While many end-use energy technologies have always been owned in a decentralized fashion (e.g., automobiles, furnaces, and electrical devices), in recent years, data show that U.S.-distributed solar generation has grown faster than utility-scale solar generation (EIA 2023).[11] Furthermore, a recent survey and interviews in New York state found that support for community or rooftop solar among rural residents is significantly higher than support for utility-scale solar (Nilson and Stedman 2022). The number of distributed solar systems is likely to continue growing nationally given the tax credits in the IRA and the potential for rooftop solar to reduce household energy bills.

Individual ownership is not the only mechanism for distributed solar systems. In the European Union, for example, energy cooperatives and neighborhood microgrids have emerged as an important strategy for enhanced public engagement and involvement in the clean energy transition (Inês et al. 2020; Lowitzsch et al. 2020). In the United States, DOE has set a target for the National Community Solar Partnership[12] of enabling community solar projects that power the equivalent of 5 million homes by 2025. Such investments can provide widespread benefits that help generate support for decarbonization by reducing long-term electricity costs and providing a means for using clean energy to strengthen other household goals (e.g., resilience to electricity grid outages). However, as noted elsewhere in this report (Chapters 2 and 6), equity is a major concern given the high up-front capital costs required for distributed energy systems and microgrids, and the challenge of ensuring affordability and reliability of system operations and maintenance. Chapter 2 describes policies to facilitate disadvantaged communities' participation in distributed and community-owned energy systems.

Tribal Energy Sovereignty

Efforts to leverage decarbonization and clean energy technologies to enhance Indigenous energy sovereignty have also emerged as an important focus of discussion, policy development, and investment (Atcitty 2021; Kinder 2021; Montoya 2022; Royster 2008; Schelly et al. 2020; Smith 2022a). This is not surprising given both the growing prevalence of distributed energy systems and the reassertion of sovereignty as a key priority for many Indigenous communities, both in the United States and around the globe (Rezaei and Dowlatabadi 2015). Tribally owned and operated community energy systems, when executed with attention to the feasibility of long-term operations and maintenance, can "multi-solve" for energy service access and climate resilience. For example, the Blue Lake Rancheria, a federally recognized tribal government

---

[11] For a study of the extent and dimensions of social preference for household and community-scale distributed solar in Puerto Rico, see Echevarria et al. (2023).

[12] See https://www.energy.gov/communitysolar/national-community-solar-partnership-targets.

and Native American community, provides an example of a successful community-led effort to exercise sovereignty and enhance climate resilience through energy systems. The community's microgrid is connected to the regional distribution system and is designed to operate autonomously—when a nearby fire in 2017 caused a grid outage, the microgrid was successfully islanded and facilities avoided a blackout. Energy savings to the Blue Lake Rancheria are estimated at nearly $200,000 annually (Carter et al. 2019). Other examples include the energy efficiency and renewable energy investments at the Navajo Nation (*Diné Bikéya* in Navajo) (Begay 2018a) and Citizen Potawatomi Nation (*Neshnabé*) (Begay 2018b).

In cases where Indigenous nations host energy infrastructure designed to export power (e.g., high-voltage transmission lines and utility-scale storage and generation facilities), there is new attention and interest in models favorable to development, in contrast to historic practices. In a landmark example in 2021, the Morongo Band of Mission Indians was approved as the first Native American tribe to be a participating transmission owner in a major system. Through the agreement with Southern California Edison (SCE), the Morongo Band secured a capital interest in the project and its returns and improved the terms of the lease, allowing access over tribally held territory (ICT News 2022). The Morongo–SCE agreement was highly complex and required many layers of regulatory approval, a barrier some have cited as one of many facing tribes that seek to use renewable energy for economic development (Zimmerman and Reames 2021). Similarly, the agreement between the Navajo Nation and Salt River Project surrounding the closure of the Navajo Generating Station coal-fired power plant and associated mine also gave the Navajo substantial access to transmission capability for future renewable energy development (Pyper 2019). The regulatory challenges associated with these agreements need to be addressed to support energy sovereignty within Indigenous nations.

Collective Benefits

Compensation and benefits schemes are critical aspects of engaging the public around large-scale energy infrastructure. According to research on social dimensions of facility siting, local stakeholders often view community or local benefits mechanisms—and the processes and negotiations associated with them—through lenses of trustworthiness and fairness. In this way, compensation emerges as an element of procedural, not just distributive, justice, which has a profound influence on the acceptability of proposed projects (Crawford et al. 2022; Hoen et al. 2019; Jørgensen et al. 2020; Knauf 2022; van Wijk et al. 2021). The rapid acceleration of renewable energy deployment is encouraging creativity in compensation models such as community benefit agreements, pooled payments to landowner collectives, and innovative state fiscal policy. These models demonstrate increased consideration of the importance of distributive justice to securing a social contract to site and host large-scale renewable energy facilities. Indeed, payments and other monetary benefits to individuals, communities, and governments *do* influence both the social acceptance and local impacts of energy developments—although public acceptance and local impacts are not always correlated in straightforward ways. However, collective payment schemes and community benefits agreements in renewable energy development are very novel tools with many potential legal issues yet to be identified and resolved (Fazio and Wallace 2017).

With vast areas capable of hosting utility-scale generation and interstate transmission lines, federal and state public lands and waters offer an important opportunity for the U.S. public to contribute to and even facilitate deployment of renewable energy (Springer and Daue 2020). While state and federal property are not taxable by local governments, their use for facility siting can generate public revenue in the form of lease and bonus payments, right of way rentals, and even generation taxes. Despite the apparent opportunities embedded in the nation's public land and water holdings as sites for renewable energy deployment, this estate will likely continue to be an underutilized decarbonization resource without necessary policy reforms.

Opportunities and Barriers in Current Policy

The current policy landscape creates new opportunities to engender public support for energy infrastructure through projects with clear local benefits that outweigh the costs. This could meaningfully comprise the growth of community-scale and energy sovereignty–focused development as well as creative public and collective benefits schemes. The primary vehicles in federal policy include the following IRA provisions listed in Table 5-1.

An additional boost for community benefits includes efforts by federal agencies to promote community benefits agreements as a new criterion in evaluating loan and grant application reviews. DOE is relying heavily on Community Benefits Plans[13] as a vehicle to meet requirements under the Justice40 Initiative, which applies to all IRA and IIJA funding opportunity announcements (FOAs) (DOE n.d.(a)). Likewise, the U.S. Department of Agriculture's New ERA program supports rural electric cooperatives

---

[13] Community Benefit Plans as defined by DOE are inclusive of community benefits agreements and include Collective Bargaining Agreements and other elements. The DOE Community Benefits Plan template can be downloaded at https://www.energy.gov/sites/default/files/2023-05/CommunityBenefitsPlan Template.docx.

TABLE 5-1 IRA Provisions Supporting Community Energy Development, Energy Sovereignty, and Collective Benefits

Provision(s)	Description
Investment and Production Tax Credits (§13101, §13102, §13103)	Provides certainty and reduces costs that have previously been prohibitive for community solar (Coalition for Community Solar Access 2022).
Investment Tax Credit and Energy Credit for Renewable Facilities Near Low-Income Communities (§13103)	Provides incentives for projects serving or located in qualified low- to moderate-income communities (Coalition for Community Solar Access 2022).
Tribal Energy Loan Guarantee Program (§50145)	Appropriates $75 million in loans for tribal investment in energy-related projects (White House 2023).
Greenhouse Gas Reduction Fund—Zero Emission Technologies Grant Program (§60103)	Appropriates $7 billion to enable low-income and disadvantaged communities to deploy or benefit from zero-emission technologies and other greenhouse gas emission reduction activities (White House 2023).
Environmental and Climate Justice Block Grants (§60201)	Appropriates $3 billion for community-led projects that address disproportionate harms related to pollution and climate change (White House 2023).

and the communities they serve to develop clean energy resources and workforce skills needed for the transition (USDA n.d.). Together, programs in the IRA create meaningful investment opportunities for communities and their advocates to design and fund new, decarbonized energy infrastructure that aligns with local priorities.

Community Energy Systems

Many households and communities face steep barriers to deploying renewable energy projects that generate significant community benefits. In the example of solar, barriers include both lower levels of home ownership and lower financial capacity to cover the high up-front costs of rooftop solar and/or batteries. However, where low- or zero-down solar opportunities are available (e.g., via leasing), the resulting arrangements generally provide significantly lower financial savings than owning the solar panels. Additionally, in cases of fraudulent or predatory behavior, these opportunities might end up costing households (Vogelsong 2022). Last, many low-income households do not own their homes or otherwise have control over what happens on their rooftops.

Community solar projects offer a potential strategy for addressing the barriers associated with rooftop solar, if designed well, but will need significant policy innovation to take off at a substantial scale (Chan et al. 2017; Grimley et al. 2022). These projects are often less expensive per watt than rooftop systems and stand-alone solar installations because they are larger in scale, involving lots of households or installations in the hundreds of kilowatts of capacity, and they do not require home ownership. When financed effectively, or granted to the community, community solar projects can deliver significant financial benefits. For example, the Canadian government has granted community ownership of solar projects to remote Indigenous communities (Government of Canada 2023). Through the ownership of energy projects, Indigenous communities have control over an energy project's planning and management, jobs, and profits (Institute for Human Rights and Business 2023). In some models, low-income households can pay for their participation over time through their savings, resulting in lower bills and part-ownership in the solar project.

Nonetheless, community solar remains a small portion of the nation's solar installations. Project-based collective benefits models are advanced and challenged by their relative flexibility and direct dependence on the capacity and will of non-regulatory actors. Community solar projects are enabled by law in fewer than half of U.S. states—and explicitly prohibited in others (DSIRE n.d.; ILSR n.d.). Even where allowed, either by law or voluntarily by utilities, community solar projects are often restricted to only one model. This limits the number and variety of communities where they can be applied and dramatically slows innovation in the sector. As the nation seeks to rapidly expand deployment of solar, especially in a future in which space for utility-scale projects is increasingly competitive, contested, and scarce, community solar projects offer a way to deploy solar and advance substantial equity goals. However, changes in federal, state, local, and utility policies are required to open opportunities for creative engagement and deployment of capital via diverse and heterogeneous community solar project models. This will be especially important to enable the historic investments anticipated in community-based solar in the IRA.

Tribal Energy Sovereignty

For tribal nations, the opportunities are also historic. The IIJA provides more than $13 billion in funding for tribal infrastructure, including $2 billion for the Tribal Broadband Connectivity Program, $200 million for climate adaptation and community relocation, and another $200 million to plug orphaned wells on tribal lands (White House 2022a). The IRA directs hundreds of millions in grants and an unprecedented $20 billion in allowable loan guarantees to support tribal climate resilience, access to clean electricity, and building electrification (i.e., §50145, §50122, §80001, §80002, §80003, and §80004). The IRA also includes elective pay and transferable credits that "allows entities with little or no tax liability—like tribes—to accelerate utilization of these credits," making renewable energy development on tribal lands "exponentially more beneficial" (Smith 2022b).

However, the IRA has been criticized for offering a "blanket solution that did not address the disparate needs of the hundreds of federally recognized tribes" (Smith 2022b) and for failing to "capture the nuances of community needs and concerns," particularly in the context of EPA and DOE funding opportunity outreach efforts (Brown 2023). Furthermore, out of the $550 million in flexible, formula-allocated funding in the Energy Efficiency and Conservation Block Grant program, DOE has encumbered $110 million for administrative and technical assistance. This comes at the expense of additional funding to organizations that need it. Critics state this will "likely do little to make the program better or easier to navigate" (Brown 2023). Moreover, "many funding opportunities require a project to be almost fully baked to be competitive" (Brown 2023)—a problem that also plagues recent federal funding programs for community solar initiatives.

In a June 2022 National Academies' information-gathering webinar, Alliance for Tribal Clean Energy[14] founder and chief executive officer Cheri Smith (2022a) noted that despite having 2 decades of experience in applying for DOE funding, she and her colleagues still need to hire someone to decipher the agency's FOAs. Three ways federal agencies can improve the grant application process are (1) reducing the amount of time needed to write grant applications; (2) standardizing the application process; and (3) giving potential funding recipients a seat at the table in the discussions leading up to the creation of the FOAs. Smith (2022a) has also noted that the majority of tribal communities will need to build capacity and technical expertise to make use of these funds. To this end, the Indigenous-led nonprofit is leveraging philanthropic and federal funding and Native experts so that tribes can build capacity to develop renewable electricity infrastructure on their homelands. See the section "Meaningful Engagement in Siting and Permitting" below for more information about utilizing funding to build community capacity.

Collective Benefits

One challenge for the deployment of utility-scale infrastructure is the variability in how public revenue policies approach renewable energy facilities (Hintz et al. 2021; Uebelhor

---

[14] Formerly the Indigenous Energy Initiative.

et al. 2021). For example, some state tax incentives for renewable energy infrastructure are less attractive to local governments than fossil fuel facilities (Haggerty and Haggerty 2015). This challenges public investment and equitable deployment of decarbonization infrastructure, hindering the progress of the energy transition. There are also situations in which states formulate fiscal policy in reactive, haphazard ways, creating confusion for developers as well as local governments (Hintz et al. 2021; Uebelhor et al. 2021). Furthermore, depending on location and jurisdiction, siting facilities on public and private land and water can be a highly complicated policy matter. In the case of local benefits that accrue via property and other taxes, multiple factors converge to affect local project "buy in": the quality of fiscal policy at the state level; the implementation of fiscal policy at the local level; and thoughtful spending decisions and associated communication by local officials (Haggerty et al. 2014; Mills et al. 2019).

In almost all cases, revenue opportunities for renewable energy are far smaller than they are for fossil fuels and mined minerals because—unlike for oil, gas, and hard rock minerals—there is no severance tax on renewable energy (Godby 2022). This does not help to engender public support for large-scale renewable energy facilities on public lands. A notable leader in addressing this policy gap is New Mexico's State Land Office, which created an Office of Renewable Energy with a mission to triple the amount of wind and solar energy generated on state trust land (Stewart 2022). Updates to auction and contractual mechanisms used by the office and investments in capacity to work with the renewable energy industry have enabled the use of state trust lands to make meaningful commitments to climate mitigation *and* diversify funds raised for beneficiaries (Stewart 2022). In contrast, Congress has yet to update federal land management guidelines to clarify key provisions regarding leasing and revenue programs, particularly for wind and solar. This regulatory gap hinders development and the delivery of public benefits from it.[15]

The policy space surrounding public revenue from private land is complicated. Regulations are underdeveloped, with many states scrambling to draft revenue policy in parallel to emerging renewable energy development. Key challenges for public revenue from renewable energy facilities on private land include depreciation and, as in the case of public land, the absence of a severance tax. The major form of public revenue

---

[15] For DOE-related lands specifically, the National Academies' Committee on Energy Resource Potential for DOE Lands conducted an inventory of the energy development potential of lands, including (1) an analysis of all oil, gas, coal, solar, wind, geothermal, and other renewable resources on the lands; and (2) an analysis of the environmental impacts associated with future development, such as mitigation actions for negative impacts. Of note, the committee recommended DOE place a higher priority on developing an inventory of lands that can be leased or sold for energy development (NASEM 2017). For more information, see https://doi.org/10.17226/24825.

comes from property taxes, from which industry advocates frequently succeed in winning relief in terms of incentives offered by state and local governments (Haggerty et al. 2014). However, scholars warn against using revenue for local tax relief (Mills 2022) because, despite being popular with voters, using new revenue to decrease local property taxes creates problems when that revenue declines—as it will in any fiscal policy regime with no counterbalance to depreciation. Elected officials must, therefore, use revenue in ways that demonstrate meaningful and sustainable value to the public.

Findings and Recommendations

Participation in energy systems offers important opportunities to engage in decarbonization by providing an economic stake in the net-zero future to more groups. The ability for individual communities to leverage the programs created in recent legislation depends on many factors, including capacity and institutional and policy environments. Sharing successful approaches will aid in the development of community energy systems, tribal energy sovereignty, and collective benefit models that support decarbonization and communities. Chapters 2, 6, and 13 offer additional insights and recommendations about building capacity for implementation. Additional opportunities to build practical expertise with community and collective benefits are noted below (see Recommendation 5-4).

> Finding 5-3: Community-scale, community-designed, and community-owned energy infrastructure can be more readily acceptable than large-scale industrial projects. The current federal policy environment encourages the expansion of community-driven energy infrastructure in places that are "ready to act" with appropriate regulations, political will, and planning capacity. Localities unable to leverage these necessary capacities will miss this historic opportunity.

> Finding 5-4: A lack of adequate expertise and institutional capacity hinders the diffusion and successful application of processes that can facilitate renewable energy development and provide collective benefits, including Community Benefits Agreements; collective leases and payments; and federal, state, and local revenue policies.

> **Recommendation 5-2:** *Accelerate the Development, Implementation, Assessment, and Sharing of Energy System Policy and Approaches That Deliver Local Benefits from Decarbonization Investments.* **State, tribal, and local governments should work in coordination with their representative coalitions and federal partners to accelerate the development, implementation, assessment, and sharing of policy and practical approaches that focus on delivering**

local benefits from energy system decarbonization investments. These benefits can include local ownership, good neighbor and collective lease payments, and community benefit agreements. Furthermore, states should review, assess, identify, and address conflicts in state fiscal policy that result in suppressing the potential for renewable energy facilities to create local benefits in the form of public revenues.

Finding 5-5: Despite the apparent opportunities embedded in the United States' vast public land and water holdings as sites for renewable energy deployment, these locations are and will continue to be underutilized as a resource in decarbonization. Among the necessary reforms, there are significant opportunities in state and federal law to improve public benefits associated with revenue payments from renewable energy facilities.

Recommendation 5-3: *Fix Policy Gaps That Limit Role of Public Land in Decarbonization.* **Congress and state legislatures should enact laws to expand the role of public land in decarbonization to facilitate long-term value creation and economic diversification.**

a. **Congress should encourage geothermal, solar, and wind energy development on public lands by establishing priority areas for development, developing associated conservation and mitigation provisions, and providing clarity about the amount and disposition of revenues from geothermal, solar, and wind development. A long-term "legacy fund" offers a preferred model for saving revenue from public land leasing for renewable energy development.**
b. **State legislatures should consider the example of New Mexico's State Land Office and reform public policies governing the use of state-owned property to enable long-term, sustainable public revenue from renewable energy.**

Finding 5-6: State-level policies often suppress the potential for renewable energy facilities to create direct local benefits in the form of public revenues. The mechanisms include aggressive tax rebates for certain types of energy, which often result in fossil fuel facilities being more lucrative than renewable energy projects, as well as less well-known limits on budget and expenditure discretion for local governments.

Recommendation 5-4: *Address Barriers to Local Benefits from Renewable Energy Facilities.* **States should review, identify, and address conflicts in state**

fiscal policy that result in suppressing local benefit for hosting renewable energy facilities. By strengthening the relationships between decarbonization and direct public benefits, fiscal policy reform has the potential to grow social acceptance for renewable energy facilities.

MEANINGFUL ENGAGEMENT IN SITING AND PERMITTING

The Challenge

Innovation in public engagement to ensure distributive and procedural justice will be essential to the deployment of *all* deep decarbonization infrastructure across the full diversity and heterogeneity of communities and landscapes. While the factors influencing the social acceptance of energy infrastructures are multi-faceted (Boudet 2019), the local public processes to develop and execute infrastructure projects are a key venue for forming social acceptance of and included support for accelerated decarbonization. The relationship between social acceptance of renewable energy infrastructure and compensation schemes is about perceptions of procedural justice, trust and communication, and the level of compensation being provided. Policy and practices that encourage projects to provide clear and meaningful benefits to local stakeholders are critical to accelerating decarbonization.

Scholars have been studying the relationship between siting policy and practice and social acceptance of energy projects since the emergence of a strong anti-nuclear movement in the 1970s (e.g., see Freudenburg 1986). They continue to produce novel and important findings in the context of new energy technologies as well as new research questions and approaches (Batel 2020; Bessette and Crawford 2022; Krupnik et al. 2022; van de Grift and Cuppen 2022). Research demonstrates that the character and quality of the process of engaging the public in the context of siting and permitting projects will affect the pace and scale of decarbonization.[16] Taken as a whole, this literature underscores that there are no perfect solutions for public engagement to deliver speedy and conflict-free industrial siting decisions in an open democratic society. In addition, even the most creative and robust public engagement is unlikely to sway ardent opponents of projects. On the other hand, shortcutting public engagement can lead to far longer delays owing to the risk of driving alienated publics to courts, alternative policy forums, and other forms of protest.

---

[16] See Chapter 6 for a discussion of how models project the impact of siting and project development.

In that context, scholars and practitioners point to key features of effective siting and permitting process that have the potential to reduce conflict and delay:

1. Public engagement in the context of project development requires inclusive, expansive, and immersive communication. This means that communication between project representatives, contractors, government officials, and local and public stakeholders is conducted in multiple languages and in diverse and accessible formats; begins early in the process and features continuous updates of project progress with easily accessible archives of past discussions; and utilizes both low- and high-tech strategies to help different groups visualize and guide changes to the natural and built environment.

2. Public engagement professionals representing developers and permitting agencies should treat local perspectives as constructive expertise in project design and give local communities the opportunity to participate in shaping the process and outcomes of important design decisions (Devine-Wright 2022; Goedkoop and Devine-Wright 2016; Sherren 2021). A corollary priority is supporting communities in the development of local and regional visions for land use and economic development when such plans are absent or neglected—and doing this *prior* to discussion of facility siting whenever possible. In this manner, the siting discussion can build on and incorporate local visions rather than the other way around.

3. Public engagement needs to be customized to unique regions, demographics, politics, economics, and social values. To every extent possible, flexibility in public engagement processes must be a priority for permitting practitioners to align with local circumstances. Clustering review processes for projects in the same geography also has merit for equitable, rapid, and intensive infrastructure deployment that acknowledges the risk of consultation fatigue (Bice 2020; Noble 2017). Zoned permitting is also noted to facilitate effective environmental impact assessment (Faconti 2013).

4. Public engagement should emphasize clarity, transparency, and accountability in all activities, particularly in the terms and conditions of engagement. That is, participants should know and see when and how their input is used through clear and accessible information with time for discussion about the implications of the findings. Every effort must be made to provide opportunities for deliberative social learning about the credibility and accuracy of estimates of how projects will affect quality of life, public health, local environments, and economics to build trust and confidence in the data used to assess siting proposals. Participatory impact assessments lead to better project design and can strengthen perceptions of procedural justice.

Stakeholders who would otherwise be excluded from participation must be prioritized for engagement as an equity measure. For example, WHEJAC, discussed above, convenes environmental and climate justice experts together to provide advice and input on policy development and implementation. Additionally, Indigenous Knowledge has recently been elevated in federal policy making. The White House Office of Science and Technology Policy and Council on Environmental Quality (CEQ) have begun to institute this practice via statements, implementation and guidance memos, and establishment of the Subcommittee on Indigenous Knowledge (OSTP and CEQ 2021, 2022a,b). Included were strategies to grow and maintain relationships to support Indigenous Knowledge, and practices and opportunities to apply Indigenous Knowledge in federal processes, including the National Environmental Policy Act (NEPA). These initiatives advance the inclusion of Indigenous people and their knowledge in impact assessment and siting processes. Additional initiatives that support innovative forms of engagement offered, including those discussed in Boxes 5-3 and 5-4 above, increase opportunities for meaningful engagement with key stakeholders.

Rapid and expansive landscape changes driven by the amount of new industrialization necessary for decarbonization will meet resistance from local and otherwise place-invested publics for a variety of complex reasons (Boudet 2019; Fergen et al. 2021; Nilson 2022; Sherren 2021). A 2023 public opinion poll (not peer reviewed) found that when biodiversity and land conservation is posed as a trade-off with rapid emissions reductions, a majority of Americans prefer a slower buildout (Meyer 2023). The growing frequency of newspaper stories about public resistance to renewable energy projects in many parts of the country suggests resistance is likely to strengthen and calcify in key landscapes as the pace and scale of development accelerates (e.g., see Roth 2023; Saul et al. 2022; Stang 2022). Continued conflicts over the appropriate use of high-value farmland and rangeland, ecosystem values, the disruption of scenic and cultural amenities, economic uses of land, and individual private property rights are to be anticipated. This is especially true in the absence of robust public engagement efforts that seek to understand local sources of resistance and local input into the design of preferred and acceptable deployment strategies.

Emerging technologies and the associated industrial infrastructure are particularly likely to meet public skepticism (Nielsen et al. 2022) as well as outright resistance from those parties with the least trust in the energy sector. For example, the environmental justice community continues to express concerns about carbon capture deployment[17] (Anchondo 2022). Where the electric grid meets the built environment in key shared

---

[17] See Chapters 2 and 3 and Appendix E for more information about the environmental justice concerns and health risks associated with carbon capture investments and other decarbonization technologies.

elements (e.g., electricity distribution lines, distributed generation, and EV charging), the pace and intensity of infrastructure additions may result in unacceptable or undesirable conditions. By extension, strong local resistance and/or inequitable outcomes may develop. A difficult feature of the contemporary environment for renewable energy deployment is exacerbation of conflicts through the rapid spread of misinformation and uncertainty via social media (Fergen et al. 2021). This emphasizes the importance of proactive and generative public dialogues prior to and during project development and of authentic and reliable investments in building interpersonal relationships and trust. Beyond known best practices, there is a pressing need to accelerate and expand social science research about how to build trust in the context of contentious decisions. (See the section "Building the Nation's Expertise in the Human Dimensions of Decarbonization" below.)

Opportunities and Barriers in Current Policy

Calls for more robust and innovative public engagement found in the social science literature on renewable energy project development seem at odds with widespread concern in public policy circles about the need to reduce permitting barriers through major policy reform. Legal scholars find that permitting processes for large-scale infrastructure are made burdensome by a lack of interjurisdictional alignment, the ensuing redundant and circular processes of both public participation and detailed environmental review, and their vulnerability to litigation by project opponents (Gerrard 2017; Ruhl and Salzman 2020). However, streamlining permitting in ways that shortcut public engagement is not a "silver bullet," and calls for permitting reform need to be weighed against scholarship and expert commentary about where the problem really lies. Permitting professionals in many levels of government emphasize that it is not permitting regulations but understaffing and resource shortages that hinders the efficiency of permitting processes (Robinson 2022; Roth 2023). The dominance of decision frameworks that focus on a single measure (i.e., cost) also impede effective national and state siting decisions by minimizing the scope of review in ways that exclude meaningful public input (Kurth et al. 2017).

If permitting reform includes significant reductions in meaningful opportunities for and forms of public engagement, then such reform would create a real risk of slowing, rather than hastening, the process of building out a net-zero infrastructure. Policy makers must simultaneously consider eliminating redundant and conflicting permitting policies and practice robust and creative engagement in project development and permitting. Whether public engagement innovation is mandated by statute or implemented as agency or private-sector priority, its efficacy will depend in large part on available resources. These resources include subject-matter expertise and the capacity of participating parties, including project developers, public-sector regulatory bodies, and local and broader publics and civil society. An effective public engagement workforce for decarbonization includes public and community engagement professionals from utilities; community-advocacy groups; tribes; clean energy demonstration projects; local, state, and federal agencies; and other relevant organizations and programs.

Federal Actions

The IRA directs funds to improve environmental review processes in multiple agency budgets, namely: $40 million for EPA to invest in more accurate and timely environmental reviews (§60115); $30 million for CEQ to improve stakeholder and community engagement (§60402); $100 million for the Federal Highway Administration to develop review documents and a process that provides for a timelier environmental review process (§60505); $350 million to accelerate and streamline the environmental review process (§70007); and nearly $500 million for the implementation of the NEPA to properly review proposed infrastructure projects (§23001, §40003, §50301, §50302, §50303). Furthermore, two important initiatives from the IRA require the incorporation of innovative public processes into siting procedures:

- *Grants to Facilitate the Siting of Interstate Electricity Transmission Lines* (§50152)—$760 million in grants for state and local governments for purposes including transmission project studies, examination of alternative siting corridors, hosting negotiations with project backers and opponents, participating in federal and state regulatory proceedings, and promoting economic development in affected communities.
- *Interregional and Offshore Wind Electricity Transmission Planning, Modeling, and Analysis* (§50153)—$100 million for expenses for convening stakeholders and conducting analysis related to interregional transmission development and development of transmission for offshore wind energy.

Because of how broadly these two sections of the IRA are written, there is potential to support creative public processes, which could be used as pilots or test cases for innovation.

Staffing and resources for environmental permitting and reviews remain inadequate. For many agencies, the additional funding for environmental permitting and reviews was only sufficient to address staffing losses that occurred under the previous administration (Gordon 2022). At the same time, the IRA and IIJA will create an enormous volume of *new* permitting and public engagement work; hence, simply returning to a previous baseline is not adequate to the task. And the resource shortage extends well beyond the federal government. As mentioned above, the federal agencies are promoting community benefits agreements as a new element in their loan and grant application reviews. Community benefits agreements must be developed using state-of-the-art engagement practices that build confidence, equity, and transparency. They also require that local governments and community-based organizations have access to legal expertise. Capacity and access falls deeply short in many companies, states, cities, and communities.

Through the NEPA, federal agencies are required to provide opportunities for meaningful public participation. CEQ has developed documents guiding individuals through engagement processes (e.g., see CEQ and DOE n.d.) and providing clarity to federal agencies about compliance (e.g., see DOE-ONPC n.d.). Recent amendments to the NEPA included in the Fiscal Responsibility Act of 2023 (FRA 2023) (P.L. 118-5) contain the requirement that one federal agency coordinate with participating agencies in the development of a single NEPA document (Diller et al. 2023).[18] Additionally, FRA 2023 allows project sponsors to prepare an Environmental Assessment (EA) or Environmental Impact Statement (EIS) with lead agencies providing guidance. Given the focus on a lead agency status for complex EIS processes and the provision that developers can develop their own EISs, the need for public engagement workforce expertise to facilitate decarbonization is likely widespread.

Non-Federal Actions

Non-governmental organizations (NGOs), including grassroots organizations and national-level nonprofits, play an essential role liaising between the federal government and specific communities, especially communities that do not have the existing capacity to apply for or appropriately utilize available funding. Engagement with civil society leaders can produce decarbonization strategies that represent the priorities and concerns of communities. For example, the Union of Concerned Scientists convened an advisory committee to develop a holistic framework for decarbonization

---

[18] The FRA 2023 is the federal agreement to suspend the debt ceiling, but the legislation impacted multiple future federal actions, including the processes associated with the NEPA. In addition to the changes mentioned above, the FRA 2023 allows federal agencies to adopt categorial exclusions, categories of projects that do not need an EA and EIA—meaning federal agencies will be able to determine which projects do not have a significant impact on the environment without seeking public input on this categorization. This removal of a public engagement opportunity has the potential of having adverse impacts on the energy transition. For more information about the changes to the NEPA included in the FRA, see Diller et al. (2023).

that is equitable and just. The advisory committee identified three core principles for holistic approaches to a transformative energy transition: effectively address the impacts of the climate crisis; advance equity and justice; and drive systemic change (Baek et al. 2021). Policy recommendations from NGOs about decarbonizing the energy transition need to be reviewed and considered by policy makers at the federal, state, and local levels. The key feature of many reports produced by NGOs is a platform that brings together stakeholders to discuss transition pathways, what challenges may arise, and how to avoid or mitigate adverse outcomes.

In addition to developing policy frameworks for state and national government, some non-federal actors are convening cross-sectoral stakeholders to develop local decarbonization strategies. For example, the Southwest Pennsylvania Decarbonization (SWPD) Forum gathers to discuss critical opportunities and challenges of regional decarbonization in 10 counties within the state. These opportunities and challenges include creating jobs and driving economic growth; developing a healthy public and environment; supporting thriving and engaged communities; and facilitating innovation in technologies and infrastructure. The convening activities of the SWPD Forum are hosted by the Pennsylvania Environmental Counsel, which aims to be a model for implementing collaborative solutions (PEC n.d.), and the Allegheny Conference on Community Development, which brings together Pittsburgh's public- and private-sector leaders to define and mobilize regional and action (ACCD n.d.). Funding for the SWPD Forum comes from the Henry L. Hillman Foundation, whose goals include funding innovative solutions that address community needs (Henry L. Hillman Foundation n.d.).

As further discussed in Chapter 2, non-federal actors are critical to the development of multi-sectoral partnerships that connect local and state action with broader federal funding and policy. Furthermore, these organizations can provide independent information about decarbonization and its trade-offs to protect the public from potential misinformation. These groups benefit from consistent, multi-year funding, and where federal funding is absent or insufficient, philanthropic foundations can provide needed support.

Findings and Recommendations

Capacity is not only a matter of having the personnel and know-how to implement state-of-the-art permitting processes or streamlining permitting for priority initiatives. Thus, enhanced permitting capacity will depend on an effort to integrate research, practice, and policy activities, and to coordinate across scales of government and within and across economic sectors—for example, the international and nationwide coordination directed to the COVID-19 public health crisis demonstrated the needed urgency and dedication (Patnaik et al. 2023). Robust community and stakeholder engagement practices need those knowledgeable about diverse social science methods of community engagement and existing inequalities and policy performance in energy equity collaborating with experts in law and public administration.

Finding 5-7: The resources currently dedicated to building and strengthening public-sector capacity for permitting and environmental review at the federal, state, and local levels are not adequate to address public resistance that may well occur in the face of the extensive infrastructure deployment anticipated. Altogether, friction in the public permitting arena has the potential to delay emissions mitigation and equity goals significantly.

Finding 5-8: The United States currently lacks a sufficiently large or well-trained professional workforce to implement the full scope of public engagement activities that public-sector, private-sector, and civil society organizations will need to undertake to achieve deep decarbonization. This is especially true for permitting and siting processes and for hosting inclusive policy dialogues and developing robust strategies for ensuring a broad and impactful distribution of benefits from deep decarbonization for households and communities. It will be critical to use available funding to develop and implement new, creative precedents and practices to support the workforce needed for public engagement activities. Furthermore, public engagement professionals are essential to the success of the transition and need to be included systematically in federal energy workforce development planning and funding.

Recommendation 5-5: *Convene a National Working Group on Siting Process Innovation with Input from State Energy Officials.* **The Department of Energy and Council on Environmental Quality, with participation from the Federal Energy Regulatory Commission, National Association of Regulatory Utility Commissioners, and National Association of State Energy Officials as appropriate, should collaborate to convene a national working group on siting process innovation. The role of this working group will be to develop innovative public engagement practices for electricity generation and transmission facility siting processes. These practices could be modeled on the International Energy Agency working groups and Canada's Impact Agency public policies dialogues. It will be critical to incorporate adaptive management into the design of these public engagement practices to ensure that insufficient processes are removed or revised. The National Working Group on Siting Process**

Innovation should provide recommendations that can inform the allocation of resources for a national public engagement workforce assessment.

Recommendation 5-6: *Mandate and Allocate Resources for a National Assessment on the Public Engagement Workforce and Gaps.* Congress should mandate and allocate resources for an interagency national assessment and subsequent Department of Energy (DOE) initiative focused on capacity gaps in the public engagement workforce.

a. Congress should mandate a workforce assessment to be overseen by the National Climate Task Force (NCTF) with participation from academic experts, industry leaders, and public-sector representatives. The assessment should focus on future workforce needs in public processes for clean energy deployment and community advocacy organizations planning and impact assessment, including health, social, economic, and environmental impacts, with particular attention to the needs of utilities and large-scale energy developers in public engagement expertise. The assessment should also include the public engagement implications of recent amendments to the National Environmental Policy Act with a focus on where workforce investments are most critical. NCTF should make recommendations for training programs to grow this workforce via multiple post-secondary pathways, with a focus on enabling current engagement professionals and students to train for and participate in clean energy deployment as quickly as possible.

b. Through appropriations, Congress should direct DOE to establish an agency-wide workforce development initiative for public engagement in the energy transition, informed by the findings from the workforce assessment (5-6a). The Regional Clean Hydrogen Hubs and the Regional Direct Air Capture Hubs may provide prime opportunities for pilot public engagement workforce initiatives that incorporate participants into its existing clean energy workforce development programs and demonstration projects. The purpose of the initiative would be to advance community-led energy and environmental justice initiatives, lead planning and organizational and cultural change for deep decarbonization, and help regions navigate the human complexities of clean energy transitions. Furthermore, workforce initiatives should include opportunities for place-based internships to deliver capacity for planning and federal program access in under-resourced areas, potentially using DOE's Oak Ridge Institute for Science and Education fellowship program and utilizing the AmeriCorps model. These internships would generate non-technical career opportunities that address the climate crisis for young professionals.

c. **DOE should fund legal clinics at public institutions to provide technical assistance for Community Benefit Agreement and other collective benefits negotiations. This would advance community-level engagement in decarbonization by providing equitable access to programs providing local benefits.**

Finding 5-9: The limited number of dedicated efforts to promote deployment by credible multi- and cross-sectoral partnerships—for example, between environmental NGOs, industry, finance, and government—is another notable capacity gap that is creating friction for clean energy deployment and openings for misinformation and disinformation.

Recommendation 5-7: *Develop Collaborative Regional Renewable Energy Deployment Plans.* **Civil society leaders should use available public and private resources to develop collaborative regional deployment plans for renewable energy.**

a. **The philanthropic sector should immediately support the establishment of a set of pilot regional planning efforts, each focused on single renewable energy technology and other relevant social and economic choices along the region's path to net-zero emissions. The efforts should model robust, sustained, creative engagement and discourse around regional energy futures that include dimensions of the energy transition most salient to local stakeholders and publics.**

b. **Civil society leaders should work with industry and government to determine the best use of available land and resources for renewable energy deployment opportunities. This process will involve difficult trade-offs; engaging with and arriving at consensus about those trade-offs is a much-needed public exemplar of the spirit of compromise and determination necessary to generate progress on climate mitigation.**

Finding 5-10: While legislative progress on statutes that enshrine "meaningful engagement" into the NEPA is stalled, there are opportunities to integrate these approaches as standard practice in private- and public-sector activities. Working groups and programs can be modeled after the Interagency Working Group on Indigenous Traditional Ecologic Knowledge to support the inclusion of specialized expertise in government policy and guidance.

Recommendation 5-8: *Address the Priorities of Native American and Environmental Justice Communities.* Congress and federal agency leads should address the priorities of Native American and environmental justice communities through legislation and, in the interim, purposeful adoption of best practices in meaningful engagement.

a. Congress should pass legislation to codify "meaningful engagement" in environmental review practices. Furthermore, key federal actors in renewable energy and transmission deployment should include "meaningful engagement" practices in existing public engagement and environmental review processes, including providing many points of engagement (e.g., in time and across social groups) and materials in accessible forms (e.g., diverse languages), and requiring the consideration of alternative actions.

b. Federal program designers should involve social and behavioral researchers in the appropriate design of the social, behavioral, and other non-financial elements of deployment programs to enable communities to make informed technology adoption decisions and effectively use technologies to decarbonize, reduce energy consumption, save money, and obtain other additional benefits.

BUILDING THE NATION'S EXPERTISE IN THE HUMAN DIMENSIONS OF DECARBONIZATION

The Challenge

Effectively engaging U.S. publics in clean energy transitions will require upgrading the nation's expertise in the human dimensions of deep decarbonization. Doing public engagement well entails not only listening to people's voices and concerns but also facilitating an informed dialogue about the aspects of the issues that are important to them (Reed et al. 2018). For energy transitions, this means developing a rich and contextualized understanding of the ways that decarbonization matters to people, impacts their lives and livelihoods, and intersects with other aspects of society, the economy, and the environment that they care about. The capability of the public, decision makers, and institutions to effectively understand these issues, assess their significance, and integrate them into decision-making at multiple scales will be crucial for the success of public engagement. Building the nation's capacity for development and deployment will also require the ability to conduct credible, strategic assessments of outcomes for adaptive management (see Chapter 1).

Energy Literacy

Energy is one of the most important elements of modern economies, yet also one of the least well understood by the public. This is true even with regard to knowledge about energy sources or how to conserve energy (Bodzin 2012; DeWaters and Powers 2011; Murphy 2002), let alone the more complex challenges of navigating sustainable energy transitions (Martins et al. 2020). In this context, energy literacy[19] goes well beyond basic knowledge of scientific and engineering principles of energy taught in K–12 classrooms and science museums. Few people in the United States have even a rudimentary understanding of energy sources, infrastructures, or security (van den Broek 2019). Frequently, the only source of public understanding of energy systems is often simplified news coverage of exciting new technology developments or controversies over power plant or infrastructure siting.

Federal investment has prioritized improved public understanding and engagement in science and technology for non-energy topics, such as the National Aeronautics and Space Administration's fiscal year (FY) 2023 $144 million budget for educational programs for the public at large. In comparison, the United States has invested relatively little in ensuring that people have the energy literacy needed to participate effectively in energy decisions. It should not be surprising, therefore, that U.S. consumers significantly underinvest in technologies that could considerably improve their household energy economics (Brent and Ward 2018) or that misinformation pervades public understanding of energy technologies and their ability to contribute to decarbonization (Sovacool 2009). Misinformed understandings of the energy sector and systems undermine robust public engagement and the development and implementation of effective energy transition policies. Choices will need to be made to upgrade the efficiency of homes and businesses, electrify heating and transportation systems, and perhaps adopt dietary changes or new distributed energy technologies. Chadwick et al. (2022) show that knowledge is one of the most important factors influencing technology adoption and rejection.

Recent scholarship has highlighted the importance of adopting integrated social and technical framing of energy systems for decarbonization policy making (Miller et al. 2015). Interesting examples of this are Richard Scarry's well-read children's books about Busytown, which contain highly illuminative illustrations and stories about a coal mine and power plant and the people they serve with electricity (*What Do People Do All Day?* [1968]) and our automobile-intensive society (*Cars and Trucks and Things*

---

[19] The understanding of the role and nature of energy in daily lives accompanied by the application of this understanding to solve problems (DOE n.d.(b)).

That Go [1974]). These books portray a rich picture of how people's everyday lives and work are interdependent with energy technologies and infrastructures—and things that might be at stake in energy transitions, from the sector jobs to the organization of communities. Energy literacy education needs to follow this lead, not only for children but also in public engagement initiatives and for energy transition leaders across diverse sectors and organizations.

Enhanced understanding among consumers—as well as the array of contractors, technicians, salespeople, and influencers they interact with—will be crucial to effective household decision-making on decarbonized energy systems. States, cities, tribes, and communities will also benefit from improved energy literacy among residents and leaders as they face increasingly consequential choices about complex regional energy transitions (Miller et al. 2022). To make sense of the choices, decisions, and trade-offs entailed and their societal implications requires rich understandings of energy systems: who and what they serve, how they work, and their constraints in serving regional economies. For example, the National Science Foundation's (NSF's) Directorate for Social and Behavioral Sciences (SBE) Sciences, which had a budget of $286 million in FY 2022, supports research on human behavior and societal factors (NSF n.d.). Experiences from incorporation of social science and community engagement in interdisciplinary NSF research centers, funded through the SBE, could usefully inform energy programs (see Radatz et al. [2019]). Similar lessons might be drawn from the integration of ethical, legal, and social research into the National Institutes of Health's Human Genome Project (see Hilgartner et al. [2016] and McEwen et al. [2014]).

Anticipatory Methods

Recent scholarship has demonstrated the value of using anticipatory methods to examine the potential unanticipated impacts of new and emerging technologies (Guston 2014). Such methods use participatory public engagement to inform technology assessment, policy deliberation, and organizational decision-making (Kaplan et al. 2021) alongside other forward-looking analytic methods, such as responsible innovation (Stilgoe et al. 2013) and anticipatory and social life-cycle analysis (Fortier et al. 2019; Wender et al. 2014). Anticipatory methods expand insights into new technologies beyond the limits of market-based technology adoption studies. This will be especially valuable for informing energy transitions because technology adoption studies alone miss broader aspects of technology deployment that can slow decarbonization and lead to a range of risks and adverse social or economic outcomes. For example, strategic energy and environmental assessments using an anticipatory approach have recently illuminated pathways for institutional reform and coordination that can facilitate renewable energy deployment (Nwanekezie et al. 2022).

Anticipatory methods attend to the dynamics created by new technologies that ripple outward from their construction and use via complex social and technological systems. The consequences of these ripples are not intuitively obvious either from the perspective of the technology's intended function and use or when used differently than their inventors and designers initially imagined (Oudshoorn and Pinch 2005). These human complexities are particularly significant for decarbonization planning, which anticipates rapid and near-universal adoption of new technologies by "average" people. However, these narrow assumptions fail to account for obvious asymmetries between users and contexts, between urban and rural users (Kline and Pinch 1996), or among users with and without disabilities (Wolbring 2008, 2011).

Anticipatory analysis can also inform systems-level elements of the energy transition. Two examples in the electrification of light-duty transportation illustrate this phenomenon:

- A lack of anticipatory analysis in technology development may result in the need to redesign the technology after deployment. For example, hybrid and electric vehicles are inherently nearly silent when operating at low speeds; a fact that many early EV purchasers appreciated but that created potential safety risks for pedestrians and other road users who could not hear them moving. Redesign was necessary to adapt vehicles to real-world human contexts that initial designs had failed to consider by adding audible external sounds for the safety of pedestrians (e.g., see P.L. 111-373). Few assessments—especially involving robust public engagement—have rigorously explored how the heterogeneity of vehicle use (among different kinds of users, as well as day-to-day for an individual user) matches the capabilities of EVs (e.g., see He et al. 2016).
- The use of anticipatory analysis can allow for design of systems with preferred properties. For example, EV adoption is transforming vehicle supply chains resulting in new social and environmental consequences in the automobile manufacturing and repair sector, a major contributor to the U.S. and global economy. Anticipatory analyses are increasing being implemented to understand and predict how a shift to EVs will change the environmental, employment, and consumer aspects of mineral and material resource requirements and manufacturing, and the servicing of vehicles (e.g., see Colato and Ice 2023; EIA 2021; Shrestha et al. 2022).

A lack of anticipatory assessment and planning can lead to a slow pace of learning, incremental redesign, or less-than-near-universal adoption of key technologies, none of which support the accelerated decarbonization of energy systems. Ambitious anticipatory assessment and engagement is critical to inform and modify technology design, development, and markets, as well as to help diverse people and communities learn about new technologies and understand their implications.

Research and Inquiry Capabilities

Developing new capabilities for research and inquiry into the complexities of energy transitions is important to inform inclusive policy deliberation and infrastructure siting (Sovacool et al. 2020). Areas where research capabilities can inform decarbonization planning include

- *Mobilizing and supporting people and organizations in implementing key decarbonization strategies.* The scale of effort required to achieve a net-zero economy is unprecedented—much of it will require significant organizational, workforce, and even behavioral change from individuals and households to entire industries. Research can identify the human and organizational changes needed and the strategies to advance them, evaluate outcomes, and enhance the sharing of good practices. Research can also identify and suggest strategies for reducing workforce shortages, inflation in the pricing of materials, and backlashes against social and environmental goal setting.
- *Evaluating the societal and economic implications of deep decarbonization.* Many aspects of U.S. society and economy are organized around the ways energy is produced, distributed, and consumed. As a result, the consequences of adopting clean energy technologies and reconfiguring their manufacture and supplies will ripple outward into other areas of social and economic life and work. Research can help anticipate and comparatively evaluate trends and their potential implications for different groups, communities, and regions— especially for equity and justice considerations.
- *Anticipating vulnerabilities in interdependent infrastructure systems.* People depend on energy systems to provide essential services for an array of critical infrastructures and systems, including food, water, transport, communication, manufacturing, and the built environment. While climate and disaster interdependencies among critical infrastructures receive careful attention, significantly less attention has been paid to vulnerabilities that might arise owing to energy transitions. Research can identify how the human and organizational dimensions of interdependent systems may exacerbate or reduce vulnerabilities created by technological dependencies.

Opportunities and Barriers in Current Policy

The committee's first report identified opportunities for Congress to invest in educational and research programs focused on the knowledge and skills needed to implement and manage the transition (NASEM 2021). The IIJA, IRA, and CHIPS and Science Act direct nearly $18 billion[20] to career and skills training programs located at institutions of higher education. Several programs recommended in the first report, including $5 billion per year for the 10-year, GI Bill–type program and $100 million per year for the creation of innovative new degree programs, could be realized by the combined efforts of these three laws. However, support for workforce development in recent legislation focuses almost exclusively on applied science and engineering and less on the skills needed for an equitable and just energy transition. Outcomes will also likely be uneven given the heavy reliance on states to implement education and training programs. Workforce training will require additional support to drive innovation (see Chapter 4).

Missing from current legislation is a key element in the committee's prior recommendations: the explicit recognition that the United States needs to develop substantial knowledge, expertise, and workforces focused on higher-level understanding, analysis, and management of energy transitions, including among disciplines and sectors, and across research, application, and decision-making. This includes use-inspired research and training that intersects with technology development and deployment but focuses on the effective and equitable integration of technology into diverse societal, organizational, and market contexts. Such research areas include public and community engagement; the human and social dimensions of energy transitions; organizational change; interdisciplinary collaboration and convergence; energy policy and economics; the social and environmental impacts of technology; and energy and environmental equity and justice. Additionally, it is critical to assess methods in interdisciplinary convergence and co-production of knowledge among researchers and diverse knowledge-users in industry, government, and society.

To build this expertise, the committee's first report recommended $50 million per year for interdisciplinary doctoral and postdoctoral training programs, similar to those funded by NIH; $375 million per year to support doctoral and postdoctoral fellowships in energy transitions, with at least 25 fellowships per state; and support for lowering barriers to non-U.S. researchers. In principle, such investments might be made via the new NSF Directorate for Technology, Innovation, and Partnerships (TIP), established

---

[20] IIJA §40503, §40512, §40513, and §40521; IRA §60201; CHIPS and Science Act §10113, §10303, §10316, §10322, §10392, §10393, §10601, and §10745.

to "accelerate breakthrough technologies and solutions that address national-scale societal and economic challenges" with multidisciplinary, use-inspired research and collaboration that includes traditional and nontraditional players (NSF 2022; U.S. Senate Committee on Commerce, Science, and Transportation 2022). TIP is unlikely to serve this goal, however, due to its focus on advancing breakthrough technologies rather than tackling the broader challenges of integrating technologies into diverse social and economic contexts to advance national goals, including decarbonization, social and economic inclusion, and equity and justice.

Similarly, although the CHIPS and Science Act authorizes more than $13 billion in funding over 5 years for programs that include scholarships and fellowships,[21] this investment focuses on science, technology, engineering, and mathematics and entrepreneurship rather than the social science research and education needed to facilitate improved transition management. Thus, despite substantial new investments in clean energy R&D, recent legislation and executive action continues to significantly underinvest in efforts to understand and build knowledge and capacity relative to navigating the human complexities of the energy transition. This underinvestment risks replicating the misperception that the energy transition is a technological problem with social and economic dimensions rather than an integrated technological, social, and economic challenge.

Findings and Recommendations

Upgrading the nation's expertise in the human dimensions of deep decarbonization will require innovative action by the federal agencies that invest in and regulate the energy sector. Fortunately, much of this will require only modest shifts in and intentional implementation of already appropriated funding. To date, however, federal agencies are largely unprepared to do this work, and while recent legislation has provided extensive funding that could be leveraged for these purposes, the IRA, IIJA, and CHIPS and Science Act have not prioritized them.

> Finding 5-11: The United States has not yet implemented the expanded program of research into the human dimensions and complexities of energy transitions needed to inform effective decarbonization and public engagement strategies. This area represents a persistent gap in research portfolios. The committee highlighted in its first report and recommended that Congress appropriate $25 million per year. Neither it nor an alternative is included in current policies.

---

[21] CHIPS and Science Act §10113, §10303, §10316, §10322, §10392, §10393, §10601, and §10745.

Recommendation 5-9: *Invest in and Integrate Social Science Research into Transition Decision-Making.* **The federal agencies whose research and development efforts impact the clean energy transition should invest in and integrate robust human dimensions and social science research into energy transition decision-making.**

a. **The Department of Energy (DOE) Office of Science should establish an independent research and graduate training program focused on the basic human and social sciences of energy, including economics and behavioral sciences, anthropology, sociology, and political science. This program would help develop a robust foundation of knowledge and expertise necessary to navigate the human complexities of energy transitions in the United States. The initial budget for this program should be $25 million annually and grow to $200 million by 2030. Representatives from the nation's energy social sciences research community should design and lead the program's research agenda.**

b. **DOE, Department of Transportation, Department of Defense, Environmental Protection Agency, and National Science Foundation (NSF) should integrate human dimensions research and graduate training into clean energy technology research, innovation, and deployment programs. This should include the NSF Directorate for Technology, Innovation, and Partnerships; DOE technology offices; and the new DOE Office of Clean Energy Demonstrations. Lessons from NSF's prior experience integrating social science research into major science and engineering research centers should be used to guide this effort, as should research on anticipatory assessment and governance of emerging technologies.**

c. **NSF and DOE should establish a network of 10 regional, university-led research centers to develop and apply fundamental new strategies for managing the social and technical dynamics of energy transitions. The research centers would draw together interdisciplinary teams of science, engineering, and social science researchers with government, industry, and community stakeholders to apply anticipatory methods to the energy transition.**

Finding 5-12: The U.S. public is under-prepared and insufficiently educated to fully carry out the work required of them for the nation to achieve deep decarbonization or to participate and engage effectively in deep decarbonization planning processes.

Recommendation 5-10: *Establish an Energy Systems Education Network.* **The Department of Energy and the Department of Education should establish**

a 5-year, $50 million national energy science, technology, engineering, and mathematics (STEM) network for informal education and a parallel $50 million annual national energy STEM education program for K–12 schools. The focus of these initiatives should introduce students to (a) the organization, development, and operation of the energy cultures, infrastructures, and systems that underpin the U.S. economy; (b) the ways in which those infrastructures and systems are changing and will need to change to achieve deep decarbonization; (c) the opportunities and challenges that decarbonization might pose; and (d) the ways that people can effectively participate in envisioning and guiding energy transitions. These initiatives should draw lessons from other large-scale, public STEM education initiatives, such as the National Science Foundation's Nanoscale Informal Science Education Network and the recent National Aeronautics and Space Administration's SciAct STEM Ecosystems project.

CONCLUSION

This chapter raises and attempts to address how to engage and mobilize the U.S. people in the project of deep decarbonization, which has to date received far too little attention from Congress, the White House, and federal agencies. The social contract for decarbonization is the shared understanding among all sectors and groups in society about the necessity of decarbonization, the willingness to deliberation and follow steps to get there, and the agreed-upon character of the transition. It hinges on decisions and actions taken now and over the next decade to enable people to meaningfully participate in envisioning, planning, and implementing the transition in ways that they judge fair, equitable, and beneficial. This includes strategies about how to imagine, design, and build energy systems with the public as well as policies that affect when, who, where, and how people will experience the everyday material realities of decarbonization and its impacts on their livelihoods and their access to energy services.

Although many of the features of recent policy initiatives create opportunities to engage and invest various publics in clean energy futures, there is a persistent mismatch between the scale of the decarbonization endeavor and the resources, capacity, and vision currently dedicated to mobilizing all the people of the United States to achieve deep decarbonization. Without additional resources and determined strategies, current public engagement efforts will be inadequate to preempt substantial public resistance to the pace and scale of systemic change necessary. Inadequate public engagement also curtails opportunities to advance creative, collaborative, and place-based energy system designs and their many potential advantages for equitable deep decarbonization.

Rather than being derailed by the complexity and enormity of the public engagement challenge, proponents of deep decarbonization can turn to the policies, practices, and investments reviewed and recommended here as actionable steps toward building a social contract. Recommendations in this chapter include commitments to a growth mindset about public engagement—including through significant investments in applied social science research and a determination to engage the nation's youth in the search for climate solutions. We can also turn to our history: in crucial moments in the past, determined and robust efforts helped the U.S. public understand the gravity of existential problems, our critical roles in tackling the challenges, and the benefits that we can achieve together as a nation. Public engagement for deep decarbonization is a task no less significant than that undertaken by President Franklin D. Roosevelt via his fireside chats to help the nation navigate the challenges of the Great Depression, prepare for the prospect of war, and come together as a nation to fight for freedom and democracy. The present challenge is no less existential, and the gravity of the public engagement task no less important. Table 5-2 summarizes the committee's recommendations to support innovative public engagement in decarbonization.

SUMMARY OF RECOMMENDATIONS ON PUBLIC ENGAGEMENT TO BUILD A STRONG SOCIAL CONTRACT FOR DEEP DECARBONIZATION

TABLE 5-2[a] Summary of Recommendations on Public Engagement to Build a Strong Social Contract for Deep Decarbonization

Short-Form Recommendation	Actor(s) Responsible for Implementing Recommendation	Sector(s) Addressed by Recommendation	Objective(s) Addressed by Recommendation	Overarching Categories Addressed by Recommendation
5-1: Encourage Prospective, Inclusive Dialogue at National and Regional Levels	National Climate Task Force (NCTF), Department of Energy (DOE), and Environmental Protection Agency (EPA)	• Non-federal actors	• Equity • Employment • Public engagement	Ensuring Procedural Equity in Planning and Siting New Infrastructure and Programs Building the Needed Workforce and Capacity
5-2: Accelerate the Development, Implementation, Assessment, and Sharing of Energy System Policy and Approaches That Deliver Local Benefits	Subnational governments, elected officials and their representative coalitions, federal partners	• Non-federal actors	• Equity • Public engagement	Ensuring Equity, Justice, Health, and Fairness of Impacts Ensuring Procedural Equity in Planning and Siting New Infrastructure and Programs
5-3: Fix Policy Gaps That Limit Role of Public Land in Decarbonization	Congress and state legislatures	• Electricity • Non-federal actors • Land use	• Equity • Public engagement	A Broadened Policy Portfolio Ensuring Procedural Equity in Planning and Siting New Infrastructure and Programs

[a] The text in this table was changed during editorial review to improve clarity and alignment with information in other sections of the report.

TABLE 5-2 Continued

Short-Form Recommendation	Actor(s) Responsible for Implementing Recommendation	Sector(s) Addressed by Recommendation	Objective(s) Addressed by Recommendation	Overarching Categories Addressed by Recommendation
5-4: Address Barriers to Local Benefits from Renewable Energy Facilities	State legislatures	• Non-federal actors	• Equity • Public engagement	Ensuring Equity, Justice, Health, and Fairness of Impacts Ensuring Procedural Equity in Planning and Siting New Infrastructure and Programs
5-5: Convene a National Working Group on Siting Process Innovation with Input from State Energy Officials	DOE, Council on Environmental Quality, Federal Energy Regulatory Commission, National Association of Regulatory Utility Commissioners, and National Association of State Energy Officials	• Non-federal actors • Electricity	• Equity • Public engagement	Ensuring Procedural Equity in Planning and Siting New Infrastructure and Programs Siting and Permitting Reforms for Interstate Transmission

continued

TABLE 5-2 Continued

Short-Form Recommendation	Actor(s) Responsible for Implementing Recommendation	Sector(s) Addressed by Recommendation	Objective(s) Addressed by Recommendation	Overarching Categories Addressed by Recommendation
5-6: Mandate and Allocate Resources for a National Assessment on the Public Engagement Workforce and Gaps	Congress, DOE, NCTF	• Electricity • Non-federal actors	• Equity • Employment • Public engagement	Ensuring Procedural Equity in Planning and Siting New Infrastructure and Programs Siting and Permitting Reforms for Interstate Transmission Building the Needed Workforce and Capacity
5-7: Develop Collaborative Regional Renewable Energy Deployment Plans	Civil society leaders and philanthropic organizations	• Non-federal actors	• Equity • Public engagement	Ensuring Procedural Equity in Planning and Siting New Infrastructure and Programs Siting and Permitting Reforms for Interstate Transmission

TABLE 5-2 Continued

Short-Form Recommendation	Actor(s) Responsible for Implementing Recommendation	Sector(s) Addressed by Recommendation	Objective(s) Addressed by Recommendation	Overarching Categories Addressed by Recommendation
5-8: Address the Priorities of Native American and Environmental Justice Communities	Congress and federal program designers	• Electricity	• Equity • Public engagement	A Broadened Policy Portfolio Ensuring Equity, Justice, Health, and Fairness of Impacts Ensuring Procedural Equity in Planning and Siting New Infrastructure and Programs
5-9: Invest in and Integrate Social Science Research into Transition Decision-Making	DOE, Department of Transportation, Department of Defense, EPA, and National Science Foundation	• Non-federal actors	• Equity • Employment • Public engagement	Building the Needed Workforce and Capacity Research, Development, and Demonstration Needs
5-10: Establish an Energy Systems Education Network	DOE and Department of Education	• Electricity • Buildings • Transportation • Industry • Non-federal actors	• Public engagement	Building the Needed Workforce and Capacity

REFERENCES

ACCD (Allegheny Conference on Community Development). n.d. "Who We Are." https://www.alleghenyconference.org. Accessed May 11, 2023.

ACUS (Administrative Conference of the United States). 2021. "Managing Mass, Computer-Generated, and Falsely Attributed Comments." June 30. https://www.acus.gov/recommendation/managing-mass-computer-generated-and-falsely-attributed-comments.

Aidun, H., J. Elkin, R. Goyal, K. Marsh, N. McKee, M. Welch, L. Adelman, and S. Finn. 2022. "Opposition to Renewable Energy Facilities in the United States: March 2022 Edition." Sabin Center for Climate Change Law, Columbia Law School, March 2022. https://scholarship.law.columbia.edu/sabin_climate_change/186.

Anchondo, C. 2022. "White House CCS Guidance Exposes Environmental Justice Rifts." *E&E News*, February 16. https://www.eenews.net/articles/white-house-ccs-guidance-exposes-environmental-justice-rifts.

Atcitty, S. 2021. "Native American Energy Sovereignty: Energy Storage and Power Electronic Benefits." In *Proposed for Presentation at the IEEE Applied Power Electronics Conference (APEC) 2021 Held June 14–17, 2021 in Phoenix, AZ*. Washington, DC: Department of Energy. https://doi.org/10.2172/1868001.

Batel, S. 2020. "Research on the Social Acceptance of Renewable Energy Technologies: Past, Present and Future." *Energy Research and Social Science* 68(October):101544. https://doi.org/10.1016/j.erss.2020.101544.

Begay, S.K. 2018a. "How Citizen Potawatomi Nation Utilizes Energy Efficiency and Renewable Energy to Address Its High Energy Burden." *Electricity Journal* 31(6):16–22. https://doi.org/10.1016/j.tej.2018.07.005.

Begay, S.K. 2018b. "Navajo Residential Solar Energy Access as a Global Model." *Electricity Journal* 31(6):9–15. https://doi.org/10.1016/j.tej.2018.07.003.

Bessette, D., and J. Crawford. 2022. "All's Fair in Love and WAR: The Conduct of Wind Acceptance Research (WAR) in the United States and Canada." *Energy Research and Social Science* 88(June):102514. https://doi.org/10.1016/j.erss.2022.102514.

Bice, S. 2020. "The Future of Impact Assessment: Problems, Solutions and Recommendations." *Impact Assessment and Project Appraisal* 38(2):104–108. https://doi.org/10.1080/14615517.2019.1672443.

Biswas, S., A. Echevarria, N. Irshad, Y. Rivera-Matos, J. Richter, N. Chhetri, M.J. Parmentier, and C.A. Miller. 2022. "Ending the Energy-Poverty Nexus: An Ethical Imperative for Just Transitions." *Science and Engineering Ethics* 28(4):36. https://doi.org/10.1007/s11948-022-00383-4.

Bodzin, A. 2012. "Investigating Urban Eighth-Grade Students' Knowledge of Energy Resources." *International Journal of Science Education* 34(8):1255–1275. https://doi.org/10.1080/09500693.2012.661483.

Boudet, H.S. 2019. "Public Perceptions of and Responses to New Energy Technologies." *Nature Energy* 4(6):446–455. https://doi.org/10.1038/s41560-019-0399-x.

Brent, D.A., and M.B. Ward. 2018. "Energy Efficiency and Financial Literacy." *Journal of Environmental Economics and Management* 90(July):181–216. https://doi.org/10.1016/j.jeem.2018.05.004.

Brown, A. 2023. "An Open Letter to the Department of Energy: Enough with the Technical Assistance." Utility Dive. January 30. https://www.utilitydive.com/news/an-open-letter-to-the-department-of-energy-enough-with-the-technical-assis/641491.

Brown, C. 2021. "PA Senate and House Dems Host Hearing on 'A People's Budget: The Environment.'" *Pennsylvania Senate Democrats* (blog), February 26. https://pasenate.com/pa-senate-house-dems-host-hearing-on-a-peoples-budget-the-environment.

Burke, M.J., and J.C. Stephens. 2018. "Political Power and Renewable Energy Futures: A Critical Review." *Energy Research and Social Science* 35:78–93. https://doi.org/10.1016/j.erss.2017.10.018.

Cameron, J., and K. Gibson. 2005. "Participatory Action Research in a Poststructuralist Vein." *Geoforum* 36(3):315–331. https://doi.org/10.1016/j.geoforum.2004.06.006.

Carter, D., J. Zoellick, M. Marshall, G. Chapman, P. Lehman, D. Saucedo, S. Shoemaker, C. Chamberlin, J. Ganion, and P. Singh. 2019. "Demonstrating a Secure, Reliable, Low-Carbon Community Microgrid at the Blue Lake Rancheria." CEC-500-2019-011. California Energy Commission. https://www.energy.ca.gov/sites/default/files/2021-05/CEC-500-2019-011.pdf.

Catenacci, T. 2022. "Green Energy Projects Face Stark Environmental, Local Opposition Nationwide." *Fox News*, October 13. https://www.foxnews.com/politics/green-energy-projects-face-stark-environmental-local-opposition-nationwide.

CCSA (Coalition for Community Solar Access). 2022. "Federal Inflation Reduction Act Key Community Solar Provisions." https://www.maine.gov/energy/sites/maine.gov.energy/files/inline-files/CCSA_InflationReductionAct_FactSheet_Final.pdf.

CEQ (Council on Environmental Quality). 2022. "CEQ Response to the WHEJAC March 2022 Letter (Submitted May 2022)." https://www.epa.gov/system/files/documents/2022-05/CEQ%20Response%20to%20the%20WHEJAC%20March%202022%20Letter%20%28submitted%20May%202022%29.pdf.

CEQ and DOE (Department of Energy). n.d. *Citizens Guide to NEPA.* https://ceq.doe.gov/get-involved/citizens_guide_to_nepa.html. Accessed May 15, 2023.

Chadwick, K., R. Russell-Bennett, and N. Biddle. 2022. "The Role of Human Influences on Adoption and Rejection of Energy Technology: A Systematised Critical Review of the Literature on Household Energy Transitions." *Energy Research and Social Science* 89(July):102528. https://doi.org/10.1016/j.erss.2022.102528.

Chan, G., I. Evans, M. Grimley, B. Ihde, and P. Mazumder. 2017. "Design Choices and Equity Implications of Community Shared Solar." *Electricity Journal*, Energy Policy Institute's Seventh Annual Energy Policy Research Conference, 30(9):37–41. https://doi.org/10.1016/j.tej.2017.10.006.

Chilvers, J., H. Pallett, and T. Hargreaves. 2018. "Ecologies of Participation in Socio-Technical Change: The Case of Energy System Transitions." *Energy Research and Social Science* 42(August):199–210. https://doi.org/10.1016/j.erss.2018.03.020.

Clifford, C. 2022. "Fierce Local Battles Over Power Lines Are a Bottleneck for Clean Energy." CNBC, June 26. https://www.cnbc.com/2022/06/26/why-the-us-has-a-massive-power-line-problem.htm.

Colato, J., and L. Ice. 2023. "Charging into the Future: The Transition to Electric Vehicles." *Beyond the Numbers* 12(4). https://www.bls.gov/opub/btn/volume-12/charging-into-the-future-the-transition-to-electric-vehicles.htm.

CPUC (California Public Utilities Commission). 2023. "Equity Initiatives and Clean Energy Access Program Webinar." https://www.cpuc.ca.gov/events-and-meetings/webinar-2023-02-15.

Crawford, J., D. Bessette, and S.B. Mills. 2022. "Rallying the Anti-Crowd: Organized Opposition, Democratic Deficit, and a Potential Social Gap in Large-Scale Solar Energy." *Energy Research and Social Science* 90(August):102597. https://doi.org/10.1016/j.erss.2022.102597.

CUB (Citizens Utility Board) Minnesota. n.d. https://cubminnesota.org. Accessed May 11, 2023.

Cyrs, T., and C. Elliot. 2018. "Insider: Expand the Role of Subnational Actors in Climate Policy." World Resources Institute. https://www.wri.org/technical-perspectives/insider-expand-role-subnational-actors-climate-policy.

Dawson, D. 2023. "New Law Removes Local Control in Siting Solar, Wind Farms." *Jacksonville Journal-Courier*, February 1. https://www.myjournalcourier.com/news/article/local-boards-losing-control-siting-illinois-17754277.php.

Devaney, L., D. Torney, P. Brereton, and M. Coleman. 2020. "Ireland's Citizens' Assembly on Climate Change: Lessons for Deliberative Public Engagement and Communication." *Environmental Communication* 14(2):141–146. https://doi.org/10.1080/17524032.2019.1708429.

Devine-Wright, P. 2011. "Public Engagement with Large-Scale Renewable Energy Technologies: Breaking the Cycle of NIMBYism." *WIREs Climate Change* 2(1):19–26. https://doi.org/10.1002/wcc.89.

Devine-Wright, P. 2022. "Place and Participation: Principles for Public Engagement with Large Scale Net Zero Infrastructures." Presentation at the National Academies Webinar on Public Responses to Large-Scale, Net-Zero Infrastructure: Research Perspectives. June 6. https://www.nationalacademies.org/event/06-06-2022/accelerating-decarbonization-in-the-united-states-technology-policy-and-societal-dimensions-public-responses-to-large-scale-net-zero-infrastructure-research-perspectives.

Devine-Wright, P., and A. Moseley. 2019. "Developing a Net Zero Citizens' Assembly for Devon: A Rapid Review of Evidence and Best Practice Prepared for the Devon Climate Emergency Response Group and the Devon Net Zero Task Force." University of Exeter. https://devonclimateemergency.org.uk/devon-carbon-plan/citizens-assembly.

DeWaters, J.E., and S.E. Powers. 2011. "Energy Literacy of Secondary Students in New York State (USA): A Measure of Knowledge, Affect, and Behavior." *Energy Policy* 39(3):1699–1710. https://doi.org/10.1016/j.enpol.2010.12.049.

Diller, E., R. Walter, and J. Hansel. 2023. New Amendments to NEPA in the Fiscal Responsibility Act of 2023. *Inner City Fund Insights.* https://www.icf.com/insights/environment/new-nepa-amendments-fiscal-responsibility-act-2023.

DOE (Department of Energy). 2022. "Consent-Based Siting." https://www.energy.gov/ne/consent-based-siting.

DOE. 2023. "Office of Energy Justice Policy and Analysis." https://www.energy.gov/diversity/office-energy-justice-policy-and-analysis.

DOE. n.d.(a). "About Community Benefits Plans." https://www.energy.gov/infrastructure/about-community-benefits-plans. Accessed May 15, 2023.

DOE. n.d.(b). "Energy Literacy: Essential Principles for Energy Education." https://www.energy.gov/energysaver/energy-literacy-essential-principles-energy-education. Accessed May 15, 2023.

DOE. n.d.(c). "What Challenges Are Associated with Shale Gas Production?" https://www.energy.gov/fecm/articles/challenges-associated-shale-gas-production. Accessed May 15, 2023.

DOE-OCED (Department of Energy Office of Clean Energy Demonstrations). n.d. "OCED EXCHANGE: Funding Opportunity Exchange." https://oced-exchange.energy.gov/Default.aspx#Foald3ec25bcf-a385-4b5a-87d2-2a0b8fa4ca5a. Accessed May 11, 2023.

DOE-ONPC (Department of Energy Office of NEPA Policy and Compliance). n.d. CEQ Guidance Documents. DOE Office of NEPA Policy and Compliance. https://www.energy.gov/nepa/ceq-guidance-documents. Accessed May 15, 2023.

DSIRE. n.d. "Database of State Incentives for Renewables and Efficiency®." DSIRE. https://www.dsireusa.org. Accessed June 13, 2023.

Duvic-Paoli, L.A. 2022. "Re-Imagining the Making of Climate Law and Policy in Citizens' Assemblies." *Transnational Environmental Law* 11(2):235–261. https://doi.org/10.1017/S2047102521000339.

Echevarria, A., Y. Rivera-Matos, N. Irshad, C. Gregory, M. Castro-Sitiriche, R.R. King, and C.A. Miller. 2023. "Unleashing Sociotechnical Imaginaries to Advance Just and Sustainable Energy Transitions: The Case of Solar Energy in Puerto Rico." *IEEE Transactions on Technology and Society* 4(3):255–268. https://doi.org/10.1109/TTS.2022.3191542.

EIA (Energy Information Administration). 2023. "Electric Power Monthly Table 1.17.A. Net Generation from Solar Photovoltaic by State, by Sector, February 2023 and 2022 (Thousand Megawatthours)." https://www.eia.gov/electricity/monthly.

Elmallah, S., and J. Rand. 2022. "'After the Leases Are Signed, It's a Done Deal': Exploring Procedural Injustices for Utility-Scale Wind Energy Planning in the United States." *Energy Research and Social Science* 89(July):102549. https://doi.org/10.1016/j.erss.2022.102549.

EPA (Environmental Protection Agency). 2010. "State Climate and Energy Technical Forum Background Document: An Overview of PUCs for State Environment and Energy Officials." https://www.epa.gov/sites/default/files/2016-03/documents/background_paper.pdf.

EPA. 2021. "White House Environmental Justice Advisory Council." Overviews and Factsheets. https://www.epa.gov/environmentaljustice/white-house-environmental-justice-advisory-council.

EPA. 2023. "Climate Pollution Reduction Grants: Other Policies and Guidance." https://www.epa.gov/inflation-reduction-act/climate-pollution-reduction-grants.

Eschrich, J., and C.A. Miller. 2019. *The Weight of Light: A Collection of Solar Futures.* Center for Science and the Imagination. Arizona State University. https://csi.asu.edu/books/weight.

Eschrich, J., and C.A. Miller. 2021. *Cities of Light: A Collection of Solar Futures.* Center for Science and the Imagination. Arizona State University. https://csi.asu.edu/books/cities-of-light.

Faconti, M.C. 2013. "How Texas Overcame California as Renewable State: Look at the Texan Renewable Energy Success." *Vermont Journal of Environmental Law* 14(3):411–434.

Fazio, C., and J. Wallace. 2017. "Legal and Policy Issues Related to Community Benefits Agreements." *Fordham Environmental Law Review* 21(3):543. https://ir.lawnet.fordham.edu/elr/vol21/iss3/2.

FERC (Federal Energy Regulatory Commission). 2022. "Office of Public Participation (OPP)." https://www.ferc.gov/OPP.

Fergen, J.T., J.B. Jacquet, and R. Shukla. 2021. "'Doomscrolling' in My Backyard: Corrosive Online Communities and Contested Wind Development in Rural Ohio." *Energy Research and Social Science* 80(October):102224. https://doi.org/10.1016/j.erss.2021.102224.

Fischer, D., P. Lochner, and H. Annegarn. 2020. "Evaluating the Effectiveness of Strategic Environmental Assessment to Facilitate Renewable Energy Planning and Improved Decision-Making: A South African Case Study." *Impact Assessment and Project Appraisal* 38(1):28–38. https://doi.org/10.1080/14615517.2019.1619389.

Forester, J. 1999. *The Deliberative Practitioner: Encouraging Participatory Planning Processes*. Cambridge, MA: MIT Press. https://mitpress.mit.edu/9780262561228/the-deliberative-practitioner.

Fortier, M.-O.P., L. Teron, T.G. Reames, D.T. Munardy, and B.M. Sullivan. 2019. "Introduction to Evaluating Energy Justice Across the Life Cycle: A Social Life Cycle Assessment Approach." *Applied Energy* 236:211–219. https://doi.org/10.1016/j.apenergy.2018.11.022.

French, R. 2023. "With Growing Backlash to Wind Energy, Michigan Turns to Solar Power." *Bridge Michigan*, January 16. https://www.bridgemi.com/michigan-environment-watch/growing-backlash-wind-energy-michigan-turns-solar-power.

Freudenburg, W.R. 1986. "Social Impact Assessment." *Annual Review of Sociology* 12(1):451–478. https://doi.org/10.1146/annurev.so.12.080186.002315.

Gastil, J. 2018. "The Lessons and Limitations of Experiments in Democratic Deliberation." *Annual Review of Law and Social Science* 14(1):271–291. https://doi.org/10.1146/annurev-lawsocsci-110316-113639.

Gearino, D. 2022. "In the End, Solar Power Opponents Prevail in Williamsport, Ohio." *ABC News*, December 24. https://abcnews.go.com/US/end-solar-power-opponents-prevail-williamsport-ohio/story?id=95752351.

Gelles, D. 2022. "The U.S. Will Need Thousands of Wind Farms. Will Small Towns Go Along?" *The New York Times*, December 30. https://www.nytimes.com/2022/12/30/climate/wind-farm-renewable-energy-fight.html.

Gerrard, M. 2017. "Legal Pathways for a Massive Increase in Utility-Scale Renewable Generation Capacity." *Environmental Law Reporter* 47(January):10591. https://scholarship.law.columbia.edu/faculty_scholarship/2045.

Godby, R. 2022. "Panel Discussion." Presented at Accelerating Decarbonization in the United States: Technology, Policy, and Societal Dimensions | Local Benefits and Compensation Strategies for Deep Decarbonization Infrastructure Webinar. October 3. https://www.nationalacademies.org/event/10-03-2022/accelerating-decarbonization-in-the-united-states-technology-policy-and-societal-dimensions-local-benefits-compensation-strategies-for-deep-decarbonization-infrastructure.

Goedkoop, F., and P. Devine-Wright. 2016. "Partnership or Placation? The Role of Trust and Justice in the Shared Ownership of Renewable Energy Projects." *Energy Research and Social Science* 17(July):135–146. https://doi.org/10.1016/j.erss.2016.04.021.

Gordon, A. 2022. "Why Doesn't America Build Things?" *Vice*. https://www.vice.com/en/article/93a39e/why-doesnt-america-build-things.

Government of Canada. 2017. "Generation Energy." Natural Resources Canada. https://natural-resources.canada.ca/climate-change/canadas-green-future/generation-energy/20093.

Government of Canada. 2018. "Generation Energy Council." Natural Resources Canada. https://www.nrcan.gc.ca/20380.

Government of Canada. 2023. "Clean Energy for Rural and Remote Communities Funded Projects." Natural Resources Canada. https://natural-resources.canada.ca/reducingdiesel/clean-energy-for-rural-and-remote-communities-funded-projects/22524.

Grimley, M., V. Shastry, D.G. Kânoğlu-Özkan, E. Blevins, A.L. Beck, G. Chan, and V. Rai. 2022. "The Grassroots Are Always Greener: Community-Based Organizations as Innovators of Shared Solar Energy in the United States." *Energy Research and Social Science* 90(August):102628. https://doi.org/10.1016/j.erss.2022.102628.

Gustafson, A., S.A. Rosenthal, M.T. Ballew, M.H. Goldberg, P. Bergquist, J.E. Kotcher, E.W. Maibach, and A. Leiserowitz. 2019. "The Development of Partisan Polarization Over the Green New Deal." *Nature Climate Change* 9(12):940–944. https://doi.org/10.1038/s41558-019-0621-7.

Guston, D.H. 2014. "Understanding 'Anticipatory Governance.'" *Social Studies of Science* 44(2):218–242. https://doi.org/10.1177/0306312713508669.

Haggerty, J.H., M. Haggerty, and R. Rasker. 2014. "Uneven Local Benefits of Renewable Energy in the U.S. West: Property Tax Policy Effects." *Western Economics Forum* 13(1). https://headwaterseconomics.org/wp-content/uploads/Uneven_Local_Renewable_Tax_Benefits.pdf.

Haggerty, M.N., and J.H. Haggerty. 2015. "Energy Development as Opportunity and Challenge in the Rural West." Pp. 161–191 in *The Rural West: Common Regional Issues*, D. Danbom, ed. Salt Lake City, UT: University of Utah Press.

He, X., Y. Wu, S. Zhang, M.A. Tamor, T.J. Wallington, W. Shen, W. Han, L. Fu, and J. Hao. 2016. "Individual Trip Chain Distributions for Passenger Cars: Implications for Market Acceptance of Battery Electric Vehicles and Energy Consumption by Plug-In Hybrid Electric Vehicles." *Applied Energy* 180:650–660. https://doi.org/10.1016/j.apenergy.2016.08.021.

Henry L. Hillman Foundation. n.d. https://henrylhillmanfoundation.org. Accessed May 11, 2023.

Hilgartner, S., B. Prainsack, and J. Hurlbut. 2016. Chapter 28, "Ethics as Governance in Genomics and Beyond." In *Handbook of Science and Technology Studies*, U. Felt, R. Fouché, C.A. Miller, L. Smith-Doerr, eds. Cambridge, MA: MIT Press.

Hintz, O., E. Uebelhor, and E. Gold. 2021. "Inventory of State Solar Property Tax Treatments." CLOSUP Working Paper Series Number 55. University of Michigan Center for Local, State, and Urban Policy. https://closup.umich.edu/research/working-papers/inventory-state-solar-property-tax-treatments.

Hoen, B., J. Firestone, J. Rand, D. Elliott, G. Hübner, J. Pohl, R.H. Wiser, E. Lantz, R. Haac, and K. Kaliski. 2019. "Attitudes of U.S. Wind Turbine Neighbors: Analysis of a Nationwide Survey." *Energy Policy* 134. https://emp.lbl.gov/publications/do-wind-turbines-make-good-neighbors.

H.R.2021. 2021. "Environmental Justice for All Act." https://www.congress.gov/bill/117th-congress/house-bill/2021.

ICT News. 2022. "Landmark Federal Energy Regulatory Commission Decision Advances Morongo Transmission, LLC." https://ictnews.org/the-press-pool/landmark-federal-energy-regulatory-commission-decision-advances-morongo-transmission-llc.

IEA (International Energy Agency). 2021. "The State of Play." Pp. 23–42 in *The Role of Critical Minerals in Clean Energy Transitions*.

IHRB (Institute for Human Rights and Business). 2023. "Community Ownership of Renewable Energy: How it Works in Nine Countries." *Just Transitions*. https://www.ihrb.org/focus-areas/just-transitions/community-ownership-of-renewable-energy-how-it-works-in-nine-countries.

ILSR (Institute for Local Self-Reliance). n.d. "Community Power Map." Institute for Local Self-Reliance. https://ilsr.org/community-power-map. Accessed June 13, 2023.

Impact Assessment Agency of Canada. 2022. "Funding Programs." Government of Canada. https://www.canada.ca/en/impact-assessment-agency/services/public-participation/funding-programs.html.

Inês, C., P.L. Guilherme, M. Esther, G. Swantje, H. Stephen, and H. Lars. 2020. "Regulatory Challenges and Opportunities for Collective Renewable Energy Prosumers in the EU." *Energy Policy* 138(March):111212. https://doi.org/10.1016/j.enpol.2019.111212.

Ingram, G. 2020. Civil Society: An Essential Ingredient of Development. *Brookings*. https://www.brookings.edu/articles/civil-society-an-essential-ingredient-of-development.

IWG (Interagency Working Group on Coal and Power Plant Communities and Economic Revitalization). 2021. "Initial Report to the President on Empowering Workers Through Revitalizing Energy Communities." https://netl.doe.gov/sites/default/files/2021-04/Initial%20Report%20on%20Energy%20Communities_Apr2021.pdf.

IWG. 2023. "Revitalizing Energy Communities: Two-Year Report to the President." https://energycommunities.gov/wp-content/uploads/2023/04/IWG-Two-Year-Report-to-the-President.pdf.

Jørgensen, M.L., H.T. Anker, and J. Lassen. 2020. "Distributive Fairness and Local Acceptance of Wind Turbines: The Role of Compensation Schemes." *Energy Policy* 138(March):111294. https://doi.org/10.1016/j.enpol.2020.111294.

Kaplan, L.R., M. Farooque, D. Sarewitz, and D. Tomblin. 2021. "Designing Participatory Technology Assessments: A Reflexive Method for Advancing the Public Role in Science Policy Decision-Making." *Technological Forecasting and Social Change* 171(October):120974. https://doi.org/10.1016/j.techfore.2021.120974.

Kinder, J.B. 2021. "Solar Infrastructure as Media of Resistance, or Indigenous Solarities Against Settler Colonialism." *South Atlantic Quarterly* 120(1):63–76. https://doi.org/10.1215/00382876-8795718.

Kline, R., and T. Pinch. 1996. "Users as Agents of Technological Change: The Social Construction of the Automobile in the Rural United States." *Technology and Culture* 37(4):763–795. https://doi.org/10.2307/3107097.

Knauf, J. 2022. "Can't Buy Me Acceptance? Financial Benefits for Wind Energy Projects in Germany." *Energy Policy* 165(June):112924. https://doi.org/10.1016/j.enpol.2022.112924.

Kok, M.T.J., and K. Ludwig. 2022. "Understanding International Non-State and Subnational Actors for Biodiversity and Their Possible Contributions to the Post-2020 CBD Global Biodiversity Framework: Insights from Six International Cooperative Initiatives." *International Environmental Agreements: Politics, Law, and Economics* 22(1):1–25. https://doi.org/10.1007/s10784-021-09547-2.

Krupnik, S., A. Wagner, O. Vincent, T.J. Rudek, R. Wade, M. Mišík, S. Akerboom, et al. 2022. "Beyond Technology: A Research Agenda for Social Sciences and Humanities Research on Renewable Energy in Europe." *Energy Research and Social Science* 89(July):102536. https://doi.org/10.1016/j.erss.2022.102536.

Kurth, M.H., S. Larkin, J.M. Keisler, and I. Linkov. 2017. "Trends and Applications of Multi-Criteria Decision Analysis: Use in Government Agencies." *Environment Systems and Decisions* 37(2):134–143. https://doi.org/10.1007/s10669-017-9644-7.

Lacelle-Webster, A., and M. Warren. 2021. "Citizens' Assemblies and Democracy." *Oxford Research Encyclopedias, Politics*, 25 March, 2021. https://doi.org/10.1093/acrefore/9780190228637.013.1975.

Lilienthal, D.E. 1944. *TVA: Democracy on the March.* Penguin Books.

Longley, R. 2022. "Civil Society: Definition and Theory." *ThoughtCo.* https://www.thoughtco.com/civil-society-definition-and-theory-5272044.

Lopez, A., and A. Levine. 2022a. "U.S. Solar Siting Regulation and Zoning Ordinances." DOE Open Energy Data Initiative (OEDI), National Renewable Energy Laboratory. https://doi.org/10.25984/1873867.

Lopez, A., and A. Levine. 2022b. "U.S. Wind Siting Regulation and Zoning Ordinances." DOE Open Energy Data Initiative (OEDI), National Renewable Energy Laboratory. https://doi.org/10.25984/1873866.

Lotfi, M., J.P.S. Catalão, and H.A. Gabbar. 2020. "The Rise of Energy Prosumers and Energy Democracy: History and Future Prospects." *IEEE Smart Grid Bulletin* July. https://smartgrid.ieee.org/bulletins/july-2020/the-rise-of-energy-prosumers-and-energy-democracy-history-and-future-prospects.

Lowitzsch, J., C.E. Hoicka, and F.J. van Tulder. 2020. "Renewable Energy Communities Under the 2019 European Clean Energy Package—Governance Model for the Energy Clusters of the Future?" *Renewable and Sustainable Energy Reviews* 122(April):109489. https://doi.org/10.1016/j.rser.2019.109489.

Martins, A., M. Madaleno, and M.F. Dias. 2020. "Energy Literacy: What Is Out There to Know?" *Energy Reports.* 6th International Conference on Energy and Environment Research—Energy and Environment: Challenges Towards Circular Economy 6(February):454–459. https://doi.org/10.1016/j.egyr.2019.09.007.

McCoy, M.L., and P.L. Scully. 2002. "Deliberative Dialogue to Expand Civic Engagement: What Kind of Talk Does Democracy Need?" *National Civic Review* 91(2):117–135. https://doi.org/10.1002/ncr.91202.

McEwen, J.E., J.T. Boyer, K.Y. Sun, K.H. Rothenberg, N.C. Lockhart, and M.S. Guyer. 2014. "The Ethical, Legal, and Social Implications Program of the National Human Genome Research Institute: Reflections on an Ongoing Experiment." *Annual Review of Genomics and Human Genetics* 15(1):481–505. https://doi.org/10.1146/annurev-genom-090413-025327.

Measham, T., A. Walton, and S. Felton. 2021. "Mining Heritage and Community Identity in the Social License of Proposed Renewed Mining." *Extractive Industries and Society* 8(3):100891. https://doi.org/10.1016/j.exis.2021.02.011.

Meeker, A. 2021. "Making Laws WITH the People." *POPVOX* (blog), January 29. https://medium.com/popvox/making-laws-with-the-people-d619b0aa84f1.

Meyer, R. 2023. "Protecting Nature Is More Important Than 'Quickly' Building Renewables, Most Americans Say." https://heat-map.news/climate/protecting-nature-is-more-important-than-quickly-building-renewables-most-americans-say.

Miller, C.A. 2022. "13 Redesigning Political Economy: The Promise and Peril of a Green New Deal for Energy." Pp. 270–292 in *The Green New Deal and the Future of Work*, C. Calhoun and B. Fong, eds. New York: Columbia University Press. https://doi.org/10.7312/calh20556-013.

Miller, C.A., and S. Moore. 2011. "Arizona's Energy Future." Phoenix, AR: Arizona Town Hall. https://www.aztownhall.org/resources/Documents/99%20Arizona's%20Energy%20Future%20Final%20Report.pdf.

Miller, C.A., J. O'Leary, E. Graffy, E.B. Stechel, and G. Dirks. 2015a. "Narrative Futures and the Governance of Energy Transitions." *Futures* 70(June):65–74. https://doi.org/10.1016/j.futures.2014.12.001.

Miller, C.A., J. Richter, and J. O'Leary. 2015b. "Socio-Energy Systems Design: A Policy Framework for Energy Transitions." *Energy Research and Social Science* 6(March):29–40. https://doi.org/10.1016/j.erss.2014.11.004.

Miller, C.A., L.W. Keeler, J. Loughman, and B. Davis. 2022a. "Pathways to a Carbon-Neutral Arizona Economy." Tempe, AR: Arizona State University, School for the Future of Innovation in Society, College of Global Futures. Report. https://doi.org/10.17605/OSF.IO/DRT7X.

Miller, C.A., Y. Rivera-Matos, A. Echevarria, and G. Dirks. 2022b. "Intentional and Responsible Energy Transitions: Integrating Design Choices in the Pursuit of Carbon-Neutral Futures." In *Routledge Handbook of Energy Transitions*, K.M. Araújo, ed. Milton Park, Abingdon, Oxon, UK, and New York: Routledge.

Mills, S. 2022. "Community-Wide Benefits via Property Taxes." Presented at Accelerating Decarbonization in the United States: Technology, Policy, and Societal Dimensions | Local Benefits and Compensation Strategies for Deep Decarbonization Infrastructure Webinar. October 3. https://www.nationalacademies.org/event/10-03-2022/accelerating-decarbonization-in-the-united-states-technology-policy-and-societal-dimensions-local-benefits-compensation-strategies-for-deep-decarbonization-infrastructure.

Mills, S.B., D. Bessette, and H. Smith. 2019. "Exploring Landowners' Post-Construction Changes in Perceptions of Wind Energy in Michigan." *Land Use Policy* 82(March):754–762. https://doi.org/10.1016/j.landusepol.2019.01.010.

Montoya, T. 2022. "Stockpile: From Nuclear Colonialism to 'Clean' Energy Futures." Society for Cultural Anthropology. https://culanth.org/fieldsights/stockpile-from-nuclear-colonialism-to-clean-energy-futures.

Murphy, T.P. 2002. "The Minnesota Report Card on Environmental Literacy: A Benchmark Survey of Adult Environmental Knowledge, Attitudes and Behavior." Minnesota Office of Environmental Assistance. https://eric.ed.gov/?id=ED474505.

Nadesan, M., M.J. Pasqualetti, and J. Keahey. 2023. "Introduction to Collection." Pp. xxxvii–xlviii in *Energy Democracies for Sustainable Futures*, M. Nadesan, M.J. Pasqualetti, and J. Keahey, eds. Academic Press. https://doi.org/10.1016/B978-0-12-822796-1.09986-1.

NASEM (National Academies of Sciences, Engineering, and Medicine). 2017. *Utilizing the Energy Resource Potential of DOE Lands.* Washington, DC: The National Academies Press.

NASEM. 2021. *Accelerating Decarbonization of the U.S. Energy System.* Washington, DC: The National Academies Press. https://doi.org/10.17226/25932.

New York State. 2022. "Climate Justice Working Group." New York's Climate Leadership and Community Protection Act. https://climate.ny.gov/resources/climate-justice-working-group.

Nielsen, J.A.E., K. Stavrianakis, and Z. Morrison. 2022. "Community Acceptance and Social Impacts of Carbon Capture, Utilization and Storage Projects: A Systematic Meta-Narrative Literature Review." *PLOS ONE* 17(8):e0272409. https://doi.org/10.1371/journal.pone.0272409.

Nilson, R. 2022. "Utility-Scale Solar in New York State: An Exploration of Public Response, Policy, and Justice." August. https://doi.org/10.7298/mc5p-2c69.

Nilson, R.S., and R.C. Stedman. 2022. "Are Big and Small Solar Separate Things?: The Importance of Scale in Public Support for Solar Energy Development in Upstate New York." *Energy Research and Social Science* 86(April):102449. https://doi.org/10.1016/j.erss.2021.102449.

Noble, B.F. 2000. "Strategic Environmental Assessment: What Is It? And What Makes It Strategic?" *Journal of Environmental Assessment Policy and Management* 02(02):203–224. https://doi.org/10.1142/S146433320000014X.

Noble, B.F. 2017. "Getting the Big Picture: How Regional Assessment Can Pave the Way for More Inclusive and Effective Environmental Assessments." https://doi.org/10.13140/RG.2.2.30630.52809.

NSF (National Science Foundation). 2022. "NSF's Convergence Accelerator Releases 2022 Portfolio Guide." https://beta.nsf.gov/funding/initiatives/convergence-accelerator/updates/nsfs-convergence-accelerator-releases-2022.

NSF. n.d. "NSF and Congress: Final Action Completed on Appropriations for FY22." https://www.nsf.gov/about/congress/119/highlights/cu22.jsp. Accessed May 11, 2023.

Nwanekezie, K., B. Noble, and G. Poelzer. 2022. "Strategic Assessment for Energy Transitions: A Case Study of Renewable Energy Development in Saskatchewan, Canada." *Environmental Impact Assessment Review* 92(January):106688. https://doi.org/10.1016/j.eiar.2021.106688.

NYPA (New York Power Authority). n.d. "Environmental Justice." https://www.nypa.gov/communities/nypa-engagement/environmental-justice. Accessed May 11, 2023.

OSTP and CEQ (White House Office of Science and Technology Policy and Council on Environmental Quality). 2021. "Indigenous Traditional Ecological Knowledge and Federal Decision Making." https://www.whitehouse.gov/wp-content/uploads/2021/11/111521-OSTP-CEQ-ITEK-Memo.pdf.

OSTP and CEQ. 2022a. "Guidance for Federal Departments and Agencies on Indigenous Knowledge." https://www.whitehouse.gov/wp-content/uploads/2022/12/OSTP-CEQ-IK-Guidance.pdf.

OSTP and CEQ. 2022b. "Memorandum on Implementation of Guidance for Federal Departments and Agencies on Indigenous Knowledge." https://www.whitehouse.gov/wp-content/uploads/2022/12/IK-Guidance-Implementation-Memo.pdf.

Oudshoorn, N., and T.J. Pinch, eds. 2005. *How Users Matter: The Co-Construction of Users and Technology,* Inside Technology. Cambridge, MA, and London: MIT Press.

Patnaik, S., J. Kunhardt, and R.G. Frank. 2023. "Why an Interdepartmental Coordination Group Should Be Part of the CDC's Reforms for Future Pandemics." https://www.brookings.edu/articles/why-an-interdepartmental-coordination-group-should-be-part-of-the-cdcs-reforms-for-future-pandemics.

Paullin, C. 2022. "State Work Group on Solar Development Achieves Little Consensus on New Regulations." *Virginia Mercury*. https://www.virginiamercury.com/2022/12/20/state-work-group-on-solar-development-unable-to-reach-consensus.

PEC (Pennsylvania Environmental Council). n.d. "Pennsylvania Environmental Council—Home." https://pecpa.org. Accessed May 11, 2023.

P.L.111-373. 2011. "Pedestrian Safety Enhancement Act of 2010." https://www.congress.gov/bill/111th-congress/senate-bill/841.

P.L.117-167. 2022. "Chips and Science Act." https://www.congress.gov/bill/117th-congress/house-bill/4346.

P.L.118-5. 2023. "H.R.3746—Fiscal Responsibility Act of 2023." https://www.congress.gov/bill/118th-congress/house-bill/3746.

Pyper, J. 2019. "The Navajo Generating Station Coal Plant Officially Powers Down. Will Renewables Replace It?" Greentech Media. November 20. https://www.greentechmedia.com/articles/read/navajo-generating-station-coal-plant-closes-renewables.

Rabe, B.G. 2014. "Shale Play Politics: The Intergovernmental Odyssey of American Shale Governance." *Environmental Science and Technology* 48(15):8369–8375. https://doi.org/10.1021/es4051132.

Radatz, A., M. Reinsborough, E. Fisher, E. Corley, and D. Guston. 2019. "An Assessment of Engaged Social Science Research in Nanoscale Science and Engineering Communities." *Science and Public Policy* 46(6):853–865. https://doi.org/10.1093/scipol/scz034.

Reed, M.S., S. Vella, E. Challies, J. de Vente, L. Frewer, D. Hohenwallner-Ries, T. Huber, et al. 2018. "A Theory of Participation: What Makes Stakeholder and Public Engagement in Environmental Management Work?" *Restoration Ecology* 26(S1):S7–S17. https://doi.org/10.1111/rec.12541.

Reeves, J.M., T. Baumgartl, D. Morgan, V. Reimers, and M. Green. 2022. "Community Capacity to Envisage a Post-Mine Future: Rehabilitation Options for Latrobe Valley Brown Coal Mines." M. Tibbett, A.B. Fourie, and G. Boggs, eds. Pp. 173–185. Perth, Australia: Australian Centre for Geomechanics (ACG). https://doi.org/10.36487/ACG_repo/2215_09.

ReImagine Appalachia. 2021. "A New Deal That Works for Us." https://reimagineappalachia.org/wp-content/uploads/2021/03/ReImagineAppalachia_Blueprint_042021.pdf.

Rezaei, M., and H. Dowlatabadi. 2015. "Off-Grid: Community Energy and the Pursuit of Self-Sufficiency in British Columbia's Remote and First Nations Communities." *Local Environment* 21(7):789–807. https://doi.org/10.1080/13549839.2015.1031730.

Richter, J., M.J. Bernstein, and M. Farooque. 2022. "The Process to Find a Process for Governance: Nuclear Waste Management and Consent-Based Siting in the United States." *Energy Research and Social Science* 87(May):102473. https://doi.org/10.1016/j.erss.2021.102473.

Rittman, B. 2023. "Alternative Energy Being Rejected in Central Illinois." *CIProud.com*. https://www.centralillinoisproud.com/news/local-news/alternative-energy-being-rejected-in-central-illinois.

Robinson, M. 2022. "Panel Discussion." Presented at Accelerating Decarbonization in the United States: Technology, Policy, and Societal Dimensions | Public Engagement Across the Transmission Development Lifecycle: From Planning to Permitting Webinar. September 30. https://www.nationalacademies.org/event/09-30-2022/accelerating-decarbonization-in-the-united-states-technology-policy-and-societal-dimensions-public-engagement-across-the-transmission-development-lifecycle-from-planning-to-permitting.

Roth, S. 2023. "How Can We Speed Up Solar and Wind Energy? Here Are Some Ideas." *Los Angeles Times*. https://www.latimes.com/environment/newsletter/2023-05-11/how-can-we-speed-up-solar-and-wind-energy-here-are-some-ideas-boiling-point.

Royster, J. 2008. "Practical Sovereignty, Political Sovereignty, and the Indian Tribal Energy Development and Self-Determination Act." *Lewis and Clark Law Review* 12:1065. University of Tulsa Legal Studies Research Paper No. 2011-01. https://papers.ssrn.com/abstract=1174472.

Ruhl, J.J., and Salzman, J. 2020. "What Happens When the Green New Deal Meets the Old Green Laws?" *Vermont Law Review* 44(4):693–722. https://scholarship.law.vanderbilt.edu/cgi/viewcontent.cgi?article=2180&context=faculty-publications.

Sandover, R., A. Moseley, and P. Devine-Wright. 2021. "Contrasting Views of Citizens' Assemblies: Stakeholder Perceptions of Public Deliberation on Climate Change." *Politics and Governance* 9(2):76–86. https://doi.org/10.17645/pag.v9i2.4019.

Saul, J., N. Malik, and D. Merril. 2022. "The Clean-Power Megaproject Held Hostage by a Ranch and a Bird." *Bloomberg*.

Scarry, R. 1968. *Richard Scarry's What Do People Do All Day?* 2015 edition. New York: Golden Books.

Scarry, R. 1974. *Richard Scarry's Cars and Trucks and Things That Go*. New York: Golden Press.

Schelly, C., D. Bessette, K. Brosemer, V. Gagnon, K.L. Arola, A. Fiss, J.M. Pearce, and K.E. Halvorsen. 2020. "Energy Policy for Energy Sovereignty: Can Policy Tools Enhance Energy Sovereignty?" *Solar Energy* 205(July):109–112. https://doi.org/10.1016/j.solener.2020.05.056.

Shehabi, A., and M. Al-Masri. 2022. "Foregrounding Citizen Imaginaries: Exploring Just Energy Futures Through a Citizens' Assembly in Lebanon." *Futures* 140(June):102956. https://doi.org/10.1016/j.futures.2022.102956.

Sherren, K. 2021. "From Climax Thinking Toward a Non-Equilibrium Approach to Public Good Landscape Change." Pp. 17–44, in *Energy Impacts: A Multidisciplinary Exploration of North American Energy Development*, J.B. Jacquet, J.H. Haggerty, and G.L. Theodori, eds. University Press of Colorado. http://www.jstor.org/stable/j.ctv19t41pj.4.

Shrestha, R., J. Neuberger, and D. Saha. "Table 4." *Federal Policy Building Blocks*. World Resources Institute. https://www.greenpolicyplatform.org/sites/default/files/downloads/resource/Federal-policy-building-blocks-support-just-prosperous-new-climate-economy-united-states_World_Resource_Institute.pdf.

Sides, J., C. Tausanovitch, and L. Vavreck. 2022. *The Bitter End: The 2020 Presidential Campaign and the Challenge to American Democracy*. Princeton, NJ: Princeton University Press. https://press.princeton.edu/books/hardcover/9780691213453/the-bitter-end.

Smith, C. 2022a. "Empowering Native Americans: Self-Determined Regenerative Energy Programs Restoring Sovereignty and Hope to Indigenous Communities." Presented at Accelerating Decarbonization in the United States: Technology, Policy, and Societal Dimensions | Public Responses and Engagement with Distributed Energy Infrastructure Webinar. June 28. https://www.nationalacademies.org/event/06-28-2022/accelerating-decarbonization-in-the-united-states-technology-policy-and-societal-dimensions-public-responses-and-engagement-with-distributed-energy-infrastructure.

Smith, C. 2022b. "IRA—Does It Stand for Something Good for Indian Country This Time?" *LinkedIn*. https://www.linkedin.com/pulse/ira-does-stand-something-good-indian-country-time-ch%C3%A9ri-smith.

SMUD (Sacramento Municipal Utility District). n.d. "Sustainable Communities." https://www.smud.org/en/Corporate/Landing/Sustainable-Communities. Accessed May 11, 2023.

Sobczyk, N. 2020. "Dems Roll Out Sweeping Environmental Justice Bill." *E&E News*. February 27. https://www.eenews.net/articles/dems-roll-out-sweeping-environmental-justice-bill.

Sovacool, B.K. 2009. "The Cultural Barriers to Renewable Energy and Energy Efficiency in the United States." *Technology in Society* 31(4):365–373. https://doi.org/10.1016/j.techsoc.2009.10.009.

Sovacool, B.K., D.J. Hess, S. Amir, F.W. Geels, R. Hirsh, L.R. Medina, C. Miller, et al. 2020. "Sociotechnical Agendas: Reviewing Future Directions for Energy and Climate Research." *Energy Research and Social Science* 70(December):101617. https://doi.org/10.1016/j.erss.2020.101617.

Springer, N., and A. Daue. 2020. "Key Economic Benefits of Renewable Energy on Public Lands." Yale Center for Business and the Environment and Wilderness Society. https://cbey.yale.edu/research/key-economic-benefits-of-renewable-energy-on-public-lands.

SRP (Salt River Project). n.d. "Coal Community Transition." https://www.srpnet.com/grid-water-management/grid-management/improvement-projects/coal-communities-transition. Accessed May 11, 2023.

Stang, J. 2022. "Washington Wind Power Farms May Conflict with Habitat Preservation Projects." *Energy News Network*. https://energynews.us/2022/02/04/washington-wind-power-farms-may-conflict-with-habitat-preservation-projects%EF%BF%BC.

State of Wyoming Legislature. 2023. "SJ0004—Phasing Out New Electric Vehicle Sales by 2035." https://wyoleg.gov/Legislation/2023/SJ0004.

Stewart, S. 2022. "Panel Discussion." Presented at Accelerating Decarbonization in the United States: Technology, Policy, and Societal Dimensions | Local Benefits and Compensation Strategies for Deep Decarbonization Infrastructure Webinar. October 3. https://www.nationalacademies.org/event/10-03-2022/accelerating-decarbonization-in-the-united-states-technology-policy-and-societal-dimensions-local-benefits-compensation-strategies-for-deep-decarbonization-infrastructure.

Stilgoe, J., R. Owen, and P. Macnaghten. 2013. "Developing a Framework for Responsible Innovation." *Research Policy* 42(9):1568–1580. https://doi.org/10.1016/j.respol.2013.05.008.

Szulecki, K. 2018. "Conceptualizing Energy Democracy." *Environmental Politics* 27(1):21–41. https://doi.org/10.1080/096 44016.2017.1387294.

TAMEST (The Academy of Medicine, Engineering, and Science of Texas). 2017. *Environmental and Community Impacts of Shale Development in Texas.* https://doi.org/10.25238/TAMESTstf.6.2017.

Trutnevyte, E., M. Stauffacher, and R.W. Scholz. 2011. "Supporting Energy Initiatives in Small Communities by Linking Visions with Energy Scenarios and Multi-Criteria Assessment." *Energy Policy, Clean Cooking Fuels and Technologies in Developing Economies* 39(12):7884–7895. https://doi.org/10.1016/j.enpol.2011.09.038.

UCS (Union of Concern Scientists). 2021. "A Transformative Climate Action Framework: Putting People at the Center of Our Nation's Clean Energy Transition." https://www.ucsusa.org/sites/default/files/2021-08/A-Transformative-Climate-Action-Framework.pdf.

Uebelhor, E., O. Hintz, and E. Gold. 2021. "Inventory of State Wind Property Tax Treatments." CLOSUP Working Paper Series Number 55. University of Michigan Center for Local, State, and Urban Policy. https://closup.umich.edu/research/working-papers/inventory-state-wind-property-tax-treatments.

U.S. Senate Committee on Commerce, Science, and Transportation. 2022. "CHIPS and Science Act of 2022 Section-by-Section Summary." https://www.commerce.senate.gov/services/files/1201E1CA-73CB-44BB-ADEB-E69634DA9BB9.

USDA (U.S. Department of Agriculture). n.d. "Empowering Rural America New ERA Program. U.S. Department of Agriculture—Rural Development." https://www.rd.usda.gov/programs-services/electric-programs/empowering-rural-america-new-era-program. Accessed May 11, 2023.

van de Grift, E., and E. Cuppen. 2022. "Beyond the Public in Controversies: A Systematic Review on Social Opposition and Renewable Energy Actors." *Energy Research and Social Science* 91(September):102749. https://doi.org/10.1016/j.erss.2022.102749.

van den Broek, K.L. 2019. "Household Energy Literacy: A Critical Review and a Conceptual Typology." *Energy Research and Social Science* 57(November):101256. https://doi.org/10.1016/j.erss.2019.101256.

van Wijk, J., I. Fischhendler, G. Rosen, and L. Herman. 2021. "Penny Wise or Pound Foolish? Compensation Schemes and the Attainment of Community Acceptance in Renewable Energy." *Energy Research and Social Science* 81(November):102260. https://doi.org/10.1016/j.erss.2021.102260.

Vogelsong, S. 2022. "Predatory Residential Solar Installers Could Sow Mistrust, Advocates Fear." *Virginia Mercury.* https://www.virginiamercury.com/2022/05/25/as-residential-solar-grows-more-popular-advocates-worry-about-consumer-protections.

Washington State Department of Ecology. n.d. "Climate Commitment Act." https://ecology.wa.gov/Air-Climate/Climate-Commitment-Act. Accessed May 11, 2023.

Wender, B.A., R.W. Foley, V. Prado-Lopez, D. Ravikumar, D.A. Eisenberg, T.A. Hottle, J. Sadowski, et al. 2014. "Illustrating Anticipatory Life Cycle Assessment for Emerging Photovoltaic Technologies." *Environmental Science and Technology* 48(18):10531–10538. https://doi.org/10.1021/es5016923.

WHEJAC (White House Environmental Justice Advisory Council). 2022a. "RE: Request for Resources for the White House Environmental Justice Advisory Council; Request for Timelines for Key Deliverables; Request for Key Agency Contacts to Attend Public Comment Period During Public Meetings; Recommendation for Increase in the Council on Environmental Quality Budget and Staff." White House Environmental Justice Advisory Council. https://www.epa.gov/system/files/documents/2022-04/whejac-ceq-final-letter-february-24-2022-public-meeting_.pdf.

WHEJAC. 2022b. "White House Environmental Justice Advisory Council May 2022 Meeting Summary." https://www.epa.gov/system/files/documents/2022-09/WHEJACPublicMeetingMay112022Final.pdf.

White House. 2022. "Bipartisan Infrastructure Law Tribal Playbook." https://www.whitehouse.gov/build/resources/bipartisan-infrastructure.

White House. 2023. *Guidebook to the Inflation Reduction Act's Clean Energy and Climate Investments in Indian Country.* https://www.whitehouse.gov/wp-content/uploads/2023/04/Inflation-Reduction-Act-Tribal-Guidebook.pdf.

Wolbring, G. 2008. "The Politics of Ableism." *Development* 51(2):252–258. https://doi.org/10.1057/dev.2008.17.

Wolbring, G. 2011. "Ableism and Energy Security and Insecurity." *Studies in Ethics, Law, and Technology* 5(1). https://doi.org/10.2202/1941-6008.1113.

Zimmerman, M.G., and T.G. Reames. 2021. "Where the Wind Blows: Exploring Barriers and Opportunities to Renewable Energy Development on United States Tribal Lands." *Energy Research and Social Science* 72(February):101874. https://doi.org/10.1016/j.erss.2020.101874.

Zullo, R. 2023. "Across the Country, a Big Backlash to New Renewables Is Mounting." *Idaho Capital Sun*, February 16. https://idahocapitalsun.com/2023/02/16/across-the-country-a-big-backlash-to-new-renewables-is-mounting.

6

The Essential Role of Clean Electricity

ABSTRACT

Reducing and eventually eliminating greenhouse gas (GHG) emissions from power generation is essential to reducing overall GHG emissions in the U.S. energy system. Power-sector emissions are currently the second-highest source of emissions. Equally important, electrification of buildings', vehicles', and other uses of energy can support decarbonization only if their supply of electricity contributes no net carbon emissions. Recent federal legislation and executive branch action, combined with policy support by many state governments, commitments by private companies, and favorable energy market and technology conditions, are helping to put the U.S. electric system on track to eliminate power-sector GHG emissions by midcentury, if not a decade earlier.

Although further developments are needed for ensuring post-2030 commercial readiness and availability of net-zero carbon generating and storage technologies that provide dispatchable, around-the-clock capabilities, significant and tenacious challenges remain in many non-technical factors necessary to staying on track.

These latter challenges for actors in federal agencies, state and local governments, the private sector, and civil society include

- Public engagement and decision-making support for siting the new transmission facilities and other electric projects and for the expansion of regional wholesale markets necessary to support an electric system that depends increasingly on renewable energy and other carbon-free resources;
- Regulatory support for investment in distribution-system infrastructure and communications, controls, interconnection policies, and other technologies necessary for supporting much-more expanded distributed energy resources at the grid edge;
- Regulatory support for innovative pricing of electricity, utility business model changes, and the provision of related retail services to enable dynamic interactions between customer-sited technologies and equipment and grid operations;

- Support from leaders, boards and stakeholders for the organizational, behavioral, and cultural changes needed in the electric sector;
- Regional coordination of transition planning in the electricity sector with other critical infrastructure sectors and effective engagement with diverse stakeholders and publics around equity and innovation in electricity systems; and
- Policy makers' support for ensuring that low-income electricity consumers and disadvantaged communities get access to clean and affordable electricity, energy bill savings, and reliable power.

In addition to continued investment of federal funds in research on and development of advanced technologies, greater attention to these non-technological issues is essential for the electric-system transitions needed for affordable, reliable, and equitable decarbonization outcomes.

Other chapters address the important changes that need to occur in the built environment, transportation, and industrial sectors as they rely increasingly on electrical energy, and in public engagement and financial systems to support these transitions.

INTRODUCTION

Clean electricity is essential to decarbonizing the U.S. economy. The many chapters of this report—along with the committee's first report (NASEM 2021a)—highlight the role that electrification will play in an affordable and effective path toward a net-zero economy. Increased electrification of vehicles, buildings, and industrial processes will be a fundamental element of achieving net-zero GHG emissions by 2050. But if the electricity powering these sectors does not come from net-zero carbon-emitting sources, this approach will not succeed. This chapter addresses issues on the way to a net-zero electricity system, focusing on the supply and delivery of clean power.[1] Table 6-1, at the end of the chapter, summarizes all the recommendations that appear in this chapter to support decarbonizing the electricity system.

---

[1] It is common for discussions of methods to reduce emissions from the power sector to focus on the role of avoiding emissions through such things as energy efficiency and other demand-side measures. Such actions are essential to a decarbonized economy and are discussed in the chapters on end uses of energy (i.e., Chapter 7 on the Built Environment, Chapter 9 on Transport, and Chapter 10 on Industrial Decarbonization) rather than in this chapter on the electric system itself.

The committee's first report included several recommendations to promote and facilitate a low-carbon electricity system, with particular attention to federal action:[2]

- Double the share of electricity generated by non-carbon-emitting sources, through
  - Setting a clean energy standard for electricity generation designed to reach 75 percent clean electricity by 2030 and net-zero emissions by 2050.
  - Expanding wind and solar capacity so that it would supply 45–55 percent of electricity nationwide.
  - Retiring coal plants (or retrofitting them to capture more than 90 percent of CO_2 emissions).
  - Preserving all safely operating existing nuclear power plants.
  - Maintaining existing gas-fired capacity (with modest reductions in new capacity).
  - Increasing electric transmission capacity by 40 percent to open up access to regions with strong renewable resources and deliver that power to distant regions of the United States.
  - Conducting research, development, and demonstration for long-duration energy storage technologies and on advanced electric generation technologies capable of providing around-the-clock power supply with no carbon emissions.
  - Reinforcing local electricity distribution networks to accommodate increasing peak demand from electric vehicles, heat pumps, and other new loads.
  - Adopting several new authorities to enable the Federal Energy Regulatory Commission (FERC) and the Department of Energy (DOE) to improve the design and functioning of wholesale markets and expansion of electric transmission.[3]

---

[2] Additional recommendations to support the electrification necessary for decarbonization and a just transition can be viewed in NASEM (2021c). These recommendations can be filtered, with terms most-relevant to clean electricity including "Produce carbon free electricity"; "Electrify energy services in transportation, buildings, and industry"; "Invest in energy efficiency and productivity"; and "Plan, permit, and build critical infrastructure."

[3] As summarized in Table S.1 in the first report (and repeated in Table C-1 in Appendix C in this publication), the committee's first report's recommendations included that
- FERC take the following actions:
  - "work with regional transmission organizations (RTOs) and independent system operators (ISOs) to ensure that markets in all parts of the country are designed to accommodate the shift to 100 percent clean electricity on the relevant timetable" (p. 199).

◦ Establishing and funding a new national Green Bank to support a wide variety of investments (e.g., in the power sector, buildings, and other eligible project targets).

Recommendations also included financial and technical support for public engagement in infrastructure planning and siting, and for communities and regions that have or will face challenges associated with the energy transition.

The committee made those recommendations in early 2021 with an understanding that the nation's electricity grid has been undergoing significant change in response to factors beyond the drive to equitably decarbonize the energy system. These other influences include the growth of distributed energy resources and behind-the-meter technology, evolution of the electricity system's institutional and market structure, new roles for consumers in the electricity system, and growing complexities in operational and electric-system security issues. Although implemented for economic, social, and other reasons, many of these other influences help to drive decarbonization outcomes as well.

◦ "direct [the North American Electricity Reliability Corporation] to establish and implement standards to ensure that grid operators have sufficient flexible resources to maintain operational reliability of electric systems" (p. 199).
- Congress enact the following:
  ◦ establish that it is the "National Transmission Policy to rely on the high-voltage transmission system to support the nation's" goals to achieve net-zero carbon emissions in the power sector (p. 211).
  ◦ authorize and direct FERC to require transmission companies and RTOs "to analyze and plan for economically attractive opportunities to build out the interstate electric system to connect regions that are rich in renewable resources with high-demand regions" (p. 211).
  ◦ assign to FERC "the responsibility to designate any new National Interest Electric Transmission Corridors and to clarify that it is in the national interest for the United States to achieve net-zero climate goals as part of any such designations" (p. 211).
  ◦ "authorize FERC to issue certificates of public need and convenience for interstate transmission lines (along the lines now in place for certification of gas pipelines), with clear direction to FERC that it should consider the location of renewable and other resources to support climate-mitigation objectives, as well as community impacts and state policies as part of the need determination (i.e., in addition to cost and reliability issues) and that FERC should broadly allocate the costs of transmission enhancements designed to expand regional energy systems in support of decarbonizing the electric system" (p. 211).
  ◦ "clarify that the Federal Power Act does not limit the ability of states to use policies (e.g., long-term contracting with zero-carbon resources procured through market-based mechanisms) to support entry of zero-carbon resources into electric utility portfolios and wholesale power markets . . . [and] direct FERC to exercise its rate-making authority over wholesale prices in ways that accommodate state action to shape the timing and character of the transitions in their electric resource mixes" (p. 199).

Finding 6-1: The transition to a net-zero power supply must occur alongside progress on other needs for the electric system—that is, that electricity supply is affordable, safe, secure, sustainable, equitable, and resilient (NASEM 2021b)— all of which pose challenges of their own and complicate the challenge of decarbonization.

DEVELOPMENTS SINCE EARLY 2021

A lot has happened since the committee issued its first report in February 2021. Changes have occurred in energy markets, in federal policy support for investment in clean power resources, and in subnational-government and private-sector actions.

Market Conditions

Although the U.S. economy overall remains highly dependent on fossil fuels (with coal, oil, and natural gas providing nearly four-fifths of all primary energy consumption [Chapter 12]), the electricity sector's reliance on fossil fuels continues to decline, with nuclear, hydro, wind, solar, and other zero-carbon energy sources accounting for 41 percent of total power supply (Rivera et al. 2023). In fact, virtually all of the zero-carbon energy used anywhere in the U.S. economy was used to produce electricity.

With increases in wind and solar capacity additions (EIA 2022a), renewable resources have accounted for a steadily growing share of power supply and are expected to account for 22 percent of electricity in 2022 (EIA 2022a), with much of those renewables occurring in the West, Southwest, and Plains states, where there is high renewable resource potential. Over the past decade, wind generation grew by 2.7 times, and solar photovoltaic generation at utility-scale and behind-the-meter sources increased from a minimal amount in 2012 to 160,799 GWh in 2021 (EIA 2022b). Coal-fired electricity generation has declined and now accounts for one-fifth of the nation's electricity, after decades during which it accounted for half of the country's power supply.[4] Power sector CO_2 emissions dropped by over one-third from 2005 through the end of 2022, with emissions slightly increasing from 2020 through 2022 (EIA 2023).

Based on analyses that incorporate the impacts of recent federal legislation, the Congressional Budget Office (CBO) anticipates that over the upcoming decade,

---

[4] Coal-fired generation provided between 45 percent to 57 percent of U.S. electricity for the period from 1950 to 2010, with most years having higher than 50 percent. Since then, coal-fired power production decreased gradually as gas-fired and renewable generation have increased and power demand has remained relatively flat. EIA (2023, Tables 7.2a and 7.2b), https://www.eia.gov/electricity/data.php#generation.

power-sector CO_2 emissions can be expected to decrease further, largely because of the introduction of new wind and solar projects and with the pace and magnitude of reductions depending on changes in costs of wind, solar and battery technologies, fossil fuel prices, and the siting of new renewable projects and transmission lines (CBO 2022).

> Finding 6-2: Total power-sector CO_2 emissions dropped by 36 percent between 2005 and 2022, and the carbon intensity of electric supply dropped 42 percent since 2005.[5] While progress has slowed in recent years, reductions are expected to accelerate again in upcoming years.

Federal Policy

Progress ahead is anticipated in light of several major federal actions adopted in the executive branch and in Congress since early 2021.

Executive Branch Action

In the earliest days of his administration, President Biden issued multiple executive orders (EOs) committing to put "the climate crisis at the center of United States foreign policy and national security," take "a government-wide approach to the climate crisis," use "the federal government's buying power and real property and asset management" to advance a clean energy transition, empower "workers through rebuilding our infrastructure for a sustainable economy," and secure environmental justice and economic opportunity (White House 2021a). The President announced that the United States would rejoin the Paris Agreement.

In March 2021, President Biden proposed a $2 trillion infrastructure plan, including hundreds of billions of investment dollars for the electric grid and clean energy and climate technology (Parlapiano and Tankersley 2021). By April 2021, the President had set a goal of reaching 100 percent clean electricity by 2035 (White House 2021c). On the eve of the United States' participation in the United Nations Framework Convention on Climate Change's (UNFCCC's) Conference of the Parties (COP26) in Glasgow, Scotland, in late 2021, the President announced the climate components of his proposed "Build Back Better" framework, including tax credits, grants, loans, and other incentives for clean energy investment (including in the power sector), and consumer rebates and tax credits to help families shift to electric equipment and install energy efficiency measures (White House 2021b).

---

[5] Carbon intensity (pounds of CO_2 per MWh), as of Q2 2022 (Samaras 2022).

Agency actions further supported the clean power agenda. In the summer of 2021, FERC asked for comments on what changes were needed in interstate transmission planning and cost support in order to plan more holistically for the nation's needs as the electric system shifted to greater reliance on renewable resources often located far from consumers (FERC 2021). In April 2022, FERC proposed new regulations to modify its transmission planning, cost-allocation, and generator interconnection policies so as to address "some perceived shortcomings in current regional transmission planning processes" and its ability to keep pace with changing electric-system needs (FERC 2022b).

In the month prior to FERC's issuance of its proposed transmission-planning rule, the Securities and Exchange Commission (SEC) proposed rules to enhance publicly traded companies' disclosure of climate-related risks and GHG emissions (SEC 2022). This was important for the power sector in light of the large number of electric companies that had made commitments to reduce GHG emissions, and the role that transparency and reporting plays in maintaining corporate commitment to decarbonization. (See Chapter 11 for further discussion.)

Power-plant emissions regulation by the Environmental Protection Agency (EPA) was addressed by the Supreme Court in its June 2022 decision in the *West Virginia v. U.S. EPA* case, which held that without a clear statement of congressional intent that delegated authority to an agency for certain "major" regulations, the agency was limited in adopting and implementing regulations in that domain (Bergman et al. 2022; RFF 2022; Supreme Court of the United States 2021). In May 2023, EPA issued new proposed standards for regulating CO_2 emissions from new and existing fossil-fueled power plants, with EPA analyses indicating that the proposal would avoid 617 million metric tons of CO_2 through 2042 and that the reduced air pollutants would lead to up to $85 billion in climate and public health benefits over the next 2 decades (EPA 2023). Complying with these proposed standards would require emissions-control through carbon capture and storage/sequestration (CCS), or co-firing coal plants with natural gas and co-firing natural gas plants with low-emissions hydrogen (Lashof 2023). Enacting these emissions-control strategies will require testing and adaptation by the power sector, which could influence the electricity system's trajectory along its energy transition pathway.

Congressional Action

The President's proposals and commitments ultimately found their way into two congressional laws: the Infrastructure Investment and Jobs Act (IIJA) enacted in late 2021 and the Inflation Reduction Act (IRA) of August 2022. Although neither law included a binding target for reducing total CO_2 or GHG emissions in either the power sector or the economy at large, both laws provided billions of dollars of incentives to encourage clean energy investment and project development to enter the market over the upcoming decade.

Power-Sector Provisions of the IIJA

The IIJA authorizes more than $50 billion for clean energy for the power sector (and other large-scale facilities), grid hardening and resilience, retention of existing nuclear and hydropower facilities, and building out the nation's electric grid (McLaughlin and Bird 2021). Figure 6-1 shows funding amounts for relevant sections (not counting the support for electrifying transportation [shown in the yellow rectangle] and reducing energy costs and improving efficiency [shown in the dark green rectangle]):

- Support the clean energy economy and innovation (shown in the dark orange rectangle): $28 billion
  - Carbon capture and storage for industrial and power-sector purposes: $10 billion
  - Battery supply chains used in energy storage and electric vehicle (EV) applications: $7 billion

FIGURE 6-1 Funding for Clean Energy in the Infrastructure Investment and Jobs Act.
SOURCE: McLaughlin and Bird (2021), https://www.wri.org/insights/implementing-clean-energy-investments-us-bipartisan-infrastructure-law. CC BY 4.0.

- Hydrogen hubs, electrolysis cost reduction, and recycling: $8 billion
  - Advanced modular nuclear reactors: $3 billion
- Support grid hardening and resilience (shown in the blue rectangle): $11 billion
  - Grid reliability and resilience research and development: $6 billion
  - Competitive grants for grid hardening and weatherization: $5 billion
- Retain existing nuclear and hydro generation (shown in the light green rectangle): $6.7 billion
  - New civil nuclear credit program to retain existing nuclear fleet: $6 billion
  - Retain efficiency and improve operations of existing hydropower facilities: $0.7 billion
- Build the transmission system (shown in the purple rectangle): $5.5 billion
  - Expansion of the Smart Grid Investment Matching Grant Program: $3 billion
  - Revolving loan fund to support nationally significant transmission lines: $2.5 billion.

In August 2022, DOE established a new Grid Deployment office to lead the $17 billion in IIJA programs and projects related to the nation's electric transmission, distribution, and power generation needs, as well as the programs to retain carbon-free power from nuclear and hydroelectric facilities (DOE 2022). The IIJA also included a requirement that state utility regulators consider establishing electricity rate mechanisms that allow utilities to recover the costs of promoting customer demand-response practices (BIL Summary 2021). Such flexible demand has the potential to help grid managers operate a grid reliably and economically with greater penetration of intermittent resources like solar and wind generation (NASEM 2021b).

Power-Sector Provisions of the IRA

Although President Biden and many congressional leaders had hoped to see greater financial incentives for clean power in 2021, it was not until August 2022 that such incentives were put into law through the IRA. Considered the largest climate bill ever enacted by Congress, the IRA includes numerous additional incentives for a low-carbon electric system on top of those in the IIJA.

For example, the IRA increases DOE's lending and loan-guarantee authority for power-system and other energy infrastructure;[6] authorizes $27 billion in funding for state,

---

[6] The Bipartisan Policy Center (BPC) observes that the IRA gave the DOE Loan Programs Office "$40b in new Title 17 loan authority available through 2026 with $3.6 billion for credit subsidies" and creates the Energy Infrastructure Reinvestment Financing program with "$5 billion to carry out program authorities and $250 billion in loan authority through 2026" (BPC 2022a, p. 17).

local, tribal, and nonprofit financial institutions to invest in equitable access to clean energy through a new Greenhouse Gas Reduction Fund (§60103) under authority given to EPA; and provides significant tax-credit incentives for private investment in clean energy. As described by analysts from the Rhodium Group in their analysis of IRA impacts, the new law includes a suite of long-term, full-value, flexible clean energy tax credits and other programs in the IRA focus on the "4 Rs" of electric generation decarbonization:

- *Reinvigorate* new clean capacity additions: production and investment tax credits (PTC and ITC);
- *Retain* existing clean capacity: zero-emitting nuclear PTC;
- *Retire* fossil capacity: U.S. Department of Agriculture (USDA) investments in rural electric cooperatives (coops) and [DOE] loan programs; and
- *Retrofit* remaining fossil capacity: section 45Q carbon capture tax credit (Larsen et al. 2022, p. 4).

More specifically, the production and investment tax credits (e.g., for clean hydrogen production, for renewable power investment and/or electrical output, for new advanced manufacturing for clean energy equipment components) either extend existing ones that are slated to expire or add new ones available to investors over the next decade. Many of the tax credits include direct pay provisions (in lieu of a tax credit) that open up greater investment opportunities for tax-exempt entities (e.g., electric coops) and enable efficient use of federal incentives (BPC 2022b). Legal analysts (Eversheds Sutherland 2022; Schurle et al. 2022; Sidely 2022) have pointed to other notable changes in the IRA's clean-energy tax credits, including providing bonus credits where facilities are constructed on brownfield sites or in an area where there have been closures of coal mines or coal-fired generating facilities; tying full availability of the credit value to such things as near-term commencement of construction and prevailing wage and apprenticeship requirements; relying on product inputs (e.g., steel, iron, other manufactured products) produced in the United States. The Department of the Treasury plays an important role in specifying and clarifying these tax credit provisions and qualifications.

Subnational Government Policy

Meanwhile, states and localities have also taken steps to lower power-sector emissions. Most states have some sort of mandate to add increasing amounts of renewable energy to their electricity supply, and many states have further requirements to significantly reduce or eventually eliminate electricity sources with GHG emissions in the decades ahead (Figure 6-2). Notably, states accounting for more than 50 percent of the nation's

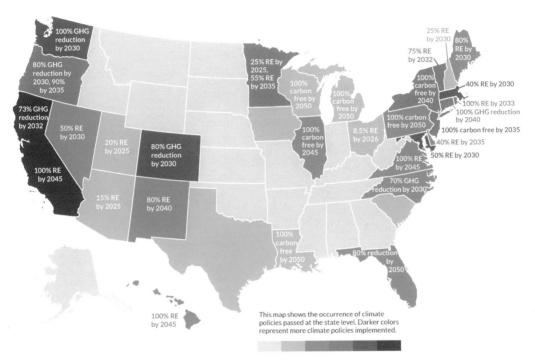

FIGURE 6-2 States with climate, renewable, or GHG-reduction mandates for the electricity sector as of September 2023. NOTE: Map annotated by the committee on September 12, 2022. SOURCES: Adapted from Climate XChange (C2ESab). Data for *Louisiana and Michigan* from Micek (2020). Data for *Pennsylvania* from Pennsylvania DEP (2021).

power demand have adopted clean-power goals (EIA 2021).[7] These actions by states provide additional heft in the nation's push toward decarbonizing the power system.

Private-Sector Commitments

Many other public and many private entities have also made substantial climate commitments affecting clean-power transitions. As of 2019, subnational and private-sector entities representing 71 percent of U.S. Gross Domestic Product (GDP), 68 percent of the population, and 51 percent of GHG emissions had pledged to reduce GHG emissions (see Figure 6-3). The federal statutory support from the IIJA and the IRA can be expected to strengthen the chances that these emissions reductions will occur

---

[7] These states include California, Colorado, Connecticut, Florida, Hawaii, Illinois, Louisiana, Maryland, Massachusetts, Michigan, Minnesota, Nevada, New Jersey, New Mexico, New York, North Carolina, Oregon, Pennsylvania, Virginia, and Washington.

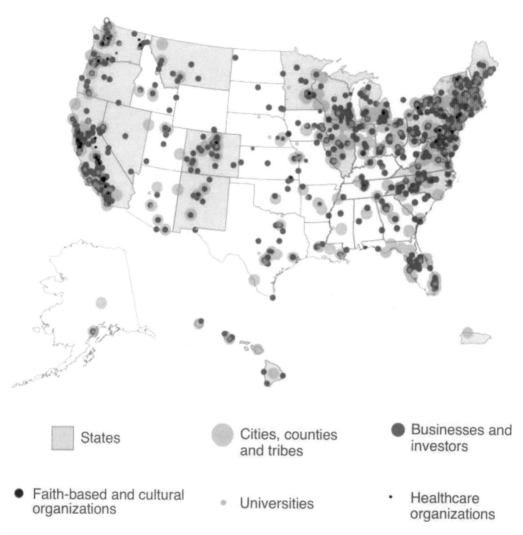

States

Cities, counties and tribes

Businesses and investors

Faith-based and cultural organizations

Universities

Healthcare organizations

FIGURE 6-3 Subnational and private-sector climate commitments. SOURCE: Reproduced with permission from Hultman et al. (2019), Bloomberg Philanthropies, with modifications from Hultman et al. (2020), https://doi.org/10.1038/s41467-020-18903-w. CC BY 4.0.

through private-sector commitments and do so earlier and at lower cost than without the new federal financial incentives in place.

Notably, electric utilities serving 84 percent of the nation's electricity customer accounts have made commitments to have either 100 percent renewable/clean power or net-zero emissions by no later than 2050 (Figure 6-4). In all, 497 individual utilities are preparing to meet their state's 100 percent carbon-reduction mandates (SEPA 2023).

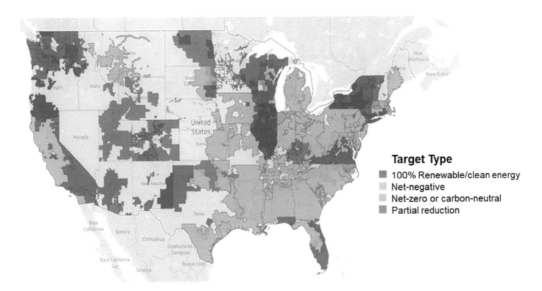

FIGURE 6-4 Service territories of electric utilities with commitments made to reduce power-sector emissions. SOURCE: Smart Electric Power Alliance, generated by ORNL, ©2023 Mapbox ©OpenStreetMap.

Power-sector actions emerging from these subnational-government and private-sector commitments include investment in and contractual procurements of onshore and offshore wind, and utility-scale and distributed solar capacity (MA Department of Public Utilities 2017; McGovern 2022; Penrod 2022; Peretzman 2022); retirements of coal plant capacity (Brown 2022); programs to compensate zero-carbon generating resources (like existing nuclear plants) for their value in avoiding carbon emissions (Illinois Power Agency 2021; Lopez 2022; Trabish 2021); electric transmission project plans (Grid North Partners 2022; Willson 2022); adopting carbon prices in state electricity markets (McCarthy 2022; RGGI 2023); and corporate efforts (e.g., the 300-member Clean Energy Buyers Alliance[8]) to accelerate the achievement of carbon-free electricity through procurements of clean energy (CEBA 2023). Figure 6-5 depicts carbon-reduction targets by utility companies compared to state requirements.

> Finding 6-3: A significant share of actors with decision-making responsibility for power-sector developments has made climate commitments. States accounting for more than 50 percent of the nation's power demand have adopted clean-power goals. Subnational and private-sector entities representing

---

[8] Since 2014, "energy customers have voluntarily procured over 60 gigawatts (GW) of clean energy," which is "equivalent to 26 percent of the total capacity added to the U.S. grid 2014–2022" and includes "16.9 GW of new clean energy deals announced by energy customers in 2022" CEBA (2023).

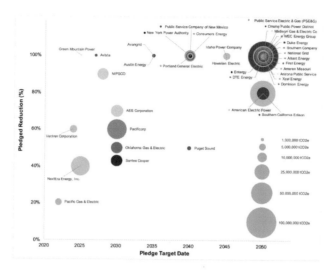

FIGURE 6-5 Plot of voluntary electric utility company emissions reduction pledges compared with state requirements. SOURCE: This article was published in D. Godlevskaya, C. Galik, and N. Kaufman, 2021, "Major US Electric Utility Climate Pledges Have the Potential to Collectively Reduce Power Sector Emissions by One-Third," *One Earth* 4(12):1741–1751. Copyright Elsevier (2021).

71 percent of U.S. GDP, 68 percent of the population, and 51 percent of GHG emissions have pledged to reduce GHG emissions. Electric utilities representing 84 percent of the nation's electricity customer accounts have made commitments to have either 100 percent renewable/clean power or net-zero emissions by no later than 2050. The federal support from the IIJA and the IRA can be expected to strengthen the chances that these emissions reductions will occur and do so earlier and at lower cost than without the new federal financial incentives in place.

HOW FAR DO CURRENT/NEW POLICY AND MARKET CONDITIONS GET US?

From a technology-deployment and investment point of view, the types of clean-power incentives and commitments taking shape in recent years align with the recommendations offered by the committee in its first report, which indicated that non-carbon-emitting sources of power would need to account for 75 percent of the electric power supply by 2030 to put the nation on a path to net zero by midcentury. (Figure 1-3 in Chapter 1 indicates that such a trajectory of emissions reductions is consistent with the White House's Long-Term Strategy and its outlook for the power sector [DOS and EOP 2021].)

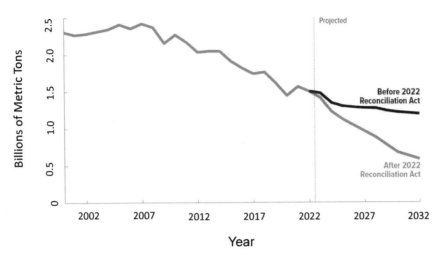

FIGURE 6-6 Congressional Budget Office's anticipated trends in carbon dioxide emissions in the U.S. power sector. SOURCE: Courtesy of CBO (2022).

Early reporting on (Ewing and Penn 2022) and estimates of (CBO 2022; Jenkins et al. 2022b; Larsen et al. 2022; Mahajan et al. 2022) the impacts of these federal, state, local, and private efforts show that they will help to make important progress toward decarbonizing the electric system.

For example, CBO has estimated that the IRA (also known as the 2022 Reconciliation Act) alone could lead to significant carbon emission reductions in the power sector over the next decade (as shown in Figure 6-6), with potential uncertainty related to factors such as the cost of new wind, solar, and battery capacity; future fossil fuel prices; and the availability of transmission expansion potentially leading to higher or lower emissions levels by 2032 (CBO 2022).

The REPEAT modeling team (Jenkins et al. 2022b) estimates that the combination of IIJA and IRA provisions and other existing federal and state policy is likely to accelerate progress toward the goal of zero-emitting resources accounting for 75 percent of power supply by 2030. By substantially reducing investors', owners', and consumers' cost of adding and maintaining zero-emitting generation technologies, the IRA's tax-credit provisions, for example,

> could spur record-setting growth in wind and solar capacity, with annual additions increasing from 15 GW of wind and 10 GW of utility-scale solar PV in 2020 to an average of 39 GW/year of wind additions in 2025–2026 (~2× the 2020 pace) and 49 GW/year of solar (~5× the 2020 pace), with solar growth rates increasing thereafter. The bill will also incentivize deployment of carbon capture at new and existing natural gas power plants and retrofits of existing coal plants, owing to the enhanced 45Q tax credit. (Jenkins et al. 2022b, p. 11)

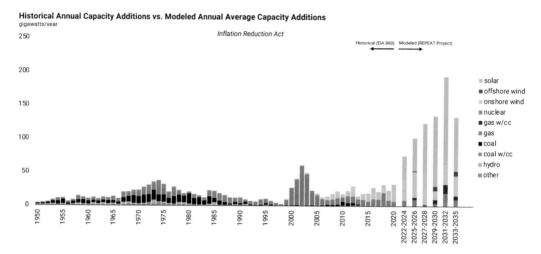

FIGURE 6-7 Historical annual generating capacity additions and modeled annual average generating capacity additions (GW/yr) assuming the IIJA and IRA. NOTE: See footnote 9 and/or source for description of modeling assumptions. SOURCE: Jenkins et al. (2022b), https://doi.org/10.5281/zenodo.7106218. CC BY 4.0.

Figure 6-7 shows the projected annual increases in solar, onshore wind, and offshore wind capacity additions that REPEAT estimates could enter the system after 2022.[9]

The REPEAT Project examined the sensitivity of different carbon-reduction outcomes in the power sector to changes in assumptions about the pace of build-out of the nation's high-voltage transmission grid. Zero-emitting resources would accelerate and expand so as to contribute from approximately 50 percent to 80 percent of power supply by 2030 and between 60 percent and 85 percent by 2035 (see Figure 6-8). The availability of increased transmission would allow for faster reductions in fossil generation and their GHG emissions.

In April 2023, Jenkins et al. previewed a revised analysis, which reflected supply chain and other limiting constraints slowing policy implementation, estimating the

---

[9] The REPEAT project authors add the following caveat to this analysis: "Several constraints that are difficult to model may limit these growth rates in practice, including the ability to site and permit projects at requisite pace and scale, expand electricity transmission and CO_2 transport and storage to accommodate new generating capacity, and hire and train the expanded energy workforce to build these projects. Modeled results should thus be taken as indicative that IRA establishes strong financial incentives to build capacity at the modeled pace, while non-financial challenges may constrain the pace of real-world deployment relative to modeled results. Several policies in IRA and the Bipartisan Infrastructure Law, as well as proposed permitting reforms to be considered by Congress this Fall, can reduce these non-financial barriers (e.g., reforms to transmission siting and funding for CO_2 transport and storage in IIJA; funding to expedite NEPA review in IRA; transmission investment funding in both bills)" (Jenkins et al. 2022, p. 11).

Electricity Generation
terawatt-hours (TWh)

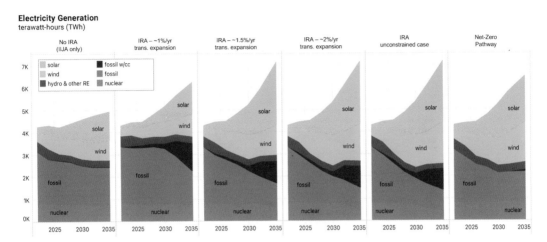

FIGURE 6-8 Generation shares by resource by 2035, under various transmission-expansion assumptions.
SOURCE: Jenkins et al. (2022a), https://doi.org/10.5281/zenodo.7106176. CC BY 4.0.

IRA's effects to result in approximately 37–41 percent of emissions reductions below 2005 levels, compared with approximately 43 percent in the preliminary analysis (Jenkins et al. 2023).

Estimates by several other analytic teams—Rhodium Group, Resources for the Future (RFF), the National Renewable Energy Laboratory (NREL), Energy Innovation, and ClearPath—reach similar conclusions about the impact of new federal policy on changes in the power sector. Figure 6-9 summarizes the results of power-sector estimates of clean generation shares in 2030 as conducted by the Rhodium Group (Larsen et al. 2022); Figure 6-10 shows Energy Innovation's range of estimates of 2030 GHG emission reductions associated with investments and operations, with most of the reductions occurring in the electricity sector (Mahajan et al. 2022).

Analysis by the NREL team indicates that shares of clean electricity[10] would increase from 41 percent in 2022 to 71–90 percent of total generation by 2030 (a 25–38 percentage point increase relative to a "no new policy" scenario) with the implementation of key provisions from the IRA and the IIJA (also referred to as the Bipartisan Infrastructure Law) (Steinberg et al. 2023). In turn, this modeling indicates that by 2030, annual power sector CO_2 would be between 72–91 percent of the 2005 baseline along with a 5–13 percent reduction in average annual bulk power system costs.

---

[10] Includes nuclear, fossil energy with CCS, wind, solar, hydroelectric, geothermal, landfill gas, and biomass.

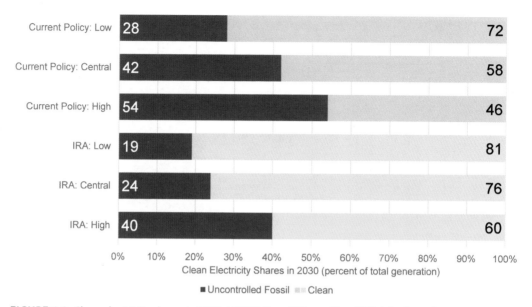

FIGURE 6-9 Clean electricity shares in 2030. NOTES: "Low," "Central," and "High" reflect Rhodium Group's three core emissions scenarios: "A central case that relies on central energy market prices, central clean technology costs, and baseline economic growth. A low emissions case that uses continued rock-bottom prices for clean energy technologies paired with high oil and gas prices and baseline economic conditions. A high emissions case that considers the inverse: expensive clean technologies, cheap oil and gas, and a high economic growth rate" (King et al. 2022, p. 9). SOURCE: Larsen et al. (2022), *Rhodium Group*.

RFF analysts have modeled electric-industry transitions in light of the new federal statutes (and in particular, their tax incentives for clean energy supply) (Roy et al. 2022). Their analysis compared "the average annual change in retail prices in the no-policy baseline and in the policy scenarios under expected natural gas prices, as well as with alternative natural gas price scenarios from [EIA's] *AEO 2021*. The IRA is projected to have a deflationary effect on retail electricity prices under all of the alternative scenarios we modeled" (Roy et al. 2022, p. 2).[11] The RFF researchers concluded that electricity costs to consumers could be expected to decline 5.2–6.7 percent over the next decade, saving them $209 billion–$278 billion, and smaller electricity bills and lower costs of other goods and services would mean $170–$220 in annual savings to the average household (Roy et al. 2022).[12] The RFF analysis estimated that by 2030,

---

[11] "Even if natural gas prices are higher than expected, as they have been in recent months owing to global shocks in fuel prices, electricity rates are still projected to decline under the legislation" (Roy et al. 2022, p. 3).

[12] The study notes other outcomes of the federal legislation, as well. "There are several provisions in the IRA subsidizing the domestic manufacturing and production of inputs to electricity generation that also can contribute to a reduction in the capital costs of qualifying generation, and which our modeling does not

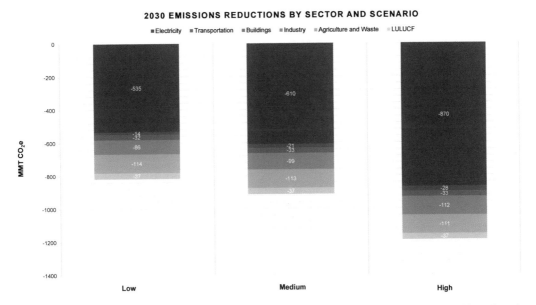

FIGURE 6-10 2030 emissions reductions by sector and scenario. SOURCE: Tallackson and Baldwin (2022), https://energyinnovation.org/wp-content/uploads/2022/11/Implementing-The-Inflation-Reduction-Act-A-Roadmap-For-State-And-Federal-Buildings-Policy.pdf. CC BY 4.0.

power-sector emissions would drop by 70–75 percent relative to 2005, as compared to approximately a 48.5 percent reduction without the policy (Roy et al. 2022).[13]

ClearPath researchers modeled the implications of the combination of utility climate commitments and the provisions of the IRA, as shown in Figure 6-11. The model results indicated that together, these commitments combined with the IRA's federal financial incentives would lead to deeper near-term reductions than would be the

represent. Additionally, domestic manufacturing can reduce bottlenecks in the energy supply chain that could contribute to inflationary pressures. . . . Lower electricity prices under the IRA also can be expected to accelerate electrification of transportation and buildings, which would likely complement the nation's climate policy goals and provide additional savings to households. The IRA provides substantial incentives for energy efficiency and electrification that are expected to provide substantial additional savings to consumers" (Roy et al. 2022, p. 4).

[13] Another study that modeled the impacts of the IRA's clean electricity tax credits found that they "are projected to cut the average residential bill by 3.4 percent in 2030 and 4.6 percent in 2035, relative to business as usual (Figure 3). This amounts to annual electricity bill savings of $37 and $52 in 2030 and 2035 (2021$), respectively, for the average U.S. household. The total this comes to $60 billion in electricity bill savings for U.S. households over the next 15 years (2021$)" (Levin and Ennis 2022). Additionally, Bistline et al. (2023) observe that IRA incentives "have implications for electricity prices and affordability" and "can have large impacts on electricity markets by lowering wholesale prices and increasing prevalence of negative-priced periods, which can alter operational, investment, and retirement decisions."

Projected Power Sector CO₂ Emissions vs Net-Zero by 2050

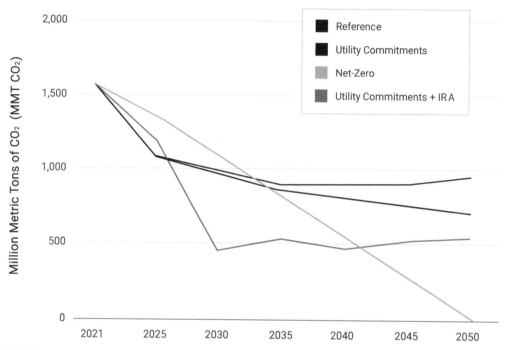

FIGURE 6-11 ClearPath modeled CO_2 emissions reductions associated with electric utility commitments, utility commitments plus IRA incentives, and a net-zero pathway for the power sector.
SOURCE: ClearPath (2023).

case with the corporate commitments alone and helped to put the power sector on a trajectory of 2040 emissions close to what is needed to head toward a net-zero pathway as of that year.

A recent nine-model intercomparison published in *Science* shows the IRA resulting in wind and solar growth rates ranging from 10 to 99 GW/year between 2021 to 2035, with particularly wide variation in predicted increases in energy storage, ranging from 1 to 18 GW/year. Despite this variation, generation and emissions outcomes for the power sector are more closely aligned across models by 2035, with key provisions from the law resulting in emissions reductions of 43 percent and 48 percent below 2005 levels by 2035 (Bistline et al. 2023).

> Finding 6-4: From a technology point of view, the combination of new federal policies (e.g., the IIJA and the IRA), state policy adoption, and other subnational and private-sector commitments put the United States on track to reduce power-sector emissions in 2030/2035 by a percentage consistent with achieving net-zero emissions by midcentury.

Finding 6-5: Power-sector clean-energy transitions enabled by recent federal legislation can help to lower electricity bills and produce energy bill savings for consumers.

BARRIERS, IMPEDIMENTS, AND WHAT STILL NEEDS TO BE DONE TO ACHIEVE NET-ZERO POWER SUPPLY

As the committee stated in its first report, "decarbonization studies find that reaching net-zero emissions is technically feasible (and relatively low cost) provided that significant proactive effort is invested over the next decade to drive the maturation and improvement of a range of more nascent technologies and solutions needed to reach net-zero emissions" (NASEM 2021a, pp. 59–60).

The positive outcomes that could result from technically feasible and increasingly economical deployment of clean-power technologies—especially in light of federal financial support introduced by the IIJA and the IRA—depend to a large degree on resolving persistent non-technological challenges. Examples of non-technological issues are institutional and jurisdictional complexities in the power sector—barriers that were summarized by an expert workshop hosted by RFF and that include

> market structures that disfavor renewable energy resources, backed-up interconnection queues, local siting opposition, policy uncertainty, and challenges in arranging for efficient procurement of clean power . . . institutional mismatch between state agencies and Regional Transmission Operators (RTOs) with authority over transmission, difficulty agreeing on cost allocation for interstate and interregional transmission, insufficient state government capacity for studying and engaging with the planning process, and local opposition. Demand management is hampered by lack of access to energy efficiency for low-income households and renters, inadequate metrics for energy efficiency, inadequate price incentives for consumers, incomplete incentives for utilities and transmission investors, and inequitable and confusing rate structures. (Domeshek et al. 2022, p. ii)

Many of these same issues were discussed during panels hosted by the committee and in publications, government convenings, and articles of experts in the industry (see, e.g., FERC 2021; NASEM 2021a,b). Chapters 5 and 11 of the committee's report also discuss new, reformed, and innovative aspects of public engagement and activities of subnational governments as critical to addressing non-technical issues associated with transitions in the power sector.

Many of these challenges relate to the adoption of government policies that were recommended in our first report but not subsequently enacted in the IIJA or the IRA, including a cap on power-sector emissions and/or a clean power standard. Additional policies are still needed to build out the electric grid to accommodate more renewables, expand regional power markets, improve the design of electricity rates, deploy distributed energy resources, and modernize local distribution grids. Equally, such governmental actions depend on public acceptance of policy changes and regulatory decisions, which often does not happen under traditional decision-making approaches.

Finding 6-6: There is broad consensus in the published literature that rapidly decarbonizing the electric system is critical to the success of the nation's decarbonization transition. Decarbonization needs to happen while still meeting the objective of an affordable, equitable, reliable, and resilient electricity system that reduces its other environmental impacts. In spite of clear progress to date (including with the IIJA, the IRA, state policy, and private-sector commitments), additional policies will be needed to successfully address persistent barriers to decarbonizing the electricity sector during the next decade and for the years beyond 2030/2035.

Finding 6-7: Business-as-usual approaches to government decision-making on such things as transmission and distribution system build-out, regional power markets, retail rate design, and distributed energy resource deployment can impede the type and pace of actions needed for decarbonizing the power system consistent with a net-zero transition. Changes in public engagement and governmental leadership on these issues will be necessary to enable equitable emissions reductions and affordable decarbonization opportunities offered by technological change. Without such changes, electric system transitions will be more expensive, slower to accomplish, and less equitable than otherwise could occur.

The remainder of this chapter addresses these persistent challenges and makes recommendations related to resolving them. These recommendations are intended to complement and amplify the recommendations the committee makes in other chapters—including with respect to equity, workforce, public engagement, subnational action, financial markets, and end-use electrification and efficiency improvements in the built environment and transportation sectors.

Note that this chapter concludes with one recommendation that relates to ensuring the readiness of technology options for a decarbonized electric system for the period beyond 2030/2035.

Adoption of Limits on Power-Sector Emissions

Given the trajectory of power-sector emissions reductions and the expectation that policy and market forces will lead to the entry of additional zero-carbon power sources, it might seem unnecessary to introduce a cap on GHG emissions from power production. But electrification of buildings, vehicles, and other activities will increase the nation's overall demand for electricity. And depending on the extent to which the system's increase in size to meet new loads is modulated by such things as flexible demand, energy efficiency, storage, other on-demand capacity, transmission expansion, and other factors, the electric system could grow, and with it, GHG emissions could only slowly go down—or even rise (Jenkins et al. 2022b; Larson et al. 2021; Mai et al. 2018).[14] (See further discussion below.) Thus, even with the powerful incentives provided by federal, state, and private-sector policies and commitments, actual success in accomplishing a net-zero electricity system in time to achieve a net-zero economy by midcentury requires an additional national policy: a formal cap on power-sector emissions or a clean energy standard (as originally recommended by this committee in its first report). The absence of either one remains a key gap in policy.

Recommendation 6-1: *Adopt National Policy to Limit Power-Sector Greenhouse Gas (GHG) Emissions.* **As recommended in the first report, Congress should adopt a cap on power-sector GHG emissions, or clean electricity standard for the power sector, which would be designed to reach roughly 75 percent clean electricity share by 2030 and a declining emissions intensity reaching net-zero emissions by 2050.**

Build-Out of the Electric Transmission Grid

Given the essential role that the nation's electric transmission system will need to play in enabling decarbonization of the grid in a relatively affordable and timely way, it is imperative that the nation moves forward with planning, siting, building out, and paying for expansion of the high-voltage transmission system. Particularly because many of the best locations for high-quality, utility-scale wind and solar generation are far from population centers, the current grid is insufficient to enable extensive expansion of those resources (ESIG 2021; Goggin et al. 2021). Various analysts have concluded that by 2050 the nation will need from two to five times the amount of transmission capacity that is in place today (Larson et al. 2021; Reed et al. 2021). Figure 6-12a shows the U.S. high-voltage transmission grid in 2020, while Figure 6-12b shows the grid as of 2035 in a net-zero

---

[14] NREL's Electrification Futures Study (Mai et al. 2018) analyzed the implications for electricity demand of various electrification scenarios and found that "electrification has the potential to significantly increase overall demand for electricity" and "to significantly shift load shapes." The Net-Zero America and REPEAT studies analyses of different scenarios indicate that capacity additions could be as high as four times current levels, with varying degrees of coal-fired and gas-fired generation and GHG emissions between now and 2030–2040, depending on energy efficiency, transmission expansion, and firm-generating capacity build-outs (Jenkins et al. 2022; Larson et al. 2021).

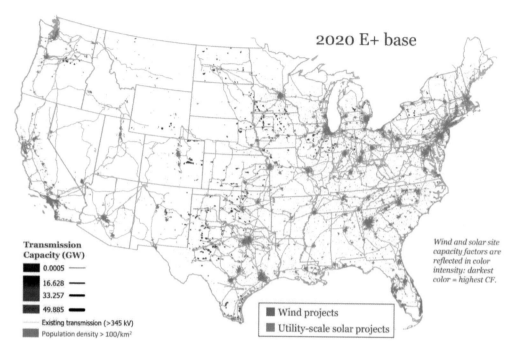

FIGURE 6-12 (a) Existing transmission grid (2020). SOURCE: Larson et al. (2021), https://netzeroamerica. princeton.edu/the-report. CC BY 4.0.

scenario (the "E+" scenario in Larson et al. [2021][15]). Much of the additions to the transmission grid would be needed to bring new renewable electricity to centers of demand.

Timely expansion of transmission will make a significant difference in the total amount of generating capacity needed to serve demand and attain national decarbonization goals, as illustrated by the recent 2022 REPEAT Project Report:

> [The] IRA could cut U.S. greenhouse gas emissions by roughly one billion tons per year in 2030 and reduce cumulative greenhouse gas emissions by 6.3 billion tons of CO_2-equivalent over the decade (2023–2032). That outcome depends on more than doubling the historical pace of electricity transmission expansion over the last decade in order to interconnect new renewable resources at sufficient pace and meet growing demand from electric vehicles, heat pumps, and other electrification. . . . Failing to accelerate transmission expansion beyond the recent historical pace (~1%/year) increases 2030 U.S. greenhouse emissions. . . . Over 80% of the potential emissions reductions delivered by IRA in 2030 are lost if transmission expansion is constrained to 1%/year, and roughly 25% are lost if growth is limited to 1.5%/year. (Jenkins et al. 2022, pp. 3–4)

---

[15] The E+ scenario "assumes aggressive end-use electrification, but energy-supply options are relatively unconstrained for minimizing total energy-system cost to meet the goal of net-zero emissions in 2050" (Larson et al. 2021, p. 9).

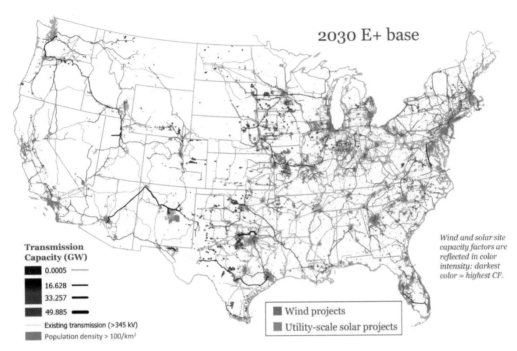

2030 E+ base

Transmission Capacity (GW)
- 0.0005
- 16.628
- 33.257
- 49.885
- Existing transmission (>345 kV)
- Population density > 100/km²

Wind and solar site capacity factors are reflected in color intensity: darkest color = highest CF.

- Wind projects
- Utility-scale solar projects

FIGURE 6-12 (b) Transmission grid as of 2035 (estimates to support wind and solar generation in E+ scenario with base siting availability). SOURCE: Larson et al. (2021), https://netzeroamerica.princeton.edu/the-report. CC BY 4.0.

Siting long-distance transmission has historically been a fraught and difficult process involving stakeholders ranging from federal to local actors (NASEM 2021b). Historically, some of the most significant barriers—such as the "Five Ps" (Reed et al. 2021) of transmission planning, permitting, paying, participation, and process—are quite tenacious. These transmission issues are being addressed by actions at FERC[16] and DOE, but—however well intended and critically important to decarbonization—each of those agency's processes are time-consuming and their outcomes are uncertain. Many states and other stakeholders do not support expanded federal action to accelerate

---

[16] In December 2022, FERC issued proposed rules to implement its newly expanded backstop siting authority under the IIJA (FERC 2022a). The proposed rules explain that under the Federal Power Act section 216(b)(1)(c) and the Proposed Rule, FERC's backstop siting authority is triggered for a project: "(i) where the state does not have the authority to approve the siting of the facilities or to consider the interstate or interregional benefits expected to be achieved from the project; (ii) the applicant is not eligible for state siting approval because it does not serve end-use customers within the state; (iii) a state has not made a determination on an application within the specified timeframes; (iv) the state siting authority conditioned its approval such that the proposed facilities would not significantly reduce transmission capacity constraints or congestion in interstate commerce or in such a way that is not economically feasible; or (v) the state has denied an application" (Lipinski et al. 2022).

build-out of the interstate transmission system (Klein 2022; NARUC to FERC 2022). Differing positions on the siting of high-voltage transmission—like other infrastructure (Gelles 2022)—also arise from concerns about cost, local land-use, and environmental impacts versus regional benefits; different views about the need for new transmission in general and specific lines in particular; polarization and "not in my back yard" sentiments among members of the public; and other private versus public interests (Mohl 2021; NASEM 2021a,b; NASEM Committee 2022).

The amount of transmission capacity needed will depend on decarbonization policies and actions, with certain technologies (e.g., CCS and nuclear) presenting the potential to decrease inter-regional transmission build-out (Blanford and Bistline 2021), and with some modeling suggesting that the power sector can attain 100 percent clean energy by 2035 under a constrained technology and infrastructure scenario limiting transmission build-out (Denholm et al. 2022). DOE's Transmission Facilitation Program now allows the agency to act as an "anchor tenant"—buying up to 50 percent of a planned line's capacity rating for up to 40 years—for transmission projects supporting new lines of 1,000 MW and greater, or upgrading or replacing lines rated at least 500 MW (Grid Deployment Office 2022). An anchor tenant agreement from DOE provides a guaranteed customer, improving the financial case for transmission expansion, and allows for the sale of this contract to recover costs. Improving public engagement in transmission planning, siting, and permitting is broadly viewed as necessary (see Chapters 2, 5, and 13), but not necessarily sufficient to expansion of the grid. Benefit sharing is also essential, as is expansion of distributed energy resources and the modernization of local electric distribution systems to support them (see further discussion below). The committee reaffirms the importance of transmission expansion that was a finding and recommendation in its first report.

> Finding 6-8: Expansion of the nation's electric transmission system is a key enabler of and critical success factor for economical and accelerated decarbonization of the grid.

> Finding 6-9: The nation needs a grand compromise on transmission—one that relies on a more inclusive and participatory set of processes to develop national-interest plans for needed expansion of the grid and that results in decisions that support actual expansion of the grid—as integral to accelerating decarbonization of the U.S. energy system. That compromise on transmission is also necessary to accomplish the public health, equity, affordability, reliability, and resilience benefits of the clean-energy transformation of the nation's electric system.

Recommendation 6-2: *Support the Expansion of the Transmission Grid.* **With regard to the siting of new high-voltage transmission infrastructure needed**

to provide for an economical, reliable, resilient, and equitable transition to a net-zero electric system in regions where such transmission expansion is needed,

a. The Federal Energy Regulatory Commission (FERC) should expeditiously update its transmission policy statements (i.e., its Notice of Proposed Rulemakings on transmission planning and cost allocation, and on interconnection of generating facilities).

b. The Department of Energy (DOE), FERC, the states, transmission companies, grid operators, and public stakeholders should expeditiously implement the new provisions of the Infrastructure Investment and Jobs Act (IIJA) and the Inflation Reduction Act (IRA), in conjunction with other aspects of Federal Power Act requirements and state statutes and regulations, so that

- DOE conducts its analyses and identifies corridors where transmission expansion is in the national interest and supports states' goals for decarbonizing their electric systems;
- Transmission companies and/or, as appropriate, Regional Transmission Organization plan for projects to fulfill the need(s) identified in those and in other regional planning studies;
- DOE expeditiously disburses its $760 million under the IRA to fund states and local government's technical assistance and financial support for communities' meaningful participation;
- DOE collaborates with potential project applicants to use its "anchor tenant" authority to facilitate progress on high-value projects (in particular, interregional projects);
- DOE and the Department of the Interior (where appropriate on federal lands and offshore waters) conduct community impact analyses on transmission projects;
- Transmission providers work proactively with potentially affected and diverse stakeholders to discuss options for addressing routing for needed transmission infrastructure;
- State regulatory agencies establish expeditious procedural schedules to review and approve transmission project proposals and take into consideration both quantitative and qualitative information about the economic, reliability, resilience, decarbonization, regional interdependence, and other aspects of proposals; and
- FERC acts under its IIJA backstop siting authority to review and approve transmission projects that are in the national interest and do so on expedited procedural schedules.

Expansion of Regional Power Markets and Design to Accommodate Clean Energy Resources

In its first report, the committee previously identified the expansion of regional power markets as an important element of an efficient clean-energy transition of the electric industry. In conjunction with expansion of regional and inter-regional transmission systems and implementation of federal financial incentives for investment in clean energy, the expansion of wholesale markets will help with economical and reliable decarbonization of the electric system.

Researchers at RFF estimate that these new policies—including expansion of the interstate transmission grid—will produce billions in savings for consumers as well as cost-effective emissions reductions (Shawhan et al. 2022). Such outcomes result from efficient dispatch of a broad set of diverse power supplies and customers across broad regions, the ability of customers to gain access to green purchasing options, the leveraging of local distributed energy resources in conjunction with more centralized technology options, and the ability to achieve power-sector emissions caps at lowest cost.

States, too, have begun examining in earnest the potential implications of new, enhanced, expanded wholesale power markets. In December 2021, the Colorado Public Utilities Commission, for example, found significant consumer, environmental, and economic benefits of a Western RTO, which could reduce electricity costs for consumers by up to 5 percent annually while accelerating the state's clean energy goals (CO PUC 2021). A 2023 study by NREL that synthesized other analyses of the potential benefits to California of expanded regional power-market cooperation found that both California and other states would realize cost savings, and reliability and common energy policy goals, with the size and character of these outcomes depending on the structure of regional cooperation approaches (Hurlbut et al. 2023). Another recent study found regional savings of $1.2 billion per year for a scenario in which all parts of the Western electrical interconnection join a West-Wide Enhanced Day-Ahead Market (Moyer and Ramirez 2022). Chang et al. (2020) observed that numerous studies have pointed to the role of regional markets in "reducing the cost of power, increasing reliability, reducing the need for new generation and transmission facilities, aiding the integration of renewable energy resources, and improving system balancing and operations." Chen and Bardee (2020, pp. 1, 7) concluded that the benefits of "enhanced trading between utilities and resource sharing across broader regions can be substantial," with greater net benefits potentially "derived from a platform transparently operated by an independent entity."

Recommendation 6-3: *Expand Regional Wholesale Power Markets Consistent with Decarbonization Objectives.* **To achieve the economic, environmental,**

greenhouse gas emissions, and consumer benefits of a clean electricity transition:

a. **The Federal Energy Regulatory Commission (FERC) should work with Regional Transmission Organizations (RTOs) and regional energy imbalance markets to ensure that wholesale power markets in all parts of the country are designed to accommodate the shift to 100 percent clean power.**
b. **Congress should provide incentives for transmission companies to join regional power systems administered by RTOs or participate in other wholesale market coordination approaches.**

Design of Retail Electricity Rates and Pricing Options

In its first report, the committee also highlighted the role of the demand side of retail and wholesale electricity markets and systems in accomplishing electric-system decarbonization transitions that avoid massive build-outs of the supply side of the system that might otherwise be needed to meet the needs of an electrified economy. As explained in Chapters 7 and 9, electrification will produce economic efficiency, energy savings, and lower emissions, but it also enables opportunities to operate the electric grid devices in new and more efficient ways. For example, managing the timing of energy use of appliances or equipment (e.g., water heating, refrigeration and cooling, and EV batteries) can take advantage of the thermal and/or electrical storage capabilities of such resources and in so doing avoid installation of supply-side capacity. Designing electricity pricing and service options to encourage customers and suppliers of smart equipment, appliance-management applications, and other communications and control systems to deploy such resources can provide customer and system savings. In light of the many and very challenging issues in ratemaking and other utility policies and practices, doing so will require regulatory innovation and creativity beyond that shown in most utility rate cases at present—or even allowed in some jurisdictions (Madduri et al. 2022; Matisoff et al. 2020). Ensuring that electricity regulation provides both creative rate and service options to all customers while also ensuring basic electricity service rates for low-income customers is both technically possible (from a utility ratemaking point of view) and essential (to ensure equity as a core element of the decarbonization transition).

Recommendation 6-4: *Provide Rate Options to Encourage Flexible Demand While Ensuring Affordable Electricity.* **Decision makers on utility rates**

(i.e., state utility regulators and boards of publicly owned utilities) should direct utility management to:

a. **Prepare plans for the deployment of advanced meters for retail electricity customers where such do not already exist;**

b. **Pilot and/or directly move to implementation of time-varying rate options (and where possible, dynamic rate options) for residential, commercial, and industrial customer classes, with communications strategies and control systems to enable flexible demand to meet customer and electric system needs cost-effectively; and**

c. **Make sure that there are also options to ensure basic electricity service at affordable (and where possible, discounted) rates for low-income customers.**

Deployment of Distributed Energy Resources on the Electric System

For decades, many large electricity customers have been able to take advantage of installing on their own premises various small-scale energy technologies (e.g., cogeneration systems) and energy management systems so as to manage their electricity usage and bills. In recent years, new cost-effective technology developments (such as rooftop solar, small-scale electricity storage systems, smart appliances, smart meters), combined with ratemaking and other regulatory tools (such as net energy metering, time-of-use rates, programmable controls on appliances, utility hosting studies showing locations where on-site generation can be added without utility distribution-system upgrades), have opened up the possibility for large and small customers to manage their electricity usage and in some cases inject power into the grid when it is not being used on site. These technologies present the opportunities to improve economic and equity outcomes for some consumers, but access to these benefits may also be limited owing to structural factors, including difficulties integrating these technologies with multi-family dwellings and/or in situations where renters may not control or capture the long-term benefits of investments in and on buildings. (See Chapter 5 of NASEM [2023] for further discussion on distributed generation and equity.) In the past decade or so, community solar farms and large-scale, individual solar projects that can be (and in many cases are) interconnected to the distributed system have experienced cost reductions so that these technologies have become relatively attractive for meeting renewable energy goals and for providing customers and the utility itself with more options. (DOE has a goal of seeing another 20 GW of community solar deployed commercially by 2025—for example, DOE [2021].)

Collectively known as "distributed energy resources" (DERs), such technologies have experienced notable growth, with 78 GW of capacity installed between 2017 through 2021, and the pace of DER deployment anticipated to increase in the near future (Hertz-Shargel 2022). They are seen in many states and localities as critical components of the technology option set needed for decarbonization and for meeting myriad other private objectives (e.g., customer bill savings, energy management, resilience from back-up supplies—which may enable improved energy justice outcomes; see Chapter 2) and public ones (e.g., renewable energy deployment; see Chapter 7). In many cases, DERs like energy efficiency, rooftop solar, and on-site battery storage can serve needs locally and avoid some generating and delivery capacity additions (although such outcomes are quite location-specific and in some cases the addition of DERs can cause additional investment in local distribution system capacity and capabilities above and beyond what might otherwise occur on the local grid; see further discussion below). It should be noted that increased amounts of DER on the electricity system motivates the need for improved planning and integration between wholesale and retail markets, as well as for planning across timescales (ranging from seconds to day-ahead). (See NASEM [2021b] for an in-depth discussion of these topics.)

The IRA provides consumer and corporate tax incentives for the installation of DERs, especially solar and storage systems. The IRA also supports expanded investment in energy efficiency and other weatherization measures, including for switching heating systems so as to use electricity. (See further discussion in Chapter 7.) Significantly, the IRA authorizes EPA to award $27 billion in funding from a new GHG Reduction Fund,[17] a substantial portion of which may end up supporting the financing of various types of eligible projects, which could include DERs that reduce GHG emissions in

---

[17] The IRA requires that the recipients of EPA's awards of funding from the GHG Reduction Fund direct more than half of the funding to low-income and disadvantaged communities. EPA explains (in its October 21, 2022, Request for Information for the GHG Reduction Fund) that:

- Section 134(a)(1) of the IRA makes available "$7 billion to EPA to make competitive grants to States, municipalities, tribal governments, and eligible recipients . . . to provide subgrants, loans, or other forms of financial assistance as well as technical assistance to enable low-income and disadvantaged communities to deploy or benefit from zero-emission technologies, including distributed technologies on residential rooftops, and to carry out other greenhouse gas emission reduction activities" (p. 1).
- Section 134(a)(3) of the IRA "makes available $8 billion to EPA to make competitive grants to eligible recipients [i.e., certain non-profit institutions] for the provision of financial and technical assistance to projects that reduce or avoid greenhouse gas emissions and other forms of air pollution in low-income and disadvantaged communities" (p. 2).
- Section 134(a)(2) of the IRA "makes available $11.97 billion to EPA to make competitive grants to eligible recipients [i.e., certain non-profits] for the provision of financial and technical assistance to projects that reduce or avoid greenhouse gas emissions and other forms of air pollution" (p. 2). These funds may be applied in low-income and disadvantaged communities among other places.

low-income and disadvantaged communities (EPA 2022). Growing consumer interest in DERs, combined with new federal incentives and financing opportunities offered in the IRA and cost reductions in DER options, create opportunities for much-faster deployment as part of decarbonization efforts over the next decade. States, localities, and tribes that move expeditiously to help their constituents—especially those that otherwise lack the financial, informational, or other resources to install and use DERs—take advantage of such opportunities can achieve near-term emissions reductions and other economic and equity benefits. This can happen by using technical assistance to assist such constituents in ways that incorporate the practical needs of recipients (Brown 2023) and also to learn about and make use of lessons from other jurisdictions about the important roles of aligning various enabling policies such as utility rate design (described above) and local distribution system enhancements (described below) to facilitate the integration of DERs. Just as importantly as these direct benefits, accelerating adoption of DERs can also help invest households and communities in the clean energy transition and recognize and reap some of the rewards of the transition, both of which are important elements in building the social contract for the transition (see Chapters 2 and 5).

> **Recommendation 6-5:** *Support Equitable Deployment of Distributed Energy Resources (DERs).* **States, localities, and tribal governments should use the technical assistance, funding, and financing opportunities in the Inflation Reduction Act (especially, e.g., Environmental Protection Agency Greenhouse Gas Reduction Fund) and other programs (e.g., Department of Energy National Community Solar Partnership, state or utility-based net energy metering programs, community solar programs) to facilitate the deployment of distributed energy resources and to enable low-income and disadvantaged households and communities to have access and ownership opportunities for such DERs. Such technical assistance should take advantage of lessons learned about actions in other jurisdictions that facilitate DER deployment and access—especially in terms of the benefits that can arise if policies are well designed versus poorly designed. Such technical assistance should also be provided in ways that serve the practical needs of recipients.**

Modernization of Local Electric Grids

The role of electric distribution systems has become increasingly important as part of innovative approaches to decarbonization. While local distribution systems have always played an essential role in connecting end-use electricity consumers to the grid that provides them with around-the-clock access to reliable electricity supply, a decarbonized energy system amplifies the importance of a highly well-functioning local grid. Depending on the ways in which electrification occurs—for example, with substantial energy efficiency as part of building electrification; with rate designs that enable much-more flexible demand; with DERs located in particular in places that meet customer needs *and* provide grid services; with different pathways to electric vehicle charging buildout; with different levels or patterns of electrification in industry or heavy transport; with different levels of demand for firming hydrogen generation or powering CCS—the local distribution system will need very substantial investment and expansion.

At a minimum, the local grid will require different types of local-distribution system planning and operation than in the past, so that it accommodates two-way flows of power on circuits, more DER installations, more technologies to visualize and in some cases manage and control flows on the local system, and so forth. Although some of these activities can be supported by third-party actions and investments, much of these local-distribution system enhancements will need to be done by the local utility. Modernizing the local grid is an essential part of the transition to a decarbonized electric system (Gillis and Norris 2022), and additional research, analytics, other tools, and investment are needed to make the lower-voltage electric grid fit the needs of a highly electrified economy with much more complicated interactions with the electricity system (McDermont et al. 2022; NASEM 2021b). Investment in technical improvements to the local grid, its communication and control systems, and its delivery capacity are as essential to an equitable transition as are the DER technologies and bulk-power-system transformations more commonly identified as key to the needs of the future power system. This is especially true when local grid upgrades are piecemeal and follow customer investment in new electrical technologies (as is widely practiced currently) rather than flowing from a broad and participatory grid upgrade planning process designed to meet the long-term needs of decarbonization (see Chapters 2, 5, and 7). Attention to ensuring efficiencies in such investment and recovery of it from electricity users is also an essential piece of the decarbonization equation.

> **Recommendation 6-6:** *Support Planning, Public Participation, and Investment in Modernizing Local Grids.* **Decision makers on utility service provision (i.e., state utility regulators for jurisdictional investor-owned utilities and boards of cooperatives, municipal electric utilities and other publicly owned utilities) should direct their utilities to carry out planning, public participation, investment projects, rate proposals, and other actions necessary to modernize and ready local distribution systems for increased deployment of**

distributed energy resources; for new loads driven by electrification actions in buildings, vehicles, and industry; for ensuring that all customers have equitable access to resilient and reliable power; and for operating and maintaining the local grid under much more complicated conditions than in the past. Such plans should anticipate the types of changes needed on the local distribution system to be ready as these changes occur, and to be equitable across customers, consistent with timely transitions to a net-zero economy. In jurisdictions where such decision makers lack existing authorities to require such changes, state legislatures should authorize and direct utilities to conduct such planning and grid modernization and to require the equitable recovery of costs related to them. Congress should continue to fund research and analysis on these topics through the Department of Energy.

Research, Development, Demonstration, and Deployment

Commercially available power technologies are already capable of decarbonizing electricity supply to a substantial degree. But with increasing and deep penetration of variable generating resources—which are unable to provide around-the-clock, dispatchable generation—reliable system operations will require a combination of on-demand generating technologies, storage that allows for flexible generation, long-duration storage to sustain supply during periods of high demand and/or constraints on wind and solar availability, flexible demand, and power-delivery technologies. Many of these resources—including flexible demand and transmission—are the subject of other recommendations. But other such resources—including advanced nuclear, hydrogen technologies, and long-duration storage—will require additional research, development, demonstration, and deployment (Kelly and Nelson 2023). As the committee concluded in its first report (and as recommended by the National Academies' Committee on the Future of Electric Power in the United States [NASEM 2021b]), the federal government is uniquely positioned to provide such support. The committee reiterates its first report recommendation that Congress provide such funding as a high national priority.

Additionally, deployment of various technologies (including commercially ready technologies like wind, solar, and battery technologies) to advance net-zero energy transitions depend on critical minerals whose supply chains are subject to numerous domestic challenges and international risks (DOE 2022a; Kelly and Nelson 2023). These have been analyzed in detail by DOE, as part of executive branch efforts to identify and address technical, economic, and other policy drivers of such challenges (DOE 2022a).

Recommendation 6-7: *Invest in Research, Development, and Demonstration (RD&D) of On-Demand Electric-Generating Technologies and Long-Duration Storage Technologies.* **Congress should continue to increase inflation-adjusted funding of RD&D for innovation on and commercialization of on-demand/firm electric generating technologies and long-duration storage technologies. Continued work in this area needs to support early-stage deployment, as well as research and analysis to ensure that commercial developments and applications, including siting of such technologies, addresses potential local adverse impacts to host communities. Additionally, Congress should support the Department of Energy's work to secure resilient supply chains that will be critical in harnessing emissions outcomes and capturing the economic opportunity inherent in the energy-sector transition.**

Table 6-1 summarizes all the recommendations in this chapter to support decarbonizing the electricity system.

SUMMARY OF RECOMMENDATIONS ON THE
ESSENTIAL ROLE OF CLEAN ELECTRICITY

TABLE 6-1 Summary of Recommendations on the Essential Role of Clean Electricity

Short-Form Recommendation	Actor(s) Responsible for Implementing Recommendation	Sector(s) Addressed by Recommendation	Objective(s) Addressed by Recommendation	Overarching Categories Addressed by Recommendation
6-1: Adopt National Policy to Limit Power-Sector Greenhouse Gas Emissions	Congress	• Electricity	• Greenhouse gas (GHG) reductions • Health	A Broadened Policy Portfolio
6-2: Support the Expansion of the Transmission Grid	Federal Energy Regulatory Commission (FERC), Department of Energy (DOE), states, transmission companies, public stakeholders, and Department of the Interior	• Electricity • Non-federal actors	• GHG reductions • Health • Public engagement	Siting and Permitting Reforms for Interstate Transmission
6-3: Expand Regional Power Markets Consistent with Decarbonization Objectives	Congress, FERC, regional transmission organizations	• Electricity • Non-federal actors	• GHG reductions • Equity • Health	Siting and Permitting Reforms for Interstate Transmission
6-4: Provide Rate Options to Encourage Flexible Demand While Ensuring Affordable Electricity	Decision makers on utility rates (i.e., state utility regulators for jurisdictional investor-owned utilities and boards of cooperatives, municipal electric utilities, and other publicly owned utilities)	• Electricity • Non-federal actors	• Equity	Ensuring Equity, Justice, Health, and Fairness of Impacts Siting and Permitting Reforms for Interstate Transmission

TABLE 6-1 Continued

Short-Form Recommendation	Actor(s) Responsible for Implementing Recommendation	Sector(s) Addressed by Recommendation	Objective(s) Addressed by Recommendation	Overarching Categories Addressed by Recommendation
6-5: Support Equitable Deployment of Distributed Energy Resources	States, localities, and tribal governments	• Electricity • Non-federal actors	• GHG reductions • Equity • Health • Public engagement	Ensuring Equity, Justice, Health, and Fairness of Impacts Siting and Permitting Reforms for Interstate Transmission
6-6: Support Planning, Public Participation, and Investment in Modernizing Local Grids	Decision makers on utility service provision (i.e., state utility regulators for jurisdictional investor-owned utilities and boards of cooperatives, municipal electric utilities, and other publicly owned utilities)	• Electricity • Non-federal actors	• GHG reductions • Equity • Health • Public engagement	Ensuring Procedural Equity in Planning and Siting New Infrastructure and Programs Siting and Permitting Reforms for Interstate Transmission
6-7: Invest in Research, Development, and Demonstration of On-Demand Electric-Generating Technologies and Long-Duration Storage Technologies	Congress	• Electricity	• GHG reductions • Equity	Research, Development, and Demonstration Needs

REFERENCES

Bergman, A., Burtraw, D., Richardson, N. 2022. "The Consequences of the Supreme Court's *West Virginia v. EPA* Ruling for Environmental Regulation." Resources. https://www.resources.org/common-resources/the-consequences-of-the-supreme-courts-west-virginia-v-epa-ruling-for-environmental-regulations.

BIL (Bipartisan Infrastructure Law) Summary. 2021. "Bipartisan Infrastructure Investment and Jobs Act Summary: A Road to Stronger Economic Growth." https://www.cantwell.senate.gov/imo/media/doc/Infrastructure%20Investment%20and%20Jobs%20Act%20-%20Section%20by%20Section%20Summary.pdf.

Bistline, J., G. Blanford, M. Brown, D. Burtraw, M. Domeshek, J. Farbes, A. Fawcett, et al. 2023a. "Emissions and Energy Impacts of the Inflation Reduction Act." *Science* 380(6652):1324–1327. https://doi.org/10.1126/science.adg3781.

Bistline, J., C. Raney, G. Blanford, and D. Young. 2023b. "Impacts of Inflationary Drivers and Updated Policies on U.S. Decarbonization and Technology Transitions." #3002026229. Palo Alto, CA: EPRI. https://www.epri.com/research/products/000000003002026229.

Blanford, G., and J. Bistline. 2021. *Powering Decarbonization: Strategies for Net-Zero CO$_2$ Emissions.* #3002020700. Palo Alto, CA: EPRI.

BPC (Bipartisan Policy Center). 2022a. "Inflation Reduction Act Summary: Energy and Climate Provisions." https://bipartisanpolicy.org/blog/inflation-reduction-act-summary-energy-climate-provisions/#.

BPC. 2022b. "3 Reasons Why 'Direct Pay' Is Crucial for Clean Energy Tax Policy." https://bipartisanpolicy.org/wp-content/uploads/2022/06/Energy-Direct-Pay-Infographic.pdf.

Brown, A. 2023 "An Open Letter to the Department of Energy: Enough with the Technical Assistance." *Utility Dive.* January 30. https://www.utilitydive.com/news/an-open-letter-to-the-department-of-energy-enough-with-the-technical-assis/641491.

Brown, M.T. 2022. "Nearly a Quarter of the Operating U.S. Coal-Fired Fleet Scheduled to Retire by 2029." https://www.eia.gov/todayinenergy/detail.php?id=54559.

C2ES. 2022. "U.S. State Climate Action Plans." *Center for Climate and Energy Solutions.* https://www.c2es.org/document/climate-action-plans.

CBO (Congressional Budget Office). 2022. "Emissions of Carbon Dioxide in the Electric Power Sector." December. https://www.cbo.gov/system/files/2022-12/58419-co2-emissions-elec-power.pdf.

CEBA (Clean Energy Buyers Alliance). 2023. https://cebuyers.org/programs/market-policy-innovations and https://cebuyers.org/deal-tracker.

Chang, J., J. Pfeifenberger, and J. Tsoukalis. 2019. "Potential Benefits of a Regional Wholesale Power Market to North Carolina's Electricity Customers." Brattle Group. https://www.brattle.com/wp-content/uploads/2021/05/16092_nc_wholesale_power_market_whitepaper_april_2019_final.pdf.

Chen, J., and Bardee, M. 2020. "How Voluntary Electricity Trading Can Help Efficiency in the Southeast." https://www.rstreet.org/wp-content/uploads/2020/08/No.-201-Energy-Trade-in-the-Southeast.pdf.

CO PUC (Colorado Public Utilities Commission). 2021. "Colorado Transmission Coordination Act: Investigation of Wholesale Market Alternatives for the State of Colorado §§ 40-2.3-101 to 102, C.R.S." https://www.dora.state.co.us/pls/efi/efi_p2_v2_demo.show_document?p_dms_document_id=961152.

Denholm, P., P. Brown, W. Cole, M. Trieu, B. Sergi, M. Brown, P. Jadun, et al. 2022. *Examining Supply-Side Options to Achieve 100% Clean Electricity by 2035.* Golden, CO: National Renewable Energy Laboratory. NREL/TP6A40-81644. https://www.nrel.gov/docs/fy22osti/81644.pdf.

DOE (Department of Energy). 2021. "DOE Sets 2025 Community Solar Target to Power 5 Million Homes." https://www.energy.gov/articles/doe-sets-2025-community-solar-target-power-5-million-homes.

DOE. 2022a. "America's Strategy to Secure the Supply Chain for a Robust Clean Energy Transition." https://www.energy.gov/policy/articles/americas-strategy-secure-supply-chain-robust-clean-energy-transition.

DOE. 2022b. "Biden-Harris Administration Launches New Offices to Lower Energy Costs and Deploy Clean Power Nationwide." https://www.energy.gov/articles/biden-harris-administration-launches-new-offices-lower-energy-costs-and-deploy-clean-power.

Domeshek, M., T. Bowen, C. Ivanova, K. Palmer, and B. Shobe. 2022. "State-Level Planning for Decarbonization: Critical Elements of State Action." Resources for the Future. https://media.rff.org/documents/Report_22-10_v4.pdf.

DOS and EOP (Department of State and Executive Office of the President). 2021. "The Long-Term Strategy of the United States: Pathways to Net-Zero Greenhouse Gas Emissions by 2050." Washington, DC. https://www.whitehouse.gov/wp-content/uploads/2021/10/US-Long-Term-Strategy.pdf.

EIA (Energy Information Administration). 2021. "U.S. Electricity Profile 2021." https://www.eia.gov/electricity/state.

EIA. 2022a. "EIA Expects Renewables to Account for 22% of U.S. Electricity Generation in 2022." *Today in Energy*, August 16. https://www.eia.gov/todayinenergy/detail.php?id=53459.

EIA. 2022b. "The U.S. Power Grid Added 15 GW of Generating Capacity in the First Half of 2022." *Today in Energy*, August 3. https://www.eia.gov/todayinenergy/detail.php?id=53299.

EIA. 2023. "Total Energy: Monthly Energy Review." https://www.eia.gov/totalenergy/data/monthly.

EPA (Environmental Protection Agency). 2022. "Request for Information: Greenhouse Gas Reduction Fund (RFI GHGRF)." EPA-HQ-OA-2022-0859. https://www.regulations.gov/docket/EPA-HQ-OA-2022-0859.

EPA. 2023. "EPA Proposes New Carbon Pollution Standards for Fossil Fuel-Fired Power Plants to Tackle the Climate Crisis and Protect Public Health." https://www.epa.gov/newsreleases/epa-proposes-new-carbon-pollution-standards-fossil-fuel-fired-power-plants-tackle.

ESIG (Energy Systems Integration Group). 2021. "Transmission Planning for 100% Clean Electricity." Energy Systems Integration Group. https://www.esig.energy/wp-content/uploads/2021/02/Transmission-Planning-White-Paper.pdf.

Eversheds Sutherland. 2022. "Inflation Reduction Act of 2022: The Energy Tax Provisions You Need to Know About." https://us.eversheds-sutherland.com/NewsCommentary/Legal-Alerts/252759/Inflation-Reduction-Act-of-2022-The-energy-tax-provisions-you-need-to-know-about.

Ewing, J., and I. Penn. 2022. "Clean Energy Projects Surge After Climate Bill Passage." *The New York Times*, September 7. https://www.nytimes.com/2022/09/07/business/energy-environment/clean-energy-climate-bill.html.

FERC (Federal Energy Regulatory Commission). 2021. "Advance Notice of Proposed Rulemaking: Building for the Future Through Electric Regional Transmission Planning and Cost Allocation and Generator Interconnection." 176 FERC ¶ 61,024. Docket No. RM2-1-17-000. July 15. https://www.ferc.gov/news-events/news/advance-notice-proposed-rulemaking-building-future-through-electric-regional.

FERC. 2022a. "Applications for Permits to Site Interstate Electric Transmission Facilities." Federal Energy Regulatory Commission. Proposed Rules. 181 FERC ¶ 61,205. Docket No. RM22-7-000. https://www.ferc.gov/media/e-1-rm22-7-000.

FERC. 2022b. "Explainer on the Transmission Notice of Proposed Rulemaking." https://www.ferc.gov/explainer-transmission-notice-proposed-rulemaking.

Gelles, D. 2022. "The U.S. Will Need Thousands of Wind Farms. Will Small Towns Go Along?" *The New York Times*, December 30. https://www.nytimes.com/2022/12/30/climate/wind-farm-renewable-energy-fight.html?action=click&module=Well&pgtype=Homepage&section=US%20News.

Gillis, J., and T. Norris. 2022. "Here Is What Is Really Strangling the Energy Transition." *The New York Times*, December 17. https://www.nytimes.com/2022/12/16/opinion/solar-wind-electricity.html.

Goggin, M., R. Gramlich, and M. Skelly. 2021. "Transmission Projects Ready to Go: Plugging into America's Untapped Renewable Resources." Americans for a Clean Energy Grid. https://cleanenergygrid.org/wp-content/uploads/2019/04/Transmission-Projects-Ready-to-Go-Final.pdf.

Grid Deployment Office. 2022. "Transmission Facilitation Program (TFP)." Department of Energy. https://www.energy.gov/sites/default/files/2022-11/11.18.22%20TFP%20Fact%20Sheet_final.pdf.

Grid North Partners. 2022. "Grid North Partners Support MISO's Long Range Transmission Plan." https://gridnorthpartners.com/grid-north-partners-support-misos-long-range-transmission-plan.

Grid Strategies. 2023. "The Benefits of New Regional Transmission Planning Entities in the U.S. West and Southeast Regions." https://cebi.org/wp-content/uploads/2023/02/CEBI-The-Benefits-of-New-Regional-Transmission-Planning-Entities-in-The-U.S.-West-And-Southeast-Regions.pdf.

Hertz-Shargel, B. 2022. "Distributed Energy Is Poised to Take Center Stage in 2022, But Policymakers and Regulators Must Step Up." *Utility Dive*. https://www.utilitydive.com/news/distributed-energy-is-poised-to-take-center-stage-in-2022-but-policymakers/618331.

Howland, E. 2022. "Accelerating Renewable Energy Buildout Faces Big Hurdles, Even with Inflation Reduction Act: Developers." *Utility Dive*, September 6. https://www.utilitydive.com/news/clean-energy-challenges-ira-inflation-reduction-wind-solar-storage/630988.

Hultman, N., C. Frisch, L. Clarke, K. Kennedy, P. Bodnar, P. Hansel, T. Cyrs, et al. 2019. *Accelerating America's Pledge: Going All-In to Build a Prosperous, Low-Carbon Economy for the United States*. America's Pledge Initiative on Climate Change. New York: Bloomberg Philanthropies with University of Maryland Center for Global Sustainability, Rocky Mountain Institute, and World Resources Institute. https://assets.bbhub.io/dotorg/sites/28/2019/12/Accelerating-Americas-Pledge.pdf.

Hultman, N., L. Clarke, C. Frisch, K. Kennedy, H. McJeon, T. Cyrs, P. Hansel, et al. 2020. "Fusing Subnational with National Climate Action Is Central to Decarbonization: The Case of the United States." *Nature Communications* 11:5255. https://doi.org/10.1038/s41467-020-18903-w.

Hurlbut, D., Greenfogel, M. and Speetles, B. 2023. *The Impacts on California of Expanded Regional Cooperation to Operate the Western Grid (Final Report)*. Golden, CO: National Renewable Energy Laboratory. NREL/TP-6A20-84848. https://www.nrel.gov/docs/fy23osti/84848.pdf.

Illinois Power Agency. 2021. *Carbon Mitigation Credit Procurement Plan*. https://ipa.illinois.gov/content/dam/soi/en/web/ipa/documents/ipa-final-cmc-procurement-plan-(-dec-16-2021).pdf.

Jenkins, J.D., J. Farbes, R. Jones, N. Patankar, and G. Schivley. 2022a. *Electricity Transmission Is Key to Unlock the Full Potential of the Inflation Reduction Act*. Princeton, NJ: REPEAT Project. https://repeatproject.org/docs/REPEAT_IRA_Transmission_2022-09-22.pdf.

Jenkins, J.D., E.N. Mayfield, J. Farbes, R. Jones, N. Patankar, Q. Xu, and G. Schivley. 2022b. *Preliminary Report: The Climate and Energy Impacts of the Inflation Reduction Act of 2022*. Princeton, NJ: REPEAT Project. https://repeatproject.org/docs/REPEAT_IRA_Prelminary_Report_2022-08-04.pdf.

Jenkins, J.D., G. Schively, E.N. Mayfield, N. Patankar, J. Farbes and R. Jones. 2023. *Preview: Final REPEAT Project Findings on the Emissions Impacts of the Inflation Reduction Act and Infrastructure Investment and Jobs Act*. Princeton, NJ: REPEAT Project.

Kelly, C., and Nelson, S. 2023. "Clear Path to a Clean Energy Future 2022." https://static.clearpath.org/2023/02/CPCEF22-2-23.pdf.

King, B., H. Kolus, N. Dasari, E. Wimberger, W. Herndon, E. O'Rear, A. Rivera, J. Larsen, and K. Larsen. 2022. "Taking Stock 2022: US Greenhouse Gas Emissions Outlook in an Uncertain World." Rhodium Group.

Klein, E. 2022. "All Biden Has to Do Now Is Change the Way We Live." *The New York Times*, September 11. https://www.nytimes.com/2022/09/11/opinion/biden-climate-congress-infrastructure.html.

Larsen, J., B. King, H. Kolus, N. Dasari, G. Hiltbrand, and W. Herndon. 2022. "A Turning Point for US Climate Progress: Assessing the Climate and Clean Energy Provisions in the Inflation Reduction Act." Rhodium Group. https://rhg.com/research/climate-clean-energy-inflation-reduction-act.

Larson, E., C. Greig, J. Jenkins, E. Mayfield, A. Pascale, C. Zhang, J. Drossman, et al. 2021. "Net-Zero America: Potential Pathways, Infrastructure, and Impacts, Final Report Summary." https://netzeroamerica.princeton.edu/img/Princeton%20NZA%20FINAL%20REPORT%20SUMMARY%20(29Oct2021).pdf.

Lashof, D. 2023. "EPA's Proposed Rules for Power Plant Emissions: 6 Key Questions, Answered." https://www.wri.org/insights/epa-power-plant-rules-explained.

Levin, A., and Ennis. J. 2022. "Clean Electricity Tax Credits in the Inflation Reduction Act Will Reduce Emissions, Grow Jobs, and Lower Bills." Issue Brief 22-09-2. Natural Resources Defense Council. https://www.nrdc.org/sites/default/files/clean-electricity-tax-credits-inflation-reduction-act-ib.pdf.

Lipinski, R.L., T. Ellis, J.B. Nelson, and J.D. Simon. 2022. "Backstop Siting: FERC Issues Notice of Proposed Rulemaking." *Van Ness Feldman* (blog), December 22. https://www.vnf.com/backstop-siting-ferc-issues-notice-of-proposed-rulemaking.

Lopez, N. 2022. "Diablo Canyon: Nuke Plant a Step Closer to Staying Open Longer." CalMatters, September 1. http://calmatters.org/environment/2022/09/diablo-canyon-legislature-california.

MA (Massachusetts) Department of Public Utilities. 2017. 220 CMR 23.00: St. 2016, c. 188, §12. https://www.mass.gov/regulations/220-CMR-2300-competitively-solicited-long-term-contracts-for-offshore-wind-energy.

Madduri, A., M. Foudeh, and P. Phillips. 2022. "Advanced Strategies for Demand Flexibility Management and Customer DER Compensation: Energy Division White Paper and Staff Proposal." California Public Utilities Commission. https://www.cpuc.ca.gov/-/media/cpuc-website/divisions/energy-division/documents/demand-response/demand-response-workshops/advanced-der—demand-flexibility-management/ed-white-paper—advanced-strategies-for-demand-flexibility-management.pdf.

Mahajan, M., O. Ashmoore, J. Rissman, R. Orvis, and A. Gopal. 2022. "Update Inflation Reduction Act Modeling Using the Energy Policy Simulator." Energy Innovation. https://energyinnovation.org/wp-content/uploads/2022/08/Updated-Inflation-Reduction-Act-Modeling-Using-the-Energy-Policy-Simulator.pdf.

Mai, T., P. Jadun, J. Logan, C. McMillan, M. Muratori, D. Steinberg, L. Vimmerstedt, R. Jones, B. Haley, and B. Nelson. 2018. *Electrification Futures Study: Scenarios of Electric Technology Adoption and Power Consumption for the United States*. Golden, CO: National Renewable Energy Laboratory. NREL/TP-6A20-71500. https://www.nrel.gov/docs/fy18osti/71500.pdf.

Matisoff, D., D. Beppler, G. Chan, and S. Carley. 2020. "A Review of Barriers in Implementing Dynamic Electricity Pricing to Achieve Cost-Causality." *Environmental Research Letters* 15. https://iopscience.iop.org/article/10.1088/1748-9326/ab9a69/pdf.

McCarthy, E. 2022. "Pennsylvania Reaches the Regional Greenhouse Gas Initiative Starting Line." *Utility Dive*.

McDermott, T.E., K. McKenna, M. Heleno, B.A. Bhatti, M. Emmanuel, and S.P. Forrester. 2022. "Distribution System Research Roadmap: Energy Efficiency and Renewable Energy." Pacific Northwest National Laboratory, prepared for the Department of Energy. February 1. https://www.osti.gov/servlets/purl/1843579.

McGovern, J. 2022. "FERC Accepts PJM, NJBPU Agreement to Further Offshore Wind Goals." *PJM Inside Lines* (blog), April 19.

McLaughlin, K., and L. Bird. 2021. "Implementing the Clean Energy Investments in US Bipartisan Infrastructure Law." World Resources Institute. https://www.wri.org/insights/implementing-clean-energy-investments-us-bipartisan-infrastructure-law.

Micek, K. 2020. "Commodities 2021: States Racing to Set Goals Toward Net-Zero Emission, 100% Renewable Electricity." *S&P Global*, December 24. https://www.spglobal.com/commodityinsights/en/market-insights/latest-news/electric-power/122420-commodities-2021-states-racing-to-set-goals-toward-net-zero-emission-100-renewable-electricity.

Mohl, B. 2021. "Maine Voters Tell Mass. to Stick Its Transmission Line." *Commonwealth Magazine*. https://commonwealthmagazine.org/energy/maine-voters-tell-mass-to-stick-its-transmission-line.

Moyer, K., and Ramirez, D. 2022. "California EDAM Benefits Study: Estimating Savings for California and the West Under EDAM Market Scenarios." Energy Strategies, prepared for the California ISO. November 4. http://www.caiso.com/Documents/Presentation-CAISO-Extended-Day-Ahead-Market-Benefits-Study.pdf.

NARUC to FERC (National Association of Regulatory Utility Commissioners to Federal Energy Regulatory Commission). 2022. "Comments of the National Association of Regulatory Utility Commissioners." *Federal Energy Regulatory Commission*. Docket No. RM21-17-000. August 17. https://pubs.naruc.org/pub/F3AF556A-1866-DAAC-99FB-BFE2357A9443?_gl=1*1n13vec*_ga*MTE2OTkwMDYwNS4xNjcwOTM0OTI0*_ga_QLH1N3Q1NF*MTY3MDkzNDkyMy4xLjEuMTY3MDkzNDkzOS4wLjAuMA.

NASEM (National Academies of Sciences, Engineering, and Medicine). 2021a. *Accelerating Decarbonization of the U.S. Energy System*. Washington, DC: The National Academies Press. https://doi.org/10.17226/25932.

NASEM. 2021b. *The Future of Electric Power in the United States*. Washington, DC: The National Academies Press. https://doi.org/10.17226/25968.

NASEM. 2021c. "Recommended Policies for Reaching Net-Zero Carbon Emissions." https://nap.nationalacademies.org/resource/25932/interactive/table/index.html#top.

NASEM. 2023. *The Role of Net Metering in the Evolving Electricity System*. Washington, DC: The National Academies Press. https://doi.org/10.17226/26704.

NASEM Committee. 2022. "Public Engagement Across the Transmission Development Lifecycle: from Planning to Permitting." NASEM Committee on Accelerating Decarbonization of the U.S. Energy System. https://vimeo.com/757193592.

Parlapiano, A., and J. Tankersley. 2021. "What's in Biden's Infrastructure Plan?" *The New York Times*, March 31.

Pascale, A., and Jenkins, J.D. 2021. "Princeton's Net-Zero America Study—Annex F: Integrated Transmission Line Mapping and Costing." https://netzeroamerica.princeton.edu/the-report.

Pennsylvania DEP (Department of Environmental Protection). 2021. "Pennsylvania Climate Action Plan 2021: Strategies for government, business, agriculture, and community leaders—and all Pennsylvanians." *Pennsylvania DEP, ICF, Penn State University* and *Hamel Environmental Consulting.* http://www.depgreenport.state.pa.us/elibrary/GetDocument?docId=3925177&DocName=2021%20PENNSYLVANIA%20CLIMATE%20ACTION%20PLAN.PDF%20%20%3cspan%20style%3D%22color:green%3b%22%3e%3c/span%3e%20%3cspan%20style%3D%22color:blue%3b%22%3e%28NEW%29%3c/span%3e%209/21/2023.

Penrod, E. 2022. "Corporate Clean Energy Procurement on Track for Another Record Year After Adding 11 GW in 2021." *Utility Dive.* https://www.utilitydive.com/news/corporate-clean-energy-procurement-ceba-report/623926.

Peretzman, P. 2022. "Board of Public Utilities | New Jersey Advances Offshore Wind Transmission Proposal at Federal Energy Regulatory Commission." Newsroom and Public Notices. https://www.nj.gov/bpu/newsroom/2022/approved/20220127.html.

Reed, L., L. Abrahams, A. Cohen, J. Majkut, B. Phillips, A. Place, and J. Prochnik. 2021. "Report: How Are We Going to Build All That Clean Energy Infrastructure?" Niskanen Center and Clean Air Task Force. August 24. https://www.niskanencenter.org/report-how-are-we-going-to-build-all-that-clean-energy-infrastructure.

RFF (Resources for the Future). 2022. "Climate Change and the Supreme Court: *West Virginia v. EPA.*" Resources for the Future. https://www.rff.org/events/rff-live/climate-change-and-the-supreme-court-west-virginia-v-epa.

RGGI (Regional Greenhouse Gas Initiative). n.d. "The Regional Greenhouse Gas Initiative: An Initiative of Eastern States of the US." https://www.rggi.org/program-overview-and-design/elements. Accessed January 11, 2023.

Rivera, A., B. King, J. Larsen, and K. Larsen. 2023. "Preliminary US Greenhouse Gas Emissions Estimates for 2022." Rhodium Group. January 10. https://rhg.com/research/us-greenhouse-gas-emissions-2022.

Roy, N., D. Burtraw, and K. Rennert. 2022. "Retail Electricity Rates Under the Inflation Reduction Act of 2022." *Resources for the Future Issue Brief* 22-07. https://media.rff.org/documents/IB_22-07_HcKDycO.pdf.

Samaras, C. 2022. "Power Sector Carbon Index—2022 Q2 Update." *Power Sector Carbon Index.* Scott Institute for Energy Innovation, Carnegie Mellon University. https://emissionsindex.org/news-events/power-sector-carbon-index-2022-q2-update.

SEC (Securities and Exchange Commission). 2022. "SEC Proposes Rule to Enhance and Standardize Climate-Related Disclosures for Investors." https://www.sec.gov/news/press-release/2022-46.

SEPA (Smart Electric Power Alliance). 2023. "Utility Carbon Reduction Tracker." Smart Electric Power Alliance. https://sepapower.org/utility-transformation-challenge/utility-carbon-reduction-tracker.

Schurle, A., and T. Roessler. 2022. "The Inflation Reduction Act: Key Provisions Regarding the ITC and PTC." Foley and Lardner. https://www.foley.com/en/insights/publications/2022/08/inflation-reduction-act-key-provisions-itc-ptc.

Shawhan, D., S. Witkin, and C. Funke. 2022. "Pathways Toward Grid Decarbonization: Impacts and Opportunities for Energy Customers from Several US Decarbonization Approaches." Resources for the Future. https://media.rff.org/documents/Shawhan_Pathways_II_final_pres.pdf.

Sidley. 2022. "Inflation Reduction Act: Overview of Energy-Related Tax Provisions—An Energy Transition 'Game Changer.'" August 18. https://www.sidley.com/en/insights/newsupdates/2022/08/inflation-reduction-act-an-energy-transition-game-changer.

Steinberg, D.C., M. Brown, R. Wiser, P. Donohoo-Vallett, P. Gagnon, A. Hamilton, M. Mowers, C. Murphy, and A. Prasana. 2023. "Evaluating Impacts of the Inflation Reduction Act and Bipartisan Infrastructure Law on the U.S. Power System." NREL. Technical Report NREL/TP-6A20-85242. https://www.nrel.gov/docs/fy23osti/85242.pdf.

Supreme Court of the United States. 2021. *West Virginia et al. v. Environmental Protection Agency et al.* https://www.supremecourt.gov/opinions/21pdf/20-1530_n758.pdf.

Tallackson, H., and S. Baldwin. 2022. "Implementing the Inflation Reduction Act: A Roadmap for Federal and State Buildings Policy." Energy Innovation: Policy & Technology. https://energyinnovation.org/wp-content/uploads/2022/11/Implementing-The-Inflation-Reduction-Act-A-Roadmap-For-State-And-Federal-Buildings-Policy.pdf.

Trabish, H.K. 2021. "State, Federal Actions Show Growing Push for a Nuclear Role in Reaching Net Zero Emissions." *Utility Dive*. September 28. https://www.utilitydive.com/news/state-federal-actions-show-growing-push-for-a-nuclear-role-in-reaching-net/606107.

White House. 2021a. "Executive Order on Tackling the Climate Crisis at Home and Abroad." https://www.whitehouse.gov/briefing-room/presidential-actions/2021/01/27/executive-order-on-tackling-the-climate-crisis-at-home-and-abroad.

White House. 2021b. "President Biden Announces the Build Back Better Framework." https://www.whitehouse.gov/briefing-room/statements-releases/2021/10/28/president-biden-announces-the-build-back-better-framework.

White House. 2021c. "President Biden Sets 2030 Greenhouse Gas Pollution Reduction Target Aimed at Creating Good-Paying Union Jobs and Securing U.S. Leadership on Clean Energy Technologies." https://www.whitehouse.gov/briefing-room/statements-releases/2021/04/22/fact-sheet-president-biden-sets-2030-greenhouse-gas-pollution-reduction-target-aimed-at-creating-good-paying-union-jobs-and-securing-u-s-leadership-on-clean-energy-technologies.

Willson, M. 2022. "MISO Approves Large Grid Expansion, Paving Way for Renewables." *E&E News*. July 26. https://www.eenews.net/articles/miso-approves-large-grid-expansion-paving-way-for-renewables.

The Built Environment

ABSTRACT

The planning, design, construction, maintenance, and disposition of the built environment creates more than 40 percent of U.S. energy demand and resulting greenhouse gas (GHG) emissions. From our individual homes, workplaces, and institutions through neighborhoods and greater metropolitan and regional geographies, the built environment's demand for energy continues to evolve in complex ways. A broad range of public policies and private actors make decisions fundamental to the built environment that, in turn, define how, how much, and what kind of energy is consumed. The overall reduction of total energy demand in the built environment through energy-efficiency improvements and advanced energy management is as critical for U.S. decarbonization as the elimination of carbon-based energy directly used by them through building electrification and property-level renewable energy installations. The execution of these integrated building-level decarbonization actions alongside community-scale interventions such as mixed-use, transit-oriented redevelopment, and shared renewable power generation could be transformational.

Most of these actions are well within the current technological capacity and knowledge of the industries that support new and existing buildings. Movement along the decarbonization path has indeed accelerated since the committee's first report owing to new funding and supportive policy at all levels of government, not the least of which are the 2021 Infrastructure Investment and Jobs Act (IIJA) and the Inflation Reduction Act of 2022 (IRA). However, further opportunity for decarbonization exists in the built environment, well beyond the goals that underpin these policies, and more aggressive targets for the built environment's decarbonization are feasible.

The immediate challenges lie in these policies' implementation. This holds particularly true of their primary use of voluntary, financial incentives that will require extensive coordination with existing private service providers, owners and occupants of buildings that must be harnessed and sustained beyond their course over the next decade. To succeed, there needs to be clarity and deep coordination in federal, state, and local building programs, with a commitment to ensuring uniform access, minimal burden on both providers and households, and diligent monitoring of take-up across all states. The current incentives should catalyze the institutional changes—the technologies, workforce, industrial organization, owner demand, and, where needed, sufficient household assistance—that will facilitate the move to more regulatory policies for eliminating carbon-based demands in the built environment.

The committee identified immediate actions regarding the implementation of recent investments to ensure that the next decade's targets are met, including promoting equity and access to the benefits availed under the incentives within the IRA and working to improve and expand programs such as the Weatherization Assistance Program. Deep conservation and innovations at the building to community level are needed to support the strategic integration of the building sector with the emerging renewable electric grid. For this to succeed, there needs to be clarity and deep coordination in federal, state, and local building programs with increased regulatory rigor for both efficiency and electrification in buildings, appliances, and equipment. There is an equally critical need for national investment in increased research, development, demonstration (RD&D), and commercialization of building technologies for homes, commercial buildings, and grid-interactive efficient buildings (GEBs) with distributed renewables and storage. Last, there is the transformational pursuit of community-level decarbonization of buildings and community infrastructures that integrate buildings more comprehensively with transportation systems and the grid.

INTRODUCTION

The committee's first report highlighted the built environment of the United States—the millions of individual homes, offices, schools, and other facilities as well as the communities and infrastructure that serve and connect them. Although not separately analyzed for its decarbonization potential, the sector was the subject of numerous recommendations. Movement along that path has accelerated since the first report's publication owing to new public funding and supportive policy at all levels of government, not the least of which are the IIJA, the IRA, and their combined federal investment in the electrification and energy-efficient improvement of our residential and commercial building stocks, the largest component of the stationary built environment's contribution to GHGs by far. Combined with the current administration's presumed targets for the built environment's decarbonization by 2050, simply put, the goal posts have moved. Consequently, the committee provides three overarching observations in relation to these recent changes in addition to specific recommendations and the individual findings that are enumerated in this report.

First, the committee is very conscious of the pivotal moment in building decarbonization in which the nation finds itself, but also that the transformation that is asked of our buildings is largely voluntary and market based. It requires extensive coordination with existing private providers of building services. However, that coordination must be sustained and formalized beyond the next decade. The practices established now must be based on solid principles of equity, efficiency, and effectiveness. Therefore, the committee's first recommendations center on the implementation of recent investments to ensure that the next decade's targets are met.

- **Recommendation 7-1:** Ensure Clarity and Consistency for the Implementation of Building Decarbonization Policies.
- **Recommendation 7-2:** Promote an Equitable Focus Across Building Decarbonization Policies.
- **Recommendation 7-3:** Expand and Evaluate the Weatherization Assistance Program.
- **Recommendation 7-4:** Coordinate Subnational Government Agencies to Align Decarbonization Policies and Implementation.

Second, the committee explicitly calls for approaching the built environment's decarbonization through both the reduction of carbon-based fuels that are directly consumed in building (a shift to electrification powered by renewables) as well as the range of interventions that reduce the overall demand for energy production. Reducing overall demand ensures that energy production decreasingly relies on any carbon-based sources, but also promotes management of a clean electricity grid.

This approach begins with improving all buildings' energy efficiency. The committee referenced this multi-pronged strategy in the first report but provides more detail about the range of building interventions that are possible and whose sum are greater than the individual parts that are proposed in contemporary efforts. Additional recommendations ensure that this range of interventions are central to the mid- and long-term plans for building decarbonization.

- **Recommendation 7-5:** Build Capacity for States and Municipalities to Adopt and Enforce Increased Regulatory Rigor for Buildings and Equipment.
- **Recommendation 7-6:** Increase Research, Development, Demonstration, and Deployment for Built Environment Decarbonization Interventions.

Third, the committee believes that more aggressive actions are needed and possible from buildings than what is projected in the original report's targets, what is enabled with IRA funds, and what is stipulated even under current White House ambitions. The building industry's proven capacity to adapt, combined with the urgent co-benefits to occupant health and reduced cost burdens as well as community economic development, demand that our nation's decarbonization strategy center the built environment more squarely.

To that end, one additional strategy is needed to ensure that the lessons of this first decade in comprehensive building carbonization are sustained.

- **Recommendation 7-7:** Extend Current Decarbonization Incentives Beyond the Next Decade While Scaling Up Mandates.

Given the significant portion of our GHG emissions that come from both the direct combustion of natural gas and other carbon fuels in our buildings and in the electricity sources that power them, the committee expands on the earlier recommendations by identifying the gaps between current conditions and the first report's targets as well as the estimated contributions from recent policy. This report describes the range of physical interventions in the built environment that present decarbonization opportunities, arguing for a nuanced approach to policy and program priorities between them. Last, the report sheds light on the physical, industrial, and social parameters of the built environment's energy demand that shape the implementation of recently passed legislation as well as the committee's recommendations, as these structures will define how midcentury decarbonization goals will be met in this sector. Table 7-8, at the end of the chapter, summarizes all the recommendations that appear in this chapter to support decarbonizing the built environment.

SUMMARY OF CURRENT GOAL SETTING

Several proposals have been put forth regarding the built environment's capacity to decarbonize and, consequently, quantified targets for this vision, over the past decade. Two are most relevant to the committee's current considerations: (1) the committee's last set of recommendations in 2021 that are mirrored in recently enacted public policy, and (2) the more recent goals defined by the Biden administration in 2022 for keeping with international promises for the nation's GHG reductions. Both included the built environment as sectors to be decarbonized within a larger economic and material transition, but to variable degrees of magnitude and urgency.

2021 Committee Goals

The committee's first report addressed the contribution from the built environment to the national decarbonization vision at different levels of policy magnitude. These included the broad goal of reductions in total energy consumption (not exclusively carbon-based energy consumption) as well as detailed heat pump adoption targets (as the largest technological opportunity for both electrification and efficiency gains in the building sector). These goals and targets included variably quantified outcomes.

Specifically in relation to overarching goals at the time, the report recommended the reduction of overall energy consumption in existing buildings by 3 percent per year over the next decade to achieve a 30 percent total reduction, and in new construction by 50 percent total, both by 2030. As described in the report, the strategy for meeting these targets combines incentives for private-sector voluntary actions as well as mandates for new construction and equipment (the latter primarily for buildings supported with public funding). Beyond the first decade, the committee provided a broad range of long-term goals that also differentiate between new buildings (designated to be carbon-neutral by 2050) and existing ones (to have 50 percent reductions in overall energy demand by 2050). In short, the higher-order goals target overall energy demand as opposed to the carbon-based energy sources that currently fill that demand.

Last, the committee provided specific measurable targets for one strategy toward these decarbonization goals. The primary technological vehicle for reaching the report's targets is the electrification of space and water heating; the report set a target of increasing electric heat pumps' share of heating and hot water equipment to 25 percent in residential buildings and 15 percent in commercial buildings by 2030 (NASEM 2021a, p. 6). This specific intervention is particularly strategic given that heat pumps bring important benefits from reducing carbon-based fuel dependence and indoor air pollution for occupant health and from enhancing electrical grid management and optimization (owing to the digitalization of electrical information and energy-efficiency of new equipment). The recommended enabling policy to support this target is in the form of national equipment and appliance standards with increasing requirements for their electrification.

The committee also recommended a few other policy actions to assist in meeting the 2030 targets, including expanding funding for the Weatherization Assistance Program (WAP) and including electrification requirements within the program's eligible activities; enforcing efficiency standards on federally owned buildings; and encouraging more RD&D across a range of building interventions, such as building benchmarking, district and combined building energy, building-level energy storage, and embodied carbon in construction materials.

When taken in aggregate, the first report encouraged building electrification policies that reduce and eventually eliminate carbon-based fuel consumption directly in U.S. buildings, while also acknowledging the critical role that increasing energy efficiency requirements of new buildings and expanding energy-efficient retrofits in existing ones have had for reducing the overall demand for energy and, in turn, reducing the magnitude of carbon-based energy needs. However, although the report addressed a range of technological interventions that fall within the built environment's realm, it does not prioritize or assign quantitative targets for each intervention's contributions to decarbonization.

Finding 7-1: The first report's recommendations for the built environment's decarbonization were multipronged, with overarching goals and specific activity targets aimed at different challenges and variable actions. The overall reduction of buildings' total energy demand is as critical for comprehensive decarbonization as the elimination of carbon-based fuels that are directly used in buildings—the former generating benefits in equitable building transitions, grid management in the built environment, and overall energy demand beyond buildings.

2022 Federal Goals

The reentry of the United States into the global climate treaties collectively known as the Paris Agreement in January 2021 launched a reconsideration of the magnitude of ambition in the national decarbonization efforts and each sector's contribution to it.[1] The goal posts moved significantly. In April of 2021, the nation committed to reducing net GHG emissions to 50 to 52 percent below 2005 levels by the year 2030 and to zero by 2050 (DOS and EOP 2021).

For the building sector, the policy commitment combines the multi-pronged strategies of energy efficiency and electrification recommended by the committee with consequent, critical interventions such as digitalization. Equity, mainly with regard to building decarbonization's costs, is noted as part of this vision and moved forward through the implementation of Justice40 across the federal government (see Chapter 2 for more details on Justice40). Policy actions like RD&D funding, support for increasingly rigorous codes and standards, and investments in the public building stock add to the approaches for realizing the vision.

However, the ambition for the built environment's decarbonization is more muted than for other sectors, and less ambitious than the committee's prior (and achievable) 30 percent reduction goal. Emissions from buildings have already been falling since 2005 owing largely to increases in energy efficiency, even though both the number and size of residential and commercial buildings have continued to increase. The potential for much deeper emissions reductions is large. More aggressive actions are needed to ensure that such reductions are tapped.

> Finding 7-2: The White House's 2022 projections for each sector's decarbonization understate buildings' decarbonization magnitude and pace. As a leading contributor to GHG emissions from both direct carbon-based fuel use and as an end use for utility-provided electricity that relies on carbon-based fuels, the built environment has shown demonstrable transformation over the past half-decade and could, technologically and economically, transform further.

---

[1] This chapter considers the current administration's proposal as the overarching update to the committee's 2021 goals as described in *The Long-Term Strategy of the United States: Pathways to Net-Zero Greenhouse Gas Emissions by 2050* (DOS and EOP 2021).

GAPS

The built environment directly consumes a significant amount of energy, most of which continues to derive from carbon-based sources. The aggregation of buildings and their spatial placement has historically relied on carbon-based energy sources directly (50 percent of residences and 47 percent of commercial buildings use oil or natural gas energy directly) and through electric power production (adding another 25 percent of residences and 30 percent of commercial facilities that rely on electricity produced through fossil fuel or coal combustion). When including the electricity transmission for their consumption, buildings account for the largest end-use share (35 percent) of total U.S. GHG emissions of any sector (EIA 2022a,b). (See Chapter 8 for a discussion on emissions from construction material extraction and production.)

Current Energy Use

Consequently, the residential and commercial building sector accounted for more than 40 percent of total energy end uses last year, including electrical generation, transmission, and losses, as well as direct energy uses in buildings, or almost 23 quadrillion BTUs, according to the Energy Information Administration (EIA 2022c). Through direct combustion alone, buildings consumed an aggregate 7.2 quadrillion BTUs in energy from natural gas, heating oils, and other carbon-based fuels according to recent consumption surveys, almost two-thirds of it coming from the housing stock and the remainder from commercial buildings (Table 7-1). Buildings, in short, matter.

Comparing the White House's 2021 long-term national targets (DOS and EOP 2021) with the current energy demands of the built environment provides important perspective. Projections of building decarbonization to support the national goals suggest that the sector's emissions would be reduced to 0.4–0.5 gigatons from the 0.6 gigatons at 2005 levels by 2035 and to 0.1–0.3 gigatons by 2050 (DOS and EOP 2021, Figures 4 and 9). The reductions stem in part from the reductions in carbon-based fuels used directly in buildings, as they electrify and increase efficiency, presumed to decrease from approximately 11 exajoules in 2005 to 7.5–9 exajoules (an 18–32 percent reduction, or the equivalent of saving 2–3 quadrillion BTUs) by 2035 and to 2–6 exajoules by 2050 (or 45–82 percent reductions, or 5–8.5 quadrillion BTUs saved).

If the lower range of these values were achieved, however, fully 2.2 quadrillion BTUs of carbon-based fuels would still be used directly in buildings by 2050—almost one-quarter of what is currently used. This sizable, persistent use of carbon fuels in buildings is neither aligned with the transformations proposed for other sectors, nor is

TABLE 7-1 Recent Annual Total and Building Average Primary Energy Consumption

	Number of Buildings (thousands)	Energy Sources (millions of BTUs)			
		All Energy Sources	Electricity	Natural Gas and Fuel Oil	Other Source
2015 Residential	118,200	9,114,000,000	4,324,000,000	4,790,000,000	—
Building average	118,200	77	37	34	—
Recent average	3,800	67	40	27	—
2018 Commercial	5,918	6,807,000,000	4,090,000,000	2,401,000,000	316,000,000
Building average	5,918	1,150	691	406	53
Recent average	537	1,359	885	439	35

NOTES: Averages for energy source types in buildings derived across the entire building stock (not by buildings that use an individual type solely). Most recent = 2010–2018 commercial buildings; 2010–2015 residential. Other sources include district heating. Primary electricity and all energy consumption numbers do not include transmission losses. SOURCE: Tabulations of 2015 Residential Energy Consumption Surveys and 2018 Commercial Buildings, courtesy of the U.S. Energy Information Administration, May 2018 and December 2022 (EIA 2018, 2022a).

it compatible with the usual functional life cycle of energy-consuming equipment and appliances in buildings or the turnover in building ownership.[2]

To reach a more aggressive outcome for transforming individual buildings, each with its own owners and occupants, construction qualities, and local building markets, significant implementation challenges must be overcome. Reducing buildings' energy use requires both eliminating their direct use of fossil fuels toward electric appliances and equipment alongside deep efficiency improvements and energy management innovations.

For the former objective, the gap between current consumption and the desired reduction is formidable. Over half of all commercial buildings and almost two-thirds of homes currently combust natural gas, fuel oil, or propane solely or in some combination directly in buildings. In absolute terms, this translates to almost 3 million commercial buildings and almost 75 million homes using just natural gas currently that would need to be electrified to meet national net-zero decarbonization goals by midcentury (Table 7-2). Like other sectors, fortunately, buildings

---

[2] For example, the average life cycle of a natural gas space-heating furnace is 15 to 20 years, and the current median duration of home occupancy is 13 years, per DOE technical monitoring (Energy Star Undated) and census estimates (NAR 2020). In theory, homes could have two system replacements and two new owners in the 3 decades before 2050.

TABLE 7-2 Building Counts by Energy Source and Year of Construction (millions of buildings)

	All Buildings	Electricity	Natural Gas	Fuel Oil	Propane	Wood, Coal, Other	District Heat	District Chilled Water
Residential	123.53	123.53	74.65	5.72	11.68	10.83	—	—
Before 1950	20.26	20.26	15.28	1.87	1.83	1.75	—	—
1950 to 1959	12.48	12.48	9.09	1.06	0.78	0.80	—	—
1960 to 1969	12.76	12.76	8.72	0.72	0.98	1.06	—	—
1970 to 1979	18.34	18.34	9.76	0.76	1.65	2.01	—	—
1980 to 1989	16.30	16.30	8.10	0.55	1.37	1.89	—	—
1990 to 1999	17.16	17.16	9.30	0.42	2.08	1.68	—	—
2000 to 2009	16.16	16.16	8.97	0.31	1.91	1.20	—	—
2010 to 2015	5.53	5.53	2.93	0.03	0.58	0.24	—	—
2016 to 2020	4.56	4.56	2.49	—	0.49	0.19	—	—
Commercial	5,918	5,613	2,974	583	676	180	86	55
Before 1920	329	323	184	54	49	—	—	—
1920 to 1945	379	368	172	52	65	—	11	3
1946 to 1959	517	496	310	56	52	—	20	—
1960 to 1969	685	673	420	57	61	—	14	10
1970 to 1979	831	787	400	87	115	25	8	6
1980 to 1989	794	752	372	69	85	—	6	4
1990 to 1999	921	844	412	73	113	—	8	10
2000 to 2009	924	860	454	94	91	—	6	6
2010 to 2018	537	508	250	41	45	—	4	4

NOTES: Rows do not sum, as multiple sources may be used in individual buildings. Approximately 3.7 percent of residential buildings and 1.8 percent of commercial buildings have property-level solar installed as part of their electric source (EIA 2022a,b). SOURCE: Tabulations of 2018 Commercial Buildings and 2020 Residential Energy Consumption Surveys, courtesy of the U.S. Energy Information Administration, December 2022 and May 2022 (EIA 2022a,b).

have undergone a significant although gradual internal transformation over the past half-century.

Increasing electrification and consequent reductions in carbon-based fuels have been the long-term trend, albeit a slowly realized one. For example, 75 percent of homes built before 1950 currently rely on natural gas compared to 55 percent of the homes built more recently. Although 60 percent of commercial buildings built before the mid-1900s rely on natural gas, that share is about 47 percent in the buildings built in the past decade (see Table 7-2). Continuing the trend toward full electrification in new buildings along with a massive transformation for electrification within existing buildings is needed to reach net-zero carbon goals solely within their operations—that is, before considering electrical sources and transmission.

Regarding the second objective of decreased total energy consumption, new commercial and residential buildings are built to stricter energy performance requirements, using more efficient equipment and appliances and often with their own energy production potential via distributed, building-level renewable energy installations. Consequently, the average energy consumption (and certainly the energy intensity) for the most recent additions to the building stock tend to be lower than their older counterparts. (See Table 7-3 for average residential energy consumption for different energy sources by year of construction.)

Efficiency improvements in newer building construction over time, however, are offset by two other realities that affect overall energy use and GHG emissions. First, the number of buildings continues to increase, mirroring national demographic and economic growth. Second, the average size of new buildings has been increasing in both commercial and residential construction. Consequently, both primary energy use and total consumption from the sector's two building stocks have leveled over the past 2 decades (Figure 7-1).

Because neither of these trends is expected to change, the ability to further decarbonize buildings requires an intricate surgery to (1) gradually remove the primary use of natural gas, heating fuels, and other carbon-based sources—that is, electrify; and (2) further reduce the overall energy demand through efficiency improvements, grid management, distributed electricity production, and other electric grid management strategies. These approaches are required for both existing and new construction. Because of the magnitude of these challenges, targeting the largest end uses in buildings of both primary carbon-based fuel consumption as well as the overall largest uses of electric energy provide strategic opportunities for immediate reductions.

TABLE 7-3 Average Annual Residential Energy Consumption by Year of Construction, 2015

Year of Construction	All Fuels	Electricity	Natural Gas (millions of BTUs)	Propane	Fuel Oil/ Kerosene
Before 1950	88.7	30.1	65.3	34.5	68.7
1950 to 1959	84.4	31.7	60.3	26.9	79.9
1960 to 1969	75.0	32.5	53.9	26.3	63.2
1970 to 1979	70.3	36.7	52.2	28.0	64.3
1980 to 1989	65.7	37.5	48.0	25.8	58.6
1990 to 1999	78.3	42.3	60.2	29.8	62.0
2000 to 2009	78.2	43.8	59.1	40.9	63.6
2010 to 2015	67.0	39.8	51.7	31.2	

NOTES: The "All Fuels" category includes consumption for biomass (wood), coal, district steam, and solar thermal, which are not represented in individual columns in this table. Electricity consumption from on-site solar photovoltaic generation (i.e., solar panels) is included. SOURCE: Calculated using data from Table CE2.1, "Annual Household Site Fuel Consumption in the United States—Totals and Averages," in the 2015 Residential Energy Consumption Surveys, courtesy of the U.S. Energy Information Administration, May 2018 (EIA 2018).

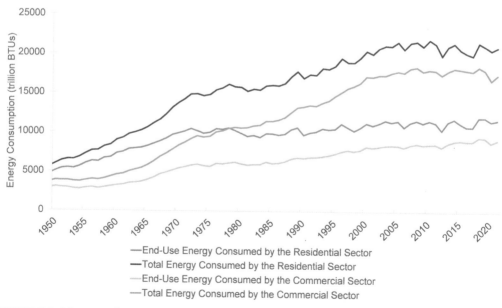

FIGURE 7-1 Primary end use and total energy consumption for residential and commercial buildings, 1949–2021 (trillion BTUs). SOURCE: Data from EIA, October 2022 (EIA 2022d) and August 2023 (EIA 2023c).

Technology Gaps and Transitions: The Example of Heat Pump Adoption

The first report's recommendation to focus on space and water heating as the largest carbon-based fuel end uses in commercial and residential buildings and its subsequent integration into national policy as noted later in this chapter, then, is well founded.[3] Approximately 69 percent of all carbon-based energy consumed in homes is used for space heating, 24 percent for water heating, and another 3 percent for cooking (Table 7-4). These end uses dominate consumption of most energy sources except for fuel oil (which includes kerosene), 86 percent of which is used for space heating and 13 percent for water heating. The shares of carbon energy sources in commercial buildings for space and water heating are comparable—for example, 70 percent of the 2,974 commercial buildings that use natural gas use it for space heating.

The committee's focus on heat pumps as the vehicle for electrification of current space and water heating's energy consumption is also supported by the evidence, given current technological options for immediate electrification and the financial feasibility of their installation. The gap between heat pumps' diffusion and the rates of this technology's adoption that would be necessary to meet the stated decarbonization goals is significant. For example, approximately 673,000 commercial buildings currently use heat pumps as primary heating equipment (or 11.4 percent of commercial buildings) (EIA 2022a). This leaves less than a 4 percentage-point gap between current conditions and the committee's original targets for commercial building diffusion. If 2022 construction rates hold, half of all new buildings would need heat pumps to meet the 15 percent target by 2030.

The residential sector demonstrates similar challenges. At last count, slightly more than 14.4 percent of homes use electric heat pumps, posing a more than 10 percentage-point gap from committee targets by 2030 (EIA 2022b) (Table 7-5).[4] This share represents a significant increase of residential buildings from 5 years prior, when only 10 percent of homes contained heat pumps (EIA 2018). In fact, new homes constructed in the past 5 years for which data are available show a diffusion rate of almost 24 percent—or only 1 percentage-point away from committee targets (NASEM 2021).

---

[3] The White House goals do not specify the technologies that would need to be adopted in order to meet the reduction targets in buildings. However, the IRA incentivizes a range of energy-efficiency and electrification technologies, with electric space heating and cooling heat pumps receiving the largest incentives by dollar value.

[4] This count does not include ductless heat pumps, commonly called "mini-splits," which would increase the current share to 15.3 percent of all homes. It does not include natural gas heat pumps.

TABLE 7-4 Aggregate Residential Energy Consumption by Source and End Use, 2015

	Total Consumption	Electricity	Natural Gas	Fuel Oil	Propane
	(trillion BTUs)				
Total	4,790	4,324	3,965	464	361
Space heating	3,307	638	2,678	397	234
Water heating	1,154	590	1,019	59	76
Space cooling	731	731	—	—	—
Cooking	132	62	113	—	19
Clothes drying	39	196	36	—	3
Other	156	1,804	119	8	30
	(million BTUs)				
Average	77.1	36.6	57.8	67.3	31.2
Space heating	—	11.1	45.0	59.9	31.0
Water heating	—	10.7	18.1	20.7	17.8
Space cooling	—	7.1	—	—	—
Cooking	—	0.8	2.9	—	3.1
Clothes drying	—	2.6	2.1	—	2.4
Other	—	16.6	0.9	24.4	1.9

NOTE: Averages are based on population of homes using each energy type as a primary source for an end use and therefore do not sum across energy types. SOURCE: Tabulations of 2015 Residential Energy Consumption Surveys, courtesy of the U.S. Energy Information Administration, May 2018 (EIA 2018).

If all new homes were required to have heat pumps, the housing stock would meet the committee's heat pump diffusion rate targets by 2028, assuming the average construction rates of approximately 1.2 million new homes built each year continues (U.S. Census Bureau 2023). However, requirements that all new construction be built to certain specifications are not always viable financially or logistically, nor are mandates without incentives politically viable. They also do not catalyze the industry to retrofit the existing building inventory—an industrial transformation that is needed to meet long-term decarbonization goals beyond 2030.

TABLE 7-5 Share of Homes with Electric Heat Pumps by Year of Home Construction over Time

Year of Observation	Total Share	Year of Home Construction (percentage of home age group)										
		Before 1950	1950 to 1959	1960 to 1969	1970 to 1979	1980 to 1989	1990 to 1999	2000 to 2009	2010 to 2015	2016 to 2020	2020 to 2030	2030 to 2040
2009	8.6%	3.1%	3.7%	5.3%	8.7%	11.8%	13.4%	14.1%	—	—	—	—
2015	10.0%	1.9%	7.1%	7.8%	9.8%	14.4%	14.3%	14.1%	13.2%	—	—	—
2020	14.4%	5.5%	8.4%	11.3%	14.2%	18.4%	19.1%	18.8%	21.0%	23.7%	—	—
2030 (est.)	20.6%	7.7%	12.7%	16.8%	19.1%	24.4%	24.2%	23.1%	40.0%	45.2%	25%	—
2040 (est.)	25.8%	9.9%	17.0%	22.2%	24.1%	30.5%	29.3%	27.4%	59.1%	66.7%	25%	25%

NOTES: The committee's estimates assume identical annual increases in rates of replacement for homes built before 2015 based on the 2009–2020 average and a consistent adoption rate of 25 percent for new homes built after 2020. New home counts after 2020 assume 1.2 million completions annually. SOURCES: Data from 2009, 2015, and 2020 Residential Energy Consumption Surveys, courtesy of the U.S. Energy Information Administration, May 2013, 2018, and 2022 (EIA, 2013, 2018, 2022b). Data for estimates of 2030 and 2040 heat pump adoption rates and housing units were committee generated.

Existing buildings' retrofitting will be the critical vehicle for the sector's decarbonization. As such, another important consideration from the review of past diffusion rates is the fact that replacement rates of heating equipment across the inventory of older building stock are also trending toward electric heat pumps—that is, not all gains in heat pumps' diffusion are from new construction. For example, almost 12 percent, or about 2 million homes, built in the 1980s had heat pumps over a decade ago. Now, 18.4 percent of the same homes have heat pumps—or about 3 million homes (see Tables 7-2 and 7-5). The transformation is notable because equipment in homes built in that decade will have completed its functional life and require replacing.

If electrification of space heating equipment in existing housing persists at the same acceleration in the future, the residential stock could hit 25 percent total adoption rates—that is, including existing and new housing barring any new policy intervention—by 2039. Assuming a 50 percent reduction in space heating energy consumption in these homes owing to heat pump installation, this also translates to about a 7.4 percent reduction in total residential energy demand by 2038 from heat pump adoption alone—below the committee's 30 percent reduction goal or the administration's more varying 18–32 percent goal (Energy Saver 2022).[5] Furthermore, assuming that all existing homes that transition to electric heat pumps were using carbon-based fuels for space heating beforehand, this would result in a 14.4 percent reduction in total demand for those fuels. In short, although the heat pump revolution is already happening, and adoption rates are increasingly meeting the committee's original targets or the new national projections, it is not happening fast enough to meet the broader goals of reductions in carbon-based energy demands or overall energy demand.

A second consideration regarding the characteristics of households that are likely to replace current equipment with heat pumps and to live in the new homes in which heat pumps are installed during construction is also worth noting (Table 7-6). Currently, the rates of heat pump adoption do not vary significantly across either income or housing tenure groups—for example, 14.5 percent of households with annual incomes lower than $40,000 have heat pumps compared to 14.3 percent of households with incomes greater than $40,000 (EIA 2022b). Similarly, 14.8 percent of owner-occupied homes have heat pumps compared to 13.5 percent of renters' homes. When factoring in the size of these groups, however, wide differences are seen in where heat pumps are found. Very low income households make up a much smaller

---

[5] A reduction of 50 percent of space heating energy consumption (or about 20 mBTUs per home annually) in 25 percent of the existing housing stock (or 33.9 million homes by 2030) could result in approximately 678 tBTUs reduction in primary overall energy demand from residential buildings.

TABLE 7-6 Share of Households in Homes with Heat Pumps by Income and Tenure, 2020

	Total U.S.	Less Than $5,000	$5,000–$9,999	$10,000–$19,999	$20,000–$39,999	$40,000–$59,999	$60,000–$99,999	$100,000–$149,999	$150,000 or More
Income group	14.4%	12.7%	14.0%	14.2%	15.1%	15.7%	14.6%	13.3%	13.0%
Heat pump homes	100.0%	3.2%	3.2%	8.2%	20.5%	17.4%	22.8%	12.4%	12.4%

	Total U.S.	All Owners	All Renters	SF Owners	SF Renters	All SF	MF Owners	MF Renters	All MF
Income group	14.4%	14.8%	13.5%	15.1%	14.7%	15%	9.5%	12.9%	12.5%
Heat pump homes	100.0%	69.1%	30.9%	66.8%	10.6%	77.4%	2.3%	20.3%	22.6%

NOTE: MF = multifamily; SF = single family. SOURCE: Tabulations of 2020 Residential Energy Consumption Surveys, courtesy of the U.S. Energy Information Administration, May 2022 (EIA 2022b).

share of heat pump home occupants than their moderate-income counterparts (although wealthier counterparts lag as well). Heat pumps are also more than three times more likely to be found in single-family homes than multifamily apartment buildings. Despite modest differences across household demographic groups, then, there is still sufficient disparity in adoption to warrant a policy intervention to close the gaps.

Equity Gaps: The Role of the Weatherization Assistance Program

Along this same vein, a review of the committee's remaining quantified target regarding funding for WAP is also in order. However, identifying the gap between the committee's recommendations is more complicated because eligibility is based on income qualifications (specifically, households earning less than 200 percent of the federal poverty level) rather than the energy consumption, efficiency or performance, or energy source of the home. Further complicating this calculation is the fact that energy benefits, including natural gas savings, from WAP have been estimated variably (Fowlie et al. 2018; Tonn et al. 2018).[6]

Barring more refined monitoring, outcome evaluation, and tracking of implementation across grantees, rudimentary calculations estimate the committee's recommendations as minimally contributing to decarbonization goals.[7] Assuming an accurate estimate lies between currently measured ranges, the increase in appropriations recommended by the committee would lead to about 105,000 homes being weatherized per year, for an additional nearly 0.95 million homes by 2030 since the first report. The committee's desired program budgets could lead to total reductions of approximately 23.8 tBTUs in overall energy demand (or about a 0.3 percent reduction from the current total residential demand of 9.1 qBTUs) and a reduction of 19 tBTU in natural gas demand (the equivalent of 0.4 reduction).[8] Even when added to the reductions estimated for the heat pump goals, then, WAP contributes very modestly to reductions in

---

[6] Tonn et al. (2018) estimated a 29.3 mBTU/unit reduction in 2008 and a 26.6 mBTU/unit reduction in 2010; Fowlie et al. (2018) calculated annual energy savings of 17.2 mBTU/unit. For gas savings, Allcott and Greenstone (2012) calculate 20–25 percent natural gas savings, while Brown et al. (1994) estimated first-year savings of 17.3 mBTU.

[7] The Department of Energy (DOE) estimates that 35,000 households are served per year based on an annual budget of approximately $300 million, or an average $8,571 per home (EERE 2022a). However, the most recent accounting of WAP grantees in 2019 found that the $1.12 billion in funding for the program year resulted in 85,200 weatherized homes, averaging $13,146 per home (NASCSP 2019). For the purposes of this exercise, the average of $10,858.50 is assumed.

[8] The estimate assumes 950,000 weatherized homes with an average 25 mBTU total energy savings and a 20 mBTU natural gas savings per unit by 2030.

both overall and specifically carbon-based energy demand given its current statutory rules and implementation framework.

In addition to its energy ambitions, WAP is widely varied in its ability to meet its social impacts. Approximately 38.8 million households (about 30 percent of the United States) are currently at or below twice the U.S. poverty level and therefore eligible for WAP assistance—a need that would take centuries to serve even with the committee's ambitious budget recommendation. Other indicators for identifying energy-specific needs may help better target decarbonization efforts. For example, recent data suggest a growth in overall energy insecurity. Considering just energy cost hardship, high energy burden (defined by Drehobl et al. [2020] and Fisher Sheehan & Colton [2021][9] as paying more than 6 percent of income for energy bills) remains a persistent national challenge for at least a quarter of the nation's households, and two-thirds of WAP-eligible households face this burden (Drehobl et al. 2020). Median energy costs as a share of a household's total housing costs are more than one-third lower for wealthier counterparts than WAP-eligible households, although lower-income households consume a third less energy per household and per household member.[10]

Other measures of household energy hardships such as housing inadequacy may provide a narrower target of opportunity for WAP than solely financial hardship. The physical energy performance of homes could be one obvious consideration. Such an overlay could better serve the 4.5 million households that experience uncomfortable cold or heat for extended periods owing to equipment breakdown, utility interruption, inadequate heating system capacity, or inadequate insulation among other reasons—568,000 of which experience severe housing-based heating inadequacy (data from AHS 2021; U.S. Census Bureau 2022). Alternatively, if the committee's recommendation to include electrification as a WAP activity could be used to help further identify the most burdened households, then serving the 25.7 million WAP-eligible households that live in homes that rely on natural gas or heating oils may be a preferable target.[11] There are also 13.1 million households with incomes between $40,000–$60,000 living

---

[9] This affordability percentage is based on the assumption that an affordable housing burden is less than 30 percent of income spent on energy, and no more than 20 percent of housing costs should be allocated to energy bills.

[10] For households with incomes under 200 percent of poverty thresholds, median energy costs across all energy types are 32.8 percent of median housing costs compared to 22.3 percent of households with incomes above per 2021 AHS housing cost data. Per 2015 Residential Energy Consumption Survey (RECS) consumer expenditure data, households with incomes less than $40,000 consumed 63.4 mBTUs compared to the 92.2 mBTUs for those with higher incomes (EIA 2018).

[11] Using the 25.7 million households in homes with carbon-based energy uses and the $13,146/unit average, the total budget needed is $337.9 billion, or $33.8 billion annually for the next decade.

in homes that rely on carbon-based fuels that are not WAP eligible but may not be able to afford electrification (AHS 2021).

Furthermore, the range of housing inadequacies that need to be addressed such that weatherization interventions are beneficial for decarbonization and energy reductions—as well as for the recipient households' health and finances—are greater than the program currently has capacity to address (Kresowik and Reeg 2022). Given that the predominate beneficiaries of positive energy improvements continue to be wealthier households, a more nuanced eligibility metric that includes households' energy burdens and their housing's energy performance and decarbonization needs is needed (Frank and Nowak 2015; Hernández et al. 2014). Ultimately, targeted definitions of WAP eligible populations, services, and allowable expenditures per household would provide analysis to assess gaps more fully between need and committee recommendations.

> Finding 7-3: The committee's original recommendations for increased investments in WAP are meant to prioritize decarbonization efforts for households experiencing a range of energy hardships. However, defining the population and articulating outcomes (in both removal of carbon-based energy sources as well as improving energy efficiency to reduce overall energy burden) remains a work in progress, as does WAP's capacity to meet the demand for eligible households.

Workforce Gaps

A major implementation challenge will be ensuring an adequate workforce to accomplish the necessary building, retrofitting, and construction. In 2022, the construction industry experienced its highest recorded level of job openings combined with an industry-low unemployment rate, and the Associated Builders and Contractors estimated that in 2023, the construction industry would need to attract an estimated 546,000 workers in addition to the normal pace of hiring to meet the demand for labor (ABC 2023). The federal and subnational actions described in the following section will further increase spending on construction. McKinsey modeling of the Bipartisan Infrastructure Law found that it could create 3.2 million new jobs across the nonresidential construction value chain, ~30 percent increase in the overall U.S. nonresidential construction workforce, or 300,000 to 600,000 new workers entering the sector every year (Hovnanian et al. 2022). However, these major legislative actions primarily include labor provisions that would grow the workforce (e.g., prevailing wage and apprenticeship programs) not as requirements, but as opportunities for "bonus" rates on tax incentives. The major legislation passed includes workforce training as an option for several funding opportunities but does not directly invest in the education and training of tradespeople. (See Chapter 4 for a more detailed discussion of the labor provisions in recent legislative actions and on the challenge of attracting and retaining the necessary workforce to accomplish decarbonization.)

Summary of Gap Analysis

In sum, the committee's targets for specific technological interventions (e.g., heat pump adoption) and program budgets (for WAP) lead to relatively modest levels of decarbonization. The short-term targets established in the first report are likely to be achieved independent of external intervention, although additional policies are needed to ensure that desired outcomes are met in future buildings and more tenuously for existing ones. Given the modest gap to meet the short-term targets, voluntary financial incentives would be reasonable to meet the committee's original milestones. Yet, the sum of these individual targets is not likely to meet the committee's more aggressive goals for energy reductions over the same time.

As precedent for the net-zero decarbonization visions for 2050, the targets are even more modest. For outcomes beyond the next decade, more concrete targets for the range of building interventions could be established in addition to clearer goals for both primary energy decarbonization and overall reductions in energy consumption. Like the committee's first recommendations, these should be specified for new and existing buildings distinctly. Short-term targets such as heat pump adoption rates would need to expand to meet the magnitude of those goals in combination with other grid-level decarbonization and building-level direct interventions.

Significant intervention is needed to make up the gaps between current conditions, the committee's more detailed intervention targets, and the committee's broad goals for 2030 as stated in the first report. This holds true for all the quantified gaps in overall energy efficiency (particularly for the least efficient older building stock), heat pump diffusion, weatherization funding, and need. Yet, it holds even more true for what is needed beyond 2030. The interventions that could be employed for the first decade of the decarbonization transition must also set the stage for the decades after.

Finding 7-4: The committee's original technological targets were modest and could be accomplished within a longer timeframe, although not one ending at 2030. Even new national goals established by the current administration, further, are more modest than the sector is capable of achieving. Meeting the overarching goals for reductions in the building sector's energy demands and the removal of carbon-based fuels that are currently used directly in building operations requires more aggressive action. Both policy incentives and regulatory intervention are needed to secure current trends and to move the sector toward the committee's and the administration's broader goals of comprehensive building-level electrification, overall energy demand reductions, and equitable transitions to both.

Policy Updates

Major policy adoption since the first report's publication have set us on a few of these pathways. Indeed, federal, state, and municipal policy support for building decarbonization have increased during and since the publication of the committee's first report. Each provides some achievement against the committee's original 2030 energy reduction goals, although none as much as the federal investments in the past 2 years.[12]

National Policy Changes

In particular, the IIJA authorized $3.5 billion to WAP until expended (§40551)—which the White House estimates will aid 700,000 income-eligible households over the next decade based on past WAP mean expenditures per unit.[13] However, a likely scenario given past expenditure rates is more than 322,000 homes weatherized, resulting in an additional 8 tBTUs reduced in total energy consumption (6.5 tBTUs in natural gas consumption). Assuming these funds get expended before 2030, this is only a fraction of $1.2 billion annual budget increases, or $10.8 billion by 2030, recommended by the committee in its first report (NASEM 2021).

Other IIJA authorizations that will shape the outcomes projected in the committee's goals include the Energy Efficiency and Conservation Block Grant Program (§40552), State Energy Program (§40109), and Energy Efficiency Revolving Loan Fund Capitalization Grant Program (§40502), although their individual contributions are not readily measurable given the lack of evidence of reductions for the share of these

---

[12] It should be noted that the IRA and the IIJA are not equivalent in funding mechanisms. The IRA primarily consists of spending programs (appropriations) and tax expenditures. Spending programs can allocate federal resources to projects and activities up to the amount of their appropriation. By contrast, tax expenditures, such as the production tax credits in the IRA, typically have no limit on the amount that could be claimed by taxpayers. The IIJA consists of a mix of authorizations and appropriations. Authorizations are laws that establish or continue a federal program or agency and are typically passed by Congress for a set period of time, but authorizations require appropriations before funds can be spent. Appropriations are laws that actually provide the money for government programs and must be passed by Congress every year in order for the government to continue to operate.

[13] DOE assumes an average $4,695/unit (EERE 2022b).

programs that already exist and the implementation questions for the new programs. The IIJA authorizations comes with no extensive statutory changes to the program beyond the removal of Davis-Bacon requirements for the weatherization of buildings of fewer than five units, which make up the bulk of interventions. This change will likely increase the number of those building types given the added program resources that can be used for their weatherization, although at the expense of prevailing wage benefits. This segment of the residential stock is also the least efficient.[14] Consequently, this targeting could result in greater efficiency gains than estimated previously (e.g., closer to 25 percent natural gas reductions).

The IRA, however, is the most relevant federal policy change for decarbonizing the built environment due its combination of tax credits and direct rebates for a wide range of technologies and energy-performance actions in new and existing stocks of both residential and commercial buildings (ACEEE 2022; Azerbegi 2022; DOE Office of Policy 2022; Evans 2023; Guidehouse Insights 2022; Philadelphia Energy Authority 2023; Smedick et al. 2022). Estimates of the act's possible building outputs for electrification and energy efficiency range widely.

Taking heat pump adoption projections alone, the IRA is projected in one scenario to directly install 7.2 million in retrofits via the 25C tax credit.[15] Additional opportunities that are likely to add to heat pump adoption counts in the existing housing stock include the low-income rebate and home energy retrofits programs, respectively projected to touch 2.4 million and 1.2 million households (Smedick et al. 2022). Presuming that one-half of these later programs' recipients opt to utilize the incentives for heat pump installation, 1.8 million more heat pumps will be added. Current projections for new homes benefiting from the 45L tax credits can also assume to add 650,000 heat pumps at a minimum to the housing stock (Smedick et al. 2022), bringing the total to 10 million existing homes with new heat pump installations by 2032.

With these new tax credits and direct rebates from the IRA, the committee estimates increased heat pump adoption rates starting in 2023, which would accomplish the committee heat pump adoption goals by 2029, 10 years before the currently projected industrial diffusion. By the IRA benefits' closeout in 2033, furthermore, the residential heat pump adoption will have surpassed the committee's goals by an additional 2.3 percent and 27.3 percent of the entire housing stock (or 38 million homes) will have electric heat pumps. The IRA's heat pump incentives alone would

---

[14] According to WAP American Recovery and Reinvestment Act evaluations and RECS consumption data for 2015, 2–4 unit residential buildings consumed 52.5 kBTUs/SF compared to 37.1 kBTUs for single-family attached and 38.8 kBTUs for buildings of more than five units. See also Martín et al. (2023).

[15] Among the most extensive estimates are those produced by RMI (Smedick et al. 2022).

reduce overall residential energy consumption by 8.3 percent and carbon-based fuels specifically by 15.9 percent by 2033—or one-third of the committee's 2030 goal.[16]

The range of other electrification and energy-efficiency improvements incented by the IRA would produce additional modest reductions. For example, based on analysis of realized project savings in annual consumption averages at current rates, the approximately 1.2 million households who could benefit from Home Owner Managing Energy Savings (HOMES) rebate programs[17] would produce 32.4 tBTUs aggregate reduction in overall energy consumption (or about 0.7 percent), assuming all households attempt the aggressive 35 percent reduction rebate threshold. If 115 million square feet of commercial building space are retrofitted at 25 percent efficiency improvement owing to the IRA's extended 179C tax credit, 2 tBTUs are reduced.

Appropriations for federal buildings—another opportunity targeted by the committee—could result in additional gains.[18] Other IRA provisions, such as local government incentives to adopt and enforce more rigorous building energy codes, would lead to additional reductions for new construction (Tyler et al. 2021). Of important note is the IRA's Greenhouse Gas Reduction Fund (§60103), the $27 billion set of competitive grants for "green banks" that could finance distributed community energy projects as well as household efficiency and electrification programs.

> Finding 7-5: The new federal policy terrain helps to fill the gaps or, more accurately, expedite the achievement of several of the committee's original targets. IIJA funding for WAP will provide an opportunity to target eligible households and energy-reducing improvements, including electrification, more effectively. IRA tax credits and rebates are also significant: IRA incentives for heat pump installations will meet the committee's targets by 2029 (a decade ahead of business-as-usual adoption rates). These installations alone could help reduce energy in 38 million homes by 8.3 percent of current consumption (15.9 percent of carbon-based energy consumption). Combined with other building improvements funded, recent federal laws could move the country significantly toward the committee's 2030 goals and the sector's expected national greenhouse gas reductions.

---

[16] See Table 7-5 for more on the committee's estimates for heat pump diffusion, and NASEM (2021a) for more on the committee's 2030 goals.

[17] HOMES rebate programs are implemented by state energy offices, providing rebates to cover a percentage of costs for retrofit projects that achieve 15 to 35 percent energy system savings. For more details on the $4.3 billion appropriated in the IRA to DOE to distribute as grants to these programs, see P.L. 117-169, title V, §50121, August 16, 2022, 136 Stat. 2033.

[18] The administration has moved forward with actions related to the federal stock as well (White House 2022a).

There are several caveats to these projections. This analysis assumes that eligible property owners would likely choose the technologies offering the largest credit or rebate (e.g., heat pumps with a $8,000 rebate for the lowest-income eligible households). This may not be additive—that is, many of the households and homebuilders that take the tax credits or rebates may have been purchasing heat pumps anyway. Regardless of motivation, the IRA incentives certainly help secure those purchases and produce the same decarbonization results.

A second concern is ensuring that the lower-income households that are eligible for larger rebates can actually access the incentives, which require them to own property and assume they have sufficient information and resources to initiate a retrofit (or that they are renters in multifamily buildings with willing property owners). Households that are WAP-eligible but have not benefited from the program—as well as the larger pool of low–moderate income households that are not WAP-eligible but eligible for the IRA rebates—will require extensive outreach and engagement programs for which most current programs have not had the capacity to experiment successfully (Cluett et al. 2016). Harnessing the marketing potential of product vendors, retailers, and service providers such as building contractors and remodelers will be critical for the IRA's building decarbonization ambitions among this population.

Consequently, the third and most critical concern about achieving the IIJA's and the IRA's projected outcomes rests in their implementation across various building sector programs. Fundamental challenges in relation to the kinds of information that property owners and occupants receive regarding the various incentives, the potential confusion between incentive benefit duplication, and their eligibility will need to be established via federal program rules (particularly the Departments of the Treasury and Energy) as well as state rebate programs and their expected plans. The burdens associated with proving eligibility and compliance could inhibit take-up as well. The availability of local technical assistance to help households make complicated financial decisions and the gaps in assistance providers' capacity for both WAP agencies and state and local government energy offices could undermine implementation over the next decade.

> Finding 7-6: Recent national laws such as the IIJA and IRA may not have sufficiently streamlined implementation plans and resulting execution to meet those targets. Capacity gaps at all levels of government and service providers may further complicate these policies' achievements—and will define perceptions of further public appropriation and programming for efforts after the next decade and through the 2050 goals.

Subnational Policy Changes

A significant reason for the increasing take-up of heat pumps prior to the IRA has been the combined mandates and incentives established by state, tribal, and local governments. Most of these laws and programs have focused on the reduction or near-elimination of carbon-based fuels in new construction—that is, electrification—though energy efficiency improvements by regulation as largely represented by state adoptions of model energy codes for new buildings have been the much longer policy vehicle (Berg 2022; Berg et al. 2020). A few states have also incented or mandated distributed renewable energy production.[19] For example, California's goal of achieving zero net energy for new residential buildings by 2020 and by 2030 for commercial buildings has resulted in codes with prescriptive requirements for heat pumps (CPUC n.d.). Other states, including Massachusetts and New York, have also adopted multiple intervention points to meet net-zero building energy legislation, including requiring all-electric new home construction.[20]

The focus on new construction provides the avoidance of inefficient energy consumption; increased energy codes have resulted in an estimated 45 percent reduction in new home energy consumption and almost 55 percent reduction in new commercial buildings since the first codes were introduced a half-century ago (PNNL and DOE 2022). A mandated gradual increase in the number of new homes with heat pumps during the IRA timeframe leading to a complete requirement by 2033 would yield an additional 180 tBTUs in energy reductions (more than 10 percent of total residential consumption when combined with the IRA incentive reductions).

Several states are also leading the "net-zero" charge. For example, Massachusetts established a Stretch Building Energy Code (i.e., beyond the national model energy code) focused on carbon performance rather than prescriptive construction specifications (State of Massachusetts 2022). The challenge with focusing exclusively on new construction is the extensive increase in construction costs, which, in turn, yield unaffordable housing overall and the transfer of lower-income households to the more inefficient existing stock.

Furthermore, the overwhelming share of building sector energy use comes from existing stock, not new buildings. Yet only a handful of subnational policy

---

[19] For example, in 2019 and 2022, respectively, California passed new ordinances requiring all new homes to have a solar photovoltaic system and all new commercial buildings to have a solar photovoltaic array and an energy storage system.

[20] For examples of mandates, see New York State Climate Action Council Scoping Plan (New York State Climate Action Council 2022). Incentive programs include California's Building Initiative for Low-Emissions Development Program (California Energy Commission 2023).

interventions have focused on the existing building stock, and they have typically focused on larger commercial and multifamily buildings (BDC 2023; Sobin 2021). For example, building "benchmarking," or measurement and public reporting of large buildings' energy use, has become the norm in most larger cities over the past decade (IMT 2022).[21] The specific requirements of these laws vary, but typically involve annual reporting of energy consumption and GHG emissions, energy audits, and retrofits. Several of these cities have since used benchmarking of larger properties to set building performance standards to match the cities' long-term GHG reduction goals or, more prescriptively, to conduct mandatory audits with retrofit or retro-commissioning reporting as well (Hart et al. 2022). New York City's Local Law 97 is among the most aggressive of these mandates (NYC 2022). However, small cities such as Ithaca, New York, and Menlo Park, California, have committed to the carbon neutrality of their entire building stocks by 2030 (Harding 2021; Woody 2022). While some stakeholders have raised concerns about the costs and burdens of monitoring and evaluation, building data disclosure laws are increasingly recognized as a useful tool for driving energy efficiency and reducing emissions in the built environment (ACEEE 2014; C40 Cities Climate Leadership Group and C40 Knowledge Hub 2019; Palmer and Walls 2015, 2017; Shang et al. 2019).

For existing buildings, state mandates can focus on replacement of energy-consuming equipment (Berg et al. 2022). Although federal law typically leads equipment standards, several states have moved forward with their own increased efficiency requirements and, in some cases, electrification requirements. For example, California recently prohibited natural gas heating equipment beginning in 2030. California has also set goals for the installation of heat pumps and set aside funds for this effort. Other jurisdictions are following suit, although often relying on direct funding or financial incentives to encourage property owners to convert in addition to strict mandates for the equipment (Cohn and Esram 2022). More evidence is needed to determine the effect of these subnational interventions on the building energy consumption within these jurisdictions and to the overall reductions for the nation.

> Finding 7-7: State, tribal, and local governments play a major role in decarbonizing the nation's building stock, particularly for new construction through mandated codes. However, returns on these policies in overall energy reductions and the removal of direct carbon-based fuel are diminishing. There are signs that

---

[21] Single-family housing, however, does employ common energy labels, although these are typically used only for new construction mandates (NASEO 2022).

governments are turning their attention to retrofits for electrification and effi-ciency in their existing buildings, although there are only a few cases to date. Most of these, further, are awaiting enaction, so there are few clear implementation plans from which to learn and report upon.

In short, recent adoption of local, state, and national policies are projected to move the nation slowly toward decarbonization goals for 2050. These recently enacted policies are a work in progress and, in cases like the IRA, require signifi-cant implementation design and action—not just for the public-sector officials charged with designing and launching programs but also for the building indus-try and owners (including individual households in the latter) to act, particularly when the policy supports are voluntary. Implementation will involve private industry actors and individual building owners and households—the same stake-holder groups that built and maintain the current built environment. A deep understanding of these stakeholders and their role in manifesting the built envi-ronment is needed.

Consequently, after considering all the findings presented in this chapter, the committee is most focused on the implementation challenges of this first decade after IIJA and IRA passage into law and during which the more aggressive state and local policies will have taken shape. A particular concern is the integration of equity and burden during implementation.

Recommendation 7-1: *Ensure Clarity and Consistency for the Implementa-tion of Building Decarbonization Policies.* **The Department of Energy should develop rules for the implementation of energy programs (particularly those funded through the Inflation Reduction Act [IRA]) that increase access for households and other property owners of various incomes while decreasing their burden for participation. Such rules include simplifying the definitions of points-of-sale and income verification protocols for rebates, which might benefit from universal screening for all federal assistance programs. These terms should align with other federal agencies' definitions for eligibility of services and should also serve to set up an evaluation of building program outcomes (such as across state IRA plans and outputs).**

Recommendation 7-2: *Promote an Equitable Focus Across Building Decarbonization Policies.* **The Department of Energy (DOE) should establish requirements for states to market and provide appropriate levels of service to the owners and occupants of rental, multifamily, and low-to-moderate-income occupied buildings as a prerequisite for approval for state energy rebate plans. These plans should include educating households and**

commercial building occupants about all the incentives and resources available to them, as the rebate plans will be a gateway for additional education needed to support further decarbonization after 2033. Ensuring that all states' processes for creating awareness of contemporary financial incentives are consistent, fair, and include authentic local community engagement will ensure that a wider population can access them. DOE's frequent and comprehensive monitoring of the execution of these processes after approvals of state plans will also allow department staff to assign additional technical assistance and related guidance to states to revise and improve their implementation.

Recommendation 7-3: *Expand and Evaluate the Weatherization Assistance Program (WAP).* As the predominant national vehicle for energy improvements among low-income households, WAP should be enhanced in its purpose, scope, and activity in addition to the appropriations increases the committee supported in its first report. Given questions about the program's capacity, the Department of Energy should fund an independent set of evaluations of WAP with an eye toward reform that provides a range of decarbonization interventions discussed in this report that are appropriate to each housing unit. An eligibility analysis should derive alternative population targets that ensure that WAP covers any household not covered by other programs such as rebates and tax credits. An engineering study should determine an appropriate increase in the assistance per household. An implementation study and monitoring program should ensure that decarbonization targets are met, and an outcome study should assess strategies that ensure positive household financial and health outcomes.

Recommendation 7-4: *Coordinate Subnational Government Agencies to Align Decarbonization Policies and Implementation.* States should focus on utilizing internal governmental expertise to de-silo the knowledge and training required to expand all building-level energy interventions' implementation within their state government offices and, in turn, county, tribal, municipal, and other subnational government offices. For example, state energy offices should coordinate with community services, housing, and community development agencies to better target weatherization assistance funds as well as state rebate programs. Furthermore, local building inspection and housing departments should be utilized to better support building improvements in new and existing residential and commercial buildings.

Recommendation 7-5: *Build Capacity for States and Municipalities to Adopt and Enforce Increased Regulatory Rigor for Buildings and Equipment.* **Congress should require and provide resources for the Department of Energy's continued upgrading of building appliance standards (including phasing in electrification of appliances, as applicable). Congress should also require the adoption of increasingly rigorous energy building codes by state and local governments in conjunction with state and municipal block grant programs (also including mandatory electrification) over the next decade. State and local governments should also require that (a) local governments resource, staff, and enhance building departments' capacity to enforce codes, as well as assist in state and local building benchmarking and audit services for existing buildings; and (b) subnational governments consider adopting mandatory appliance and equipment electrification in new buildings by 2035, after which national standards could increasingly require electric-only alternatives. Relevant regulatory initiatives such as the Infrastructure Investment and Jobs Act's Cost-Effective Codes Implementation for Efficiency and Resilience Program ($225 million, §40511) and the Inflation Reduction Act's Technical Assistance for Latest and Zero Building Energy Code Adoption Program ($1 billion, §50131) should dedicate a portion of their available funds to monitoring the effectiveness of this technical assistance as well as tracking its outcomes in energy savings or electrification rates.**

BROADER DECARBONIZATION STRATEGIES FOR THE BUILT ENVIRONMENT

The opportunities for decarbonization in the built environment rest on more than just addressing buildings' energy consumption. Taking into account the life cycle of energy systems, almost twice as much energy is produced *for* buildings than is used *by* them—making the building sector alone one of the biggest end use sectors for carbon-based fuels. Electricity used within buildings account for 74 percent of total retail electricity sales and the emissions related to producing and delivering that power—well beyond the "Scope 1" emissions associated solely with buildings' direct use of energy. Improving the efficiency and managing building energy use through tools like digitalization and demand management, distributed energy generation, and energy storage are as critical to decarbonizing the built environment as are the actions to clean up the electricity grid that supplies buildings with power.

But the GHG implications of the built environment go beyond just the buildings themselves. Where physical infrastructure is located and how it is managed may contribute negatively to local environmental impacts such as air quality and the health and equity outcomes of neighboring communities, and affect such things as land values, congestion, and other community outcomes. Energy infrastructure shapes the built environment just as the built environment affects demands for energy. These patterns are shaped not only by natural phenomena (such as topography and waterways) but also by societal structures, like land use policies and building regulations—the governing frameworks of the built environment. These policies and regulations vary across and within states. Although complicated and challenging to accomplish, changes in such frameworks may provide opportunities for reducing GHG emissions from the overall built environment.

> Finding 7-8: Given the various patterns of development in cities, towns, and rural communities, there is no one size fits all for decarbonization strategies across buildings and communities and regions. Decarbonization of the built environment requires tailored, place-based approaches.

Like increasing building requirements, federal incentives for subnational governments to implement community-level decarbonization strategies should involve appropriate implementation transition periods. Because the process of adopting and implementing community and land use changes in the built environment takes longer than retrofitting of individual buildings, it would be helpful to use modest incentives across a wider range of federal programs for subnational governments (including transportation, water, broadband, housing, and energy block-grant resources) to encourage community decarbonization efforts as early as reasonably possible for outcomes to show up as early as 2033.

Given the range of potential decarbonization interventions in the built environment and the likely diversity of their combinations that might be applied in settings across the nation, issues might arise related to ownership, regulation, and governance structures of different local infrastructure (including such things as electric delivery systems, district heating, microgrids), where current structures stand in the way of efficient and effective development and use of distributed energy resources (DERs). Better coordination among states' energy offices, public utilities commissions, and housing and community-development agencies could support the provision of broader access for all households and communities to such technology and infrastructure options. Further experimentation, technical resources, funding, and financing resources are needed for community distributed renewable projects. Financing alternatives may include revolving loan funds and green banks, to be paid back from production revenues and household energy savings. Financing from federal seed monies from the new IRA Greenhouse Gas Reduction Fund (§60103) could support such outcomes. Civic and public assistance for lower-income communities should also support sustained maintenance and operation.

There are important research questions for the future that relate to approaches affecting the broader built environment, their potential to reduce GHG emissions in the long term, and the matching of alternative decarbonization strategies with different local conditions. These questions go beyond building science, engineering, and construction research—all of which have been underfunded by the national research agencies.[22] Funding for research on community-level decarbonization interventions in the built environment is generally even more limited and less realized than those for individual buildings. DOE, the National Science Foundation, the Department of Housing and Urban Development (HUD), and the General Services Administration (GSA) all have a role to play in identifying important social science, engineering, economic, and legal research and policy analysis questions.

Building Policies and Actions for Rapid Decarbonization in the United States

Policies, incentives, and investments in new and existing buildings and community systems can improve individuals' quality of life while offering the deep carbon savings needed to reach carbon neutral targets. Technology already exists to pursue greater emissions reductions than current targets specify (DOS and EOP 2021), which would take pressure off other sectors. For example, increased effort in the building sector would reduce the need for the risky reliance in the administration's plan on large deployments of technologies that are not yet ready, such as direct air capture (see Figure 1-3). A more aggressive target for increased energy efficiency in buildings and the built environment would reduce demand for heating, cooling, and transport electricity. This would reduce perhaps the greatest execution risk the nation faces in decarbonization, by allowing a less daunting pace of deployment for new transmission and renewables infrastructure (see Chapter 6). A number of published emissions scenarios include much more aggressive decreases in emissions from buildings and the built environment (IPCC 2023). The Biden administration explains the targets for the built environment in their Long-Term Strategy by pointing to the longevity of existing buildings and the high average cost and disruption of retrofits (DOS and EOP 2021). This is one sector in which the policy instruments available at subnational scales may provide stronger targets and incentives than those available to the federal government. In large part, this is because building codes and standards in the United States are set at state and local scales.

The following sections take a fresh look at what design, engineering, and innovation can do to reduce emissions from the building environment and conclude that the

---

[22] Housing-related technological research was estimated at 0.6 percent of all federal non-defense RD&D (Hassel et al. 2001).

federal administration's goals could be significantly tightened through a variety of national and subnational actions. It is in the nation's interest to significantly accelerate reductions in energy demand and associated GHG emissions from the built environment by implementing these actions wherever practical and politically possible to do so. In addition, subnational policies in this sector offer an unusually large opportunity to reduce decarbonization risks by broadening the nation's climate and energy policy portfolio, by adding standards to a portfolio dominated by tax incentives.

The following prioritized actions build on the strategies laid out in NASEM (2021b) and outline what could be achieved through a more aggressive set of policies emphasizing increased efficiency and reduced demand for energy in the built environment, alongside decarbonization opportunities that emerge when building technologies are more tightly coupled to the energy system and the grid. The built environment has the technical potential to reduce emissions by 900 MMtCO$_2$/year by 2050 (Ungar and Nadel 2019), supporting a net-zero future in the built environment with the expansion of renewables, while improving quality of life. In addition to strategically implementing the IIJA and IRA, and extending commitments beyond the next 10 years, the United States needs to increase its portfolio of decarbonization innovations to fully engage the built environment, in keeping with the leading nations around the world.

1. Accelerate Appliance and Equipment Efficiency. 5.6 quads of reduced annual energy demand, 210 MMtCO$_2$ of reduced annual emissions.

In 2019, the American Council for an Energy-Efficient Economy (ACEEE) identified 5.6 quads per year of energy efficiency that could be achieved through specific appliance and equipment upgrades, leading to 210 MMtCO$_2$/y of reductions or an equivalent of 1.64 trillion kWh (Ungar and Nadel 2019, p. 54). Of these savings, 70 percent come from a dozen products that could be accelerated to achieve Energy Star performance in the top 25 percent of their market, including residential water heaters, heat pumps/central air conditioners, boilers and furnaces, refrigerators, as well as commercial/industrial fans, electric motors, transformers, air compressors, and packaged unitary air conditioners and heat pumps. An NREL study found that efficiency improvements in the range of 0.5 to 2 percent per year for electric building technologies could completely offset the electricity load growth associated with building electrification for decarbonization (Steinberg et al. 2017). The Biden administration estimates that increased stringency of appliance and equipment standards have the potential to reduce emissions by 2.4 billion metric tons by 2050

(White House 2022b). With the expansion of the recommendations from the committee's first report, the further acceleration of appliance efficiency goals for the following appliances will be critical for decarbonization:

1. Heating (efficiency with electrification)
2. Domestic hot water (DHW) (efficiency with electrification)
3. Lighting (even more efficiency possible)
4. Refrigerator-freezers (more efficiency possible)
5. Air conditioning (with more innovation and decarbonized refrigerants)
6. Cooking and clothes drying (efficiency with electrification)
7. Miscellaneous electric loads now 30 percent of building electric demand

The committee's first report and Recommendations 7-5 and 10-2 of this report include energy efficiency and emissions manufacturing standards for appliances as a backstop to the incentives offered in the IIJA and IRA. These standards would include rigorous support for Energy Star certifications and national (or, as historically, state by state) requirements that codify the top 20 percent of performers (receiving Energy Star designation; ASAP 2023) to become the mandatory minimum within 5–10 years, with continuously updated minimums to reflect advances in appliance efficiency.

The most rapid acceleration for appliance and equipment upgrades may occur in response to mandates for building electrification, especially in the installation of heat pumps for heating and hot water. For building electrification to contribute to significant carbon savings, however, renewable electricity sources and demand efficiency measures would need to be fully in place. In addition, heat pump electrification priorities would be needed to ensure carbon benefits and to protect occupants from higher energy bills. Priorities for heat pump installation for these outcomes may necessitate a critical path (Deetjen et al. 2021; DOE 2016; Pantano 2020; Waite and Modi 2020) that includes air conditioner upgrades to heat pumps for heating and cooling; oil heating system upgrades; electric resistance heating system upgrades; gas heating system upgrades in mild climates; and, only then, gas heating system upgrades in cold climates with hybrid fuel capability for extremely cold days—each of these with cooling as needed and potentially integrated hot water. DOE's Residential Cold Climate Heat Pump Challenge is intended for rapidly improving cold climate heat pumps so that hybrid heat may not be needed in a decade, and reducing the installed cost of geothermal heat pumps may be even more impactful for reducing the large carbon footprint of heating and cooling in the United States (DOE n.d.).

2. Mandate and/or Incentivize Zero-Energy New Homes and Commercial Buildings. 5.7 quads of reduced annual energy demand, 265 MMtCO$_2$ of reduced annual emissions.

The design, engineering, and construction professions have been delivering net-zero new buildings wherever clients or codes mandate, by combining very low energy use per square foot requirements with on-site or purchased renewable energy sources. However, the vast majority of new buildings today are not net zero. The Architecture 2030 Challenge calls for all new construction to be net zero by 2030 and for all major retrofits to achieve a 50 percent emissions reduction by 2030 (Architecture 2030 2023). The 2030 Challenge goals have been adopted by 1,200 architecture firms, 15 cities, and the U.S. Conference of Mayors. If zero-energy goals for new homes and commercial buildings were to be achieved nationally, then the United States would save more than 5.7 quads of energy and 265 MMtCO$_2$/y by 2050 (Ungar and Nadel 2019).

To achieve net-zero energy use intensity in all new residential and commercial buildings by 2030, a full suite of subnational actions would need to be advanced state by state: (1) ASHRAE, Zero Code, and IECC2021 code adoption; (2) tax credits; (3) qualified allocations for passive house construction (especially for low-income housing); and (4) distributed renewable targets and incentives, and stretch code mandates that reduce site Energy Use Intensities (EUIs) below 30 kBtu/ft^2/year (e.g., State of Massachusetts 2022). Together, these actions would reduce energy use on site by 70 percent in new residential and commercial buildings by 2030, with the remaining 30 percent met by on-site or purchased renewable sources to achieve net zero (NBI 2019; USGBC MA 2019). Figure 7-2 displays one modeling exercise showing the impact these actions could have on the energy use intensity of new construction compared to the current commercial building stock.

The design expertise, technologies, and standards for net-zero new construction have been demonstrated but not enacted beyond a few leading states (e.g., California, New York, Massachusetts, and Colorado). These states have demonstrated that the incremental financial cost of purchasing and operating a new net-zero building offers some of the most cost competitive carbon savings, with 0–10 percent increased first costs even for low-income projects, offset by operational cost savings (Leach et al. 2014; NBI 2012). The Zero Energy Buildings targets for low-income housing in New York City and Massachusetts address inequities in the built environment, including energy security and resiliency for all citizens during power outages (Cleveland et al. 2019; NASEM 2021b; NBI 2019). The IIJA includes modest related funding for housing, including $225 million for DOE's Building Technologies Office to offer state grants for advancing sustained, cost-effective implementation of updated building energy codes (fiscal years 2022–2026) (§40522).

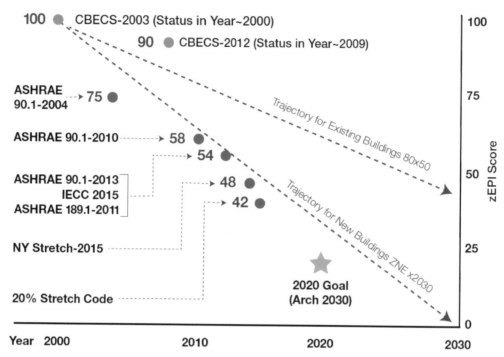

zEPI: The Path to Zero Energy in Building Energy Codes

FIGURE 7-2 Model stretch code provisions for additional performance improvements in new commercial construction. NOTES: The Zero Energy Performance Index (zEPI) is a relative scale that allows various levels of building energy performance to be compared against each other. zEPI sets an energy use intensity (EUI) target for building type and is adjusted for climate. This graph charts zEPI scores for the current national model energy codes and standards. SOURCE: Reproduced with permission of NBI (2017).

3. Incentivize Retrofits for Existing Homes and Commercial Buildings. 3.8 quads of reduced annual energy demand, 148 MMtCO$_2$ of reduced annual emissions (not including the reductions in plug loads).

Even by 2050, existing buildings will still dominate the residential and commercial portfolio. The largest energy demands in residential and commercial buildings are for heating, cooling, ventilation, and lighting—each of which can be measurably reduced through thermal and air tightness improvements in building roofs, walls, windows, and foundations, which directly impact the sizing and performance of mechanical and electrical equipment. Retrofitting the nation's existing buildings to improve their energy efficiency would require significant expansion of relevant manufacturing and training, and would need to address the barriers and disruptions that accompany most retrofits, as well as the up-front capital cost, even when reduced operating costs would more than compensate over time. Many households lack access to financing or even ownership, especially in underserved communities (Kirk 2021; see Chapter 11). On the other hand, the technologies are proven, unlike the atmospheric carbon removal technologies that the nation will need if it cannot do more to reduce emissions from buildings and industry (Chapter 10). To advance a national plan for retrofitting the nation's current portfolio of buildings, the first step would be the energy use benchmarking, followed by annual reporting of progress. The Environmental Protection Agency's (EPA's) Portfolio Manager is the nation's repository for energy benchmarking and needs to be strengthened to ensure that the data are robust, transparent, and analyzed with the most advanced expertise. Benchmarking would be followed with aggressive goals and funding for prioritized investments. In 2022, President Biden launched the National Building Performance Standards Coalition, a nationwide group of approximately 40 state and local governments that have committed to inclusively design and implement existing building performance policies and programs in their jurisdictions, with shared goals and solutions (National BPS Coalition n.d.). Leaders including New York City, Boston, Denver, and others have matched benchmarking and aggressive goals with prioritized investments in existing buildings based on the detailed understanding of where the energy is lost through end use load breakdowns (Figure 7-3; USGBC MA 2019).

A national program to retrofit existing buildings would perhaps do more to advance environmental justice during the energy transition than any other action, with the possible exception of reduced air pollution exposure from conventional fossil pollutants (Chapter 2). The IIJA and IRA provide a significant start by a commitment to retrofitting 1.3–2.5 million buildings inhabited by low-income households with heat pumps. These residential expenditures will address only a portion of the urgent need for investments in the energy efficiency of both the residential and commercial buildings that house, educate, employ, and service the poorest U.S. residents.

Not only have disadvantaged communities been locked out of home ownership, they are also often fated to live, study, and work in substandard buildings with the highest energy costs, relying on subsidy or sacrifice to pay the bills. Low-income households bear an energy burden that is 3 times greater than that of non-low-income households, with the national average standing at 8.6 percent (DOE 2020a), and even higher for households in the lowest decile of income. While these costs may be classified as an energy burden (unaffordable), they often are accompanied by energy insecurity (reliability and outage risk) as well as full energy poverty (no electricity or gas) with serious risk to health and life (Biswas et al. 2022; see Chapter 2). The federal program known as LIHEAP—the Low Income Home Energy Assistance Program—has been the primary federal instrument aimed at reducing energy burdens and energy

Energy Consumption—Existing Office

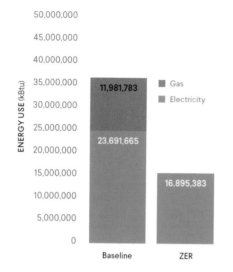

EUI Breakdown and PV—Existing Office

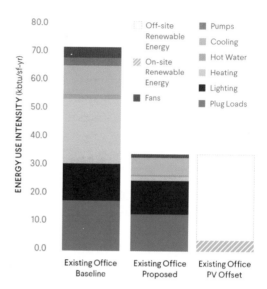

FIGURE 7-3 Zero Energy Buildings in MA: Saving Money from the Start | 2019 Report.
SOURCE: Courtesy of BE+, USGBC MA (2019).

poverty (HHS 2023). These annual subsidies keep the lights and heat on for many low-income households, but do not address the underlying challenge of excessive energy demands of inadequately built and maintained buildings. Thirty percent of U.S. households (38.6 million homes) meet the income requirement for WAP assistance (Drehobl 2020). A national effort to retrofit all existing housing stock would necessarily make the greatest difference in the homes that need upgrades the most and would thus provide the largest benefits to disadvantaged communities. Also, a comprehensive retrofitting effort would cover all eligible households, not just owner-occupied homes (Kirk 2021). Earlier studies estimated the impact could be at least 35 percent residential energy savings and 20 new jobs per million invested (ACEEE 2011; DOE 2011).

The potential of existing building retrofits, both residential and commercial, across all socioeconomic spectrums, can be more than 50 percent savings of the present energy demand in buildings (RMI n.d.). The IRA's $4.3 billion for grants from DOE to state energy offices to develop and implement a whole-house energy savings rebate program (§50121) and $0.2 billion for states to establish training and education programs for contractors who install home energy efficiency and electrification improvements (§50123) is a critical start and could be expanded to address existing commercial buildings as well.

4. Accelerate the RD&D for Smart Technologies and Systems for Homes and Commercial Buildings. 3.2 quads of reduced annual energy demand, 125 MMtCO$_2$ of reduced annual emissions.

In addition to increased use of proven technologies, an effort to significantly tighten and accelerate national targets for building decarbonization (i.e., Figure 1-3) would benefit from additional investments in RD&D of innovations in the built environment. Four major areas of innovations for buildings are described here to illustrate critical needs: engaging the Internet of Things (IoT) for building energy use controls; developing hydrofluorocarbon (HFC) alternatives; innovating to reduce embodied carbon; and advancing GEBs.

The IoT has created the potential for a transformational change in society but has experienced a surprisingly slow pace of integration into buildings. With each piece of mechanical equipment and every appliance, light fixture, and window control having an IP address, the ability to monitor and control energy consumption while ensuring a high level of occupant service provides unprecedented opportunities for increasing the energy efficiency of buildings. At the residential building level, smart thermostats had been installed in more than 19 million homes at the end of 2021, reducing national energy consumption for space heating and cooling by 1.4 percent and saving residents an average of 8 percent on their heating and cooling bills (Barbour 2021; EnergyStar n.d.; Walton 2022). At the commercial building level, BAS can monitor and control heating, ventilation, cooling, lighting, elevators, and multiple energy-intensive devices, reducing commercial building energy consumption by 10 to 30 percent (EIA 2022a; Fernandez et al. 2017). Sixty percent of large commercial buildings (>50,000 square feet) in the United States have a BAS to control heating, ventilating, and air conditioning; lighting; and more. But only 13 percent of small- to medium-size (<50,000 square feet) buildings have adopted the technology, leaving more than 75 percent of all commercial buildings in the United States primed for opportunity (Trenbath et al. 2022).

The challenge is learning from the growing repository of monitored sensor and controller settings to provide a high level of occupant service with low energy and carbon demand. Given the volume of data and the complexity of optimal control for the significant variations in building types and climates, the addition of an IoT with smart controls for carbon savings in buildings is an ideal challenge for new private and public investment in artificial intelligence and machine learning to be rapidly developed with the building sector.

HFC Alternatives for All Heat Pumps, Refrigerators, and Air Conditioners

Air conditioning is a triple threat to climate—from its operational energy to its impact on peak demand, to its use of refrigerants. The HFC refrigerants in heat pumps, refrigerators, and air conditioners are major sources of fugitive HFC emissions. HFCs have a 100-year global warming potential 3790 times larger than CO_2, and the use of air conditioning is growing at a rate of 10–15 percent per year worldwide. HFCs also persist in the atmosphere for an average of 29 years (Climate and Clean Air Coalition 2020). With the Kigali Amendment to the Montreal Protocol, nearly 200 countries have committed to reduce the production and consumption of HFCs by more than 80 percent over the next 30 years to avoid more than 70 billion metric tons of CO_2 equivalent emissions by 2050—and to reduce up to 0.5°C of warming by the end of the century (UN 2023). The Senate ratified and President Biden signed this amendment in 2022 (Department of State 2022). The IRA includes a commitment to the American Innovation and Manufacturing Act of 2020 that requires EPA to implement an HFC phasedown plan to reduce 2011–2013 levels 85 percent by 2036—meeting the goals of this agreement. Local and state codes will need to be modified to support HFC-free equipment. Increased federal funding to develop and deploy the next generation of refrigerants would help the United States to remain a leader in the manufacturing of heat pumps, refrigerators, and air conditioners, while helping the United States meet or exceed its target for reducing refrigerant emissions.

Embodied Carbon

As building energy demands continue to drop through significant design and engineering improvements, the energy and carbon costs of building material extraction, production, transportation, installation, and end-of-life disposal become more significant. In the most energy efficient buildings, the embodied carbon in building material selection can be equivalent to 30 years of operational energy (see Figure 7-4; Carbon Leadership Forum 2020). The largest contributors to embodied carbon in buildings are the extensive use of concrete, steel, aluminum, petroleum-based insulation, plastics, and disposable technologies. Consequently, the most significant strategies for reducing embodied carbon in buildings include (1) minimizing concrete and ensuring CarbonStar certification for what is used (CSA Group 2021); (2) 100 percent recycled content in steel and aluminum; (3) shifting to wood construction; (4) design for disassembly so that all steel, aluminum, glass and other materials can be reused without down-cycling; and (5) reusing existing buildings rather than new construction. Today, operational carbon and embodied carbon

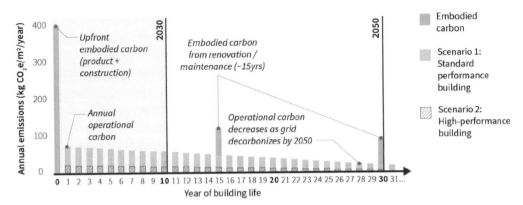

FIGURE 7-4 With the most energy efficient buildings, the embodied carbon in building material selection can be equivalent to 30 years of operational energy. SOURCE: © Copyright 2020, Carbon Leadership Forum.

need to be on the same balance sheet, with cutting-edge tools emerging for comprehensive carbon accounting. New research and development are needed to advance design, engineering, and manufacturing of building assemblies and systems that can be produced with reduced GHG emissions (or even negative emissions in the case of some alternatives to concrete and steel).

Innovation for Building Electrification with Grid Integration

The electrification of existing heating, hot water, and gas appliances offers only modest reductions in energy demand, estimated at 0.9 quads in savings by 2050 (76 MMtCO$_2$ reductions) (Ungar and Nadel 2019). However, if electrification in buildings is combined with high energy efficiency, on-site energy storage, and smart technologies with grid integration, the energy and GHG emissions reductions can be substantial. In new residential and commercial buildings, electrification often reduces the first cost of construction, with records of 27 percent lower upfront costs for an all-electric single-family new home (Group14 Engineering 2020). With the addition of smart technologies and on-site energy storage, peak loads are also reduced, supporting greater grid stability. The addition of time-of-day pricing and equipment controls, as well as batteries, can both reduce and align the "camel curves" of conventional electricity demand (high late afternoon demand) and the "duck curves" of renewable energy sources (photovoltaic [PV] displaced late afternoon demand). Increased federal and industry investments in RD&D in building-level thermal and electric energy storage, utility integration of distributed PV and energy storage, and integrated heating/DHW/cooling/ventilation technologies can accelerate the relatively weak federal target for decarbonizing the built environment (see Figure 1-3). Success may also depend on significant input from the social sciences, beyond economics and behavioral analysis, to fully understand how to improve technology use and impact (Dietz et al. 2013; Gromet et al. 2013; Shove 2021; Stern et al.1986; Sussman and Chikumbo 2016; and see Chapter 5).

Community Policies and Actions for Rapid Decarbonization in the United States

In addition to modifications to buildings for decarbonization, federal, state, and local policies could promote land use policies that substantially increase the energy efficiency of buildings and infrastructure as described in the next four sections on mixed-use, transit-oriented development; community renewables with micro-grids; smart surfaces; and innovation, research, and rapid development of district energy through GEBs with thermal energy distribution and storage systems.

5. Incentivize Mixed-Use Walkable Infill Instead of Sprawl Communities.

Sprawl is accompanied by infrastructure growth with both a significant carbon footprint and a long-term maintenance cost. A 2020 *Transportation for America* report identified that the nation's largest 100 urban areas added 30,511 new freeway lane-miles of roads between 1993 and 2017, a 42 percent increase, while population only increased 32 percent (Bellis 2020). Each lane-mile of road costs between $4.2 million and $15.4 million to build and approximately $24,000 per year to maintain. States alone spent $500 billion annually to expand roads between 1993 and 2017, and the *public* road infrastructure grew by almost 224,000 miles between 2009 and 2017. The expanding road infrastructure was accompanied by a 50 percent increase in vehicle miles traveled over 25 years (AFDC 2022), with compounded impacts for air quality, health, equity, and family life. Moreover, with every mile of new roadway there are equivalent miles of electric, gas, water, and cable infrastructures, which have embodied carbon and long-term maintenance costs.

The CoolClimate Network has captured California's household Carbon Footprints in 2010 and 2020 with a vision for 2030 and 2050 that relies heavily on rethinking land use (Figure 7-5), to include low-energy new housing, shifts in household diets, and rethinking transportation with walkability and transit (CoolClimate

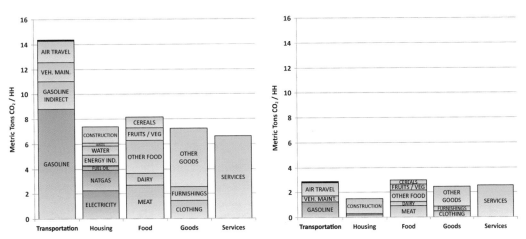

FIGURE 7-5 Carbon footprint of average California household (HH) in 2010 and 2050 under deep GHG abatement. NOTE: Green-colored bars are indirect emissions from the life cycle of products and services that are not typically covered in production-based inventories (unless produced locally).
SOURCE: Adapted from Jones et al. (2018), https://doi.org/10.17645/up.v3i2.1218. CC BY 4.0.

Network n.d.). These actions could reduce California's household total carbon footprint 11–45 tCO$_2$e/y per household through a full menu of actions that significantly feature urban infill in transit-centric low-carbon zones (Figure 7-5; Jones et al. 2018).

A Review of Physical Planning and Transportation Impacts

Feng and Gauthier (2021) identified the range of environmental consequences of today's subsidized sprawl beyond the growing demand for fossil or electric energy sources—including atmospheric pollution; hydrographic system alteration, increased impervious surfaces and flooding; and loss of biodiversity, forests, and agriculture. Additional social costs include increased spatial segregation, commuting time, and demands for automobile ownership, as well as diminished access to jobs and amenities.

The IRA and IIJA contain a number of programs that would modify urban and suburban land use policies, including $1.893 billion in a Neighborhood Access and Equity Grants Program (IRA §60501) to reconnect communities separated by highways and other infrastructure, and $1.5 billion for grants under the Urban and Community Forestry Assistance Program (IRA §23002) to promote tree planting in communities, both with priority given to underserved populations. However, these programs aim to fill particular needs, and would not be sufficient if the nation were to decide to transform the built environment to reduce emissions and achieve objectives related to environmental justice (see Chapter 2), health (see Chapter 3), and quality of life (see Chapter 3). A shift in federal and state subsidies that presently incentivize sprawl alongside changes in zoning laws to support mixed-use infill in transit-oriented developments could transform our cities and towns into walkable, bikeable, transit-serviced, low-energy, landscape-rich communities with well-designed and maintainable public infrastructures.

6. Bundle Retrofits for Improved Energy Efficiency, Electrification, and On-Site Power Generation and Storage, at Both Building and Whole Community Scales.

The shift to all-electric buildings and transportation will lead to significant new demands on the electric system (Chapter 6), especially if we continue to rely on personal vehicle–centric mobility and inefficient buildings. Without more intentional policies, the new demands for electricity will increase the needed size of the electric system, including the power generation and transmission infrastructures that will impact rural and urban neighborhoods. Chapters 5 and 6 identify difficulties in siting electricity infrastructure as the single most likely point of failure for the climate and energy provisions in the IRA and IIJA. Delays in siting new transmission lines to support the increased demand for electricity from the shift to electric transport and heat pumps could result in fossil emissions that increase through the 2020s, making decarbonization look like a complete failure (Chapter 6). Thus, accelerating and tightening targets for increased building energy efficiency and energy load management, along with changes in the built environment that will reduce demand for automotive travel, would pay significant dividends beyond their direct effects on emissions (and overall costs of a decarbonized economy). By reducing the amount of new electricity infrastructure needed, these building energy efficiency and load management measures would reduce the number of new electricity infrastructure projects and increase the number of trained people available to facilitate each siting (Chapter 5). Combining electrification with 50 percent reduction in energy demand in both buildings and transportation would significantly improve health and quality of life (Chapter 3).

Moreover, bundling electrification with energy efficiency retrofits reduces the needed size and operational energy demands of the new electrical equipment such as heat pumps. This provides both direct and synergistic reductions in demand for new electricity. Adding rooftop solar and battery storage to the mix would provide additional synergies because the efficiency upgrades and reduced electrical demand reduces the size and expense of the rooftop system and batteries. All three would interact synergistically to decrease the amount of new generating capacity and distribution and transmission investments needed locally and regionally.

While the IRA, IIJA, and Justice40 Initiative provide significant resources to address community infrastructure improvements, they do not explicitly bundle these improvements into integrated community action plans. Bundling would provide additional synergies if deployed in community-scale retrofit packages—which could combine new community solar power projects with rooftop solar installations; heat pump DHW heaters with water saving fixtures; and heat pump heating and cooling with insulation, reglazing, and air tightness—in municipal as well as privately owned buildings. The community could also decrease automotive travel by offering tax and other incentives for employers and people to locate close to the municipal center, and by improving electric mass transit and corridors for walking and biking. These upgrades could reduce both directly and synergistically the new transmission needed by the municipality, provide direct revenue from the solar power generation, and provide a new source of employment. The synergies created by bundled community-scale retrofitting would provide significant benefits, not evident in separate analyses of each option alone.

The Lawrence Berkeley National Laboratory (LBNL) has analyzed how trade-offs and synergies among building efficiency, electrification, and the various options for electric power generation affect energy demand and greenhouse emissions. LBNL concluded that the most aggressive strategies would bundle building efficiency, electrification, and grid decarbonization to offer approximately 90 percent reductions in carbon emissions from the built environment by 2050 (Figure 7-6; Langevin et al. 2022).

A final advantage of bundling distributed photovoltaics, battery storage, and increased efficiency is that the bundled system could provide sufficient on-site electricity during short-duration power outages, which are increasingly frequent (Bowen et al. 2019; EIA 2021, 2023b). Concerns about reliability have contributed to the explosive increase in fossil fuel–powered standby generators (low-, medium-, and high-power gensets) with health and carbon consequences (CARB n.d.). DERs including rooftop and community solar and microgrids can make buildings more self-reliant and resilient to disruptions of service, although DERs also require significant coordination between customers and grid operators (NASEM 2018, 2021b, 2023).

7. Ensure Smart Surfaces for Carbon Reduction and Equity.

More than 25 percent of land in U.S. urban areas is impervious today, combining the areas of dark and impervious roofs, parking lots, streets, and sidewalks (Center for Sustainable Systems 2021; Nowak and Greenfield 2018). The most effective way to reduce the intensity of the urban heat island is to decrease the area of dark,

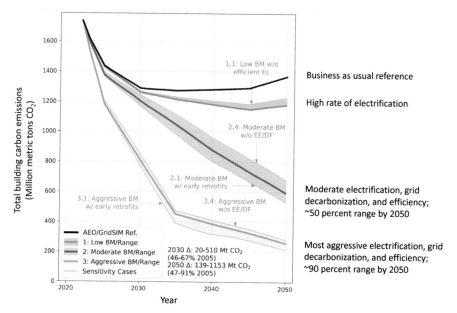

FIGURE 7-6 The most aggressive building efficiency, electrification, and grid decarbonization offer 90 percent reductions in carbon emissions from the built environment. SOURCE: Published in Langevin et al., 2022, "The Role of Buildings in U.S. Energy System Decarbonization by Mid-Century." *One Earth*, Copyright Elsevier (2022).

paved, concrete surfaces, and increase the number of reflective, vegetated, and porous surfaces, which reduce absorption of shortwave solar radiation, improve rainwater management, and support evaporative cooling (Strohbach et al. 2019; Taha 2021). Decreasing the use of concrete also reduces the associated carbon emissions. The integrated deployment of green, porous, and reflective surfaces, as well as trees and solar PV, allows cities to save money, mitigate and adapt to climate change, reduce public health and equity risks, and cut urban heat and flooding. The Smart Surfaces Coalition has quantified the major contributions of smarter surfaces to neighborhood quality of life, reducing urban heat island temperatures over 2°C, reducing flooding devastation, while also reducing or sequestering carbon and generating electricity (SSC 2022; Table 7-7).

Expansion of the electric grid and providing sufficient charging points for vehicles will force substantial changes in the built environment that will be challenging everywhere, but particularly in areas with high population density. The urban scape would be most resilient, walkable, and aesthetic if transmission lines were to move underground (out of the way of storms). The need for millions of new charge points

TABLE 7-7 CO_2e Reduction Potential of Integrated Deployment of Smart Surfaces in Baltimore, Maryland

Intervention	Adoption Scenario	Climate Mitigation Mechanism	CO_2e Reduction (metric tons)	CO_2e Reduction as a Percent of Total City Emissions (from 2017)
Reflective Roofs	80% low-slope	Negative radiative forcing	2,860,000	1.9%
	20% steep-slope	Negative radiative forcing	715,000	0.5%
Reflective Roads	15% of road area	Negative radiative forcing	748,000	0.5%
Reflecting Parking Lots	50% of parking area	Negative radiative forcing	216,000	0.1%
Trees	40% tree coverage (11% increase)	Carbon sequestration and reduced cooling demand from shading	345,000	0.2%
Solar PV	40% low-slope	Electricity generation	8,907,000	6%
	20% steep-slope	Electricity generation	3,492,000	2.3%

NOTE: Integrated deployment of porous and reflective surfaces as well as trees and solar PV would reduce urban heat island temperatures more than 2°C, reduce flooding, and generate electricity to offset as much as 8 percent of Baltimore's CO_2 emissions. SOURCE: Data from Kats et al. (2022), https://smartsurfaces coalition.org/analysis/baltimore-report.

could be coordinated with other upgrades to streets and sidewalks, including the removal or repurposing of natural gas infrastructures; upgrades or repair of data, water, and waste infrastructures; storm water management; greater walkability; added greenspace; and improved electric mass transit.

8. Innovate and Deploy Community and District Thermal Energy Systems Including Combined Heating, Cooling, and Power. 4.0 quads of reduced annual energy demand, 150 MMtCO₂ of reduced annual emissions.

There are abundant alternatives to the use of electricity for space heating, hot water, and even cooling loads that combined are responsible for 37 percent to 46 percent of the total energy consumed in the United States (EIA 2022a). Campus and city district thermal energy systems distribute "waste" energy through shallow, insulated networks of hot water and chilled water, with a wide array of thermal conditioning sources, including

- Waste heat utilization from high cooling demand buildings and data center combined heat and power (CHP) for electric uninterrupted power supply and heating;
- Low-temperature geothermal (with heat pumps) and geo-exchange (closed loop exchange with the earth or bodies of water for conditioning);
- High-efficiency gas and biogas district heating systems;
- High-efficiency heat pump cooling, open and closed loop aquifer cooling;
- Waste to energy (highly managed waste, addressing a parallel national challenge);
- Regional CHP (reducing the 70 percent source to site thermal waste of power generation to 30 percent or less; Litjens et al. 2018); and
- Sewer mining.

District heating and cooling systems that combine ground source heat pumps with PV and building energy storage systems have resulted in major residential carbon savings in Europe and the United Kingdom, under climatic conditions present in many regions of the United States (Litjens et al. 2018). In a detailed study, Andrés et al. (2018) concluded that local, urban, and regional district energy systems have evolved significantly in the past 50 years, and can now supply low-energy thermal conditioning with substantially lower and more effective infrastructure costs. Real-world examples include waste heat recovery from data centers (Brunswick, Germany), sewage water (Nice, France), cooling systems in tertiary buildings (Madrid, Spain), and waste heat recovery underground railway stations (Bucharest, Romania). Despite examples in Minneapolis and other locations, relatively few district and community energy systems have been installed in the United States (DOE 2020b). Björnebo et al. (2018) concluded that small-scale district heating that replaces oil-fired heating in cool and cold climates would provide the largest GHG emissions reductions per dollar in the United States.

The United Nations Environment Programme (UNEP) concluded that modern district energy systems can achieve a 50 percent reduction in building conditioning energy demand through the full range of thermal energy sources, with corresponding reductions in CO_2, SO_x, NO_x, particulates, mercury, and other pollutants from conventional fossil sources of heat and electric power (UNEP 2019). With data gathered from cities around the world, they quantified the CO_2 emissions reductions from a variety of options, including geothermal (with heat pumps) and geo-exchange (no heat pumps), solar thermal, industrial waste heat recovery, data center heat recovery, waste to energy, district boilers, district chillers, combined heat and power and combined heating, cooling (absorption chillers), and power (expanding on Riahi [2015]). District energy systems are predominantly piped infrastructures, and the Building Decarbonization Coalition, in public commentary to the committee,[23] suggests that these new thermal infrastructures can provide a just transition for gas workers, using existing rights-of-way, and reengaging labor with state training.

Grid Actions for Integrating the Built Environment in Rapid Decarbonization

More than 70 percent of the nation's electricity is consumed by or within a building (Figure 7-7a; DOE 2015; EIA 2023a; EPA 2023), and buildings offer the opportunity to play a critical role in supporting grid stability. Buildings create the largest peak demand on the electric grid, challenging both capacity during peak hours and effective renewable energy use during non-peak hours (Figure 7-7b; Hale et al. 2018). Moreover, peak demands that typically emerge in summer heat waves will be outpaced by peak demands in winter cold snaps as heating loads are shifted to electric sources.

In addition to ensuring that new transmission infrastructures improve communities, there are two key opportunities that depend on integration of the grid and the built environment beyond a simple utility service: advancing buildings (and their cars) as batteries and peak load managers for grid resilience; and incorporating distributed renewables and district energy with GEBs into an interconnected grid of grids.

9. Advance Buildings as Batteries and Peak Load Managers.

Buildings can be critical partners in reducing and shifting electricity demand, supporting distributed renewables with load shifting and storage, and extracting low-temperature energy to replace electricity. Because of the size of their peak and total demand for electricity, buildings disproportionately shape the overall needs of the electric system. However, with deep efficiency, the growth of rooftop and community renewables and storage, and replacement of some uses of electricity with low-quality thermal energy, buildings could become critical active partners in managing peak electrical loads, rather than being simply one-way users of the grid. All of the

---

[23] Joanna Partin, Building Decarbonization Coalition, October 21, 2022; https://www.nationalacademies.org/event/10-20-2022/accelerating-decarbonization-in-the-united-states-technology-policy-and-societal-dimensions-perspectives-on-priority-actions-for-the-built-environment-and-building-technologies-rd-d-needs.

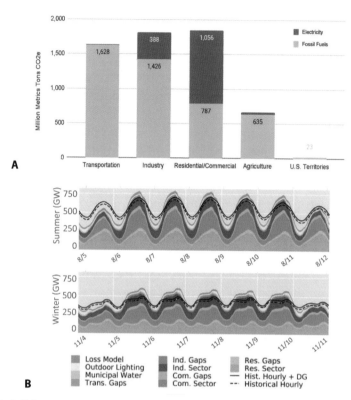

FIGURE 7-7 (a) Buildings are 70 percent of electricity sector GHG and (b) the variable load on the grid. SOURCES: (a) Courtesy of Carnegie Mellon CBPD, based on January 11, 2022, EPA data: https://www.epa.gov/ghgemissions/inventory-us-greenhouse-gas-emissions-and-sinks; (b) Hale et al. (2018), NREL, EERE.

innovations needed for fully engaging buildings in grid management are not yet on the shelf. (See NASEM 2021b.) To realize the potential of buildings as batteries and peak load managers for grid resilience, the United States would need to increase long-term RD&D investments in a number of technologies, including

- Heat pumps for integrated heating, cooling, and hot water (buildings);
- Elimination of "parasitic or vampire" miscellaneous electric loads (buildings);
- Solar DHW and energy storage systems (buildings);
- Grid-connected smart building systems—equipment, appliances, controls (buildings);
- Geothermal heat pumps for heating, cooling, hot water (communities);
- Grid-connected Building Electric and Thermal Energy Storage systems (communities);

- Grid-integrated distributed batteries (i.e., car batteries and building thermal and electric batteries) that can run homes or feed the grid (communities);
- Building and community distributed renewables with deeply efficient buildings; and
- New grid infrastructures that meet multiple performance and decarbonization goals.

10. District, Community, and Building Renewables with GEBs.

The IRA dedicates $65 billion to funds for grid reliability and resiliency actions and supports a Grid Deployment Office for critical minerals and supply chains for clean energy technology; key technologies including carbon capture, hydrogen, direct air capture, and energy efficiency; and energy demonstration projects outlined in the bipartisan Energy Act of 2020. These federal funding opportunities do not include a focused RD&D program to develop grids that fully engage distributed renewables, storage, and GEBs as an integral player in the grid. Such a program would encompass a range of technological advances: V2G (vehicle to grid) bi-directional charging to store up to 80 to 100 kilowatt-hours of electricity, sufficient to run a U.S. home for 2–3 days (Blumsack 2022); ice storage, water storage, and phase change materials that can provide long-term cooling for buildings with off-peak power; smart time-of-day controls for appliances and equipment; and new approaches to reducing AC to DC conversion losses through strategic management or replacement with direct DC from PV. The potential of these technologies would be invaluable to shave and shift peak loads and avoid brownouts and blackouts. A grid that flexibly integrates building and community renewable energy supplies would help the nation meet its renewable energy goals and advance resiliency.

The barriers to rapid advances in **building-scale PV systems** are both regulatory and profit driven. In the United States, soft costs (including installation labor, customer acquisition through sales and marketing, and permitting/inspection/interconnection) are more than 65 percent of the overall costs of rooftop solar panels (EERE 2023). Reducing soft costs and increasing incentives for distributed PV will spur U.S. manufacturing and installation jobs and bring the United States in line with international gains in DERs (Birch 2018).

Community-owned renewables are emerging across the United States, led by special-purpose community entities, the local utility, or a third party (which might be a private developer or nonprofit organization). Minnesota's community solar comprises more than 70 percent of the state's total solar photovoltaic capacity of

1,057 megawatts, followed by 18 percent of New York State's 3,950 megawatts, and 15 percent of Massachusetts's total solar capacity of 2,805 megawatts (Heeter et al. 2021; McCoy 2022). HUD has updated guidance to support community solar on public and assisted multifamily housing (HUD 2023), and multiple states have launched low- and moderate-income solar programs to facilitate community solar by helping to overcome barriers such as lack of access to capital, insufficient tax burden to take advantage of tax credits, the large fraction of renters who are often ineligible for incentives, frustrating interconnection policies, and lack of familiarity with solar products (Mai et al. 2018).

Given their growing popularity, building- and community-scale renewables can be critical partners in the expansion of the electric grid. To date, the United States has been focused on utility-scale renewables with the lowest cost and legislative challenges, rather than those likely to be easiest to site (Chapter 5). With national and utility leadership, Europe and Australia have far more successfully integrated building and community renewables into the grid, as shown for Germany in Figure 7-8.

In the United States, 20 states and the District of Columbia (DC) have laws ensuring that consumers can participate in community solar power generation, 28 states authorize or allow third-party Power Purchase Agreements for solar PV, and 39 states (plus DC, American Samoa, Guam, Puerto Rico, and the U.S. Virgin Islands) have mandatory net metering rules (Heeter et al. 2019, pp. 28 and 30, 2021). Extensions of these programs to all states would facilitate deployment of DERs and keep the United States from falling behind at a time in which the future architecture of the grid is in development. Interconnected building- and community-owned renewable electricity, from wind and photovoltaic installations, can be designed to act as a microgrid to operate both with the regional grid or independently, as needed. Microgrids can increase reliability in the event of power outages owing to hurricanes and other natural and human-made disasters, and can provide electricity access, jobs, and revenues in poor, rural, and isolated communities.

Fully engaging the electricity industry and its regulatory bodies in the critical frontier of GEBs is a strategic opportunity for decarbonizing the United States. DOE, ACEEE, GSA, and RMI have released a national roadmap that illustrates the potential and the urgency of integrating buildings into the future grid, moving well beyond grid-communicating buildings to grid interactive for reducing, shifting, storing, and offsetting energy loads with innovation (Figure 7-9; Dean et al. 2021; Satchwell et al. 2021).

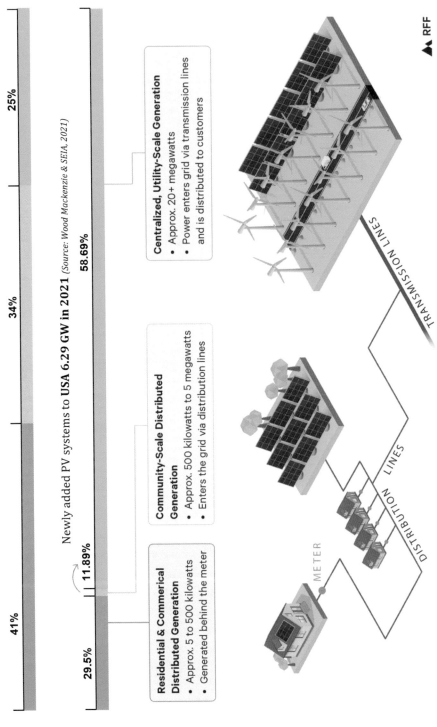

Newly added PV systems to **Germany 5.25 GW in 2021** *(Source: Bundesverband Solarwirtschaft e. V., 2022)*

41%   34%   25%

Newly added PV systems to **USA 6.29 GW in 2021** *(Source: Wood Mackenzie & SEIA, 2021)*

29.5%   11.89%   58.69%

Residential & Commerical Distributed Generation
- Approx. 5 to 500 kilowatts
- Generated behind the meter

Community-Scale Distributed Generation
- Approx. 500 kilowatts to 5 megawatts
- Enters the grid via distribution lines

Centralized, Utility-Scale Generation
- Approx. 20+ megawatts
- Power enters grid via transmission lines and is distributed to customers

METER

DISTRIBUTION LINES

TRANSMISSION LINES

▲▲ RFF

FIGURE 7-8 Accelerating U.S. investments in building and community PV is key to competitiveness. SOURCE: Modified with permission from Cleary and Palmer, 2022, *Resources for the Future*, with data from Bundesverband Solarwirtschaft e.V. (2022) and Wood Mackenzie and Solar Energy Industries Association (2021).

Grid Integrated Building: Load Profiles

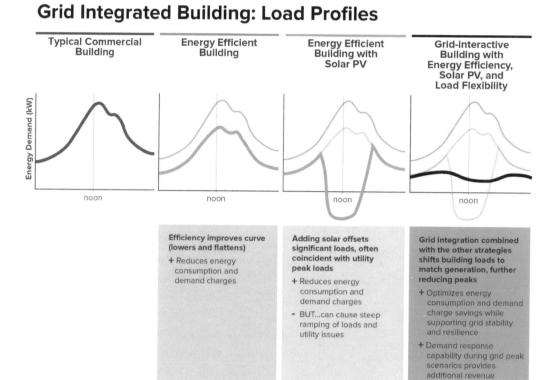

FIGURE 7-9 Efficiency, distributed PV, load flexibility, and thermal storage can flatten peak electricity demand and ensure resiliency. SOURCE: Courtesy of M. Jungclaus, C. Carmichael, and P. Keuhn, *Value Potential for Grid-Interactive Efficient Buildings in the GSA Portfolio: A Cost Benefit Analysis*, Rocky Mountain Institute, 2019. http://www.rmi.org/GEBs_report.

Summary of 10 Actions

Consistent long-term federal and subnational leadership and incentives, combined with subnational mandates, could significantly accelerate the pace of emissions reductions associated with the built environment. For this to succeed, there would need to be clarity and coordination across federal, state, and local building programs, with a commitment to ensure equity across all programs and policies. Significant reductions in demand could be ensured by (1) increased regulatory rigor for buildings and equipment, including accelerated federal appliance and equipment efficiency standards, especially with electrification; (2) subnational mandates and incentives for zero energy new homes and commercial buildings; and (3) accelerated retrofits of existing homes and commercial buildings, with benchmarking and reporting. Research has also demonstrated that the ability of incentives and other strategies to increase adoption of technologies and their effective use to reduce energy consumption can be significantly enhanced through appropriate use of social and behavior insights to inform user-centered and community-centered design (Dietz et al. 2013; Gromet et al. 2013; Stern et al. 1986; Sussman and Chikumbo 2016). Federal and subnational entities adopting new incentives or rules should be sensitive to and consult social and behavioral experts about the variety of non-financial design elements that can dramatically improve long-term efficiency gains for the same financial investment.

There is an equally critical need for national investment in RD&D of low-carbon and smart technologies, and systems for homes, commercial buildings, and GEBs with distributed renewables and storage. Beyond the building level, there is the challenging but transformational pursuit of community-level decarbonization of buildings and community infrastructures, including incentivizing mixed-use walkable infill instead of sprawl communities; accelerating building and community renewables; and innovating in addressing low-temperature thermal (heating and cooling demands) to preserve electricity for higher and better uses. The extent of the carbon savings in the built environment and the cost-effectiveness among decarbonization strategies should further elevate the ongoing commitment to investment, policy, and innovation at the building, community, and grid-levels.

> **Recommendation 7-6:** *Increase Research, Development, Demonstration, and Deployment for Built Environment Decarbonization Interventions.* **Congress should provide funding in support of a Department of Energy (DOE)-led research agenda on the wide-ranging operational and embodied energy savings; greenhouse gas emissions reductions, including refrigerants; and socioeconomic outcomes of different strategies for implementing advanced building and community decarbonization technologies. DOE and state research and development agencies and public utility commissions should expand their budgetary and regulatory support (respectively) for community-level pilots and demonstrations for distributed renewable electric and thermal installations, including district energy systems and grid-integrated efficient buildings. These should include multi-disciplinary socioeconomic strategies to inform the implementation approaches best suited for different communities, including through the potential of alternative ownership strategies over distributed energy and transmission.**

BEYOND THE NEXT 10 YEARS: EVALUATION AND GROWTH

The built environment is composed of a range of buildings and infrastructure, and is designed, built, and maintained by a diverse set of service providers such as developers, builders, and remodelers, along with the range of energy auditors, product manufacturers, retail and supplier distributing building inspectors, real estate agents, lenders, appraisers, and title agents, to name a few of the stakeholders in addition to building owners, occupants, and tenants that are vested in current building practices (Hassel et al. 2003; Martín and McCoy 2019). Many of these same actors would need to be trained, educated, or otherwise made aware of the long-term decarbonization goals for individual buildings and properties—a challenge that is further discussed in Chapter 4.

The planning, design, construction, maintenance, and disposition of human settlement and the built environment defines a significant portion of energy demand. Yet this sector collectively retains a unique social, technological, and economic role in the United States in addition to their energy and environmental consequences. Construction and real estate transactions have averaged 8.4 percent of U.S. gross domestic product over the past half-century (U.S. Bureau of Economic Analysis 2022). From our individual homes and workplaces, through our neighborhoods, and to the greater metropolitan and regional geographies that define our collective sense of place and daily social and economic activities, the built environment's demand for energy has evolved in complex way.

A broad range of public policies and private decisions associated with the built environment significantly define how, how much, and what kind of energy is consumed in the United States. For example, improvements in energy efficiency within new buildings have modestly offset the sheer growth in these buildings' numbers. National decarbonization strategies should undoubtedly consider the multifaceted role that the built environment plays in defining the speed, quality, equity, and consequences of current and future energy transitions in the history of this sector's carbon-based development. In the same way that the transformation for electrification and energy provision in buildings over a century ago required substantial coordination across these groups, so is its decarbonization (Hughes 1993). Indeed, as this chapter describes, there is a much greater potential for emissions reductions from the built environment with even more ambitious strategies for reducing energy demand and more tightly weaving building technologies into the broader energy system.

The IRA, the IIJA, and state and local policies have initiated this transition and suggest ways in which the committee's first recommendations need to be revised.

Indeed, the first report's recommendations for individual buildings remain relevant, but require

1. More rigor in the form of advanced targets for both the current decade and through 2050;
2. Attention to the market transformations, incentives, and other voluntary transitions that must occur among current built environmental sectors, particularly by the end of the first decade to ensure sufficient evidence of their outcomes and feasibility;
3. Articulation of the pivot from voluntary incentives such as those established in the IIJA and IRA for the first decade to mandatory decarbonization across new and existing development leading up to 2050; and
4. Consideration of the interactions across intervention types and their resulting contributions to both comprehensive decarbonization of the sector and the inequitable access and impacts across populations—particularly for low-income, energy-burdened, energy-insecure households; households in severely inadequate housing; and other disadvantaged groups.

For the next decade, investments in market transformation across all building and community intervention points are needed to meet the goal of reducing the built environment's contribution to carbon-based energy demand and its resulting GHG emissions by 2050. Public-sector investments are most needed for workforce training, for fomenting of building company specialization and property fuel provider transition, and in consumer awareness and incentives for existing building retrofits. The market transformation must be conscious of but not neglect the need to increasingly require all intervention types for new development over the same time. Increasing national appliance and equipment standards for efficiency and electrification goals are included in this immediate regulatory framework, as are requirements of federally funded public works, federal facilities, and federally assisted housing.

After the next decade, increasing requirements by subnational governments on developments—and consequently sequenced national regulations with technical assistance and other resources for subnational governments—will be needed to complete transformations for all existing construction in the built environment using all intervention types by 2050. However, the rollout of the newest federal, state, and local policies along with the continued innovation and experimentation of the building and land use industries over the next decade are certainly setting us on the course. The question then becomes: what happens in 2033?

In both the market transformations and regulations sequenced to reach 2035 and 2050 targets, more resources—including direct grants and service provision—will be needed to assist low- and moderate-income households and those living in historically disadvantaged communities. These resources must be allocated simultaneous to market incentives for ensuring that these communities benefit from the transformations, and to regulations for ensuring that they are not punitively burdened by lack of compliance. The resources can come through existing assistance programs for the lowest-income households, such as WAP, if they are armed with ample resources, more flexible eligibility requirements, broader intervention action allowances, improved outreach and community engagement protocols, and safeguards to minimize the adverse effects of improvements such as housing unaffordability or displacement. Expanded product and service rebates and tax credits with similar changes but for a wider population of low- and moderate-income households will ensure comprehensive diffusion of all building decarbonization intervention types for these communities.

> **Recommendation 7-7:** *Extend Current Decarbonization Incentives Beyond the Next Decade While Scaling Up Mandates.* **By 2033, Congress should continue appropriation for rebates and tax credits of residential and commercial decarbonization to incentivize voluntary upgrades for energy efficiency and electrification for an additional decade. Larger incentives must be allotted to building audit services and to owner and occupant service management to ensure the technologically appropriate selection and installation order of decarbonization technologies that are suited to the housing unit and household or owner capacity. For example, the creation of additional targets for 2040–2043 could ensure a higher adoption rate of heat pumps beyond the 25 percent and 15 percent targets laid out in the committee's first report for residential and commercial buildings by 2030. The total appropriation could be equivalent to the funding provided in the Inflation Reduction Act (IRA), but the per household rebate caps and eligible activities should be redesigned based on evaluation of the IRA's outcomes. Extensions to tax credits for both residential and commercial improvements could also be extended just as the IRA extended these from their previous enactment. Furthermore, Congress should increase support for Weatherization Assistance Program (WAP) services by an order of magnitude equivalent to meet the needs of decarbonizing all WAP-eligible households living in homes that use carbon-based energy in the next decade (approximately $33.8 billion annually).**

Table 7-8 summarizes all the recommendations in this chapter to support decarbonizing the built environment.

SUMMARY OF RECOMMENDATIONS ON THE BUILT ENVIRONMENT

TABLE 7-8 Summary of Recommendations on the Built Environment

Short-Form Recommendation	Actor(s) Responsible for Implementing Recommendation	Sector(s) Addressed by Recommendation	Objective(s) Addressed by Recommendation	Overarching Categories Addressed by Recommendation
7-1: Ensure Clarity and Consistency for the Implementation of Building Decarbonization Policies	Department of Energy (DOE)	• Buildings	• Greenhouse gas (GHG) reductions • Equity • Public engagement	Rigorous and Transparent Analysis and Reporting for Adaptive Management Ensuring Procedural Equity in Planning and Siting New Infrastructure and Programs Tightened Targets for the Buildings and Industrial Sectors and a Backstop for the Transport Sector
7-2: Promote an Equitable Focus Across Building Decarbonization Policies	DOE	• Buildings • Non-federal actors	• GHG reductions • Equity • Public engagement	Ensuring Procedural Equity in Planning and Siting New Infrastructure and Programs Tightened Targets for the Buildings and Industrial Sectors and a Backstop for the Transport Sector

continued

TABLE 7-8 Continued

Short-Form Recommendation	Actor(s) Responsible for Implementing Recommendation	Sector(s) Addressed by Recommendation	Objective(s) Addressed by Recommendation	Overarching Categories Addressed by Recommendation
7-3: Expand and Evaluate the Weatherization Assistance Program	DOE	• Buildings	• GHG reductions • Equity • Health	Rigorous and Transparent Analysis and Reporting for Adaptive Management Ensuring Procedural Equity in Planning and Siting New Infrastructure and Programs Tightened Targets for the Buildings and Industrial Sectors and a Backstop for the Transport Sector
7-4: Coordinate Subnational Government Agencies to Align Decarbonization Policies and Implementation	State and municipal government offices	• Buildings • Non-federal actors	• Equity • Employment	Tightened Targets for the Buildings and Industrial Sectors and a Backstop for the Transport Sector Building the Needed Workforce and Capacity

TABLE 7-8 Continued

Short-Form Recommendation	Actor(s) Responsible for Implementing Recommendation	Sector(s) Addressed by Recommendation	Objective(s) Addressed by Recommendation	Overarching Categories Addressed by Recommendation
7-5: Build Capacity for States and Municipalities to Adopt and Enforce Increased Regulatory Rigor for Buildings and Equipment	Congress	• Buildings • Non-federal actors	• GHG reductions • Equity • Health • Employment	A Broadened Policy Portfolio Rigorous and Transparent Analysis and Reporting for Adaptive Management Tightened Targets for the Buildings and Industrial Sectors and a Backstop for the Transport Sector Building the Needed Workforce and Capacity
7-6: Increase Research, Development, Demonstration, and Deployment for Built Environment Decarbonization Interventions	Congress	• Buildings • Non-federal actors	• Equity • Health • Employment • Public engagement	Siting and Permitting Reforms for Interstate Transmission Research, Development, and Demonstration Needs
7-7: Extend Current Decarbonization Incentives Beyond the Next Decade While Scaling Up Mandates	Congress; DOE	• Buildings	• GHG reductions • Equity • Health • Employment	A Broadened Policy Portfolio Tightened Targets for the Buildings and Industrial Sectors and a Backstop for the Transport Sector

REFERENCES

ACEEE (American Council for an Energy-Efficient Economy). 2011. "How Does Energy Efficiency Create Jobs?" Fact Sheet. https://www.aceee.org/files/pdf/fact-sheet/ee-job-creation.pdf.

ACEEE. 2014. "Residential Energy Use Disclosure: A Guide for Policymakers." https://www.aceee.org/toolkit/2014/01/residential-energy-use-disclosure-guide-policymakers.

ACEEE. 2022. "Home Energy Upgrade Incentives: Programs in the Inflation Reduction Act and Other Recent Federal Laws." Policy Brief. https://www.aceee.org/sites/default/files/pdfs/home_energy_upgrade_incentives_9-27-22.pdf.

AFDC (Alternative Fuels Data Center). 2022. "Annual Vehicle Miles Traveled in the United States." Updated June 2022. Department of Energy. https://afdc.energy.gov/data/10315.

Allcott, H., and M. Greenstone. 2012. "Is There an Energy Efficiency Gap?" *Journal of Economic Perspectives* 26(1):3–28. https://doi.org/10.1257/jep.26.1.3.

Andrés, M., M. Regidor, A. Macía, A. Vasallo, and K. Lygnerud. 2018. "Assessment Methodology for Urban Excess Heat Recovery Solutions in Energy-Efficient District Heating Networks." *Energy Procedia*. 16th International Symposium on District Heating and Cooling, DHC2018, 9–12 September 2018, Hamburg, Germany, 149(September):39–48. https://doi.org/10.1016/j.egypro.2018.08.167.

Architecture 2030. 2023. "The 2030 Challenges." *Architecture 2030*. https://architecture2030.org/the2030challenges.

ASAP (Appliance Standards Awareness Project). 2023. "National." *Appliance Standards Awareness Project*. https://appliance-standards.org/national.

Azerbegi, R. 2022. "The Inflation Reduction Act and Its Impact on Your Next Building Project." Mead and Hunt. https://meadhunt.com/ira-impact-building-projects.

Barbour, N. 2021. "Smart Thermostats Gain Traction in US, Point to Modest Electricity Savings." *S&P Global Market Intelligence* (blog), July 20. https://www.spglobal.com/marketintelligence/en/news-insights/blog/smart-thermostats-gain-traction-in-us-point-to-modest-electricity-savings.

BDC (Building Decarbonization Coalition). 2023. "Zero Emission Building Ordinances." *Building Decarbonization Coalition*. https://buildingdecarb.org/zeb-ordinances.

Bellis, R. 2020. "The Congestion Con: How More Lanes and More Money Equals More Traffic." Transportation for America. https://t4america.org/wp-content/uploads/2020/03/Congestion-Report-2020-FINAL.pdf.

Berg, W. 2022. "Leading States Chart Path for Cutting Emissions with Electrification, Pointing Way for Peers." *ACEEE*, July 21. https://www.aceee.org/blog-post/2022/07/leading-states-chart-path-cutting-emissions-electrification-pointing-way-peers-1.

Berg, W., S. Vaidyanathan, B. Jennings, E. Cooper, C. Perry, M. DiMascio, and J. Singletary. 2020. *The 2020 State Energy Efficiency Scorecard*. Washington, DC: ACEEE. https://www.aceee.org/research-report/u2011.

Berg, W., E. Cooper, and M. DiMascio. 2022. *State Energy Efficiency Scorecard: 2021 Progress Report*. Washington, DC: ACEEE. aceee.org/research-report/u2201.

Birch, A. 2018. "How to Halve the Cost of Residential Solar in the US." Greentech Media, January 5. https://www.greentechmedia.com/articles/read/how-to-halve-the-cost-of-residential-solar-in-the-us.

Biswas, S., A. Echevarria, N. Irshad, Y. Rivera-Matos, J. Richter, N. Chhetri, M.J. Parmentier, and C.A. Miller. 2022. "Ending the Energy-Poverty Nexus: An Ethical Imperative for Just Transitions." *Science and Engineering Ethics* 28(4):36. https://doi.org/10.1007/s11948-022-00383-4.

Björnebo, L., S. Spatari, and P.L. Gurian. 2018. "A Greenhouse Gas Abatement Framework for Investment in District Heating." *Applied Energy* 211:1095–1105. https://doi.org/10.1016/j.apenergy.2017.12.003.

Blumsack, S. 2022. "Can My Electric Car Power My House? Not Yet for Most Drivers, But Vehicle-to-Home Charging Is Coming." *Fast Company*, March 31. https://www.fastcompany.com/90736485/can-my-electric-car-power-my-house-not-yet-for-most-drivers-but-vehicle-to-home-charging-is-coming.

Bowen, T., I. Chernyakhovskiy, and P.L. Denholm. 2019. "Grid-Scale Battery Storage: Frequently Asked Questions." NREL/TP-6A20-74426, 1561843. https://doi.org/10.2172/1561843.

Brown, M., L. Berry, and L. Kinney. 1994. *Weatherization Works: Final Report of the National Weatherization Evaluation.* Oak Ridge, TN: Oak Ridge National Laboratory. ORNL/CON-395. https://technicalreports.ornl.gov/cppr/y2001/rpt/109939.pdf.

Bundesverband Solarwirtschaft e.V. 2022. "2021: Zehn Prozent mehr Solarmodule installiert." January 3. https://www.solarwirtschaft.de/2022/01/03/2021-zehn-prozent-mehr-solarmodule-installiert.

C40 Cities Climate Leadership Group and C40 Knowledge Hub. 2019. "How to Use Reporting and Disclosure to Drive Building Energy Efficiency." C40. April. https://www.c40knowledgehub.org/s/article/How-to-use-reporting-and-disclosure-to-drive-building-energy-efficiency?language=en_US.

California Energy Commission. 2023. "Building Initiative for Low-Emissions Development Program." https://www.energy.ca.gov/programs-and-topics/programs/building-initiative-low-emissions-development-program.

CARB (California Air Resources Board). n.d. "Use of Back-Up Engines for Electricity Generation During Public Safety Power Shutoff Events." https://ww2.arb.ca.gov/resources/documents/use-back-engines-electricity-generation-during-public-safety-power-shutoff. Accessed March 31, 2023.

Carbon Leadership Forum. 2020. "1—Embodied Carbon 101." https://carbonleadershipforum.org/embodied-carbon-101.

Carmichael, C., M. Jungclaus, P. Keuhn, and K.P. Hydras. 2019. "Value Potential for Grid-Interactive Efficient Buildings in the GSA Portfolio: A Cost-Benefit Analysis." *RMI and GSA.* https://rmi.org/wp-content/uploads/2019/07/value-potential-grid-integrated-buldings-gsa-portfolio.pdf.

Center for Sustainable Systems, University of Michigan. 2021. "U.S. Cities Factsheet." Pub. No. CSS09-06.

Cleveland, C.J., P. Fox-Penner, M.J. Walsh, M. Cherne-Hendrick, S. Gopal, J.R. Castigliego, and T. Perez, et al. 2019. "Carbon Free Boston Summary Report." Boston University Institute for Sustainable Energy. https://www.bu.edu/igs/2019/01/29/carbon-free-boston-summary-report.

Climate and Clean Air Coalition. 2020. *HydroFlurocarbons (HFCs).* https://www.ccacoalition.org/en/slcps/hydrofluorocarbons-hfcs.

Cluett, R., J. Amann, and S. Ou. 2016. *Building Better Energy Efficiency Programs for Low-Income Households.* Washington, DC: American Council for an Energy-Efficient Economy. Report No. A1601. https://www.aceee.org/sites/default/files/publications/researchreports/a1601.pdf.

Cohn, C., and N.W. Esram. 2022. *Building Electrification: Programs and Best Practices.* Washington, DC: American Council for an Energy-Efficient Economy. https://www.aceee.org/research-report/b2201.

CoolClimate Network. n.d. "California Carbon Footprints Under Climate Targets v2.0." https://cal.maps.arcgis.com/apps/MapSeries/index.html?appid=2c0d0d53239c45ba980211842837cc32. Accessed March 31, 2023.

CPUC (California Public Utilities Commission). n.d. "Zero Net Energy." Cpuc.ca.Gov. https://www.cpuc.ca.gov/industries-and-topics/electrical-energy/demand-side-management/energy-efficiency/zero-net-energy. Accessed August 10, 2023.

CSA Group. 2021. "CSA SPE-112:21 | CarbonStar®: Technical Specification for Concrete Carbon Intensity Quantification and Verification." CSA Group. https://www.csagroup.org/store/product/CSA%20SPE-112%3A21.

Dean, J., P. Voss, D. Gagne, D. Vasquez, and R. Langner. 2021. "Blueprint for Integrating Grid-Interactive Efficient Building (GEB) Technologies into U.S. General Services Administration Performance Contracts." NREL/TP-7A40-78190, 1784273, MainId:32099. https://doi.org/10.2172/1784273.

Deetjen, T.A., L. Walsh, and P. Vaishnav. 2021. "US Residential Heat Pumps: The Private Economic Potential and Its Emissions, Health, and Grid Impacts." *Environmental Research Letters* 16(8):084024. https://doi.org/10.1088/1748-9326/ac10dc.

Department of State. 2022. "U.S. Ratification of the Kigali Amendment." https://www.state.gov/u-s-ratification-of-the-kigali-amendment.

DOE (Department of Energy). 2011. "Saving Money by Saving Energy (Fact Sheet)." DOE/GO-102011-3280, 1024064. https://doi.org/10.2172/1024064. Accessed August 10, 2023.

DOE. 2015. Chapter 5, "Increasing Efficiency of Building Systems and Technologies." In *Quadrennial Technology Review 2015.* Washington, DC.

DOE. 2016. "10 CFR Parts 429 and 430 [Docket No. EERE-2016-BT-TP-0029] RIN 1904-AD71 Energy Conservation Program: Test Procedures for Central Air Conditioners and Heat Pumps." Office of Energy Efficiency and Renewable Energy, Department of Energy. Final Rule.

DOE. 2019. "Low-Income Household Energy Burden Varies Among States—Efficiency Can Help in All of Them." https://www.energy.gov/sites/prod/files/2019/01/f58/WIP-Energy-Burden_final.pdf.

DOE. 2020a. "Low-Income Energy Affordability Data (LEAD) Tool." https://lead.openei.org/docs/LEAD-Factsheet.pdf.

DOE. 2020b. "Minnesota Implementation Model—Combined Heat and Power Action Plan." https://www.energy.gov/scep/slsc/articles/minnesota-implementation-model-combined-heat-and-power-action-plan.

DOE. n.d. "Residential Cold Climate Heat Pump Challenge." https://www.energy.gov/eere/buildings/residential-cold-climate-heat-pump-challenge.

DOE Office of Policy. 2022. "The Inflation Reduction Act Drives Significant Emissions Reductions and Positions America to Reach Our Climate Goals." Department of Energy. DOE/OP-0018. www.energy.gov/sites/default/files/2022-08/8.18%20InflationReductionAct_Factsheet_Final.pdf.

DOS and EOP (Department of State and Executive Office of the President). 2021. *The Long-Term Strategy of the United States: Pathways to Net-Zero Greenhouse Gas Emissions by 2050.* Washington, DC. https://www.whitehouse.gov/wp-content/uploads/2021/10/US-Long-Term-Strategy.pdf.

Drehobl, A. 2020. "Weatherization Cuts Bills and Creates Jobs But Serves Only a Tiny Share of Low-Income Homes." American Council for an Energy-Efficient Economy. https://www.aceee.org/blog-post/2020/07/weatherization-cuts-bills-and-creates-jobs-serves-only-tiny-share-low-income.

Drehobl, A., L. Ross, and R. Ayala. 2020. *How High Are Household Energy Burdens?* Washington, DC: American Council for an Energy-Efficient Economy. https://www.aceee.org/research-report/u2006.

EERE (Department of Energy Office of Energy Efficiency and Renewable Energy). 2022a. "How Historic Weatherization Investments Will Make Life Better for Low-Income Families." https://www.energy.gov/eere/articles/how-historic-weatherization-investments-will-make-life-better-low-income-families.

EERE. 2022b. "Weatherization Assistance Program Fact Sheet." DOE/EE-2124. https://www.energy.gov/sites/default/files/2022-06/wap-fact-sheet_0622.pdf.

EERE. 2023. "Soft Costs: Solar Energy Technologies Office." https://www.energy.gov/eere/solar/soft-costs.

EIA (U.S. Energy Information Administration). 2013. "2009 RECS Survey Data." https://www.eia.gov/consumption/residential/data/2009.

EIA. 2018. "2015 RECS Survey Data." May. https://www.eia.gov/consumption/residential/data/2015.

EIA. 2021. "U.S. Electricity Customers Experienced Eight Hours of Power Interruptions in 2020." https://www.eia.gov/todayinenergy/detail.php?id=50316.

EIA. 2022a. "2018 CBECS Survey Data." https://www.eia.gov/consumption/commercial/data/2018/index.php?view=consumption.

EIA. 2022b. "2020 RECS Survey Data." May. https://www.eia.gov/consumption/residential/data/2020/hc/pdf/HC%206.1.pdf.

EIA. 2022c. *Monthly Energy Review: 2. Energy Consumption by Sector.* December. https://www.eia.gov/totalenergy/data/monthly/pdf/sec2.pdf.

EIA. 2022d. "Today in Energy: Homes and Buildings in the West and Northeast Have the Largest Share of Small-Scale Solar." https://www.eia.gov/todayinenergy/detail.php?id=54379.

EIA. 2023a. "Electric Power Monthly—Table 2.1A Energy Consumption: Residential, Commercial, And Industrial Sectors." https://www.eia.gov/electricity/monthly/epm_table_grapher.php.

EIA. 2023b. "Electric Power Monthly—Table B.1 Major Disturbances and Unusual Occurrences, Year-to-Date 2023." https://www.eia.gov/electricity/monthly/epm_table_grapher.php.

EIA. 2023c. "Energy Consumption by Sector." https://www.eia.gov/totalenergy/data/monthly/pdf/sec2.pdf.

Energy Saver. 2022. "Heat Pump Systems." Department of Energy Office of Energy Efficiency and Renewable Energy. https://www.energy.gov/energysaver/heat-pump-systems.

EnergyStar. n.d.(a). "Energy Efficiency Program Sponsor Frequently Asked Questions About ENERGY STAR Smart Thermostats." https://www.energystar.gov/products/heating_cooling/smart_thermostats/smart_thermostat_faq. Accessed August 10, 2023.

EnergyStar. n.d.(b). "When Is It Time to Replace?" www.energystar.gov/saveathome/heating_cooling/replace.

Energymag. n.d. "Daily Energy Demand Curve." https://energymag.net/daily-energy-demand-curve.

EPA (Environmental Protection Agency). 2023. "Inventory of U.S. Greenhouse Gas Emissions and Sinks: 1990–2021." EPA 430-D-23-001. https://www.epa.gov/ghgemissions/draft-inventory-us-greenhouse-gas-emissions-and-sinks-1990-2021.

Evans, B. 2023. "IRA Update: The New Economics for Allowing Public Sector Buildings to Go Green." U.S. Green Building Council. April 25. https://www.usgbc.org/articles/ira-update-new-economics-allowing-public-sector-buildings-go-green.

Feng, Q., and P. Gauthier. 2021. "Untangling Urban Sprawl and Climate Change: A Review of the Literature on Physical Planning and Transportation Drivers." *Atmosphere* 12(5):547. https://doi.org/10.3390/atmos12050547.

Fernandez, N., S. Katipamula, W. Wang, Y. Xie, M. Zhao, and C. Corgin. 2017. *Impacts of Commercial Building Controls on Energy Savings and Peak Load Reduction.* PNNL Report to DOE. May. PNNL-25985. https://buildingretuning.pnnl.gov/publications/PNNL-25985.pdf.

Fisher Sheehan & Colton. 2021. "What Is the Home Energy Affordability Gap?" *Fisher Sheehan & Colton Public Finance and General Economics.* http://www.homeenergyaffordabilitygap.com/01_whatIsHEAG2.html.

Fowlie, M., M. Greenstone, and C. Wolfram. 2018. "Do Energy Efficiency Investments Deliver? Evidence from the Weatherization Assistance Program." *Quarterly Journal of Economics* 133(3):1597–1644. https://doi.org/10.1093/qje/qjy005.

Frank, M., and S. Nowak. 2015. "Who Is Participating, and Who Is Not? An Analysis of Demographic Data from California Residential Energy-Efficiency Programs, 2010–2012." "Behavior, Energy and Climate Change." Presented at BECC Conference, October 2015. https://beccconference.org/wp-content/uploads/2015/10/presentation_frank.pdf.

Group14 Engineering. 2020. "Electrification of Commercial and Residential Buildings: An Evaluation of the System Options, Economics, and Strategies to Achieve Electrification of Buildings." Denver, CO: Group 14 Engineering, PCB. Prepared for Community Energy Inc., November 2020. https://www.communityenergyinc.com/wp-content/uploads/Building-Electrification-Study-Group14-2020-11.09.pdf.

Guidehouse Insights. 2022. "The IRAs Impact on Building Efficiency and Decarbonization Markets." *Latest Insights,* September 26. https://guidehouseinsights.com/news-and-views/the-iras-impact-on-building-efficiency-and-decarbonization-markets.

Hale, E., H. Horsey, B. Johnson, M. Muratori, E. Wilson, B. Borlaug, C. Christensen, et al. 2018. "The Demand-Side Grid (Dsgrid) Model Documentation." NREL/TP-6A20-71492, 1465659, MainId:16928. https://doi.org/10.2172/1465659.

Harding, T. 2021. "Ithaca Becomes First City in U.S. to Try and Electrify All Buildings." *Ithaca Times,* November 9. https://www.ithaca.com/news/ithaca/ithaca-becomes-first-city-in-u-s-to-try-and-electrify-all-buildings/article_03c6e998-41bb-11ec-9a84-47a7c90ee120.html.

Hart, Z., C. Majersik, and J. Eagles. 2022. *Leveling Up Building Performance Regulations: How Governments Can Craft Equitable, Effective Building Performance Standards to Drive Widespread Market Transformation.* Institute for Market Transformation. https://www.imt.org/wp-content/uploads/2022/08/SummerStudy22-BPS-Equity-Hart.pdf.

Hassell, S., A. Wong, A. Houser, D. Knopman, and M.A. Bernstein. 2003. "Building Better Homes: Government Strategies for Promoting Innovation in Housing." RAND Corporation.

Heeter, J., K. Xu, and E. Fekete. 2020. "Community Solar 101 [Slides]." NREL/PR-6A20-75982, 1602184. https://doi.org/10.2172/1602184.

Heeter, J., K. Xu, and G. Chan. 2021. *Sharing the Sun: Community Solar Deployment, Subscription Savings, and Energy Burden Reduction.* National Renewable Energy Laboratory. NREL/PR-6A20-80246. https://www.nrel.gov/docs/fy21osti/80246.pdf.

Hernández, D., Y. Aratani, and Y. Jiang. 2014. *Energy Insecurity Among Families with Children.* New York: National Center for Children in Poverty (NCCP), Columbia University Mailman School of Public Health. https://www.nccp.org/wp-content/uploads/2020/05/text_1086.pdf.

HHS (Department of Health and Human Services). 2023. "Low Income Home Energy Assistance Program (LIHEAP)." https://www.acf.hhs.gov/ocs/programs/liheap.

HUD (Department of Housing and Urban Development). 2023. "HUD Publishes Updated Public and Assisted Housing Guidance for Treatment of Solar Programs for Residents to Benefit from President Biden's Investing in America Agenda." https://www.hud.gov/press/press_releases_media_advisories/hud_no_23_162.

Hughes, T.P. 1993. "Networks of Power: Electrification in Western Society, 1880–1930." Baltimore, MD: Johns Hopkins University Press. https://www.press.jhu.edu/books/title/2031/networks-power.

IMT (Institute for Market Transformation). 2022. "U.S. City, County, and State Policies for Existing Buildings: Benchmarking, Transparency, and Beyond." Institute for Market Transformation. https://www.imt.org/wp-content/uploads/2022/06/IMT-Benchmarking-Map-1.pdf.

IPCC (Intergovernmental Panel on Climate Change). 2023. "Sixth Assessment Report." Intergovernmental Panel on Climate Change. https://www.ipcc.ch/assessment-report/ar6.

Jones, C.M., S.M. Wheeler, and D.M. Kammen. 2018. "Carbon Footprint Planning: Quantifying Local and State Mitigation Opportunities for 700 California Cities." *Urban Planning* 3(2:35–51). https://doi.org/10.17645/up.v3i2.1218.

King, J. 2018. "Energy Impacts of Smart Home Technologies." American Council for an Energy-Efficient Economy. https://www.aceee.org/research-report/a1801.

Kirk, C. 2021. "Los Angeles Building Decarbonization: Tenant Impact and Recommendations." Strategic Actions for a Just Economy (SAJE). https://www.saje.net/wp-content/uploads/2021/12/LA-Building-Decarb_Tenant-Impact-and-Recommendations_SAJE_December-2021-1.pdf.

Kresowik, M., and L. Reeg. 2022. "Funding Our Future: Creating a One-Stop Shop for Whole-Home Retrofits." RMI. https://rmi.org/creating-a-one-stop-shop-for-whole-home-retrofits.

Langevin, J., A. Satre-Meloy, A.J. Satchwell, R. Hledik, J. Olszewski, K. Peters, and H.C. Putra. 2022. "The Role of Buildings in U.S. Energy System Decarbonization by Mid-Century." SSRN Scholarly Paper. https://doi.org/10.2139/ssrn.4253001.

Leach, M., S. Pless, and P. Torcellini. 2014. "Cost Control Best Practices for Net Zero Energy Building Projects." National Renewable Energy Laboratory. NREL/CP-5500-61365. https://www.nrel.gov/docs/fy14osti/61365.pdf.

Litjens, G.B.M.A., E. Worrell, and W.G.J.H.M. van Sark. 2018. "Lowering Greenhouse Gas Emissions in the Built Environment by Combining Ground Source Heat Pumps, Photovoltaics and Battery Storage." *Energy and Buildings* 180:51–71.

Mai, T., P. Jadun, J. Logan, C. McMillan, M. Muratori, D. Steinberg, L. Vimmerstedt, R. Jones, B. Haley, and B. Nelson. 2018. *Electrification Futures Study: Scenarios of Electric Technology Adoption and Power Consumption for the United States.* Golden, CO: National Renewable Energy Laboratory. NREL/TP-6A20-71500. https://www.nrel.gov/docs/fy18osti/71500.pdf.

Martín, C., and A. McCoy. 2019. *Building Even Better Homes: Strategies for Promoting Innovation in Home Building.* Washington, DC: Department of Housing and Urban Development. https://www.huduser.gov/portal/sites/default/files/pdf/BuildingEvenBetterHomes.pdf.

Martín, C., M. Bueno, M. Johnson, F. Montes, and R. Frost. 2022. "Targeting Weatherization: Supporting Low-Income Renters in Multifamily Properties Through the Infrastructure Investment and Jobs Act's Funding of the Weatherization Assistance Program and Beyond." Harvard University Joint Center for Housing Studies. www.jchs.harvard.edu/research-areas/working-papers/targeting-weatherization-supporting-low-income-renters-multifamily.

McCoy, M. 2022. "The State(s) of Distributed Solar—2021 Update." Institute for Local Self-Reliance. https://ilsr.org/the-states-of-distributed-solar.

NAR (National Association of Realtors). 2020. "How Long Do Homeowners Stay in Their Homes?" *Economists' Outlook* (blog), January 8. www.nar.realtor/blogs/economists-outlook/how-long-do-homeowners-stay-in-their-homes.

NASCSP (National Association for State Community Services Programs). 2019. *Weatherization Assistance Program PY2019 Funding Report.* Washington, DC: National Association of State Community Service Programs, WAP. https://nascsp.org/wp-content/uploads/2021/01/NASCSP-2019-WAP-Funding-Survey_Final.pdf.

NASEM (National Academies of Sciences, Engineering, and Medicine). 2017. *Enhancing the Resilience of the Nation's Electricity System.* Washington, DC: The National Academies Press. https://doi.org/10.17226/24836.

NASEM. 2021a. *Accelerating Decarbonization of the U.S. Energy System.* Washington, DC: The National Academies Press. https://doi.org/10.17226/25932.

NASEM. 2021b. *The Future of Electric Power in the United States.* Washington, DC: The National Academies Press. https://doi.org/10.17226/25968.

NASEM. 2023. *The Role of Net Metering in the Evolving Electricity System.* Washington, DC: The National Academies Press. https://doi.org/10.17226/26704.

NASEO (National Association of State Energy Officials). 2022. "Home Energy Labeling." National Association of State Energy Officials. November 2022. https://naseo.org/issues/buildings/home-energy-labeling.

National BPS (Building Performance Standards) Coalition. n.d. "White House National Building Performance Standards Coalition." National BPS Coalition. https://nationalbpscoalition.org. Accessed August 10, 2023.

NBI (New Buildings Institute). 2012. "Getting to Zero 2012 Status Update: First Look at the Costs and Features of Zero Energy Commercial Buildings." https://newbuildings.org/wp-content/uploads/2015/11/GettingtoZeroReport_01.pdf.

NBI. 2017. "Model Stretch Code Provisions for a 20% Performance Improvement in New Commercial Construction." https://newbuildings.org/wp-content/uploads/2017/11/20percent_code_provisions_SummaryDoc_FINAL.pdf.

NBI. 2019. "2019 New York Getting to Zero Status Report." New Buildings Institute for NYSERDA. https://newbuildings.org/wp-content/uploads/2019/04/NY-GTZ-Status-Report_0419.pdf.

New York State Climate Action Council. 2022. "New York State Climate Action Council Scoping Plan." https://climate.ny.gov/-/media/project/climate/files/NYS-Climate-Action-Council-Final-Scoping-Plan-2022.pdf.

Nowak, D.J., and E.J. Greenfield. 2018. "Declining Urban and Community Tree Cover in the United States." *Urban Forestry and Urban Greening* 32:32–55. https://doi.org/10.1016/j.ufug.2018.03.006.

NYC (New York City). 2022. "Sustainable Buildings: Compliance." *City of New York.* https://www.nyc.gov/site/sustainablebuildings/requirements/compliance.page.

Palmer, K., and M. Walls. 2015. "Can Benchmarking and Disclosure Laws Provide Incentives for Energy Efficiency Improvements in Buildings?" Discussion Paper RFF-DP-15-09. Resources for the Future. https://media.rff.org/documents/RFF-DP-15-09.pdf.

Palmer, K., and M. Walls. 2017. "Using Information to Close the Energy Efficiency Gap: A Review of Benchmarking and Disclosure Ordinances." *Energy Efficiency* 10(3):673–691. https://doi.org/10.1007/s12053-016-9480-5.

Pantano, S. 2020. "Heating Electrification: The Next Opportunity for Coordinated Climate Action." CLASP. https://www.clasp.ngo/updates/heating-electrification-the-next-opportunity-for-coordinated-climate-action.

Philadelphia Energy Authority. 2023. "The Inflation Reduction Act (IRA) and Commercial Properties." Philadelphia Energy Authority. https://philaenergy.org/the-inflation-reduction-act-ira-and-commercial-properties.

PNNL and DOE (Pacific Northwest National Laboratory and Department of Energy). 2022. "Historical Model Energy Code Improvement." *Tableau Public.* December 21. https://public.tableau.com/app/profile/doebecp/viz/HistoricalModelEnergyCodeImprovement/CombinedHistoricalCodeImprovement_1.

Riahi, L. 2015. "District Energy in Cities: Unlocking the Potential of Energy Efficiency and Renewable Energy." United Nations Environment Programme. http://www.unep.org/resources/report/district-energy-cities-unlocking-potential-energy-efficiency-and-renewable-energy.

RMI. n.d. "The Retrofit Depot." https://rmi.org/our-work/buildings/deep-retrofit-tools-resources/deep-retrofit-case-studies.

Shang, L., H.W. Lee, S. Dermisi, and Y. Choe. 2020. "Impact of Energy Benchmarking and Disclosure Policy on Office Buildings." *Journal of Cleaner Production* 250(March):119500. https://doi.org/10.1016/j.jclepro.2019.119500.

Shove, E. 2021. "Time to Rethink Energy Research." *Nature Energy* 6(2):118–120. https://doi.org/10.1038/s41560-020-00739-9.

Smedick, D., R. Golden, and A. Petersen. 2022. "The Inflation Reduction Act Could Transform the US Buildings Sector." RMI. https://rmi.org/the-inflation-reduction-act-could-transform-the-us-buildings-sector.

Sobin, R. 2021. "State and Local Building Policies and Programs for Energy Efficiency and Demand Flexibility." NASEO. https://www.naseo.org/data/sites/1/documents/publications/NASEO%20BldgPolicies%20EE%20and%20DF%20Feb%202021.pdf.

SSC (Smart Surfaces Coalition). 2022. "Cooling Cities, Slowing Climate Change, and Enhancing Equity: Costs and Benefits of Smart Surfaces Adoption for Baltimore." https://smartsurfacescoalition.org/baltimore-report.

State of Massachusetts. 2022. "Stretch Energy Code Development 2022." https://www.mass.gov/info-details/stretch-energy-code-development-2022.

Steinberg, D., D. Bielen, J. Eichman, K. Eurek, J. Logan, T. Mai, C. McMillan, A. Parker, L. Vimmerstedt, and E. Wilson. 2017. "Electrification and Decarbonization: Exploring U.S. Energy Use and Greenhouse Gas Emissions in Scenarios with Widespread Electrification and Power Sector Decarbonization." NREL/TP—6A20-68214, 1372620. https://doi.org/10.2172/1372620.

Strohbach, M.W., A.O. Döring, M. Möck, M. Sedrez, O. Mumm, A.K. Schneider, S. Weber, and B. Schröder. 2019. "The 'Hidden Urbanization': Trends of Impervious Surface in Low-Density Housing Developments and Resulting Impacts on the Water Balance." *Frontiers in Environmental Science* 29. https://doi.org/10.3389/FENVS.2019.00029.

Taha, H. 2021. "Development of an Urban Heat Mitigation Plan for the Greater Sacramento Valley, California, a Csa Koppen Climate Type." *Sustainability* 13(17):9709. https://doi.org/10.3390/su13179709.

Tonn, B., E. Rose, and B. Hawkins. 2018. "Evaluation of the US Department of Energy's Weatherization Assistance Program: Impact Results." *Energy Policy* 118:279–290. https://doi.org/10.1016/j.enpol.2018.03.051.

Tyler, M., D. Winiarski, M. Rosenberg, and B. Liu. 2021. *Impacts of Model Building Energy Codes—Interim Update*. USDOE. PNNL-31437. https://doi.org/10.2172/1808877.

UN (United Nations). 2023. "2.f. Amendment to the Montreal Protocol on Substances That Deplete the Ozone Layer." United Nations Treaty Collection. https://treaties.un.org/Pages/ViewDetails.aspx?src=IND&mtdsg_no=XXVII-2-f&chapter=27&clang=_en.

UNEP (United Nations Environment Programme). 2019. *District Energy in Cities Initiative—[Factsheet]*. https://wedocs.unep.org/20.500.11822/31588.

Ungar, L., and S. Nadel. 2019. *Halfway There: Energy Efficiency Can Cut Energy Use and Greenhouse Gas Emissions in Half by 2050*. U1907. Washington, DC: American Council for an Energy-Efficient Economy. https://www.aceee.org/research-report/u1907.

U.S. Bureau of Economic Analysis. 2022. "Gross Domestic Product (Third Estimate), Corporate Profits (Revised Estimate), and GDP by Industry, Third Quarter 2022: Table 18. Price Indexes for Gross Output by Industry Group: Percent Change from Preceding Period." https://www.bea.gov/sites/default/files/2022-12/gdp3q22_3rd.pdf.

U.S. Census Bureau. 2022. "American Housing Survey (AHS)." Department of Housing and Urban Development. https://www.census.gov/programs-surveys/ahs.html.

USGBC MA (U.S. Green Building Council Massachusetts). 2019. *Zero Energy Buildings in Massachusetts: Saving Money from the Start 2019 Report*. USGBC MA and Integral Group. https://builtenvironmentplus.org/wp-content/uploads/2019/09/ZeroEnergyBldgMA2019.pdf.

Waite, M., and V. Modi. 2020. "Electricity Load Implications of Space Heating Decarbonization Pathways." *Joule* 4(2): 376–394. https://doi.org/10.1016/j.joule.2019.11.011.

Walton, R. 2022. "Slow Adoption of Smart Thermostats in the US Misses Big Potential Energy Savings: S&P." *Utility Dive*, August 31. https://www.utilitydive.com/news/smart-thermostats-us-slow-adoption-misses-energy-savings/630901.

White House. 2022a. "Fact Sheet: Biden-Harris Administration Announces First-Ever Federal Building Performance Standard, Catalyzes American Innovation to Lower Energy Costs, Save Taxpayer Dollars, and Cut Emissions." Briefing Room, Statements and Releases. https://www.whitehouse.gov/briefing-room/statements-releases/2022/12/07/fact-sheet-biden-harris-administration-announces-first-ever-federal-building-performance-standard-catalyzes-american-innovation-to-lower-energy-costs-save-taxpayer-dollars-and-cut-emissions.

White House. 2022b. "Fact Sheet: Biden-Harris Administration Takes More Than 100 Actions in 2022 to Strengthen Energy Efficiency Standards and Save Families Money." https://www.whitehouse.gov/briefing-room/statements-releases/2022/12/19/fact-sheet-biden-harris-administration-takes-more-than-100-actions-in-2022-to-strengthen-energy-efficiency-standards-and-save-families-money.

Wood Mackenzie and Solar Energy Industries Association. 2021. *US Solar Market Insight, Executive Summary, Q4 2021*. https://www.woodmac.com.

Woody, T. 2022. "Silicon Valley's Menlo Park Plans to Electrify 10,000 Buildings." *Bloomberg*, June 17. https://www.bloomberg.com/news/articles/2022-06-17/silicon-valley-s-menlo-park-plans-to-electrify-10-000-buildings.

8

Land Use

ABSTRACT

Land used for agriculture and forestry plays a significant role in U.S. decarbonization strategies. It serves as a potential terrestrial carbon sink (in forest biomass and agricultural soils), as a major source of the greenhouse gases (GHGs) CH_4 and N_2O, and—through bioenergy production capacity—as a partial replacement for fossil fuels and a potential source for carbon capture and storage. In addition, land availability constrains components of decarbonization within the energy sector (e.g., wind and solar facility siting, transmission grid expansion).

The available "safe" land carbon sink capacity in the United States is more than sufficient to support the negative emissions needed from the land sector consistent with 2050 net-zero goals. Funding from the Inflation Reduction Act (IRA) to incentivize terrestrial carbon sinks and non-CO_2 emission reductions is technically capable of generating an annual land sink plus CH_4 and N_2O abatement of 211 Mt CO_2e/y in 2030, and a total net emission (carbon sinks plus non-CO_2 abatement) reduction of 845 Mt CO_2e over the 8-year funding cycle from the IRA, at implementation costs ≤$50/tonne CO_2e. This exceeds the estimated land sector contribution needed to follow emission reduction trajectories in current comprehensive U.S. decarbonization scenarios. However, there is substantial uncertainty with respect to the extent that IRA investments will actually achieve these potential emissions reductions, and monitoring, learning, and adaptive management will be necessary for lands to play a meaningful role in decarbonization.

Land use sinks and sources of GHGs are among the most difficult to accurately quantify because they are non-point source, highly variable both spatially and temporally, and subject to many influencing factors. Hence, Recommendations 8-1, 8-3, 8-4, and 8-5 are for the Secretary of Agriculture to direct funding to improve measurement and monitoring of forest and agriculture soil carbon sinks and CH_4 and N_2O emissions, and to better track practice adoption rates, barriers to adoption, and overall performance of the land use–related climate mitigation initiatives in the IRA and related legislation. Demand-side factors—including development of artificial meat and dairy food products, reduced food waste, and shifts toward a more plant-based diet—could significantly reduce land needed for agricultural production. Sufficient consumer acceptance of and demand for these products (see Recommendation 8-8) could cut N_2O

emissions associated with fertilizer use, reduce livestock-sourced CH_4 and N_2O emissions, and free up substantial land area for expanding perennial forest and grassland systems with high carbon sink capacity.

INTRODUCTION

Positive emissions of carbon dioxide (CO_2) from terrestrial ecosystems to the atmosphere occur whenever photosynthetic uptake of atmospheric CO_2 is smaller than emissions from plant respiration, biomass combustion, and microbial decomposition of dead biomass. Positive emissions cause an ecosystem to lose carbon mass because of a net transfer from terrestrial organic carbon to atmospheric CO_2.[1] Carbon dioxide emissions are net negative whenever losses are smaller than gains, which causes a net gain in ecosystem carbon stocks (a carbon sink) and a net reduction in atmospheric CO_2. In addition, terrestrial ecosystems—particularly those managed for agricultural production—emit two other potent GHGs: methane (CH_4, primarily from livestock production) and nitrous oxide (N_2O, primarily from synthetic fertilizer). This chapter covers four kinds of GHG mitigation measures involving land—specifically, forests croplands, and pastures, all of which are targeted by the IRA[2]:

1. Increasing negative CO_2 emissions through reforestation and afforestation and changes in forest management.
2. Reducing positive emissions caused by wildfire through changes in forest management.
3. Increasing negative CO_2 emissions by promoting agricultural practices that increase carbon stores in soils.
4. Reducing CH_4 and N_2O emissions from agriculture and animal husbandry.

Additionally, this chapter discusses land requirements for siting energy infrastructure (e.g., wind turbines, solar panels, and transmission lines); land requirements for production of biofuels feedstocks; and the land use impacts of animal agriculture and the potential implications of demand-side changes.

The 2021 National Academies' report *Accelerating Decarbonization of the U.S. Energy System* deferred discussion of the policies to create and manage agricultural and forestry carbon sinks to this report, and hence a full chapter has been allocated here. This chapter is unusual, in that many of the technical and scientific methods for achieving

---

[1] While carbon transport occurs between land and water (rivers, oceans) in terrestrial ecosystems, this chapter will focus only on land.

[2] In contrast to the IRA, neither the Infrastructure Investment and Jobs Act (IIJA) nor the CHIPS and Science Act includes programs or funding for terrestrial carbon sinks.

negative CO_2 emissions in the land use sector were addressed comprehensively in the 2019 National Academies' report *Negative Emissions Technologies and Reliable Sequestration: A Research Agenda.* That report covered strategies and achievable CO_2 removals for (1) and (3) above as well as biofuels used for Biomass Energy with Carbon Capture and Storage (BECCS), in which biomass is used to produce electricity, hydrogen, or a carbo-hydrate fuel, with CO_2 capture and geologic storage.[3] However, mitigating agricultural emissions of non-CO_2 GHGs were not covered in the 2019 report, and so here an analysis is provided—including available practices and magnitude of mitigation potential—of N_2O and CH_4 reductions that play a key role in any net-zero GHG emissions pathway. The committee will not recapitulate everything in the 2019 report but will summarize and update its findings about terrestrial GHG mitigation. In the 4 years since the release of the 2019 report, the most important relevant new developments are:

a. Terrestrial ecosystem GHG mitigation has become a more prominent topic in the public and scientific literature under the rubric of "Nature-Based Climate Solutions" (NBCS).[4] Several state governments and the federal government are giving increased emphasis on NBCS approaches (California Natural Resources Agency n.d.; White House 2022). Several comprehensive and peer-reviewed analyses of the potential of NBCS have been published, including some that caution against previous estimates that NBCS could supply more than one-third of needed climate mitigation (e.g., Cook-Patton et al. 2021a,b; Fargione et al. 2021; Seddon 2022; Seddon et al. 2020).

b. There has been rapid growth in private markets for carbon offsets, many from forestry and agricultural practices (see Box 8-2 below). Some private companies have established large-scale experiments intended to verify the negative emissions they sell, like those described later in this chapter. However, these experiments are not particularly useful for safeguarding the public expenditures for carbon sinks in the IRA, because the information is proprietary. Also, concerns persist about the permanence, additionality, and leakage of carbon offsets. Permanence is a concern because a negative emission that is subsequently returned to the atmosphere, resulting in only short-term temporary storage, largely negates the intended mitigation benefit.[5] Additionality means that the offset would not have occurred without the payment to adopt C sequestering practices; incentives are wasted when provided where practice

---

[3] Note: Not all biomass is converted to electricity, and biomass removal can potentially degrade the land sink.

[4] "Nature-based solutions" (NBS) is also frequently used in the same context.

[5] There is still value in temporary storage that spans a few decades, as it could buy time for other more technical or costly sequestration solutions to be developed and deployed.

changes would have occurred anyway (non-additionality). Leakage occurs when an offset in one location simply shifts the emissions to another—for example, when reduced deforestation causes forest clearance in another location because of unabated demand for new cropland and/or timber products. To accurately quantify carbon removal for a given offset, one must know how much additional carbon a project removes from the atmosphere and the magnitudes and timing of any subsequent re-emissions owing to lack of permanence.

c. The market for plant-based artificial meat has grown substantially. Meat substitutes might offer the potential to significantly reduce constraints on GHG mitigation caused by limited arable land, and by competing demands for food and fiber production and biodiversity preservation. The National Academies (2019) reviewed the abundant literature showing that global managed land mainly comprises croplands, grasslands, and forests. Today's croplands and grasslands will be needed to feed increasingly wealthy and numerous humans through midcentury and beyond, given current demand for animal protein. Thus, the land for large-scale additional deployment of land-hungry methods of carbon mitigation, such as forest planting or biofuels feedstock production, must come either from agricultural land or forest. If this land comes from agricultural land, then this could cause food price increases or shortages, which have repeatedly caused political unrest and violence in the past (Bellemare 2015; Calvin et al. 2014; Kreidenweis et al. 2016; Powell and Lenton 2012; Rosegrant 2008; Smith et al. 2013). If it comes from forests, then this will harm biodiversity, particularly in tropical forests, and release carbon currently sequestered in the forests (Law 2022). While the use of degraded agricultural lands would partially mitigate these concerns and could provide additional benefits, the National Academies (2019) were unable to identify a large body of these degraded lands. Meat substitutes have the potential to free up cropland currently devoted to animal feed production to directly feed more people or to provide services in reforestation or sequestration (Santo et al. 2020). Meat substitutes are discussed at greater depth later in this chapter.

d. The National Academies (2019) reviewed the possibility of adding crushed mafic and ultramafic rocks (which are silicate minerals high in base [e.g., Mg, Ca] cations) to agricultural lands, but only as a frontier negative emissions technology. These rocks react with CO_2 from the atmosphere, yielding stable carbonate minerals and providing co-benefits by increasing soil pH, which improves the productivity of acidic soils (Beerling et al. 2020). Many start-up companies have emerged to bring agricultural carbon mineralization to market, and the technology now appears much closer to deployment than it did 4 years ago.

In what follows, the committee analyzes gaps and barriers between the terrestrial carbon sinks (biomass and soils) needed for the United States to reach net-zero emissions by midcentury, and those that might be created by the IRA. Next, the committee analyzes mitigation measures in the IRA that target agricultural-based emissions of methane and nitrous oxide. The committee offers specific recommendations to the Secretary of Agriculture about the implementation of programs in the bill that target carbon sinks through forestry and agricultural soils, and through abatement of methane and nitrous oxide. The committee then addresses the total amount of land needed for the net-zero transition, including with the role of biofuels feedstock production. Last, the committee discusses the potential role of meat and dairy substitutes as a demand-side approach to reducing net GHG emissions from the agricultural sector. Table 8-3, at the end of the chapter, summarizes all the recommendations that appear in this chapter regarding land use in support of decarbonization efforts.

TERRESTRIAL CARBON SINKS NEEDED TO REACH NET-ZERO EMISSIONS AT MIDCENTURY

How large do terrestrial carbon sinks need to be for the United States to reach net-zero emissions by 2050, and for the United States to reach the Biden administration's goal of a 50–52 percent emissions reduction from 2005 levels by 2030? These two goals are nearly the same for a linearly decreasing emissions trajectory, and are treated as equivalent in the rest of this chapter.

The ~6.0 Gt carbon dioxide equivalents (CO_2e) per year of U.S. GHG emissions in 2020 was about 78 percent CO_2, 12 percent methane, and 6 percent N_2O (EPA 2022). Here, CO_2e is calculated using 100-year global warming potentials (GWPs), which means that methane has 25 times and N_2O has 298 times the radiative impact of an equivalent mass of CO_2 (EPA 2022). These GWPs are from the Intergovernmental Panel on Climate Change (IPCC) AR4, which is the current convention for national reporting. Updated 100-year GWP estimates from the IPCC AR6 are 27 for CH_4 and 273 for N_2O.

The current U.S. emissions of 6.0 Gigatonnes CO_2 equivalent per year ($GtCO_2e/y$) do not include a net CO_2 sink of 0.8 $GtCO_2$ from the Land Use, Land-Use Change, and Forestry (LULUCF) sector, which is caused primarily by forest regrowth, probably amplified by CO_2 fertilization of tree growth (EPA 2022). This sink counts as part of U.S. net anthropogenic emissions under United Nations Framework Convention on Climate Change

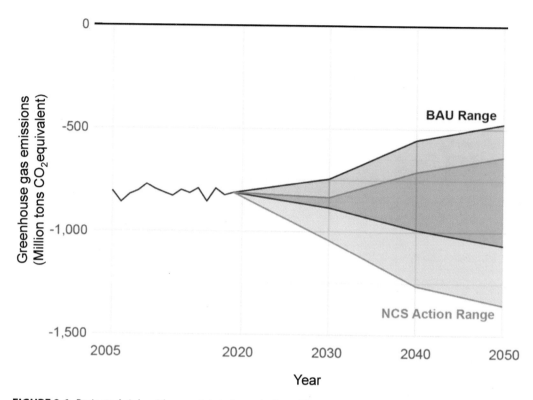

FIGURE 8-1 Projected sink, with uncertainty bounds, from *The Long-Term Strategy of the United States: Pathways to Net-Zero Greenhouse Gas Emissions by 2050*. NOTE: BAU = business as usual; LULUCF = land use, land use change, and forestry; NCS = U.S. National Climate Strategy. SOURCE: Courtesy of Department of State (2021).

(UNFCCC) guidelines.[6] The LULUCF net CO_2 sink has declined from 892 Megatonnes CO_2 per year ($MtCO_2$/y) in 1990 to 812 $MtCO_2$/y today because regrowing forests in the eastern United States are beginning to reach maturity and because of growing losses from wildfire and insect pests primarily in the west (EPA 2022). Furthermore, the current net CO_2 sink from LULUCF is uncertain; EPA (2022) reports 95 percent confidence limits of 648 and 1076 $MtCO_2$e/y. There is also considerable scientific uncertainty about the future sink because of the difficulty of balancing losses that are linked to climate change with gains owing to CO_2 fertilization and recovery from historic land cover change, as well as unknown future forest harvesting, deforestation, reforestation, and afforestation (Figure 8-1). These many uncertainties are also reflected in projections from the peer-reviewed literature (Hurtt et al. 2002; USGCRP 2018).

---

[6] See United Nations Climate Change, "Reporting of the LULUCF Sector Under the Convention," https://unfccc.int/topics/land-use/workstreams/land-use—land-use-change-and-forestry-lulucf/reporting-of-the-lulucf-sector-by-parties-included-in-annex-i-to-the-convention.

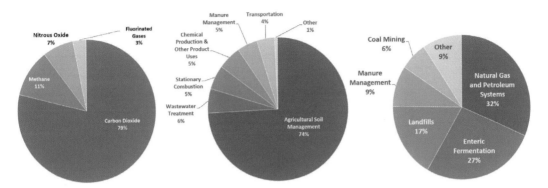

FIGURE 8-2 Greenhouse gas emissions in the United States in 2020. Overview by gas (left) and break-down by source for nitrous oxide (center) and methane (right). For methane, the "Other" category includes rice cultivation, responsible for 3 percent of emissions. SOURCE: Data from EPA (2022).

The net LULUCF GHG sink is estimated to be 759 $MtCO_2e$ because the 812 $MtCO_2/y$ net carbon sink is partially offset by CH_4 and N_2O emissions on forested land of 38 and 15 $MtCO_2e/y$, respectively (EPA 2022). However, the bulk of managed land emissions of N_2O and CH_4 are from agriculture (which is a separate sector in national GHG inventory reporting), comprising almost 80 percent of total N_2O emissions (total of 426 $MtCO_2e/y$; 95 percent confidence range: 342–551) and 39 percent of total CH_4 emissions (total of 650 $MtCO_2e/y$; 95 percent confidence range: 596–724) (Figure 8-2).

Net-zero emissions require a CO_2 sink equal in magnitude to remaining CO_2e positive emissions of all GHGs covered by the UNFCCC: CO_2, CH_4, N_2O and the fluorinated gases. Most global net-zero scenarios assume negative emissions equal to 10–20 percent of today's emissions by the year in which net zero is achieved, which is usually 2050 in 1.5-degree scenarios (Figure 8-3). It is possible to envision extreme scenarios in which the needed sink is much smaller because of curtailed energy demand or much larger because of more residual fossil fuel use and/or higher non-CO_2 GHG emissions. Large energy demand reductions like those assumed in the left-most panel in Figure 8-4 might be difficult to sustain politically in the United States, while a larger sink requirement would entail a large deployment of BECCS, like in the right-most panel, taking more land from food production and biodiversity preservation.

Most net-zero emissions budgets for the United States assume midcentury sinks totaling 1 $GtCO_2/y$ plus or minus several hundred $MtCO_2/y$ from a mix of forestry, agriculture, and industrial carbon removal methods like BECCS and direct air capture (DAC)

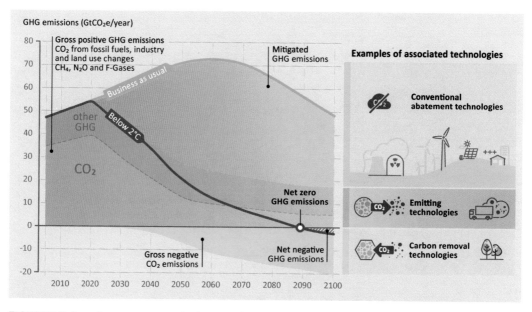

FIGURE 8-3 Sample net-zero scenario showing the role of negative emissions in climate change mitigation. SOURCE: UNEP (2017)/Jérôme Hilaire, Mercator Research Institute on Global Commons and Climate.

(Figures 8-5 and Figure 8-6; Larson et al. 2021; Lempert et al. 2019; White House 2021; Williams et al. 2019).

Because technological options like BECCS and DAC are unlikely to be deployed at levels that would materially affect 2030 emissions, carbon sinks in the United States through 2030 should be thought of as the "business as usual" (BAU) land sink augmented by new policies designed to enhance it. In Figure 8-5, a 2030 land sink of

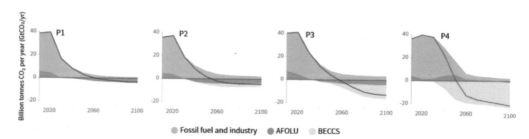

FIGURE 8-4 Breakdown of contributions to global net CO_2 emissions from AFOLU, BECCS, and reduced emissions from fossil fuel and industry in four illustrative model pathways. NOTE: AFOLU = Agriculture, Forestry, and Other Land Use; BECCS = Biomass Energy with Carbon Capture and Storage. SOURCE: IPCC (2018), https://doi.org/10.1017/9781009157940. CC BY-NC-ND 4.0.

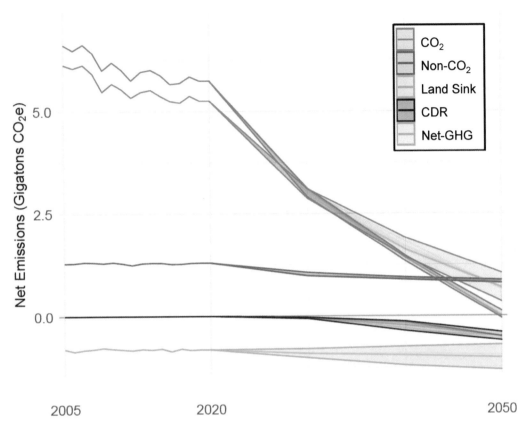

FIGURE 8-5 Net-zero pathway for the United States. NOTE: CDR means carbon dioxide removal by BECCS and other technological methods like direct air capture. SOURCE: Courtesy of Department of State (2021).

940 MtCO$_2$/y is required to keep the United States on the assumed net-zero trajectory (defined here as the midpoint of the range shown, which is the midpoint of the "NCS Action Range" in Figure 8-1). This implies that the 2030 BAU land sink of 750–875 MtCO$_2$/y in Figure 8-1 must increase by 65–190 MtCO$_2$/y, with most 2030 values from other recent analyses falling within this range. For example, in Larson et al. (2021), the 2030 land sink is 750 MtCO$_2$/y on the net-zero trajectory and 600 MtCO$_2$/y on the BAU trajectory, which means that policies must increase the land sink by 150 MtCO$_2$/y. Orvis and Mahajan (2021) produce a 2030 estimate of 112 MtCO$_2$/y. The highest value that the committee has found is an increase of 336 MtCO$_2$/y from Larsen et al. (2021), which combines optimistic assumptions about forestry and agricultural options, and assumptions about the decline of the BAU land sink. Corresponding

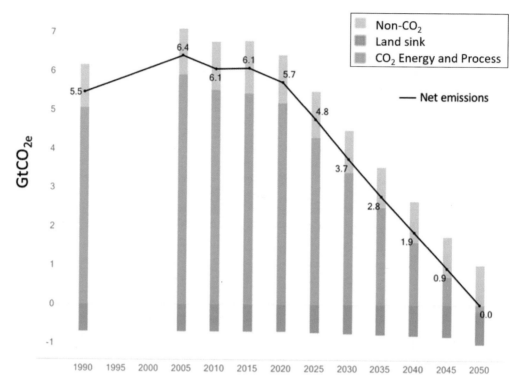

FIGURE 8-6 The net-zero emissions scenario modeled in the Net Zero America Project (NZAP). SOURCE: Larson et al. (2021), https://netzeroamerica.princeton.edu/the-report. CC BY 4.0.

numbers for new LULUCF carbon sinks needed in 2050 range from zero to 495 $MtCO_2/y$, because of the large range of uncertainty in the BAU trajectory in Figure 8-1. In the rest of this chapter, the range from the White House strategy paper is used because it encompasses most published values and reflects current U.S. policy goals.

The IRA covers the next 8 years of the 28 years to midcentury, and so it is useful to consider a linear ramp from zero in 2023 to 65–190 $MtCO_2/y$ at the close of 2030 (sequestering an additional 9.3–27.1 $MtCO_2$ in each year), implying a total 8-year sink of 260–760 $MtCO_2$.

Potential Size of Safe Terrestrial Carbon Sinks in the United States

The National Academies (2019, p. 351) estimated the "safe" capacity for U.S. terrestrial negative emissions, where "safe" was defined as: no "large adverse societal, economic, and environmental impacts." This chapter offers summary statistics about feasible and safe land sinks. Interested readers should consult the National Academies (2019) for a detailed description of forestry and agricultural options on which these statistics are based (including rates of uptake, total capacity and cost for each option, and available U.S. land). Forestry options include avoided deforestation, afforestation, reforestation, and changes in forest management, including tree replanting after natural disturbance or fire, pesticide application, and increased harvest rotation length. Agricultural options include cropland and grazing land management practices that increase the rate of organic matter addition into soils and/or reduce the rate of carbon losses from soil. Relevant cropland practices include use of cover crops, reduced tillage, conservation buffers, perennial crops, rewetting of peat soils, and restoration of marginal lands. For grassland, improved grazing systems and/or amendments to boost plant productivity can increase soil organic carbon stocks. Application to soils of ground silicate minerals, having high base cation (Ca, Mg) content, can sequester inorganic carbon on croplands and grazing land.

The safe limit for new forests from afforestation and reforestation is primarily constrained by judgment about how much agricultural land can be converted to forest before food demand creates destabilizing price increases or deforestation elsewhere (leakage). The safe limit for sinks caused by an increased harvest interval (as a forest management option) is also constrained by the need to prevent leakage, which in this case means reducing harvest in one area can increase harvesting elsewhere to meet ongoing demand for forest products. Other forest management options, like pesticide application to reduce mortality from insect attack, have limited total capacity. The net climate benefit of afforestation can also be affected by changes in land surface reflectance (i.e., albedo) from tree planting—for example, creating a darker vegetated surface will absorb more radiation and have a warming effect that might partially offset the climate benefits from CO_2 removal (NASEM 2019). In addition, all forestry options suffer from concerns about permanence, which limits opportunities for afforestation because of expected losses owing to fire and other factors on historically non-forested land (all landscapes have some degree of net carbon flux at baseline).[7] Usual forest sink contracts in private markets are on the order of decades, which is longer than the 8-year duration of the IRA. This is a potential concern for permanence (White et al. 2018).

The safe limit for agricultural C sequestration is less constrained by land use change concerns compared to forestry, in that most of the management practices to increase soil carbon stocks can be employed on agricultural lands that remain in production for

---

[7] Baselines are benchmarks that represent a one-time fixed benefit. For a meaningful market, additional carbon that is measured relative to the baseline and the absolute carbon flux into the landscape requires separate consideration. Carbon that is not sequestered permanently is at risk and should be considered lower value (Arcusa et al. 2022). See Arizona State University, "Carbon Removal Accounting Methodologies: How to Rethink the System for Negative Carbon Emissions," https://keep.lib.asu.edu/items/170043.

food and fiber. However, some of the highest rates of soil C sequestration do involve conversion of cropland to conservation set-asides (with perennial grass or tree cover), which can result in market-based leakage if set-aside acreage is too high. Hence, use of conservation set-asides should be limited to conversion of marginal cropland, unless other co-benefits are considered to be equally important; this could result in minimal impacts on overall food and fiber production, and reduce leakage risks. In general, there is little concern with respect to leakage related to reductions in non-CO_2 GHGs (i.e., N_2O and CH_4), in that farmers are unlikely to adopt practices to reduce emissions that would result in significantly reduced crop or livestock yields.

Safe U.S. carbon sink size limits from the National Academies (2019) are 150 $MtCO_2$/y for afforestation/reforestation at ≤20 $/$tCO_2$ and 100 $MtCO_2$/y for changes in forest management that promote carbon uptake, also at ≤20 $/$tCO_2$. For agricultural carbon sinks from improved agricultural practices, the National Academies (2019) estimated 250 $MtCO_2$/y at ≤100 $/$tCO_2$, if existing conservation management practices were widely deployed on agricultural lands; at least 50 Mt CO_2/y is achievable at ≤20 $/$tCO_2$ and an additional 100 $MtCO_2$/y at 20–50 $/$tCO_2$, based on published studies (Chambers et al. 2016; Robertson et al. 2022). These limits are also consistent with estimates in other peer-reviewed publications. For example, Fargione et al. (2018) estimated limits and costs for forestry and agricultural soils of 115 $MtCO_2$/y, each at ≤$10/$tCO_2$, and 520 $MtCO_2$/y from forestry and 260 $MtCO_2$/y from agricultural soils at ≤$50/$tCO_2$ (not including BECCS).

Reducing non-CO_2 GHG emissions lessens the size of carbon sinks required to meet overall GHG mitigation targets, and thus potential reductions in agricultural CH_4 and N_2O emission are relevant to include. Supporting material for a recent report by the Environmental Defense Fund and ICF of the technical potential for agricultural and forestry GHG mitigation in the United States contains marginal cost curves for reductions in CH_4 and N_2O emissions from agricultural sources, using current abatement technology, totaling ~90 $MtCO_2$/y at ≤100$/$tCO_2$e (Eagle et al. 2022).

In addition, a safe total for BECCS is 500 $MtCO_2$/y at 20–100 $/$tCO_2$, using only waste biomass from forestry and agriculture and feedstocks produced on lands currently devoted to corn ethanol production (NASEM 2019). Robertson et al. (2022) estimated that BECCS deployment on the 79 Mha of cropland that is marginal or abandoned or used for corn ethanol production would contribute 225–480 (mean of 349) $MtCO_2$/y from biorefinery CCS if the land were in production for cellulosic ethanol. Assuming substitution of electric vehicles for internal combustion vehicles and full biomass utilization for CCS on the same land area would yield 370–807 (mean of 581) $MtCO_2$/y.

Forestry and Agricultural Programs in the IRA

The agriculture and forestry programs in the IRA provide $24.5 billion to help farmers, ranchers, and forest managers adopt conservation practices that enhance landscape resilience and confer other climate benefits.

The bill directs $5 billion to forestry programs, with $2.15 billion for National Forest System restoration and fuels reduction projects (§23001), $2.2 billion for state and private forestry conservation programs (§23003), $0.55 billion for competitive grants for non-federal forest landowners (§23002), and $0.1 billion for administration (§23005). A substantial fraction of the funding is devoted to objectives other than carbon sequestration, although increased carbon sequestration could be a benefit as well. The $2.15 billion program is intended primarily to reduce the risk of wildfire. The $2.2 billion fund is divided into $0.7 billion for the acquisition of land for forest programs, and $1.5 billion for tree planting and related activities.

Tree cover, especially in urban areas and at the urban–forest interface promotes healthy living spaces. Furthermore, better land use planning, including avoiding urban development on mature forest lands and productive croplands, can reduce loss of biomass and soil carbon stocks through both direct and indirect land use change impacts (see Chapter 7). Urban reforestation can also reduce energy costs in cities through shading and evaporative cooling, as well as contributing to the forest C sink. The $1.5 billion spent on tree planting would yield a sink of 15 Mt CO_2e/y, assuming the upper end of the cost estimate ($100/t$CO_2$) for urban reforestation (Fargione et al. 2018); this could also yield energy savings for residential and business heating and cooling. Last, $0.3 billion of the $0.55 billion fund is for either climate mitigation or forest resilience. In addition, a fund of $0.1 billion for the hauling and use of material cleared for fire suppression may create a net carbon sink, but its magnitude and even its sign will depend on details of implementation. The $0.1 billion for administration is a legitimate portion of any expenditure to create carbon sinks or manage fire.

The IRA directs $19.45 billion to agricultural conservation, with the majority as additional funding to four existing U.S. Department of Agriculture (USDA) conservation programs: $8.45 billion for the Environmental Quality Incentives Program (EQIP), $3.25 billion for the Conservation Stewardship Program (CSP), $1.4 billion for USDA conservation easements (ACEP), and $4.95 billion for the Regional Conservation Partnership Program (RCPP) (§21001), with the requirement to prioritize the buildup of soil carbon stocks, the reduction of nitrogen losses and N_2O emissions, and the reduction, avoidance, or capture of CH_4 emissions. These programs could benefit from further assessment of environmental advantages and constraints to achieving a tighter coupling between rural and urban waste streams and potential carbon removals, such as producing compost from food and agricultural waste and then applying as a soil amendment. The IRA also provides the USDA National Resources Conservation Service (NRCS) with $1 billion to provide conservation technical assistance, $0.3 billion to engage in a program quantifying carbon sequestration and GHG emissions in agricultural lands, and $0.1 billion for administrative expenses (§21002). It is possible that during negotiations for the 2023 or 2028 Farm Bills, the IRA funding to the programs could be rescinded and reallocated to Farm Bill baseline funding, meaning they and their climate focus get carried over more permanently in future Farm Bills. This, however, carries a risk that the rescinded money gets allocated away from climate or conservation focused programs entirely.

On top of all of this new funding, the IRA also extends the working lands conservation programs' authorities beyond their current 2018 Farm Bill expiration in 2023 to 2031, annually at $2.025 billion for EQIP, $1 billion for CSP, $0.45 billion for ACEP, and $0.3 billion for RCPP. This, could make them more resilient in case the 2023 or 2028 Farm Bills do not pass on time, as the conservation programs would still be in place.

POTENTIAL TERRESTRIAL CARBON SINKS AND CH_4 AND N_2O ABATEMENT CREATED BY THE IRA

Forestry

The $2.15 billion for fire suppression may not create net carbon sinks. Current U.S. policy to reduce the risk of wildfire focuses on thinning to reduce fuel loads. Thus, there are two opposing forces. Thinning reduces forest carbon stocks, which represents a positive CO_2 emission, but may reduce future fire emissions. As a result, the science is currently unsettled about the sign of the carbon flux caused by thinning treatments to reduce wildfire. Some studies show that thinning reduces forest carbon more than the reduced burning increases it (Ager et al. 2010; Campbell et al. 2012; Chiono et al. 2017; North et al. 2009), while others offer evidence that the reverse is possible (Foster et al. 2020; Hurteau and Brooks 2011).[8]

In contrast, all of the $2.2 billion and $0.55 billion funds target enhanced forest carbon net sinks, with the proviso that $1 billion could also be used to target related

---

[8] Prescribed burning is another option that mimics a low-severity fire more than thinning. Prescribed burns may have fewer consequences than thinning owing to the preservation of large-diameter trees, which are often removed in unregulated thinning practices (Birdsey 2023; Hudiburg 2011; Mitchell 2012). Although there are potential benefits to prescribed burns, in some areas they create wildfire risk, especially when compounded with extreme weather.

properties such as resilience (e.g., to disaster, or severe weather). Also $100 million is directed to the Wood Innovation Grant Program (§23002), whose impact on carbon sequestration is uncertain. Investments to develop wooden replacements for structural steel and concrete in buildings is a promising option for mitigation in the buildings sector (see Chapter 7). If all of this $2.75 billion is used effectively to create net carbon sinks at $20/tCO$_2$, then it will sequester a total of 138 MtCO$_2$ over 8 years. A linear 8-year ramp, like the accumulating sink that would be created by planting a constant amount of new forest each year, would mean a total sink of 34 MtCO$_2$/y in 2030. This is well below the safe total of 250 MtCO$_2$/y from the National Academies (2019) for afforestation, reforestation, and forest management. The implication is that reluctance of landowners to enroll in the program is not the primary concern that it would be with a larger program. Only a small fraction of landowners need respond to the incentives offered by the forestry incentives in the IRA to exhaust all funds.

Agriculture

The largest proportion of the IRA funds for agriculture is devoted to increased funding to USDA programs that incentivize farmers and ranchers to adopt more conservation-oriented management practices that increase soil organic C stocks and lower N$_2$O and CH$_4$ emissions, as well as providing co-benefits such as reduced soil loss, greater biodiversity, and improved water quality, with its associated positive impacts on human health in rural areas. Incentives for these programs come largely in the form of cost share or direct payments for conservation practice adoption as well as technical assistance. A key issue for the longer-term efficacy of these incentives is that they be of sufficient duration to get the practice changes to "stick"—that is, to maximize continuance of the conservation management practice even if subsidies are eventually discontinued. Experience with early adopters show that it can take several years for new management practices to be "perfected" by farmers, for the improved soil and ecosystem functions to be manifested, and for increased profitability for the new practices to be sustained (Bagnall et al. 2021). Once management systems have successfully transitioned, the practice changes can be quite durable along with the increased soil carbon storage, reducing concerns over C sink permanence. Hence, incentive programs need to have sufficient duration to make it through that transition. In July 2023, USDA announced that the Biden administration will support $300 million to quantify emissions from agriculture, to include evaluating some of these new farming and soil management practices.

Currently, adoption rates of conservation practices (e.g., cover crops, nutrient management, tillage reduction, conservation buffers) are increasing, attributable in part to recent increased government subsidies (Zhou et al. 2022). However, the rate and scale of adoption resulting from the planned increase in support payments from the IRA is uncertain. In addition to direct financial incentives, the importance of technical service providers and peer-to-peer farmer networks that can assist farmers and ranchers in adopting locally appropriate conservation practices cannot be overstated.

USDA programs, including incentives to promote conservation practice adoption, have historically underserved smaller producers, producers from Indigenous populations and communities of color, and other disadvantaged producer groups (Horst and Marion 2019). Past discriminatory practices and program design, including lack of access to finance and insufficient provision of technical services (FBLE 2022), need to be addressed to provide equitable opportunities for marginalized producers to participate in USDA programs in general, and specifically in IRA initiatives, and to contribute to improved agricultural sustainability and climate mitigation efforts.

Background on CH$_4$ and N$_2$O Emissions from Agriculture

In the National Academies (2019) report, forestry and agricultural carbon sinks were examined in detail, but the important non-CO$_2$ greenhouses gases CH$_4$ and N$_2$O—of which the agricultural sector is a major source—were not covered. Hence, here the committee provides more background on the specific sources of these gases within agriculture, management practices to mitigate their emissions (see Box 8-1), emissions reduction potential using current technologies, and estimated costs of abatement, in order to include CH$_4$ and N$_2$O in the analysis of potential outcomes from the IRA and related legislation.

Methane emissions are a product of microbial metabolism in anaerobic (low oxygen) environments (e.g., the rumen and animal digestive tract, flooded soils). Agriculture is the second-largest overall emission source (behind the energy sector), accounting for nearly 40 percent of total U.S. methane emissions, mostly from livestock production, including enteric fermentation (27 percent) and manure management (9 percent), and a smaller amount (3 percent) from rice cultivation (Figure 8-2; EPA 2022). Within livestock, ruminant animals (primarily cattle and sheep) contribute the majority of emissions via enteric (digestive tract) emissions as well from the manure they produce (depending on how it is stored and managed).

Soil N$_2$O emissions are a product of the metabolism of microorganisms present in soils, and thus they occur to some extent in all terrestrial (and aquatic) ecosystems. However, the high rates of nitrogen additions to agricultural systems used to boost plant productivity, including synthetic N fertilizers, manure additions, and N$_2$-fixing

---

BOX 8-1
MANAGEMENT PRACTICES TO MITIGATE AGRICULTURAL EMISSIONS OF CH$_4$ AND N$_2$O

Improving feed quality, animal genetics, and health are important for decreasing CH$_4$ emissions per unit product (meat, milk) produced. However, most livestock production in the United States is already quite efficient and there is less room for further reductions in CH$_4$ emission intensity (emission per unit product) with conventional nutritional improvements. Thus, current research to reduce CH$_4$ release from enteric fermentation focuses on the development of new vaccines or feed additives, such as 3-nitrooxypropan, asparagopsis (seaweed), and selective breeding of low-emission animals (Waite et al. 2022). However, these are still emerging technologies with concerns over efficacy, health, and safety. Moreover, some feed additives show initial short-term reductions in CH$_4$ production that later attenuate as the microflora in the rumen adjust to the presence of the additive. How livestock manure is managed (largely linked to whether it is stored in an aerobic or anaerobic environment) affects both CH$_4$ and N$_2$O emissions from microbial processes in the waste. Various storage methods and handling can increase or decrease either CH$_4$ or N$_2$O emissions, but the most effective system to largely eliminate both CH$_4$ and N$_2$O is anaerobic digestion in which the methane is captured and then taken from a sealed reactor vessel and volatile N compounds are fully reduced to N$_2$ gas. The biogas production also represents a renewable fuel that can substitute for fossil-derived methane, thus adding additional value to use of anaerobic digestors to manage manure. Caution should be exercised, however, if use of biodigestors only serves to enable larger concentrated animal feeding operations (CAFOs) that generate more waste. Such trade-offs require further analysis.

Reductions in N$_2$O emissions primarily involve reducing the total amount of reactive nitrogen added to agricultural soils while increasing the nitrogen use efficiency to the plant, thus maintaining yields and also reducing other losses to the environment such as N leaching. This can be achieved by reducing nitrogen-based fertilizer and manure application through methods such as variable rate technology, using nitrogen inhibitors and "slow-release" fertilizers to suppress soil microbial activity, and improving nitrogen nutrient management through methods such as "precision farming" to optimize N efficiency (EPA 2022; Winiwarter et al. 2018).

---

crops (e.g., legumes), have greatly increased N$_2$O emissions in agricultural soils compared to native ecosystems. Currently, about 74 percent of total U.S. N$_2$O emissions are associated with agricultural soil management (Figure 8-2; EPA 2022).

Agricultural emissions of nitrous oxide and methane have been remarkably stable over the past 30 years (averaging 336 MtCO$_2$e/y for N$_2$O and 235 MtCO$_2$e/y for methane), with 2020 emissions of 336 MtCO$_2$e of N$_2$O and 251 MtCO$_2$e of methane (EPA n.d.(a)). Minor increases in emissions are projected by 2030, to 341 MtCO$_2$e of N$_2$O from agricultural soils and 265 MtCO$_2$e of methane and N$_2$O from livestock (EPA 2019). Methane and nitrous oxide have high GWPs, and a rapid reduction of methane emissions has been identified as the single most effective strategy to reduce warming over the next 30 years (White House 2021). As such, agriculture presents a large opportunity to reduce emissions. However, in practice, while we can eliminate the majority of coal- and natural gas–associated methane emissions by substituting for energy from non-emitting sources (see Chapters 10 and 12), agricultural N_2O and methane emissions result from natural biogeochemical processes occurring in soils and in the digestive tract of livestock that are impossible to fully eliminate and difficult to substantially reduce with current mitigation practices without affecting food production systems. Agricultural emissions will therefore come to dominate non-CO_2 GHG emissions, and reducing agricultural emissions will be the major bottleneck in mitigating non-CO_2 GHGs emissions. This highlights the important potential role of a dietary shift away from animal products in enabling a reduction of emissions beyond the technical methods described here (see the section "Meat and Dairy Alternatives" below).

A recent report on a 2030 mitigation pathway for U.S. agriculture identified a technically feasible 2030 target of reducing agricultural methane emissions by 63 $MtCO_2e/y$ (a 25 percent reduction from 2020) and agricultural N_2O emissions by 32 $MtCO_2e/y$ (a 10 percent reduction from 2020), for a net reduction of 95 $MtCO_2e/y$ (Eagle et al. 2022), as part of a total targeted reduction of emissions from the agriculture and forestry sectors of 560 $MtCO_2e/y$. The U.S. Long-Term Strategy uses a similar technical mitigation potential of 70 $MtCO_2e/y$ at ≤$100/tonne for livestock emissions, but just 1.7 $MtCO_2e/y$ for cropland and rice production methane and 8.8 $MtCO_2e/y$ for agricultural nitrous oxide emissions, totaling 80.5 $MtCO_2e/y$ (White House 2021).

Eagle et al. (2022) did a comprehensive meta-analysis of 16 recent peer-reviewed studies on the impact of carbon pricing on the abatement of all U.S. agricultural and forestry emissions via different approaches. They defined marginal abatement cost curves (MACCs) for each approach, which they used to identify cumulative emissions reductions per year at different prices. For compatibility with the soil carbon calculations, abatement potentials are shown for methane and nitrous oxide at ≤$20/t CO_2e, at $20–$50/t CO_2e, and the total at ≤$50/t CO_2e, as well as the percent reduction from 2020 emissions (EPA 2022) and the total cost of said reduction (Table 8-1).

For estimating the total potential soil carbon sink plus N_2O and CH_4 abatement through IRA expenditures for agriculture, the committee uses the potential amounts and per tonne costs from the National Academies (2019) described above for soil carbon, and the amounts and costs from Table 8-1 for N_2O and CH_4 reductions. To be conservative, the abatement costs for amounts available are assumed at <$20/tonne CO_2e, from the marginal abatement curves, to be priced at $20/tonne, and the costs for abatement available at $20–$50/tonne to be priced at $50/tonne. The $19.45 billion expenditure for agriculture in the IRA is more than sufficient to

TABLE 8-1 Abatement Potentials for Methane and Nitrous Oxide

	At ≤$20/t CO$_2$e			At $20–$50/t CO$_2$e			Total at ≤$50/t CO$_2$e		
	Abatement (MtCOe/y)	Reduction from 2020 (%)	Total Cost (billions 2020 USD)	Abatement (MtCOe/y)	Reduction from 2020 (%)	Total Cost (billions 2020 USD)	Abatement (MtCOe/y)	Reduction from 2020 (%)	Total Cost (billions 2020 USD)
CH$_4$	32	12.7	0.64	11	4.4	0.55	43	17.1	1.19
N$_2$O	13	3.9	0.26	6	1.8	0.30	19	5.7	0.56
Total	45	7.7	0.9	17	2.9	0.85	62	10.6	1.75

NOTE: To be conservative, for cost estimates, abatement options at ≤$20/t were priced as $20/t and abatement options in the $20–$50/t class were priced at $50/t. SOURCES: Adapted from Eagle et al. (2022). Copyright © 2023 Environmental Defense Fund. Used by permission. The original material is available at https://www.edf.org/sites/default/files/documents/climate-mitigation-pathways-us-agriculture-forestry.pdf and EPA (2022).

obtain the 50 Mt CO_2/y from increased soil C sinks and 45 $MtCO_2e$/y from CH_4 and N_2O reductions available at <\$20/tonne, which for the 8-year period of the IRA programs would total to 760 Mt CO_2e (95 × 8), at a total cost of \$15.2 billion (760 million × \$20). This suggests that a relatively even balance of incentive funding toward C sink enhancement and toward CH_4 and N_2O abatement is merited for a least-cost option. The estimated opportunity for higher cost (\$20–\$50/tonne) C sink increases (100 $MtCO_2$/y) and CH_4+N_2O abatement (17 $MtCO_2e$/y) substantially exceeds what would remain of IRA funds after investing in the least-cost options, but that remaining funding of \$4.25 billion (19.45 minus 15.2) would be sufficient for an additional 85 $MtCO_2e$ at \$50/tonne, bringing the total abatement potential for the 8-year period to 845 $MtCO_2e$. If ramped up linearly across the 8 years (like that created by annual payments for adopting cover crop or nutrient management improvements), then the annual sink and GHG reduction totals in 2030 would be 211 $MtCO_2$/y. This is well under the "safe" CO_2 removal potential that would not require changes in land use; hence, food production/food security issues would be minimally impacted in the short term and leakage owing to reduced agricultural production should not be a concern.

Adding the expected forestry and agricultural carbon sinks plus N_2O and CH_4 reductions, the IRA could create a net 8-year net sink of 983 $MtCO_2e$ (138 + 845) and a 2030 net sink of 245 $MtCO_2e$/y (34 + 211, assuming a linear ramp) if available funding were effectively deployed at ≤\$50/tCO_2e. These are larger than the new sinks needed for the nation to stay on a linear net-zero trajectory: 260–760 $MtCO_2$ for the 8-year total and 65–190 $MtCO_2$/y in 2030. These estimates are consistent with the one previously published estimate for which the source calculations were available to the committee (Mahajan et al. 2022; 95 $MtCO_2$/y in 2030).

Last, the upper bound of the 0–495 $MtCO_2$/y range for the 2050 LULUCF carbon sink, from the U.S. National Climate Strategy analysis shown in Figure 8-1 (i.e., from the midpoint of the BAU range to the upper and lower limits of the NCS Action Range) is very close to the 500 $MtCO_2$/y safe and practically achievable limit reported in the National Academies' report (2019). Box 8-2 expands on how the markets for forest and agricultural carbon have developed thus far and how they may progress.

In addition to the IRA funding, it is also expected that the USDA Partnerships for Climate-Smart Commodities projects,[9] currently funded at more than \$3.1 billion, will produce learnings about cost-effective methods to reduce methane and nitrous oxide emissions, including improved nutrient management, enhanced efficiency fertilizers,

---

[9] See U.S. Department of Agriculture, "Partnerships for Climate-Smart Commodities," https://www.usda.gov/climate-solutions/climate-smart-commodities.

BOX 8-2
DEVELOPMENT AND TRAJECTORY OF FOREST AND AGRICULTURE CARBON MARKETS

Forestry and agriculture GHG mitigation projects are predominantly associated with voluntary markets,[a] in which buyers (e.g., companies seeking to offset emissions) purchase units of CO_2 removals to (biomass or soil) sinks and/or non-CO_2 GHG reductions, from projects in which farmers/foresters adopt new management practices that increase C sinks and/or reduce GHG emissions. Currently, there are very few agriculture and forestry project types that are eligible for inclusion in compliance (cap-and-trade) markets[b] (e.g., California and the Regional Greenhouse Gas Initiative [RGGI] in the United States).

Recent rapid growth[c] in the use of offsets has been driven by companies pursuing "net zero"/"net negative" emissions policies—in particular, to offset their Scope 3 emissions.[d] Voluntary markets include both registry-certified offsets, with project-specific accounting protocols that account, to varying degrees, for the effects of leakage, additionality, uncertainty, and permanence on the actual abatement achieved, and climate change risks compound uncertainty (Novick 2022; Wu 2023). However, a substantial proportion of voluntary market activity involves direct transactions between buying and selling parties, without registry-certified protocols, and likely less rigor in inventory and accounting procedures. Furthermore, even for carbon market participants that base their projects on registry-certified protocols and measurement and monitoring methodologies, there is no formal standardization across different protocols, which reduces confidence in integrity of the claimed carbon removals (or GHG reductions) and opens the opportunity for selection bias for which protocols project developers might choose.

The rationale for promoting a vigorous voluntary carbon market within the land sector is that adoption of different practices that increase CO_2 removals/emissions reductions incur additional costs and risks to forest and agricultural producers (e.g., delayed forest harvest, added costs for cover crop seed and management on cropland, etc.) that are a key barrier to practice adoption, which revenues from offset sales can help to overcome. Within the IRA, the major funding to forestry and agriculture is through increases in USDA programs to incentivize mitigation practice adoption through direct payments and cost-sharing. However, funding that is available to finance expanded C stock and GHG emission inventory methods and on-farm soil monitoring has the capacity to substantially improve quantification methods and reduce uncertainty in quantifying net CO_2 removals. Moreover, the proposed investments in monitoring systems in the IRA will provide ongoing data on the durability of practice adoption and on practice outcomes, which is critical in addressing permanence and leakage issues particularly in the agriculture sector. Improved carbon and GHG accounting with reduced uncertainties would increase offset buyer confidence and reduce the size of uncertainty "discounts" in offset value that are widely applied by registries, increasing the value of the offset for both buyers and sellers. The expanded inventory and monitoring capabilities recommended in this chapter would contribute significantly toward developing a common set of standards for offset accounting that could allow for

continued

BOX 8-2 Continued

harmonization across different registry systems and offset market participants. More immediately, increased transparency in the tracking and reporting of sources and protocols for offset use by the private sector can encourage these developments. (See the Chapter 11 discussion on strengthening corporate climate disclosures.)

---

[a] Prominent registries that support voluntary markets include Verra Registry, Climate Action Reserve (CAR), American Carbon Registry (ACR), the Gold Standard, and others (Stubbs et al. 2021).

[b] For compliance cap-and-trade markets (e.g., California and RGGI in the United States), some forest practices are allowed to generate CO_2 offsets, whereas only CH_4 abatement practices (from manure management and rice) are currently included as agricultural-based options.

[c] Globally, annual offset issuances in voluntary markets (including all energy and emission sectors) grew sixfold (46 $MtCO_2e$ in 2017 to 239 $MtCO_2e$ for the first 8 months of 2021), with forestry and land use making up 45 percent of the total market volume in 2021 (Ecosystem Marketplace 2021).

[d] Scope 3 emissions are emissions that are outside the direct control of a company, resulting from the use/consumption of the company's product.

---

manure management, feed management to reduce enteric emissions, and soil amendments like biochar. These learnings would then need to be promoted through current USDA conservation programs, or through new policies or laws, to be broadly applied and result in emissions reductions at scale.

Forestry Carbon Sink Findings and Recommendations

Finding 8-1: The $5 billion of forestry funding in the IRA is theoretically sufficient to create additional terrestrial carbon sinks as large as those in simple net-zero trajectories. There is sufficient land available to stay well within the safe limits proposed by the National Academies (2019), and to avoid likely impediments caused by low landowner adoption rates. However, the actual performance of the terrestrial sink provisions in the IRA is still highly uncertain for several reasons, including unpredictable climate, demand for fuel and food, and changes in afforestation, reforestation, and development.

The size of the terrestrial sink needed to reach net zero is uncertain by more than a factor of 2 because of uncertainty about both the magnitude of residual emissions that will need to be offset and the fate of the current (BAU) U.S. forest sink. Moreover, the realized price per tonne of CO_2 that will be sequestered by the constellation of programs in the IRA is also highly uncertain.

Afforestation, reforestation, and reduced development-driven deforestation are the options most certain to create large net carbon sinks locally. While afforestation and reforestation may promote leakage, this is less likely if marginal land is converted to forest. Also, the relatively modest size of the $2.75 billion in funding that could promote these options reduces the danger of a reduction in agricultural land that would promote leakage. Changes in forest management such as increasing the length of the interharvest interval may create local sinks. However, an important recent paper used remote sensing methods to compare 37 forest management projects in California designed to enhance carbon sinks with matched controls and found no additional carbon storage (Coffield et al. 2022). Additionally, longer harvest rotations may reduce wood supply and increase incentives for additional harvest elsewhere.

A substantial fraction of the funding in the IRA that could promote carbon storage can also be directed toward other ecosystem benefits like "resilience." Moreover, the IRA does not specify how landowners will be compensated or the length of contracts. Its funds must be spent within 8 years, which is shorter than the 20 years or more in most forestry offset contracts. These factors add to the uncertainty about how much carbon the bill will cause to be sequestered.

For these reasons, the committee has two recommendations for the Secretary of Agriculture, who is tasked with implementing the forestry provisions in the IRA.

Recommendation 8-1: *Convene an Expert Group to Recommend Ways to Measure Additional Forest Sinks.* **Using the $100 million in administrative funding in the Inflation Reduction Act (IRA) for forestry efforts, the Secretary of Agriculture should convene an expert group of foresters, ecologists, and accounting experts to recommend a way to measure the sizes of forest sinks created by the IRA funding and how they change over time. Their recommendations should then be implemented. Most important is to establish appropriate control areas for a statistically representative sample of treatment areas within the first 2 years. Treatment and control areas should be resurveyed 5 years later. This, together with the Forest Inventory and Analysis census, should inform any needed revisions to management recommendations for increasing the U.S. forest sink beyond 2030 (pending subsequent legislation).**

Recommendation 8-2: *Prioritize Ecosystem-Level Carbon Storage.* **The Secretary of Agriculture should prioritize increased ecosystem-level carbon storage when allocating the $2.75 billion of non-fire-suppression funding in the Inflation Reduction Act, and especially reduced gross deforestation at the urban–forest interface, as well as increased reforestation and afforestation. When sites are selected for reforestation and afforestation, priority should be given to productive but agriculturally marginal land and degraded land, which will reduce the potential to cause deforestation elsewhere; lands with a high risk of non-permanence should be avoided. A committee of experts should review new expenditures annually to evaluate effectiveness.**

Agricultural Carbon Sink and CH$_4$ and N$_2$O Findings and Recommendations

Finding 8-2: If allocated efficiently, the $19.5 billion of funding in the IRA for agricultural conservation directed at GHG mitigation appears to be adequate to provide the needed contribution by 2030 from soil C sinks on cropland and grazing land and from reduced agricultural CH$_4$ and N$_2$O emissions, for the net-zero trajectories described above. However, the actual performance of proposed measures for soil C sink enhancement and GHG abatement is uncertain for a number of reasons, and the amount of agricultural methane and nitrous oxide emissions that the IRA will avoid is highly uncertain and depends on priorities for which conservation practices are incentivized, the level of farmer/rancher participation and practice adoption, and the actual cost and efficacy of emission reduction practices. Additional soil C sink and GHG abatement could come from transitioning grain-based biofuels to cellulosic feedstocks from perennial biomass plantings. There is sufficient land available upon which to adopt new practices, and a number of farmers are successfully transitioning to more regenerative/conservation practices.

Improved monitoring and GHG inventory capabilities will be crucial to track adoption rates and durability of carbon sequestering/GHG reducing practices and whether the projected C sinks and N$_2$O and CH$_4$ abatement are achieved. Different management practices to mitigate agricultural emissions of CH$_4$ and N$_2$O vary in their effectiveness. Similarly, the efficacy of different conservation practices to store C and achievable per area rates varies by climate and soil type, the baseline management system, and other local factors. Thus, the results from providing subsidies to incentivize agricultural producers to adopt new practices will depend highly on the geographic distribution of the land area involved and the types of incentives. Additionally, rates and uncertainty bounds of soil C accrual for some practices in certain systems and geographies are poorly known. Last, the long-term durability of practices change is key to ensuring long-term maintenance of increased soil C stocks. Better understanding of the economic sustainability of regenerative/conservation agricultural practices is needed, along with improved outreach and training to assist farmers and ranchers in successfully adopting regionally tailored technologies.

The committee has four recommendations for the Secretary of Agriculture, who is tasked with implementing these programs.

Recommendation 8-3: *Establish a Permanent, National-Scale, High-Quality Soil Monitoring Network.* **The U.S. Department of Agriculture (USDA) should establish a permanent, national-scale, high-quality soil monitoring network, as an augmentation of the National Resources Inventory (NRI) system**

maintained by USDA. This would allow for periodic (re)measurement of soil C stocks and soil heath indicators within a subset (e.g., 10,000–20,000 points) of the 400,000 NRI points on U.S. agricultural lands. The current NRI system collects valuable data on land use and management practices at these locations, but does not include the on-site soil measurements that are needed for a robust monitoring program. This system would be analogous in function to the Forest Inventory and Analysis system for the nation's forests, which entails periodic measurement of forest biomass and soils, carbon, and forest condition. Measurements of soil carbon stocks and other key soil indications at agricultural NRI points would reflect actual on-farm conditions, and could be done on a rotating basis (e.g., 1,000 points per year, on a 10-year remeasurement cycle), at low cost (e.g., $5 million–$10 million per year), and would provide the single most valuable data source to improve accuracy and reduce uncertainty in regional to national-scale quantification of soil C stocks.

Recommendation 8-4: *Build Out Long-Term Agricultural Field Experiments.* The U.S. Department of Agriculture (USDA) should provide funding to further build out the USDA-funded network of agroecosystem sites at land grant universities and USDA Agricultural Research Service locations, or similar long-term field experiments, to fill the gap in the existing long-term experiment site network to include agroecosystems in underrepresented cropping systems, regions, climates, and soil types, and to expand the measurement of non-CO_2 greenhouse gases (GHGs), particularly soil N_2O and enteric CH_4 emissions on both existing and new sites. These field sites are designed to quantify impacts of adopting conservation/regenerative management practices and contrasting their performance to conventional systems. The expanded network would have a coordinated set of measurement protocols and methods, including whole-system carbon balance, utilizing eddy covariance instrumentation, and other GHG (e.g., N_2O and CH_4) flux measurements. The data provided would be of great value in improving capabilities to accurately quantify decarbonization and GHG emission reduction capabilities on the nation's agricultural lands.

Recommendation 8-5: *Fund Research to Quantify Indicators That Influence Adoption of Regenerative Agriculture Practices.* The U.S. Department of Agriculture should fund research and survey efforts to quantify economic cost/benefit, social barriers, and equity issues influencing adoption of conservation or regenerative agricultural management practices. Considerable anecdotal evidence suggests that there are reduced input costs and higher net returns for many farmers adopting conservation/regenerative practices. However, broader and more targeted economic and behavioral analysis is needed to project the potential for up-scaling of soil-based carbon removal and greenhouse gas reduction approaches as climate mitigation strategies. In addition, economic and behavioral research need to investigate approaches to improve policy design and address inequities in participation of disadvantaged communities in conservation incentive programs.

Recommendation 8-6: *Incentivize the Abatement of CH_4 and N_2O Emissions and Improve Soil Carbon Sequestration.* **The U.S. Department of Agriculture (USDA) should split the $19.45 billion in Inflation Reduction Act agricultural conservation funding between incentive payments to abate CH_4 and N_2O emissions and incentive payments to improve soil C sequestration. Incentive payments should also address inequities in participation by disadvantaged communities, in line with stated USDA policies for expanding the delivery of conservation assistance to low-income and marginalized racial/ethnic communities. To maximize climate and environmental benefits, incentives should target performance (and not strictly acres enrolled) and prioritize practices that can be adopted at scale and that have the potential to continue longer term, by improving soil health, improving yield stability, and reducing input costs. Additionally, practices should be encouraged that can achieve both C sequestration and reduced CH_4 and N_2O emissions, as well as improve other environmental co-benefits. Such practices include conservation buffers, within field set-asides of unprofitable "patches" and precision nitrogen management.**

Recommendation 8-3 could be supported under the $300 million funding in the IRA that specifically targets improvement in carbon and GHG inventory on managed lands, and Recommendation 8-4 could also be supported through the $300 million GHG inventory funds and/or via USDA's National Institute of Food and Agriculture (NIFA) funding that supports current agroecosystem sites. Funding for economic and behavioral studies in elucidating constraints and outcomes for adoption of regenerative agricultural management practices in Recommendation 8-5 could be prioritized within current funding of USDA's Economic Research Service and in research funding through USDA's NIFA and possibly be included in other land use related funding lines in the IRA.

LAND REQUIREMENTS

The 915 million hectares (Mha) in the United States (765 Mha outside Alaska and Hawaii) is divided primarily among grazing lands (265 Mha), cropland (159 Mha), and forests (256 Mha), with the rest amounting to only a few percent outside developed land and Alaskan taiga woodlands and tundra (USDA 2017). This shows why large-scale reforestation or afforestation would have to come from food-producing lands and would thus bring with it the dual risks of destabilizing food price shocks abroad and increasing deforestation elsewhere to meet ongoing demand for food (leakage) (NASEM 2019; White House 2021). Similarly, any significant increase in biofuels feedstock production would incur these same risks because it must come at the expense of lands currently devoted to food or forest products. Moreover, clearing forest to produce biofuels feedstocks would cause an initial emission in the United States of approximately 100–300 tCO_2/ha from the decomposition or combustion of forest trees (Fargione et al. 2008; Hoover 2021; Law 2018). This "carbon debt" would require decades to be repaid by the production of new biofuels that displace fossil fuels, implying decades of net positive emissions, which are not congruent with a 30-year approach to net zero.

For these reasons, this report follows the recommendation of the National Academies (2019) to limit (1) biofuels feedstocks to crop and forestry residues, biomass waste, and marginal cropland and croplands already devoted to biofuels (16 Mha of corn devoted to grain-based ethanol production); and (2) changes in forest area to a few percent of the total. As benchmarks, from 1977 to 2012, U.S. forests increased in area by approximately 10 Mha (USDA 2014), which is about 3 percent of the current U.S. forest total; similarly, croplands currently kept out of production by the Conservation Reserve Program total approximately 10 Mha, which is about 6 percent of the U.S. cropland area.

With biofuels restricted in this way, the large new land use areas needed for the energy transition are for renewable electricity infrastructure and new forest carbon sinks (reforestation and afforestation). Sinks from changes in forest or agricultural management do not require land use change. Engineered sinks, such as direct air capture (DAC) would require large amounts of land, but this is a problem for the final part of the energy transition, if DAC becomes cheap enough to deploy at scale. And if it is deployed at scale, most of the land required for a DAC installation could still be used for agriculture, just like in wind farms on agricultural land. To first order then, the challenge during the next 30 years is to site all the needed wind and solar electricity and carbon sinks, while retaining the nation's capacity to produce essential food and fiber and to preserve forest biodiversity.

Terrestrial Carbon Sinks

Technically achievable "safe" totals from the National Academies (2019) are again 150 $MtCO_2$/y from new forests, 100 $MtCO_2$/y from changed forest management, 250 $MtCO_2$/y from agricultural soils and 500 $MtCO_2$/y from BECCS. Using the midpoint of the range of new forest sequestration rates from this study (2.6–23.5 tCO_2/ha/y), attaining the "safe" upper bound for afforestation/deforestation would require conversion of 11.5 Mha of agricultural land, which is 2.7 percent of the total area (of cropland plus grazing land). Changed forest management would be needed on 70–90 Mha of existing forest lands (NASEM 2019).

Renewable Electricity

There are a significant number of studies on the land required for the wind turbines, solar panels, transmission lines, and support infrastructure of a net-zero energy system in the United States. An excellent entry to this vast literature is Saunders (2020): *Land Use Requirements of Solar and Wind Power Generation: Understanding a Decade of Academic Research*. Peer-reviewed studies virtually all conclude that the United States (Jacobson et al. 2015; Li et al. 2018) and most other countries (Calvert 2018; Capellán-Pérez et al. 2017; Jacobson and Delucchi 2011) have sufficient suitable land for renewable electricity infrastructure, including supporting roads and other equipment, even with the complex existing legal restrictions in the United States at federal, state, and local levels. Published studies also generally conclude that public resistance to new renewable infrastructure will significantly limit the pace of deployment and represents the most serious impediment to a 30-year energy transition (Saunders 2020). For example, only one-ninth of new renewables projects currently reach completion in the United States, after an average completion time of 5–8 years (Jenkins et al. 2021). These statistics are inconsistent with the pace of deployment needed in net-zero scenarios. Chapters 5 and 6 itemize barriers to renewable infrastructure and offers policy recommendations to speed deployment.

The most complete and fine-grained analysis of the land required by renewable electricity and transmission is offered by the Net-Zero America study (Larson et al. 2021), which planned a half-dozen net-zero electricity generation and transmission systems on a 4×4 km grid over the United States, using constrained least-cost algorithms. Deployment was limited by two different sets of constraints, including more than 50 factors such as high population density, legal prohibitions, Native American jurisdiction, high conservation value, low site suitability (i.e., wetlands, too steep, too little wind), incompatible prior infrastructure (i.e., airports), high agricultural productivity, and high to moderate sensitivity under the Bureau of Land Management's designation system (Figure 8-7). The net-zero energy systems ranged from one with 100 percent renewable electricity, through a least-cost system allowing nuclear and fossil with CCS, and to a system that limited the pace of renewables deployment to the maximum historic rate. The study used National Renewable Energy Lab estimates

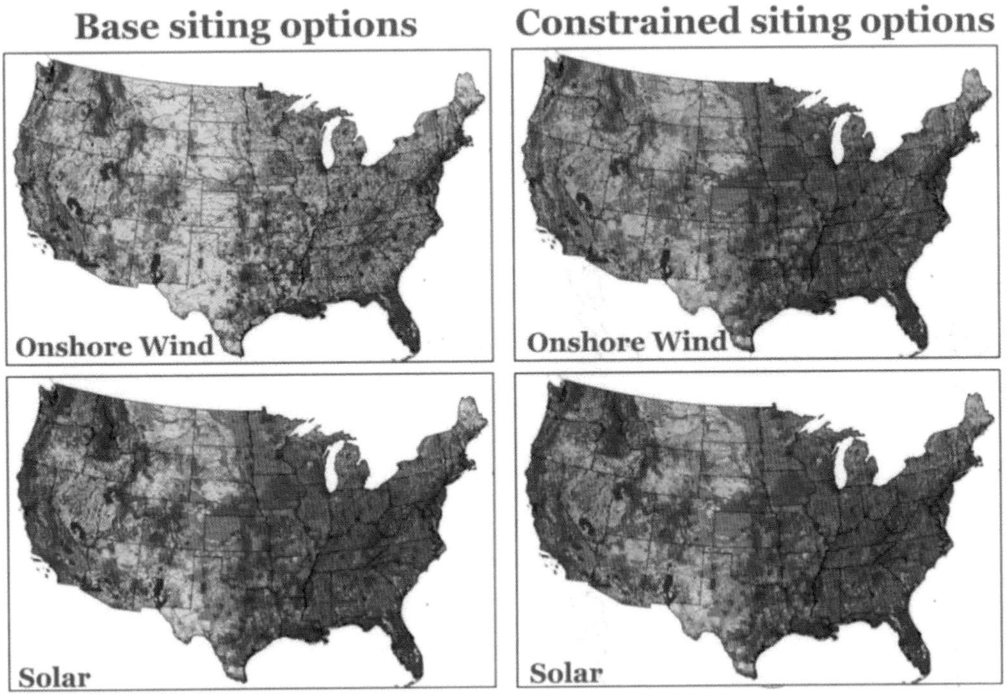

Base siting options

Constrained siting options

FIGURE 8-7 Land required by renewable electricity and transmission in NZAP. NOTES: The green regions are excluded from development. The light gray regions are suitable for siting projects (candidate project areas). SOURCE: Larson et al. (2021), https://netzeroamerica.princeton.edu/the-report. CC BY 4.0.

of average power density: 45 MW/km^2 for solar, and 2.7 MW/km^2 for onshore wind. It separately estimated the land area that would be unavailable for joint use (i.e., the area actually occupied by the wind turbines and associated access roads) and the area that could also be used for other purposes (i.e., croplands beneath a windfarm), under the assumption that the visual transformation of the landscape is the most likely source of public resistance to deployment. The study found that there is technically enough land for any of the scenarios considered, including the one with 100 percent renewables (Table 8-2), although that scenario required 90 percent of available land in the case with the strongest constraints on deployment. The large amount of visually transformed land from wind and solar in 2050—58.9 Mha for the cheapest scenario and 105.9 for 100 percent renewables—indicates a large potential for public resistance to deployment, as discussed extensively in Chapter 5.

Increased solar and wind deployment will create conflicts in land use, conservation acts, and land ownership status, affecting local communities. Promoting and

TABLE 8-2 Land Use Scenarios in NZAP

Scenario	Source	Year	GW	Million Hectares Direct Use	Mha Visually Transformed	Percent of Available Base (Stronger Constraints)
Cheapest	Wind	2030	414	0.2	15.7	4 (14)
Cheapest	Wind	2050	1,479	0.6	55.1	12 (45)
Cheapest	Solar	2030	320	0.7	0.8	<1 (1)
Cheapest	Solar	2050	1,495	3.5	3.8	1 (4)
100% Renewable	Wind	2030	462	1.7	17.4	4 (16)
100% Renewable	Wind	2050	2,700	10.0	100.3	22 (90)
100% Renewable	Solar	2030	402	2.5	2.5	<1 (1)
100% Renewable	Solar	2050	2,750	5.6	5.6	2 (6)

SOURCE: Larson et al. (2020), https://netzeroamerica.princeton.edu/img/Princeton_NZA_Interim_Report_15_Dec_2020_FINAL.pdf. CC BY 4.0.

incentivizing dual use of land for distributed energy resource (DER) infrastructure can help to utilize formerly contaminated areas or existing infrastructures, reducing the extra burden of construction, and can even economically revitalize communities. This can be implemented through various approaches such as "Solar pollinator"—for example, Oregon Community Solar Program (Oregon Solar + Storage Industries Association n.d.); "Agrivoltaics" (Fraunhofer Institute for Solar Energy Systems ISE 2020); "Floatovoltaics" (YSG Solar 2022); "Canal top" (DOI 2016); "Bridge-top"—for example, Blackfriars Bridge, London (Thameslink Programme n.d.); and many more.

Required land for electrical transmission lines is reviewed in Chapter 6. In the NZAP interim report, this was zero because the study restricted transmission to existing rights-of-way (Larson et al. 2020).

Biofuels Feedstocks

Biofuel production in the United States currently consists of approximately 15 billion gallons per year of corn ethanol, which is about 10 percent of gasoline consumption, and 2.5 billion gallons of biodiesel (including renewable diesel), which is about 5 percent of diesel consumption (USDA 2023). Corn ethanol was once subsidized but is now supported by a 15-billion-gallon annual mandate (Energy Policy Act of 2005, P.L. 109-58; Energy Independence and Security Act of 2007, P.L. 110-140), while biodiesel is supported by a $1/gallon tax credit (DOE n.d.(a)). Corn ethanol production currently requires fossil energy and yields substantial N_2O emissions from fertilizer use, such that ethanol from corn grain produces 56 percent the emissions of gasoline per unit energy (Scully et al. 2021). Biodiesel is more greenhouse efficient, with 74 percent lower emissions than conventional diesel (DOE n.d.(b)). Biofuels are marginally more expensive than fossil fuels in most locations, but within the historic range of fossil prices (DOE 2022a).

Advanced cellulosic fuels, which use most of a plant's biomass instead of just the starches and sugars in corn kernels, can have emissions as low as 10–15 percent of fossil fuels (DOE n.d.(c); Rinke Dias De Souza et al. 2021). U.S. production of these advanced "cellulosic" fuels is only about 10 million gallons per year (EIA 2018). Such cellulosic fuels can also be produced from high-yield perennial crops like switchgrass or *Miscanthus* and from annual crop residues (Schmer 2008). The amount of biofuel that can be produced from an acre varies considerably depending on the crop—the ability to utilize cellulose could raise the per-acre yields from ~100–600 gallons to more than 1,000 gallons.

Perennial bioenergy crops provide co-benefits including enhanced nitrogen and water use efficiency, and some studies have reported soil carbon sequestration of 0.5–1.0 tCO_2/ha/y (Glover et al. 2010; NASEM 2019). Perennials also offer the highest yields. A hectare of corn grain ethanol produces about 3,000 liters of fuel (~3,000 l/ha/y, Achinas et al. 2019), whereas advanced biofuels can have considerably higher yields—perhaps up to 10,000 l/ha/y (Coyle 2007).

As discussed in the previous section, biofuels are constrained by available land and are assumed in this report to be limited to feedstocks from biomass waste and feedstocks produced on the 16 million hectares (DOE 2022b) currently used for corn ethanol. Note that at the upper limit of cellulosic fuel yields, these lands could produce 40 billion gallons per year, which is over twice U.S. aviation fuel consumption (EIA 2019).

Biomass energy with carbon capture and storage (BECCS) can produce electricity or fuels with negative emissions (NASEM 2019). When used to produce hydrogen, BECCS provides (1) a source of industrial heat; (2) hydrogen transport fuels for fuel cell–powered heavy trucks or perhaps for aviation; and (3) a carbon sink. The combined greenhouse benefit of a carbon sink and a zero-carbon fuel might approach double that from direct biofuels combustion (NASEM 2019). With yields of 10–20 tons of perennial biomass per hectare per year on the 16 million hectares currently devoted to corn ethanol and 90 percent CO_2 capture, BECCS would produce a carbon sink of 264–528 tCO_2/y (assuming half the biomass is carbon). The National Academies (2019) estimated an upper bound of 500 $MtCO_2$/y for the practically achievable U.S. carbon sink that could be produced using only biomass waste in BECCS. Thus, the upper bound including both biomass waste and dedicated energy crops on current corn ethanol lands ranges from 0.75 $GtCO_2$/y to just over 1 $GtCO_2$/y. This implies that BECCS alone might fill the need for carbon dioxide removal from industrial (e.g., BECCS, DAC) and other non-LULUCF methods (White House 2022) (gray CDR range center in Figure 8-5, which has a midpoint of about 500 $MtCO_2$/y).

Before electric land transport became economically competitive with vehicles powered by internal combustion engines, biofuels were seen as a primary means of decarbonizing road vehicle transport (Pacala and Socolow 2004). However, biofuels are now seen primarily as an energy source for low-carbon aviation and other sectors that require the high energy density of hydrocarbons. Most current net-zero scenarios assume little emissions reduction from biofuels before 2030 (Larsen et al. 2021; Larson et al. 2021; White House 2021) and heavy deployment of BECCS in the 2030s and 2040s (IPCC 2022). The optimization algorithms in NZAP (Larson et al. 2021) would deploy substantial BECCS hydrogen in the 2040s because the energy economy requires larger carbon sinks than are available from agricultural soils or forestry to reach net-zero emissions by 2050, and also needs zero-carbon fuels. As in this report, most scenarios in NZAP limited biomass production to waste and corn ethanol land, which caps available biomass at 0.7 Gt/year. Demand for BECCS was high enough by 2050 in NZAP to consume all 0.7 Gt/year, and to consume the 1.3 Gt/year allowed in the study's high biomass scenario. This underscores the risk of an unconstrained rush to biomass that might disrupt food production and biodiversity.

Because hydrocarbon biofuels are burned in internal combustion engines or turbines and must be produced in chemicals plants, and because the agricultural production of crops for biofuels itself causes air pollution that results in approximately 2,000 air quality–related deaths in the United States each year (Domingo et al. 2021), these fuels entail the same environmental justice concerns as fossil fuels, except for their greenhouse benefits. Environmental justice concerns thus reinforce the need to electrify transport as soon as possible (including hydrogen fuel cell heavy trucks). These environmental justice concerns are discussed extensively in Chapters 3 and 9. On the other hand, biofuels offer employment and revenue in rural communities, many of which struggle economically. However, BECCS hydrogen or electricity could provide much the same benefits.

The IIJA includes $500 million for biofuels infrastructure (Section 11401), while the IRA contains about $6 billion of biofuels incentives (BPC 2022), a $6.5 billion total investment mostly devoted to biofuels production. The IRA extends the current $1/gallon tax credit for biodiesel until 2024, and then replaces it with a credit of $0.2/gallon for biofuels with 50 percent greenhouse efficiency, like current biodiesel (where greenhouse efficiency equals the percent of the GHG emissions of an energy-equivalent amount of fossil fuel). This credit increases linearly with the greenhouse efficiency of the fuel, until it reaches $1/gallon for fuels with zero emissions. The IRA also includes a tax credit for aviation biofuels that starts at $1.25/gallon for 50 percent greenhouse efficiency and increases to $1.75 gallon for fuels with zero greenhouse emissions (BPC 2022).

BECCS is also implicitly subsidized by the Carbon Capture and Sequestration Tax Credit (45Q) incentives for CCUS ($85/tCO$_2$ for CCS and $60/tCO$_2$ for CCU), and biohydrogen by the New Clean Hydrogen Production Tax Credit (45V) incentives (ranging from $0.6/kgH$_2$ to $3/kgH$_2$ depending on the greenhouse intensity of the fuel). These incentives are large enough when combined with other existing incentives, such as California's low carbon fuel subsidy, to make some biofuels highly cost competitive (Cheng et al. 2023). However, other analyses of the bills see relatively little expansion of biofuels production during the 2020s because of the time and capital required to scale up production, the uncertainty of future demand given the progressive electrification of transport, and uncertainty about federal and state incentives (Jenkins et al. 2022; Mahajan et al. 2022).

The IRA, IIJA, and CHIPS do not include funding for BECCS demonstration projects, and DOE has not yet announced a specific plan to develop BECCS. This is a critical shortcoming, given the system complexity of BECCS and the potential need for relatively cheap and efficient BECCS after 2030. Also, Recommendation 9-6 in Chapter 9 calls for increased research on advanced liquid fuels in general, given the long-term needs of aviation and other models of transport that prove difficult to electrify economically.

> Finding 8-3: The $6.5 billion of biofuels-related funding in the IRA and IIJA is sufficient for biofuels production in the 2020s and for research and development (R&D) so that advanced land transport and aviation biofuels are ready after 2030. However, there is not currently a sufficient comprehensive plan to develop BECCS.

> Finding 8-4: The need for carbon sinks and net-zero chemical fuels during 2030–2050 would likely cause a rush to biomass production that would decrease agricultural and forest land. This could impact food prices around the world, with implications for political stability, and could cause deforestation with adverse effects on biodiversity and increased carbon emissions.

> **Recommendation 8-7:** *Release a Comprehensive Research, Development, Demonstration, and Deployment (RDD&D) Program for Biomass Energy with Carbon Capture and Storage (BECCS).* **The Department of Energy should complete and release a comprehensive RDD&D program for BECCS based on the recommendations in the 2019 National Academies' report** *Negative Emissions Technologies and Reliable Sequestration: A Research Agenda.* **This should include both biomass to electricity and biomass to fuels.**

EMISSIONS AND LAND USE RELATED TO FOOD AND DIETS

Reducing Food Waste

Food waste is also a major environmental and economic challenge, with approximately one-third of U.S. food never eaten, representing a significant waste of resources. The GHG emissions embodied in this food waste amount annually to 170 Mt CO_2e (excluding landfill emissions, where food is the single most common material and decomposes into methane) (Read et al. 2020). The majority of these emissions occur during the process of growing and harvesting food, so source reduction (preventing food overproduction and subsequent diversion to another use) is a more effective method to reduce emissions than methods that recycle food into another use such as animal feed, compost, or anaerobic digestion (EPA 2021). Because roughly half of food waste in the United States occurs at the consumption stage (in households and restaurants), and because embodied emissions accumulate as food progresses through the supply chain, reducing food waste at the consumption stage can result in the greatest emissions reductions. Pathways to reduce food waste at the consumer level include (1) changing the U.S. food environment to discourage waste, (2) strengthening consumers' ability and motivation to reduce food waste, and (3) leveraging and applying research and technology to support consumers in food waste reduction (NASEM 2020). Additionally, reducing waste of animal products such as dairy and beef—whose production emits a relatively greater amount of GHGs per unit of food—can result in greater reductions of emissions than reducing a similar amount of other foods (EPA 2021). EPA has identified addressing food waste as a priority area for reducing GHG emissions, with a national goal to halve food waste by 2030, which could reduce food sector emissions by roughly 10 percent. While the nation has not yet made significant progress toward this goal, if it were achieved, emissions from the food system could be reduced by 92 $MtCO_2e/y$ (EPA 2021; Read et al. 2020), so efforts to make progress on reducing food waste could be a part of national net-zero emissions strategy.

> Finding 8-5: Reducing food waste could be an important way to reduce food sector GHG emissions more broadly. If food waste were halved, up to 10 percent of emissions from the food system could be mitigated.

Land Use Requirements and Emissions Related to Animal Agriculture

Multiple indicators need to be considered in assessing dietary sustainability (e.g., overall nutritional quality, GHG emissions, water and land use, economic cost, health), and most of these indicators are dependent on the type of food produced and the efficiency of production (Chen et al. 2019). It follows that the composition of the human diet is inextricably linked to all of these indicators, and significantly, environmental impact. Even if fossil fuel emissions were to be eliminated entirely, current trends in food systems might prevent the achievement of the 1.5 degree Paris Agreement target (Clark et al. 2020). Animal agriculture specifically accounts for 5 percent of global carbon dioxide emissions, 37–44 percent of global methane emissions, 44 percent of nitrous oxide emissions, and 75–80 percent of total agricultural emissions (FAO et al. 2021). Furthermore, approximately 50 percent of global habitable land is currently used for agriculture; however, that percentage varies by country depending on dietary patterns, which are strongly correlated to a country's gross domestic product (Alexander et al. 2016). Of the finite amount of land available for agriculture, 77 percent (~4 Bha) is in use today for animal agriculture, either for animal feed production or grazing (Figure 8-8). This percentage will only grow as the global population continues to grow, so reducing per capita consumption of livestock, meat, and dairy both globally and in the United States will be critical to stay within both the arable land and emissions budgets for the agriculture sector (Willet et al. 2019).

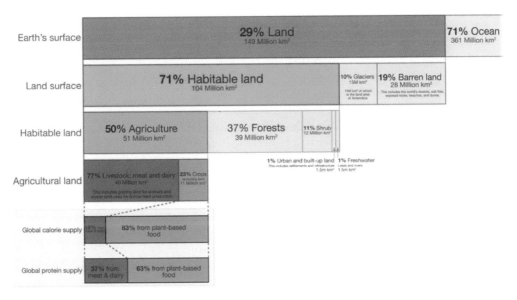

FIGURE 8-8 Global land use for food production. NOTE: Of land available for agriculture, 77 percent (~4 billion hectares) is in use for animal feed production or grazing. SOURCE: Ritchie (2017), https://ourworldindata.org/agricultural-land-by-global-diets. CC BY 4.0.

In the United States, approximately 75–80 percent of protein consumption comes from animal sources, making up only 18 percent of total calorie consumption. Furthermore, within diets in the United States, meat and dairy constitute 75 percent of all emissions from food production and consumption (Heller et al. 2018). In addition to the emission profile of animal products, conventional meat supply chains pose other environmental and dietary health impacts, discussed in Chapter 3.

Alternative ways to produce protein could reduce environmental and health impacts in addition to addressing the disproportionate level of land and resource use required by animal agriculture. Eighty-five percent of harvested soybean and grain is fed to animals each year; if soybeans and grains were directly fed to humans, farm acreage could be dramatically reduced, and land could be freed up for biofuels, sequestration, and other purposes (Cassidy et al. 2013). Hayek et al. (2021) mapped the magnitude of the carbon opportunity cost of resource-intensive diets and found that a global shift to more plant-based diets could lead to the sequestration of 332–547 $GtCO_2$ by 2050, providing evidence that shifts to plant-based diets could lead to greater mitigation options. Eliminating the cycling of calories through animals represents a huge opportunity to reinvent the agricultural supply chain in the United States, which is based disproportionally on supporting livestock. Directly converting protein from crops to human food redefines the value proposition of the agricultural supply chain.

> Finding 8-6: If plant-based substitutes were to substantially replace meat and dairy, this would fundamentally change net-zero strategies, as a significant portion of arable land could then be used for other purposes (e.g., carbon sinks, biofuels).

Meat and Dairy Alternatives and Market Potential

Meat Substitutes

Today, the plant-based meat industry in the United States totals about $1.4 billion (Grand View Research 2023). With many meat substitutes already on the market and looking to grow, other innovative technologies are paving the way to producing "real" meat without the animal. There are three main alternative protein technologies: (1) protein isolates of traditional foods (e.g., legumes, grains); (2) precision fermentation proteins (e.g., fungal and yeast); and (3) cell-based and cultured proteins ("tissue engineering"). Hybrid products that leverage several of these processes are also being explored. The products that result from these technologies require regulatory evaluation by the Food and Drug Administration and USDA to establish safety, and investments in R&D as well as supply chain and manufacturing.

Protein isolates and fermentation products have a long history in the food supply chain, and many meat substitutes made from these materials have been on the market for decades. Recent years have seen an uptick in interest for meat-mimicking foods made by these processes, with companies like Beyond® and Impossible® creating commercial products that are already close to price parity with conventional meat.[10] This, however, has not significantly altered the market share of conventional meat. While the market for plant-based meat substitutes is projected to grow by 11 percent by 2029 (Fortune Business Insights 2022), the much larger conventional meat market ($3 trillion, globally) is also expected to grow by 5.7 percent. The effects of a growing meat-substitute market on cattle production will likely be small unless meat substitutes can capture much more of the market than these projections indicate (Lusk et al. 2022).[11]

Cell-based and cultured meat products are still a developing but potentially disruptive technology with cost curves not yet amenable to widespread commercialization. However, there are currently more than 70 companies working on this technology (Bandoim 2022), and prices have fallen precipitously over the past decade, similar to cost and performance advances in computing technologies governed by Moore's law (Shigeta 2020). Achieving cost parity with conventional meat production could be realized by 2030 with sufficient R&D and investments in supply chain (Vergeer et al. 2021).

Should cultivated meat products reach price parity, their impact on the market share of conventional meat will still depend on a variety of sociological and cultural factors. Consumer ideology and culture will greatly affect interest or aversion to new types of food, even foods that replicate traditional products down to the DNA. Global protein needs will increase substantially by 2050, and policies and incentives that take consumer ideologies, anxieties, and taste preferences into account will be necessary to

---

[10] A recent scenario-driven assessment of the plant-based meat industry's future suggests that manufacturing capacity (not raw ingredients) is expected to be rate-limiting, and the industry will need to invest $27 billion in global capital expenditure to meet projected global production needs and a minimum of $17 billion in annual operating costs (Troya et al. 2022).

[11] An economic model estimating how a reduction in price or increase in demand for plant-based meat in the United States affects cattle production found that increases in U.S. demand alter trade patterns, leading to a reduction of beef imports and an increase in beef exports, which reduces emissions and land use given the relative efficiency of U.S. beef production. The study found that for every 10 percent reduction in price, the global reduction in emissions is equivalent to 0.34 percent of U.S. emissions from beef production and 1.14 percent when including reduced land use change emissions, such that even substantial reductions in prices are unlikely to have substantive impacts on the U.S. cattle population and emissions (Lusk 2022). This suggests the need to also pursue alternative mitigation strategies, such as innovations to reduce methane emissions per head.

facilitate a transition to more sustainable diets (Meybeck and Gitz 2017; Yin et al. 2020). For example, although unprocessed plant-based foods are often cheaper than animal foods, most consumers today prefer to keep meat in their diet despite higher costs, higher environmental impact, and potential higher health impacts.

Dairy Substitutes

While the GHG and resource footprint is highest for beef cattle than any other animal product, dairy is a large emissions contributor that cannot be overlooked. The dairy and beef industries are intertwined, but the mitigation potential of dairy alone is likely only about one-third that of beef. However, alternative dairy products are currently viewed differently (and more favorably) by consumers than meat substitutes, making dairy substitutes an important mitigation opportunity (NASEM 2023). Plant-based milks are currently the largest plant-based category of foods in North America, and currently account for 10 percent of the market (Hale 2021).

> Finding 8-7: Owing to a variety of cultural and sociological factors, demand growth for meat substitutes may be too slow to have major impact on land use and GHG emissions before 2050, especially considering what is known about current consumer behavior. There is potentially more opportunity but less mitigation benefit with dairy substitutes.

Despite the current animal protein–oriented food system's high GHG emissions, potential adverse health impacts, and high demand for arable land, research shows that a fundamental change in the food animal production system will be difficult, requiring technological innovation, policies that make sustainable food more accessible and affordable, and buy-in from consumers. There is no investment in alternative protein technologies in the IRA. There is an opportunity in the Farm Bill to invest in R&D, and in business and consumer incentives for meat and dairy substitutes, as well as to increase incentives for whole foods such as fruits, vegetables, and legumes.

> **Recommendation 8-8:** *Convene an Expert Group to Recommend Policies That Could Encourage Sustainable Diets.* **The Secretary of Agriculture should quickly convene an expert group to recommend policies that could facilitate societal acceptance and adoption of sustainable diets with lower consumption of animal products and increased consumption of plant-based foods.**

All options should be considered, including some potentially politically difficult options—for example, the following:

- Including food in a comprehensive tax on greenhouse gas (GHG) emissions (while addressing the impacts on consumers with lower incomes).
- Creating federal incentives for plant-based meat substitutes and cell-cultivated products that reflect their life-cycle GHG and land use benefits relative to conventional meat.
- Including sustainability in U.S. Department of Agriculture dietary guidelines, and supporting data collection and availability on the life-cycle environmental impacts of emerging food products.
- Including sustainability in federal food procurement requirements.
- Phasing-out programs that incentivize the production or promotion of high-emission conventional meat products, while supporting just transitions for farms and workers.

Table 8-3 summarizes all of the recommendations in this chapter regarding land use in support of decarbonization efforts.

SUMMARY OF RECOMMENDATIONS ON LAND USE

TABLE 8-3 Summary of Recommendations on Land Use

Short-Form Recommendation	Actor(s) Responsible for Implementing Recommendation	Sector(s) Addressed by Recommendation	Objective(s) Addressed by Recommendation	Overarching Categories Addressed by Recommendation
8-1: Convene an Expert Group to Recommend Ways to Measure Additional Forest Sinks	Secretary of Agriculture	• Land use	• Greenhouse gas (GHG) reductions	Rigorous and Transparent Analysis and Reporting for Adaptive Management
8-2: Prioritize Ecosystem-Level Carbon Storage	Secretary of Agriculture	• Land use	• GHG reductions	A Broadened Policy Portfolio
8-3: Establish a Permanent, National-Scale, High-Quality Soil Monitoring Network	U.S. Department of Agriculture (USDA)	• Land use	• GHG reductions	Rigorous and Transparent Analysis and Reporting for Adaptive Management
8-4: Build Out Long-Term Agricultural Field Experiments	USDA	• Land use • Non-federal actors	• GHG reductions	Rigorous and Transparent Analysis and Reporting for Adaptive Management Research, Development, and Demonstration Needs
8-5: Fund Research to Quantify Indicators That Influence Adoption of Regenerative Agriculture Practices	USDA	• Land use	• GHG reductions • Equity • Public engagement	Research, Development, and Demonstration Needs

TABLE 8-3 Continued

Short-Form Recommendation	Actor(s) Responsible for Implementing Recommendation	Sector(s) Addressed by Recommendation	Objective(s) Addressed by Recommendation	Overarching Categories Addressed by Recommendation
8-6: Incentivize the Abatement of CH_4 and N_2O Emissions and Improve Soil Carbon Sequestration	USDA	• Land use	• GHG reductions • Equity	A Broadened Policy Portfolio
8-7: Release a Comprehensive Research, Development, Demonstration, and Deployment Program for Biomass Energy with Carbon Capture and Storage	Department of Energy	• Land use	• GHG reductions	Research, Development, and Demonstration Needs
8-8: Convene an Expert Group to Recommend Policies That Could Encourage Sustainable Diets	Secretary of Agriculture	• Land use	• GHG reductions • Health	A Broadened Policy Portfolio

REFERENCES

Achinas, S., J. Horjus, V. Achinas, and G.J.W. Euverink. 2019. "A PESTLE Analysis of Biofuels Energy Industry in Europe." *Sustainability* 11(21):5981. https://doi.org/10.3390/su11215981.

Ager, A.A., M.A. Finney, A. McMahan, and J. Cathcart. 2010. "Measuring the Effect of Fuel Treatments on Forest Carbon Using Landscape Risk Analysis." *Natural Hazards and Earth System Sciences* 10(12):2515–2526. https://doi.org/10.5194/nhess-10-2515-2010.

Alexander, P., C. Brown, A. Arneth, J. Finnigan, and M.D.A. Rounsevell. 2016. "Human Appropriation of Land for Food: The Role of Diet." *Global Environmental Change*. http://www.sciencedirect.com/science/article/pii/S0959378016302370.

Bagnall, D.K., J.F. Shanahan, A. Flanders, C.L.S. Morgan, and C.W. Honeycutt. 2021. "Soil Health Considerations for Global Food Security." *Agronomy Journal* 113(6):4581–4589. https://doi.org/10.1002/agj2.20783.

Bandoim, L. 2022. "Making Meat Affordable: Progress Since the $330,000 Lab-Grown Burger." *Forbes*, March 8. https://www.forbes.com/sites/lanabandoim/2022/03/08/making-meat-affordable-progress-since-the-330000-lab-grown-burger.

Beerling, D.J., E.P. Kantzas, M.R. Lomas, P. Wade, R.M. Eufrasio, P. Renforth, B. Sarkar, et al. 2020. "Potential for Large-Scale CO_2 Removal via Enhanced Rock Weathering with Croplands." *Nature* 583(7815):242–248. https://doi.org/10.1038/s41586-020-2448-9.

Bellemare, M.F. 2015. "Rising Food Prices, Food Price Volatility, and Social Unrest." *American Journal of Agricultural Economics* 97(1):1–21. https://doi.org/10.1093/ajae/aau038.

Birdsey, R.A., D.A. DellaSala, W.S. Walker, S.R. Gorelik, G. Rose, and C.E. Ramírez. 2023. "Assessing Carbon Stocks and Accumulation Potential of Mature Forests and Larger Trees in US Federal Lands." *Frontiers in Forests and Global Change* 5(277).

BPC (Bipartisan Policy Center). 2022. "Inflation Reduction Act Summary." https://bipartisanpolicy.org/download/?file=/wp-content/uploads/2022/08/Energy-IRA-Brief_R04-9.26.22.pdf.

California Natural Resources Agency. n.d. "Expanding Nature-Based Solutions." https://resources.ca.gov/Initiatives/Expanding-Nature-Based-Solutions. Accessed February 17, 2023.

Calvert, K.E. 2018. "Measuring and Modelling the Land-Use Intensity and Land Requirements of Utility-Scale Photovoltaic Systems in the Canadian Province of Ontario: Solar Energy Land Use." *Canadian Geographer/Le Géographe Canadien* 62(2):188–199. https://doi.org/10.1111/cag.12444.

Calvin, K., M. Wise, P. Kyle, P. Patel, L. Clarke, and J. Edmonds. 2014. "Trade-Offs of Different Land and Bioenergy Policies on the Path to Achieving Climate Targets." *Climatic Change* 123(3):691–704. https://doi.org/10.1007/s10584-013-0897-y.

Campbell, J.L., M.E. Harmon, and S.R. Mitchell. 2012. "Can Fuel-Reduction Treatments Really Increase Forest Carbon Storage in the Western US by Reducing Future Fire Emissions?" *Frontiers in Ecology and the Environment* 10(2):83–90. https://doi.org/10.1890/110057.

Capellán-Pérez, I., C. De Castro, and I. Arto. 2017. "Assessing Vulnerabilities and Limits in the Transition to Renewable Energies: Land Requirements Under 100% Solar Energy Scenarios." *Renewable and Sustainable Energy Reviews* 77:760–782. https://doi.org/10.1016/j.rser.2017.03.137.

Cassidy, E.S., P.C. West, J.S. Gerber, and J.A. Foley. 2013. "Redefining Agricultural Yields: From Tonnes to People Nourished per Hectare." *Environmental Research Letters* 8(3):034015. https://doi.org/10.1088/1748-9326/8/3/034015.

Chambers, A., R. Lal, and K. Paustian. 2016. "Soil Carbon Sequestration Potential of US Croplands and Grasslands: Implementing the 4 per Thousand Initiative." *Journal of Soil and Water Conservation* 71(3):68A–74A. https://doi.org/10.2489/jswc.71.3.68A.

Chen, C., A. Chaudhary, and A. Mathys. 2019. "Dietary Change Scenarios and Implications for Environmental, Nutrition, Human Health and Economic Dimensions of Food Sustainability." *Nutrients* 11(4):856. https://doi.org/10.3390/nu11040856.

Cheng, F., H. Luo, J.D. Jenkins, and E.D. Larson. 2023. "The Value of Low- and Negative-Carbon Fuels in the Transition to Net-Zero Emission Economies: Lifecycle Greenhouse Gas Emissions and Cost Assessments Across Multiple Fuel Types." *Applied Energy* 331:120388. https://doi.org/10.1016/j.apenergy.2022.120388.

Chiono, L.A., D.L. Fry, B.M. Collins, A.H. Chatfield, and S.L. Stephens. 2017. "Landscape-Scale Fuel Treatment and Wildfire Impacts on Carbon Stocks and Fire Hazard in California Spotted Owl Habitat." *Ecosphere* 8(1):e01648. https://doi.org/10.1002/ecs2.1648.

Clark, M.A., N.G.G. Domingo, K. Colgan, S.K. Thakrar, D. Tilman, J. Lynch, I.L. Azevedo, and J.D. Hill. 2020. "Global Food System Emissions Could Preclude Achieving the 1.5° and 2°C Climate Change Targets." *Science* 370(6517):705–708. https://doi.org/10.1126/science.aba7357.

Coffield, S.R., C.D. Vo, J.A. Wang, G. Badgley, M.L. Goulden, D. Cullenward, W.R.L. Anderegg, and J.T. Randerson. 2022. "Using Remote Sensing to Quantify the Additional Climate Benefits of California Forest Carbon Offset Projects." *Global Change Biology* 28(22):6789–6806. https://doi.org/10.1111/gcb.16380.

Cook-Patton, S.C., R.C. Drever, K. Hamrick, H. Hardman, T. Kroeger, S. Yeo, P.W. Ellis, et al. 2021a. "Protect, Manage and Then Restore Lands for Climate Mitigation." *Nature Climate Change* 11(12):1027–1034. https://doi.org/10.1038/s41558-021-01198-0.

Cook-Patton, S.C., D. Shoch, and P.W. Ellis. 2021b. "Dynamic Global Monitoring Needed to Use Restoration of Forest Cover as a Climate Solution." *Nature Climate Change* 11(5):366–368. https://doi.org/10.1038/s41558-021-01022-9.

Coyle, W.T. 2007. "The Future of Biofuels: A Global Perspective." U.S. Department of Agriculture Economic Research Service. November 1. https://www.ers.usda.gov/amber-waves/2007/november/the-future-of-biofuels-a-global-perspective.

DOE (Department of Energy). 2022a. "Clean Cities Alternative Fuel Price Report." DOE Office of Energy Efficiency and Renewable Energy. https://afdc.energy.gov/files/u/publication/alternative_fuel_price_report_july_2022.pdf.

DOE. 2022b. "U.S. Corn Production and Portion Used for Fuel Ethanol." DOE Office of Energy Efficiency and Renewable Energy. https://afdc.energy.gov/data/10339.

DOE. n.d.(a). "Biodiesel Production and Blending Tax Credit." Alternative Fuels Data Center. https://afdc.energy.gov/laws/5831. Accessed February 17, 2023.

DOE. n.d.(b). "Biodiesel Vehicle Emissions." Alternative Fuels Data Center. https://afdc.energy.gov/vehicles/diesels_emissions.html. Accessed February 17, 2023.

DOE. n.d.(c). "Biofuels and Greenhouse Gas Emissions: Myths Versus Facts." https://www.energy.gov/sites/prod/files/edg/media/BiofuelsMythVFact.pdf. Accessed February 17, 2023.

DOI (Department of the Interior). 2016. "Fundamental Considerations Associated with Placing Solar Generation Structures at Central Arizona Project Canal."

Domingo, N.G.G., S. Balasubramanian, S.K. Thakrar, M.A. Clark, P.J. Adams, J.D. Marshall, N.Z. Muller, et al. 2021. "Air Quality–Related Health Damages of Food." *Proceedings of the National Academy of Sciences* 118(20):e2013637118. https://doi.org/10.1073/pnas.2013637118.

DOS and EOP (Department of State and Executive Office of the President). 2021. *The Long-Term Strategy of the United States: Pathways to Net-Zero Greenhouse Gas Emissions by 2050.* https://www.whitehouse.gov/wp-content/uploads/2021/10/US-Long-Term-Strategy.pdf.

Eagle, A.J., A.L. Hughes, N.A. Randazzo, C.L. Schneider, C.H. Melikov, E. Puritz, K. Jaglo, and B. Hurley. 2022. *Ambitious Climate Mitigation Pathways for U.S. Agriculture and Forestry: Vision for 2030.* New York: Environmental Defense Fund and Washington, DC: ICF International. https://www.edf.org/sites/default/files/documents/climate-mitigation-pathways-us-agriculture-forestry.pdf.

Ecosystem Marketplace. 2021. *Markets in Motion: State of Voluntary Carbon Markets 2021, Installment 1.* Washington, DC: Forest Trends Association. https://www.ecosystemmarketplace.com/publications/state-of-the-voluntary-carbon-markets-2021.

EIA (U.S. Energy Information Administration). 2018. "EPA Finalizes Renewable Fuel Standard for 2019, Reflecting Cellulosic Biofuel Shortfalls." https://www.eia.gov/todayinenergy/detail.php?id=37712.

EIA. 2019. "EIA Projects Energy Consumption in Air Transportation to Increase Through 2050." *Today In Energy*, November 6. https://www.eia.gov/todayinenergy/detail.php?id=41913.

EPA (Environmental Protection Agency). 2019. "Global Non-CO_2 Greenhouse Gas Emission Projections and Mitigation 2015–2050." EPA-430-R-19-010. https://www.epa.gov/global-mitigation-non-co2-greenhouse-gases/global-non-co2-greenhouse-gas-emission-projections.

EPA. 2021. "From Farm to Kitchen: The Environmental Impacts of US Food Waste." EPA 600-R21 171. https://www.epa.gov/system/files/documents/2021-11/from-farm-to-kitchen-the-environmental-impacts-of-u.s.-food-waste_508-tagged.pdf.

EPA. 2022. "Inventory of U.S. Greenhouse Gas Emissions and Sinks: 1990–2020." EPA-430-R-22-003. https://www.epa.gov/system/files/documents/2022-04/us-ghg-inventory-2022-main-text.pdf.

EPA. n.d.(a). "Greenhouse Gas Inventory Data Explorer." https://cfpub.epa.gov/ghgdata/inventoryexplorer/#iagriculture/entiresector/allgas/gas/all. Accessed February 15, 2023.

EPA. n.d.(b). "Overview of Greenhouse Gases." https://www.epa.gov/ghgemissions/overview-greenhouse-gases. Accessed February 15, 2023.

FAO (Food and Agriculture Organization), IFAD (International Fund for Agricultural Development), UNICEF (United Nations Children's Fund), WFP (World Food Programme), and WHO (World Health Organization). 2021. *The State of Food Security and Nutrition in the World 2021*. Rome, Italy: Food and Agriculture Organization. https://doi.org/10.4060/cb4474en.

Fargione, J., J. Hill, D. Tilman, S. Polasky, and P. Hawthorne. 2008. "Land Clearing and the Biofuel Carbon Debt." *Science* 319(5867):1235–1238. https://doi.org/10.1126/science.1152747.

Fargione, J., S. Bassett, T. Boucher, S.D. Bridgham, R.T. Conant, S.C. Cook-Patton, P.W. Ellis, et al. 2018. "Natural Climate Solutions for the United States." *Science Advances* 4(11):eaat1869. https://doi.org/10.1126/sciadv.aat1869.

Fargione, J., D.L. Haase, O.T. Burney, O.A. Kildisheva, G. Edge, S.C. Cook-Patton, T. Chapman, et al. 2021. "Challenges to the Reforestation Pipeline in the United States." *Frontiers in Forests and Global Change* 4(February):629198. https://doi.org/10.3389/ffgc.2021.629198.

FBLE (Farm Bill Law Enterprise). 2022. "Equity in Agricultural Production and Governance." https://www.farmbilllaw.org/wp-content/uploads/2022/10/Equity-Report.pdf.

Fortune Business Insights. 2022. "Meat Substitutes Market Size, Share and COVID-19 Impact Analysis, by Source (Soy-Based Ingredients, Wheat-Based Ingredients, Other Grain-Based Ingredients, and Textured Vegetable Proteins), by Distribution Channel (Mass Merchandisers, Specialty Stores, Online Retail, Other Retail Channels, and Food Service), and Regional Forecast, 2022–2029." FBI100239. https://www.fortunebusinessinsights.com/industry-reports/meat-substitutes-market-100239.

Foster, D.E., J.J. Battles, B.M. Collins, R.A. York, and S.L. Stephens. 2020. "Potential Wildfire and Carbon Stability in Frequent-Fire Forests in the Sierra Nevada: Trade-Offs from a Long-Term Study." *Ecosphere* 11(8). https://doi.org/10.1002/ecs2.3198.

Fraunhofer Institute for Solar Energy Systems ISE. 2020. *Agrivoltaics: Opportunities for Agriculture and the Energy Transition: A Guideline for Germany*. Freiburg, Germany.

Glover, J.D., J.P. Reganold, L.W. Bell, J. Borevitz, E.C. Brummer, E.S. Buckler, C.M. Cox, et al. 2010. "Increased Food and Ecosystem Security via Perennial Grains." *Science* 328(5986):1638–1639. https://doi.org/10.1126/science.1188761.

Grand View Research. 2023. "Plant-Based Meat Market." https://www.grandviewresearch.com/industry-analysis/plant-based-meat-market.

Hale, M. 2021. "Exploring the Growth of Plant-Based Milk." *Food Manufacturing*, September 23. https://www.foodmanufacturing.com/consumer-trends/article/21723117/exploring-the-growth-of-plantbased-milk.

Hayek, M.N., H. Harwatt, W.J. Ripple, and N.D. Mueller. 2021. "The Carbon Opportunity Cost of Animal-Sourced Food Production on Land." *Nature Sustainability* 4(1):21–24. https://doi.org/10.1038/s41893-020-00603-4.

Heller, M.C, A. Willits-Smith, R. Meyer, G.A. Keoleian, and D. Rose. 2018. "Greenhouse Gas Emissions and Energy Use Associated with Production of Individual Self-Selected US Diets." *Environmental Research Letters* 13(4):044004. https://doi.org/10.1088/1748-9326/aab0ac.

Hoover, C.M., and J.E. Smith. 2021. "Current Aboveground Live Tree Carbon Stocks and Annual Net Change in Forests of Conterminous United States." *Carbon Balance and Management* 16(1):17. https://doi.org/10.1186/s13021-021-00179-2.

Horst, M., and A. Marion. 2019. "Racial, Ethnic and Gender Inequities in Farmland Ownership and Farming in the U.S." *Agriculture and Human Values* 36(1):1–16. https://doi.org/10.1007/s10460-018-9883-3.

Hudiburg, T.W., B.E. Law, C. Wirth, and S. Luyssaert. 2011. "Regional Carbon Dioxide Implications of Forest Bioenergy Production. *Nature Climate Change* 1:419–423. https://doi.org/10.1038/nclimate1264.

Hudiburg, T.W., B.E. Law, W.R. Moomaw, M.E. Harmon, and J.E. Stenzel. 2019. "Meeting GHG Reduction Targets Requires Accounting for All Forest Sector Emissions. *Environmental Research Letters* 14:095005. https://doi.org/10.1088/1748-9326/ab28bb.

Hurteau, M.D., and M.L. Brooks. 2011. "Short- and Long-Term Effects of Fire on Carbon in US Dry Temperate Forest Systems." *BioScience* 61(2):139–146. https://doi.org/10.1525/bio.2011.61.2.9.

Hurtt, G.C., S.W. Pacala, P.R. Moorcroft, J. Caspersen, E. Shevliakova, R.A. Houghton, and B. Moore. 2002. "Projecting the Future of the U.S. Carbon Sink." *Proceedings of the National Academy of Sciences* 99(3):1389–1394. https://doi.org/10.1073/pnas.012249999.

IPCC (Intergovernmental Panel on Climate Change). 2022. *Global Warming of 1.5°C: IPCC Special Report on Impacts of Global Warming of 1.5°C Above Pre-Industrial Levels in Context of Strengthening Response to Climate Change, Sustainable Development, and Efforts to Eradicate Poverty*, 1st ed., V. Masson-Delmotte, P. Zhai, H.-O. Pörtner, et al., eds. Cambridge, UK, and New York: Cambridge University Press. https://doi.org/10.1017/9781009157940.

Jacobson, M.Z., and M.A. Delucchi. 2011. "Providing All Global Energy with Wind, Water, and Solar Power, Part I: Technologies, Energy Resources, Quantities and Areas of Infrastructure, and Materials." *Energy Policy* 39(3):1154–1169. https://doi.org/10.1016/j.enpol.2010.11.040.

Jacobson, M.Z., M.A. Delucchi, M.A. Cameron, and B.A. Frew. 2015. "Low-Cost Solution to the Grid Reliability Problem with 100% Penetration of Intermittent Wind, Water, and Solar for All Purposes." *Proceedings of the National Academy of Sciences* 112(49):15060–15065. https://doi.org/10.1073/pnas.1510028112.

Jenkins, J.D., E.N. Mayfield, E.D. Larson, S.W. Pacala, and C. Greig. 2021. "Mission Net-Zero America: The Nation-Building Path to a Prosperous, Net-Zero Emissions Economy." *Joule* 5(11):2755–2761. https://doi.org/10.1016/j.joule.2021.10.016.

Jenkins, J.D., E.N. Mayfield, J. Farbes, R. Jones, N. Patankar1, Q. Xu, and G. Schivley. 2022. "Preliminary Report: The Climate and Energy Impacts of the Inflation Reduction Act of 2022." Princeton, NJ: REPEAT Project, p. 17. https://repeatproject.org/docs/REPEAT_IRA_Prelminary_Report_2022-08-04.pdf.

Jennings, R., A.D. Henderson, A. Phelps, K.M. Janda, and A.E. Van Den Berg. 2023. "Five U.S. Dietary Patterns and Their Relationship to Land Use, Water Use, and Greenhouse Gas Emissions: Implications for Future Food Security." *Nutrients* 15(1):215. https://doi.org/10.3390/nu15010215.

Kreidenweis, U., F. Humpenöder, M. Stevanović, B.L. Bodirsky, E. Kriegler, H. Lotze-Campen, and A. Popp. 2016. "Afforestation to Mitigate Climate Change: Impacts on Food Prices Under Consideration of Albedo Effects." *Environmental Research Letters* 11(8):085001. https://doi.org/10.1088/1748-9326/11/8/085001.

Larsen, J., B. King, E. Wimberger, H. Pitt, H. Kolus, A. Rivera, N. Dasari, C. Jahns, K. Larsen, and W. Herndon. 2021. "Pathways to Paris: A Policy Assessment of the 2030 US Climate Target." Rhodium Group. https://rhg.com/research/us-climate-policy-2030.

Larsen, J., B. King, H. Kolus, N. Dasari, G. Hiltbrand, and W. Herndon. 2022. "A Turning Point for US Climate Progress: Assessing the Climate and Clean Energy Provisions in the Inflation Reduction Act." Rhodium Group. https://rhg.com/research/climate-clean-energy-inflation-reduction-act.

Larson, E., C. Greig, J. Jenkins, E. Mayfield, A. Pascale, C. Zhang, J. Drossman, et al. 2020. *Net-Zero America: Potential Pathways, Infrastructure, and Impacts, Interim Report*. Princeton, NJ: Princeton University. https://netzeroamerica.princeton.edu/img/Princeton_NZA_Interim_Report_15_Dec_2020_FINAL.pdf.

Larson, E., C. Greig, J. Jenkins, E. Mayfield, A. Pascale, C. Zhang, J. Drossman, et al. 2021. *Net-Zero America: Potential Pathways, Infrastructure, and Impacts, Final Report*. Princeton, NJ: Princeton University. https://netzeroamerica.princeton.edu/the-report.

Law, B.E., T.W. Hudiburg, L.T. Berner, J.J. Kent, P.C. Buotte, and M.E. Harmon. 2018. "Land Use Strategies to Mitigate Climate Change in Carbon Dense Temperate Forests." *Proceedings of the National Academy of Sciences* 115(14):3663–3668.

Law, B.E., W.R. Moomaw, R.W. Hudiburg, W.H. Schlesinger, J.D. Sterman, and G.M. Woodwell. 2022. "Creating Strategic Reserves to Protect Forest Carbon and Reduce Biodiversity Losses in the United States." *Land* 11:721.

Lempert, R., B.L. Preston, J. Edmonds, L. Clarke, T. Wild, M. Binsted, E. Diringer, and B. Townsend. 2019. "Pathways to 2050: Scenarios for Decarbonizing the U.S. Economy." Center for Climate and Energy Solutions. https://www.c2es.org/document/pathways-to-2050-scenarios-for-decarbonizing-the-u-s-economy.

Li, Y., C.K. Miskin, and R. Agrawal. 2018. "Land Availability, Utilization, and Intensification for a Solar Powered Economy." *Computer Aided Chemical Engineering* 44:1915–1920. https://doi.org/10.1016/B978-0-444-64241-7.50314-1.

Lusk, J.L., D. Blaustein-Rejto, S. Shah, and G.T. Tonsor. 2022. "Impact of Plant-Based Meat Alternatives on Cattle Inventories and Greenhouse Gas Emissions." *Environmental Research Letters*, January. https://doi.org/10.1088/1748-9326/ac4fda.

Mahajan, M., O. Ashmoore, J. Rissman, R. Orvis, and A. Gopal. 2022. "Updated Inflation Reduction Act Modeling Using the Energy Policy Simulator." Energy Innovation. https://energyinnovation.org/wp-content/uploads/2022/08/Updated-Inflation-Reduction-Act-Modeling-Using-the-Energy-Policy-Simulator.pdf.

Meybeck, A., and V. Gitz. 2017. "Sustainable Diets Within Sustainable Food Systems." *Proceedings of the Nutrition Society* 76(1):1–11. https://doi.org/10.1017/S0029665116000653.

Mitchell, S.R., M.E. Harmon, and K.E.B. O'Connell. 2012. "Carbon Debt and Carbon Sequestration Parity in Forest Bioenergy Production." *GCB Bioenergy*.

NASEM (National Academies of Sciences, Engineering, and Medicine). 2019. *Negative Emissions Technologies and Reliable Sequestration: A Research Agenda*. Washington, DC: The National Academies Press. https://doi.org/10.17226/25259.

NASEM. 2020. *A National Strategy to Reduce Food Waste at the Consumer Level*. Washington, DC: The National Academies Press. https://doi.org/10.17226/25876.

NASEM. 2021. *Accelerating Decarbonization of the U.S. Energy System*. Washington, DC: The National Academies Press. https://doi.org/10.17226/25932.

NASEM. 2023. *Alternative Protein Sources: Balancing Food Innovation, Sustainability, Nutrition, and Health: Proceedings of a Workshop*. Washington, DC: The National Academies Press. https://doi.org/10.17226/26923.

Network Rail. 2023. "Thameslink Programme." https://www.networkrail.co.uk/running-the-railway/railway-upgrade-plan/key-projects/thameslink-programme.

North, M., M. Hurteau, and J. Innes. 2009. "Fire Suppression and Fuels Treatment Effects on Mixed-Conifer Carbon Stocks and Emissions." *Ecological Applications* 19(6):1385–1396. https://doi.org/10.1890/08-1173.1.

Novick, K.A., S. Metzger, W.R.L. Anderegg, M. Barnes, D.S. Cala, K. Guan, K.S. Hemes, et al. 2022. Informing Nature-Based Climate Solutions for the United States with the Best-Available Science. *Global Change Biology* 28:3778–3794.

Oregon Solar 1 Storage Industries Association. n.d. "Oregon Community Solar Program." https://www.orssia.org/communitysolar. Accessed February 21, 2023.

Orvis, R., and M. Mahajan. 2021. "A 1.5°C NDC for Climate Leadership by the United States." Energy Innovation. https://energyinnovation.org/wp-content/uploads/2021/04/A-1.5-C-Pathway-to-Climate-Leadership-for-The-United-States_NDC-update-2.pdf.

Pacala, S., and R. Socolow. 2004. "Stabilization Wedges: Solving the Climate Problem for the Next 50 Years with Current Technologies." https://www.science.org/doi/10.1126/science.1100103.

Powell, T.W.R., and T.M. Lenton. 2012. "Future Carbon Dioxide Removal via Biomass Energy Constrained by Agricultural Efficiency and Dietary Trends." *Energy and Environmental Science* 5(8):8116. https://doi.org/10.1039/c2ee21592f.

Read, Q.D., S. Brown, A.D. Cuéllar, S.M. Finn, J.A. Gephart, L.T. Marston, E. Meyer, K.A. Weitz, and M.K. Muth. 2020. "Assessing the Environmental Impacts of Halving Food Loss and Waste Along the Food Supply Chain." *Science of the Total Environment* 712:136255. https://doi.org/10.1016/j.scitotenv.2019.136255.

Rinke Dias De Souza, N., B.C. Klein, M.F. Chagas, O. Cavalett, and A. Bonomi. 2021. "Towards Comparable Carbon Credits: Harmonization of LCA Models of Cellulosic Biofuels." *Sustainability* 13(18):10371. https://doi.org/10.3390/su131810371.

Ritchie, H. 2017. "How Much of the World's Land Would We Need in Order to Feed the Global Population with the Average Diet of a Given Country?" Our World in Data. October 3, 2017. https://ourworldindata.org/agricultural-land-by-global-diets.

Robertson, G.P., S.K. Hamilton, K. Paustian, and P. Smith. 2022. "Land-Based Climate Solutions for the United States." *Global Change Biology* 28(16):4912–4919. https://doi.org/10.1111/gcb.16267.

Rosegrant, M.W. 2008. *Biofuels and Grain Prices: Impact and Policy Responses*. Washington, DC: International Food Policy Research Institute.

Santo, R.E., B.F. Kim, S.E. Goldman, J. Dutkiewicz, E.M.B. Biehl, M.W. Bloem, R.A. Neff, and K.E. Nachman. 2020. "Considering Plant-Based Meat Substitutes and Cell Based Meats: A Public Health and Food Systems Perspective." *Frontiers*. https://www.frontiersin.org/articles/10.3389/fsufs.2020.00134/full.

Saunders, P.J. 2020. "Land Use Requirements of Solar and Wind Power Generation: Understanding a Decade of Academic Research." Energy Innovation Reform Project. https://www.innovationreform.org/2020/10/12/land-use-requirements-of-solar-and-wind-power-understanding-a-decade-of-academic-research.

Schmer, M.R., K.P. Vogel, R.B. Mitchel, R.K. Perrin. 2008. "Net Energy of Cellulosic Ethanol from Switchgrass." *Proceedings of the National Academy of Sciences* 105(2)464. https://www.pnas.org/doi/10.1073/pnas.0704767105.

Scully, M.J, G.A. Norris, T.M. Alarcon Falconi, and D.L. MacIntosh. 2021. "Carbon Intensity of Corn Ethanol in the United States: State of the Science." *Environmental Research Letters* 16(4):043001. https://doi.org/10.1088/1748-9326/abde08.

Seddon, N. 2022. "Harnessing the Potential of Nature-Based Solutions for Mitigating and Adapting to Climate Change." *Science* 376(6600):1410–1416. https://doi.org/10.1126/science.abn9668.

Seddon, N., A. Chausson, P. Berry, C.A.J. Girardin, A. Smith, and B. Turner. 2020. "Understanding the Value and Limits of Nature-Based Solutions to Climate Change and Other Global Challenges." *Philosophical Transactions of the Royal Society B: Biological Sciences* 375(1794):20190120. https://doi.org/10.1098/rstb.2019.0120.

Shigeta, R. 2020. "Lab-Grown Meat Is Scaling Like the Internet." *The Spoon*, October 26. https://thespoon.tech/lab-grown-meat-is-scaling-like-the-internet.

Smith, L.J., and M.S. Torn. 2013. "Ecological Limits to Terrestrial Biological Carbon Dioxide Removal." *Climatic Change* 118(1):89–103. https://doi.org/10.1007/s10584-012-0682-3.

Stubbs, M., K. Hoover, J.L. Ramseur. 2021. "Agriculture and Forestry Offsets in Carbon Markets: Background and Selected Issues." R46956. Washington, DC: Congressional Research Service. https://crsreports.congress.gov/product/pdf/R/R46956.

Swartz, E., A. Ravi, A. Reeber, J. Levink, T. Huang, and B. Smith. 2023. *Anticipated Growth Factor and Recombinant Protein Costs and Volumes Necessary for Cost-Competitive Cultivated Meat*. Washington, DC: The Good Food Institute.

Troya, M., H. Kurawadwala, B. Byrne, R. Dowdy, and Z. Weston. 2022. "Plant-Based Meat: Anticipating 2030 Production Requirements." Good Food Institute. https://gfi.org/resource/anticipating-plant-based-meat-production-requirements-2030.

UNEP (United Nations Environment Programme). 2017. "Emissions Gap Report 2017." http://www.unep.org/resources/emissions-gap-report-2017.

USDA (U.S. Department of Agriculture). 2014. "U.S. Forest Resource Facts and Historical Trends." FS-1035. https://mff.forest.mtu.edu/TreeBasics/USFSfacts&trends2014.pdf.

USDA. 2017. "Major Land Uses: Summary Table 1: Major Uses of Land, by Region, State, and United States, 2012." https://www.ers.usda.gov/data-products/major-land-uses/major-land-uses/#Summary%20tables.

USDA. 2023. "U.S. Bioenergy Statistics." Economic Research Service. https://www.ers.usda.gov/data-products/u-s-bioenergy-statistics.

USGCRP (U.S. Global Change Research Program). 2018. *Second State of the Carbon Cycle Report (SOCCR2): A Sustained Assessment Report*, N. Cavallaro, G. Shrestha, R. Birdsey, M.A. Mayes, R.G. Najjar, S.C. Reed, P. Romero-Lankao, and Z. Zhu, eds. Washington, DC. https://doi.org/10.7930/SOCCR2.2018.

Vergeer, R., P. Sinke, I. Odegard. 2021. "TEA of Cultivated Meat: Future Projections for Different Scenarios." CE Delft.

Waite, R., T. Searchinger, J. Ranganathan, and J. Zionts. 2022. "6 Pressing Questions About Beef and Climate Change, Answered," World Resources Institute. https://www.wri.org/insights/6-pressing-questions-about-beef-and-climate-change-answered.

White, A.E., D.A. Lutz, R.B. Howarth, and J.R. Soto. 2018. "Small-Scale Forestry and Carbon Offset Markets: An Empirical Study of Vermont Current Use Forest Landowner Willingness to Accept Carbon Credit Programs," C.T. Bauch, ed. *PLOS ONE* 13(8):e0201967. https://doi.org/10.1371/journal.pone.0201967.

White House Council on Environmental Quality, White House Office of Science and Technology Policy, White House Domestic Climate Policy Office. 2022. *Opportunities for Accelerating Nature-Based Solutions: A Roadmap for Climate Progress, Thriving Nature, Equity, and Prosperity*. Report to the National Climate Task Force. Washington, DC. https://www.whitehouse.gov/wp-content/uploads/2022/11/Nature-Based-Solutions-Roadmap.pdf.

Willett, W., J. Rockström, B. Loken, M. Springmann, T. Lang, S. Vermeulen, T. Garnett, et al. 2019. "Food in the Anthropocene: The EAT–Lancet Commission on Healthy Diets from Sustainable Food Systems." *Lancet* 393(10170):447–492.

Williams, J.H., R.A. Jones, B. Haley, G. Kwok, J. Hargreaves, J. Farbes, and M.S. Torn. 2021. "Carbon-Neutral Pathways for the United States." *AGU Advances* 2(1):21. https://doi.org/10.1029/2020AV000284.

Winiwarter, W., L. Höglund-Isaksson, Z. Klimont, W. Schöpp, and M. Amann. 2018. "Technical Opportunities to Reduce Global Anthropogenic Emissions of Nitrous Oxide." *Environmental Research Letters* 13(1):014011. https://doi.org/10.1088/1748-9326/aa9ec9.

Wu, C., S.R. Coffield, M.L. Goulden, J.T. Randerson, A.T. Trugman, and W.R.L. Anderegg. 2023. "Uncertainty in US Forest Carbon Storage Potential Due to Climate Risks." *Nature Geoscience* 16:422–429.

Yin, J., D. Yang, X. Zhang, Y. Zhang, T. Cai, Y. Hao, S. Cui, and Y. Chen. 2020. "Diet Shift: Considering Environment, Health and Food Culture." *Science of the Total Environment* 719:137484. https://doi.org/10.1016/j.scitotenv.2020.137484.

YSG Solar. 2022. "3 Largest Floating Solar Farms in the United States in 2022." https://www.ysgsolar.com/blog/3-largest-floating-solar-farms-united-states-2022-ysg-solar.

Zhou, Q., K. Guan, S. Wang, C. Jiang, Y. Huang, B. Peng, Z. Chen, et al. 2022. "Recent Rapid Increase of Cover Crop Adoption Across the U.S. Midwest Detected by Fusing Multi-Source Satellite Data." *Geophysical Research Letters* 49(22). https://doi.org/10.1029/2022GL100249.

9

Transport

ABSTRACT

Transportation emissions represent nearly one-third of greenhouse gas (GHG) emissions in the United States, the majority of which will be reduced through vehicle electrification. The costs to produce, purchase, and operate electric vehicles (EVs) have fallen significantly, due primarily to reduced battery costs and total costs of ownership, and are now reaching parity with comparable internal combustion engine models. The Infrastructure Investment and Jobs Act (IIJA)[1] (2021) and the Inflation Reduction Act (IRA) (2022) have provided historic levels of funding and tax credits to address climate change. Despite this legislation, there remain barriers to reaching zero-emission vehicle (ZEV) sales goals for light-duty vehicles (LDVs), including consumer reticence about current EV initial cost premiums over internal combustion engine vehicles (ICEVs), lack of awareness about available incentives for EV purchase and home chargers, insufficient overall funding for public chargers to enable EV use for drivers without home charging and for those making trips longer than their vehicle range, and constraints on critical minerals for EV batteries. Requirements in the IRA itself regarding battery minerals sourcing and North American manufacturing have reduced the number of models qualifying for tax credits, although it is not clear how quickly automakers will adapt in the near term. Midpoints of projections suggest that the United States may not achieve its goal of 50 percent ZEV sales by 2030. Even if it does, long-distance, heavy-duty (HD) land transport; aviation; and marine vessels will require development and large-scale production of net-zero carbon liquid fuels for successful decarbonization by 2050 and beyond.

Given the current technological and policy situation for transportation decarbonization, the committee recommends actions to help achieve ZEV sales goals, including continued tightening of federal fuel economy and emissions standards; federal and state adoption of California ZEV sales mandates; additional incentives for vehicle purchase and charger installation; and local funding and policies preferencing EVs and chargers (Recommendation 9-1). The committee also recommends cost-effective electrification of port and airport operations (Recommendation 9-2); cost-effective state and local policies to reduce vehicle emissions through enhanced

---

[1] Also referred to as the Bipartisan Infrastructure Law (BIL).

traffic management and operational efficiency; substitution of information technology for travel (especially via aircraft); expansion of transit and non-motorized travel; and land use changes to enhance the density of development (Recommendation 9-3). To decarbonize the embodied carbon in infrastructure, the committee recommends state and private standards and procurement policies to reduce the carbon content of infrastructure materials and carbon emissions during construction and maintenance (Recommendation 9-4). To enhance equity, the committee recommends state and local efforts to support EV purchase by low-income households and equitable distribution of chargers; targeted expansion of transit, car sharing, and other modal options for those unable to afford EVs; and representation of low-income residents on public planning, zoning, and transportation decision-making boards (Recommendation 9-5). Last, the committee recommends that targeted federal investments in research, development, and demonstration (RD&D) be made to improve battery and fuel cell design and performance and production of net-zero liquid fuels for hard-to-decarbonize modes such as aviation, ocean shipping, freight rail, and long-distance heavy trucks (Recommendation 9-6).

INTRODUCTION

Transportation is the nation's largest source of GHG emissions, accounting for 29 percent of total GHG emissions, edging ahead of electric power generation (25 percent) in 2019 (EPA 2021). On-road vehicles (automobiles and trucks) dominate transport GHG emissions (82 percent) (Figure 9-1 and Table 9-1).[2] Within that group, LDVs (58 percent) emit the largest share of GHGs because of the large stock of 260 million LDVs driving more than 3 trillion vehicle miles per year.[3] Medium- and heavy-duty vehicles (MHDVs) (24 percent), including trucks moving about half of the nation's freight (BTS 2021b), are the second largest emitters from a stock of 13.2 million vehicles moving almost 2.5 trillion ton-miles per year. Aviation for freight and passenger transportation (10 percent) is the third largest, although aviation's GHG emissions alone considerably understate its adverse climate forcing effects (Lee et al. 2021).

The majority of GHG emissions from transportation results from combustion of fossil fuels onboard vehicles in internal combustion engines. A primary target for deep decarbonization is vehicle electrification, which eliminates onboard vehicle GHG

---

[2] Note that these estimates of total transportation emissions do not include estimated methane leakages from pipelines, which recent studies (Von Fischer et al. 2017; Weller et al. 2020) are finding to be much more substantial than previously believed.

[3] LDV and MHDV stock estimates from EIA (2022) Tables 39 and 49. LDV vehicle miles and truck ton-miles from Bureau of Transportation Statistics (2021a,c), Tables 1-35 and 1-50.

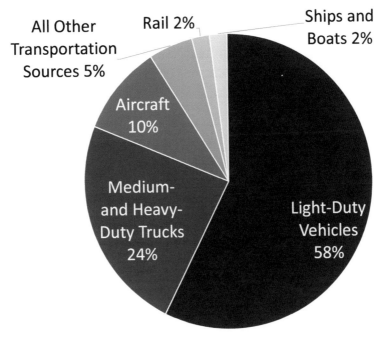

FIGURE 9-1 Share of transportation GHG emissions by source (2019). SOURCE: Courtesy of EPA (2021).

emissions, uses energy more efficiently to move the vehicle, and allows energy use from lower-emitting sources of electricity that reduces life-cycle emissions relative to use of fossil fuels. Electrification takes on many forms for efficiency and emissions reduction. Hybrid electric vehicles (HEVs) are powered by combusting liquid fuels in an efficient powertrain that harvests electrical energy from regenerative braking, which is stored in a battery and used to supplement engine power. Plug-in hybrid electric vehicles (PHEVs) are similar to HEVs, but they have larger batteries, motors, and other electrical equipment that allow for the vehicle to be charged with electricity from the grid and travel on stored power for a typical range of up to 30–50 miles in addition to efficiently operating with an internal combustion engine. Battery electric vehicles (BEVs) are of most interest for deep decarbonization because their entire motive power comes from grid electricity stored in an onboard battery. Fuel cell electric vehicles (FCEVs) are powered via fuel cells, which oxidize fuels (usually hydrogen) stored onboard the vehicle and can aid deep decarbonization if the hydrogen source has low-carbon emissions.[4] This chapter refers to plug-in electric vehicles (PEVs) as

---

[4] This report considers low-carbon hydrogen as defined in the IIJA §40315: hydrogen with a carbon intensity of less than 2 kilograms of CO_2 equivalent per kilogram of hydrogen at the site of production. The Department of Energy (DOE) has also released a Clean Hydrogen Production Standard with a life-cycle-based target of well-to-gate carbon intensity of less than 4 kg CO_2 equivalent per kilogram of hydrogen (DOE 2023b).

TABLE 9-1 Transportation GHG Emissions (2019)

Vehicle Type	Total GHG Emissions (Tg CO$_2$e)	Percent of Transportation Emissions
On-Road Vehicles[a]		
Passenger cars	762.3	40.6
Light-duty trucks	323.1	17.2
Motorcycles	3.6	0.2
Buses	22.2	1.2
Medium- and heavy-duty trucks	444.4	23.7
Aircraft		
Commercial	135.4	7.2
Military	12	0.6
General aviation	33.7	1.8
Ships and Boats	40.4	2.2
Rail	37.6	2.0
Pipelines	53.7	2.9
Lubricants	8.9	0.5

[a] Off-road vehicles, such as those used in construction and agriculture, add 11 percent more GHGs from fuel combustion in internal combustion engines (Ledna et al. 2022). SOURCE: EPA (2021).

the combination of BEVs and PHEVs. It refers to ZEVs as any vehicle type that has zero carbon tailpipe emissions—for example, BEVs and FCEVs.

This chapter begins with a focus on transportation electrification, describing the committee's 2030 ZEV sales goals, supportive policies (including describing the measures in the IIJA and IRA that support the committee's first report goals for electrifying roadway transportation by 2030), and the barriers that remain in achieving them. The other sections go on to identify and describe other important transportation decarbonization opportunities and challenges, including reducing GHG emissions through increased efficiency of travel and reduced carbon content of infrastructure materials, construction, and maintenance (see the section "GHG Reduction Through Transport Efficiency"); crosscutting issues such as equity, PEV demand on the electric grid, and agricultural and carbon sink constraints on bio-fuel production (see the section "Equity and Other Crosscutting Issues"); and innovation priorities for RD&D (see the section "Actions to Expand the Innovation Toolkit"). Table 9-5, at the end of the chapter, summarizes all the recommendations that appear in this chapter regarding decarbonizing the transportation system.

TRANSPORTATION ELECTRIFICATION INCLUDING ZEV SALES GOALS, SUPPORTIVE POLICIES, AND BARRIERS

Light-Duty PEVs and Charging Infrastructure

LDV Goals and Scenarios

LDVs are increasingly being electrified owing to market demand, corporate offerings, and incentives offered for electrification and its attributes, primarily fuel efficiency and reduced emissions. In its first report, the committee recommended specific goals for decarbonizing transportation (NASEM 2021a):

- A national standard for a 50 percent sales share of ZEVs by 2030 and 100 percent by 2050. (The U.S. national long-term decarbonization strategy also sets a 2030 goal of 50 percent EV sales [DOS and EOP 2021].)
- Deployment of public charging infrastructure to meet charging needs, which it estimated to be at least 3 million Level 2 chargers and 120,000 fast direct current (DC Fast) chargers by 2030.[5]
- An investment goal of $5 billion for intercity charging infrastructure.

Based on its updated findings presented in this report, the committee continues to endorse these goals for light-duty (LD) ZEVs and charging infrastructure from its first report, and provides further goals, especially described in Finding 9-1 and Recommendation 9-1. Note that while the committee endorses ZEVs (BEVs and FCEVs) as the appropriate goal for deep decarbonization of road transportation, it may be appropriate to incorporate or even encourage PHEV deployment in some limited amount, especially in the early stages of decarbonization (Foster et al. 2022). Most federal and state incentives and regulations currently incorporate PHEVs, with consideration of their remaining emissions, into ZEV regulations.

Long-term investment in RD&D by the public and private sectors, especially in batteries, provided breakthroughs in EV technologies that are driving down their costs and helped stimulate more than $1.2 trillion in North American and European original equipment manufacturer (OEM) investment commitments to PEVs (Leinert 2022). Well before passage of the IRA, automakers announced multiple new EV models, including LD pick-up trucks and sport utility vehicles (SUVs) (*Car and Driver* 2022). Many automakers have announced corporate decarbonization or electrification goals and commitments. (See Table 9-4 below.) Similarly, major investments in manufacturing

---

[5] Level 2 charging provides roughly 25 miles of driving per hour of charging. DC Fast charging provides roughly 100 to 200 miles of driving per half hour of charging (DOE n.d.(a)).

capacity for EVs were planned before passage of the IRA, including adding 13 new EV battery manufacturing plants (DOE 2021; Voelker 2021). The appeal of personal PEVs to high-income households has already been proven; hence, the emphasis in the IRA on stimulating PEV purchases by low- and moderate-income households to accelerate and broaden penetration of PEVs into the vehicle fleet.

Demand for PEVs has increased sharply since 2020, reaching 6.8 percent of LDV sales in 2022 (Table 9-2), with corresponding reduction in GHG emissions (Figure 9-2). Meeting the LD ZEV 2030 sales goal would require a roughly 130 percent annual growth rate from 2023 to 2030 (Table 9-2)—an ambitious but achievable goal. LD PEV sales in China and Europe reached more than 15 percent of total LDV sales in 6 years by 2021 (IEA 2022a), which is comparable to what the United States would need to accomplish in the 6 years starting in 2023 to be on a trajectory to achieve a 50 percent sales share of ZEVs by 2030 (Table 9-2).

TABLE 9-2 Estimated Zero-Emission LDV Sales Growth Required to Reach 50 Percent of LDV Sales by 2030, and the Resulting Stock of Zero-Emission LDVs

Year	Yearly ZEV Sales (000s)	ZEV Share of LDV Sales (%)	ZEV Stock (000s)	ZEV Share of LDV Stock (%)
2021	608	4.08	2,350	0.90
2022	935	6.76	3,285	1.26
2023	1,215	7.82	4,500	1.71
2024	1,579	10.19	6,079	2.30
2025	2,052	13.28	8,131	3.06
2026	2,667	17.16	10,800	4.04
2027	3,465	22.39	14,260	5.30
2028	4,503	29.38	18,770	6.95
2029	5,852	38.33	24,620	9.08
2030	7,606	50.00	32,220	11.84

NOTES: ZEV sales and stock estimated based on growth rates needed to achieve the committee's ZEV sales goal. The initial 2021 and 2022 data are the actual sum of PEV sales in the United States—a mix of PHEVs and BEVs—and the very small number of FCEV sales, including the share of total sales and total LDV stock based on EIA (2022), Table 39. To reach the committee's ZEV sales goal, future projections assume a continued trend of declining share of PHEVs toward BEVs and FCEVs. SOURCE: Based on data from the U.S. Energy Information Administration, March 2020 (EIA 2022).

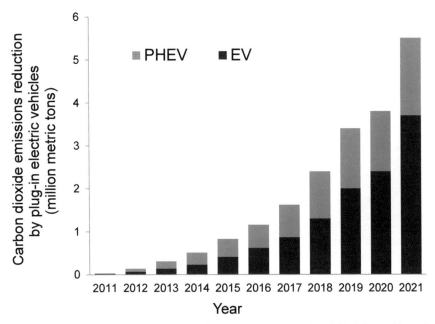

FIGURE 9-2 Carbon dioxide emissions reduction by PEVs, 2011–2021. SOURCE: Adapted from Gohlke et al. (2022).

The LD PEV and FCEV and charger tax credits in the IRA and the funding authorized and appropriated for charging infrastructure and other transport electrification in the IIJA will accelerate demand for PEVs.[6] The IRA also provides substantial tax credits to North American manufacturers of vehicles and batteries through 2032, and for production of low-carbon and net-zero liquid fuels, which will stimulate supply. Production of low-carbon and net-zero liquid fuels may help reduce future GHG emissions from the large post-2030 stock of LD ICEVs, long-distance heavy freight vehicles and vessels, and aircraft.

---

[6] It should be noted that federal legislation, such as the IIJA, IRA, and CHIPS and Science (CHIPS), contains different funding mechanisms. The IRA primarily consists of spending programs (appropriations) and tax expenditures. Spending programs can allocate federal resources to projects and activities up to the amount of their appropriation. By contrast, tax expenditures, such as the production tax credits in the IRA, typically have no limit on the amount that could be claimed by taxpayers. The IIJA consists of a mix of authorizations and appropriations, while CHIPS contains primarily authorizations. Authorizations are laws that establish or continue a federal program or agency and are typically passed by Congress for a set period of time, but authorizations require an appropriation before funds can be spent. Appropriations are laws that actually provide the money for government programs and must be passed by Congress every year in order for the government to continue to operate.

Whether demand will grow as fast as required to reach a 50 percent 2030 ZEV sales goal is uncertain and unknowable at this point. An analysis compared several economy-wide models of the impacts of the IRA versus reference scenarios, with five models including projections of EV sales share for LDVs. The cross-model comparison found that provisions in the IRA resulted in a projected 32–52 percent (41 percent average) of new LDV sales being EVs by 2030, as compared to 22–43 percent (31 percent average) in the reference scenarios (Bistline et al. 2023). The Energy Innovation and Princeton Zero Lab research groups each produced reports estimating LD and MHD EV sales based on implementation of the IRA alongside other mid-range policies, including the California emissions standard adoption by 17 states and the District of Columbia, and find, respectively, 48–61 percent LD and 39–48 percent HD EV sales (Slowik et al. 2023) and 52 percent LD and 58 percent MHD EV sales in 2030 (Jenkins et al. 2023). State and local policies discussed later in this section and in the section "GHG Reductions Through Transport Efficiency" can further stimulate ZEV demand toward achieving 2030 sales goals.

In addition to providing incentives for EV deployment, provisions in the IRA change eligibility of vehicles and buyers for tax credits, in some cases restricting eligibility, and in other cases expanding it. The IRA's requirements for North American LDV assembly; sources of battery minerals; mineral processing facility locations; and battery manufacturing facilities; as well as limits on the price of qualifying vehicles (§13401)[7] have reduced the number of PEV models eligible for federal tax credits. However, the law also expands the number of eligible vehicles by removing limits based on manufacturer volume and allowing used vehicle sales to qualify for tax credits. The law limits the buyers eligible for tax credits based on an income cap, aligned with a recent recommendation by the National Academies (NASEM 2021b), but also expands who is eligible, allowing the tax credit to be transferred to the dealer, who can pass along savings to buyers who do not have tax liability, or prefer an immediate cost reduction. All of the above IRA provisions, however, only apply to individual vehicle sales governed by Internal Revenue Code section 30D. Commercial vehicle sales, which include leased vehicles, are covered under different tax provisions (section 45W) and do not have any mineral sourcing, battery assembly, or income/price eligibility limits. It is not clear how the IRA's changes to tax credit eligibility for individual and commercial vehicle sales will affect vehicle supply and

---

[7] Eligibility for federal LDV tax credits require vehicle assembly in North America; 40 percent of battery minerals sourced or processed in the United States or U.S. trading partners with free-trade agreements in 2023 and 80 percent by 2026; and 50 percent of battery assembly in North America in 2023 and 100 percent by 2028. Income eligibility is capped at $300,000 adjusted gross income (AGI) for couples filing jointly and $150,000 AGI for individual filers (CRS 2022).

demand in the next few years, as automakers develop supply chains, build new battery factories, and change vehicle offerings so that their customers can capture tax credits. More details on the impact of critical minerals and materials on EV supply chains and tax credits are covered in the section "Barriers and Supportive Policies to Electrify Roadway Vehicles."

Charging Infrastructure

The $7.5 billion in funding authorized and appropriated for chargers in the IIJA (§11401 and Division J, Title VIII) will provide roughly 500,000 chargers (mostly on intercity highways) and make significant progress toward the total chargers needed to reach the committee's 2030 goals. Although most PEV charging in early deployment markets like Europe and California has been done either at home (50–80 percent) or workplaces (15–25 percent for owners that have a vehicle commute) (Hardman et al. 2018; NASEM 2015), prospective U.S. BEV buyers express concern about the lack of public charging stations (*Consumer Reports* 2022a). As of the first quarter of 2023, there were about 104,000 Level 2 workplace and public charging ports (at 46,000 locations), and about 30,000 DC Fast charging ports (at 7,000 locations) (Alliance for Automotive Innovation 2023)—far short of the 3 million the committee estimates will be needed by 2030.

Medium- and Heavy-Duty PEVs and Charging Infrastructure

The committee's ZEV sales goals include MHD trucks reaching 30 percent of sales by 2030. The market for PEVs is much more established for LDVs than for commercial MHD trucks. Roughly 2.3 million LD PEVs were registered in the United States by the end of 2021 (Alliance for Automotive Innovation 2022b, p. 7), compared with only 1,200 MHD BEV or FCEV trucks (Al-Alawi et al. 2022). Although lagging behind the personal vehicle EV market, several OEMs have MHD truck PEV models in development, and Tesla has produced a limited number of long-distance HD PEVs with capabilities that could prove consequential for HD truck PEV demand if proven in early use (Sriram 2022).

Commercial PEVs should be attractive to operators of local MHD trucks that can run a full day on a single charge and recharge at home-base depots overnight. The IRA (§13403) provides tax credits for purchase of commercial PEVs and FCEVs, representing the lesser of up to $40,000 or 30 percent of the purchase price for MHD trucks. The 30 percent subsidy of the purchase price of MHDs in the IRA could drive a 40 to 50 percent PEV sales share of MHD trucks by 2035 (Linn and Look 2022; Slowik et al. 2023).

Expansion of existing state efforts to subsidize fleet owner purchases of BEV trucks and chargers would also accelerate demand, as will the Environmental Protection Agency's (EPA's) 2023 approval of the California Air Resources Board's (CARB's) Advanced Clean Trucks (ACT) regulation. Among other provisions, this rule requires roughly half of new truck sales (40 percent of HD trucks) to be ZEVs by 2035. In April 2023, CARB petitioned EPA to allow enactment of a more expanded MHD truck ZEV rule that would phase out use of ICEVs by 2045 (CARB 2023).

Recent scenarios for needed MHD PEV chargers suggest a total investment requirement of $21 billion to $79 billion by 2030–2035 based on varied scenarios of fleet penetration and charger type installed.[8] The IRA's tax credits for commercial chargers (§13404) cover a maximum of 30 percent of cost, or $30,000 at each separate location, which expand to up to $100,000 if prevailing wage and apprenticeship criteria are met. Given these tax credits and the fuel and maintenance cost savings that truck BEVs are expected to provide commercial owners, the cost of chargers should not be a barrier for short-haul plug-in commercial MHD trucks.

Although representing about 10 percent of MHD trucks, long-distance HD trucks account for about half of MHD truck GHG emissions (Ledna et al. 2022). Large-scale electrification of long-range HD trucks, and associated charging requirements, remains uncertain at this time owing to the large batteries required with associated high demand for battery minerals, and heavy power and energy demand on the electric grid (Katsh et al. 2022; Slowik et al. 2023). Slower charging may also be an alternative to reduce power demands, but may be a deterrent for adoption of electric freight vehicles if they need to charge in the middle of their trips, impacting drivers' time on the road. Slowik et al. (2023) project that if these requirements could be met, the IRA could increase the sales share of HD, long-range trucks up to 17 percent by 2035 (see Slowik et al. 2023, Figure 7). Burnham et al. (2021) estimate that with aggressive assumptions about technology advancement, the total cost of ownership of HD PEV trucks could be competitive with ICEVs by 2035 (see Burnham et al. 2021; Figure 4-8).

Importance of Achieving 2030 ZEV Sales Goals

Achieving or approaching the committee's 2030 sales goal of 50 percent ZEVs is a central element of decarbonizing transportation by 2050. Doing so would quickly achieve the scale economies necessary to bring down the cost of ZEVs and create sustainable markets for private providers of charging infrastructure. More importantly,

---

[8] Estimated from McKenzie et al. (2021) and Phadke et al. (2021).

illustrative stock turnover scenarios for LDVs indicate that meeting a sales goal of 100 percent ZEVs by 2050 with no interim sales goal leads to a fleet in 2050 that still has 46 percent ICEVs (Figure 9-3), which would require significant volumes of fossil fuels and produce large amounts of GHG and other emissions. In contrast, meeting an interim 2030 goal of 50 percent ZEV sales as well as a 2050 goal of 100 percent ZEV sales would lead to a vehicle stock in 2050 that is about 10 percent ICEVs and 90 percent EVs. If the goal of 100 percent ZEV sales is further advanced to 2035, then there will be close to zero ICEVs operating, requiring fossil fuels, and producing emissions, in 2050. Any legacy ICE LDVs—combined with hard-to-electrify-transport vehicles such as aircraft, ships, locomotives, and long-distance heavy trucks—would continue to demand liquid fuels at a scale that would make it very challenging to fully decarbonize the transport sector by 2050 for reasons discussed in the sections "Equity and Other Crosscutting Issues" and "Actions to Expand the Innovation Toolkit" below, as well as in Chapter 8. Thus, it may be necessary to reduce the legacy stock of ICEVs even faster than would be accomplished by achieving the 50 percent 2030 ZEV sales share. One option for reducing the legacy stock of ICEVs is to accelerate scrappage of vehicles; however, as noted by the National Academies (2021b), the effectiveness and impacts of accelerated scrappage programs are not well understood and should be

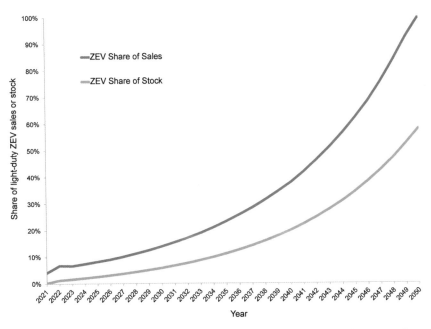

A

FIGURE 9-3 Illustrative model of ZEV share of LDV sales (blue) and stock (gray) if various sales goals are met. NOTES: Image (a) illustrates a scenario where a goal of 100 percent of sales is met in 2050 with no interim goal. *continued*

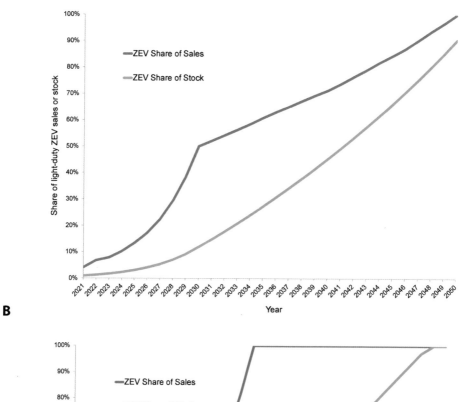

B

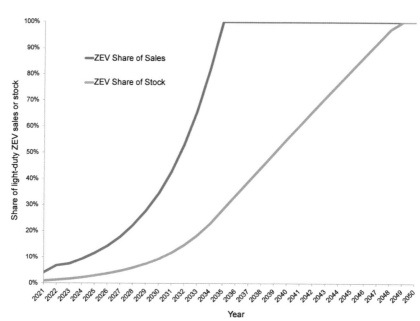

C

FIGURE 9-3 *Continued* Image (b) illustrates a scenario where the 2030 goal of 50 percent of sales is met as well as the 2050 100 percent sales goal; and image (c) illustrates the scenario where a sales goal of 100 percent of ZEVs is met in 2035. NOTE: The stock of ICEVs remaining in 2050 (represented by the gap between the ZEV stock and total stock) is largest if only a 100 percent ZEV sales goal in 2050 is met, and smallest if a 100 percent ZEV sales goal is met in 2035. SOURCES: Data from ANL (2023), DOE (2022), DOT (2021), and EIA (2022).

studied with respect to their benefits of emissions reduction, increasing ZEV sales, and addressing equity considerations as well as program costs.

Low and net-zero carbon liquid fuels are a possible route to decarbonization of vehicles, both for older legacy vehicles during the phase-out of combustion engines as well as for harder to electrify transportation subsectors. Cost-competitive net-zero carbon fuels at scale, while feasible, do not yet exist (NASEM 2021b). With innovation and either an explicit or implicit carbon price or regulation, cost-competitive net-zero carbon fuels may become available in the future, but today's options for low-carbon fuels are currently lacking. The actual carbon content of current low-carbon fuels on a full life-cycle basis varies considerably across feedstocks and is subject to uncertainties especially in the case of biofuels (NASEM 2022b). Sustained research and development (R&D), innovation, and demonstration at scale is required in order for truly net-zero carbon liquid fuels to contribute substantially to transport GHG reductions in the future (see the sections "Equity and Other Crosscutting Issues" and "Actions to Expand the Innovation Toolkit"). Such fuels will still be limited in use owing to concerns about biofuel production competing for land with agriculture and forest and marginal lands needed for carbon sinks (Chapter 8), and their harmful air pollution impacts, especially for environmental justice communities that are disproportionately impacted by vehicle emissions.

Barriers and Supportive Policies to Electrify Roadway Vehicles

Three important barriers to accelerated LD PEV market penetration remain: consumer discounting of vehicle operating cost savings, current lack of public charger availability, and manufacturer access to critical minerals and materials, especially within the context of IRA incentives. These barriers, and supportive policies to overcome barriers, are discussed next.

Barriers in Consumer Cost and Valuation of Electric Vehicles

Cost is one of the main considerations for vehicle market decision-making. Production costs[9] for PEVs and ICEVs are important inputs to the decisions of vehicle producers and sellers. Purchase and operating costs are important inputs to decisions of vehicle

---

[9] This section discusses vehicle costs, rather than vehicle prices, because vehicle costs offer a more fundamental comparison between PEVs and ICEVs. Vehicle costs are fundamental expenses, such as the cost of materials, labor, and capital to produce a vehicle, or the cost of fuel, supplies, and labor to operate and maintain a vehicle. Vehicle prices are impacted indirectly by the same fundamentals that underlie vehicle costs, as well as by additional variables such as automaker and vehicle dealer market strategies to influence sales of different models.

buyers and users. PEV production costs have been falling in comparison to rising costs of ICEVs owing to technology developments in BEVs, particularly the dramatically falling costs of batteries, and efficiency, safety, and other regulations impacting both ICEVs and PEVs (NASEM 2021b).

Two cost-parity points may be particularly salient for consumer decision-making: (1) first-cost parity, when purchase prices of PEVs and ICEVs are equal, and (2) total cost of ownership (TCO) parity, when the costs to purchase and operate PEVs and ICEVs are equal. The most recent National Academies' study of LDV efficiency technology estimates first-cost parity by 2025–2030 for manufacturers producing high volumes of EVs (NASEM 2021b). TCO parity is likely already present for some PEV models relative to comparable ICEVs and will be present by 2025 for additional models (Lutsey and Nicholas 2019). Going forward, PEVs will begin showing considerable cost savings owing to electric drivetrains that require 70–80 percent less energy to operate, less expensive fuel (electricity), and reduced electric drivetrain maintenance compared to ICEVs (DOE n.d.(c)). PHEVs are not projected to reach first cost parity with ICEVs in the 2030s because they contain significant aspects of both BEV and ICE powertrains, so have comparatively higher costs than BEVs (Lutsey and Nicholas 2019; NASEM 2021b).

Despite cost reductions, consumers do not show consistent behavior in purchasing vehicles based on TCO. It is not clear why this is the case, but it may be because consumers are unfamiliar with or inattentive to vehicle operating costs, that they value other vehicle attributes like acceleration or vehicle size, that they see predicted fuel savings as a "risky bet" that may not come to fruition with their vehicle purchase, or that TCO it is not communicated in a way that consumers can use, such as monthly cost of ownership comparisons (Dumortier et al. 2015; Greene 2011, 2019; Leard 2018; NASEM 2021b). Past battery reliability issues with certain vehicle models, and low reliability ratings for new electric SUV models, likely contribute to concerns about capturing maintenance cost savings (*Consumer Reports* 2022b). Consumer perception of first-cost parity is also impacted by the availability of purchase incentives from federal and state governments, which effectively reduce PEV purchase price for many consumers.

Even when faced with PEVs that are less expensive to produce and purchase than ICEVs, consumers may still see lack of charging infrastructure, cost of installing Level 2 chargers at home, or lack of model and product diversity as barriers to individual purchase decisions (*Consumer Reports* 2022a; NASEM 2021b). Passenger safety of electric vehicles appears comparable to conventional vehicles (IIHS-HILDI 2021), and thus does not appear to be a major consumer vehicle purchase consideration.

Charging Infrastructure Barriers

As discussed above, IIJA funding authorized and appropriated for public chargers and IRA tax credits for chargers may fall short of the amount required to achieve 2030 sales goals. Whether private investment in charger installation, operation, and maintenance will close the gap in time to meet 2030 sales share goals is unclear.

Shared charging stations for individuals without dedicated parking spaces will be needed to enable widespread electric vehicle adoption. About one-third of home-owners do not have a garage or carport that could be used for charging, but when renters are included 44 percent of households lack residential charging capability (*Consumer Reports* 2022a; DOE-VTO 2022). IRA §13404 provides substantial tax credits (estimated to total $1.7 billion through 2032) for home installation of chargers and to encourage siting of commercial charging stations in low-income or non-urban areas (CRS 2022).

Estimates of the funding required by 2030–2035 for LD charging infrastructure of all types (residential, workplace, and public) range from roughly $73 billion to $87 billion.[10] The estimated investment requirements for single-family and multi-unit residential buildings alone range from $39 billion to $45 billion. Tax credits for charger installation in the IRA fall far short of this amount, although strong demand for BEVs so far indicates that higher income households are not limited by tax credits. Cost estimates for local and intercity shared charging infrastructure range from $28 billion to $53 billion. The IIJA §11401 and Division J, Title VIII authorization and appropriation of $7.5 billion for intercity and other priority charging is but a fraction of this amount; however, all states have submitted plans to fund charging infrastructure improvements using this program, and the Federal Highway Administration has approved these plans (Joint Office of Energy and Transportation n.d.).

State, utility, and commercial funding sources will likely close some of this gap, with an expanding list of rebates and incentives being offered (AFDC n.d.(a)). Another state funding source is from the Volkswagen diesel emissions violation settlement agreement, which provides $2 billion to states for national ZEV enhancing investments, including charging and purchases of electric vehicles, primarily school buses (NASEO n.d.). Moreover, private sustainable business models for charger installation and operation are beginning to be demonstrated, partly through

---

[10] Range estimated from lowest to highest cost scenarios in McKenzie and Nigro (2021) and Phadke et al. (2021).

manufacturer-supported local and intercity networks, such as Tesla's network of chargers for its vehicle owners (Jockims 2022; Tesla n.d.). Recently, several large automakers have chosen to adopt Tesla's charging adapter and buy in to its charging network, which is reported to be the most reliable network by consumers (J.D. Power 2023). Public charging infrastructure is limited, but new programs are providing grants to extend charging networks. For example, investor-owned utilities have been approved to invest $3.3 billion as of early 2022 for charging infrastructure, education, and, in limited cases, vehicle purchase (Lepre 2022). Many local jurisdictions are adding chargers in public spaces using public funds. However, PEV owners report experiencing one in five public chargers out of service when they attempt to use them, adding consumer concerns about charger reliability to concerns about overall charger availability (J.D. Power 2023).

Critical Minerals and Materials Supply Barriers

The growth in decarbonization technologies (including batteries, motors, electronics, and other components) is expected to dominate future global needs for various critical materials (including lithium, cobalt, and nickel). All mineral demand for clean energy is expected to increase by 2–6 times by 2040, based on stated national policies ($2\times$) and a global net-zero scenario ($6\times$). The majority of the minerals demand in these scenarios is for EVs and battery electricity storage (IEA 2022b). The growth in demand for energy transition minerals will occur alongside a decrease in extraction of fossil fuels, particularly coal mining. Notably for transportation applications, EVs are more materials-intensive to produce than ICEVs,[11] and require larger amounts of critical minerals than ICEVs.

Critical minerals needs for EVs are dominated by the materials required to produce high-capacity batteries, especially for BEVs (IEA 2021b). DOE identified several important materials for the PEV and FCEV industries described in Table 9-3.

The U.S. Geological Survey also produces a list of critical minerals relevant to the entire U.S. economy, which is broader than those required for energy technologies (USGS 2022). The IRA defines critical minerals for EVs to include aluminum, cobalt, graphite (natural and synthetic), lithium, manganese, and nickel.

---

[11] Although they are more materials-intensive to build, EVs have lower life-cycle energy and GHG emissions than ICEVs. Representative life-cycle GHG emissions from BEVs are significantly less than ICEVs because the total emissions are dominated by the operational phase, where ICEV combustion emissions are very high (EPA n.d.; IEA 2021a).

TABLE 9-3 Important Minerals and Materials for Electric and Other Vehicles as Identified by DOE

Mineral or Material	Primary Vehicle Applications	Short-Term Criticality, 2020–2025	Medium-Term Criticality, 2025–2035
Silicon carbide	Power electronics	Near critical	Critical
Manganese	Lightweighting (EVs and ICEVs), batteries, fuel cells	Not critical	Not critical
Magnesium	Lightweighting (EVs and ICEVs)	Near critical	Critical
Aluminum	Lightweighting (EVs and ICEVs), batteries	Not critical	Near critical
Nickel	Lightweighting (EVs and ICEVs), batteries, fuel cells	Near critical	Critical
Silicon	Lightweighting (EVs and ICEVs)	Not critical	Near critical
Neodymium	Magnets	Critical	Critical
Praseodymium	Magnets	Near critical	Critical
Dysprosium	Magnets	Critical	Critical
Boron	Magnets	Not listed	Not listed
Iron	Magnets, batteries	Not listed	Not listed
Lithium	Batteries	Near critical	Critical
Cobalt	Batteries, fuel cells	Critical	Critical
Graphite	Batteries, fuel cells	Critical	Critical
Phosphorus	Batteries	Not critical	Not critical
Light rare-earth elements	Batteries	Not listed	Not listed
Electrical steel	Motors (EVs and ICEVs)	Near critical	Near critical
Copper	Motors, wiring (EVs and ICEVs)	Not critical	Near critical

continued

TABLE 9-3 Continued

Mineral or Material	Primary Vehicle Applications	Short-Term Criticality, 2020–2025	Medium-Term Criticality, 2025–2035
Platinum	Fuel cells, ICEV emissions catalysts	Near critical	Critical
Lanthanum	Fuel cells	Not listed	Not listed
Strontium	Fuel cells	Not listed	Not listed
Yttrium	Fuel cells	Not listed	Not listed
Zirconium	Fuel cells	Not listed	Not listed
Palladium	ICEV emissions catalysts	Not listed	Not listed
Rhenium	ICEV emissions catalysts	Not listed	Not listed

SOURCE: DOE (2023a).

Supply of critical materials is important both for automakers' ability to produce electric and conventional vehicles and also because of the critical minerals sourcing requirements in the IRA. The rapid global increases in EV production and vehicle range are major challenges to production of components for lithium-ion batteries and neodymium iron boron magnets, a key component of highly efficient electric motors. The chemical composition of lithium-ion EV batteries is changing as automakers improve their technologies to reduce cost; improve energy density, charging capability, and range; and reduce supply chain risks. The most common battery type has graphite anodes with nickel-manganese-cobalt cathodes of varying composition. Also common are nickel-cobalt-aluminum batteries. Lithium-iron-phosphate batteries are growing in use, especially in the Chinese market. In the past several years, cobalt use in batteries has dramatically decreased with increased nickel content, and that change in chemistry, along with growth in lithium-iron-phosphate batteries will further reduce supply chain constraints related to critical minerals.

The growth in minerals needs will require increases in production from existing mines and other mineral resources, development of new mines, and development of recycling technologies and facilities. Near-term supplies of minerals are expected to be able to meet demand; however, medium- and longer-term supply chain risks may slow the energy transition as soon as 2030–2035, if current production and investment trends continue (DOE 2023a; IEA 2022b). Current production of many critical minerals is concentrated geographically by location of mineral deposits, and of production or processing facilities, which is in part associated with the cost of mineral extraction and processing. Some sites of mineral production and processing have low labor and environmental standards and regulations, and so mineral and material production is sometimes associated with child and slave labor as well as environmental destruction, notably cobalt production in the Democratic Republic of the Congo. Companies are under pressure to end sourcing of materials produced under such conditions, which is a major motivator to produce batteries with limited cobalt content, for example. China is a dominant producer and processor of both critical minerals and materials, as well as finished battery components, cells, and packs, and is the largest market for electric vehicles. For example, China produces 60 percent of rare earth elements, and refines approximately 35 percent of nickel, 50–70 percent of lithium and cobalt, and nearly 90 percent of rare earth elements (IEA 2022b). Other major producers of critical minerals and materials include Australia, Chile, the Democratic Republic of the Congo, and Indonesia.

Concern for U.S. economic competitiveness and security in critical mineral and EV battery production led to provisions in the IRA to encourage domestic production and processing of battery minerals, domestic assembly of batteries, and sourcing from countries with free trade agreements with the United States (IEA 2022b). A recent study of the feasibility of the IRA's critical minerals requirements for EVs found that even with maximum availability of minerals from U.S. or free trade sources on the market value basis required by law, available materials are just shy of the requirements for lithium-iron phosphate and nickel cobalt-aluminum chemistries and reach only one-quarter to one-half of the requirements for nickel-manganese cobalt chemistries (Trost et al. 2023). Development of mines from discovery of a resource to first production takes more than 15 years on average, so in the medium term, there is a risk of critical minerals supply constraining electric vehicle production (IEA 2022b).

The IRA materials sourcing and EV battery assembly requirements also reduce the availability of electric vehicles eligible for tax credits in the near and possibly medium term; however, there are signs that automakers are adjusting their production of both minerals as well as batteries to capture higher tax credit value for their customers (Schwartz 2023). Countries are also considering signing new trade agreements to garner the higher value that their mineral exports would have if eligible to qualify under the IRA requirements, so more qualifying resources may become eligible over time (Bond et al. 2023). Recycling of used batteries and other components is another possible source of critical materials in the future. There is considerable commercial interest and authorized and appropriated funding support from the IIJA for facilities and processes (DOE n.d.(b)).

Supportive Policies

Federal, state, and local governments can help institute regulations, policies, and programs to overcome barriers and promote ZEV purchases toward achieving 2030 sales goals beyond the authorized and appropriated funding provided through the IIJA and IRA. As discussed in Chapter 5, innovative public engagement strategies may also be required to help people navigate the transition to new vehicles and technologies and the adjustments required to expectations and practices. Additionally, private sector actors including fleet owners and operators as well as manufacturers are taking actions that advance decarbonization owing to their own business interests. Some common strategies include

- *Federal Vehicle Fuel Economy and Emissions Standards.* The National Highway Traffic Safety Administration (NHSTA) and EPA have released new vehicle fuel economy and GHG emissions standards, respectively, under their existing legislative authorities. In April 2023, EPA proposed more stringent, performance-based GHG and criteria pollutant standards under the Clean Air Act for model year 2027–2032 light-, medium-, and heavy-duty vehicles. EPA projected that in model year (MY) 2032, the standards could result in nearly 70 percent BEV sales in the LD fleet, 40 percent in the medium-duty van and pickup fleet, 50 percent ZEV sales in vocational vehicles, 34 percent ZEV sales in day cab tractors, and 25 percent ZEV sales for sleeper cab tractors in MY 2032 (EPA 2023a,b,c,d). Upon a review mandated in Executive Order (EO) 13990, the Department of Transportation (DOT) revised the fuel economy standards for MY 2024–2026, which would result in a fleet-wide average fuel economy of 49 miles per gallon for MY 2026, and, according to DOT projections, yield an 8 percent reduction in CO_2 emissions from passenger cars and light trucks between 2021 and 2100 compared to the alternative of leaving the less stringent Safer Affordable Fuel Efficient Vehicles Rule in place (EO 13990 2021; NHTSA 2022). Under the same regulatory review required by EO 13990, in 2022 EPA restored its waiver of preemption of California's GHG and ZEV standards, allowing their Advanced Clean Cars (ACC) program to continue as well as allowing other states to adopt the California standards pursuant to Clean Air Act Section 117 (EPA 2022a). In July 2023, NHTSA continued to update its regulations under its existing authority from the Energy Policy and Conservation and Energy Independence and Security Acts, proposing an 18 percent increase in fuel economy from MY 2027–2032, with trucks requiring greater yearly fuel economy increases than cars (NHTSA 2023). The 2021 National Academies' report on setting national CAFE and GHG standards recommended that federal agencies "use all their delegated authority to drive the development and deployment of zero-emission vehicles (ZEVs)," especially EPA's continued setting of GHG standards based on growing availability of ZEVs and efforts to inform and educate consumers about EV fuel and maintenance cost savings (NASEM 2021b, p. 6).

- *State ZEV Purchase Mandates.* California ZEV regulation enacted in 2022, known as Advanced Clean Cars II (ACC II), requires 100 percent of LDV sales to be either BEV, FCEV, or PHEV by 2035. More than 40 percent of the LDV market may follow California's ZEV policy if the 17 states that currently adopt other California emissions standards also adopt the ZEV policy (Tal et al. 2022).[12] As noted earlier, California has added the ACT mandate that would require all new MHD trucks to be ZEVs by 2045. Slowik et al. (2023) estimate that if the 17 states that have previously adopted California's emissions policies adopt California's 2035 LDV and 2050 truck ZEV sales share mandates, LD PEV sales share would reach 63 percent and MHD PEV sales share would reach 56 percent by 2035, even with the phaseout of IRA incentives in 2032.

- *State Low Carbon Fuel Standards.* California includes a low carbon fuel standard (LCFS) as one of the main pillars of its efforts to promote ZEVs (CARB n.d.(b)). Oregon, Washington, and British Columbia have adopted similar standards. LCFSs set a gradually more stringent requirement for fuel providers to reduce the carbon intensity of liquid fuels brought to market. LCFSs do not dictate which fuels or technologies should be adopted; rather, they rely on market mechanisms to achieve a performance standard.

- *State Vehicle Purchase and Charger Installation Incentives.* As described in the previous section, the IRA has numerous tax credits to enhance supply and demand for PEVs and low carbon liquid fuel production. Several states (e.g., California, Colorado, Pennsylvania, Vermont, and Washington) and the District of Columbia offer additional tax credits or rebates for PEV purchase (AFDC n.d.(a)). Several states also provide tax credits for home charger installation, and/or state regulators allow utilities to provide discounts and charge the cost to their entire rate bases. Additional states could adopt these policies. State incentives for purchase of electric vehicles or charging infrastructure are countered by PEV-specific registration, charging, and other fees being instituted in more than 30 states. In many cases, these fees are described as an attempt to replace the gas tax revenue from BEVs that do not use gasoline; however, they are often set at levels much larger than the equivalent gas tax

---

[12] As of June 2023, 12 states have adopted (Maryland, Massachusetts, New York, Oregon, Vermont, Virginia, and Washington), partially adopted (Colorado), or plan to adopt (Delaware, New Jersey, and Rhode Island) ACC II. Six states have adopted California's previous emissions standards but have not adopted ACC II (Connecticut, Maine, Minnesota, Nevada, New Mexico, and Pennsylvania).

(Lee and Aton 2023; Preston 2022). The gap in transportation funding from decreasing gas tax revenues is only minorly owing to PEVs because they are a very small portion of the fleet, but will need to be addressed comprehensively at the state and federal levels as PEV deployment becomes widespread (TRB and NRC 2015).

- *Improvements in Charging Infrastructure.* Building code requirements that include home charging capability in new structures are being enacted by local jurisdictions and some states (Salcido et al. 2021). Standardization of plug-in connectors and open consumer search and reservation capability across charger providers and networks would also be helpful (Alliance for Transportation Electrification n.d.). Provision of charging infrastructure requires the capability of the local electric grid to provide sufficient power (see the section "Actions to Expand the Innovation Toolkit" and Chapter 6), which is something state utility regulators can consider when reviewing utility capital investment plans.

- *Innovation for Electric Vehicles.* Government support for battery research has been essential to make PEVs commercially competitive (DOE-VTO n.d.). Further areas of government R&D to improve vehicle performance and reduce cost are described in the section "Actions to Expand the Innovation Toolkit."

- *Other Supportive Policies and Actions, Including by Private Sector Actors.*
  - *Fleet Electrification.* Fleet electrification can be a useful additional strategy. Numerous private firms have announced plans to electrify part of their fleets, including Amazon, FedEx, and UPS (Domonosky 2021). Government agencies and nonprofit organizations that own fleets are also beginning to require purchases of electric vehicles, in part for operating cost and emissions reductions, including vehicle fleets from the postal service and the Department of Defense (DoD) (Department of the Army 2022; EO 14057 2021; USPS 2023). DoD and the National Aeronautics and Space Administration have used their procurement power to drive innovation in clean energy technologies, and continue to do so for vehicles (ITIF 2022). While rail transit generally is already electrified, transit and school buses can be electrified, especially as battery range improves and electric bus costs decline (Tong et al. 2017). The IIJA (§11101, §30017, §71101) includes $10 billion in authorization and appropriations for low- or no-emissions transit buses as well as electrification of school buses and associated charging infrastructure. These funds will address only a small share of the existing fleets of buses, but should help motivate a shift in demand, especially if ramping up production to achieve scale economies

can reduce initial cost premiums over diesel vehicles. Fleet operators can often centralize necessary charging infrastructure, an area where IRA tax credits and additional discounts provided by utilities could accelerate demand.

- o *Business Commitments from Manufacturers and Others.* Automakers are committing to produce EVs and other clean vehicles to meet customer demand, lower production costs, comply with regulations in many jurisdictions, meet corporate sustainability commitments, and compete in the global marketplace. In particular, major U.S. and global automaker electrification targets are produced in Table 9-4 (IEA 2023). As noted in IEA (2023), these announced automaker targets are often more ambitious than regulatory requirements or stated government pledges, but are generally non-binding. Some of the most ambitions pledges are for full electrification by 2025–2030, and less ambitious pledges are for a smaller percent of sales, a fixed number of models, or a mix of technologies including non-ZEV technologies.

Many other state and local actions can also boost demand for PEVs (Baldwin et al. 2021). Included among them are giving preferences in road and parking space allocation to PEVs and facilitating zoning and siting of charging infrastructure, including on public property such as municipal parking lots and garages and street parking spaces. Tax credits provided through the IRA and authorized and appropriated funding for chargers provided through the IIJA are market-pull strategies designed to incentivize consumers. Comparable, but limited, state tax credits for vehicle purchases and home charging serve the same function. California's experience indicates that its ZEV sales mandate, a market-push strategy, is its single most important ZEV policy, but it is complemented by its LCFS and several other state programs (Sperling et al. 2020). Federal fuel economy and GHG emissions standards, although less direct, can also serve to push manufacturers to produce and market PEVs. These market-pull and market-push strategies complement one another and will have even greater influence if the up to 17 states that follow California's ZEV purchase mandate implement comparable policies (Slowik et al. 2023; Tal et al. 2022).

Electrification of Railroads, Ships, and Aircraft

Electrifying vehicles coupled with net-zero electricity generation is a decarbonization strategy that can be applied to all types of vehicles. However, range, power, and weight issues make battery electric approaches challenging for trains, ships, and aircraft.

TABLE 9-4 Automakers' Electrification Targets for LDV Since 2022

Automaker	Target	Region	Group/Brand
Ford	600,000 BEV sales by 2026	Europe	Group
General Motors	400,000 EV sales from 2022–2024; 1 million EV production capacity in 2025	North America	Group
Volkswagen	Targets fully electric production by 2033 (brought forward by 2 years)	Europe	Brand
Toyota	1,500,000 BEV sales; introduce 10 additional models by 2026; committed to a multi-pathway approach to reduce CO_2, including continuing development of FCEVs and PHEVs	Global	Group
Mazda	Expects at least 25% of sales globally to be BEV in 2030	Global	Group
Honda	Aims to launch 30 EV models globally by 2030, with production volume of more than 2 million units annually	Global	Group
Nissan	Updated global target to 44% EV sales by 2026 (with regional subtargets for Europe, Japan, China, and the United States) and to 55% EV sales by 2030	Global	Group
Mitsubishi	Plans for 100% of EV sales by 2035 and 50% EV sales by 2030 in their Environmental Targets 2030	Global	Group
Porsche	80% of sales to be electric by 2030	Europe	Brand
BMW Group	Cumulative sales of more than 2 million EVs by the end of 2025; EV sales shares of 30% by 2025, 50% by 2030	Global	Group
MINI and Rolls-Royce	Aims to have fully electric line-up by 2030	Global	Brand
Lancia	All new model launches from 2026 to be electric; to sell 100% EVs by 2028	Global	Brand
Jaguar	Aims to go all-electric by 2025	Global	Brand
Land Rover	Aims to go all-electric by 2036	Global	Brand

TABLE 9-4 Continued

Automaker	Target	Region	Group/Brand
BYD	Ceased ICE vehicle production; has produced only EVs since March 2022	Global	Brand
Geely	600,000 EV sales over this year	Global	Group
SAIC-GM-Wuling	Annual sales of 1 million NEVs by 2023 including small EVs; 40% NEVs in total sales by 2025	China	Group
BAIC Group	NEVs to make up 1 million of 3 million in total sales in 2025	China	Group
FAW Group	Half of its total 1 million sales target by 2025 to be NEVs; 1.5 million vehicles (mostly NEVs) sold by 2030	China	Group

NOTE: Note that most all-electric automakers such as Tesla are not represented in this table. SOURCE: IEA (2023), https://iea.blob.core.windows.net/assets/dacf14d2-eabc-498a-8263-9f97fd5dc327/GEVO2023.pdf. CC BY 4.0.

Locomotives and Rail Vehicles

Because locomotives and rail vehicles operate on fixed routes with dedicated infra-structure, electric power can be provided to moving vehicles, rather than being stored on the vehicle as for roadway vehicles. Locomotives and rail vehicles can use catenary or third rail for electricity, but the catenary is expensive, vulnerable to failures, and unaesthetic to many. The alternative of third rail power is also expensive, may require power distribution upgrades, and raises safety concerns for maintenance and yard workers and public trespassers across and along rail lines. While catenary and third rail can be used effectively in some situations, they are unlikely to prove practical for all rail lines. One area where electrification may be adopted more quickly is in rail yards, analogous to port operations, where electrification, including electric locomotives, can provide emissions reductions in and near urban areas as well as benefit from easier access to infrastructure and no requirement for long-distance travel.

Battery electric locomotives are now commercially available and are being used in demonstrations (Popovich et al. 2021). Cost, range, and weight limit their wide-spread adoption. Fuel cells are also feasible, but also currently have significant cost penalties and safety concerns. Until technology and IRA incentives significantly reduce the cost of fuel cells and low-carbon hydrogen (see the section "Actions to Expand the Innovation Toolkit"), widespread adoption of battery electric or fuel cell locomotives will likely be slow.

Ships

As with locomotives, battery electric and fuel cell ships are available and in use for limited applications. Large marine vessels are good candidates for fuel cells, but cost-effectively producing and providing low-carbon hydrogen or ammonia produced from hydrogen at port locations requires further innovation. Nuclear propulsion is used for submarines and large military vessels but higher cost limits widespread adoption of these alternative propulsion systems. Providing net-zero shore power for ships in port is one strategy that can be cost effective and reduce emissions in urban areas, but requires substantial port investments (EPA 2022b).

Aircraft

Aviation represents 10 percent, and growing, of transportation GHG emissions (EPA 2021; see Figure 9-1). Aircraft, however, require high power and are severely constrained by weight, so the prospect of electrifying air travel with batteries is daunting. Battery electric technologies can be employed for short flights on small aircraft (air taxis) or drones. Operational improvements and aircraft design changes can also provide emissions reductions. For example, airports and passenger and freight airlines could increase use of electric vehicles in multiple airport operations, including to tow aircraft to and from runways rather than aircraft taxiing (NREL 2017). However, the largest improvement will likely be from low-carbon fuels, which will not be available in large volumes for several decades (see the section "Actions to Expand the Innovation Toolkit"). Figure 9-4 shows a possible scenario for reduction of emissions from flights within the United States and international flights by U.S. carriers as developed by the Federal Aviation Administration (FAA n.d.). The history of emissions from passenger and freight aviation operations from 2020–2023 are plotted in white. A future scenario shows the trajectory of emissions from 2019 levels to zero by 2050. Emissions assuming frozen 2019 technology is the base case, and the emissions reductions from various technology, operations, and fuel improvements are shown as different-colored wedges.

In addition to on-road vehicles, other transportation activities may also be usefully electrified, and in some cases are already being electrified, driven by local air quality or cost of ownership considerations. Examples include pipeline processes (e.g., pumping), port operations including drayage vehicles, and off-road vehicles.

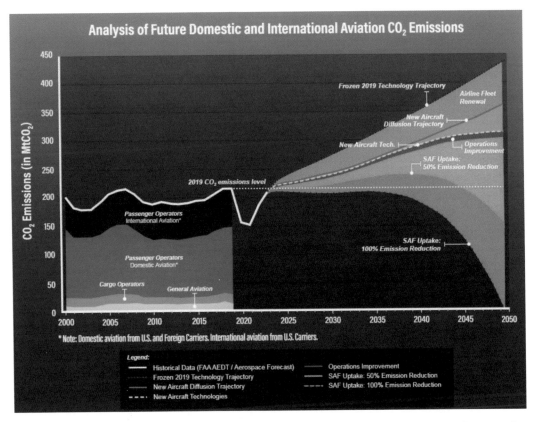

FIGURE 9-4 Analysis of future domestic and international aviation CO_2 emissions. SOURCE: Courtesy of FAA (n.d.).

Findings and Recommendations

Finding 9-1: The IIJA and IRA authorized and appropriated funding and tax credits to electrify road vehicles offer the largest, most cost-effective opportunity to decarbonize the transportation sector and will make substantial progress toward achieving the 2030 ZEV sales share goals of the committee's first report and the nation's long-term strategy for decarbonization. However, further efforts are needed to ensure that 2030 sales goals are not only met but exceeded in order to substantially reduce the legacy stock of ICEVs by 2050.

Finding 9-2: Barriers to achieving 2030 sales goals include consumer purchase decisions for LDVs that are more driven by initial vehicle prices rather than the total cost of ownership and operation, concerns about charging capability, and barriers to sourcing critical minerals. Expanded public sector support beyond that available from the IIJA and IRA for building out charger infrastructure over the next decade or so will help overcome consumer reticence about battery vehicle range and provide charging options for residents without home charging capability. Commercial MHD truck investments are more likely to be made on a life-cycle economic basis than are consumer purchases of LDVs, but short-term incentives to accelerate commercial operator adoption of PEVs would have valuable environmental and public health benefits.

Finding 9-3: Critical materials production will need to grow in multiples of current levels to accommodate the growth of electric vehicle manufacture in the United States and globally. This may constrain electric vehicle development in the medium term. Concentration of both mineral deposits and especially mining and processing facilities presents a risk to secure supply chains for both companies seeking to produce decarbonization technologies, as well as countries relying on these technologies for their national decarbonization plans.

Finding 9-4: Boosting demand for personal and commercial PEVs beyond what the IIJA and IRA are expected to achieve will encourage faster development of PEV manufacturing and achievement of scale economies that will further reduce PEV costs to consumers. Achieving the higher growth rates required to reach or exceed the 2030 50 percent sales share goals of ZEVs, rather than simply reaching the 2050 100 percent sales goal, would substantially reduce the number of vehicles dependent on net-zero-emission liquid fuels after 2030 as well as after 2050.

Recommendation 9-1: *Accelerate the Adoption of Battery Electric Vehicles.* **Federal, state, and local government policies should build on the provisions of the Inflation Reduction Act and the Infrastructure Investment and Jobs Act to accelerate the cost-effective adoption of battery electric roadway vehicles, through**

a. **Continued ratcheting up of federal fuel economy and greenhouse gas vehicle emissions standards by the National Highway Traffic Safety Administration and Environmental Protection Agency to achieve a lower bound of 50 percent zero-emission vehicle (ZEV) sales by 2030;**
b. **Federal and state adoption of ZEV sales mandates in line with California's ZEV goals and supportive policies to achieve 100 percent new light-duty ZEV sales by 2035 and 100 percent new medium- and heavy-duty (MHD) ZEV sales by 2045;**
c. **Enactment of a carbon tax by the U.S. Congress to facilitate decarbonization of the whole transportation fleet;**
d. **Expanded state funding for vehicle purchase incentives and rebates for home charging infrastructure targeted to low- and moderate-income households and, through state utility regulation and**

oversight, allowing utilities to offer incentives for home charger installation and to cover the cost from their rate bases;

e. Expanded state and utility incentives for MHD truck purchase and charger installation designed to accelerate conversion to electric drive;

f. Expanded state and local support (funding, permitting, and allocation of public infrastructure) for build-out of public chargers;

g. State and local funding for conversion of public vehicle fleets, including transit and school buses, to electric drive; and

h. Expanded public engagement programs to help consumers better understand and navigate the changes entailed in adopting and adapting their practices and household infrastructures to the capabilities and requirements of electric vehicles.

Recommendation 9-2: *Promote Vehicle Electrification at Ports and Airports.* Applications for vehicle electrification should be promoted by ports and airports (and their state and local government owners) beyond the incentives available in the Inflation Reduction Act and the Infrastructure Investment and Jobs Act as the plug-in electric vehicle fleet expands. Examples include providing shore power for ships, converting port equipment and drayage trucks to electric power, towing aircraft to and from runways with electric vehicles, and converting other airport ground operations vehicles for baggage movement and other logistics to electric drive.

GHG REDUCTION THROUGH TRANSPORT EFFICIENCY

Improving the efficiency of energy used to provide transportation of goods and people generally results in lower impacts from transportation systems, including fewer GHG emissions. A wide range of measures to improve travel efficiency can reduce transport GHG emissions. Although these efficiency improvements would make only modest contributions to reaching 2030 and 2050 decarbonization goals, they could have co-benefits such as reducing travel costs and increasing overall economic efficiency.

Internal Combustion Engine Vehicle Efficiency Improvements

Fuel economy of LD ICE vehicles has improved significantly over the past 50 years. This efficiency improvement resulted in a drop in total fuel use from 2005 to 2020 even though number of vehicle registrations increased (NASEM 2021b) (Figure 9-5). Regulatory requirements such as the federal CAFE standards for passenger vehicles and

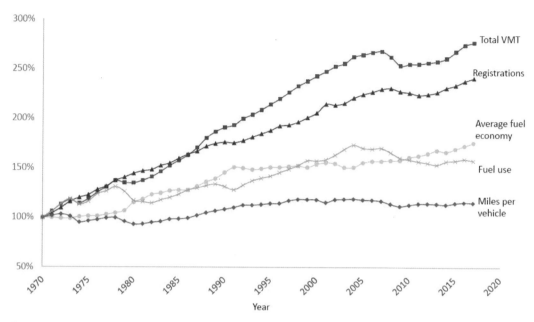

FIGURE 9-5 LDV transportation characteristics, including total vehicle miles traveled, vehicle registrations, average fuel economy, and miles per vehicle. SOURCE: NASEM (2021b), generated with data from Davis and Boundy (2020).

trucks, which was established in 1975 and has tightened over time, have provided a major incentive for this efficiency improvement and will remain important because of the tens of millions of fossil fuel vehicles that are likely to continue to be in operation for decades to come. Numerous technical means to improve fuel economy exist, so efficiency improvements are likely to continue in the next few decades. For example, vehicle lightweighting can dramatically improve energy efficiency for either ICEs or EVs (Lovins 2020). Fuel economy regulations require consideration of not only technical opportunities but also their costs and benefits to the consumer, manufacturers, and the economy as a whole, including aspects such as safety[13] and national energy security.

---

[13] Various factors affect vehicle safety, including factors associated with the driver, transportation system, and vehicles. Implications of fuel economy regulations have been small, relative to primary determinants of vehicle safety; however, they are important to understand and address. The National Academies' Committee on Assessment of Technologies for Improving Fuel Economy of Light-Duty Vehicles—Phase 3 examined the implications of a future mixed fleet with various technologies, especially a mix of different vehicle weights, sizes (especially a shift from sedans to crossovers, SUVs, and trucks), and safety technologies, and recommended that NHTSA study potential changes in mass disparity and societal safety risk (NASEM 2021b).

For MHD roadway vehicles, regulation of fuel economy has only occurred over the past 10 years (NASEM 2020). As a result, fuel use efficiency improvements in this transport subsector have been scant. Over the next 2 decades, only a 13 percent improvement (from 6.9 miles per gallon to 7.8 miles per gallon) is anticipated for freight trucks. As a result, other means to reduce GHG emissions are priorities for these vehicles, such as alternative fuels or freight system improvements, as discussed below.

Airplanes, locomotives, and ships also present opportunities for fuel efficiency improvements (Lovins 2021). For example, from 1960 to 2020, revenue passenger kilometers per kilogram of CO_2 emitted from airplanes grew eightfold (Lee et al. 2021). Although these transport sectors have strong economic incentives to reduce fuel consumption, there are no regulatory requirements for fuel economy applied to them. As discussed previously, alternative fuels or electrification, where possible, are primary means to achieve net-zero vehicle movements for airplanes, locomotives, and ships. Development of sustainable alternative fuels for these sources—which represent 14 percent of transportation GHG emissions (see Figure 9-1)—is a high priority of innovation and federal and private RD&D.

Traffic Flow Improvements

Improving efficiency with better traffic flow management is a widely held goal for agencies and firms. Operational improvements are also being pursued, such as adaptive traffic signal systems or optimized airplane routing and operations. Tracking vehicle locations and using vehicle connectivity aid these operational strategies. Avoiding vehicle crashes can also improve traffic flow by avoiding the congestion resulting from incidents. Partial vehicle automation, including positive train control for freight and passenger rail, and active roadway vehicle braking can help avoid such crashes.

Road traffic flow management improvements will typically not be the result of federal actions, as there is no current federal role in managing on-road traffic, unlike air traffic, which is a fully federal responsibility. Urban roadway networks are a state and local responsibility. Rail networks are managed and owned by private corporations for the most part, except for Amtrak's ownership of the Northeast Corridor rail lines and state ownership of limited other mileage used for passenger rail. The federal government can help by funding research on improvements and participating in standard setting.

As travel demand grows, the GHG emissions reduction potential of improved traffic flow management will erode. In essence, greater traffic volumes without infrastructure capacity expansions will reduce the efficiency of traffic flow owing to congestion (NASEM 2019d). While improved traffic flow management may reduce congestion and avoid bottlenecks, it likely will not achieve significant GHG emissions reductions. Even so, it can mitigate GHG emission increases from ICEVs during the decades in which they will continue to operate, and pricing lane additions through electronic tolls that vary with demand can mitigate induced travel (Milam et al. 2017, p. 14).

Rail and Marine Freight Efficiency Improvements

While non-truck freight movements are only a fraction of overall transport GHG emissions, there is potential for efficiency improvements in other modes, including rail and marine. Freight railroads have made considerable reductions in fuel expended per ton-mile in the past decades by rebuilding their tracks, carrying heavier weights, pulling more cars, and more carefully managing speed and acceleration. Large ships have also reduced fuel per ton-mile, largely through scale economies associated with larger container ships. However, infrastructure constraints limit the amount of further improvement achievable in a cost-effective fashion. For example, double tracking railroads would improve movement efficiency, but obtaining the required rights-of-way is difficult and raising bridges is expensive. Nevertheless, infrastructure investments to remove major freight bottlenecks can be pursued for situations such as congested land-side access to ports and heavily congested interstate highway interchanges.

Modal shifts in freight transportation offer another means to improve efficiency. While routing on rail or inland waterways may be longer in both distance and time, and short-sea shipping along coastal routes has so far failed to gain substantial market share in the United States, these modes have lower fuel use and GHG emissions per ton-mile of freight movement (Corbett et al. 2008). Shifting freight from trucks to rail or water for long-distance movements can reduce overall GHG emissions. The carbon tax endorsed in the committee's first report and in Recommendation 1-1 of this report, or increased federal and state motor fuels taxes on trucks, would encourage such mode shifting. Efficiencies may also be achieved through information technology. Delivery loads and routes can be optimized. Greater consolidation of freight to fill combination truck trailers (within size and weight limits) can improve efficiency.

Freight transport efficiency can help reduce GHG emissions and reduce overall costs. For example, data sharing, communication, and more efficient routing of vehicles, both within fleets and also among all vehicles, ports, and other origins and destinations can reduce congestion, wait times, and emissions. While GHG emission reductions for modes other than trucks will be modest, the improvements can be cost effective, and may be important for modes like marine freight in reducing port congestion and associated emissions.

Automation and Connectivity

Vehicle automation and connectivity is already appearing in new vehicles to some extent. Partial automation and driver warning systems have reduced the numbers of collisions (Flannagan and Leslie 2020). Adaptive cruise control has improved fuel efficiency with smoother driving. Highly automated vehicles could have greater savings, including aerodynamic savings from truck platooning. Safer automated vehicles might also be smaller and lighter than vehicles in a comparable non-automated fleet. Depending on public policies, in a scenario with greater automation, travel demand may increase owing to traveler shifts from transit into automated vehicles, and automated vehicle trips without passengers. Automation and connectivity can increase or decrease energy efficiency of vehicle operation in both scenarios with limited automation as well as scenarios with nearly full penetration of highly automated vehicles (NASEM 2021b). Changes to vehicle ownership models and increased vehicle electrification to facilitate automation may add to the complexity of predicting efficiency outcomes.

Information and Communications Technology Substitutes for Transport

In many cases, information and communications technologies (ICT) can substitute for travel and thus reduce GHG emissions. Telework can reduce or eliminate commuting travel for those workers able to use this option. E-commerce can replace some consumer shopping trips (Matthews et al. 2001). Video conferencing can reduce some travel for meetings, including long-distance trips by air. School and medical trips can be reduced through online learning and virtual visits with medical practitioners, respectively. Of course, these substitutions often result in only a partial reduction of motor vehicle trips and travel and will not always replace the quality of in-person interactions. For example, teleworkers generally travel more overall than workers who do not telecommute (Speroni and Taylor 2023). E-commerce purchases increase delivery vehicle travel while also reducing personal vehicle shopping trips (Matthews et al. 2001). Most advantageous in reducing GHG emissions is use of enhanced videoconferencing technologies that became available during the COVID-19 pandemic to substitute for energy-intensive trips by aircraft. A variety of federal, state, and local policies can affect ICT travel substitution. For example, public support for Internet access in rural and less-developed urban areas would support wider use of ICT and provide equity benefits.

Biking, Shared Rides, Transit, and Walking Coupled with Development Density

Reducing single-person trips in LDVs by shifting to bicycling, shared rides, transit, or walking can reduce vehicle miles traveled and vehicle emissions. Biking and walking also offer the advantage of healthy exercise, as discussed in Chapter 3 (Public Health).[14] The impact of biking, walking, and transit on transportation GHG emissions, especially in the near term, is limited by the nature of transportation needs in the United States today, which are themselves heavily influenced by land use patterns. Trips tend to be relatively long, and thus accomplished by personal vehicles, which therefore results in the majority of emissions. The majority of annual average person trips (about 83 percent) is in personal motor vehicles, with walking at about 10 percent, public transit at about 2.5 percent, and biking at about 1 percent (ORNL 2017). Passenger miles of travel are the more important comparison for GHG emissions, and are even more skewed toward personal motor vehicles. Passenger miles (excluding aviation and intercity rail) are dominated by personal motor vehicles (86 percent), with public transit at 1 percent, walking at 0.6 percent, and biking at 0.15 percent (BTS 2021b). A variety of policies can affect these modal shares somewhat, such as increasing transit service frequency or providing dedicated bike lanes and sidewalks (NASEM 2021e), but it would take very large shifts away from personal motor vehicles to significantly reduce total passenger GHG emissions. For example, doubling walking, biking, and transit trips, and assuming that this doubling replaced trips by personal motor vehicles, would reduce personal motor vehicle miles of travel from 86 to 82.5 percent, and the GHG emission benefits would be lessened to the extent that substituted trips were made in PEVs.

Changes to denser, mixed-use, and active-transportation friendly land use patterns could also reduce transportation GHGs, but on a smaller scale and more slowly than policies aimed at electrifying vehicles. Making communities more walkable and bikeable through density increases, mixed-use development, and improving transit service could reduce the reliance on personal vehicle travel, and serve important public health and equity policy goals (DOT 2022). Such policies are being actively pursued in many communities; however, fragmented regional governance of land use, entrenched zoning policies, and public preferences that determine residential and commercial development patterns can be slow to change (Cervero 2003; Savitch and Adhikari 2017; Schuetz 2022).[15] Also, the turnover of the LDV

---

[14] Cycling, like other transportation modes, has applications where it is more or less effective and accessible. Cycling tends to be seasonal, limited by topography, and not equally accessible across age and ability groups. Electric assisted bicycles are a growing aspect of cycling for transportation that can better serve different ages, ability groups, and types of trips.

[15] See also A. Downs, 1992, *Stuck in Traffic*, Brookings, https://www.brookings.edu/book/stuck-in-traffic, and A. Downs, 2004, *Still Stuck in Traffic*, Brookings, https://www.brookings.edu/book/still-stuck-in-traffic.

fleet is inherently faster, on the order of 1–2 decades, relative to land development, which, during periods of slow economic and population growth, turns over on the timescale of a century or more.

State policies can have an influence on development patterns. States have the power under their constitutions to override local autonomy on zoning and land use, but doing so to increase density has been rare to date. As important exceptions, Oregon (2019) and California (2021) passed state laws that would override local zoning restrictions on density increases that prohibit conversion of single-family lots to duplexes, and up to quadruplexes in limited cases (*California City News* 2021; Shumway 2021). However, density increases that double the density of single-family residential development, and assuming mixed-use zoning to reduce motor vehicle trips, would likely have modest impacts on reducing auto trips (TRB 2009). Although increases in residential density have been accepted in cities such as Portland, Oregon, and are promoted by the "yes in my backyard" movement, strong resistance to changes in residential zoning has a long history in the United States (Downs 1992, 2004).

California has gone further to comprehensively address housing availability and affordability. In 2022, the state enacted a bevy of new and modified laws to streamline permitting, increase development density, restrict parking, and other measures that could have greater impact on motor vehicle trips (Maclean et al. 2022). To reach California's goal to fully decarbonize transportation by 2045, Brown et al. (2021) assumed that 15 percent of California's transport GHG reduction could be achieved by reducing per-capita vehicle miles traveled through pricing roads and parking (−5.5 percent); Transit Oriented Development, Active Transportation, and Transit (−4.7 percent); telework (−2.5 percent); and other strategies (−2.3 percent).

Transport Infrastructure Construction and Maintenance

Relative to direct vehicle emissions, the GHG emissions from infrastructure construction and maintenance are relatively modest but still important. For example, including them would result in a roughly 10 percent increase over passenger vehicle emissions in the case of highways (Chester and Horvath 2009). Emissions from infrastructure are associated with purchases of carbon-intensive materials such as cement. The largest supply chain inputs into new highways, bridges, and other horizontal construction are engineering services, wholesale trade, trucking, concrete, asphalt, stone, concrete products, asphalt felts, petroleum, steel, and fabricated metal (Hendrickson and Horvath 2000). A variety of approaches can be used to reduce emissions from these inputs. Reducing direct emissions from carbon intensive inputs can be a very effective approach. Cement and steel production can use innovative processes to reduce GHG emissions in their production, as discussed in Chapter 10 (Industrial Decarbonization). Vehicles used for construction and maintenance can be electrified.

The IRA includes several new programs to help speed decarbonization of infrastructure materials (Margolies 2022). Industries can apply for grants from the $5.8 billion Advanced Industrial Facilities Deployment Program (§50161) as well as receive tax credits to speed decarbonization (§13502). The IRA included $4.5 billion for federal procurement of low-carbon materials for projects (§60116, §60502, §60503, §60506) and designated another $250 million to aid in producing Environmental Product Declarations that document carbon intensity of materials (§60112).

Infrastructure design, construction, and maintenance processes also provide opportunities for GHG emission reductions (Rangelov et al. 2022; Santero et al. 2011). For design, material use can be reduced, or less GHG intensive materials employed. For example, scrap tire material and asphalt shingles are used in pavements to reduce the need to produce new materials (TRB 2013), and asphalt is commonly recycled in repaving projects. For construction, GHG emissions may be reduced through procurement, contracting, and operational changes such as work zone controls (NASEM 2019e). Maintenance equipment and vehicles can be electrified. Standards and best practice guides can be formulated at the federal or state levels for such changes.

Findings and Recommendations

Finding 9-5: Efficiency improvements can reduce transportation GHG emissions between now and 2050. GHG emissions from ICE vehicles, vessels, and aviation operations not easily electrified can be reduced cost effectively through enhanced fuel economy, traffic flow management, freight operational efficiencies and mode shift, enhanced mode choices, and land use and zoning policies. Enhancing efficiency also has direct economic and non-GHG environmental benefits even though the marginal GHG reductions from these efforts may be modest. Aside from federal fuel economy standards (covered in Recommendation 9-1a) and more efficient management of air traffic, other efficiency improvements depend on the actions of the private sector and state and local governments. The efficiency recommendations listed below build on existing efforts and policies.

Recommendation 9-3: *Pursue Cost-Effective Efficiency Improvements to Reduce Greenhouse Gas (GHG) Emissions.* **Private companies and state and local governments should pursue cost-effective transportation efficiency**

improvements as a means to further reduce GHG emissions and advance other social goals.

   a. In the absence of federal action on a carbon tax or increased fossil fuel taxes, states should incentivize private-sector freight efficiency and mode-shift to less carbon-intensive modes through fuel taxes based on the carbon content of motor fuels and with a share of the revenues allocated for equity-enhancing strategies.
   b. States and local governments should enhance mode choice wherever feasible and environmentally cost effective through transit expansion, expanded sidewalks and separated bike lanes, and zoning for mixed uses and densification of development to reduce distances between origins and destinations.

Finding 9-6: Transportation infrastructure construction and maintenance represents a small share of transportation GHG emissions relative to combustion from operation of ICE vehicles, but they can be further reduced in cost-effective ways by reducing the full life-cycle carbon content of input materials, enhanced use of recycled materials, electrification of construction equipment and vehicles, and low-carbon materials procurement standards.

Recommendation 9-4: *Pursue Infrastructure Designs, Standards, Specifications, and Procedures That Effectively Reduce Transportation Carbon Emissions.* **State Departments of Transportation, the American Association of State Highway and Transportation Officials, the American Road and Transportation Builders Association, and other specialized transportation infrastructure materials and construction associations should pursue infrastructure designs, consensus standards, specifications for materials and construction, and procurement procedures that cost-effectively reduce carbon emissions over the life cycle of transportation infrastructure.**

EQUITY AND OTHER CROSSCUTTING ISSUES

Energy and Climate Justice and Equity

For decades, transportation policies have focused on technical aspects such as optimizing the performance and efficiency of the transport system. This focus has had broad economic benefits but has also imposed deep inequities. Some U.S. populations have enjoyed the fruits of the improving system, while others—such as low-income and rural populations and people of color—have experienced fewer benefits, greater economic burdens, and increased health risks from exposure to vehicle emissions and noise (Chapter 3). Racially segregated neighborhoods, attributable to a history of housing discrimination and redlining, are disproportionately located in central cities with poorer access to employment and amenities (Blumenberg 2017). They have also had their communities divided or destroyed by transportation infrastructure (Martens 2016; NASEM 2021e). Electrification of transportation vehicles will provide substantial benefits to low- and moderate-income residents living near transportation infrastructure and ports by reducing vehicle emissions harmful to public health. Although they will benefit from lower emissions exposures, without additional policies and programs, the initial growth in EVs will have limited benefits for the lower-income drivers or owners of vehicles who will find it hard to purchase a new EV, nor will EV growth result in expanded mobility options for those who cannot afford to own vehicles. As discussed in the section "2030 ZEV Sales Goals and Barriers," the IRA includes specific efforts to make PEVs and charging accessible to low-income households, but broad penetration of PEV ownership in this income group will require additional efforts.

Although roughly 20 percent of low-income households depend on public transportation to reach jobs and other destinations, about 80 percent live in households with a least one vehicle and rely on vehicle sharing and other LDV strategies for access to employment and amenities (Blumenberg 2017; Figure 13-2). Despite the considerable variation in employment accessibility across metro areas, households with automobiles have access to far more jobs than transit-dependent households, as well as higher earnings and job tenure (Smart and Klein 2020). Moreover, as jobs and poverty have increasingly suburbanized over time, and given the difficulty fixed-route transit has in serving suburban and exurban geographies, automobile access has joined transit service to low-income communities as an equally important equity issue (Romero-Lankao et al. 2022). As the personal vehicle fleet shifts to PEVs, equitable access to these vehicles will also loom large.

Equity Policies and Programs

Equity policies and programs can improve access to mobility for underserved populations as the transportation system decarbonizes. As noted above, 20 percent of low-income households depend on public transit. Transit service to low-income, transit-dependent populations is provided across the country by thousands of transit agencies, but service is typically limited and infrequent outside of a few transit-rich urbanized areas. Funding could be supplemented for this purpose and could be drawn from sources that also serve to reduce ICEV demand, such as carbon taxes, congestion fees, or highway tolls imposed to manage auto demand. Access to transit could be expanded through subsidies for car-, bike-, and scooter-sharing services in low-income areas (NASEM 2021e).

The IRA and the IIJA include important equity-enhancing policies to make LD PEVs more affordable to low- and moderate-income households. Up until 2020 or so, relatively few EVs had filtered outside of higher-income areas (Tal et al. 2021). The IRA for the first time includes a federal tax credit for the purchase of used EVs with a price cap of $25,000 (§13402). Used cars represent the major source (66 percent) of vehicle purchases for low-income households (Board of Governors of the Federal Reserve System 2016). The IRA (§13401) also sets price limits for new EVs to qualify for federal tax credits for sedans ($55,000) and SUVs, pickups, and vans ($80,000), which will encourage manufacturers to offer PEVs at a wider range of price points than their recent emphasis on luxury models (Hardman et al. 2021). Moreover, beginning in 2024, the IRA allows consumers to transfer their tax credit to auto dealers, who would then provide buyers with an equal price discount, which would not require buyers to wait for a tax return for reimbursement, nor require them to have tax liability. This should encourage purchase of PEVs by some low- and moderate-income households. Although new PEVs and new cars generally are still beyond the reach of most low- and moderate-income households, EVs at the price ceilings set in the IRA for tax credit eligibility will begin filtering into the used car market within a few years of sale. Moreover, several compact sedans priced well below these ceilings are being introduced by OEMs. The tax credits available in the IRA for home and commercial installation of charging infrastructure target rural and low-income census tracts, which will facilitate access to public charging by low-income households and renters less able to charge at home.

Enhancing clean automobile access for low-income households could also be pursued through programs such as those that CARB offers for low-income households to purchase ZEV vehicles: scrappage of older vehicles for $9,500; cash assistance of up to $7,000 for qualified households to buy or lease a ZEV; and special financing assistance of up to $5,000 for ZEV vehicle down payments (CARB 2022b). CARB is also pioneering subsidized carsharing and ridesharing programs for low-income households (CARB n.d.(a)). The expanding supply of used PEVs will make them more affordable through cost reductions and as the secondary market develops. Local programs to increase access to used vehicles and to provide counsel on vehicle and insurance decisions and avoiding predatory lending practices would be particularly helpful for first-time, low-income vehicle purchasers (Pendall et al. 2016). Used PEVs entail other issues, such as the risk of owning PEV batteries beyond warranty and accessibility of affordable charging (Hardman et al. 2021). Expanded battery warranty programs and equitable distribution of recharging infrastructure supported with public funds would help address these issues. States and local governments can learn from the many aforementioned policies and programs that are being experimented with and implemented in California.

Workforce Needs, Opportunity, and Support for Transportation Decarbonization

Transitioning from ICEVs to PEVs will have broad positive consequences for society and consumers, but some transport sector workers will face diminished employment opportunities as a result of EVs that are simpler to produce and maintain than ICEVs. For example, a recent set of decarbonization scenarios finds net increases in employment across the economy but declines in fossil fuel and transportation employment by 2035 (WRI 2022). In these WRI projections, most of the transportation employment decline is owing to reduced ICEV manufacturing employment (a loss of about 5.5 million jobs in the net-zero scenario), whereas employment growth in manufacturing for PEVs and in charging infrastructure would fall 2 million jobs short of replacing these losses. PEVs, having more integrated designs with fewer parts, can be produced with fewer workers per unit of output than ICEVs, and new factories are expected to be more reliant on automation than existing ones.

Domestic semiconductor and battery manufacturing and mining may be stimulated by the Creating Helpful Incentives for Producing Semiconductors and Science (CHIPS and Science) Act and the IRA. For example, the North American assembly and minerals sourcing requirements of the IRA will provide new domestic demand for vehicle battery suppliers and their employment needs. The WRI scenarios do not account for local repair and maintenance shops, which are expected to have reduced demand in the future because PEVs have fewer moving parts and electric motors are more reliable than internal combustion engines. CARB estimates that its new LD ZEV mandate (100 percent PEV and FCEV new sales by 2035) will reduce auto repair and maintenance jobs in the state by 13.8 percent (CARB 2022a). If that same percentage is applied to the current U.S. auto repair and maintenance workforce (BLS 2022), it implies a loss of roughly 127,000 jobs, although this reduction would occur slowly over the next 3 decades owing to the very large and slowly declining stock of ICEVs. Recommendations addressing any future employment losses appear in Chapter 4.

Engaging the Public in the Transportation Decarbonization Transition

Special efforts are needed to involve low-income and rural populations in transportation infrastructure planning and decision-making and in researching, developing, and implementing more effective ways of doing so (NASEM 2021d). For more than

3 decades, public participation has been mandated for federally funded transportation projects (NASEM 2019c). Title VI of the Civil Rights Act of 1964 and the Americans with Disabilities Act of 1990 formed the foundations of this requirement. The electric vehicle charging grants authorized by the IIJA require public participation and include 50 percent of funds set aside for community grants that prioritize projects for rural areas, low- and moderate-income neighborhoods, and communities with a low ratio of private parking spaces.

Requirements for public participation in transportation planning in the past, however, have not proven effective in participants' perceptions of being heard, improvement in the decisions made, or inclusiveness of the full spectrum of the public (Innes and Booher 2004). More meaningful processes require active participation of adversely affected or underserved communities in defining goals, resource allocation, and metrics by which to measure progress (Karner and Marcantonio 2018). In addition to improving opportunities for more meaningful participation, a useful step would be to expand the representation of discriminated-against, low-income communities on the planning, zoning, and transportation agency decision-making boards that plan and provide for transportation infrastructure and services.

Certain transportation-related technologies implemented for decarbonization will introduce new or heightened interest and concern from the public and require special consideration for public engagement. Carbon capture and sequestration, net-zero GHG emissions synthetic fuels, and other carbon management strategies may require investment in new or modified pipelines to transport carbon dioxide or ammonia (Larson et al. 2021; NASEM 2019b). Permitting new pipelines is a lengthy process requiring considerable government and public participation. Thus, making decisions on such pipelines is a priority for net-zero emissions planning, as is ensuring meaningful participation in siting decisions by the public and low-income and minority communities in particular.[16] Chapters 2 and 12 discuss pipeline challenges and needs in greater depth.

Health and Environmental Justice in Transportation

The transport sector is the second largest source of U.S. air pollution illness and death next to electric power (Chapter 3). Roughly 20,000 premature deaths in 2017 were attributed to ICEV emissions, with roughly one-third of those deaths resulting from

---

[16] Construction and design of new or modified pipelines provides an opportunity to reduce GHG emissions from these processes. For example, recycled materials and electrified construction equipment could be employed.

heavy-truck emissions (Choma et al. 2021). Transport pollutants disproportionately harm low-income and historically marginalized people, primarily in urban areas (Chapter 2). Although net-zero biofuels or synfuels could be employed to eliminate some or all transport-related GHG emissions, they would still emit conventional pollutants at approximately the same levels per distance traveled as fossil fuels. The need to dramatically reduce health impacts of fuel combustion is one of the reasons why combustion of net-zero liquid fuels is expected be a limited solution for transportation decarbonization. More details on the health impacts of transportation decarbonization are found in Chapter 3.

Clean Electricity for Transportation Electrification

Vehicle electrification will result in greater demands on the power grid, in aggregate power demand, and potentially in peak hour demands. For example, California's overall power demand would increase 5 percent with 50 percent penetration of electric vehicles by 2030, but peak hour demand could increase up to 25 percent with uncontrolled charging times (Powell et al. 2022). Achieving the committee's 2050 ZEV sales could increase average demand on the electric grid by as much as 28 percent (Oke et al. 2022). The power demands from intercity charging at large passenger/truck stops could reach the magnitude of demand of a small town (approximately 20 MW) by 2035 (National Grid et al. 2022). A variety of strategies can be employed to reduce peak demand by 2030, including local charging from solar panels, switching to off-peak charging (often at the workplace), smart charging systems responding to off-peak tariffs, or even two-way power flows with electric vehicles discharging at peak hours. The overall impact of EV charging on the grid is analogous to the introduction of widespread air conditioning (NASEM 2021c), but there is more opportunity to manage charging demand on the grid from PEVs.

Transportation Fuel Impact on Agriculture, Forestry, and Nature-Based Solutions

Even if the committee's ZEV sales goals are reached, there will still be substantial demand for low-carbon fuels for industrial heat, aviation, marine shipping, and perhaps heavy road transport. Some of this demand could be met by expanded production of biofuels, but all available land is already claimed for food, wood, and fiber production; biodiversity preservation, land carbon sinks; human settlements, or current biofuels production (16 million hectares, Chapter 8). Constraints are likely to limit carbohydrate biofuels use to aviation and other applications without likely alternatives. Hydrogen is another alternative to use of biofuels. By midcentury, bioenergy with carbon capture and storage hydrogen may compete with carbohydrate liquid biofuels because this process also produces a sizable and much-needed negative emission (Larson et al. 2021). Heavy trucking and trains may ultimately have an economic hydrogen fueling option, while marine shipping might be economically powered by ammonia produced with low-carbon hydrogen. The IRA provides substantial incentives for biofuels (reviewed in Chapter 8). The largest of these (§13203) appropriately targets aviation (up to $1.75/gallon), but the other major program offers a $1 per gallon tax credit for any second-generation biofuel (§13202).

Findings and Recommendations

Finding 9-7: Although technocratic arguments support continued public investment in zero-emission fuels for ICE land transport, including biofuels and synfuels, their use would continue to harm people with conventional air pollutants. Environmental justice and health impacts argue strongly against ICE land transport whenever there are economically viable battery-electric or fuel cell alternatives. Moreover, competition for land will or should restrict feedstock production for carbohydrate biofuels. However, aviation will almost certainly require liquid fuel combustion for the foreseeable future. Recommended actions to address future liquid fuel demand are made in the section "Actions to Expand the Innovation Toolkit."

Finding 9-8: Reducing ICEV emissions and noise through electrification offers greater health benefits to low-income communities of color than other groups because such communities tend to be located near major highways, freight depots, and ports. As noted above, the IRA has made notable efforts to enhance equity through targeted tax credits to low- and moderate-income households for EV purchase and tax credits for locating charging infrastructure in low-income areas. Both the IRA and IIJA provide authorized and appropriated funds to help communities separated or displaced by transportation infrastructure to develop plans and programs to ameliorate these effects as well as funding to reduce ICEV emissions at ports. Even so, a just transition to decarbonized transportation will require additional efforts to enhance the equitable access to PEVs by low-income households and equitable allocation of charging infrastructure, as well as additional efforts to ameliorate past injustices to low-income and minority communities.

Recommendation 9-5: *Enhance Transportation Equity and Environmental Justice Through Programs, Planning, and Services.* **States and local governments should enhance transportation equity and environmental justice through**

a. **New programs to assist low-income households in purchasing, owning, leasing, and insuring new and used plug-in electric vehicles;**
b. **Assurance that public charging locations are equitably allocated, accessible, and affordable by low-income residents unable to rely on home charging;**
c. **Improved mobility for low-income residents unable to afford vehicles by subsidizing car-sharing and ride-sharing programs and through location and subsidy of micro-mobility and micro-transit programs to improve connections to fixed-route public transportation;**
d. **Expanded transit services funded through a carbon tax or increased carbon-based fuel taxes or highway tolls to reduce highway trips;**
e. **Greater targeting of transit services to communities adversely affected by past infrastructure location, redlining, and housing discrimination; and**
f. **Increased representation of low-income residents of communities historically discriminated against on public regional and local planning, zoning, and transportation decision-making boards that plan for and provide transportation infrastructure and services and transparency about the proportion of such representation by reporting on the websites of such organizations.**

ACTIONS TO EXPAND THE INNOVATION TOOLKIT

Although EVs and other transportation decarbonization technologies are commercially available, others are still in research and demonstration status. Even those that are commercially available would benefit from continued innovation to reduce costs and improve effectiveness. Investment in innovation is essential to aid the process of deep decarbonization in transportation. As with all innovation processes, the technical and market successes of developing technologies are uncertain. For example, it is unclear whether FCEVs can be cost-competitive in the marketplace, as discussed below (NASEM 2021b). Pursuing innovation requires flexibility as results are obtained or conditions change. In this section, some priorities for innovation investment are outlined, but innovation investment should shift as results accrue.

Innovation investments can be made by a wide range of individuals and entities. The IRA and IIJA authorize and appropriate substantial RD&D funding to advance decarbonization goals. State governments also play a role, such as the long-standing RD&D support for ZEVs including fuel cell development by the State of California (McConnell et al. 2019). Private companies often pioneer innovation, such as ride-hailing companies, micro-mobility (bike and scooter sharing), and mobility as a service (NASEM 2021e). University research programs such as the DOT-supported university transportation centers often partner with local and state agencies or private firms to research and deploy innovations (NASEM 2019f). Innovation investment is also global in nature, with many innovations pioneered outside of the United States. Motor vehicle manufacturers are a good example of global entities in which designs and technologies are developed for new vehicles sold around the world.

Additional Innovation for Electrification

The EVs available today are the result of focused, long-term investments in basic and applied R&D over decades by automakers, suppliers, and federal and state governments. New motor, power electronics, and battery technologies; manufacturing process improvements, design innovations; and other R&D investments have all contributed to improved power density, reduced costs, and improved range for BEVs and other types of EVs. Further innovation is needed in these areas to continue these trends and increase the attractiveness of BEVs in the marketplace.

As BEVs become more prevalent in the marketplace, other enabling or supporting technologies need innovation investment. Ensuring adequate supply of mineral resources for batteries and electric motors is a concern, as well as developing batteries less reliant on minerals such as cobalt. Recycling processes for batteries and BEVs will be needed, building on the extensive recycling infrastructure for conventional vehicles. End-of-life recycling of electric vehicles and batteries is in its infancy (Chokshi and Browning 2022) and needs innovation. Innovations in support for EV supply chains, especially for domestic manufacturers, are critical to address minerals sourcing constraints imposed by the IRA and reduce costs. Improved recycling and increased recovery of critical minerals and materials requires design of the battery materials, cells, modules, and packs that considers ease of end-of-life recycling and recovery, in addition to R&D of improved processes and systems for battery recycling. Also related to BEV, and particularly battery technology: first responders to crashes (and battery fires) and vehicle mechanics need training to deal with BEVs.

Charging technology, operation practices, and infrastructure for BEVs also need innovation investment, including for shared and wireless charging. Currently, several different types of connector plugs are used for charging, reflecting different choices and proprietary incentives. Standardization of connector plugs has significant benefits for BEV users. Provisions in the IIJA will encourage this standardization because any recharging infrastructure funded through this program will have to be interoperable across proprietary designs. Automakers have recently been partnering to develop or adopt new standards, including several major automakers buying into the Tesla network and standard. Operating procedures for sharing chargers could also be improved, such as across real-time reservation systems and open standards that facilitate this process. Facilitating "smart" charging to optimize the charging cycles, time-shift electricity demand to periods of excess supply (and low prices), and avoid peak electricity demand periods would be beneficial. The IIJA provides substantial R&D funding for DOE to advance technologies for these purposes. Further innovation in bidirectional charging systems can aid resiliency by providing emergency power supply when the grid is unavailable, as already demonstrated commercially by the Ford F-150 BEV (Zhou et al. 2021).

Developing appropriate incentives for using PEVs and discouraging petroleum-based vehicles also require public funding and policies. Much can be learned from other countries taking different approaches to promoting PEVs and doing so equitably.

Although hydrogen fuel cells are a proven technology and in limited commercial use in FCEVs, they currently require a significant capital premium relative to BEVs as well as development of a low-carbon hydrogen supply infrastructure (NASEM 2020). FCEVs generally have the advantage of faster refueling time and longer range, so they could be most competitive for long-distance trucking and potentially for locomotives and ships. Innovations to reduce the cost of fuel cells, improve durability, ensure safety, and improve supplies of low-carbon hydrogen could make the technology competitive (NASEM 2021b). The alternative fuel subsidies in the IRA may accelerate the latter. Innovation for fuel cells might focus on cost reductions and safety assurance (NASEM 2021b), as well as innovations in how hydrogen can be distributed to points of demand in a cost-competitive way.

Other important areas of innovation, as described next, include development at scale of net-zero-carbon liquid and gaseous fuels to provide further options to decarbonize heavy-duty, long-distance road transport, aviation, and shipping and in estimating and verifying the carbon intensities of these fuels on a full life-cycle basis (NASEM 2022a,b).

Innovation in Net-Zero-Carbon Liquid Fuels

Assuming that the committee's 2030 and 2050 ZEV sales goals are met for LDVs, gasoline consumption by LDVs could be reduced by 80 percent in 2050 in line with the reduction in LDV stock. However, as noted earlier (see Figure 9-3), conversion of ICEVs to EVs may languish and thereby require provision of liquid fuels for decades to come. In this case, some use of biofuels in land transport may need to continue, which supports continued investment in biofuels R&D. Synthetic low-carbon liquid fuels could also be commercially available by 2050. For the hard-to-electrify transportation applications, especially aviation, true net-zero-carbon liquid fuels, with energy density comparable to current fossil fuels, provide the most likely option for decarbonization despite their harmful emissions of conventional pollutants. Such fuels are hydrocarbons where the carbon source and all other inputs result in zero net emissions of GHGs to the atmosphere on a life-cycle basis. Net-zero-carbon fuels can be developed from a variety of carbon sources, such as biomass, recycled carbon-based materials, and carbon dioxide captured from emissions sources or the atmosphere, and net-zero emitting inputs like clean hydrogen and electricity from renewables (Figure 9-6). Net-zero fuel composition may be tailored for standard operation in existing vehicles with no modification of the vehicle required, or for improved operation with optimized vehicle-fuel combinations.

Biomass such as corn, agricultural wastes, or algae could be used as a carbon source for net-zero-carbon liquid fuels, within the limits imposed by land constraints described in Chapter 8, but all GHG emissions with all aspects of fuel recovery and use would need to be eliminated or balanced by negative emissions such as carbon capture and sequestration. Areas of research for other biomass-based fuel processes include biomass-to-gasoline, involving the gasification of biomass and subsequent chemical conversion to fuel, and thermochemical conversion of biomass via pyrolysis or hydrothermal liquefaction followed by chemical refining steps (Phillips et al. 2011; Royal Society 2019).

Captured carbon dioxide is an alternative carbon source for synthesizing net-zero-carbon liquid fuels; however, at present there are no large-scale, low-carbon synthetic fuels available for LDV transportation. Existing gas-to-liquid processes like Fischer–Tropsch synthesis, methanol synthesis, and the methanol-to-gasoline process could be modified to utilize non-fossil carbon and low-carbon hydrogen, or direct chemical conversion of CO_2 may develop, as it is being explored at fundamental research and benchtop-proof-of-concept stages (Basic Energy Sciences Roundtable 2019; NASEM 2019a, 2022a).

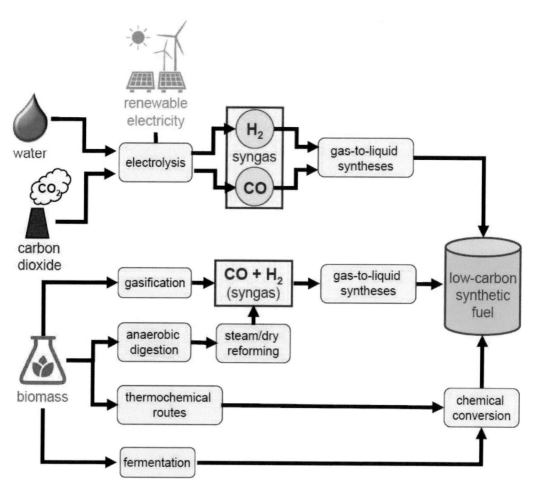

FIGURE 9-6 Pathways for production of low-carbon synthetic fuel. SOURCES: NASEM (2021b), inspired by Royal Society (2019), org/-/media/policy/projects/synthetic-fuels/synthetic-fuels-briefing.pdf. CC BY 4.0.

Production of low-carbon synthetic fuels is currently limited by high costs and inefficiencies (Cai et al. 2018; Li et al. 2016; Royal Society 2019). High costs and inefficiencies may be acceptable for low-volume, high-value commodities, but they are untenable for very high volume, low-margin products like mass market motor fuels, especially in comparison to inexpensive and readily available gasoline, diesel, and electricity. There are more near-term options for commercial drop-in, diesel-like fuels, as compared to lighter, gasoline-like spark-ignition engine fuels (AFDC n.d.(b)). Net-zero carbon synthetic and biofuels will likely be first introduced as blends with existing fossil fuels (Farrell et al. 2018). Examples of this are already available for diesel blends (Neste 2016; Renewable Energy Group 2020). Low- and net-zero-carbon liquid fuels require robust life-cycle analysis methods to be incorporated into transportation decarbonization policy (NASEM 2022b).

In summary, there are a variety of R&D and innovation investments that need to be undertaken to improve deep decarbonization processes for transportation, especially for hard to electrify transportation modes, and to improve electrification for the majority of vehicles.

Findings and Recommendations

Finding 9-9: Innovation supported by public and private RD&D remains essential for achieving GHG reduction goals for 2030 and beyond. Considerable additional innovations are needed in development of batteries less dependent on minerals not economically available domestically or from U.S. free-trade partners; extension in battery range and life-cycle performance; improvement of battery recycling to reuse minerals such as cobalt and lithium; and cost reductions in FCEVs and production of low-carbon hydrogen as well as in hydrogen distribution.

Finding 9-10: Despite successful electrification of LDVs and most MHD trucks, net-zero-carbon liquid fuels will still be required to decarbonize high-power, high-energy-consumption applications such as aviation and perhaps also heavy long-distance land transport and marine vessels, which could amount to 25 percent of current demand for fossil fuels in transport. There are no current commercially available truly net-zero-carbon liquid fuels, but technologies in development include carbon capture paired with either biofuels or synthetic low-carbon fuels from CO_2.

Recommendation 9-6: *Support Advances in Battery Design and Recycling, Fuel Cell Electric Vehicles (FCEVs), and Net-Zero Liquid Fuels.* **The Department of Energy (DOE) and the National Science Foundation (NSF) should continue to support advances in battery design and recycling and FCEVs to reduce their associated environmental and social costs and to make battery electric vehicles and FCEVs more cost-effective. In order to address hard to electrify aircraft, ships, locomotives, and long-distance heavy trucks, DOE and NSF should target their investments in research, development, and demonstration on technologies that produce liquid fuels that use feedstocks and energy inputs efficiently to reduce costs, reduce the life-cycle greenhouse gas emissions of the fuels to approach zero, and make possible the scale-up of these fuels on the order of tens of percentage points of current fuel volumes.**

Table 9-5 summarizes all of the recommendations in this chapter regarding decarbonizing the transportation system.

SUMMARY OF RECOMMENDATIONS ON TRANSPORT

TABLE 9-5 Summary of Recommendations on Transport

Short-Form Recommendation	Actor(s) Responsible for Implementing Recommendation	Sector(s) Addressed by Recommendation	Objective(s) Addressed by Recommendation	Overarching Categories Addressed by Recommendation
9-1: Accelerate the Adoption of Battery Electric Vehicles	Federal, state, and local governments	• Transportation • Finance • Non-federal actors	• Greenhouse gas (GHG) reductions • Equity • Health • Public engagement	A Broadened Policy Portfolio Tightened Targets for the Buildings and Industrial Sectors and a Backstop for the Transport Sector
9-2: Promote Vehicle Electrification at Ports and Airports	Ports and airports and their state and local government owners	• Transportation • Non-federal actors	• GHG reductions • Health	Tightened Targets for the Buildings and Industrial Sectors and a Backstop for the Transport Sector
9-3: Pursue Cost-Effective Efficiency Improvements to Reduce Greenhouse Gas Emissions	Private companies and state and local governments	• Buildings • Transportation • Fossil fuels • Non-federal actors	• GHG reductions • Equity • Health	A Broadened Policy Portfolio Tightened Targets for the Buildings and Industrial Sectors and a Backstop for the Transport Sector
9-4: Pursue Infrastructure Design, Standards, Specifications, and Procedures That Effectively Reduce Transportation Carbon Emissions	State Departments of Transportation, American Association of State Highway and Transportation Officials, American Road and	• Transportation • Industry • Non-federal actors	• GHG reductions	Tightened Targets for the Buildings and Industrial Sectors and a Backstop for the Transport Sector

TABLE 9-5 Continued

Short-Form Recommendation	Actor(s) Responsible for Implementing Recommendation	Sector(s) Addressed by Recommendation	Objective(s) Addressed by Recommendation	Overarching Categories Addressed by Recommendation
	Transportation Builders Association, and other specialized transportation infrastructure materials and construction associations			
9-5: Enhance Transportation Equity and Environmental Justice Through Programs, Planning, and Services	States and local governments	• Buildings • Transportation • Finance • Non-federal actors	• GHG reductions • Equity • Health • Public engagement	Rigorous and Transparent Analysis and Reporting for Adaptive Management Ensuring Procedural Equity in Planning and Siting New Infrastructure and Programs Ensuring Equity, Justice, Health, and Fairness of Impacts Tightened Targets for the Buildings and Industrial Sectors and a Backstop for the Transport Sector

continued

TABLE 9-5 Continued

Short-Form Recommendation	Actor(s) Responsible for Implementing Recommendation	Sector(s) Addressed by Recommendation	Objective(s) Addressed by Recommendation	Overarching Categories Addressed by Recommendation
9-6: Support Advances in Battery Design and Recycling, Fuel Cell Electric Vehicles, and Net-Zero Liquid Fuels	Department of Energy and National Science Foundation	• Land use • Transportation • Industry	• GHG reductions	Ensuring Equity, Justice, Health, and Fairness of Impacts Research, Development, and Demonstration Needs

REFERENCES

AFDC (Alternative Fuels Data Center). n.d.(a). "All Laws and Incentives Sorted by Type." DOE Office of Energy Efficiency and Renewable Energy. https://afdc.energy.gov/laws/matrix?sort_by=incentive. Accessed May 12, 2023.

AFDC. n.d.(b). "Renewable Hydrocarbon Biofuels." https://afdc.energy.gov/fuels/emerging_hydrocarbon.html. Accessed September 29, 2022.

Al-Alawi, B.M., O. MacDonnell, R. McLane, and K. Walkowicz. 2022. "Zeroing in on Zero-Emission Trucks: The Advanced Technology Truck Index: A U.S. ZET Inventory Report." CALSTART. https://calstart.org/zeroing-in-on-zero-emission-trucks.

Alliance for Automotive Innovation. 2022. "Get Connected Electric Vehicle Report: Fourth Quarter 2021." https://www.autosinnovate.org/posts/papers-reports/Get%20Connected%202021%20Q4%20%20Electric%20Vehicle%20Report.pdf.

Alliance for Automotive Innovation. 2023. "Get Connected Electric Vehicle Report: First Quarter 2023." https://www.autosinnovate.org/posts/papers-reports/Get%20Connected%20EV%20Quarterly%20Report%202023%20Q1.pdf, p. 8.

Alliance for Transportation Electrification. n.d. "Open Standards Move Technology." https://evtransportationalliance.org/about-us/open-standards-move-technology. Accessed January 31, 2023.

ANL (Argonne National Laboratory). 2023. "Light Duty Electric Drive Vehicles Monthly Sales Updates." https://www.anl.gov/esia/light-duty-electric-drive-vehicles-monthly-sales-updates.

Baldwin, S., A. Myers, M. O'Boyle, and D. Wooley. 2021. "The 2035 Report 2.0: Accelerating Clean, Electrified Transportation by 2035: Policy Priorities, a 2035 2.0 Companion Report." Energy Innovation and University of California, Berkeley, Goldman School of Public Policy. https://energyinnovation.org/wp-content/uploads/2021/04/Energy-Innovation_2035-2.0-Accelerating-Clean-Transportation-Policy-Report.pdf.

Basic Energy Sciences Roundtable. 2019. https://science.osti.gov/-/media/bes/pdf/reports/2020/Liquid_Solar_Fuels_Report.pdf.

Bistline, J., G. Blanford, M. Brown, D. Burtraw, M. Domeshek, J. Farbes, A. Fawcett, et al. 2023. "Emissions and Energy Impacts of the Inflation Reduction Act." *Science* 380:1324–1327. https://doi.org/10.1126/science.adg3781.

BLS (Bureau of Labor Statistics). 2022. "May 2021 National Industry-Specific Occupational Employment and Wage Estimates, NAICS 811100—Automotive Repair and Maintenance." https://www.bls.gov/oes/current/naics4_811100.htm#00-0000.

Blumenberg, E. 2017. "Social Equity and Urban Transportation." In *The Geography of Urban Transportation*, 4th ed., G. Giuliano and S. Hanson, eds. New York: Guilford Press.

Board of Governors of the Federal Reserve System. 2016. "Report on the Economic Well-Being of U.S. Households in 2015." https://www.federalreserve.gov/2015-report-economic-well-being-us-households-201605.pdf.

Bond, D.E., I. Macvay, C. Thomas, J. Marssola, and I. Saccomanno. 2023. "Will the United States' New Critical Minerals Agreements Shape Electric Vehicle Investments?" https://www.whitecase.com/insight-alert/will-united-states-new-critical-minerals-agreements-shape-electric-vehicle.

Brown, A.L., D. Sperling, B. Austin, J.R. DeShazo, L. Fulton, T. Lipman, C. Murphy, et al. 2021. "Driving California's Transportation Emissions to Zero." UC-ITS-2020-65. University of California Institute of Transportation Studies. https://doi.org/10.7922/G2MC8X9X.

BTS (Bureau of Transportation Statistics). 2017. "National Household Travel Survey Daily Travel Quick Facts." Department of Transportation. https://www.bts.gov/statistical-products/surveys/national-household-travel-survey-daily-travel-quick-facts.

BTS. 2021a. "National Transportation Statistics, Table 1-35, U.S. Vehicle Miles." Department of Transportation. https://www.bts.gov/content/us-vehicle-miles

BTS. 2021b. "National Transportation Statistics, Table 1-40, U.S. Passenger Miles." Department of Transportation. https://www.bts.gov/content/us-passenger-miles.

BTS. 2021c. "National Transportation Statistics, Table 1-50: U.S. Ton-Miles of Freight." Department of Transportation. https://www.bts.gov/content/us-ton-miles-freight.

Burnham, A., D. Gohlke, L. Rush, T. Stephens, Y. Zhou, M. Delucchi, A. Birky, et al. 2021. "Comprehensive Total Cost of Ownership Quantification for Vehicles with Different Size Classes and Powertrains." ANL/ESD-21/4, 1780970, 167399. https://doi.org/10.2172/1780970.

CAA (Clean Air Act). Title 42—The Public Health and Welfare. 42 U.S.C. §§7507, 7543.

Cai, H., J. Markham, S. Jones, P.T. Benavides, J.B. Dunn, M. Biddy, and L. Tao. 2018. "Techno-Economic Analysis and Life-Cycle Analysis of Two Light-Duty Bioblendstocks: Isobutanol and Aromatic-Rich Hydrocarbons." *ACS Sustainable Chemistry and Engineering* 6(7):8790–8800. https://doi.org/10.1021/acssuschemeng.8b01152.

California City News. 2021. "California Says Goodbye to Single-Family Zoning." https://www.californiacitynews.org/2021/09/california-says-goodbye-single-family-zoning.

Car and Driver. 2022. "Future EVs: Every Electric Vehicle Coming Soon." https://www.caranddriver.com/news/g29994375/future-electric-cars-trucks.

CARB (California Air Resources Board). 2022a. "Advanced Clean Cars II Proposed Amendments to the Low Emission, Zero Emission, and Associated Vehicle Regulations: Standardized Regulatory Impact Assessment (SRIA)." https://dof.ca.gov/wp-content/uploads/sites/352/Forecasting/Economics/Documents/ACCII-SRIA.pdf.

CARB. 2022b. "California Moves to Accelerate to 100% New Zero-Emission Vehicle Sales by 2035." https://ww2.arb.ca.gov/news/california-moves-accelerate-100-new-zero-emission-vehicle-sales-2035.

CARB. 2023. "California Approves Groundbreaking Regulation That Accelerates the Deployment of Heavy-Duty ZEVs to Protect Public Health." https://ww2.arb.ca.gov/news/california-approves-groundbreaking-regulation-accelerates-deployment-heavy-duty-zevs-protect.

CARB. n.d.(a). "Funding Opportunities for Individuals and Families." Moving California: Cleaner Transportation for All Communities. https://ww2.arb.ca.gov/sites/default/files/movingca/opportunities.html. Accessed January 31, 2023.

CARB. n.d.(b). "Low Carbon Fuel Standard." https://ww2.arb.ca.gov/our-work/programs/low-carbon-fuel-standard/about. Accessed March 2, 2023.

Casady, C.B., J.A. Gómez-Ibáñez, and E. Schwimmer. 2020. "Toll-Managed Lanes: A Simplified Cost Benefit Analysis of Seven U.S. Projects." *Transport Policy* 89:38–53.

Cervero, R. 2003. "Coping with Complexity in America's Urban Transport Sector." University of California, Berkeley: Institute of Urban and Regional Development. https://escholarship.org/uc/item/4wf4n16r.

Chester, M.V., and A. Horvath. 2009. "Environmental Assessment of Passenger Transportation Should Include Infrastructure and Supply Chains." *Environmental Research Letters* 4(2):024008. https://doi.org/10.1088/1748-9326/4/2/024008.

Chokshi, N., and K. Browning. 2022. "Electric Cars Are Taking Off, But When Will Battery Recycling Follow?" *The New York Times*, December 21. https://www.nytimes.com/2022/12/21/business/energy-environment/battery-recycling-electric-vehicles.html.

Choma, E.F., J.S. Evans, J.A. Gómez-Ibáñez, Q. Di, J.D. Schwartz, J.K. Hammitt, and J.D. Spengler. 2021. "Health Benefits of Decreases in On-Road Transportation Emissions in the United States from 2008 to 2017." *Proceedings of the National Academy of Sciences* 118(51):e2107402118. https://doi.org/10.1073/pnas.2107402118.

Consumer Reports. 2022a. "Battery Electric Vehicles and Low Carbon Fuel Survey: A Nationally Representative Multi-Mode Survey." https://article.images.consumerreports.org/image/upload/v1657127210/prod/content/dam/CRO-Images-2022/Cars/07July/2022_Consumer_Reports_BEV_and_LCF_Survey_Report.pdf.

Consumer Reports. 2022b. "Insights for More Reliable Electric Vehicles." https://data.consumerreports.org/wp-content/uploads/2022/01/Consumer-Reports-Insights-for-More-Reliable-Electric-Vehicles-Jan-2022.pdf.

Corbett, J.J., J.J. Winebrake, J. Hatcher, and A.E. Farrell. 2008. "Emissions Analysis of Freight Transport Comparing Land-Side and Water-Side Short-Sea Routes: Development and Demonstration of a Freight Routing and Emissions Analysis Tool (FREAT)." Department of Transportation Research and Special Programs Administration. https://www.transportation.gov/sites/dot.gov/files/docs/emissions_analysis_of_freight.pdf.

CRS (Congressional Research Service). 2022. Tax Provisions in the Inflation Reduction Act of 2022 (H.R. 5376). CRS Report R47202. https://crsreports.congress.gov/product/pdf/R/R47202.

Davis, S.C., and R.G. Boundy. 2020. *Transportation Energy Data Book: Edition 38.* Oak Ridge National Laboratory. https://TEDB.ORNL.GOV.

Department of the Army, Office of the Assistant Secretary of the Army for Installations, Energy and Environment. 2022. *United States Army Climate Strategy.* Washington, DC.

DOE (Department of Energy). 2021. "Thirteen New Electric Vehicle Plants Are Planned in the U.S. Within the Next Five Years." https://www.energy.gov/eere/vehicles/articles/fotw-1217-december-20-2021-thirteen-new-electric-vehicle-battery-plants-are.

DOE. 2022. "Energy Saver: New Plug-in Electric Vehicle Sales in the United States Nearly Doubled from 2020 to 2021." https://www.energy.gov/energysaver/articles/new-plug-electric-vehicle-sales-united-states-nearly-doubled-2020-2021.

DOE. 2023a. "Critical Materials Assessment." https://www.energy.gov/sites/default/files/2023-07/doe-critical-material-assessment_07312023.pdf.

DOE. 2023b. "U.S. Department of Energy Clean Hydrogen Production Standard (CHPS) Guidance." https://www.hydrogen.energy.gov/pdfs/clean-hydrogen-production-standard-guidance.pdf.

DOE. n.d.(a). "Alternative Fuels Data Center: Developing Infrastructure to Charge Electric Vehicles." https://afdc.energy.gov/fuels/electricity_infrastructure.html#terms. Accessed February 28, 2023.

DOE. n.d.(b). "Electric Drive Vehicle Battery Recycling and 2nd Life Apps." https://www.energy.gov/infrastructure/electric-drive-vehicle-battery-recycling-and-2nd-life-apps. Accessed July 1, 2023.

DOE. n.d.(c). "Where the Energy Goes." https://www.fueleconomy.gov/feg/atv-ev.shtml. Accessed August 3, 2023.

DOE-VTO (Department of Energy Vehicle Technologies Office). 2022. "FOTW #1268, December 12, 2022: As of 2021, Two-Thirds of U.S. Housing Units Had a Garage or Carport, Improving Opportunities for EV Adoption." https://www.energy.gov/eere/vehicles/articles/fotw-1268-december-12-2022-2021-two-thirds-us-housing-units-had-garage-or.

DOE-VTO. n.d. "Batteries." https://www.energy.gov/eere/vehicles/batteries. Accessed March 1, 2023.

Domonosky, C. 2021. "From Amazon to FedEx, the Delivery Truck Is Going Electric." NPR. https://www.npr.org/2021/03/17/976152350/from-amazon-to-fedex-the-delivery-truck-is-going-electric.

DOS and EOP (Department of State and Executive Office of the President). 2021. *The Long-Term Strategy of the United States: Pathways to Net-Zero Greenhouse Gas Emissions by 2050.* Washington, DC. https://www.whitehouse.gov/wp-content/uploads/2021/10/US-Long-Term-Strategy.pdf.

DOT (Department of Transportation). 2021. "Highway Statistics 2019." https://www.fhwa.dot.gov/policyinformation/statistics/2019/vm1.cfm.

DOT. 2022. "Equity Action Plan." https://www.transportation.gov/sites/dot.gov/files/2022-04/Equity_Action_Plan.pdf.

Downs, A. 1992: *Stuck in Traffic: Coping with Peak Hour Traffic Congestion.* Washington, DC: Brookings Institution. https://www.brookings.edu/book/stuck-in-traffic.

Downs, A. 2004. *Still Stuck in Traffic: Coping with Peak Hour Traffic Congestion.* Washington, DC: Brookings Institution. https://www.brookings.edu/book/still-stuck-in-traffic.

Dumortier, J., S. Siddiki, S. Carley, J. Cisney, R. Krause, B. Lane, J. Rupp, and J. Graham. 2015. "Effects of Providing Total Cost of Ownership Information on Consumers' Intent to Purchase a Hybrid or Plug-in Electric Vehicle." *Transportation Research Part A: Policy and Practice* 72:71–86.

EI (Energy Innovation). 2022. "Updated Inflation Reduction Act Modeling Using the Energy Policy Simulator." https://energyinnovation.org/publication/updated-inflation-reduction-act-modeling-using-the-energy-policy-simulator.

EIA (U.S. Energy Information Administration). 2022. "Annual Energy Outlook 2022—Reference Case Projections Tables." https://www.eia.gov/outlooks/aeo/tables_ref.php.

EPA (Environmental Protection Agency). 2021. "Fast Facts: U.S. Transportation Sector Greenhouse Gas Emissions 1990–2019." EPA-420-F-21-076. https://nepis.epa.gov/Exe/ZyPDF.cgi?Dockey=P1013NR3.pdf.

EPA. 2022a. "California State Motor Vehicle Pollution Control Standards: Advanced Clean Car Program; Reconsideration of a Previous Withdrawal of a Waiver of Preemption; Notice of Decision." *Federal Register* 87(49):14332. https://www.govinfo.gov/content/pkg/FR-2022-03-14/pdf/2022-05227.pdf.

EPA. 2022b. "Shore Power Technology Assessment at U.S. Ports." Reports and Assessments. https://www.epa.gov/ports-initiative/shore-power-technology-assessment-us-ports.

EPA. 2023a. "Greenhouse Gas Emissions Standards for Heavy-Duty Vehicles—Phase 3." *Federal Register* 88(81):25926–26161. https://www.govinfo.gov/content/pkg/FR-2023-04-27/pdf/2023-07955.pdf.

EPA. 2023b. "Multi-Pollutant Emissions Standards for Model Years 2027 and Later Light-Duty and Medium-Duty Vehicles." EPA-420-F-23-009 Program Announcement. https://nepis.epa.gov/Exe/ZyPDF.cgi?Dockey=P1017626.pdf.

EPA. 2023c. "Multi-Pollutant Emissions Standards for Model Years 2027 and Later Light-Duty and Medium-Duty Vehicles." *Federal Register* 88(87):29184. https://www.govinfo.gov/content/pkg/FR-2023-05-05/pdf/2023-07974.pdf.

EPA. 2023d. "Proposed Standards to Reduce Greenhouse Gas Emissions from Heavy-Duty Vehicles for Model Year 2027 and Beyond." EPA-420-F-23-011 Program Announcement. https://nepis.epa.gov/Exe/ZyPDF.cgi?Dockey=P101762L.pdf.

EPA. n.d. "Electric Vehicle Myths. Myth #2: Electric Vehicles Are Worse for the Climate Than Gasoline Cars Because of Battery Manufacturing." https://www.epa.gov/greenvehicles/electric-vehicle-myths#Myth2.

EO (Executive Order) 13990. 2021. 86 Fed. Reg. 14 (January 25, 2021). https://www.energy.gov/sites/default/files/2021/02/f83/eo-13990-protecting-public-health-environment-restoring.pdf.

EO 14507. 2021. 86 Fed. Reg. 70935 (December 8, 2021). https://www.federalregister.gov/documents/2021/12/13/2021-27114/catalyzing-clean-energy-industries-and-jobs-through-federal-sustainability.

FAA (Federal Aviation Administration). n.d. "Working to Build a Net Zero Sustainable Aviation System by 2050." https://www.faa.gov/sustainability.

Farrell, J.T., J. Holladay, and R. Wagner. 2018. "Co-Optimization of Fuels and Engines: Fuel Blendstocks with the Potential to Optimize Future Gasoline Engine Performance; Identification of Five Chemical Families for Detailed Evaluation." NREL/TP—5400-69009, DOE/GO—102018-4970, 1434413. https://doi.org/10.2172/1434413.

FHWA (Federal Highway Administration). 2021. "Highway Statistics 2019 Table VM-1." Office of Highway Policy Information. https://www.fhwa.dot.gov/policyinformation/statistics/2019/vm1.cfm.

Flannigan, C., and A. Leslie. 2020. *Crash-Avoidance Technology Evaluation Using Real-World Crash Data*. Report DOT HS 812 841. Washington, DC: National Highway Traffic Administration.

Foster, D., J. Koszewnik, W. Wade, and W. Winer. 2022. "Pathways to More Rapidly Reduce Transportation's Climate Change Impact." *Issues in Science and Technology*, November 17.

Gohlke, D., Y. Zhou, X. Wu, and C. Courtney. 2022. "Assessment of Light-Duty Plug-in Electric Vehicles in the United States, 2010–2021." ANL-22/71, 1898424, 178584. https://doi.org/10.2172/1898424.

Greene, D.L. 2011. "Uncertainty, Loss Aversion, and Markets for Energy Efficiency." *Energy Economics* 33(4):608–616. Special Issue on The Economics of Technologies to Combat Global Warming. https://doi.org/10.1016/j.eneco.2010.08.009.

Greene, D.L. 2019. "Implications of Behavioral Economics for the Costs and Benefits of Fuel Economy Standards." *Current Sustainable/Renewable Energy Reports* 6(4):177–192. https://doi.org/10.1007/s40518-019-00134-3.

Gulley, A.L. 2023. "China, the Democratic Republic of the Congo, and Artisanal Cobalt Mining from 2000 Through 2020." *Proceedings of the National Academy of Sciences* 120(26):e2212037120. https://doi.org/10.1073/pnas.2212037120.

Hardman, S., A. Jenn, G. Tal, J. Axsen, G. Beard, N. Daina, E. Figenbaum, et al. 2018. "A Review of Consumer Preferences of and Interactions with Electric Vehicle Charging Infrastructure." *Transportation Research Part D: Transport and Environment* 62(July):508–523. https://doi.org/10.1016/j.trd.2018.04.002.

Hardman, S., K. Fleming, E. Kare, and M. Ramadan. 2021. "A Perspective on Equity in the Transition to Electric Vehicle." B. Neyhouse and Y. Petri, eds. *MIT Science Policy Review* (August):46–54. https://doi.org/10.38105/spr.e10rdoaoup.

Hendrickson, C., and A. Horvath. 2000. "Resource Use and Environmental Emissions of U.S. Construction Sectors." *Journal of Construction Engineering and Management* 126(1):38–44. https://doi.org/10.1061/(ASCE)0733-9364(2000)126:1(38).

IEA (International Energy Agency). 2021a. "Comparative Life-Cycle Greenhouse Gas Emissions of a Mid-Size BEV and ICE Vehicle." https://www.iea.org/data-and-statistics/charts/comparative-life-cycle-greenhouse-gas-emissions-of-a-mid-size-bev-and-ice-vehicle.

IEA. 2021b. "Minerals Used in Electric Cars Compared to Conventional Cars." https://www.iea.org/data-and-statistics/charts/minerals-used-in-electric-cars-compared-to-conventional-cars.

IEA. 2022a. *Global EV Outlook 2022: Securing Supplies for an Electric Future.* Paris. https://iea.blob.core.windows.net/assets/ad8fb04c-4f75-42fc-973a-6e54c8a4449a/GlobalElectricVehicleOutlook2022.pdf, p. 15.

IEA. 2022b. *The Role of Critical Minerals in Clean Energy Transitions.* Paris. https://iea.blob.core.windows.net/assets/ffd2a83b-8c30-4e9d-980a-52b6d9a86fdc/TheRoleofCriticalMineralsinCleanEnergyTransitions.pdf.

IEA. 2023. *Global EV Outlook 2022: Catching Up with Climate Ambitions.* Paris. https://iea.blob.core.windows.net/assets/dacf14d2-eabc-498a-8263-9f97fd5dc327/GEVO2023.pdf.

IIHS-HILDI (Insurance Institute for Highway Safety, Highway Loss Data Institute). 2021. "With More Electric Vehicles Comes More Proof of Safety." https://www.iihs.org/news/detail/with-more-electric-vehicles-comes-more-proof-of-safety.

Innes, J.E., and D.E. Booher. 2004. "Reframing Public Participation: Strategies for the 21st Century." *Planning Theory and Practice* 5(4):419–436. https://doi.org/10.1080/1464935042000293170.

ITIF (Information Technology and Innovation Foundation). 2022. "Mission, Money, and Process Makeover: How Federal Procurement Can Catalyze Clean Energy Investment and Innovation." https://www2.itif.org/2022-federal-buying-power.pdf.

J.D. Power. 2023. "Electric Vehicle Experience (EVX) Public Charging Study." https://www.jdpower.com/business/automotive/electric-vehicle-experience-evx-public-charging-study.

Jenkins, J.D., E.N. Mayfield, J. Farbes, G. Schivley, N. Patankar, and R. Jones. 2023. *Climate Progress and the 117th Congress: The Impacts of the Inflation Reduction Act and the Infrastructure Investment and Jobs Act.* Princeton, NJ: REPEAT Project. https://repeatproject.org/docs/REPEAT_Climate_Progress_and_the_117th_Congress.pdf.

Jockims, T. 2022. "How GM, Ford, and Tesla Are Tacking the National EV Charging Challenge." CNBC. https://www.cnbc.com/2022/06/20/how-gm-ford-tesla-are-tackling-the-national-ev-charging-challenge.html.

Joint Office of Energy and Transportation. n.d. "State Plans for Electric Vehicle Charging." https://driveelectric.gov/state-plans. Accessed August 21, 2023.

Karner, A., and R.A. Marcantonio. 2018. "Achieving Transportation Equity: Meaningful Public Involvement to Meet the Needs of Underserved Communities." *Public Works Management and Policy* 23(2):105–126. https://doi.org/10.1177/1087724X17738792.

Katsh, G., C. Fagan, J. Wilke, B. Wilkie, C. Lamontagne, R. Garcia Coyne, B. Mandel, et al. 2022. "Electric Highways: Accelerating and Optimizing Fast-Charging Deployment for Carbon-Free Transportation." National Grid, RMI, Calstart. https://www.nationalgrid.com/document/148616/download.

Larsen, J., B. King, H. Kolus, N. Dasari, G. Hiltbrand, and W. Herndon. 2022. "A Turning Point for US Climate Progress: Assessing the Climate and Clean Energy Provisions in the Inflation Reduction Act." Rhodium Group. https://rhg.com/research/climate-clean-energy-inflation-reduction-act, p. 7.

Larson, E., C. Greig, J. Jenkins, E. Mayfield, A. Pascale, C. Zhang, J. Drossman, et al. 2021. "Net-Zero America: Potential Pathways, Infrastructure, and Impacts, Final Report." Princeton, NJ: Princeton University. https://netzeroamerica.princeton.edu/the-report.

Leard, B. 2018. "Consumer Inattention and the Demand for Vehicle Fuel Cost Savings." *Journal of Choice Modelling* 29(December):1–16. https://doi.org/10.1016/j.jocm.2018.08.002.

Ledna, C., M. Muratori, A. Yip, P. Jadun, and C. Hoehne. 2022. "Decarbonizing Medium- and Heavy-Duty On-Road Vehicles: Zero-Emission Vehicles Cost Analysis." NREL/TP-5400-82081, 1854583, MainId:82854. https://doi.org/10.2172/1854583.

Lee, D.S., D.W. Fahey, A. Skowron, M.R. Allen, U. Burkhardt, Q. Chen, S.J. Doherty, et al. 2021. "The Contribution of Global Aviation to Anthropogenic Climate Forcing for 2000 to 2018." *Atmospheric Environment* 244(January):117834. https://doi.org/10.1016/j.atmosenv.2020.117834.

Lee, M., and A. Aton. 2023. "Electric Cars Face 'Punitive' Fees, New Restrictions in Many States." *Politico*, August 17. https://www.politico.com/news/2023/08/17/punitive-texas-other-states-roadblocks-ev-transition-00110125#.

Leinert, P. 2022. "Exclusive: Automakers to Double Spending on EVs, Batteries, to $1.2 Trillion by 2030." *Reuters*, October 25. https://www.reuters.com/technology/exclusive-automakers-double-spending-evs-batteries-12-trillion-by-2030-2022-10-21.

Lepre, N. 2022. "23 Percent of Utility Funding for Electric Vehicles Targeted for Underserved Communities." *Atlas EV Hub*, January 28. https://www.atlasevhub.com/data_story/23-percent-of-utility-funding-for-electric-vehicles-targeted-for-underserved-communities.

Li, X., P. Anderson, H.M. Jhong, M. Paster, J.F. Stubbins, and P.J.A. Kenis. 2016. "Greenhouse Gas Emissions, Energy Efficiency, and Cost of Synthetic Fuel Production Using Electrochemical CO_2 Conversion and the Fischer–Tropsch Process." *Energy and Fuels* 30(7):5980–5989. https://doi.org/10.1021/acs.energyfuels.6b00665.

Linn, J., and W. Look. 2022. "An Analysis of U.S. Subsidies for Electric Buses and Freight Trucks." Resources for the Future. https://www.rff.org/publications/issue-briefs/an-analysis-of-us-subsidies-for-electric-buses-and-freight-trucks.

Lovins, A. 2020. "Reframing Automotive Fuel Efficiency." *SAE International Journal of Sustainable Transportation, Energy, Environment, & Policy* 1(1):59–84. https://doi.org/10.4271/13-01-01-0004.

Lovins, A. 2021. "Profitably Decarbonizing Heavy Transport and Industrial Heat: Transforming These 'Harder to Abate' Sectors Is Not Uniquely Hard and Can Be Lucrative." RMI. https://rmi.org/wp-content/uploads/2021/07/rmi_profitable_decarb.pdf.

Lutsey, N., and M. Nicholas. 2019. "An Update on Electric Vehicles Costs in the United States Through 2030." International Council for Clean Transportation Working Paper 2019-06. https://www.theicct.org/publications/update-US-2030-electric-vehicle-cost.

Maclean, C., D.R. Golub, K.J. Ashe, and W.E. Sterling. 2022. "California's 2023 Housing Laws: What You Need to Know." *Holland & Knight Alert*, October 10. https://www.hklaw.com/en/insights/publications/2022/10/california-2023-housing-laws-what-you-need-to-know.

Margolies, J. 2022. "Climate Law a 'Game Changer' for Highways and Bridges." *The New York Times*, September 6. https://www.nytimes.com/2022/09/06/business/climate-law-sustainable-infrastructure.html.

Martens, K. 2016. *Transport Justice: Designing Fair Transportation Systems*. Oxfordshire, UK: Routledge.

Matthews, S., C. Hendrickson, and D. Soh. 2001. "Environmental and Economic Effects of E-Commerce: A Case Study of Book Publishing and Retail Logistics." *Transportation Research Record* 1763(1). https://journals.sagepub.com/doi/10.3141/1763-02.

McConnell, V., B. Leard, and F. Kardos. 2019. "California's Evolving Zero Emission Vehicle Program: Pulling New Technology into the Market." Resources for the Future. Working Paper 19-22. https://media.rff.org/documents/RFF_WP_Californias_Evolving_Zero_Emission_Vehicle_Program.pdf.

McKenzie, L., and N. Nigro. 2021. "U.S. Passenger Vehicle Electrification Infrastructure Assessment: Results for Light Duty Vehicles." Washington, DC: Atlas Public Policy. https://atlaspolicy.com/wp-content/uploads/2021/04/2021-04-21_US_Electrification_Infrastructure_Assessment.pdf.

McKenzie, L., J. Di Filippo, J. Rosenberg, and N. Nigro. 2021. "U.S. Vehicle Electrification Assessment: Medium- and Heavy-Duty Truck Charging." Washington, DC: Atlas Public Policy. https://atlaspolicy.com/wp-content/uploads/2021/11/2021-11-12_Atlas_US_Electrification_Infrastructure_Assessment_MD-HD-trucks.pdf.

Milam, R.T., M. Birnbaum, C. Ganson, S. Handy, and J. Walters. 2017. "Closing the Induced Vehicle Travel Gap Between Research and Practice." *Transportation Research Record* 2653(1). http://doi.org/10.3141/2653-02.

NASEM. 2019a. *Gaseous Carbon Waste Streams Utilization: Status and Research Needs*. Washington, DC: The National Academies Press. https://doi.org/10.17226/25232.

NASEM. 2019b. *Negative Emissions Technologies and Reliable Sequestration: A Research Agenda*. Washington, DC: The National Academies Press. https://doi.org/10.17226/25259.

NASEM. 2019c. *Practices for Online Public Involvement*. Washington, DC: The National Academies Press. https://doi.org/10.17226/25500.

NASEM. 2019d. *Renewing the National Commitment to the Interstate Highway System: A Foundation for the Future*. Washington, DC: The National Academies Press. https://doi.org/10.17226/25334.

NASEM. 2019e. *Sustainable Highway Construction Guidebook*. Washington, DC: The National Academies Press. https://doi.org/10.17226/25698.

NASEM. 2019f. *The Vital Federal Role in Meeting the Highway Innovation Imperative*. Washington, DC: The National Academies Press. https://doi.org/10.17226/25511.

NASEM. 2020. *Reducing Fuel Consumption and Greenhouse Gas Emissions of Medium- and Heavy-Duty Vehicles, Phase Two: Final Report*. Washington, DC: The National Academies Press. https://doi.org/10.17226/25542.

NASEM. 2021a. *Accelerating Decarbonization of the U.S. Energy System*. Washington, DC: The National Academies Press. https://doi.org/10.17226/25932.

NASEM. 2021b. *Assessment of Technologies for Improving Light-Duty Vehicle Fuel Economy 2025–2035*. Washington, DC: The National Academies Press. https://doi.org/10.17226/26092.

NASEM 2021c. *The Future of Electric Power in the United States*. Washington, DC: The National Academies Press. https://doi.org/10.17226/25968.

NASEM. 2021d. *Racial Equity Addendum to Critical Issues in Transportation*. Washington, DC: The National Academies Press. https://doi.org/10.17226/26264.

NASEM. 2021e. *The Role of Transit, Shared Modes, and Public Policy in the New Mobility Landscape*. Washington, DC: The National Academies Press. https://doi.org/10.17226/26053.

NASEM. 2022a. *Carbon Dioxide Utilization Markets and Infrastructure: Status and Opportunities: A First Report*. Washington, DC: The National Academies Press. https://doi.org/10.17226/26703.

NASEM. 2022b. *Current Methods of Life Cycle Analyses of Low-Carbon Transportation Fuels in the United States*. Washington, DC: The National Academies Press. https://doi.org/10.17226/26402.

NASEO (National Association of State Energy Officials). n.d. "Volkswagen Settlement." https://www.naseo.org/issues/transportation/volkswagen. Accessed February 28, 2023.

National Grid, CALSTART, RMI, Stable Auto, and Geotab. 2022. "Electric Highways: Accelerating and Optimizing Fast-Charging Deployment for Carbon-Free Transportation." https://www.nationalgrid.com/document/148616/download.

Neste. 2016. "Renewable Diesel Handbook." https://www.neste.com/sites/default/files/attachments/neste_renewable_diesel_handbook.pdf.

NHTSA (National Highway Traffic Safety Administration). 2022. "Corporate Average Fuel Economy Standards Model Years 2024–2026." 87 FR 25710. https://www.federalregister.gov/documents/2022/05/02/2022-07200/corporate-average-fuel-economy-standards-for-model-years-2024-2026-passenger-cars-and-light-trucks.

NHTSA. 2023. "Corporate Average Fuel Economy Standards for Passenger Cars and Light Trucks for Model Years 2027–2032 and Fuel Efficiency Standards for Heavy-Duty Pickup Trucks and Vans for Model Years 2030–2035." *Federal Register* 88(158):56128–56390. https://www.govinfo.gov/content/pkg/FR-2023-08-17/pdf/2023-16515.pdf.

NREL (National Renewable Energy Laboratory). 2017. "Electric Support Equipment at Airports." https://afdc.energy.gov/files/u/publication/egse_airports.pdf.

Oke, D., J.B. Dunn, and T.R. Hawkins. 2022. "The Contribution of Biomass and Waste Resources to Decarbonizing Transportation and Related Energy and Environmental Effects." *Sustainable Energy and Fuels* 6(3):721–735. https://doi.org/10.1039/D1SE01742J.

ORNL (Oak Ridge National Laboratory). 2017. "National Household Transportation Survey, (2017), 'Person Trips.'" https://nhts.ornl.gov/person-trips.

Pendall, R., E. Blumenberg, and C. Dawkins. 2016. "What If Cities Combined Car-Based Solutions with Transit to Expand Access to Opportunity?" Urban Institute. https://www.urban.org/sites/default/files/publication/81571/2000818-What-If-Cities-Combined-Car-Based-Solutions-with-Transit-to-Improve-Access-to-Opportunity.pdf.

Phadke, A., N. Abhyankar, J. Kersey, T. McNair, U. Paliwal, D. Wooley, O. Ashmoore, et al. 2021. *2035 Report: Plummeting Costs and Dramatic Improvements in Batteries Can Accelerate Our Clean Transportation Future*. Goldman School of Public Policy, University of California, Berkeley. https://www.2035report.com/transportation.

Phillips, S.D., J.K. Tarud, M.J. Biddy, and A. Dutta. 2011. "Gasoline from Wood via Integrated Gasification, Synthesis, and Methanol-to-Gasoline Technologies." https://doi.org/10.2172/1004790.

Polzin, S., and T. Choi. 2021. "COVID-19's Effects on the Future of Transportation." Department of Transportation, Office of the Assistant Secretary for Research and Technology. https://rosap.ntl.bts.gov/view/dot/54292.

Popovich, N.D., D. Rajagopal, E. Tasar, and A. Phadke. 2021. "Economic, Environmental and Grid-Resilience Benefits of Converting Diesel Trains to Battery-Electric." *Nature Energy* 6(11):1017–1025. https://doi.org/10.1038/s41560-021-00915-5.

Powell, S., G.V. Cezar, L. Min, I.M.L. Azevedo, and R. Rajagopal. 2022. "Charging Infrastructure Access and Operation to Reduce the Grid Impacts of Deep Electric Vehicle Adoption." *Nature Energy* 7(10):932–945. https://doi.org/10.1038/s41560-022-01105-7.

Preston, B. 2022. "High Registration Fees Hamper Adoption of Electric Vehicles." https://www.consumerreports.org/cars/hybrids-evs/high-registration-fees-hamper-adoption-of-electric-vehicles-a2396438631.

Rangelov, M., H. Dylla, B. Dobling, J. Gudimettla, N. Sivaneswaran, and M. Praul. 2022. "Readily Implementable Strategies for Reducing Embodied Environmental Impacts of Concrete Pavements in the United States." *Transportation Research Record* 2676(9):436–450. https://doi.org/10.1177/03611981221086934.

Renewable Energy Group. 2020. "REG Ultra Clean." https://www.regi.com/products/transportation-fuels/reg-ultra-clean-diesel.

Romero-Lankao, P., A. Wilson, and D. Zimny-Schmitt. 2022. "Inequality and the Future of Electric Mobility in 36 U.S. Cities: An Innovative Methodology and Comparative Assessment." *Energy Research and Social Science* 91(September):102760. https://doi.org/10.1016/j.erss.2022.102760.

Royal Society. 2019. *Sustainable Synthetic Carbon Based Fuels for Transportation: Policy Briefing*. https://royalsociety.org/-/media/policy/projects/synthetic-fuels/synthetic-fuels-briefing.pdf.

Salcido, V., M. Tillou, and E. Franconi. 2021. "Electric Vehicle Charging for Residential and Commercial Energy Codes: Technical Brief." Pacific Northwest National Laboratory. https://www.energycodes.gov/sites/default/files/2021-07/TechBrief_EV_Charging_July2021.pdf.

Santero, N.J., E. Masanet, and A. Horvath. 2011. "Life-Cycle Assessment of Pavements. Part I: Critical Review." *Resources, Conservation and Recycling* 55(9):801–809. https://doi.org/10.1016/j.resconrec.2011.03.010.

Savitch, H.V., and S. Adhikari. 2017. "Fragmented Regionalism: Why Metropolitan America Continues to Splinter." *Urban Affairs Review* 53(2):381–402. https://doi.org/10.1177/1078087416630626.

Schuetz, J. 2022. *Fixer-Upper: How to Repair America's Broken Housing Systems*. Brookings. https://www.brookings.edu/book/fixer-upper.

Schwartz, J. 2023. "VW Pauses on Europe Battery Plants, Awaiting EU Response to U.S. Inflation Reduction Act." *Reuters*, March 8. https://www.reuters.com/business/autos-transportation/awaiting-an-ira-response-volkswagen-pauses-decision-eastern-europe-battery-plant-2023-03-08/#.

Shumway, J. 2021. "White House: Oregon Single-Family Zoning Law Could Be Model for Nation." *Oregon Capital Chronicle*. https://oregoncapitalchronicle.com/2021/10/29/white-house-oregon-single-family-zoning-law-could-be-model-for-nation.

Slowik, P., S. Searle, H. Basma, J. Miller, Y. Zhou, F. Rodríguez, C. Buysse, et al. 2023. "Analyzing the Impact of the Inflation Reduction Act on Electric Vehicle Uptake in the United States." International Council on Clean Transportation and Energy Innovation Policy and Technology LLC. https://energyinnovation.org/wp-content/uploads/2023/01/Analyzing-the-Impact-of-the-Inflation-Reduction-Act-on-EV-Uptake-in-the-U.S..pdf.

Smart, M., and N. Klein. 2020. "Disentangling the Role of Cars and Transit in Employment and Labor Earnings." *Transportation* 47(3):1275–1309.

Sovacool, B.K. 2021. "When Subterranean Slavery Supports Sustainability Transitions? Power, Patriarchy, and Child Labor in Artisanal Congolese Cobalt Mining." *The Extractive Industries and Society* 8(1):271–93. https://doi.org/10.1016/j.exis.2020.11.018.

Sperling, D., L.M. Fulton, and V. Arroyo. 2020. "Accelerating Deep Decarbonization in the U.S. Transportation Sector." Zero Carbon Action Plan (2020). https://papers.ssrn.com/abstract=3725841.

Speroni, S., and B. Taylor. 2023. *The Future of Working from Home and Daily Travel: A Research Synthesis*. Los Angeles, CA: Institute of Transportation Studies, University of California. https://escholarship.org/uc/item/23v094qk.

Sriram, A. 2022. "Musk Delivers First Tesla Truck, But No Update on Output, Pricing." *Reuters*, December 2. https://www.reuters.com/business/autos-transportation/musk-delivers-first-tesla-semi-trucks-2022-12-02.

Tal, G., J.H. Lee, D. Chakraborty, and A. Davis. 2021. "Where Are Used Electric Vehicles and Who Are the Buyers?" University of California, Davis: National Center for Sustainable Transportation. https://doi.org/10.7922/G2J38QTS.

Tal, G., A. Davis, and D. Garas. 2022. *California's Advanced Clean Cars II: Issues and Implications*. https://escholarship.org/uc/item/1g05z2x3.

Tesla. n.d. "Supercharger." https://www.tesla.com/supercharger. Accessed March 1, 2023.

Tong, F., C. Hendrickson, A. Biehler, P. Jaramillo, and S. Seki. "Life Cycle Ownership Cost and Environmental Externality of Alternative Fuel Options for Transit Buses." *Transportation Research Part D: Transport and Environment* 57(2017): 287–302.

TRB (Transportation Research Board). 2009. *Driving and the Built Environment: The Effects of Compact Development on Motorized Travel, Energy Use, and CO_2 Emissions*. Washington, DC: The National Academies Press. https://doi.org/10.17226/12747.

TRB. 2013. *Recycled Materials and Byproducts in Highway Applications Scrap Tire Byproducts*. Vol. 7. Washington, DC: The National Academies Press. https://doi.org/10.17226/22546.

TRB and NRC (National Research Council). 2015. *Overcoming Barriers to Deployment of Plug-in Electric Vehicles*. Washington, DC: The National Academies Press. https://doi.org/10.17226/21725.

Trost, J.N., and J.B. Dunn. 2023. "Assessing the Feasibility of the Inflation Reduction Act's EV Critical Mineral Targets." *Nature Sustainability* 6:639–643. https://doi.org/10.1038/s41893-023-01079-8.

USGS (U.S. Geological Survey). 2022. "2022 Final List of Critical Minerals." https://d9-wret.s3.us-west-2.amazonaws.com/assets/palladium/production/s3fs-public/media/files/2022%20Final%20List%20of%20Critical%20Minerals%20Federal%20Register%20Notice_2222022-F.pdf.

USPS (United States Postal Service). 2023. "USPS Moves Forward with Awards to Modernize and Electrify the Nation's Largest Federal Fleet." https://about.usps.com/newsroom/national-releases/2023/0228-usps-moves-forward-with-awards-to-modernize-and-electrify-nations-largest-federal-fleet.htm.

Voelker, J. 2021. "Good News. Ford and GM Are Competing on EV Investments." *Car and Driver*. https://www.caranddriver.com/features/a37930458/ford-gm-ev-investments.

Von Fischer, J.C., D. Cooley, S. Chamberlain, A. Gaylord, C.J. Griebenow, S.P. Hamburg, J. Salo, R. Schumacher, D. Theobald, and J. Ham. 2017. "Rapid, Vehicle-Based Identification of Location and Magnitude of Urban Natural Gas Pipeline Leaks." *Environmental Science and Technology* 51(7):4091–4099. https://doi.org/10.1021/acs.est.6b06095.

Weller, Z.D., S.P. Hamburg, and J.C. Von Fischer. 2020. "A National Estimate of Methane Leakage from Pipeline Mains in Natural Gas Local Distribution Systems." *Environmental Science and Technology* 54(14):8958–8967. https://doi.org/10.1021/acs.est.0c00437.

WRI (World Resources Institute). 2022. "Federal Building Blocks to Support a Just and Prosperous New Climate Economy in the United States." https://files.wri.org/d8/s3fs-public/2022-09/federal-policy-building-blocks-support-just-prosperous-new-climate-economy-united-states_0.pdf, Table 4.

Zhou, Y., R.T. Muehleisen, Z. Zhou, C. Macal, and R. Oueid. 2021. *Considerations for Building the Business Cases for Bidirectional Electric Vehicle Charging*. No. ANL/ESD-21/8. Argonne, IL: Argonne National Laboratory.

10

Industrial Decarbonization

ABSTRACT

Significant reductions in industrial greenhouse gas (GHG) emissions by 2050 will require aggressive support and pursuit of key decarbonization pillars: improving energy and materials efficiency, implementing beneficial electrification, using low-carbon energy sources and feedstocks, employing mitigation options as needed, and increasing demand for low-carbon products. Recent legislation—for example, the Energy Act of 2020, the Infrastructure Investment and Jobs Act of 2021 (IIJA), the CHIPS and Science Act of 2022, and the Inflation Reduction Act of 2022 (IRA)—provides a significant infusion of seed funding to initiate the decarbonization transition in industry and other sectors. A continued drive for innovation, focus on reducing costs, development of infrastructure supporting low-carbon solutions, and supply chain engagement are vital for that seed funding to be most effective.

Analyses of the impact of these bills suggest that industrial GHG emissions will decrease 6–14 percent by 2030. However, the scenarios in the Biden administration's strategy and the funding priorities in the IRA and IIJA court risk by placing the majority of industrial CO_2 reductions on the latter decarbonization pillars. This risk could be diminished by prompt investments in faster-acting pillars (e.g., energy and materials efficiency, electrification) that could deliver substantial CO_2 reductions in the next 5–10 years. Rapid innovation, agile learning, and implementation advances to relentlessly pursue cost parity with incumbent solutions and persistent lowering of adoption barriers will be crucial to use this funding most effectively and to support future funding justification across the next 5–10 years and beyond.

INTRODUCTION

The industrial sector accounts for nearly 30 percent of U.S. energy-related CO_2 emissions—around 1,360 MMT CO_2 in 2020 (DOE 2022a). Although industry generates significant GHG emissions, it can also play a role in emissions mitigation by making products that enable low-carbon pathways in transportation, power generation, transportation, and buildings. To achieve this goal, however, will require significant reductions in the GHG emissions associated with making products, the carbon intensity of those products, the emissions across value chains where the products are transported and used, and the products' end-of-life footprint.

Industrial companies are increasingly responding to stakeholder requests for improved sustainability by setting more aggressive emissions reduction goals, considering the carbon contributions along supply chains (e.g., Scope 3 emissions), responding to customer requests by increasingly making lower-carbon products, and investing in companies developing innovative low-carbon technologies and products. The IIJA and IRA provide incentives for industrial emissions reductions over the next 5–10 years, but to be on pace with the GHG reductions needed to reach net-zero emissions by mid-century, a significant increase is required in near-, mid-, and long-term investments. Approximately half of industrial emissions reductions will likely come from emerging technologies (IEA 2020b) that are more expensive than existing technologies, and thus will require continued support for research, development, and demonstration (RD&D) to spur innovation, improve the cost position, and drive adoption (Gaster et al. 2023).

Major technology transitions in industry can take many decades (Grubler et al. 2016). To accelerate industrial decarbonization, it will be vital to address not only technical and economic hurdles that are key elements of risk reduction but also behaviors that reinforce the status quo in the face of uncertainties on market-pull, integration, and durability associated with making major technology changes. These many socioeconomic, environmental, and behavioral elements must be understood, balanced, and managed over decades with an agile approach to successfully arrive at net-zero GHG emissions by midcentury.

This chapter begins by outlining recent federal legislation related to industrial decarbonization and examining its potential impacts on emissions reductions in the sector. The chapter then describes five major pillars of industrial decarbonization, which are relevant across heavy and light industry and small, medium, and large manufacturers: (1) energy and materials efficiency, (2) beneficial electrification, (3) low-carbon energy sources and feedstocks, (4) mitigation options, and (5) demand for low-carbon products. Given that many prior analyses of industrial decarbonization have focused on large, heavy industry, this chapter additionally highlights the considerations for decarbonizing light industry and small- and medium-size manufacturers and examines approaches to tailor industrial decarbonization strategies for different states or regions. It also analyzes technical, socioeconomic, environmental, and behavioral barriers to industrial decarbonization, as well as potential policy solutions to overcome those barriers. Throughout the chapter, the committee provides findings and recommendations to facilitate industrial decarbonization efforts over the next decade and set industry on a path to net-zero emissions by midcentury. Table 10-5, at the end of the chapter, summarizes all the recommendations that appear in this chapter to support decarbonizing industry.

PACE OF INDUSTRIAL DECARBONIZATION PER RECENT LEGISLATION

The IIJA and IRA provide funding for several decarbonization initiatives that will help advance RD&D in industry, adding to the provisions in the Energy Act of 2020 that provided appropriations for carbon management, hydrogen technology, and emissions reduction programs in heavy industry and established programs for technical assistance and smart manufacturing (Energy Act of 2020; see Titles IV, V, and VI). As shown in Figure 10-1, the IRA and IIJA give strong starting support for hydrogen, carbon capture, utilization, and storage (CCUS), and transformative process technologies (e.g., the Advanced Industrial Facilities Deployment Program [AIFDP]). The $8 billion in funding appropriated for the hydrogen hubs (DOE-OCED n.d.(b)) will support 6–10 demonstrations in various regions of the country, as well as the development of networks—producers, consumers, and local infrastructure to accelerate the use of hydrogen as an energy carrier. The $6.4 billion in authorized and appropriated CCUS funding in the IIJA continues long-running support for these technologies, dating to at least 1997 (Lawson 2022). Carbon capture funding also includes $3.5 billion in appropriations for regional direct air capture (DAC) hubs (DOE 2022b). The $5.8 billion in funding appropriations for AIFDP (IRA §50161) supports demonstrations at scale of transformative low-carbon technologies directly involved in producing products in heavy industry. Figure 10-1 also indicates the absence of support for the two fast-start decarbonization pillars—electrification and

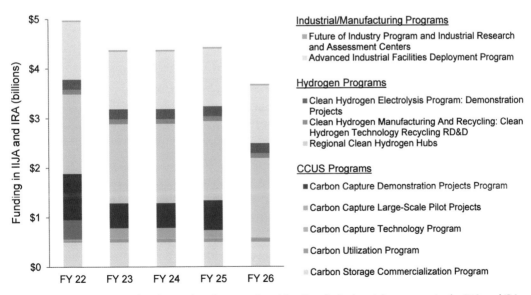

FIGURE 10-1 Summary of authorized and appropriated funding for industrial programs in the IIJA and IRA. NOTES: Program funding is shown distributed equally across the program years. The Department of Energy (DOE) funding for the programs may vary. SOURCE: Data from ITIF (2023).

energy and materials efficiency (see the section "Major Pillars of Industrial Decarbonization" below). Significant funding and program implementation are needed to accelerate the GHG reduction impact of these two pillars.

The IRA and IIJA also contain support for manufacturing decarbonization programs outside of RD&D. This includes, in the IRA, an update to the investment tax credits and storage credits for CCUS (45Q; IRA §13104), an extension of the advanced energy project investment tax credit (48C; IRA §13501), and establishment of a new clean hydrogen production tax credit (45V; IRA §13204). The IIJA appropriates $2.1 billion for CO_2 transportation infrastructure development (IIJA §40304) and $400 million for industrial energy efficiency in the form of support for industrial research and assessment centers in the Future of Industry Program (IIJA §40521). Also in the IIJA are directives for the Department of Energy (DOE) to include smart manufacturing technologies and practices within the scope of industrial assessment centers (IIJA §40532) and to study how to increase access to high-performance computing resources at National Laboratories for small- and medium-size manufacturers (IIJA §40533). (As discussed in Chapter 6, the IRA and IIJA also provide substantial incentives for decarbonizing power generation, which will indirectly affect emissions related to industrial electricity use.)

While the IIJA has support for increasing generation of clean electricity and infrastructure, it contains little direct support for industrial electrification. The relatively low levels of funding for energy efficiency and electrification in industry are major gaps considering that these decarbonization pillars are most amenable to early action and impact owing to their relatively low costs, capital requirements, and infrastructure needs (DOE 2022a).

A natural question is whether the funding provided by the IIJA and IRA for industrial emissions reductions is sufficient to set the sector on pace to reach net zero by 2050. Several groups have analyzed the potential impact of these bills on economy-wide emissions reductions (Jenkins et al. 2022a,b; Larsen et al. 2022; Mahajan et al. 2022). An analysis by the Rhodium Group, summarized in Table 10-1, separates out the impact of the industrial sector and shows that, collectively, the measures in the bills could potentially spur a 6–14 percent reduction in emissions of CO_2e by 2030 versus a 2005 baseline (King 2022).[1] The magnitude of reductions estimated in the Rhodium Group analysis is in a similar range to analyses from other groups across the economy and scenarios for CO_2e reductions in industry *if* multiple decarbonization pillars are aggressively pursued. As shown in Figure 10-2, the "net-zero" scenario

---

[1] If upstream emissions from the oil and gas sectors are included as industrial emissions, the estimated range of reduction is 3.0–16.0 percent by 2030 from a 2005 baseline, spurred by a methane emission fee and a decline in oil and gas production (Larsen et al. 2022).

TABLE 10-1 Rhodium Group Estimates of Industrial GHG Emissions Reductions Afforded by the IIJA and IRA by 2030

2030 Scenario	Percent CO_2e Reduction Versus 2005 Baseline	
	Oil and Gas Sector Emissions *Not* Included	Oil and Gas Sector Emissions Included
Low	14.0	16.0
Medium	12.9	11.0
High	6.2	3.0

NOTE: The "low" scenario achieves the greatest reductions, while the "high" scenario is the high-emissions case. SOURCES: Data for oil and gas sector emissions *not* included from King (2022). Data for oil and gas sector emissions included from Larsen et al. (2022).

within DOE's Industrial Decarbonization Roadmap estimates that an emissions reduction of 29 percent versus a 2015 baseline could be achieved by 2030 if all pillars are rigorously pursued.[2] That the Rhodium Group's estimates of industrial CO_2 emissions reductions from the IRA and IIJA are less than half of the potential reductions noted by the DOE Roadmap suggests that the reductions supported by these bills spur only a portion of the possible reductions during this time period. As noted earlier, support for energy and materials efficiency and electrification are remaining opportunities. Additional discrepancies between the emissions reductions estimates could result from the different baseline years used—2005 for the Rhodium Group analysis and 2015 for the DOE Roadmap analysis.

The Biden administration's long-term strategy document (DOS and EOP 2021) illustrates the potential impact of these major decarbonization pillars in industry as well, yet in those scenarios, major CO_2 reductions owing to electrification (via transformation of the power grid), hydrogen, and CCUS do not occur until 2030 or 2040 (e.g., see Figure 10-2). Although CO_2 reductions of 50 percent are targeted economy-wide by 2030, the Biden administration's strategy does not set a specific target for industry (DOS and EOP 2021). The infrastructure investments, deployment timeline of hydrogen and CCUS, and adoption cascade of low-carbon technologies across industry will require decades—especially considering the high investment capital, complexity, heterogeneity of industry, and cost hurdles.

---

[2] The estimated emissions reductions in the DOE Roadmap do not include upstream emissions from the oil and gas sector, only downstream decarbonization impacts at refineries. The DOE Roadmap considers only emissions from major commodity products in five industrial subsectors (chemicals, iron and steel, petroleum refining, food and beverage, and cement), which represent about 30 percent of total industrial emissions.

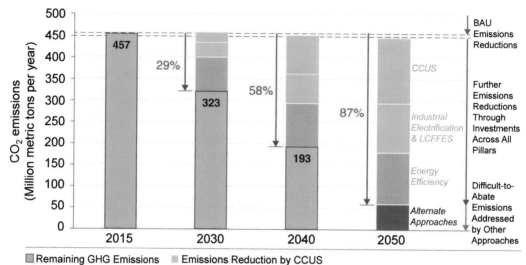

FIGURE 10-2 Potential emissions reductions in the DOE Industrial Decarbonization Roadmap's "net-zero" scenario from the application of four decarbonization pillars: energy efficiency (light pink); electrification and low-carbon fuels, feedstocks, and energy sources (green); carbon capture, utilization, and storage (blue); and alternative approaches such as negative emissions technologies (purple). NOTE: The scenario considered major commodity products in five industrial subsectors (chemicals, iron and steel, food and beverage, petroleum refining, and cement), reflecting approximately 30 percent of industrial emissions. SOURCE: DOE (2022c).

The scenarios in the Biden administration's strategy and the funding priorities in the IRA and IIJA court risk by placing the majority of industrial CO_2 reductions on these late-delivering pillars. This risk could be diminished by prompt investments in faster-acting pillars that could deliver substantial CO_2 reductions in the next 5–10 years. Accelerating the pace of reductions will require more aggressive support for the fast-start pillars of energy efficiency, materials efficiency, and electrification of process heat and key processes that are the backbone of heavy industries; pursuit of crosscutting approaches and a focus on achieving cost parity for low-carbon technologies; and decreased hurdles for implementation. Coordinated and dedicated engagement will be needed to accelerate implementation, including enhanced education and training to design, develop, demonstrate, and commission efficient energy systems.

Several energy and materials efficiency provisions that did not appear in the IRA and IIJA are good starting candidates for accelerating GHG reductions in industry. Described in Ungar et al. (2021), these programs include (1) support for audits and programs to pursue efficiency projects at large to small plants, (2) support of energy

TABLE 10-2 Cost, Energy, Emissions, Value, and Jobs Impact of Industrial Energy Efficiency Investments

Large Plant Audits and Grants for Energy Efficiency	Cost, $ Billions	Quads of Energy Reduced	CO_2 Reduced, MMT	Value, $ Billions	Jobs, Thousands
Audits and grants	2.6	10.7	458	85.0	153
Energy managers	0.3	0.74	32	5.8	9
Strategic energy management	0.2	0.41	15	4.5	9

NOTES: Estimated cumulative present value federal investments and net savings ($ billion). Estimated net job creation (thousand full-time-equivalent job-years), Quads = quadrillion British thermal units (BTUs). SOURCE: Data aggregated from Ungar et al. (2021, Tables A1, A2, and A3).

managers at small to medium plants (a workforce development and entry opportunity), (3) strategic energy management, and (4) a range of programs to catalyze energy and materials efficiency at industrial clusters. These fast-start options are attractive to industry and have relatively low technology barriers and good workforce development prospects. In particular, the programs would recognize the capital, staffing, business model, and technical capability challenges of small and medium manufacturers that differ from those of large manufacturers. This is important, as approximately 75 percent of manufacturing firms (NAICS codes 31–33) have fewer than 20 employees (U.S. Census Bureau 2022). One analysis of impacts across various metrics found that for a $3 billion investment, these programs could reduce CO_2 emissions by 500 million metric tons/year and energy use by 400 quads, provide a value return of $96 billion, and yield 171,000 full-time-equivalent job-years (see Table 10-2) (Ungar et al. 2021).

Electrification of key process technologies, process heat, and movement of goods within and outside of facilities remains a significant and largely under-pursued opportunity (Rightor et al. 2020). Hastening adoption and impact will require efforts to (1) lower the costs for known and emerging technologies (including offsetting the price difference between natural gas and electricity to spur early adopters), (2) demonstrate low-carbon technologies (e.g., industrial heat pumps, Rightor et al. 2022a) at scale, and (3) develop the community (e.g., service companies, academics) and workforce needed to support the technologies.

Process heat provides a key crosscutting opportunity, as it accounts for about 50 percent of the on-site energy use in industry and is prevalent in all heavy industries (DOE 2022c). Recognizing this, one of DOE's Energy Earthshot Initiatives™ is for process heat and targets 85 percent GHG emissions reductions in industrial heat technologies by 2035 (DOE-EERE n.d.). A greater drive for implementation and adoption across the breadth of heavy and light industry and small/medium to large manufacturers could maximize leverage and impact. For example, demonstration at scale that electrification can produce the 800–875°C temperatures needed for steam crackers (Linde Engineering 2023) would open opportunities and spur adoption in additional high-temperature applications in chemicals, as well as iron and steel, cement, and other areas (Rightor 2022).

To accelerate the deployment rate of low-carbon technologies, it is essential to decrease their cost to close to that of incumbent technologies. Demonstrations at scale of first-of-a-kind, low-carbon technologies, as supported by DOE's Advanced Industrial Technologies Deployment Program, are vital not only because they demonstrate feasibility, but also because they accelerate learning and innovation. Additional emphasis on RD&D and implementation to bring costs down will be needed so that market drivers can carry the burden of disseminating these technologies across the broad distribution of uses in multiple industrial sectors.

Increasing the pace of industrial decarbonization will require:

- *Additional support for energy efficiency, materials efficiency, and electrification*, as they are near-term opportunities that received a proportionally lower level of support in the above-mentioned bills.
- *Additional support for process technology innovation* (catalyzing changes in how materials are made). Support in the IRA for the Advanced Industrial Technologies Deployment Program provides a start, but far greater support and focus from DOE, industry, and others is needed for step-change increases in innovation, which can help integrate decarbonization pillars and yield further emissions reductions. The higher cost of low-carbon technologies compared to incumbents is a huge barrier for adoption, so innovation to reduce cost needs to be relentlessly pursued.
- *Implementation rigor to deliver the maximum possible impact* from the funding provided in the bills. While there is a tendency to add scope to initiatives in the bills, if doing so slows the rate or magnitude of emissions reductions, or hampers the acceleration of emissions reductions, then the overall potential for achieving reductions will suffer. Hence, it is vital to keep implementation the focus of industrial support provisions and to avoid the expansion of scope unnecessarily. Maximizing leverage across sectors and infrastructure projects and amplifying the how innovations are applied across opportunities in parallel will also be crucial to achieve the greatest possible emissions reductions (Rightor 2023).
- *Removal of impediments to implementation* (e.g., staffing to process permits) where possible and development of capabilities to allow major projects to proceed as fast as possible while still providing ample opportunity for community engagement.

Finding 10-1: The authorized and appropriated funding and tax credits in the IIJA and IRA will help spur CO_2 reductions in industry (estimated 6–14 percent by 2030) aligned with the long-term trajectory of climate stabilization. However, the low ambition for reductions and lack of support for early action represent a lost opportunity and an increased risk of failure to achieve 2050 net-zero targets. To accelerate industrial emissions reductions, near-term pathways (e.g., energy and materials efficiency, electrification, and low-carbon fuels substitutions) need to be supported and pursued. A focus on innovation, demonstrations at scale, cost reductions, and integration of solutions with multiple upstream and downstream processes and associated control systems is needed to relentlessly pursue price/cost equivalence for low-carbon solutions and have the market increasingly drive adoption.

Recommendation 10-1: *Develop and Enable Cost-Competitive Process and Waste Heat Solutions.* **In partnership with leading industrial companies and their supply chain partners (including small- and medium-size manu-facturers), the Department of Energy (DOE) should develop and enable cost-competitive solutions to transition 50 percent of on-site process heat use to low-carbon sources by 2035 and to increase the use of waste heat for on-site energy demands and off-site reuse applications (e.g., district or community heating). DOE should pursue the attainment of price parity for an array of low-carbon process heat options while driving the Energy Earthshot Initiative for process heat and reporting yearly on progress.**

Finding 10-2: Additional legislative support is needed for fast-start approaches (e.g., energy and materials efficiency, electrification, and low-carbon fuel substitu-tions such as biofuels) while also setting the stage for implementing low-carbon technologies that will take longer to demonstrate at scale and achieve economic parity. Support for the two fast-start decarbonization pillars—electrification and energy and materials efficiency—is largely absent in the industrial sections of the IRA and IIJA.

Recommendation 10-2: *Invest in Energy and Materials Efficiency and Indus-trial Electrification.* **Congress should increase funding for the Department of Energy (DOE) to invest in energy and materials efficiency and electrification of industrial processes, as these pillars of decarbonization were largely left out of recent legislative support. For efficiency, this could include driving optimization across entire systems and process facilities, providing partial support for energy managers at small and medium manufacturers for a limited time to pursue energy and CO_2 reduction audits and strategic energy**

management, and spurring supply chain networks to improve circular economy implementation. DOE should also support beneficial electrification via research, development, and demonstration in process heat, foundational processes, and direct use of clean energy at industrial facilities (e.g., sensing, control, demand response—connected with energy storage at scale). To support these programs, a funding level of $4 billion is recommended across 5 years or until expended.

Recommendation 10-3: *Spur Innovation to Achieve Price-Performance Parity for Low-Carbon Solutions.* To support the transition to low-carbon energy sources in industry, non-governmental organizations, associations (e.g., American Chemistry Council, American Iron and Steel Institute, Portland Cement Association, National Association of Manufacturers, and others), and industry should work with Congress to develop, propose, and adopt policies that

   a. Drive cost reductions for low-carbon technologies to achieve parity with current market solutions;
   b. Offset a portion of the price difference between incumbent energy sources and their low-carbon alternatives (e.g., low-carbon electricity versus natural gas) for a limited time to drive innovation and scale; and
   c. Initiate performance-based carbon intensity targets for major product families (e.g., steel and cement) and connect them with low-carbon product procurement and Buy Clean provisions.

These policies should also provide incentives for the manufacture and deployment of low-carbon technologies, increased use of on-site and off-site low-carbon energy, and integration of energy storage (thermal, chemical, mechanical, or electrical) with process heat generation and use. Congress should direct the Department of Energy to initiate programs to implement these policies with funding of $1 billion over 5 years or until expended.

MAJOR PILLARS OF INDUSTRIAL DECARBONIZATION

Numerous reports have described paths to deep decarbonization in the industrial sector (e.g., Bashmakov et al. 2022; DOE 2022c; USCA 2022; Williams and Bell 2022), primarily focused on decarbonizing heavy industries—chemicals, refining, iron and steel, and cement—which account for nearly 50 percent of industrial CO_2 emissions in the

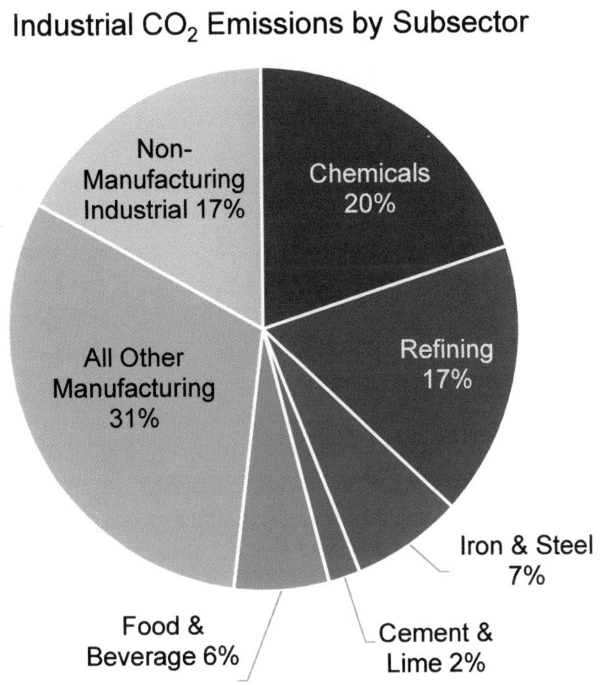

Industrial CO$_2$ Emissions by Subsector

FIGURE 10-3 Industrial CO$_2$ emissions by subsector, illustrating that nearly 50 percent of emissions come from the heavy industries of chemicals, refining, iron and steel, and cement and lime. SOURCE: Data from DOE (2022c).

United States[3] (Figure 10-3). (For a discussion of non-CO$_2$ GHG emissions from industry, see Box 10-1 below.) As one representative example, the 2022 Industrial Decarbonization Roadmap from DOE provides an overview of decarbonization pillars; RD&D needs; barriers to industrial decarbonization; and routes to accelerate deployment of technologies related to these pillars (DOE 2022c). The DOE roadmap identified four major pillars of industrial decarbonization: energy efficiency, electrification, low-carbon fuels and feedstocks, and mitigation options (e.g., CCUS and DAC). Another key pillar discussed in this report is the need to increase demand for low-carbon products. The following sections briefly summarize how each pillar can be pursued while enhancing workforce, environmental justice, and diversity objectives that are also critical to setting U.S. industry on a path to decarbonization by midcentury. A summary of opportunities for GHG emissions reductions by industry subsector and decarbonization pillar is provided in Table 10-3.

---

[3] While this report focuses on U.S. industry, the committee emphasizes that industrial decarbonization is a global challenge and refers the reader to IEA (2020a) for a discussion of technology needs and pathways for emissions reductions in heavy industry worldwide.

BOX 10-1
NON-CO$_2$ GHG EMISSIONS FROM INDUSTRY

Industrial processes accounted for approximately 16 percent of U.S. non-CO$_2$ GHG emissions in 2019 (EPA 2021). They are responsible for all emissions of fluorinated gases (F-gases), specifically hydrofluorocarbons (HFCs), perfluorocarbons (PFCs), sulfur hexafluoride (SF$_6$), and nitrogen trifluoride (NF$_3$) (EPA 2021).

The largest share of F-gas emissions (~91 percent) comes from ozone-depletant substitutes (EPA 2021) following the replacement of chlorofluorocarbons with HFCs and PFCs under the Montreal Protocol and the Clean Air Act Amendments of 1990. Indeed, since 1991, U.S. emissions owing to production and use of ozone depletant substitutes have been increasing (EPA 2021). HFCs and PFCs are not ozone depletants, but they are potent GHGs, addressed in the 2019 Kigali Amendment to the Montreal Protocol, which mandated a global phasedown of HFCs and PFCs to help avoid up to 0.5°C warming by 2100. The United States ratified the Kigali Amendment in October 2022 and adopted guidelines for the phasedown of these ozone-depletant substitutes under the American Innovation and Manufacturing (AIM) Act enacted in December 2020. The AIM Act directs EPA to implement an 85 percent reduction of the production and consumption of HFCs so that they reach 15 percent of their 2011–2013 average annual levels by 2036, as well as to minimize releases from equipment and to facilitate the transition to next-generation technologies through sector-based restrictions. The end-use sectors that contribute the most to HFC and PFC emissions are refrigeration and air-conditioning (78 percent); aerosols in metered dose inhalers, personal care, and specialty products (10 percent); and foams (9 percent) (EPA 2021). Industry stakeholders have supported the phasedown and have acted toward this goal (NRDC 2019).

Phasing down HFCs and PFCs requires developing and using alternative coolants, including hydrofluoroolefins and "natural" refrigerants such as propane, propene, ammonia, isobutene, water, and carbon dioxide. Many of these alternatives are already used; however, concerns still exist regarding their energy efficiency and safety. For example, the flammability of hydrocarbons and toxicity of ammonia call for risk assessment analyses with possible use limitations (EC 2020). In addition to substituting F-gases in equipment, monitoring and preventing leakages in existing equipment, accurately reporting F-gas emissions, and adequate disposal also help limit emissions.

Another source of F-gas emissions is semiconductor manufacturing, where various fluorinated gases are used for etching on silicon wafers and cleaning chemical vapor deposition tool chambers (EPA 2022c). The magnitude of F-gas emissions varies by the types of gas, equipment, and process used, but can range from 10 to 80 percent of the amount of input gas (EPA 2022c). Mitigation strategies—which include process improvements, source reduction, use of alternative chemicals, and destruction technologies—will become increasingly important if U.S. semiconductor production, and consequently use of F-gases, expands as anticipated with implementation of the CHIPS Act. Additional RD&D support is needed to develop alternatives for the chemicals used in these processes so that they meet performance and cost requirements with lower GHG impacts. RD&D is also needed to improve processing and mitigation until such alternative chemicals can be developed at the necessary cost and scale.

TABLE 10-3 Opportunities for GHG Emissions Reduction by Industry Subsector and Decarbonization Pillar

Industry Subsector	Decarbonization Pillar				
	Energy and Materials Efficiency	Beneficial Electrification	Low-Carbon Energy Sources and Feedstocks	Mitigation Options	Demand for Low-Carbon Products
Chemicals	• Efficiency improvements in separations, across processes, systems, and entire facilities • Materials recycling across facilities and supply chains • Improvements to catalyst conversion yields	• Clean electricity for process heat and hydrogen production • Use of variable energy from off-site, and use directly on site	• Clean hydrogen for ammonia, methanol, and ethylene syntheses • Biomass as feedstock for chemical synthesis • Low-carbon process heat from nuclear, clean electricity, solar thermal, hydrogen, and biomass	• Carbon capture • Conversion of CO_2 and other waste gases into valuable products • Incorporation of CO_2 directly into precursors and end products	• Industry accepted standards and benchmarking for reducing product carbon intensity • Shared databases of parameters used in LCAs, standards, and benchmarking
Refining	• Efficiency improvements for distillations and separations • Process conversion efficiency improvements	• Clean electricity for hydrogen production • Clean electricity to replace steam generation capacity	• Low-carbon process heat from nuclear, clean electricity, solar thermal, hydrogen, and biomass	• Carbon capture • Use of captured CO_2 for low-carbon fuels production	• Standards and benchmarking for product carbon intensity

continued

TABLE 10-3 Continued

Industry Subsector	Decarbonization Pillar				
	Energy and Materials Efficiency	Beneficial Electrification	Low-Carbon Energy Sources and Feedstocks	Mitigation Options	Demand for Low-Carbon Products
Iron and Steel[a]	• Waste heat recovery • Blast furnace optimization • Predictive maintenance, improved process control • Systems energy efficiency improvements	• Electrification of process heating pathways where viable • Direct electrolysis of iron	• Replacement of coal/petroleum coke with natural gas, biomass, biogas, or hydrogen • Use of hydrogen as reductant in DRI-EAF	• Carbon capture • Use of captured CO_2 for chemical/fuels production	• Buy Clean initiative • Standards and benchmarking for product carbon intensity
Cement	• Waste heat recovery • High-efficiency clinker cooling and grinding • Efficiency improvements for multistage preheater/precalciner kilns	• Direct and indirect calcination with electric heating	• Replacement of coal/petroleum coke with natural gas, biomass, or hydrogen • Use of supplementary cementitious materials and alternative binding materials • Use of biological routes to cement and concrete	• Capture of process-related CO_2 emissions • CO_2 use in concrete	• Buy Clean initiative • Standards and benchmarking for product carbon intensity

[a] Note that there are different solution sets for decarbonizing BF-BOFs and EAFs given their different feedstocks used, process constraints, and product markets.

SOURCES: Data from DOE (2022c), Jacoby (2023), and USCA (2022).

Energy and Materials Efficiency

Catalyzing rapid progress in the near term is vital to build momentum, develop capabilities, rally the current and future workforce to action, and drive further adoption, scale, and dispersion of low-carbon technologies. Rising energy prices and supply security concerns create strong motivation to pursue efficiency investments. Spurring innovation for further efficiency improvements with current, emerging, and future technologies and making vast improvements in materials efficiency will be important for lowering energy and resource consumption and emissions in industry. This, in turn, will minimize the power demand needed by next-generation low-carbon technologies and will help to lower the cost and resource hurdles for deploying these technologies. By accelerating investments in deep energy and materials efficiency improvements in the near term, industry can achieve 40–50 percent reductions in CO_2 emissions below a 2019 level (Nadel and Ungar 2019).

Energy efficiency (EE) is the most cost-effective option for reducing energy use and GHG emissions in the near term, as it is low cost; often provides multiple energy and non-energy benefits; and has low technical, integration, and adoption hurdles. EE can also lower the energy and resource demand for production facilities prior to implementation of more costly transformative technologies, which decreases economic hurdles and risk. Therefore, continued pursuit of EE throughout the entire course of the decarbonization transformation is critical. To meet the International Energy Agency's (IEA's) Net Zero Emissions to 2050 Scenario, the current rate of EE improvement in industry of about 1 percent per year needs to triple to 3 percent per year (IEA 2022). That level of productivity improvement could be achievable based on experience from the 250 manufacturing partners in the Better Plants program hosted by DOE, which reports an annual energy intensity improvement rate of 2.5 percent (DOE 2020a).

Considerable opportunity remains for energy efficiency improvements in heavy manufacturing. Whenever manufacturing processes are updated, altered, or replaced, additional opportunities are created for efficiency. The advent of high-speed computing, artificial intelligence, and machine learning capabilities allows efficiency optimization to occur at a higher level (e.g., across the entire production site, not just an individual process or facility). Also, as shown in Figure 10-6 below, opportunities in light manufacturing may be even larger than those in heavy manufacturing, as efficiency has received less focus historically in light manufacturing. This significant opportunity can be pursued by the audits and grants, energy managers, and strategic energy management mentioned above in the section "Pace of Industrial Decarbonization per Recent Legislation."

Materials efficiency (ME), circular economy, and related resource conservation approaches can also decrease energy demand and GHG emissions. Their impact is especially evident in cement, steel, and aluminum, where they can provide up to 30 percent of the emissions reduction targets (IEA 2019b).

As the industrial sector continues to improve its energy efficiency, recovering the estimated 20–50 percent of industrial energy input lost as waste heat will become increasingly important (DOE-EERE 2017). Waste heat can be in the form of hot exhaust gases, cooling water, and heat losses from equipment surfaces and products. Waste heat recovery provides benefits in cost and environmental impact, as well as work-flow and productivity. These technologies are not being pursued to the fullest extent possible owing to material constraints, system and process complexity, and high costs (DOE-EERE 2017).

Beneficial Electrification from Low-Carbon Sources

As the proportion of low-carbon electricity on the electric grid increases (see Chapter 6), a transformation in the way that energy is generated, stored, and used will occur. One top priority for increasing the use of low-carbon electricity and reduc-ing emissions in industry is in process heat, which currently accounts for 51 percent of the on-site energy used in manufacturing (DOE 2022c). For industry overall, some 55 percent of process heat needs are below 200°C, and more than 66 percent are below 300°C (Rightor et al. 2022a,b). For some industries such as food and textiles, the proportion of low-temperature heat used is even higher (Naegler et al. 2015). Cur-rently, heavy industries use electricity to supply <5 percent of process heat (Rightor et al. 2022b), and there is good potential for expanded use, as commercial electric technologies can provide heat at appropriate temperatures, making electrification a major near- to medium-term opportunity. For example, BASF, SABIC, and Linde are collaborating on a demonstration plant for an electric steam cracker, which could yield 90 percent reductions in CO_2 emissions by providing the 850°C heat required with electricity instead of natural gas (SABIC 2022). Boston Metal is developing a molten oxide electrolysis technology to reduce iron ore for steelmaking using electric-ity rather than coking coal (Boston Metal 2023). In addition to attainment of a target temperature, the effective transfer (and reuse where possible) of that energy to the material or process being heated (or cooled) is important. A number of low-carbon heat approaches exist, and where there are challenges, there are also numerous op-portunities for innovation, energy, and GHG reductions (Friedman et al. 2019).

Electrification of industry will face integration, control, and capital cost hurdles. There are scaling and heat transfer challenges in switching from fuels to electrification across all applications and temperature ranges, and maintaining the 24/7 capacity factors needed for some processes to be economically viable is a concern. Although these hurdles will be lower for certain application areas like low- to medium-temperature process heat, the capital cost to deliver clean electricity reliably, at the right voltage, and with 24/7 availability will be a challenge. Collaboration, negotiation, and support are needed to integrate electrical infrastructure both outside and inside the fenceline of industrial facilities; this will include addressing the cost of busbars, substations, and transformers; determining who pays for and maintains electricity generation and storage infrastructure; and negotiating an appropriate valuation of energy storage resources.

Long-duration energy storage (LDES)—whether thermal, mechanical, chemical, or electrochemical—can help mitigate the variability of renewable energy sources, translating it into energy that can be relied on 24/7 (Boyles et al. 2023). A wide variety of LDES technologies are in various stages of development (DOE 2023b; LDES Council and McKinsey & Company 2021, 2022), with several available commercially. DOE's LDES Commercial Liftoff report further describes potential use cases and technologies and highlights the need for the costs of LDES to decrease and the value allotted to LDES (i.e., compensation for its economic and reliability benefits) to increase (DOE 2023b). DOE's Energy Earthshot on LDES, which aims for 90 percent cost reductions in grid-scale storage systems delivering more than 10 hours of storage within 1 decade (by 2031), is part of the effort to decrease costs, and some $500 million is allotted for demonstration projects (DOE-EERE 2021; DOE-OCED n.d.(a)). Those demonstrations need to increasingly show the value return for LDES in a variety of end-user applications to accelerate learning, demonstrate integration aspects, and spur adoption (Boyles et al. 2023).

While electrifying industrial applications presents challenges, there are also co-benefits, such as increased reliability and resilience and the potential for improved capacity factor and load management. For example, experience in California has shown that with increased renewables, curtailment and low-capacity factors have negatively impacted utilization rates and economics of clean electricity (CAISO 2017). Increased use of clean electricity by industry could improve the utilization and economics of those low-carbon energy sources. Evaluation of these co-benefits of infrastructure upgrades to deliver and use clean electricity will be important to developing the business case for industrial electrification.

Low-Carbon Energy Sources and Feedstocks

In 2020, U.S. industrial primary energy consumption, including both fuel and feedstock energy, came primarily from natural gas (46.8 percent) and petroleum (38.6 percent), with some contributions from renewables (10.4 percent, predominantly from biomass[4]) and coal (4.2 percent) (EIA 2023). The manufacturing sector, which accounts for 81 percent of industrial energy consumption, uses predominantly fossil fuel–derived feedstocks (EIA 2021b). Switching the current fuels and feedstocks to low-carbon sources is one major pillar of industrial decarbonization. However, as illustrated in Table 2.7.1 of the committee's first report (NASEM 2021, p. 102), low-carbon fuels are typically more expensive than conventional energy carriers. This section describes opportunities for hydrogen and biomass to serve as low-carbon energy sources and feedstocks for industry, discusses co-pollutant emissions from renewable fuels, and examines opportunities for recycling carbon-based materials and using other low-carbon energy sources, such as solar thermal and nuclear, for industrial applications.

Hydrogen

As of 2020, annual U.S. hydrogen use totaled 11.4 MMT (Figure 10-4), primarily for oil refining (e.g., hydrotreatment to remove impurities and hydrocracking to upgrade crude oil), chemical production (primarily ammonia and methanol), and iron and steel production (FCHEA 2020; IEA 2019a). The demand for hydrogen in all four current use areas (see Figure 10-4) is predicted to increase through 2050, with the largest short-term increases expected for methanol and ammonia production, and the largest long-term increases for iron and steel production (IEA 2019a). Current industrial uses of hydrogen are not low carbon, however, because most hydrogen production occurs via steam reforming of natural gas, a process that also emits CO_2. The current cost and availability of low-carbon hydrogen represent substantial barriers to reducing industrial emissions from processes that use hydrogen, as discussed further in the section "Challenges for Using Hydrogen to Decarbonize Industry" below.

Hydrogen can be used in industry to replace fossil feedstocks—for example, as the reductant in iron/steel production in place of coal or in combination with captured CO_2 to synthesize hydrocarbons—or fossil fuel combustion—for example, by providing a source of high-temperature process heat or fueling furnaces for petroleum refining.

---

[4] In 2020, 97.8 percent of the renewable energy use in U.S. industry was from biomass. Solar, wind, geothermal, and hydroelectric all had minimal contributions.

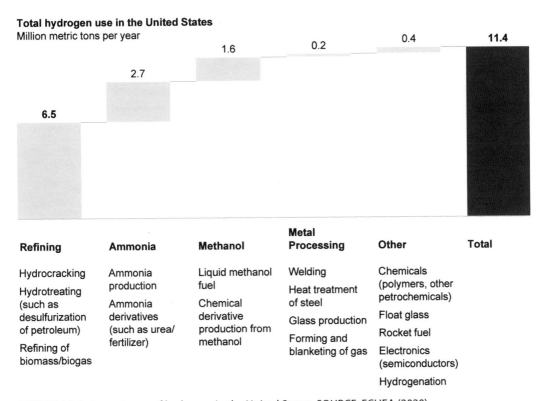

FIGURE 10-4 Current uses of hydrogen in the United States. SOURCE: FCHEA (2020).

The primary approach to decarbonize iron and steel production using hydrogen is direct reduction of iron in an electric arc furnace (DRI-EAF). The natural-gas-based DRI process can incorporate up to 30 percent hydrogen in the gas stream without any process changes, and up to 100 percent hydrogen[5] with some retrofits (Fan and Friedmann 2021). DRI-EAF is considered the most mature low-carbon emissions pathway for iron and steel production and, if operated using hydrogen and zero-carbon electricity, could reduce CO_2 emissions by around 80 percent compared to current production methods (Fan and Friedmann 2021). Bartlett and Krupnick (2020) estimated a breakeven price for hydrogen of $1.30–$1.40 per kg to make DRI-EAF cost competitive with current steelmaking processes. An analysis of the impacts of the IRA and IIJA suggests that the price range for low-carbon (e.g., "green" or "blue") hydrogen can be within or approach this range (depending on scenario) considering

---

[5] Hydrogen Breakthrough Ironmaking Technology (HYBRIT), developed in Sweden, uses green hydrogen as the sole reductant in DRI and began operation of a pilot plant in 2020, with plans for a demonstration plant to come online in 2026 (HYBRIT n.d.).

the subsidies in these bills (Larsen et al. 2022). It will be crucial to spur the technology, integration, production scale, and market innovations needed to allow low-carbon hydrogen to compete when the incentives expire.

Ammonia and methanol syntheses provide the largest volume opportunities to use low-carbon hydrogen as a feedstock to decarbonize chemical production. Each year, 31 MtH_2, or 50 percent of global hydrogen generation, is used for ammonia synthesis, a process that emits around 500 $MtCO_2$ per year, with around half of those emissions attributed to hydrogen production (Bartlett and Krupnick 2020; Sandalow et al. 2019). About 12 MtH_2 per year is used globally in the production of methanol (Bartlett and Krupnick 2020). A plethora of other chemical processes utilize hydrogen—for example, hydrogen serves as a reductant in glass manufacturing and as a hydrogenating agent in industrial food production and synthesis of olefins and BTX (DOE 2020c; FCHEA 2020)—so transitioning to low-carbon hydrogen could reduce emissions throughout the chemical industry.

Hydrogen is also being considered—along with electricity, biomass, and CCUS—as a low-carbon alternative for industrial process heat. Among these low-carbon alternatives, hydrogen is the most promising for higher-temperature applications (>2100°C), delivering sufficient heat flux, availability, and reliability (Bartlett and Krupnick 2020). However, the different combustion properties (e.g., temperature, flame speed, radiative heat transfer) of hydrogen compared to natural gas necessitate equipment modifications to accommodate its use,[6] which increases costs (Pisciotta et al. 2022; Thiel and Stark 2021). Research and development (R&D) needs include optimizing combustion controls, mitigating NO_x emissions, and improving materials compatibility (DOE 2020c; Thiel and Stark 2021).

Biomass

The *2016 Billion-Ton Report* from Oak Ridge National Laboratory indicated that the United States could produce up to 1 billion tons of biomass per year by 2030 at less than $60 per ton, although at the time of the study, biomass production was only around 400 million tons per year (DOE-BETO 2016). The 2019 National Academies' report on negative emissions technologies (NASEM 2019b), the first report of this committee (NASEM 2021), and Chapter 8 of the current report conclude that competition among alternative needs for arable land will limit annual biomass use to the amount provided by forestry and agricultural waste and feedstocks grown on lands currently devoted to corn ethanol. Together, these total about 0.7 Gt/y of biomass.

---

[6] While the specific value can vary depending on technology, hydrogen blending of up to 38 percent by volume into natural gas has been demonstrated without extensive equipment retrofits (Tisheva 2023).

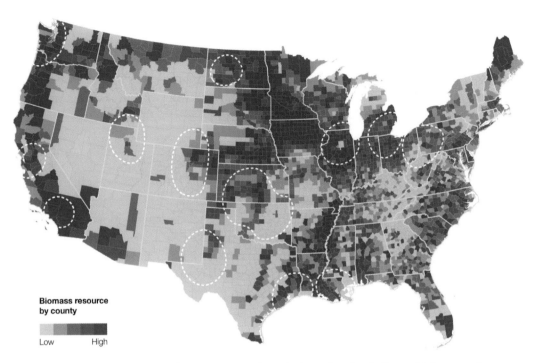

FIGURE 10-5 Biomass resources by U.S. county, where darker coloration indicates higher quantity.
SOURCES: Abramson et al. (2022), Great Plains Institute with data from NREL (2014).

Taking advantage of the full 0.7 Gt/y of biomass—in industry as well as other sectors—would require system-level considerations, such as enhanced efforts to couple the type of biomass grown and the intended use—that is, considering the crop and conversion aspects together (Abdullah et al. 2022). Other key factors to consider include the full system cost (e.g., feedstock, supply chain, refinery, conversion), feedstock availability, life-cycle GHG emissions, and related credits or incentives (Abdullah et al. 2022). As shown in Figure 10-5, biomass resources are distributed across the United States, albeit with regional differences in quantity and type; thus, regional aspects of fuel production, transportation, use, and storage need to be considered, as they impact cost and supply chains (Abdullah et al. 2022; Abramson et al. 2022). Regional availability, quality, and cost of water also need to be considered and may pose risks to biomass feedstock quality and supply (DOE-BETO 2017; Séférian et al. 2018; Stone et al. 2010).

In the industrial sector, the current primary use of biomass is as a source of process heat. Biomass accounts for about 86 percent of the renewable heat consumed by industry worldwide (IEA 2019d). The U.S. industrial sector used 2,313 TBTU of biomass energy in 2021, or 48 percent of the total biomass energy used in the United States that year (EIA 2022). Most industrial biomass use occurs in the pulp/paper and sugar/ethanol industries, which produce biomass wastes on site, and some use occurs in the cement industry (IEA 2019d). The paper and wood products industries obtain biomass primarily from wood/wood waste for consumption in combined heat and power plants (DOE-BETO 2016; EIA 2022).

Opportunities exist to expand the use of biomass as a low-carbon[7] fuel and feedstock to facilitate industrial decarbonization. Biomass in appropriate forms can be used to displace the three energy sources commonly used in industry: solid fuels (e.g., petcoke), fuel oils, and fossil-derived natural gas. Specifically, biomass pellets can substitute in applications that currently use solid fuels; bio-oils derived from pyrolysis or hydrothermal liquefaction of biomass could replace traditional fuel oils; and gas generated via biomass gasification systems can be used in place of fossil-derived natural gas (Abdullah et al. 2022). Industrial feedstock opportunities for biomass include valorizing lignocellulosic biomass to provide aromatics, anaerobically digesting food waste to generate organic acids, and incorporating biomass as a binder to reduce embodied carbon in building materials (Abdullah et al. 2022). Bio(electro)catalytic ammonia production is also being explored to displace the Haber-Bosch process, one of the most energy- and carbon-intensive industrial processes (Krietsch Boerner 2019). These industrial opportunities for biomass use would compete with the two other primary uses: as a feedstock for low-carbon transportation fuels (Chapter 9) and in Biomass Energy with Carbon Capture and Storage (BECCS), where it produces both a significant carbon sink and either energy, a carbohydrate fuel, or hydrogen.

Pollutant Emissions from Renewable Fuels

While renewable fuels may not directly release CO_2 (as in the case of hydrogen or ammonia) or may have near zero life-cycle CO_2 emissions (as in the case of biofuels or e-fuels[8]), their combustion causes emissions of conventional air pollutants, including nitrogen oxides (NO_x), carbon monoxide (CO), sulfur oxides (SO_x), and particulates (O'Connor et al. 2022).

---

[7] Or in some cases, a net-negative option, if the end of life for some of the organic carbon is sequestration underground or in long-lived materials rather than combustion.

[8] E-fuels, or electrofuels, are fuels synthetized from H_2 and captured CO_2. Depending on the emissions of the energy inputs and other upstream and downstream processes involved in their production, these fuels may emit no net CO_2 on a life-cycle basis.

NO_x is formed whenever air, composed of N_2 and O_2, is heated up to temperatures (approximately 1,800 K) at which the N_2 and O_2 begin to react. This can occur with any fuel or means of adding heat to air, regardless of its carbon emissions. Significant R&D investments over the past 2 decades have developed low-NO_x burner designs to drive NO_x emissions levels to well below Environmental Protection Agency (EPA) limits. Such technologies rely on premixing the fuel and air, maintaining combustion temperatures below levels where significant NO_x formation occurs, and minimizing residence times of combustion products at high temperatures. These technologies are often referred to as "lean, premixed" combustion technologies or "dry, low NO_x." There are several trade-offs to consider, such as the power and performance of machines, emissions of co-pollutants, economics, and complexity (Lewis 2021; Lieuwen et al. 2013).

Some studies of H_2/air combustion have shown significant increases in NO_x production relative to natural gas combustion (Therkelsen et al. 2009). However, in order to draw conclusions about the results, it is critically important to identify what is held constant in these comparisons and what type of burner technology is being used. The flame temperature of any fuel/air combination is a function of fuel/air ratio. The peak flame temperature (typically achieved at near-stoichiometric fuel/air ratios) of hydrogen is higher than that of natural gas, and NO_x production is an exponential function of temperature. As a result, if combustion occurs at stoichiometric fuel/air ratios, as in older high-NO_x technologies, then hydrogen combustion can lead to significant increases in NO_x production relative to natural gas. Modern, low-NO_x combustion systems are designed to operate in lean premixed mode that reduces the flame temperature. A hydrogen-fired system can also be operated at a set fuel/air ratio to achieve a given flame temperature. If the flame temperature is held constant, the effect of fuel switching from natural gas to hydrogen is much weaker, and NO_x emissions may actually be reduced (Breer et al. 2022).

Furthermore, NO_x numbers are typically reported in ppm (parts per million) in a dried sample, with the value corrected to some reference oxygen concentration (typically 3 percent in the industrial community and 15 percent in the gas turbine community [Douglas et al. 2022]). However, when comparing NO_x values across different fuels, the metric of primary interest is not ppm, but rather mass production rate of pollutant, normalized by the thermal or electric power of the device. Comparing ppm values rather than mass production rates can artificially inflate apparent NO_x emissions from hydrogen combustion, largely because drying the exhaust gas sample concentrates the NO_x in hydrogen (and ammonia to a lesser extent) systems because the only combustion product—water—is removed in the drying process (Douglas et al. 2022, In press).

Carbon monoxide and particulate emissions are also pollutants of concern whenever the fuel contains carbon atoms, as do biofuels and e-fuels. Carbon monoxide emissions can be managed by appropriate technologies, primarily by ensuring that sufficient time is provided for all of the fuel to burn. A serendipitous benefit of lean, premixed technologies using gaseous fuels is that particulate emissions are very low and generally not a concern. However, particulates are more of an issue with liquid-fueled systems where the fuel is not prevaporized. In this case, the composition of the fuel influences emissions—in particular, attention must be given to technologies that mitigate particulate emissions when liquid fuels are not prevaporized and contain significant aromatic content. Particulate emissions are dramatically reduced when the fuel does not contain significant aromatics, as is the case with e-fuels synthesized via the Fischer–Tröpsch process (Colket 2013). SO_x emissions occur if the fuel contains sulfur, as is the case with coal and some liquid fossil fuels, but sulfur is not typically present in most renewable fuels, such as hydrogen, ammonia, or Fischer–Tropsch-derived liquid fuels. As noted in Chapter 3, because SO_x can react with other small particles in the atmosphere and lead to particulate matter formation, its absence in renewable fuels could have additional air quality benefits.

Opportunistic Reduction of Co-Pollutants

As transformative low-carbon technologies are pursued, there will be opportunities to reduce co-pollutant emissions to air, land, and water. For example, hydrogen combustion does not cause SO_x, particulate matter, or CO emissions. Some of these health benefits are already achieved with existing technology, and others may require more investigation and investment to facilitate low-carbon, low co-pollutant technology. Trade-offs may be encountered between minimizing co-pollutants (e.g., efforts to reduce NO_x can lead to increases in CO), so rather than stipulating limits on these pollutants separately, there may be opportunities to reduce overall health impacts by providing flexibility in permitting, such as regulating weighted sums of these pollutants. Trade-offs in reduction of the co-pollutants versus project costs, complexity, and installation time may also occur.

One example of a low-carbon technology with potential co-pollutant reductions is steam crackers, which produce ethylene and hydrogen by heating ethane with natural gas until its chemical bonds break apart. This is a foundational production process in the chemical industry, as ethylene is a top commodity product. Globally, ethane crackers emit some 260 million tonnes of CO_2 per year (Sarin et al. 2021), as well as significant amounts of co-pollutants. Multiple routes for decarbonizing ethylene production via steam crackers are being pursued, including several electrification routes. Electric crackers could avoid combustion of natural gas, potentially reducing CO_2 emissions some 90 percent (SABIC 2022). For a single cracker operating at 1.5 tonnes per annum (tpa), CO_2 reductions then could approach 1.35 tpa (Rightor 2022). Because electric crackers could avoid natural gas combustion, co-pollutant emissions could also be reduced. For a single cracker, the reductions could be 40 tpa methane (a GHG that is 27 times more potent than CO_2), 160 Mtpa NO_x, 9 Mtpa SO_x, and 66 tpa particulate matter (Rightor 2022). The resulting cleaner air would provide significant benefits to the surrounding communities.

There are also potential co-pollutant reductions for processes that capture gases from process vents, capture CO_2 for CCUS, and generate alternative hydrocarbon fuels. To better understand what co-pollutant reductions are possible, evaluations of the costs, benefits, and trade-offs are needed. This is a prime opportunity for project co-funding, as additional capital, engineering, and infrastructure may be required beyond the targeted reduction of CO_2.

The reduction of refrigerants during manufacture, transportation, use, and recovery is also an area of potential co-pollutant reduction. For example, fluorinated GHGs (F-GHGs) are the most potent and long-lived GHGs and are associated with electronics manufacturing, metals, and other production processes (EPA 2022a). Where it is possible to achieve reductions in F-GHGs in parallel with CO_2 reductions, the emissions impact could be amplified greatly.

> Finding 10-3: Combustion of renewable fuels can still lead to pollutant emissions, which have to be managed by appropriate technology developments and burner upgrades. As transformative low-carbon technologies are pursued, opportunities to reduce co-pollutant emissions (e.g., refrigerants, NO_x, SO_x, particulate matter, hazardous chemicals) to air, land, and water need to be considered where feasible, and adjustments in the evaluation metrics (e.g., concentration versus mass-based metrics for NO_x from hydrogen combustion systems) may be needed to correctly evaluate reductions.
>
> **Recommendation 10-4: *Pursue Technologies That Reduce Both Greenhouse Gas (GHG) and Air Pollution Emissions.* While pursuing GHG reductions, the Department of Energy should work in parallel to reduce co-pollutants. Partners in this work could include non-governmental organizations, industry, industry associations (e.g., American Iron and Steel Institute, American Chemistry Council, Portland Cement Association, National Association of Manufacturers, and others), and engineering companies. Where co-pollutant reduction**

opportunities are viable but additional funding is required, co-funding could be lined up by the Foundation for Energy Security and Innovation. To initiate the program, a funding level of $1 billion devoted to the public match is recommended.

Recommendation 10-5: *Use Mass-Based Rather Than Concentration-Based NO$_x$ Standards.* **Regulatory and permitting organizations should eliminate all concentration-based (i.e., ppm-based) NO$_x$ standards and instead use mass output-based standards (ng/J) so that emissions can be accurately compared across different fuels. This is particularly important for hydrogen fueling, where the drying process prior to measurement and correction to a fixed O$_2$ level artificially elevates NO$_x$ levels.**

Recycled Carbon Materials

The carbon present in current waste streams represents a resource that could be repurposed as a feedstock for chemical production and gradually displace fossil fuels (Lange 2021). There is an array of options for more effectively tapping the carbon that is already in products—reuse, reprocessing (e.g., mechanical recycling), depolymerization, conversion to chemical feedstock, and energy recovery. This follows the general waste management priority pyramid (EPA 2022b), where the first step is reducing use where not necessary (e.g., superfluous packaging). One of DOE's Manufacturing USA Institutes, Reducing Embodied-Energy and Decreasing Emissions (REMADE), focuses on these challenges of reducing GHG emissions and virgin material consumption, increasing secondary material consumption, and ensuring use of waste in processes (Dyck et al. 2022).

A resurgence in RD&D on chemical recycling is adding to the options for depolymerizing polymers, even for typically difficult-to-recycle thermoset polymers (where chemical bonds need to be broken) such as polystyrene and epoxies (Li et al. 2022). The options for plastics recycling vary depending on the type of polymer; some can be depolymerized or cracked, but for others, chemical recycling methods are yet to be developed (Lange 2021). These various avenues to recover and reuse the carbon and other elements in polymers provide routes for displacing a small portion of fossil fuel feedstocks used today.

The potential for and challenges associated with repurposing the carbon in polymers and displacing fossil fuel feedstocks reflect the status of repurposing many materials. Today, less than 10 percent of materials used in manufacturing are recycled (Dyck et al. 2022; Li et al. 2022), so the volume of material available is limited, especially in regions where the local recycling rate is even lower. Mechanical recycling is largely limited to fairly clean waste streams (Barrett 2020; Lange 2021; Schyns and Shaver 2021), and products tend to end up in lower-value applications (e.g., downcycling). The material losses in recycle loops can be up to 50 percent (Lange 2021), leaving a materials gap that, in the future, will need to be filled with carbon from reused CO_2, biogenic sources, or other low-carbon pathways. There also are challenges associated with consumer awareness of what can be recycled, willingness of consumers to pay more for recycled materials, higher cost of reprocessing wastes, and waste stream purity (Collias et al. 2021). Closing materials loops with attention to the life-cycle impacts is an area where continued support, innovation, and application are needed, as they provide additional options for reducing fossil fuel use and decreasing the environmental footprint of manufacturing.

Other Low-Carbon Energy Sources

Additional opportunities to use low-carbon energy sources in industry include solar thermal for industrial process heat (IPH) and nuclear energy for IPH and other industrial applications. Solar IPH is currently deployed in 34 countries worldwide, including the United States, primarily in the food and beverage, metals, and textiles industries because of low to moderate temperature requirements (Schoeneberger et al. 2020). A 2021 NREL analysis examined the potential for three categories of solar technologies—non-concentrating collectors, concentrating collectors, and PV-connected electrotechnologies—to be used for IPH (McMillan et al. 2021). It found that most solar thermal technologies cannot meet IPH demands 100 percent of the time and would need to be augmented by fuel-based heating or grid electricity for industries that require continuous operation. Incorporating thermal or battery energy storage would enable increased use of solar IPH, as "the ability to match the temporal aspect of IPH demand is a more significant barrier than matching solar technologies to IPH temperature" (McMillan et al. 2021). Additional barriers to adopting solar IPH include temperature requirements, process integration and disruption risks, high upfront costs, and geography and land-use constraints (McMillan et al. 2021; Schoeneberger et al. 2020).

Use of nuclear-generated heat for industrial applications is an area of active R&D (Boardman et al. 2021; Rosen 2020), and some demonstrations of nuclear-generated hydrogen are under way at existing reactors (NASEM 2023b). The existing fleet of light water reactors produces heat at around 300°C, which can be used for lower-temperature processes (e.g., chemical separations, hydrogen generation via proton exchange membrane electrolysis) (NASEM 2023b). Many of the advanced reactor designs under development could generate higher-temperature heat, up to about

800°C, which would be suitable for high-temperature processes (e.g., steam methane reforming, hydrogen generation via solid oxide electrolysis, and cement and steel production) (NASEM 2023b). As one example, Dow Chemical has announced plans to partner with X-Energy to deploy a small, modular, high-temperature gas reactor to provide both electricity and process heat at Dow's Seadrift industrial site on the Gulf Coast (*World Nuclear News* 2023; X-Energy 2023). Nonetheless, key RD&D needs for industrial heat applications of nuclear remain, and include "assessing system integration, operations, safety, community acceptance, market size as a function of varying levels of implicit or explicit carbon price, and regulatory risks" (NASEM 2023b).

Mitigation Options

GHG emissions from the industrial sector will likely remain above the levels needed to reach net zero even with aggressive pursuit of the decarbonization pillars, owing to unavoidable process emissions and the high costs and technical complexity of some decarbonization solutions. Several additional strategies for mitigating or offsetting GHG emissions include CCUS, DAC, land use approaches (such as reforestation; see Chapter 8), and CO_2 mineralization. CCUS is the most recognized and developed technology following decades of research and demonstration projects (IEA 2020a). DAC is also gaining visibility (NASEM 2019b), but it is much earlier in development and has significantly higher economic costs. CCUS could be deployed directly on industrial facilities to capture and sequester their emissions, while DAC could be deployed anywhere as a means of offsetting industrial emissions at a different location. Mineralization approaches such as in the curing of cement are also being probed and could provide durable storage of CO_2 in building materials (NASEM 2019a,b, 2023a). For CCUS and DAC, pipeline networks, storage facilities, and reuse applications for CO_2, where feasible, are part of the extensive infrastructure that will be needed for industry to take full advantage of these options (NASEM 2023a).

Different deployment pathways for CCUS in net-zero scenarios have been examined, with the potential for storage approaching 1.0–1.7 billion tons of CO_2 per year by 2050 across a network of pipeline and storage facilities (Larson et al. 2021). The most favorable starting points for CO_2 capture and reuse options in industry have also been studied (Psarras et al. 2017). More concentrated CO_2 sources have substantially lower capture costs than dilute sources (IEA 2021). It is also beneficial to capture CO_2 at locations with geographically localized industry, large quantities of CO_2 available, starting pipeline infrastructure, and nearby storage options such as saline aquifers. The location of reuse options near the capture and storage areas can be important as well. Emerging reuse options include the generation of synthetic fuels, polymers, and other chemicals; mineralization; and production of elemental carbon materials and various niche products (NASEM 2023a). These options need to be further developed in regions where CCUS infrastructure is expanded.

Decreasing the cost of CCUS will be crucial to its success. Integrating CCUS with process heat may, in some applications (e.g., using heat to regenerate amine absorbents that capture the CO_2), defray some of the cost, but finding value return options that customers are willing to pay for will be vital. For example, adding CCUS to methanol or ammonia production increases the cost by 20–40 percent (IEA 2021), and further innovation is needed to bring the costs down to enable adoption. Recent approaches to capture CO_2 from industrial process vents and convert it to key intermediates such as ethanol (which can then be converted to products such as jet fuel) illustrate the potential for value return (Crumpacker 2022; IEA 2019c; NASEM 2019a). Another route for tapping co-benefits may be the reduction of co-pollutants, as discussed above.

For reuse options other than enhanced oil recovery (EOR)[9] to grow substantially, a significant degree of innovation will be needed. Policy enablers can help by providing support for RD&D and deployment of approaches to capture and repurpose CO_2 in nearby industrial applications (e.g., within the fenceline or nearby an industrial facility where the CO_2 is generated), and by improving the incentives specifically for reuse applications (other than EOR). The current 45Q tax credits provide $85 per metric ton for CO_2 storage, $60 per metric ton for EOR or other industrial uses, and $180 per metric ton for direct air capture (JDSUPRA 2022). Increasing the incentive for CO_2 reuse in industrial applications to $85 per metric ton could, depending on the process, significantly spur additional projects (Hughes and Zoelle 2022). A recent study from the National Energy Technology Laboratory showed that the cost of CO_2 capture from cement plants could reach $75 per metric ton, depending on cement plant capacity and CO_2 capture rate, although this estimate did not include transportation and storage costs (Hughes and Zoelle 2022). Establishing increased market pull for reused CO_2 is a significant opportunity to defray the costs of CCUS and argues for a renumeration for CO_2 use in industry (non-EOR applications) at least as high as that for sequestration.

Demand for Low-Carbon Products (Markets)

Numerous stakeholders—including investors, customers, supply chain partners, and non-governmental organizations—are calling for manufacturing to make materials with lower carbon intensity (i.e., low embodied carbon materials). Increased

---

[9] Enhanced oil recovery involves the injection of CO_2 into an oil reservoir to extract additional oil, and results in some long-term storage of CO_2 underground.

market pull for low-carbon products is vital to send the signal of consumer demand. The Buy Clean initiative (Federal CSO n.d.), for example, calls on large purchasers of goods (especially state and federal government entities) to increasingly request and specify materials with lower embodied carbon. The movement has started with low-carbon materials for buildings and infrastructure, specifically targeting cement and steel (Lobet 2020), with the intent of lowering the emissions footprint of buildings and infrastructure while also increasing market pull for low embodied carbon materials. Another effort to establish demand for low-carbon materials is the First Movers Coalition (FMC), which obtains purchase commitments for low-carbon products and technologies across eight sectors: aviation, shipping, steel, trucking, aluminum, carbon removal, cement/concrete, and chemicals (FMC n.d.) As of June 2023, the FMC had 106 commitments from 81 companies and 1 nonprofit organization, totaling $12 billion in demand for low-carbon products (FMC 2023). Ultimately, these initiatives should provide a strong price signal for manufacturing companies to preferentially make these materials.

Showing a viable market for these products is also important, as is demonstrating that the market will provide compensation for the likely higher-cost products. Recent work on Buy Clean proposes key elements that include (1) transparency and disclosure, (2) direct investment and RD&D in industry, and (3) establishing standards (BGA 2022). For the latter, the move toward codes and standards in building materials will help improve clarity on product carbon intensity and business case (Srinivasan et al. 2022).

Federal agencies, national laboratories, academia, industry, and non-governmental organizations are engaged in evaluating the carbon intensity of products, supporting the development of standards, and using those early standards in end-use areas such as buildings and construction (Srinivasan et al. 2022). Examples include national laboratory-developed LCA tools for energy technologies and pathways (NETL n.d.) and models for transportation fuels (ANL 2023), as well as efforts to measure embodied carbon in buildings—for example, through the university-based Carbon Leadership Forum (CLF n.d.) and the National Institute of Standards and Technology's (NIST's) metrics and tools for sustainable buildings (NIST 2020). The committee's first report recommended that Congress task EPA and DOE with establishing a library of environmental product declarations and the associated accounting and reporting infrastructure to support a Buy Clean policy (NASEM 2021). Similarly, other recent reports have recommended harmonization and standardization of LCA for CCUS projects and labeling of product carbon intensity to improve transparency for buyers (e.g., Recommendations 3-2 and 5-3 in NASEM 2023a).

Nonetheless, more work is needed to understand the carbon intensity of products as produced, as well as the addition of carbon intensity throughout the supply chain. There is a broad-based need to develop knowledge infrastructure to standardize how data on the carbon intensity of raw materials, precursors, manufactured products, and finished goods delivered to consumers are determined across the entire life cycle. This knowledge infrastructure includes the development of codes, standards, and evaluation protocols (see, e.g., Srinivasan et al. 2022), as well as the development of common data warehouses to share accepted parameters and protocols for calculating life-cycle emissions of products and energy carriers. A current effort toward this latter goal is the Federal LCA Commons, an interagency agreement for coordinating and sharing data, research, and information systems related to LCA (Federal LCA Commons n.d.). The trial and demonstration of low-carbon technologies provides an opportunity to monitor the impact on the products' carbon intensity, which could be included as a metric in DOE's and other agencies' lists of technology evaluation criteria for demonstrations. Studying reductions in carbon intensity via low-carbon technology implementation is also an opportunity to develop communities of practice that accelerate progress in low-carbon technologies. These communities can be engaged in updating product and purchasing standards, catalyzing RD&D (across industry, academia, engineering companies, etc.), and training people with the skills needed for the design, installation, and maintenance of equipment.

Finding 10-4: Market-pull mechanisms, such as Buy Clean and the First Movers Coalition, have a key role to play in helping to establish markets for low-carbon products. A number of value-chain players—from materials producers to consumers—need to be engaged in the development of codes, standards, and evaluation protocols for determining and transparently reporting the carbon intensity for products throughout their life cycle.

Recommendation 10-6: *Develop and Standardize Life-Cycle Assessment Approaches for Carbon Intensity of Industrial Products.* **The Department of Energy should lead an effort, in collaboration with the Environmental Protection Agency, National Institute of Standards and Technology, and other relevant agencies, to develop, harmonize, and standardize life-cycle assessment approaches for determining the carbon intensity of products from industry, starting with those products responsible for the largest proportion of greenhouse gas emissions. This effort should connect with related federal procurement programs for low-carbon products (e.g., Buy Clean). It should also assess the carbon impact across supply chains and develop labeling programs so that consumers can clearly evaluate the life-cycle carbon intensity of products.**

Recommendation 10-7: *Establish a Program Connecting Market-Pull Approaches to the Deployment of Low-Carbon Technologies.* **Congress should enact legislation to establish a program connecting market-pull approaches (e.g., procurement of low-embodied carbon products/Buy Clean) with the deployment of low-carbon technologies and process technology innovations to make lower-carbon products. This program could include developing protocols to quantify impacts and knowledge infrastructure to transparently report results and accelerate further improvements. It should also foster partnerships to pursue continued step-change reductions in embodied carbon across supply chains, develop "low-carbon communities" in carbon-intensive industries, and engage with states that are trailblazers in this area. The Departments of Energy (DOE), Commerce, Defense, and Transportation and the General Services Administration should be involved in this program, with DOE having responsibility for leading it.**

DECARBONIZING ACROSS THE BREADTH OF INDUSTRY

Industry is highly heterogeneous, from light industry (e.g., metal finishing, plastics processing) to heavy industry (e.g., cement, iron and steel, chemicals). As noted above, most analyses of industrial decarbonization focus on heavy industries, where strategies center around the major pillars of energy and materials efficiency, beneficial electrification, low-carbon energy sources and feedstocks, mitigation options, and demand for low-carbon products. Decarbonizing all of industry will require considering routes for light industry and engaging small and medium manufacturers (which comprise the majority of companies) in addition to the recognizable large companies in heavy industry. The pillars are applicable across industry, and this section is meant to build on the use of the decarbonization pillars across industry—while illustrating how they can be tailored for light industry and small- and medium-size manufacturers (SMMs). For example, given their lower complexity, cost sensitivity, and strong market-pull from customer demand, light industry and SMMs may see benefits from early application of low-carbon technologies in some cases (e.g., industrial heat pumps for low–moderate temperature process heat). This section highlights how opportunities and challenges for light industry and SMMs differ from those of heavy industry and large manufacturers.

Light Industry

Light industries use primary materials produced by upstream heavy industry to create final consumer products. They are characteristically smaller, more consumer-oriented, and less energy- and carbon-intensive than heavy industries (Worrell and Boyd 2022a).

Examples of light industries include food processing, consumer electronics, textiles, metal casting, and appliances.

In the United States, light industry used 2.43 exajoules (EJ) of fuel in total in 2014, and 1.69 EJ of electricity (Worrell and Boyd 2022a). According to Worrell and Boyd (2021), the primary uses of fuel are in boilers (31 percent), process heating (44 percent), and space heating and cooling (17 percent), while the primary uses of electricity are in motor systems (40 percent), facility heating ventilation and cooling and lighting (26 percent), and ovens and furnaces (13 percent). Key energy consumers include food and beverage, fabricated metals, and transportation equipment industries (EIA 2021a; Worrell and Boyd 2022a). Pursuing opportunities to reuse process heat and increasing the percentage of energy-intensive material that ends up serving customer needs are significant business opportunities (Lovins 2021).

In most heavy industries, energy use is dominated by a few processes that require high temperatures provided by fossil fuels; however, light industry uses a relatively large share of electricity because its processes generally require lower temperatures more conducive to electric technologies. Electricity currently comprises 40 percent of total site energy use in light industry—and about 60 percent when accounting for source energy—compared to less than 10 percent in heavy industry (Worrell and Boyd 2022b). There is significant opportunity for light industries to electrify further, given the wide availability of new innovative technologies (e.g., heat pumps, mechanical vapor recompression) at relatively high technology readiness levels (Worrell and Boyd 2022b). Additionally, electrification is more feasible in light industry than in heavy industry because the smaller power requirements of light industry facilities put less demand on the grid. Worrell and Boyd (2022a) further point out that light industry tends to use more electricity (and less process heat, especially that which is high temperature), have lower process integration (making it easier to change out/implement new processes), and have relatively clear early non-energy benefits (e.g., better control of temperature, minimizing maintenance, quality, product preservation). Light industry uses up to four times the amount of electricity versus fossil fuels (the opposite relationship exists for heavy industry) (Worrell and Boyd 2022b), so the heavier reliance on electricity and usage of electrical equipment suggests a lower early adoption barrier.

Looking across all of industry, the GHG emissions reduction potential of light industry is large, representing 39 percent of the total potential emissions reductions in the sector (Figure 10-6; Worrell and Boyd 2022a). Worrell and Boyd (2022a) indicate that energy efficiency could deliver upward of 40 percent of this potential. In light industry, because energy use makes up a smaller proportion of costs, improving energy efficiency has historically not been a high priority, and there have been fewer staff and resources devoted to this area.

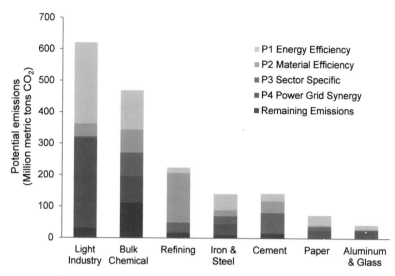

FIGURE 10-6 Potential emissions reductions in specific industries from different actions: energy efficiency, material efficiency, industry-specific, and power grid. NOTE: Remaining emissions are indicated with the darkest blue shading. SOURCE: Data from Worrell and Boyd (2022a), https://doi.org/10.1016/j.jclepro.2021.129758. CC BY 4.0.

Small- and Medium-Size Manufacturers

Of the more than 300,000 manufacturing companies in the United States, more than 90 percent have fewer than 500 employees, and most have fewer than 20 employees (U.S. Census Bureau 2022). All manufacturers and supply chain partners—whether large, medium, or small—need to participate in decarbonization efforts. While major manufacturers, strongly represented by heavy industry, use vast quantities of energy and emit large volumes of GHGs, there are also a vast number of SMMs. The SMM category includes companies that transform, combine, or customize products from earlier supply chain partners into intermediate or finished products. The supply chain partners of many heavy industrial companies (typically SMMs) use additional energy and emit additional GHGs as they prepare products for final end-use, which contributes to the Scope 3 emissions of the major manufacturers.

Key starting points for decarbonizing SMMs include dedicated energy and resource management approaches, targets, and reporting of progress. However, transmitting information to this sector has been a challenge globally. Countries like Switzerland and Germany have seen some success with networking programs that connect local SMMs to facilitate information transfer (Worrell and Boyd 2022b). Third-party aggregators may be able to play a role in coordinating small facilities and providing them with services they do not have the resources for on their own (McMillan 2022). Another approach to engaging SMMs in decarbonization efforts is by including them in supply chain partnerships with larger companies pursuing electrification, energy efficiency, and increased use of low-carbon fuels and feedstocks. Smaller companies have a lot less capital but more nimbleness, and they need programs that are cognizant of that difference (Dyck et al. 2022). Examples of SMM networks in the United States include

- Small Business Development Centers
- American Small Manufacturers Coalition
- Department of Commerce's NIST Manufacturing Extension Partnership (MEP) National Network
- National Association of Manufacturers (NAM)
- Department of Energy's Industrial Assessment Centers (IACs)

TAILORING INDUSTRIAL DECARBONIZATION APPROACHES TO SPECIFIC STATES OR REGIONS

Several factors will influence how the industrial decarbonization pillars may need to be tailored at the regional or state level, including geography, resource availability, economics of energy sources, concentration of industrial activity, infrastructure, workforce capabilities, and the policy and regulatory environment. Aligning the resources and capabilities to pursue decarbonization pillars with state incentives for accelerating reductions in energy and GHGs will help to facilitate industrial decarbonization. Recent reports show that states are taking varied approaches to supporting industrial decarbonization (Srinivasan and Esram 2022; USCA 2022) and describe best practices for industrial decarbonization at the state level (I3 2022).

The availability and delivery efficiency of the low-carbon energy sources and feedstocks needed to decarbonize industry vary across the country. For example, Figure 10-5 showed that biomass availability varies considerably by state, and likely by season as well. The capacity and delivery capability of low-carbon electricity (e.g., wind, solar, nuclear) is also variable. Texas leads the nation in wind capacity, and although transmission capabilities have been upgraded, there are still constraints considering the expected growth of both wind and solar (see LBNL 2022 for an indication of expected growth).

Electrification initiatives for industry will also need to consider the electricity/natural gas price ratio, which varies by state and region, as shown in Figure 10-7. A lower ratio decreases the economic hurdles for adopting technologies like industrial heat pumps (IHPs); in regions where the electricity/natural gas price ratio is below 3.5, simple paybacks for IHPs can be less than 2 years (Rightor et al. 2022a). Hurdles to adoption still

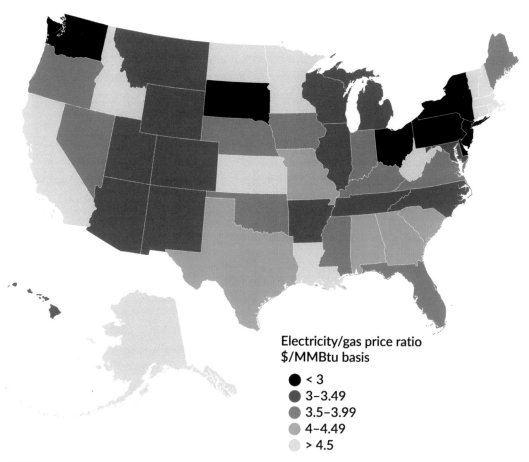

FIGURE 10-7 Electricity/natural gas price ratio variation across the United States, where lighter color represents a higher ratio. SOURCE: Rightor et al. (2022), American Council for an Energy-Efficient Economy.

exist, however, and policy incentives could help accelerate adoption of IHPs and other electric technologies. Incentives that reduce the cost of capital or electric rates would be instrumental in locations where the electricity/natural gas price ratio is higher.

BARRIERS TO INDUSTRIAL DECARBONIZATION

Common barriers to reducing emissions exist across all sizes and sectors of industry, as detailed in DOE (2022c). Such challenges need to be met with the combined efforts of state and federal policy, as well as the actions of private enterprises. The following sections briefly highlight key barriers for specific segments and technologies relevant to this work.

Challenges for Using Hydrogen to Decarbonize Industry

The primary challenges for using hydrogen to decarbonize industry are the production, distribution, and storage of low-carbon hydrogen. Low-carbon hydrogen generation, either via renewable electrolysis (i.e., "green hydrogen") or steam methane reforming (SMR) with carbon capture (i.e., "blue hydrogen"),[10] costs more than conventional hydrogen production from natural gas (Ochu et al. 2021). "Green hydrogen" is currently about 4–6 times more expensive than conventional hydrogen, and "blue hydrogen" costs about 50 percent more than conventional hydrogen on average (Ochu et al. 2021), so additional innovation is needed in both cases to reduce costs. To that end, DOE's Hydrogen Shot program, launched in June 2021, has set a target for the cost of clean hydrogen to be $1 per kg H_2 within 1 decade (DOE-HFTO n.d.(b)).

In current applications, about 85 percent of hydrogen is produced and used at the same site, minimizing costs associated with transport and storage (Bartlett and Krupnick 2020; IEA 2019a). If such co-location is not feasible for future hydrogen use cases, then infrastructure to transport and store hydrogen will need to be developed to connect sites of production and demand. Pipelines are the most efficient method to transport large quantities of hydrogen, although they may be difficult to site. Pipeline development has high capital costs, but pipelines typically have low operating costs over their 40- to 80-year lifetime (IEA 2019a). The feasibility of retrofitting natural gas pipelines to transport hydrogen or hydrogen blends is actively being explored (Blanton et al. 2021; EFI 2021; IEA 2019a). Converting gaseous hydrogen to a liquid hydrogen carrier for easier transport is also being considered (e.g., Tullo 2022) but is only cost-effective over long distances because of the high cost and energy intensity of the chemical conversion process (Bartlett and Krupnick 2020). Storing hydrogen in a gaseous, liquid, or solid form will be necessary if production and consumption are not temporally aligned. The lowest-cost option is to store large volumes of gaseous hydrogen in a geologic formation, preferably a salt cavern, but this option is geographically limited (BNEF 2020). Storing hydrogen as a liquid or solid (e.g., liquefied H_2, ammonia, liquid organic hydrogen carriers, metal hydrides) is more expensive and better suited for smaller volumes, but it does not face geographic limitations (BNEF 2020).

In addition to high costs, the transport and storage of hydrogen risk fugitive emissions. Hydrogen is an indirect GHG with a 100-year global warming potential (GWP) estimated to be between 3 and 11 (Field and Derwent 2021; Paulot et al. 2021; Warwick et al. 2022). Hydrogen leaks are much more difficult to prevent than natural gas leaks because of the

---

[10] For hydrogen generated via SMR with carbon capture to qualify as "low carbon" (≤2 kg CO_2e per kg H_2 at the site of production, per IIJA §40315), methane and CO_2 emissions need to be minimized across the supply chain and high carbon capture efficiencies must be achieved (Pettersen et al. 2022).

small size of the molecule. The possibility of future regulation of fugitive hydrogen emissions argues for use near its production sources. As the hydrogen hub demonstrations proceed, there may be an opportunity to examine technologies that most effectively monitor and reduce hydrogen leakage.

The timing of hydrogen use is a question for some applications. The carbon intensity associated with hydrogen generation affects the extent of emissions reductions achieved by incorporating hydrogen in chemical or industrial processes. For example, CO_2 emissions from methanol production would be higher if current grid-based electricity were used to make hydrogen than if traditional production methods were used; not until the grid is nearly fully decarbonized would net emissions decrease (DOE 2022c). Using clean energy to power hydrogen generation would decrease emissions, but the amount of clean energy needed at a commercial-scale facility would likely be well beyond the local supply, as methanol is one of the largest global commodity chemicals. In the near term, it would be better to focus on applications where the value proposition and market pull for low-carbon-intensity hydrogen is strongest and where the margin will support the added cost. An example is ammonia, which is used to produce fertilizer and is one of the largest commodity chemicals—at $67 billion in market value in 2020 and projected to grow to about $111 billion in 2028 (Fortune Business Insights 2021). This connection to fertilizer and hence food, which is highly visible to consumers, is apparently strong enough for an early market for green ammonia ($36 million in 2021, 0.05 percent of the overall ammonia market) and is expected to grow (Precedence Research 2022). As a result, CF Industries, Yara, and others have piloted facilities that are using hydrogen generated from renewable energy to meet customer interest for low-carbon ammonia (Jones 2022).

Identifying where market players will see the greatest value add for low-carbon H_2 (≤ 2 kg CO_2e per kg H_2 at the site of production, per IIJA §40315) remains a question. As noted above, economics is a major factor, as are location, end-customer demand for low-carbon-intensity products, and the ability to capture value from products made with low-carbon hydrogen. The market for hydrogen with various carbon intensities is at an early stage, and given the higher cost of lower-carbon options, there is uncertainty about the applications, price points, and volumes for which demand will materialize.

The rate of expansion of hydrogen production and market demand is another question. The $7 billion of DOE support for 6 to 10 hydrogen hubs (DOE-OCED n.d.(b)) authorized in the IIJA aims to catalyze production of low-carbon hydrogen, seed the market, and encourage H_2-related infrastructure development. How fast the experience in the hubs spurs developments in other areas of the country and how the learnings and scale help to reduce the cost of low-carbon hydrogen remain to be seen. Encouraging rapid reductions in the cost of hydrogen will require support for a market transformation where the low-carbon hydrogen (that likely will be at a higher price initially) can gain a foothold. The 45V Hydrogen Production Tax Credit enacted in the IRA, which gives up to a $3/kg credit,[11] is a first step. Voluntary commitments from end users, aggregation of demand by market players (perhaps inspired by bulk purchases) and potentially by governmental off-takers, and the development of standards for nomenclature, life-cycle procedures for evaluating hydrogen's carbon intensity, and energy efficiency could also help improve market confidence.

Challenges for Using Biomass to Decarbonize Industry

Three potential applications of biomass in industry require additional RD&D to be implemented at a commercial scale. First, use of biomass as a direct replacement for fossil fuels might be limited by its lower calorific value compared to some fuel types (e.g., bituminous coal) or by the fuel quality requirements and standards in some applications (Fivga and Mayer 2016). Second, bio-oils derived from pyrolysis technologies are often unstable and acidic, and current stabilization processes add cost (Abdullah et al. 2022). Third, while lignin is a promising source of aromatics, its integration into cellulose and tendency to condense into oligomers present challenges for extracting and cracking it in a manner that produces clean aromatic monomers (Abdullah et al. 2022). DOE's Bioenergy Technologies Office has several active efforts in lignin valorization aimed at solving these challenges (DOE-BETO 2021).

Realizing opportunities to use biomass for industrial decarbonization will also depend on policies and incentives, competition between biomass use cases, and consumer demand. Current U.S. policies related to biomass utilization—namely, the Environmental Protection Agency's Renewable Fuel Standard and California's Low Carbon Fuel Standard (LCFS)—center around biofuel production. Consequently, the use of biomass in industry may have to compete with an alternative use for the same biomass source that benefits from an existing policy. For example, food waste can be anaerobically digested to generate organic acids, which can serve as feedstocks in chemical syntheses. Alternatively, the same food waste could be converted to renewable natural gas, reformed into hydrogen, and then used by the petroleum industry in California to obtain LCFS credits, which is more attractive in the current policy environment (Abdullah et al. 2022). Changing consumer preferences could also influence industrial

---

[11] The exact value of the tax credit depends on the emissions associated with hydrogen production.

biomass use—for example, in recent years consumers have begun to value the "green premium" and are willing to pay more for a certifiably sustainable chemical, which could be derived from a biomass feedstock (Abdullah et al. 2022).

Barriers for Light Industry and Small- and Medium-Size Manufacturers

Light industry and SMMs face unique, additional challenges to decarbonizing compared to heavy industry—namely, a lack of sufficient resources, staffing, standardization, and coordination. McMillan (2018) identified some specific examples of these challenges:

- Limited policies exist to motivate industry to electrify; this is unlike buildings and transportation.
- Industrial electric technologies lack the public profile of electric vehicles and consumer-focused technologies for buildings.
- Researchers and policy makers face significant gaps in data (e.g., energy use, cost) and analysis tools (McMillan 2018, p. 2).

There are current and emerging opportunities to drastically lower these barriers, as well as many existing policies at various levels of jurisdiction that can be leveraged or applied at the state level. Table 10-4 describes crosscutting challenges that SMMs face in decarbonizing, state actions that can help overcome those challenges, and policy mechanisms that can enable such actions and drive emissions mitigation.

Supply Chain Challenges

Supply chain dependencies and constraints among and across industrial sectors will need to be addressed in the course of decarbonization. For example, U.S. steel production is principally based on electric arc furnaces (EAFs), which are lower carbon emitting than blast furnace–basic oxygen furnaces (BF-BOFs). EAFs typically use scrap steel, which has intertwined domestic and global supply chains, and insufficient scrap supply limits expansion of U.S. EAF production. As other countries pursue EAFs in support of decarbonization goals, increased demand for scrap could further limit capacity unless new DRI technologies are developed to provide the reduced iron for EAFs.

More generally, energy-intensive industries depend on integrated infrastructure, which today delivers power and natural gas and has overlays with transportation of raw materials and finished products. These industries are interconnected by a complex system of supply chain logistics to plan, implement, control, and optimize the movement of

TABLE 10-4 Common Barriers for Small- and Medium-Size Manufacturers Across Industry

Challenge/Barrier	Opportunities: State Perspective	Policy Connections
Energy is a smaller driver for companies	• Stimulate energy and material productivity (yield, value return, customer satisfaction, margin retention, GHG reduction)	• Expand communication, outreach, networking, and visibility of star performers • Leverage guides on energy and material efficiency (e.g., EPA Energy STAR)
Limited personnel/ resources	• Expand support, decrease hurdles/ transaction friction	• Support energy managers at company or cohort, energy assessments • Incentivize project implementation
Combined waste high, but for individual company it can be low	• Develop programs to reduce waste • Accumulate and transform, reuse where possible	• Incentivize waste reduction • Incentivize collection, reuse, and transformation of waste • Give tax breaks to companies that collect/transform waste
Limited capacity to consider/pursue decarbonization	• Provide information on decarbonization pathways • Simplify solution options • Consider working with third-party aggregators to reach, collaborate with, and serve SMMs	• Provide decarbonization roadmaps tailored to SMMs/communicate • Involve SMMs in pilots/demos of transformative technologies • Incentivize low-carbon technology choices that are commercial today • Provide support to SMMs for implementation of low-carbon technology • Expand leverage with current utility providers to reach SMMs
Limited access to emerging low-carbon infrastructure	• Ensure that SMM access needs are considered in planning	• Consider SMM needs in infrastructure planning (build experience at clusters) • Provide grants to build connections where efficient
Lack of standardization	• Work with associations and others to develop/deploy standards	• Work across jurisdictional levels to develop/convey standards

SOURCES: Data from DOE (2022c), McMillan (2022), and Worrell and Boyd (2022a,b).

materials and products. Decarbonization of industrial logistics would benefit from holistic decision-making and energy management (Miklautsch and Woschank 2022). Understanding the choices and new dependencies of clean energy and low-carbon product scenarios will require modeling to minimize constraints and avoid suboptimal solutions.

WORKFORCE, EQUITY, AND JUSTICE CONSIDERATIONS FOR INDUSTRIAL DECARBONIZATION

As discussed in Chapter 4, employment impacts will depend on the pathway taken to net zero. As new industries develop and mature and old industries decline or transform, the workforce will need to align with the developing changes in the economy. Such realignment provides an opportunity to address current racial disparities in employment and pollution risk from industrial facilities. For example, deindustrialization over the past several decades has disproportionately impacted workers of color; in the early 1990s, the Black share of the total U.S. workforce and of the manufacturing workforce were about equal, but as of 2020, Black workers comprised 10.2 percent of the manufacturing workforce, compared to 12.3 percent of the total workforce (Scott et al. 2022). Additionally, Ash and Boyce (2018) found that, on average, the share of pollution exposure risk from industrial facilities experienced by Black and Hispanic communities exceeds their share of employment at those facilities.

Manufacturing the equipment and building out the infrastructure needed for a net-zero economy will be a key part of the transition and is an opportunity to create and maintain high-quality jobs, as well as to increase global competitiveness. Studies examining job growth resulting from policies in the IRA project that manufacturing jobs could increase by about 100,000 annually, for a total of nearly 1.1 million over the 10 years of the IRA (LEP 2022; Pollin et al. 2022). However, several current trends in manufacturing that could be barriers to implementing a net-zero transition need to be addressed. Manufacturing has been hailed as a pathway to the middle class, especially for workers without a college degree, but the manufacturing wage premium has declined and disappeared in recent years (Bayard et al. 2022), and reliance on temporary workers disguises losses even further (Ruckelshaus and Leberstein 2014). The loss of manufacturing jobs has resulted in offshoring of jobs and deindustrialization of communities and has hit workers of color especially hard (Scott et al. 2022). Losses in manufacturing jobs have been in part driven by globalization, unfair trade policy, and U.S. trade deficits (Scott et al. 2020, 2022). Difficulty recruiting and retaining employees is widespread in the manufacturing sector; 76 percent of surveyed manufacturers identify attracting and retaining a quality workforce as one of the biggest problems they currently face (NAM 2022; see also DOE 2022d).

Many factors will impact industrial sector employment and workforce needs during the net-zero transition. Different segments of the industrial sector will adopt different decarbonization strategies, and decarbonization will occur along differing timelines. Decarbonization of light industry (e.g., durable goods, food and textile processing, and even mining and non-ferrous metal production) is likely to rely primarily on electrification and efficiency improvements (SDSN 2020). A WRI analysis estimates that, in a net-zero scenario, industry could add 764,000 jobs in installation of energy-efficient measures at manufacturing facilities by 2035 (Shrestha et al. 2022). Heavy industry will likely rely primarily on a switch to low-carbon fuels, feedstocks, and processes. Petrochemicals may rely on demand reduction and product substitution; semiconductors and electronics may rely on electrification, and fertilizer may utilize alternative feedstocks. Some high-carbon-intensity materials like steel and cement will have to be manufactured in a different way or integrate carbon capture (Williams and Bell 2022).

Mitigation options (e.g., CCUS and DAC) will likely also play a role in industrial decarbonization and could be another source of employment opportunities. For example, an analysis by Rhodium Group found that through 2050, capital investment in carbon capture retrofits could create 142,000 jobs, and retrofit operations could create 96,000 jobs, which would span a variety of industries, including ethanol, hydrogen, cement, refineries, steel, and power plants (Larsen et al. 2021). A DOE report estimates that supporting CCS operations of 2.0 gigatons per annum (Gtpa) by 2050 would require 35,160–155,975 jobs in operations and 236,273–1,758,000 in project/infrastructure (DOE 2022b).

Some paths to industrial decarbonization could spur new industries that may utilize skills and expertise of the existing workforce, but others may create entirely new kinds of jobs as well (see Chapter 4 for more detail on skills development). For example, jobs in CO_2 storage would require skills currently used in oil and gas industries—for reservoir characterization, well drilling, and design and operation of compression and injection facilities (LEP 2021). The expected increase in U.S. semiconductor manufacturing incentivized by the CHIPS and Science Act will require training of new skilled workers in a wide variety of fields, including materials science, electrical engineering, software development, and factory machine operation (Shivakumar et al. 2022). The hydrogen workforce, of particular interest given the significant investments being made in regional hydrogen hubs, has opportunities for both new and existing occupations across a wide range of industries (see Box 10-2).

Reshoring initiatives are also a part of the workforce landscape connected with decarbonization, as they could bring more jobs for skilled workers and help address the challenges brought about by deindustrialization (discussed above). Reshoring is creating more manufacturing jobs in the United States than foreign direct investment,

BOX 10-2
HYDROGEN: A WORKFORCE DEVELOPMENT EXAMPLE

The variety of potential uses for hydrogen in a net-zero economy, and the significant investments being made through DOE's Regional Clean Hydrogen Hubs program, provides an opportunity to examine the jobs and workforce development opportunities. Globally, hydrogen jobs span several sectors—industry, transportation, power generation—and many disciplines, including construction, manufacturing, engineering, pipeline transportation, and operations and maintenance (Bezdek 2019; DOE-HFTO n.d.(a); Hufnagel-Smith 2022; Queensland Government 2022). Other supporting industries include business and commercial development to perform financial and techno-economic analyses; environmental, social, and governance roles to understand sustainability impacts across the hydrogen value chain; and stakeholder engagement specialists to communicate with the public (Hufnagel-Smith 2022).

A study of the potential jobs connected with hydrogen estimates that 8,500 jobs could be created by 2035 related to the tax credits and other provisions in the IRA and IIJA, and 369,000 jobs could be created by 2050 in a net-zero scenario (Saha et al. 2022). The same study found that by 2050, the estimated labor income connected with hydrogen approaches $15 billion, with taxes estimated at $5.7 billion (in 2020$). This increase in hydrogen jobs would help relieve some of the job losses for fossil fuels (estimated at 1.9 million, a 44 percent decline), but clearly the increase in hydrogen jobs will be a small fraction of the jobs that could be lost across the entire fossil fuel sector.

Some hydrogen jobs can leverage skills from the existing workforce. For example, current oil and gas workers have skills in instrumentation, pipeline construction, and compression and handling of gas and liquid fuels that will be relevant for working with hydrogen (Hufnagel-Smith 2022; Queensland Government 2022). Jobs manufacturing infrastructure components like pressure vessels, piping systems, valves, and turbines will also continue to be important (Hufnagel-Smith 2022). However, a hydrogen workforce would also include jobs requiring new skills, such as fuel cell and electrolyzer technicians, technicians to maintain and repair fuel cell electric vehicles, operators of hydrogen combustion turbines, and hydrogen emergency response teams (Hufnagel-Smith 2022; Queensland Government 2022). Hufnagel-Smith (2022, p. 3) suggests that the requisite training and skill development could occur as the industry develops: "Increased demand for hydrogen will initially be marked by deployment of technology that requires retrofitting and conversion of existing systems, equipment and infrastructure to accommodate fuel switching, co-combustion and blending of hydrogen with other fuels including natural gas and diesel. This provides opportunity to augment the skills and knowledge of workers so that they can work in both systems, transitioning to full-time hydrogen roles at the same pace the industry advances."

Recognizing the need to train a future hydrogen workforce, DOE funded the Hydrogen Education for a Decarbonized Global Economy (H$_2$EDGE) program beginning in early 2021 (EPRI n.d.). H$_2$EDGE, a collaboration of the Electric Power Research Institute, Gas Technology Institute, and several universities, aims to develop training and education materials for the production, delivery, storage, and use of hydrogen (Reddoch et al. 2021). Additionally, DOE requires applicants to the Regional Clean Hydrogen Hub funding opportunity to submit a

BOX 10-2 Continued

Community Benefits Plan that "should describe the applicant's comprehensive plan for the creation and retention of high-paying quality jobs and development of a skilled workforce" (DOE-OCED 2023, p. 50).

This examination of hydrogen jobs serves as an example of how the transformations needed to achieve net-zero can provide an opportunity for job growth. Some of the fossil-fuel-related jobs lost could be transitioned to enable the growth in generation and use of hydrogen, but job training programs need to be developed to make this transition as smooth as possible.

---

a trend that has been observed for several years. In 2022, reshoring generated about 220,000 jobs, and foreign direct investment generated about 130,000 (Reshoring Initiative 2022). The COVID-19 pandemic, supply chain issues, geopolitical tensions, national security, and tariffs are major drivers for reshoring. A labor shortage, however, is putting a cap on the number of overseas manufacturing jobs that the United States can accommodate. Reshoring the ability to manufacture products that are key to supply chains, changing the perspective to total cost of ownership (versus factory price), and addressing workforce are important steps to avoid systemic trade imbalances (Moser 2022). NIST supports reshoring through its MEP program by scouting and vetting local manufacturing suppliers (NIST 2022).

A skilled workforce is crucial to industrial decarbonization efforts. Several workforce efforts and frameworks already exist at DOE, including IACs, Manufacturing USA Institutes, Renewable Energy Competency Model, and Hydrogen Education for a Decarbonized Global Economy (H_2EDGE) program (DOE 2022c). In parallel, the Department of Commerce Strategic Plan for 2022–2026 contains a number of workforce objectives, several of which cross over with the industrial sector (DOC 2022). DOE also has funding to train and provide resources for the clean energy and manufacturing energy management workforce and to provide technical assistance for implementing clean energy and efficiency practices in industry. In fiscal year 2023, the Entrepreneurial Ecosystems and Advanced Manufacturing Workforce program within the Advanced Materials and Manufacturing Technologies Office received $17.5 million, and the Technical Assistance and Workforce Development program within the Industrial Efficiency and Decarbonization Office received $45 million (DOE 2023a). The Center for Energy Workforce Development, a consortium of 120 energy companies, provides resources and training to support clean energy careers in diverse, equitable, and inclusive workplaces (CEWD 2023) and could leverage opportunities provided by DOE workforce efforts and funding. These programs provide a scale for the level of investment in jobs training programs.

A complementary level of investment is needed for expanding the range of this outreach to include the diversity of industrial sector needs and broadening training engagement at technical schools, minority serving institutions, and academia.

Despite the existing efforts, more work is required to better understand the potential employment impacts and workforce needs resulting from industrial decarbonization. Some industry-specific analyses have begun to address these questions and provide recommendations that could be generalized and applied across industries. For example, the Center for American Progress notes that understanding employment impacts of decarbonizing the steel industry will require analyzing the labor needs for existing steel-making processes (e.g., BF-BOF and EAF) compared to those for the different decarbonization options (e.g., DRI with green hydrogen or adding carbon capture to BF-BOF) (Williams and Bell 2022). DOE's Energy Storage Grand Challenge Roadmap provides recommendations for enhancing workforce development and emphasizes the need for evaluations to measure success, starting by analyzing the existing (baseline) education and workforce programs (DOE 2020b). The report also highlights the importance of stakeholder engagement to ensure that communities are aware of available programs and opportunities. Coordination and understanding workforce needs will be critical to ensuring that the transition can happen in a way that maximizes benefits and minimizes costs. Additional funding will be required to expand the scope of existing programs—for instance, by increasing outreach and training to cover additional industrial sectors and applications, and to provide crosscutting support for ongoing and future initiatives.

> Finding 10-5: While the criticality of industrial workforce development for the transition to net zero is recognized, significant training, support, and job placement needs remain to ensure the viability of this transformation.

> **Recommendation 10-8:** *Develop Effective Workforce Development Programs for Industry.* **The Department of Energy (DOE) should take the point role in convening partners in manufacturing—including labor associations, non-governmental organizations, industry leaders, and academia—to develop effective workforce development programs for industry, building on learnings from past initiatives. The programs should be piloted and improved through collaboration with state and local authorities and institutions of higher learning, including minority-serving institutions, and should stretch across small, medium, and large manufacturers. This effort should leverage and serve as crosscutting support across current/future initiatives (e.g., carbon capture, utilization, and storage; hydrogen; electrification hubs) and programs (e.g., Industrial Assessment Center, Manufacturing Extension Partnership, National Institute of Standards and Technology) with a strategy to enhance the learnings and impact of funding**

from recent legislation (the Inflation Reduction Act, the Infrastructure Investment and Jobs Act, and the Creating Helpful Incentives to Produce Semiconductors and Science Act) in order to further clarify where, how, and when workforce programs should be initiated to foster capability development for the low-carbon future. Congress should appropriate $100 million over 4 years, or until expended, for DOE to develop such programs.

POLICY ENABLERS: LANDSCAPE OF CURRENT INITIATIVES AND FUTURE NEEDS

Technologies tend to proceed from early-stage development to market commercialization along an "S-curve," in which the initial share of market penetration for a new technology is low, but then rises quickly as market adoption accelerates before slowing again at the point of market saturation. An analysis by Carey and Shepard (2022) suggested that the CHIPS and Science Act largely supports the early stages, the IIJA supports the middle stage, and the IRA invests heavily in the later stages. Box 10-3 provides select examples of technology transitions in industry.

BOX 10-3
DRIVING ADOPTION OF INNOVATIVE TECHNOLOGIES IN INDUSTRY

Wesseling et al. (2017) offers important historical insights into sociotechnical transitions in energy-intensive industries, demonstrating that adoption and diffusion of innovative technologies initially take place when a given technology has attractive benefits not necessarily related to cost. Such technologies are adopted first by a few specific industries where they fit well; as they become more broadly applicable, their diffusion across larger sections of industry becomes possible. It is therefore crucial to identify where and how those attractive features take off. McMillan (2022) provided several examples:

- **Example 1:** In 1870, the Corliss steam engine, important for the metalworking and textiles industries, enabled easier power control and greater efficiency and speed, reducing the probability of thread breakage in textiles.
- **Example 2:** The newspaper printing industry adopted the electric machine drive, a simple conversion for their machines that removed the messiness of lubricants and grease that belt drive machines required.
- **Example 3:** The pulp and paper industry, one of the last to electrify, faced difficulty in finding the right set of complementary technologies that made electrification of the paper machines possible. The industry coordinated with Westinghouse and General Electric to figure out solutions—in essence, emerging electric utilities were working directly with manufacturers to fine-tune and innovate, which ultimately gave them new electricity customers.

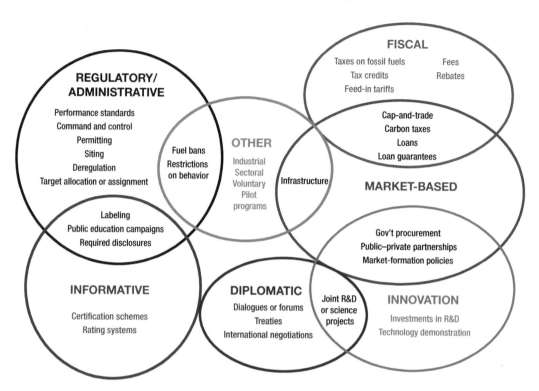

FIGURE 10-8 Policy approaches for climate mitigation. SOURCE: K.S. Gallagher and X. Xuan, forewords by J.P. Holdren and J. Zhang, *Titans of the Climate*, Fig. 5.1 (p. 106), © 2019 Massachusetts Institute of Technology, by permission of MIT Press.

Gallagher and Xuan (2019) grouped policies for decarbonization into seven categories, as shown in Figure 10-8. This typology of climate policy approaches illustrates the range of possible approaches in a policy maker's toolbox, from regulatory approaches such as performance standards to market-based approaches such as carbon taxation. Countries that have embarked on decarbonization pathways have used a variety of policy mixes, including fiscal tools such as feed-in tariffs or production tax credits, market-based tools such as emissions trading, and regulatory tools such as performance standards to achieve their objectives. Sometimes those policies have been unintentionally or deliberately sequenced to create the political conditions for new policies (Meckling et al. 2015). In China's green industrialization process, the Chinese government used a wide range of policy instruments including every type of policy depicted in Figure 10-8, such as investing in R&D, attracting human resource talent from around the world, creating pilot and demonstration programs, imposing performance standards on industry, using public education campaigns, and providing subsidies to consumers (Gallagher 2014).

The current provisions for industry in the IIJA and IRA are heavily weighted toward the "Innovation" and "Fiscal" categories of Figure 10-8. This aligns with the need to significantly decrease the cost of transformative technology, because if those technologies are not economic, adoption may not occur beyond demonstration plants (i.e., the cascade of low-carbon technologies throughout industries will be slow or nonexistent). For a low-carbon technology to be widely adopted, it needs to achieve similar or better performance as the incumbent while also delivering a similar cost/ price. A small portion of the market may be willing to pay a modest "green premium," but the larger portion will balk at paying more. The market also needs to be ready to support new technologies once installed (e.g., domestic maintenance of the technologies on site). Fiscal approaches such as production tax incentives (e.g., for low-carbon hydrogen) may not have the desired effect if economic price parity for technologies is not achieved during the time window of support for demonstrations or if market support structures (e.g., service company support, infrastructure) are not yet in place.

Some funded programs in the IRA and IIJA are connected to the "Market-Based" category in Figure 10-8, including government procurement programs such as Buy Clean. Production tax credits for hydrogen and incentives for CO_2 capture and utilization (such as 45V and 45Q, respectively) could help to establish markets for low-carbon hydrogen and aid market development for reuse of CO_2. Further proposals may be developed and implemented, such as incentives for innovations that would lower the cost of the products made by low-carbon technologies (e.g., the "green premium" for low-carbon hydrogen). Approaches tried in other countries include a Carbon Contract for Differences approach, which has been used in the United Kingdom and the European Union (Sartor and Bataille 2019), and a market auction approach such as that being used for low-carbon hydrogen in Germany (Hydrogen Europe 2022). Further study and consideration of these approaches is warranted by the market-pull they may provide.

The "Informative" category in Figure 10-8 is represented in the IRA and the IIJA to some extent by labeling programs connected with the market-based approaches. For example, connected with the Buy Clean programs, there are provisions to initiate labeling the carbon intensity of products, backed up by environmental product declarations, which would inform customers. Labeling the energy use of products (e.g., DOE's Energy Star program) has been successful for years. The confidence in metrics, transparency, and simplicity for relating numbers to customers is at an early stage, and there are numerous opportunities for future policy development.

The "Diplomatic" category in Figure 10-8 has received heightened visibility owing to discussions about developing global trade policy that seeks to address the carbon intensity of products, national competitiveness, and trade imbalances. The European Union introduced a Carbon Border Adjustment Mechanism (CBAM) where manufactured products from non-carbon-pricing countries face a similar carbon price to domestic EU products; this CBAM will start a transitional phase in October 2023 (EC 2021, 2023). The G7 established a Climate Club in December 2022 focused on international cooperation in industrial decarbonization, including strategies for mitigating carbon leakage (e.g., CBAMs) (G7 2022). As discussed by Kopp et al. (2022), the idea would be for a group of countries to align their decarbonization efforts and trade policies both to avoid shifting industrial activity and emissions to less regulated jurisdictions *and* to provide incentives for trading partners to increase their mitigation ambition.

The EU actions have sparked new interest in trade policy and industrial decarbonization in the United States. The FAIR Transition and Competition Act (H.R. 4534) was introduced in the 117th Congress with the aim of adjusting for the embodied carbon emissions from imported manufactured products. This approach would attempt to shift U.S. demand toward cleaner domestic production without seeking further domestic emission reductions. In contrast, the Clean Competition Act (S. 4355) would do both: it would set an embodied carbon benchmark and assign a fee for the carbon content above the benchmark for both domestic and imported goods.

There is continuing discussion and negotiation connected with the World Trade Organization rules to clarify if there are conflicts between these types of policies and fair-trade rules (Smith 2023). The focus of these strategies that combine both regulation and diplomatic approaches is to decarbonize heavy industry in a way that will not shift production of high-carbon-content materials to countries with weaker regulations and higher-carbon-intensity energy use (e.g., carbon leakage). Given the high degree of trade in most industrial goods, these approaches linking trade and decarbonization have high potential for further development and application in reducing GHG emissions.

Regulatory and administrative approaches can also be the subject of discordant perspectives. A carbon price on emissions has been called for, is industry supported, and would be foundational to GHG reductions (Patnaik and Kennedy 2021), yet it remains politically difficult. For the industrial sector, it would need to be paired with a CBAM, as noted above.

Approaches that decrease hurdles for siting, permitting, and other implementation aspects of IRA and IIJA initiatives can improve the effectiveness of these bills and likely will be discussed in the 118th Congress. Performance standards can advance customer acceptance and market pull for lower-carbon products and are likely to receive bipartisan support. These approaches can also serve to decrease costs for low-carbon technologies to help them compete with incumbent technologies. Combined with approaches in the "Informative" category—such as labeling, improvement of data, and transparency—"Administrative" approaches can improve consumer confidence and market-pull for low-carbon products.

Additional regulatory approaches could address issues that inhibit the turnover of capital stock to low-carbon technology—such as risk sharing/mitigation and recycling/reuse of equipment. As some industries rise, others will decline, and the best routes to phase out carbon-intensive industries and capital equipment will need to be determined (Semieniuk et al. 2020). An improved understanding of the risks, impediments, financial impacts, and paths forward to address stranded assets needs to be further developed. This is an area where U.S. trading partners will face the same challenges and where shared learnings and collaboration would be beneficial (Baron and Fischer 2015).

Regulatory approaches can force compliance costs, and, because many industries are energy intensive and trade exposed, this area needs to be approached cautiously and in concert with corresponding trade policies. With achievement of cost parity for transformative technologies and incumbent technologies prior to introducing regulations, the risk of driving manufacturing offshore and the hurdles to a cascade of adoption can be minimized. Continued development and understanding of when and where regulations can be used to spur reductions (in the last portion of GHG emissions) is needed.

The "Other" category of policy approaches includes voluntary approaches where a number of players can learn rapidly. Additional policy tools, approaches, and even additional categories will likely have to be developed during the multi-decade journey of decarbonization. An agile approach is needed to understand how well current policies are working to reduce GHG emissions while minimizing unintended consequences in parallel with developing additional policies that provide mid-course corrections as needed.

As mentioned above, provisions for industry in the IRA and IIJA largely fall into the "Innovation" or "Fiscal" categories. As those provisions make impact, complementary measures from the other categories may be useful to spur progress. For example, once innovation drives down the cost of low-carbon solutions to be competitive with incumbents and their adoption begins to accelerate, standards, benchmarking, and ultimately regulations may help to further adoption in industries that resist transitioning to low-carbon solutions (even when cost effective). "Informative" and "Market-Based" measures could be expanded to help increase customer demand for low-carbon-intensity products. Support for "Other" measures, such as pilot programs, will build experience, knowledge, and support across supply chains for low-carbon products. The timing, sequence, and reinforcement of these additional policy approaches can be important for bolstering the impacts of singular measures. Research and experience are needed in how best to sequence the various approaches to achieve maximum cost-effectiveness and rapid reductions in GHG emissions, while avoiding unintended consequences (e.g., costly technology lock-in, consumer pushback) and backsliding. Based on approaches used in the European Union and California, Meckling et al. (2017) present a "benefits-to-costs" policy sequence consisting of three steps: (1) green innovation and industrial policies, in which the government supports and invests in low-carbon energy technologies; (2) addition of carbon pricing policies; and (3) ratcheting-up the policy mix over time—for example, by tying subsidies for low-carbon energy technologies to revenues from carbon pricing policies. Any addition of carbon pricing on industrial goods would need to consider trade implications, as discussed above. Analysis, documentation, and transparent communication of the effectiveness of current and proposed measures will provide policy makers with the information required to design future policies that can complement early measures.

> **Recommendation 10-9:** *Implement a Product-Based Tradable Performance Standard for Domestic Manufacturing and Foreign Trade.* **To drive emissions to net zero in industry, Congress should task the Department of Commerce and the Department of Energy to work with a variety of stakeholders to establish declining carbon intensity benchmarks for major product families. Congress should require the Environmental Protection Agency to create a tradable performance standard for domestically produced and imported products based on these benchmarks, starting with products where there is alignment with current initiatives (e.g., Buy Clean provisions start with iron and steel, and cement in building and construction markets) to gain experience.**

Table 10-5 summarizes all of the recommendations in this chapter to support decarbonizing industry.

SUMMARY OF RECOMMENDATIONS ON INDUSTRIAL DECARBONIZATION

TABLE 10-5 Summary of Recommendations on Industrial Decarbonization

Short-Form Recommendation	Actor(s) Responsible for Implementing Recommendation	Sector(s) Addressed by Recommendation	Objective(s) Addressed by Recommendation	Overarching Categories Addressed by Recommendation
10-1: Develop and Enable Cost-Competitive Process and Waste Heat Solutions	Department of Energy (DOE) and industrial companies	• Buildings • Industry	• Greenhouse gas (GHG) reductions	Rigorous and Transparent Analysis and Reporting for Adaptive Management Tightened Targets for the Buildings and Industrial Sectors and a Backstop for the Transport Sector Research, Development, and Demonstration Needs
10-2: Invest in Energy and Materials Efficiency and Industrial Electrification	Congress and DOE	• Buildings • Industry • Finance • Non-federal actors • Transportation	• GHG reductions	A Broadened Policy Portfolio Tightened Targets for the Buildings and Industrial Sectors and a Backstop for the Transport Sector Research, Development, and Demonstration Needs

continued

TABLE 10-5 Continued

Short-Form Recommendation	Actor(s) Responsible for Implementing Recommendation	Sector(s) Addressed by Recommendation	Objective(s) Addressed by Recommendation	Overarching Categories Addressed by Recommendation
10-3: Spur Innovation to Achieve Price-Performance Parity for Low-Carbon Solutions	Congress, DOE, non-governmental organizations (NGOs), industry associations (e.g., American Chemistry Council [ACC], American Iron and Steel Institute [AISI], Portland Cement Association [PCA], National Association of Manufacturers [NAM], and others), and industry	• Industry • Finance • Non-federal actors	• GHG reductions	A Broadened Policy Portfolio Tightened Targets for the Buildings and Industrial Sectors and a Backstop for the Transport Sector
10-4: Pursue Technologies That Reduce Both Greenhouse Gas and Air Pollution Emissions	DOE, NGOs, industry, industry associations, (e.g., ACC, AISI, PCA, NAM, and others), and engineering companies	• Industry • Non-federal actors	• GHG reductions • Health	Ensuring Equity, Justice, Health, and Fairness of Impacts Tightened Targets for the Buildings and Industrial Sectors and a Backstop for the Transport Sector Research, Development, and Demonstration Needs
10-5: Use Mass-Based Rather Than Concentration-Based NO_x Standards	Regulatory and permitting organizations	• Industry • Electricity • Transportation	• GHG reductions • Health	Rigorous and Transparent Analysis and Reporting for Adaptive Management

TABLE 10-5 Continued

Short-Form Recommendation	Actor(s) Responsible for Implementing Recommendation	Sector(s) Addressed by Recommendation	Objective(s) Addressed by Recommendation	Overarching Categories Addressed by Recommendation
10-6: Develop and Standardize Life-Cycle Assessment Approaches for Carbon Intensity of Industrial Products	DOE, Environmental Protection Agency (EPA), National Institute of Standards and Technology (NIST), and other relevant agencies	• Industry • Buildings • Transportation • Non-federal actors	• GHG reductions	A Broadened Policy Portfolio Rigorous and Transparent Analysis and Reporting for Adaptive Management Tightened Targets for the Buildings and Industrial Sectors and a Backstop for the Transport Sector Research, Development, and Demonstration Needs
10-7: Establish a Program Connecting Market-Pull Approaches to the Deployment of Low-Carbon Technologies	Congress, DOE, Department of Commerce (DOC), General Services Administration, Department of Defense, and Department of Transportation	• Buildings • Transportation • Industry • Non-federal actors • Finance	• GHG reductions	A Broadened Policy Portfolio Rigorous and Transparent Analysis and Reporting for Adaptive Management Tightened Targets for the Buildings and Industrial Sectors and a Backstop for the Transport Sector

continued

TABLE 10-5 Continued

Short-Form Recommendation	Actor(s) Responsible for Implementing Recommendation	Sector(s) Addressed by Recommendation	Objective(s) Addressed by Recommendation	Overarching Categories Addressed by Recommendation
10-8: Develop Effective Workforce Development Programs for Industry	Congress, DOE, labor associations, NGOs, industry leaders, and academia	• Industry • Non-federal actors	• Employment	Building the Needed Workforce and Capacity
10-9: Implement a Product-Based Tradable Performance Standard for Domestic Manufacturing and Foreign Trade	Congress, DOE, DOC, and EPA	• Industry • Finance	• GHG reductions	A Broadened Policy Portfolio Tightened Targets for the Buildings and Industrial Sectors and a Backstop for the Transport Sector

REFERENCES

Abdullah, Z., R. Allen, L. Tao, J. Male, and B. Davison. 2022. "Industrial Decarbonization Options: Biomass as an Energy Source and Feedstock for Industry." Presentation to the Committee on Accelerating Decarbonization in the United States, online, February 8. https://www.nationalacademies.org/event/02-08-2022/accelerating-decarbonization-in-the-united-states-technology-policy-and-societal-dimensions-industrial-companies-open-session.

Abramson, E., E. Thomley, and D. McFarlane. 2022. *An Atlas of Carbon and Hydrogen Hubs for United States Decarbonization.* Minneapolis, MN: Great Plains Institute. https://scripts.betterenergy.org/CarbonCaptureReady/GPI_Carbon_and_Hydrogen_Hubs_Atlas.pdf.

ANL (Argonne National Laboratory). 2023. "GREET Model." Energy Systems and Infrastructure Analysis. https://greet.es.anl.gov.

Ash, M., and J.K. Boyce. 2018. "Racial Disparities in Pollution Exposure and Employment at US Industrial Facilities." *Proceedings of the National Academy of Sciences* 115(42):10636–10641. https://doi.org/10.1073/pnas.1721640115.

Baron, R., and D. Fischer. 2015. "Divestment and Stranded Assets in the Low-Carbon Transition." Background paper for the *32nd Round Table on Sustainable Development.* Paris: Organisation for Economic Co-Operation and Development. https://www.oecd.org/sd-roundtable/papersandpublications/Divestment%20and%20Stranded%20Assets%20in%20the%20Low-carbon%20Economy%2032nd%20OECD%20RTSD.pdf.

Barrett, A. 2020. "Mechanical vs. Chemical Recycling." *Bioplastics News,* November 20. https://bioplasticsnews.com/2020/11/20/difference-mechanical-chemical-recycling.

Bartlett, J., and A. Krupnick. 2020. "Decarbonized Hydrogen in the US Power and Industrial Sectors: Identifying and Incentivizing Opportunities to Lower Emissions." Washington, DC: Resources for the Future. https://media.rff.org/documents/RFF_Report_20-25_Decarbonized_Hydrogen.pdf.

Bashmakov, I.A., L.J. Nilsson, A. Acquaye, C. Bataille, J.M. Cullen, S. de la Rue du Can, M. Fischedick, Y. Geng, and K. Tanaka. 2022. Chapter 11 in *Climate Change 2022: Mitigation of Climate Change. Contribution of Working Group III to the Sixth Assessment Report,* P.R. Shukla, J. Skea, R. Slade, et al., eds. Intergovernmental Panel on Climate Change. Cambridge, UK, and New York: Cambridge University Press. https://doi.org/10.1017/9781009157926.013.

Bayard, K., T. Cajner, V. Gregorich, and M.D. Tito. 2022. "Are Manufacturing Jobs Still Good Jobs? An Exploration of the Manufacturing Wage Premium." Finance and Economics Discussion Series 2022-011. Board of Governors of the Federal Reserve System. https://doi.org/10.17016/FEDS.2022.011.

Bezdek, R.H. 2019. "The Hydrogen Economy and Jobs of the Future." *Renewable Energy Environmental Sustainability* 4. https://doi.org/10.1051/rees/2018005.

BGA (BlueGreen Alliance). 2022. "Buy Clean: A Tool to Create Good Jobs, Cut Pollution, and Renew American Manufacturing." https://www.bluegreenalliance.org/wp-content/uploads/2022/09/Buy-Clean-White-Paper-22-_v3_91222.pdf.

Blanton, E.M., M.C. Lott, and K.N. Smith. 2021. *Investing in the US Natural Gas Pipeline System to Support Net-Zero Targets.* New York: Columbia University Center on Global Energy Policy. https://www.energypolicy.columbia.edu/sites/default/files/file-uploads/GasPipelines_CGEP_Report_111522.pdf.

BNEF (BloombergNEF). 2020. *Hydrogen Economy Outlook: Key Messages.* London. https://assets.bbhub.io/professional/sites/24/BNEF-Hydrogen-Economy-Outlook-Key-Messages-30-Mar-2020.pdf.

Boardman, R.D., M.G. McKellar, H.C. Bryan, A.F. Wilkin, and B.D. Dold. 2021. "Process Heat for Chemical Industry." INL/JOU-21-61597-Revision-0. Idaho Falls, ID: Idaho National Laboratory. https://inldigitallibrary.inl.gov/sites/sti/sti/Sort_29023.pdf.

Boston Metal. 2023. "Molten Oxide Electrolysis for Steel Decarbonization." https://www.bostonmetal.com/green-steel-solution.

Boyles, H., E. Rightor, and D.M. Hart. 2023. "Long-Duration Energy Storage Is a Decarbonization Linchpin." *Innovation Files* (blog), June 14. https://itif.org/publications/2023/06/14/long-duration-energy-storage-is-a-decarbonization-linchpin.

Breer, B., H. Rajagopalan, C. Godbold, H. Johnson, B. Emerson, V. Acharya, W. Sun, D. Noble, T. Lieuwen. 2022 "NO$_x$ Production from Hydrogen-Methane Blends." Paper 149RKF-0019 presented at the 2022 Spring Technical Meeting of the Eastern States Section of the Combustion Institute, Orlando, FL.

CAISO (California ISO). 2017. "Impacts of Renewable Energy on Grid Operations." https://www.caiso.com/documents/curtailmentfastfacts.pdf.

Carbon Leadership Forum. n.d. "Tools for Measuring Embodied Carbon." https://carbonleadershipforum.org/tools-for-measuring-embodied-carbon.

Carey, L., and J. Ukita Shepard. 2022. "Congress's Climate Triple Whammy: Innovation, Investment, and Industrial Policy." RMI. https://rmi.org/climate-innovation-investment-and-industrial-policy.

CEWD (Center for Energy Workforce Development). 2023. "About." https://cewd.org/about.

Colket, M.B. 2013. "Particulate Formation." Pp. 123–153 in *Gas Turbine Emissions*, T.C. Lieuwen and V. Yang, eds. Cambridge Aerospace Series. Cambridge: Cambridge University Press. https://doi.org/10.1017/CBO9781139015462.009.

Collias, D.I., M.I. James, and J.M. Layman. 2021. "Introduction—Circular Economy of Polymers and Recycling Technologies." Pp. 1–21 in *Circular Economy of Polymers: Topics in Recycling Technologies.* ACS Symposium Series 1391. American Chemical Society. https://doi.org/10.1021/bk-2021-1391.ch001.

Crumpacker, L. 2022. "Twelve and LanzaTech Partner to Create Ethanol from CO$_2$." *Business Wire.* March 3. https://www.businesswire.com/news/home/20220303005328/en/Twelve-and-LanzaTech-Partner-to-Create-Ethanol-From-CO2.

DOC (Department of Commerce). 2022. *U.S. Department of Commerce Strategic Plan 2022–2026: Innovation, Equity, and Resilience: Strengthening American Competitiveness in the 21st Century.* Washington, DC. https://www.commerce.gov/sites/default/files/2022-03/DOC-Strategic-Plan-2022%E2%80%932026.pdf.

DOE (Department of Energy). 2020a. *Better Plants Progress Update.* Washington, DC. https://betterbuildingssolutioncenter.energy.gov/sites/default/files/attachments/2020%20Better%20Plants%20Progress%20Update%20-%20FINAL.pdf.

DOE. 2020b. *Energy Storage Grand Challenge Roadmap.* Washington, DC. https://www.energy.gov/sites/prod/files/2020/12/f81/Energy%20Storage%20Grand%20Challenge%20Roadmap.pdf.

DOE. 2020c. *U.S. Department of Energy Hydrogen Program Plan.* Washington, DC. https://www.hydrogen.energy.gov/pdfs/hydrogen-program-plan-2020.pdf.

DOE. 2022a. "Biden Administration Launches $3.5 Billion Program to Capture Carbon Pollution from the Air." https://www.energy.gov/articles/biden-administration-launches-35-billion-program-capture-carbon-pollution-air-0.

DOE. 2022b. *Carbon Capture, Transport, and Storage: Supply Chain Deep Dive Assessment.* Department of Energy Response to Executive Order 14017, "America's Supply Chains." Washington, DC. https://www.energy.gov/sites/default/files/2022-02/Carbon%20Capture%20Supply%20Chain%20Report%20-%20Final.pdf.

DOE. 2022c. *Industrial Decarbonization Roadmap.* DOE/EE-2635. Washington, DC: Department of Energy. https://www.energy.gov/sites/default/files/2022-09/Industrial%20Decarbonization%20Roadmap.pdf.

DOE. 2022d. *United States Energy and Employment Report 2022.* Washington, DC: Department of Energy. https://www.energy.gov/sites/default/files/2022-06/USEER%202022%20National%20Report_1.pdf.

DOE. 2023a. *Department of Energy FY 2024 Congressional Justification: Energy Efficiency and Renewable Energy, Electricity, Nuclear Energy, Fossil Energy and Carbon Management.* DOE/CF-0195, Volume 4. Washington, DC. https://www.energy.gov/sites/default/files/2023-06/doe-fy-2024-budget-vol-4-v2.pdf.

DOE. 2023b. *Pathways to Commercial Liftoff: Long Duration Energy Storage.”* Washington, DC. https://liftoff.energy.gov/wp-content/uploads/2023/05/Pathways-to-Commercial-Liftoff-LDES-May-5_UPDATED.pdf.

DOE-BETO (Department of Energy Bioenergy Technologies Office). 2016. *2016 Billion-Ton Report: Advancing Domestic Resources for a Thriving Bioeconomy. Volume I: Economic Availability of Feedstocks.* Washington, DC: Department of Energy. https://www.energy.gov/sites/default/files/2016/12/f34/2016_billion_ton_report_12.2.16_0.pdf.

DOE-BETO. 2017. *2016 Billion-Ton Report, Volume 2: Environmental Sustainability Effects of Select Scenarios from Volume 1.* Washington, DC: Department of Energy. https://www.energy.gov/eere/bioenergy/articles/2016-billion-ton-report-volume-2-environmental-sustainability-effects.

DOE-BETO. 2021. “2021 Project Peer Review—Biochemical Conversion and Lignin Utilization.” https://www.energy.gov/eere/bioenergy/2021-project-peer-review-biochemical-conversion-and-lignin-utilization.

DOE-EERE (Department of Energy Office of Energy Efficiency and Renewable Energy). 2017. “Waste Heat Recovery Resource Page.” https://www.energy.gov/eere/amo/articles/waste-heat-recovery-resource-page.

DOE-EERE. 2021. “Long Duration Storage Shot.” https://www.energy.gov/eere/long-duration-storage-shot.

DOE-EERE. n.d. “Industrial Heat Shot.” https://www.energy.gov/eere/industrial-heat-shot. Accessed February 7, 2023.

DOE-HFTO (Department of Energy Hydrogen and Fuel Cell Technologies Office). n.d.(a). “Hydrogen and Fuel Cells Career Map.” https://www.energy.gov/eere/fuelcells/hydrogen-and-fuel-cells-career-map. Accessed February 7, 2023.

DOE-HFTO. n.d.(b). “Hydrogen Shot.” https://www.energy.gov/eere/fuelcells/hydrogen-shot. Accessed February 7, 2023.

DOE-OCED (Department of Energy Office of Clean Energy Demonstrations). 2023. “DE-FOA-0002779: Bipartisan Infrastructure Law: Additional Clean Hydrogen Programs (Section 40314): Regional Clean Hydrogen Hubs Funding Opportunity Announcement.” Modified January 26. https://oced-exchange.energy.gov/Default.aspx#Foald4dbbd966-7524-4830-b883-450933661811.

DOE-OCED. n.d.(a). “Long-Duration Energy Storage Demonstrations.” https://www.energy.gov/oced/long-duration-energy-storage-demonstrations.

DOE-OCED. n.d.(b). “Regional Clean Hydrogen Hubs.” https://www.energy.gov/oced/regional-clean-hydrogen-hubs.

DOS and EOP (Department of State and Executive Office of the President). 2021. *The Long-Term Strategy of the United States: Pathways to Net-Zero Greenhouse Gas Emissions by 2050.* Washington, DC. https://www.whitehouse.gov/wp-content/uploads/2021/10/US-Long-Term-Strategy.pdf.

Douglas, C.M., S.L. Shaw, T.D. Martz, R.C. Steele, D.R. Noble, B.L. Emerson, and T.C. Lieuwen. 2022. “Pollutant Emissions Reporting and Performance Considerations for Hydrogen–Hydrocarbon Fuels in Gas Turbines.” *Journal of Engineering for Gas Turbines and Power* 144(9). https://doi.org/10.1115/1.4054949.

Douglas, C., T. Martz, R. Steele, D. Noble, B. Emerson, and T. Lieuwen. In press. “Pollutant Emissions Reporting and Performance Considerations for Ammonia-Blended Fuels in Gas Turbines.” *Journal of Engineering for Gas Turbines and Power.*

Dyck, J., I. Palou-Rivera, N. Nasr, and D. Bauer. 2022. “The Role of Manufacturing in Industrial Decarbonization.” Presentation to the National Academies’ Committee on Accelerating Decarbonization in the United States, online, September 9. https://www.nationalacademies.org/event/09-09-2022/accelerating-decarbonization-in-the-united-states-technology-policy-and-societal-dimensions-the-role-of-manufacturing-in-industrial-decarbonization.

EC (European Commission). 2020. “Report from the Commission: The Availability of Refrigerants for New Split Air Conditioning Systems That Can Replace Fluorinated Greenhouse Gases or Result in a Lower Climate Impact.” https://climate.ec.europa.eu/system/files/2020-09/c_2020_6637_en.pdf.

EC. 2021. “Carbon Border Adjustment Mechanism: Questions and Answers.” https://ec.europa.eu/commission/press corner/detail/en/qanda_21_3661.

EC. 2023. “Carbon Border Adjustment Mechanism.” https://taxation-customs.ec.europa.eu/green-taxation-0/carbon-border-adjustment-mechanism_en.

EFI (Energy Futures Initiative). 2021. "The Future of Clean Hydrogen in the United States: Views from Industry, Market Innovators, and Investors." In *Kilograms to Gigatons: Pathways for Hydrogen Market Formation in the United States*. Part of the EFI Report Series. Washington, DC. https://energyfuturesinitiative.org/wp-content/uploads/sites/2/2022/03/The-Future-of-Clean-Hydrogen-in-the-U.S._Report-1.pdf.

EIA (U.S. Energy Information Administration). 2021a. "2018 Manufacturing Energy Consumption Survey. Manufacturing Energy Consumption Survey (MECS)—Data—U.S." https://www.eia.gov/consumption/manufacturing/pdf/MECS%202018%20Results%20Flipbook.pdf.

EIA. 2021b. "Use of Energy Explained: Energy Use in Industry." https://www.eia.gov/energyexplained/use-of-energy/industry.php.

EIA. 2022. "Biomass Explained." https://www.eia.gov/energyexplained/biomass.

EIA. 2023. "Table 2.4 Industrial Sector Energy Consumption (Trillion Btu)." https://www.eia.gov/totalenergy/data/monthly/pdf/sec2_11.pdf.

Energy Act of 2020. 2020. "(Part I) Public Law 116-260—Dec. 27, 2020." In *Consolidated Appropriations Act, 2021*. https://www.congress.gov/116/plaws/publ260/PLAW-116publ260.pdf.

EPA (Environmental Protection Agency). 2021. "Inventory of U.S. Greenhouse Gas Emissions and Sinks: 1990–2019." https://www.epa.gov/sites/default/files/2021-04/documents/us-ghg-inventory-2021-main-text.pdf.

EPA. 2022a. "Overview of Greenhouse Gases." Updated May 16. https://www.epa.gov/ghgemissions/overview-greenhouse-gases.

EPA. 2022b. "Semiconductor Industry." https://www.epa.gov/f-gas-partnership-programs/semiconductor-industry.

EPA. 2022c. "Sustainable Materials Management: Non-Hazardous Materials and Waste Management Hierarchy." https://www.epa.gov/smm/sustainable-materials-management-non-hazardous-materials-and-waste-management-hierarchy.

EPRI (Electric Power Research Institute). 2022. *Taking Gas Turbine Hydrogen Blending to the Next Level*. Report No. 3002025438. Palo Alto, CA. https://www.epri.com/research/programs/113171/results/3002025438.

EPRI. n.d. "Hydrogen Education for a Decarbonized Global Economy." EPRI: The Center for Grid Engineering Education. https://grided.epri.com/H2EDGE.html. Accessed January 4, 2023.

Fan, Z., and S.J. Friedmann. 2021. "Low-Carbon Production of Iron and Steel: Technology Options, Economic Assessment, and Policy." *Joule* 5(4):829–862. https://doi.org/10.1016/j.joule.2021.02.018.

FCHEA (Fuel Cell and Hydrogen Energy Association). 2020. "Road Map to a U.S. Hydrogen Economy." Washington, DC: Fuel Cell and Hydrogen Energy Association. https://www.fchea.org/us-hydrogen-study.

Federal CSO (Office of the Federal Chief Sustainability Officer). n.d. "Federal Buy Clean Initiative." White House Council on Environmental Quality. https://www.sustainability.gov/buyclean. Accessed January 4, 2023.

Federal LCA Commons. n.d. "About Us." Federal LCA Commons. https://www.lcacommons.gov/about-us-0. Accessed January 4, 2023.

Field, R.A., and R.G. Derwent. 2021. "Global Warming Consequences of Replacing Natural Gas with Hydrogen in the Domestic Energy Sectors of Future Low-Carbon Economies in the United Kingdom and the United States of America." *International Journal of Hydrogen Energy* 46(58):30190–30203. https://doi.org/10.1016/j.ijhydene.2021.06.120.

Fivga, A., and Z.A. Mayer. 2016. "Understanding Biofuel Standards." *Biofuels Digest*. https://www.biofuelsdigest.com/bdigest/2016/10/02/understanding-biofuel-standards.

FMC (First Movers Coalition). 2023. "First Movers Coalition: Introduction and Overview of Commitments." https://weforum.ent.box.com/s/3do0gx4v0hyx3fhco7f7f8ft7wwlj66e.

FMC. n.d. "First Movers Coalition." World Economic Forum. https://www.weforum.org/first-movers-coalition. Accessed January 4, 2023.

Fortune Business Insights. 2021. "Bulk Chemicals/Ammonia Market." https://www.fortunebusinessinsights.com/industry-reports/ammonia-market-101716.

Friedmann, S.J., Z. Fan, and K. Tang. 2019. *Low-Carbon Heat Solutions for Heavy Industry: Sources, Options, and Costs Today*. New York: Columbia University Center on Global Energy Policy. https://www.energypolicy.columbia.edu/wp-content/uploads/2019/10/LowCarbonHeat-CGEP_Report_111722.pdf.

G7. 2022. "Terms of Reference for the Climate Club." G7 Germany. https://www.g7germany.de/resource/blob/974430/2153140/a04dde2adecf0ddd38cb9829a99c322d/2022-12-12-g7-erklaerung-data.pdf.

Gallagher, K.S. 2014. *The Globalization of Clean Energy Technology: Lessons from China.* Cambridge, MA: MIT Press. https://doi.org/10.7551/mitpress/9805.001.0001.

Gallagher, K.S., and X. Xuan. 2019. *Titans of the Climate: Explaining Policy Process in the United States and China.* American and Comparative Environmental Policy. Cambridge, MA: MIT Press.

Gaster, R., R. Atkinson, and E. Rightor. 2023. "Beyond Force: A Realist Pathway Through the Green Transition." Washington, DC: Information Technology and Innovation Foundation. https://www2.itif.org/2023-p3-green-transition.pdf.

Grubler, A., C. Wilson, and G. Nemet. 2016. "Apples, Oranges, and Consistent Comparisons of the Temporal Dynamics of Energy Transitions." *Energy Research and Social Science* 22(December):18–25. https://doi.org/10.1016/j.erss.2016.08.015.

Hufnagel-Smith, P. 2022. "Assessing the Workforce Required to Advance Canada's Hydrogen Economy." *Transition Accelerator Reports* 4(4):1–45. https://transitionaccelerator.ca/wp-content/uploads/2022/07/Assessing-Workforce-Required-to-Advance-Canadas-Hydrogen-Economy_V_1.pdf.

Hughes, S., and A. Zoelle. 2022. "Cost of Capturing CO_2 from Industrial Sources." Pittsburgh, PA: National Energy Technology Laboratory. https://www.netl.doe.gov/projects/files/CostofCapturingCO2fromIndustrialSources_071522.pdf.

HYBRIT. n.d. "Fossil-Free Steel—a Joint Opportunity!" HYBRIT Fossil-Free Steel. https://www.hybritdevelopment.se/en. Accessed January 4, 2023.

Hydrogen Europe. 2022. "Germany Launches H2 Global Auctions." Hydrogen Europe. https://hydrogeneurope.eu/germany-launches-h2-global-auctions.

I3 (Industrial Innovation Initiative). 2022. "State Best Practices Guide for Decarbonizing the Industrial Sector." https://industrialinnovation.org/wp-content/uploads/2022/12/i3-State-Best-Practices.pdf.

IEA (International Energy Agency). 2019a. *The Future of Hydrogen: Seizing Today's Opportunities.* Report prepared by the IEA for the G20, Japan. Paris. https://iea.blob.core.windows.net/assets/9e3a3493-b9a6-4b7d-b499-7ca48e357561/The_Future_of_Hydrogen.pdf.

IEA. 2019b. *Material Efficiency in Clean Energy Transitions.* Paris. https://iea.blob.core.windows.net/assets/52cb5782-b6ed-4757-809f-928fd6c3384d/Material_Efficiency_in_Clean_Energy_Transitions.pdf.

IEA. 2019c. *Putting CO_2 to Use: Creating Value from Emissions.* Paris. https://iea.blob.core.windows.net/assets/50652405-26db-4c41-82dc-c23657893059/Putting_CO2_to_Use.pdf.

IEA. 2019d. *Renewables 2019.* Paris. https://www.iea.org/reports/renewables-2019.

IEA. 2020a. *CCUS in Clean Energy Transitions.* Paris. https://www.iea.org/reports/ccus-in-clean-energy-transitions/a-new-era-for-ccus.

IEA. 2020b. *Energy Technology Perspectives 2020.* Paris. https://www.iea.org/reports/energy-technology-perspectives-2020.

IEA. 2021. "Is Carbon Capture Too Expensive?" https://www.iea.org/commentaries/is-carbon-capture-too-expensive.

IEA. 2022. *Energy Efficiency 2022.* Paris. https://iea.blob.core.windows.net/assets/7741739e-8e7f-4afa-a77f-49dadd51cb52/EnergyEfficiency2022.pdf.

ITIF (Information Technology and Innovation Foundation). 2023. "US Energy Department RD&D Budget: Interactive Dataviz." https://itif.org/publications/2022/05/13/energy-department-rdd-budget-interactive-dataviz.

Jacoby, M. 2023. "Brewing Building Materials." *Chemical & Engineering News* 101(19):20–25. https://doi.org/10.1021/cen-10119-cover.

JDSUPRA. 2022. "New Incentives Under the 2022 Inflation Reduction Act to Jump Start Deployment of Carbon Sequestration and Expand Clean Fuel Development." https://www.jdsupra.com/legalnews/new-incentives-under-the-2022-inflation-6891044.

Jenkins, J.D., E.N. Mayfield, J. Farbes, R. Jones, N. Patankar1, Q. Xu, and G. Schivley. 2022a. "Preliminary Report: The Climate and Energy Impacts of the Inflation Reduction Act of 2022." Princeton, NJ: REPEAT Project. https://repeatproject.org/docs/REPEAT_IRA_Prelminary_Report_2022-08-04.pdf.

Jenkins, J., J. Larsen, J. Michael, R. Orvis, K. Palmer, and K. Rennert. 2022b. "Inflation Reduction Act of 2022: Modeling Major Climate and Energy Provisions." Presented at the Resources for the Future webinar, online, August 10. https://www.rff.org/events/rff-live/inflation-reduction-act.

Jones, N. 2022. "From Fertilizer to Fuel: Can 'Green' Ammonia Be a Climate Fix?" *YaleEnvironment360*, January 20. https://e360.yale.edu/features/from-fertilizer-to-fuel-can-green-ammonia-be-a-climate-fix.

King, B. 2022. "IRA Impact Analysis Details: Industry." E-mail correspondence with committee.

Kopp, R.J., W. Pizer, and K. Rennert. 2022. "Industrial Decarbonization and Competitiveness: Building a Performance Alliance." Washington, DC: Resources for the Future. https://www.rff.org/publications/issue-briefs/industrial-decarbonization-and-competitiveness-building-a-performance-alliance.

Krietsch Boerner, L. 2019. "Industrial Ammonia Production Emits More CO_2 Than Any Other Chemical-Making Reaction. Chemists Want to Change That." *Chemical and Engineering News* 97(24). https://cen.acs.org/environment/green-chemistry/Industrial-ammonia-production-emits-CO2/97/i24.

Lange, J.-P. 2021. "Towards Circular Carbo-Chemicals—the Metamorphosis of Petrochemicals." *Energy and Environmental Science* 14(8):4358–4376. https://doi.org/10.1039/D1EE00532D.

Larsen, J., W. Herndon, G. Hiltbrand, and B. King. 2021. *The Economic Benefits of Carbon Capture: Investment and Employment Opportunities for the Contiguous United States*. New York: Rhodium Group. https://rhg.com/wp-content/uploads/2021/04/The-Economic-Benefits-of-Carbon-Capture-Investment-and-Employment-Opportunities_Phase-III.pdf.

Larsen, J., B. King, H. Kolus, N. Dasari, G. Hiltbrand, and W. Herndon. 2022. *A Turning Point for US Climate Progress: Assessing the Climate and Clean Energy Provisions in the Inflation Reduction Act*. New York: Rhodium Group. https://rhg.com/research/climate-clean-energy-inflation-reduction-act.

Larson, E., C. Greig, J. Jenkins, E. Mayfield, A. Pascale, C. Zhang, J. Drossman, et al. 2021. *Net-Zero America: Potential Pathways, Infrastructure, and Impacts, Final Report*. Princeton, NJ: Princeton University. https://netzeroamerica.princeton.edu/the-report.

Lawson, A.J. 2022. *DOE's Carbon Capture and Storage (CCS) and Carbon Removal Programs*. Washington, DC: Congressional Research Service. https://crsreports.congress.gov/product/pdf/IF/IF11861.

LBNL (Lawrence Berkeley National Laboratory). 2022. "Generation, Storage, and Hybrid Capacity in Interconnection Queues." https://emp.lbl.gov/generation-storage-and-hybrid-capacity.

LDES (Long Duration Energy Storage) Council and McKinsey & Company. 2021. *Net-Zero Power: Long Duration Energy Storage for a Renewable Grid*. Brussels, Belgium. https://www.ldescouncil.com/assets/pdf/LDES-brochure-F3-HighRes.pdf.

LDES Council and McKinsey & Company. 2022. *Net-Zero Heat: Long Duration Energy Storage to Accelerate Energy System Decarbonization*. Brussels, Belgium. https://www.ldescouncil.com/assets/pdf/221108_NZH_LDES%20brochure.pdf.

LEP (Labor Energy Partnership). 2021. *Building to Net-Zero: A U.S. Policy Blueprint for Gigaton-Scale CO_2 Transport and Storage Infrastructure*. Washington, DC: Energy Futures Initiative and AFL-CIO. https://laborenergy.org/wp-content/uploads/2021/10/LEP-Building_to_Net-Zero-June-2021-v4.pdf.

LEP. 2022. *Inflation Reduction Act Analysis: Key Findings on Jobs, Inflation, and GDP*. Washington, DC: Energy Futures Initiative and AFL-CIO. https://laborenergy.org/wp-content/uploads/2022/08/8-6-22-IRA-Impact-Analysis-V14.pdf.

Lewis, A.C. 2021. "Optimising Air Quality Co-Benefits in a Hydrogen Economy: A Case for Hydrogen-Specific Standards for NO_x Emissions." *Environmental Science: Atmospheres* 1(5):201–207. https://doi.org/10.1039/D1EA00037C.

Li, H., H.A. Aguirre-Villegas, R.D. Allen, X. Bai, C.H. Benson, G.T. Beckham, S.L. Bradshaw, et al. 2022. "Expanding Plastics Recycling Technologies: Chemical Aspects, Technology Status and Challenges." *Green Chemistry* 24(23):8899–9002. https://doi.org/10.1039/D2GC02588D.

Lieuwen, T., M. Chang, and A. Amato. 2013. "Stationary Gas Turbine Combustion: Technology Needs and Policy Considerations." *Combustion and Flame* 160(8):1311–1314. https://doi.org/10.1016/j.combustflame.2013.05.001.

Linde Engineering. 2023. "Steam Cracking Technology." https://www.linde-engineering.com/en/process-plants/petrochemical-plants/steam-cracking-technology/index.html.

Lobet, I. 2020. "The Building Industry Gets Serious About Its Embodied Carbon Problem." Greentech Media. https://www.greentechmedia.com/articles/read/climate-pressure-mounts-within-building-industry.

Lovins, A.B. 2021. "Profitably Decarbonizing Heavy Transport and Industrial Heat: Transforming These 'Harder-to-Abate' Sectors Is Not Uniquely Hard and Can Be Lucrative." Emeritus Insight Series. RMI. https://rmi.org/wp-content/uploads/2021/07/rmi_profitable_decarb.pdf.

Mahajan, M., O. Ashmoore, J. Rissman, R. Orvis, and A. Gopal. 2022. "Updated Inflation Reduction Act Modeling Using the Energy Policy Simulator." Energy Innovation. https://energyinnovation.org/wp-content/uploads/2022/08/Updated-Inflation-Reduction-Act-Modeling-Using-the-Energy-Policy-Simulator.pdf.

McMillan, C.A. 2018. "Electrification of Industry: Summary of Electrification Futures Study Industrial Sector Analysis." NREL/PR—6A20-72311, 1474033. https://doi.org/10.2172/1474033.

McMillan, C. 2022. "Industrial Decarbonization Options: Solutions for Small and Medium-Sized Manufacturers." Presentation to the National Academies' Committee on Accelerating Decarbonization in the United States, online, February 8. https://www.nationalacademies.org/event/02-08-2022/accelerating-decarbonization-in-the-united-states-technology-policy-and-societal-dimensions-industrial-companies-open-session.

McMillan, C., C. Schoeneberger, J. Zhang, P. Kurup, E. Masanet, R. Margolis, S. Meyers, M. Bannister, E. Rosenlieb, and W. Xi. 2021. "Opportunities for Solar Industrial Process Heat in the United States." NREL/TP-6A20-77760. Golden, CO: National Renewable Energy Laboratory. https://www.nrel.gov/docs/fy21osti/77760.pdf.

Meckling, J., N. Kelsey, E. Biber, and J. Zysman. 2015. "Winning Coalitions for Climate Policy: Green Industrial Policy Builds Support for Carbon Regulation." *Science* 249(6253):1170–1171. https://doi.org/10.1126/science.aab1336.

Meckling, J., T. Sterner, and G. Wagner. 2017. "Policy Sequencing Toward Decarbonization." *Nature Energy* 2(November 2017):918–922. https://doi.org/10.1038/s41560-017-0025-8.

Miklautsch, P., and M. Woschank. 2022. "Decarbonizing Industrial Logistics." *IEEE Engineering Management Review* 50(3):149–156. https://doi.org/10.1109/EMR.2022.3186738.

Moser, H. 2022. "A US Policy Roadmap for a Reshored Reality." *Force Distance Times*, November 9. https://forcedistancetimes.com/a-us-policy-roadmap-for-us-reshoring.

Nadel, S., and L. Ungar. 2019. "Halfway There: Energy Efficiency Can Cut Energy Use and Greenhouse Gas Emissions in Half by 2050." Report U1907. Washington, DC: American Council for an Energy-Efficient Economy. https://www.aceee.org/research-report/u1907.

Naegler, T., S. Simon, M. Klein, and H.C. Gils. 2015. "Quantification of the European Industrial Heat Demand by Branch and Temperature Level." *International Journal of Energy Research* 39(15):2019–2030. https://doi.org/10.1002/er.3436.

NAM (National Association of Manufacturers). 2022. "NAM Manufacturers' Outlook Survey: Third Quarter 2022." https://www.nam.org/wp-content/uploads/2022/09/Manufacturers_Third_Quarter_Outlook_Survey_September_2022.pdf.

NASEM (National Academies of Sciences, Engineering, and Medicine). 2019a. *Gaseous Carbon Waste Streams Utilization: Status and Research Needs*. Washington, DC: The National Academies Press. https://doi.org/10.17226/25232.

NASEM. 2019b. *Negative Emissions Technologies and Reliable Sequestration: A Research Agenda*. Washington, DC: The National Academies Press. https://doi.org/10.17226/25259.

NASEM. 2021. *Accelerating Decarbonization of the U.S. Energy System*. Washington, DC: The National Academies Press. https://doi.org/10.17226/25932.

NASEM. 2023a. *Carbon Dioxide Utilization Markets and Infrastructure: Status and Opportunities: A First Report*. Washington, DC: The National Academies Press. https://doi.org/10.17226/26703.

NASEM. 2023b. *Laying the Foundation for New and Advanced Nuclear Reactors in the United States*. Washington, DC: The National Academies Press. https://doi.org/10.17226/26630.

NETL (National Energy Technology Laboratory). n.d. "Life Cycle Analysis (LCA) of Energy Technology and Pathways." https://netl.doe.gov/LCA. Accessed January 4, 2023.

NIST (National Institute of Standards and Technology). 2020. "Metrics and Tools for Sustainable Buildings." https://www.nist.gov/programs-projects/metrics-and-tools-sustainable-buildings.

NIST. 2022. "Reshoring and the Pandemic: Bringing Manufacturing Back to America." NIST: Manufacturing Extension Partnership. https://www.nist.gov/mep/manufacturing-infographics/reshoring-and-pandemic-bringing-manufacturing-back-america.

NRDC (National Resources Defense Council). 2019. "U.S. States Take the Lead in HFC Phasedown." National Resources Defense Council, 41st Meeting of the Open-Ended Working Group of the Parties to the Montreal Protocol, July 2019.

NREL (National Renewable Energy Laboratory). 2014. "Biomass Resource Data, Tools, and Maps." https://www.nrel.gov/gis/biomass.html.

Ochu, E., S. Braverman, G. Smith, and J. Friedmann. 2021. "Hydrogen Fact Sheet: Production of Low-Carbon Hydrogen." https://www.energypolicy.columbia.edu/sites/default/files/pictures/HydrogenProduction_CGEP_FactSheet_052621.pdf.

O'Connor, J., B. Noble, and T. Lieuwen, eds. 2022. *Renewable Fuels: Sources, Conversion, and Utilization*. Cambridge, UK: Cambridge University Press. https://doi.org/10.1017/9781009072366.

Patnaik, S., and K. Kennedy. 2021. "Why the US Should Establish a Carbon Price Either Through Reconciliation or Other Legislation." Brookings. https://www.brookings.edu/research/why-the-us-should-establish-a-carbon-price-either-through-reconciliation-or-other-legislation.

Paulot, F., D. Paynter, V. Naik, S. Malyshev, R. Menzel, and L.W. Horowitz. 2021. "Global Modeling of Hydrogen Using GFDL-AM4.1: Sensitivity of Soil Removal and Radiative Forcing." *International Journal of Hydrogen Energy* 46(24):13446–13460. https://doi.org/10.1016/j.ijhydene.2021.01.088.

Pettersen, J., R. Steeneveldt, D. Grainger, T. Scott, L.-M. Holst, and E. Steinseth Hamborg. 2022. "Blue Hydrogen Must Be Done Properly." *Energy Science and Engineering* 10(9):3220–3236. https://doi.org/10.1002/ese3.1232.

Pisciotta, M., H. Pilorgé, J. Feldmann, R. Jacobson, J. Davids, S. Swett, Z. Sasso, and J. Wilcox. 2022. "Current State of Industrial Heating and Opportunities for Decarbonization." *Progress in Energy and Combustion Science* 91(July):100982. https://doi.org/10.1016/j.pecs.2021.100982.

Pollin, R., C. Lala, and S. Chakraborty. 2022. *Job Creation Estimates Through Proposed Inflation Reduction Act: Modeling Impacts of Climate, Energy, and Environmental Provisions of Bill.* Amherst, MA: Political Economy Research Institute. https://peri.umass.edu/publication/item/1633-job-creation-estimates-through-proposed-inflation-reduction-act.

Precedence Research. 2022. "Green Ammonia Market." https://www.precedenceresearch.com/green-ammonia-market.

Psarras, P.C., S. Comello, P. Bains, P. Charoensawadpong, S. Reichelstein, and J. Wilcox. 2017. "Carbon Capture and Utilization in the Industrial Sector." *Environmental Science and Technology* 51(19):11440–11449. https://doi.org/10.1021/acs.est.7b01723.

Queensland Government. 2022. *Hydrogen Industry Workforce Development Roadmap 2022–2032.* Queensland, Australia. https://www.publications.qld.gov.au/dataset/hydrogen-industry-workforce-development-roadmap-2022-2032/resource/5fffcbcc-7605-46ed-86b4-2c2a91e7acad.

Reddoch, T., B. Westlake, and A. Rohr. 2021. "Hydrogen Education for a Decarbonized Economy (H2EDGE)." Presented at the DOE Hydrogen Program 2021 Annual Merit Review and Peer Evaluation Meeting, online, June 9. https://www.hydrogen.energy.gov/pdfs/review21/scs028_reddoch_2021_p.pdf.

Reshoring Initiative. 2022. "Reshoring Initiative IH 2022 Data Report: Multiple Supply Chain Risks Accelerate Reshoring." Sarasota, FL. https://reshorenow.org/content/pdf/2022_1H_data_report-final5.5.pdf.

Rightor, E. 2022. "Why DOE Should Prioritize Transformational Investments in Industrial Technology to Catalyze GHG Reductions." *Innovation Files* (blog), December 19. https://itif.org/publications/2022/12/19/why-doe-should-prioritize-transformational-investments-in-industrial-technology.

Rightor, E. 2023. "Innovation Amplifiers: Getting More Bang for the Buck on GHG Reductions." Washington, DC: Information Technology and Innovation Foundation. https://www2.itif.org/2023-innovation-amplifiers.pdf.

Rightor, E., A. Whitlock, and R.N. Elliott. 2020. "Beneficial Electrification in Industry." Research Report. Washington, DC: American Council for an Energy-Efficient Economy. https://www.aceee.org/sites/default/files/pdfs/ie2002.pdf.

Rightor, E., C. McMillan, C. Arpagaus, P. Scheihing, and T. Baldyga. 2022a. "Electrifying Industry's Process Heat Supply with Industrial Heat Pumps." ACEEE Webinar, online, April 6. https://www.aceee.org/webinar/electrifying-industrys-process-heat-supply-industrial-heat-pumps.

Rightor, E., P. Scheihing, A. Hoffmeister, and R. Papar. 2022b. "Industrial Heat Pumps: Electrifying Industry's Process Heat Supply." ACEEE Report. Washington, DC: American Council for an Energy-Efficient Economy. https://www.aceee.org/sites/default/files/pdfs/ie2201.pdf.

Rosen, M.A. 2021. "Nuclear Energy: Non-Electric Applications." *European Journal of Sustainable Development Research* 5(1):em0147. https://doi.org/10.29333/ejosdr/9305.

Ruckelshaus, C., and S. Leberstein. 2014. *Manufacturing Low Pay: Declining Wages in the Jobs That Build America's Middle Class.* Washington, DC: National Employment Law Project. https://www.nelp.org/wp-content/uploads/2015/03/Manufacturing-Low-Pay-Declining-Wages-Jobs-Built-Middle-Class.pdf.

SABIC. 2022. "BASF, Sabic, and Linde Start Construction of the World's First Large-Scale Electrically Heated Steam Cracker Furnaces." https://www.sabic.com/en/news/36814-basf-sabic-and-linde-start-construction-of-the-worlds-first-large-scale-electrically-heated-steam-cracker-furnaces.

Saha, D., R. Shrestha, and P. Jordan. 2022. "How a Clean Energy Economy Can Create Millions of Jobs in the US." World Resources Institute. https://www.wri.org/insights/us-jobs-clean-energy-growth.

Sandalow, D., J. Friedmann, R. Aines, C. McCormick, S. McCoy, and J. Stolaroff. 2019. *ICEF Industrial Heat Decarbonization Roadmap*. Tokyo: Innovation for Cool Earth Forum. https://www.icef.go.jp/pdf/summary/roadmap/icef2019_roadmap.pdf.

Sarin, S., R. Verma, and R. Singh. 2021. "Reduced Carbon Intensity Ethylene Production." PEP Report 29M. IHS Markit. https://cdn.ihsmarkit.com/www/pdf/0122/RP29M_toc.pdf.

Sartor, O., and C. Bataille. 2019. Decarbonising Basic Materials in Europe: How Carbon Contracts-for-Difference Could Help Bring Breakthrough Technologies to Market. Paris: IDDRI. https://www.iddri.org/sites/default/files/PDF/Publications/Catalogue%20Iddri/Etude/201910-ST0619-CCfDs_0.pdf.

Schoeneberger, C.A., C.A. McMillan, P. Kurup, S. Akar, R. Margolis, and E. Masanet. 2020. "Solar for Industrial Process Heat: A Review of Technologies, Analysis Approaches, and Potential Applications in the United States." *Energy* 206(September):118083. https://doi.org/10.1016/j.energy.2020.118083.

Schyns, Z.O.G., and M.P. Shaver. 2021. "Mechanical Recycling of Packaging Plastics: A Review." *Macromolecular Rapid Communications* 42(3):2000415. https://doi.org/10.1002/marc.202000415.

Scott, R.E., Z. Mokhiber, and D. Perez. 2020. "Rebuilding American Manufacturing—Potential Job Gains by State and Industry." Economic Policy Institute. https://www.epi.org/publication/rebuilding-american-manufacturing-potential-job-gains-by-state-and-industry-analysis-of-trade-infrastructure-and-clean-energy-energy-efficiency-proposals.

Scott, R.E., V. Wilson, J. Kandra, and D. Perez. 2022. "Botched Policy Responses to Globalization Have Decimated Manufacturing Employment with Often Overlooked Costs for Black, Brown, and Other Workers of Color." Economic Policy Institute. https://www.epi.org/publication/botched-policy-responses-to-globalization.

SDSN (Sustainable Development Solutions Network). 2020. *America's Zero Carbon Action Plan*. New York: Sustainable Development Solutions Network. https://irp-cdn.multiscreensite.com/6f2c9f57/files/uploaded/zero-carbon-action-plan%20%281%29.pdf.

Séférian, R., M. Rocher, C. Guivarch, and J. Colin. 2018. "Constraints on Biomass Energy Deployment in Mitigation Pathways: The Case of Water Scarcity." *Environmental Research Letters* 13(5):054011. https://doi.org/10.1088/1748-9326/aabcd7.

Semieniuk, G., E. Campiglio, J.-F. Mercure, U. Volz, and N.R. Edwards. 2021. "Low-Carbon Transition Risks for Finance." *WIREs Climate Change* 12(1):e678. https://doi.org/10.1002/wcc.678.

Shivakumar, S., C. Wessner, and T. Howell. 2022. *Reshoring Semiconductor Manufacturing: Addressing the Workforce Challenge*. Washington, DC: Center for Strategic and International Studies. https://csis-website-prod.s3.amazonaws.com/s3fs-public/publication/221006_Shivakumar_Reshoring_SemiconductorManufacturing.pdf.

Shrestha, R., J. Neuberger, and D. Saha. 2022. "Federal Policy Building Blocks: To Support a Just and Prosperous New Climate Economy in the United States." Washington, DC: World Resources Institute. https://files.wri.org/d8/s3fs-public/2022-09/federal-policy-building-blocks-support-just-prosperous-new-climate-economy-united-states_0.pdf.

Smith, T. 2023. "U.S. Carbon Border Adjustment Proposals and World Trade Organization Compliance." *American Action Forum: Insight* (blog), February 8. https://www.americanactionforum.org/insight/u-s-carbon-border-adjustment-proposals-and-world-trade-organization-compliance.

Srinivasan, P., and N.W. Esram. 2022. "Guidebook Shows How States Can Tackle Industrial Emissions." *ACEEE* (blog), December 14. https://www.aceee.org/blog-post/2022/12/guidebook-shows-how-states-can-tackle-industrial-emissions.

Srinivasan, P., R.N. Elliott, E.G. Rightor, and N.W. Estam. 2022. "The Road to Industrial Buy-In for Embodied Carbon Building Standards." In *2022 ACEEE Summer Study on Energy Efficiency in Buildings*. Pacific Grove, CA. https://aceee2022.conferencespot.org/event-data/pdf/catalyst_activity_32546/catalyst_activity_paper_20220810191616919_f7fbe39c_9296_4c83_830e_ca2e76f94e2e.

Stone, K.C., P.G. Hunt, K.B. Cantrell, and K.S. Ro. 2010. "The Potential Impacts of Biomass Feedstock Production on Water Resource Availability." *Bioresource Technology* 101(6):2014–2025. https://doi.org/10.1016/j.biortech.2009.10.037.

Therkelsen, P., T. Werts, V. McDonell, and S. Samuelsen. 2009. "Analysis of NO_x Formation in a Hydrogen-Fueled Gas Turbine Engine." *Journal of Engineering for Gas Turbines and Power* 131(3). https://doi.org/10.1115/1.3028232.

Thiel, G.P., and A.K. Stark. 2021. "To Decarbonize Industry, We Must Decarbonize Heat." *Joule* 5(3):531–550. https://doi.org/10.1016/j.joule.2020.12.007.

Tisheva, P. 2023. "Constellation Achieves 38% Hydrogen Blending at Gas Plant." Renewables Now. https://renewablesnow.com/news/constellation-achieves-38-hydrogen-blending-at-gas-plant-823934.

Tullo, A.H. 2022. "Organics Challenge Ammonia as Hydrogen Carriers." *Chemical and Engineering News* 100(32). https://cen.acs.org/energy/hydrogen-power/Organics-challenge-ammonia-hydrogen-carriers/100/i32.

Ungar, L., S. Nadel, and J. Barrett. 2021. *Clean Infrastructure: Efficiency Investments for Jobs, Climate, and Consumers.* ACEEE White Paper. Washington, DC: American Council for an Energy-Efficient Economy. https://www.aceee.org/sites/default/files/pdfs/clean_infrastructure_final_9-20-21.pdf.

U.S. Census Bureau. 2022. "2019 SUSB Annual Data Tables by Establishment Industry." https://www.census.gov/data/tables/2019/econ/susb/2019-susb-annual.html.

USCA (United States Climate Alliance). 2022. "Enabling Industrial Decarbonization: A Policy Guidebook for U.S. States." https://static1.squarespace.com/static/5a4cfbfe18b27d4da21c9361/t/6399f2a93d4eb657655236c3/1671033519073/US+Climate+Alliance_2022+Industry+State+Policy+Guidebook_Industrial+Decarbonization.pdf.

Warwick, N., P. Griffiths, J. Keeble, A. Archibald, J. Pyle, and K. Shine. 2022. *Atmospheric Implications of Increased Hydrogen Use.* United Kingdom: University of Cambridge, National Centre for Atmospheric Sciences, University of Reading. https://assets.publishing.service.gov.uk/government/uploads/system/uploads/attachment_data/file/1067144/atmospheric-implications-of-increased-hydrogen-use.pdf.

Wesseling, J.H., S. Lechtenböhmer, M. Åhman, L.J. Nilsson, E. Worrell, and L. Coenen. 2017. "The Transition of Energy Intensive Processing Industries Towards Deep Decarbonization: Characteristics and Implications for Future Research." *Renewable and Sustainable Energy Reviews* 79(November):1303–1313. https://doi.org/10.1016/j.rser.2017.05.156.

Williams, M., and A. Bell. 2022. "The Pathway to Industrial Decarbonization." Center for American Progress. https://www.americanprogress.org/article/the-pathway-to-industrial-decarbonization.

World Nuclear News. 2023. "Dow's Seadrift Site Selected for X-Energy SMR Project." *World Nuclear News*, May 11. https://www.world-nuclear-news.org/Articles/Dow-s-Seadrift-site-selected-for-X-energy-SMR-proj.

Worrell, E., and G. Boyd. 2022a. "Bottom-Up Estimates of Deep Decarbonization of U.S. Manufacturing in 2050." *Journal of Cleaner Production* 330(January):129758. https://doi.org/10.1016/j.jclepro.2021.129758.

Worrell, E., and G. Boyd. 2022b. "Industrial Decarbonization Options: Light Industry." Presentation to the National Academies' Committee on Accelerating Decarbonization in the United States, online, February 8. https://www.nationalacademies.org/event/02-08-2022/accelerating-decarbonization-in-the-united-states-technology-policy-and-societal-dimensions-industrial-companies-open-session.

X-Energy. 2023. "Dow and X-Energy Advance Efforts to Deploy First Advanced Small Modular Nuclear Reactor at Industrial Site Under DOE's Advanced Reactor Demonstration Program." https://x-energy.com/media/news-releases/dow-and-x-energy-advance-efforts-to-deploy-first-advanced-small-modular-nuclear-reactor-at-industrial-site-under-does-advanced-reactor-demonstration-program.

11

Aligning the Financial Sector and Capital Markets with the Energy Transition

ABSTRACT

The financial sector directs the flow of capital and financial services to businesses and households throughout the United States and has been increasingly focused on the risks and opportunities associated with the net-zero transition. Historically, some communities have not had equal access to these services, an inequality that the energy transition must address. This chapter examines the financial sector's unique role in decarbonization, distinct from policies that change the fundamental economics of greenhouse gas (GHG) emissions.

One key role is to ensure that all households are able to equitably benefit from energy transition through targeted financial support and tracking access to government subsidies. Many communities and households lack access to the credit and financing that would allow them to participate in government subsidies for clean energy investments ranging from electric vehicles to home equipment. Targeted programs can address these inequities.

A second key role relates to the data and information that allows investors and regulators to fully understand climate-related risks and opportunities in the financial sector. Improved and standardized data collection and disclosure encourages improved risk management, facilitates the pricing of climate risk into asset values, and directs capital flows in ways that are then sensitive to climate risks.

Last, financial regulators need to improve their monitoring and supervision of climate risks. Beyond data and information collection, this includes scenario analysis and stress testing to understand the vulnerability of key financial institutions and the sector as a whole.

INTRODUCTION

The financial sector includes a wide range of financial institutions and companies and financial regulators that together are responsible for the flow of capital and financial services to businesses and households in the United States, including both large and relatively small public and private investors. This sector has become increasingly focused on climate change as both a risk and an opportunity for businesses and households that can have significant financial consequences. In recent years, discussion of capital flows associated with the nation's energy transition has taken on issues of equity. There is a long history of inequity in the processes and outcomes of actors in the financial sector. Without intentional efforts, this inequity will likely carry over and worsen with an energy transition, as disadvantaged groups will both be limited in their access to energy transition opportunities as well as face a disproportionate share of the risks.

This chapter examines a number of questions relating to decarbonization and the financial sector:

- What did the committee's first report recommend with regard to federal action needed to align the financial sector with decarbonization pathways?
- What changes have happened in the financial sector since the committee's first report in February 2021, considering actions by the federal government, state governments, and the private sector?
- What changes are still needed in the financial sector during the 2020s to address both risks and opportunities and to help move the nation toward an equitable net-zero economy by midcentury?
- What barriers to change need to be addressed to accomplish such changes?
- What are the committee's recommendations with respect to the financial sector?

As a starting point, however, the chapter's introduction explains why the committee has included a discussion of this sector in its report. Clearly, the financial decisions of investors, companies, households, and governments to invest in either conventional, GHG-emitting capital and goods and services with high embedded GHG emissions (Scope 3 emissions), versus low- or zero-emitting alternatives, can either support or impede decarbonization. Other chapters discuss policies that fundamentally alter the economics of these choices toward low- and zero-emitting alternatives. This chapter considers the additional actions that may be necessary to ensure that financial flows follow the changing economics, avoid unnecessary risks, and harness opportunities, and do all of this with a keen eye to the equity of outcomes.

Because this chapter examines current and potential financial-sector reforms, the committee focuses on the following questions: What is the financial sector's unique role in decarbonization, distinct from policies that change the fundamental economics of GHG emissions? What is the government's role in support of corporate climate information and disclosure? What further government action or regulation should be pursued in the financial sector? What changes are needed to provide better access to capital for households, small businesses, and communities that, owing to historic and structural inequalities, would otherwise not be full participants in the net-zero transition? The committee explores these questions within the framework created by prior committee recommendations, recent federal legislation and other activities, and gaps that need to be filled or barriers that need to be addressed going forward. Table 11-1, at the end of the chapter, summarizes all the recommendations that appear in this chapter regarding how the financial sector and capital markets can support decarbonization across the economy.

As noted, this chapter does *not* discuss government funding, financial incentives, and regulations that directly target the economics of conventional, GHG-emitting activities versus low- and zero-emission alternatives. Those policies focus on actions in a particular sector or system and are discussed in other chapters. This chapter also does not provide an assessment of the capital requirements for decarbonization, as this topic has been discussed in the committee's first report.

FINANCIAL-SECTOR RECOMMENDATIONS FROM THE COMMITTEE'S FIRST REPORT

The committee's first report recognized that "[f]inancial markets play an essential role in the economy by pricing risk 'to support informed, efficient capital-allocation decisions'" and recommended federal action in two areas: disclosure of financial risks associated with climate change, and creation of a national Green Bank (NASEM 2021, p. 202).

First, the committee pointed out that

> [C]limate risk still is poorly priced into financial markets, in part because there is inadequate transparency in corporate financial statements and because it is difficult to assign probabilities on government action (Litterman 2020a, 2020b). Even recognizing growing investor interest in companies with positive environmental, social, and governance (ESG) practices and outcomes (Eccles and Klimenko 2019; Fink 2020), many companies have not integrated climate risk into their governance and fiduciary responsibilities (Zaidi 2020). (NASEM 2021, p. 202)

The committee discussed the importance of private-sector actors as well as federal agencies taking climate risk into account in their own decisions. The committee recommended that Congress take several actions:

- Direct the Securities and Exchange Commission (SEC) to require public companies to formally disclose their risks from adverse impacts of climate change mitigation policies and climate change as part of their annual filings to the SEC.
- Direct the Federal Reserve System to identify climate-related financial risks, including by applying climate change policy and impact scenarios to financial stress tests.

- Direct federal agencies . . . to incorporate risks and costs from climate policies and climate change into the benefit-cost analyses required prior to the adoption of regulations or standards, or approval of public or private infrastructure investments.
- Require private firms to report their energy-related research and development investments by category (e.g., fossil, solar, wind) annually to the Department of Energy (DOE). (NASEM 2021, pp. 202–203)

Additionally, the committee recommended that the Commodity Futures Trading Commission (CFTC) "build on the recommendations of the report Managing Climate Risk in the U.S. Financial System . . . to ensure that climate risk is better reflected in the commission's and other federal financial agencies' oversight of commodities and derivative markets" (NASEM 2021, p. 203).

Second, the committee previously concluded that "the transition will be much more capital intensive than business-as-usual" and private "sources are unlikely to provide the needed capital, especially during the 2020s when the effort is new" (NASEM 2021, p. 206). The committee noted that the United States, unlike many of its economic competitor nations, does not currently have a "domestic independent development, investment, or Green Bank at the federal level," although several such green banks exist at the subnational level (NASEM 2021, p. 207).

The committee recommended the establishment of a national Green Bank "to mobilize finance for low-carbon infrastructure and business in America," with initial congressional funding of $30 billion and an additional $30 billion during this decade (NASEM 2021, pp. 206 and 208). The committee found that the Green Bank should be a non-governmental organization with that purpose and that it should use seed funding from the federal government to leverage private investment and support equitable outcomes (with emphasis added below):

> Partial financing by a Green Bank would reduce risk for private investors and encourage rapid expansion of private source capital. Such a bank would underpin the broad economic and social transitions required to achieve net-zero emissions by midcentury. The *new bank should lend, provide loan guarantees, make equity investments, cooperate with community banks to increase the availability of finance at the local level, and leverage private finance consistent with a national strategy* to compete internationally in low-carbon industries and transform the U.S. economy.
>
> It should make particular effort to be a *source of credit for innovative small and medium-size enterprises that may be locked out of commercial markets* owing to their size. The Green Bank can be a lead investor on big decarbonization projects that serve the public good, de-risking and leveraging larger commercial investors. It should *address inequities in the financing system*, working with local banks, co-ops, and rural and other marginalized communities. (NASEM 2021, p. 206)

Notably, the first report recommended that the United States adopt an economy-wide price on carbon in order to "unlock innovation in every corner of the energy economy, send appropriate signals to myriad public and private decision makers, and encourage a cost-effective route to net zero" (NASEM 2021, p. 12).

RECENT LEGISLATIVE, REGULATORY, OTHER POLICY, AND NON-GOVERNMENTAL ACTIONS RELATED TO THE FINANCIAL SECTOR

Since early 2021, the federal government has initiated action on the two topics where the committee recommended policy change: requirements for public companies to disclose climate risk and inclusion of climate risk into financial-sector risk assessments; and the seed funding for a new national Green Bank. Additionally, actors in the private sector have shown increased interest in corporate ESG accountability.

Federal Legislative Action

In Section 60103 of the Inflation Reduction Act of 2022 (IRA), Congress amended the Clean Air Act to authorize and appropriate federal funding for a new Greenhouse Gas Reduction Fund (GHG Reduction Fund). In effect, Congress has provided for the establishment of a national Green Bank, although fashioned to fit within the budget-reconciliation framework of the IRA.

The IRA appropriates $27 billion to the Environmental Protection Agency (EPA) for the GHG Reduction Fund and directs the agency to issue competitive grants to recipient entities so that they can use federal dollars to provide funding for and financing of the actions to reduce GHG emissions:

- $7 billion to states, municipalities, tribal governments, and other eligible recipients "for the purposes of providing grants, loans, or other forms of financial assistance, as well as technical assistance, to enable low-income and disadvantaged communities to deploy or benefit from zero-emission technologies, including distributed technologies on residential rooftops."
- $11.97 billion to "eligible recipients"[1]—certain nonprofit organizations—for the purposes of providing financial and technical assistance for direct and

---

[1] In this section of the IRA, an "eligible recipient" is defined as "a nonprofit organization that—(A) is designed to provide capital, including by leveraging private capital, and other forms of financial assistance for the rapid deployment of low- and zero-emission products, technologies, and services; (B) does not take deposits other than deposits from repayments and other revenue received from financial assistance provided using grant funds under this section; (C) is funded by public or charitable contributions; and (D) invests in or finances projects alone or in conjunction with other investors" (IRA §60103).

indirect[2] investment in "qualified projects"[3] that reduce GHG emissions and that would otherwise lack access to financing.

- $8 billion to "eligible recipients" for the "purposes of providing financial assistance and technical assistance in low-income and disadvantaged communities" for direct and indirect investment in "qualified projects" that reduce GHG emissions and would otherwise lack access to financing.

As of the committee's writing, EPA has taken a number of steps to implement this provision of the IRA. After issuing a request for information regarding the program guidance for the design and implementation of the awarding of funds for the GHG Fund, EPA received nearly 400 comments with disparate views about the approaches the agency should use to move the IRA funds into the financial sector (EPA 2022). On April 19, 2023, EPA announced that during the summer of 2023, the agency would hold three complementary grant competitions to distribute grant funding under the Greenhouse Gas Reduction Fund: a $14 billion National Clean Investment Fund competition to two to three national nonprofits to catalyze projects; a $6 billion Clean Communities Investment Accelerator competition to be awarded to two to seven "hub nonprofits" to provide access to financing for households and others in low-income and disadvantaged communities and to do so through networks of community lenders; and a $7 billion Solar for All competition for up to 60 grants to state, tribal, and local governments to support families' access to affordable solar installations (EPA 2023c). As of mid-July 2023, EPA announced the schedule for applications for all three funds during the second half of 2023, with awards in 2024 (EPA 2023a,b).

Federal Executive Branch Action

A 2020 report from the CFTC's Climate-Related Market Risk Subcommittee found that climate change posed complex and major risks to the U.S. financial system and urged regulators to act in a timely and decisive manner to measure, understand, and address the risks (Litterman 2020). Building on this report, the Financial Stability Oversight

---

[2] "Indirect investment" relates to the provision of "funding and technical assistance to establish new or support existing public, quasi-public, not-for-profit, or nonprofit entities that provide financial assistance to qualified projects at the State, local, territorial, or Tribal level in the District of Columbia, including community- and low-income-focused lenders and capital providers" (IRA §60103).

[3] A "qualified project" includes "any project, activity, or technology that—(A) reduces or avoids greenhouse gas emissions and other forms of air pollution in partnership with, and by leveraging investment from, the private sector; or (B) assists communities in the efforts of those communities to reduce or avoid greenhouse gas emissions and other forms of air pollution" (IRA §60103).

Council (FSOC)[4] issued a 2021 report recommending that financial regulators identify and address climate risks, laying out an agenda that includes enhancing public disclosures, addressing methodological gaps and climate data needs, as well as improving interagency coordination (FSOC 2021). After concluding that existing disclosure requirements on climate-related risks for companies and financial entities do not result in consistent, comparable, and decision-useful disclosures, FSOC identified that enhanced disclosures would increase investors' understanding of climate-related risks and allow these to be priced into markets.[5] The FSOC report acknowledged parallel efforts by individual member agencies to make progress in this area, including actions by CFTC, SEC, the Federal Reserve Board, and the Federal Housing Finance Agency (FHFA).

After signaling its intention to do so during 2021, SEC issued proposed rules in March 2022 that would require publicly traded companies to include in their public disclosure statements "certain climate-related disclosures . . . including information about climate-related risks that are reasonably likely to have a material impact on their business, results of operations, or financial condition, and certain climate-related financial statement metrics in a note to their audited financial statements" (SEC 2022b,c).[6] Furthermore, the proposed rule would require companies to disclose their GHG emissions, with the SEC's intention that providing such "GHG emissions disclosures would provide investors with decision-useful information to assess a registrant's

---

[4] FSOC is comprised of members from various federal agencies (i.e., the Department of the Treasury, the Federal Reserve Board, the Office of the Comptroller of the Currency, the Consumer Finance Protection Board, SEC, CFTC, the Federal Deposit Insurance Corporation [FDIC], FHFA, and the National Credit Union Administration) along with an independent member with insurance expertise, the head the Office of Financial Research, the Federal Insurance Office, a state insurance commissioner, a state banking supervisor, and a state securities commissioner. FSOC has responsibilities that include monitoring the financial services marketplace to identify potential threats to U.S. financial stability and to make recommendations about aspects of regulation (or gaps in it) that could pose risks or vulnerabilities to U.S. financial stability.

[5] Specific means identified for disclosures included agencies' leveraging of the Task Force on Climate-Related Financial Disclosures' existing framework (Litterman 2020), considering what constitutes appropriate information in a GHG disclosure, and coordinating disclosure data formats, comparability, and related elements of consistency.

[6] "SEC Proposes Rules to Enhance and Standardize Climate-Related Disclosures for Investors," press release, March 21, 2022, https://www.sec.gov/news/press-release/2022-46. "The proposed rule changes would require a registrant to disclose information about (1) the registrant's governance of climate-related risks and relevant risk management processes; (2) how any climate-related risks identified by the registrant have had or are likely to have a material impact on its business and consolidated financial statements, which may manifest over the short-, medium-, or long-term; (3) how any identified climate-related risks have affected or are likely to affect the registrant's strategy, business model, and outlook; and (4) the impact of climate-related events (severe weather events and other natural conditions) and transition activities on the line items of a registrant's consolidated financial statements, as well as on the financial estimates and assumptions used in the financial statements" (SEC 2022a).

exposure to, and management of, climate-related risks, and in particular transition risks" (SEC 2022c).[7]

Since SEC issued these proposed rules early in 2022, thousands of parties have filed comments that together constitute extensive legal, technical, and advocacy points—some in opposition, some in support, and many to modify SEC's proposed rule in some way (SEC 2022b). Considerable uncertainty exists with regard to future action by SEC to adopt such rules and to the legal durability of such a rule, in light of the Supreme Court's decision in *West Virginia v. EPA*. This decision limited EPA's actions to areas where Congress has authorized the agency to make decisions of economic and political significance (Zucker et al. 2022).

SEC has also proposed new guidelines regarding what may constitute an ESG investment product, in addition to providing an approach to disclosures that would allow for easier comparisons of ESG funds (SEC 2022c). The proposed SEC rule would identify three types of ESG funds, with differing disclosure requirements: "integration funds," which integrate ESG considerations along with other investment factors and would be required to describe how these elements are incorporated into investment decision making; "ESG-focused funds," which would be required to provide detailed disclosures; and "impact funds" (a subset of the former category focusing on a particular impact), which would be required to disclose how progress toward this objective will be measured (SEC 2022a).

As of this report's writing, there continues to be notable movement within elements of the federal government toward establishing means for supporting and enhancing climate-related considerations in the financial sector. This includes the development of proposals from the Federal Reserve, FDIC, and the Office of the Comptroller of the Currency for large financial institutions (with more than $100 billion in total consolidated assets) to identify and measure how climate-related risk affects them and to inform management of this exposure (Federal Reserve System 2022b). In January 2023, the Federal Reserve announced a pilot "Climate Scenario Analysis" exercise for the nation's six largest banks, in order to "learn about large banking organizations' climate risk-management practices and challenges and to enhance the ability of both large banking organizations and supervisors to identify, measure, monitor, and manage climate-related financial risks" (i.e., physical risks and transition risks on the banks' loan portfolios)

---

[7] The proposed rules also would require a registrant to disclose information about its direct GHG "emissions (Scope 1) and indirect emissions from purchased electricity or other forms of energy (Scope 2). In addition, a registrant would be required to disclose GHG emissions from upstream and downstream activities in its value chain (Scope 3) if material or if the registrant has set a GHG emissions target or goal that includes Scope 3 emissions" (SEC 2022a).

(Federal Reserve System 2023b,c).[8] In July 2023, CFTC held its second convening on voluntary markets and indicated that guidance was under development (Ellfeldt 2023).

Private-Sector Action

Although there has been interest among business-school researchers and activists in ESG efforts in the private sector for at least 20 years, the past 2 years have witnessed a number of reports by government and quasi-government agencies on climate financial risk, as well as an increasing (but still small) volume of academic research. A new academic journal, *Journal of Climate Finance*, for example, will publish its first issue in 2023 (Elsevier 2022).

ESG has begun to play an important role in climate investing.[9] The impact of "socially responsible" investing began to demonstrate strength as early as the 1960s, most notably in the form of the anti-apartheid investment campaign in which many public and private institutions holding large financial assets pressed for disinvestment in South Africa because of its racial segregation policies and practices. That campaign has been credited with influencing that country's decision to end apartheid as an official national policy in the early 1990s (Broyles 1998). Fifteen years ago, 16 national governments and financial institutions (including major public pension investment funds) representing $2 trillion in assets signed on to the United Nations' then-new Principles of Responsible Investment (PRI) (UN 2006). Since then and up through 2021, thousands of other signatories with over $120 trillion in assets have aligned themselves with these principles (PRI 2021).

ESG principles align tightly with these PRI principles and tend to reflect investor attention to corporations' attention to such factors as:

- *Environment*—for example, climate change, pollution, waste use, waste streams, and so on.

---

[8] Note that the Federal Reserve's Chair Jerome Powell explained as recently as January 2023 that the Federal Reserve's role in climate policy is extremely narrow—to ensure that financial institutions are appropriately managing their own climate-related risks (Newburger 2023).

[9] In parallel with investor focus on ESG principles (and more specifically climate financial risk), firms themselves may consider such principles alongside traditional business models. Considerable research has explored the potential for ESG to increase firm value and performance. (See, for example, Henisz et al. 2019; Young and Reeves 2020.)

Because this chapter focuses on the financial sector, it does not address the actions of entrepreneurs, start-up companies, and other private-sector entities whose core business focuses on the development, manufacture, sales, and installation of products and services consistent with a net-zero economy. These companies have seen business opportunities in doing so and are not necessarily motivated by investors pursuing ESG strategies and outcomes. Furthermore, the chapter focuses more on the actions of investors in financial markets than on the actions of corporate managers.

- *Socially responsible engagement with stakeholders*—for example, workforce safety and voice; stances on things like child protection in labor standards and product safety in supply chains; preventing sexual misconduct.
- *Governance*—for example, diversity in board membership and corporate leadership; conduct in political contributions; prohibitions against bribes and corrupt practices; attentiveness to shareholder concerns.

For decades, some nonprofit organizations have organized the voice and impact of large institutional investors around ESG types of activities, not just calling for corporate boards' attention to ESG but also urging corporations' voluntary and, more recently, mandatory disclosure of climate risk in public financial statements (NYSE 2020).

Although there is organized opposition to the ESG movement (Gelles and Tabuchi 2022; Goldstein and Farrell 2022; Read 2022; Sorkin et al. 2022) stemming, at least in part, from political polarization around the concept, the broad expectation is that investors will continue to press companies to incorporate ESG principles, including climate risks and opportunities, into their strategies (Atkins 2020; Barclays 2022; Berlin 2022). This investor interest, combined with the adoption of federal financial incentives for decarbonization discussed throughout this committee's report, is expected to help drive investments in the direction of a decarbonized economy. The actions of private-sector actors can assist and sometimes lead the policies adopted by state and federal governments.

> Finding 11-1: Investor and civil society activism to press companies to address climate change has the potential to motivate climate-friendly action by firms. This can, in turn, create additional momentum for stronger mitigation policy as firms' own financial interests become aligned with such policies. Forward-looking investors can both bet on future policies and make them more likely to occur.

More could be said about potential activism in this space, its direct and indirect mitigation consequences, and its interaction with efforts to strengthen mitigation policy. While valuable, the committee has instead chosen to focus on ensuring that financial flows can follow the changing economics of decarbonization, on addressing information and regulatory needs surrounding financial sector risks, and on providing equitable access to these flows and allocation of risk.

HOW FAR DO GOVERNMENT POLICY AND PRIVATE-SECTOR ACTIONS GET US?

The committee is not aware of modeling that captures the overall impacts of the changes in policies around financial disclosure and risk, private-sector actions affecting the alignment of ESG pressures, and the establishment of a national Green Bank on decarbonization outcomes.[10] Nor is the committee equipped with its own analytic tools to estimate the incremental impact of such activities on such outcomes.

That said, the committee views these policies and actions as supportive in moving financial markets toward better alignment with a lower-carbon energy system. To a large extent, financial sector actions create an enabling environment for an equitable decarbonization driven by *other* policies, technologies, and economics.

WHAT ISSUES AND BARRIERS TO IMPLEMENTATION NEED TO BE ADDRESSED?

Even with these recent efforts, there exist barriers to a strong alignment of financial markets with an accelerated and equitable decarbonization transition. The principal impediments are structural barriers that prevent many consumers from accessing the capital needed to buy and/or invest in low-carbon goods and services; persistent information gaps that enable decision makers to make better choices that take climate-related risks into account; and steps by financial-sector regulators to ensure adequate ongoing awareness of and ability to take action to address any adverse impacts of climate risks on financial stability.

Consumers' Access to Capital

Federal tax credits[11] and other state/federal programs[12] create financial incentives for many households and businesses to purchase and install energy-efficient and low-carbon solar systems, electric vehicles, and other home products and services. However, many such households lack access to capital, have insufficient income, or

---

[10] For example, the REPEAT Project's analysis of the impacts of the Infrastructure Investment and Jobs Act (IIJA) and IRA identifies the national Green Bank as an important element of the IRA that the project's modeling was not able to capture (Jenkins et al. 2022).

[11] The IRA includes incentives for homeowners and other households to purchase appliances and equipment that use less energy or otherwise help to reduce GHG emissions: tax credits for the purchase of certain new electric vehicles produced in the United States and certain used plug-in hybrid vehicles; expanded residential tax credits for the purchase and installation of solar panels and associated battery storage systems; rebates on households' purchases of energy-efficient appliances, electric equipment and building upgrades, with higher rebates for low-income households; tax credits for home improvements that reduce building energy use (Department of the Treasury 2022).

[12] The IIJA includes $3.5 billion in funding for weatherization assistance for low-income households and $0.55 billion for the Energy Efficiency and Conservation Block Grant Program to support, among other things, the financing of energy efficiency and other clean energy capital investments (BIL Summary 2021).

experience other situations (e.g., being renters rather than homeowners) that prevent them from taking advantage of such programs and policies.[13]

For example, 93.5 million people in the United States live in a census tract considered "disadvantaged" (CEQ 2022a,b) and 11 million households are both renters and extremely low-income (Aurand et al. 2022). A significant portion of U.S. households, especially those with low incomes, lack access to traditional forms of financing (Davidson 2018).[14]

Local lending institutions—like credit unions, Community Development Financial Institutions (CDFIs)—may exist in various parts of the nation and, as described by the Partnership on Mobility from Poverty, "attract and deliver much-needed financial services and investments in low-income and distressed communities," but these tend to be both small and far from ubiquitous (Davidson 2018). CDFIs have the mission to promote "community development in markets comprised of economically distressed people and places. CDFIs are essentially a type of public-private partnership established to advance financial inclusion, the policy goal designed to increase the accessibility of traditionally underserved populations and markets to affordable financial services and products" (Getter 2022, p. 2).

> CDFIs accomplish this goal by serving people and businesses that traditional financial institutions cannot make their predominant focus. Higher-risk clients are more likely to have weak credit histories or face above-normal levels of income volatility, making them generally more costly to serve. Consequently, traditional institutions, which must manage their liquidity and other financial risks to support public confidence in the overall financial system, often focus primarily on markets consisting of higher credit quality borrowers rather than on higher-risk borrowers. . . . CDFIs rely on a combination of public and private funding that includes grants, awards, and donations. (Getter 2022, p. 2)

---

[13] There have been efforts by federal financial regulators (e.g., Federal Reserve Board, CFPB, FHFA) and secondary finance markets (e.g., Fannie Mae, Freddie Mac) to consider the energy consumption of goods as part of these entities' financial determinations and their oversight of the primary consumer financial providers (banks, mortgage underwriters, lenders, and other creditors). Examples are "green mortgages" for individual households and "green rewards" or similar guarantee, rate discounts, or other preferred financing terms for developers. (See, e.g., https://www.fhfa.gov/PolicyProgramsResearch/Programs/Pages/Fed-Adv-Committee-AES-Housing.aspx.) State and local governments have also explored financing approaches (such as Property Assessed Clean Energy programs, or regulatory requirements related to on-bill financing of energy efficiency measures or requirements that real estate transactions on buildings include disclosure of energy use. (See, e.g., https://www.energy.gov/scep/slsc/property-assessed-clean-energy-programs and https://database.aceee.org/state/building-energy-disclosure.)

[14] As reported by the Atlanta Federal Reserve Bank, about "one in four U.S. households are either unbanked—having no relationship with a financial institution—or underbanked, meaning they have a bank account but go outside the traditional banking system for credit and other financial services. . . . [A]mong black and Hispanic households earning less than $40,000 a year (classified as low income [in 2018, when the Federal Reserve Board's survey was conducted]), 20 percent lack access to a bank account, double the proportion among all low-income households. By contrast, only 1 percent of all families with annual incomes above $40,000 lack a bank account. More than a third of low-income adults have no credit card" (Davidson 2018).

CDFIs offer numerous product lines, the majority of which are consumer, residential real estate, and small business loans, with consumer finance being the primary or secondary line of business for 41 percent of CDFI respondents and small business finance being a primary or secondary line of business for 51 percent of respondents in a 2019 survey (Carpenter et al. 2021).

> The CDFI industry represents a small percentage of the overall U.S. financial system. In 2020, the 1,271 CDFIs collectively held $151.8 billion in assets (loans). By comparison, the credit union industry in 2020 consisted of 5,099 federally insured institutions that collectively held $1.16 trillion in assets, and 4,074 small community banks, defined as having $1 billion or less in total assets, collectively held $1.158 trillion in assets. (Getter 2022, p. 4)

A recent report from the U.S. Partnership on Mobility from Poverty recommended an increase in U.S. public and private investment in CDFIs as a way to address households' and communities' access to financing resources (Elwood and Patel 2018). Notably, the IRA includes $27 billion in funding to entities that provide financial assistance for projects that reduce GHG emissions, of which at least $15 billion is targeted to projects in disadvantaged communities; CDFIs will likely participate in some fashion in such programs. (See the discussion of the national Green Bank and EPA's GHG Reduction Fund, in the section "Federal Legislative Action" above.)

> Finding 11-2: The financial sector is an important component of a just and equitable transition to a net-zero economy, because of both the financial resources and redirection required and the potential risks to the broader financial system. Key elements of addressing this topic are already under way, with private-sector initiatives among investors and advocacy groups, proposed information disclosure rules, and ongoing supervision activities by central banks. But reliance on existing private financial markets alone—without providing greater access to capital for low- and moderate-income households and other disadvantaged communities—will not ensure an equitable transition to a net-zero economy.

Recommendation 11-1: *Expand and Extend Funding and Financing Assistance for Actions Benefiting Low-Income and Disadvantaged Households and Communities.* **The federal government should support disadvantaged communities' and households' greater access to capital for greenhouse gas (GHG) emissions reductions in a number of ways:**

a. **The Environmental Protection Agency should ensure that its awards from the GHG Reduction Fund provide more than the minimum amounts of funding toward projects that benefit low-income and disadvantaged communities. In addition to the portions of the GHG Reduction Fund— that is, the $7 billion to states, municipalities, tribal governments,**

and the $8 billion to eligible nonprofit entities—where the Inflation Reduction Act specifically calls out low-income and disadvantaged communities as the beneficiaries of financing assistance, the additional $11.97 billion pot of funding to eligible nonprofit entities should also use at least 40 percent of that funding for projects that benefit low-income and disadvantaged communities.

b. **Congress should increase funding for the Community Development Financial Institution (CDFI) Fund at the Department of the Treasury, and require reporting by CDFIs on their financial assistance related to GHG emissions reductions (with the provision of technical assistance to CDFIs for complying with this requirement).**

c. **Congress should conduct hearings and support research and convenings at key consumer finance institutions—including the Consumer Financial Protection Bureau, the Federal Reserve Bank, and the Federal Housing Finance Agency—to explore how they may expand private lenders' reporting requirements regarding the energy consumption of the goods or services being financed (e.g., the mortgaged home or the auto loan) and, in turn, to incentivize finance for decarbonized alternatives and their availability to lower-income households.**

While CDFIs provide one avenue to address inequities in the access to net-zero financial incentives, others also play critical roles:

- The Consumer Financial Protection Bureau has regulatory, educational, monitoring, and enforcement authorities to protect consumers in the financial marketplace (CFPB n.d.).
- FHFA has the responsibility to oversee and regulate the nation's housing financial institutions particularly during current federal conservatorship of the government-sponsored enterprises (e.g., Fannie Mae and Freddie Mac), including through FHFA's requirements that these institutions have a duty to serve key underserved markets (e.g., manufactured housing, affordable housing, rural housing) for low- and moderate-income households and that they assist in integrating building electrification into their green financing offerings (Fannie Mae 2022).
- The Federal Reserve Board's role in regulating banks to ensure that they meet credit needs (including its oversight of the Community Reinvestment Act that encourages financial institutions to offer credit in *all* communities) could be used to identify strategies for financing inclusive energy transitions in affordable housing (Mills and Scott 2022).[15]

---

[15] Note that in October 2022, the Federal Reserve Bank of New York published a white paper with "consensus recommendations from housing and finance experts that were discussed during nine working

- The recently established Treasury Advisory Committee on Racial Equity is particularly focused on narrowing disparities faced by communities of color (Department of the Treasury n.d.).

The members of these and other groups could consider access to net-zero financial incentives and policies for decarbonization.

Distinct from addressing consumer access to capital directly and the equity of the access, it will also be important to monitor implementation of federal policies for adherence to equity principles. Major clean energy funding streams have been created by both the IRA and prior legislation, including funding to states, local governments, and nonprofits, through programs at Department of Housing and Urban Development (e.g., its Community Development Block Grant Program), DOE (e.g., its Energy Efficiency and Conservation Block Grant Program), EPA (e.g., its GHG Reduction Fund), and the Department of the Treasury (e.g., its CDFI Fund). To know whether equity is being addressed generally and with respect to the funding of energy efficiency and other decarbonization projects, these agencies will need to disclose appropriate indicators related to use of those funds in different communities.

> Finding 11-3: Many federal agencies provide funding and financial incentives for energy efficiency and other decarbonization projects for potential implementation by households, landlords, community nonprofits, and community financial institutions. These potential recipients may face barriers in accessing capital and other resources useful or necessary to take advantage of such funding programs and/or financial incentives. If such barriers are not addressed in program implementation by federal agencies, then equity goals for the transition will not be

---

sessions hosted by the Federal Reserve Bank of New York, in partnership with the New York State Energy Research and Development Authority" and the Community Preservation Corporation. Although the "recommendations in the report should not be imputed as formal recommendations from the Federal Reserve Bank of New York or Federal Reserve System, NYSERDA, or other New York state or city agencies," the report pointed to ways to help support the financing of decarbonization of buildings that serve low- and- moderate income households and neighborhoods, and included policy ideas (aimed at federal and state legislatures and other government entities) such as

- Giving tax incentives for early adopters to low-emissions heating and cooling systems;
- Providing tax relief to utility companies to induce them to reduce electricity rates to decarbonized buildings;
- Simplifying and aligning existing tax incentive programs to help owners finance retrofits to building systems;
- Recognizing the increased future value of carbon-neutral buildings in appraisals;
- Creating mortgage products that address decarbonization; and
- Charging lower interest rates for loans used to upgrade building systems to meet climate goals (Mills and Scott 2022).

met. Federal implementing agencies should also address the technical assistance needed for reporting on the outcomes of such financing.

Recommendation 11-2: *Disclose Equity Indicators for Federal Funding of Clean Energy.* **The Office of Management and Budget (OMB), with input from the White House Environmental Justice Advisory Council, should publish consolidated federal information on equity indicators related to the federal spending on energy efficiency programs and projects. OMB should then require federal agencies that issue funding for energy efficiency and other greenhouse gas (GHG) reduction projects to disclose and report on equity indicators related to the use of those funds. A non-exhaustive list of such agencies includes the Department of Housing and Urban Development (e.g., its Community Development Block Grant Program), the Department of Energy (e.g., its Energy Efficiency and Conservation Block Grant Program), the Environmental Protection Agency (e.g., its GHG Reduction Fund), and the Department of the Treasury (e.g., the Community Development Financial Institutions Fund). The analysis should also include equity outcomes based on congressional districts and an overlay of the Climate and Economic Justice Screening Tool's most disadvantaged census tracks.**

Recommendation 11-3: *Address Limited Access Faced by Low-Income and Marginalized Households.* **The Treasury Advisory Committee on Racial Equity should make recommendations to address this structural problem in the financial sector that adversely affects the ability of some Americans to access net-zero financial incentives and policies for decarbonization.**

Role of Information and Disclosure of Transition Risk

Information availability, asymmetries, and associated uncertainties all have significant impact on investor decisions about how to allocate capital and manage risk. Among institutional investors, 79 percent believe that climate-related risk is at least as important as financial risk (Ilhan et al. 2021). This climate-related risk includes financial consequences from both climate change impacts, particularly effects of droughts, wildfires, and floods on business operations, and decarbonization, including the consequences of a rapid elimination of fossil fuel use and GHG emissions through both regulation and external pressure. Investor concern about this latter transition risk motivates the need to consider and understand risk stemming from accelerating deep decarbonization. As noted by the International Panel on Climate Change

(IPCC, 15.6.2, AR6, WGIII) (Shukla et al. 2022) and TCFD (2017), such information can help increase climate financing.

A wide variety of firms are exposed to transition risk that accompanies the shift toward a decarbonized U.S. economy. Some companies are negatively exposed to risk (e.g., direct emissions impacts and business models) and some are positively exposed (e.g., renewables firms that stand to see increased demand with the energy transition) or both. The committee also notes the risk of firms *over*-investing in the transition, if policies fail to support the mitigation activities that firms provide and/or the economics of such activities fail to be profitable. This may not be a risk for the environment but remains a financial risk for investors. These issues motivate risk management by firms and investors—which requires information.

There is also a role for investors and companies that hope to benefit from the energy transition to push beyond immediate profits. The financial sector has an important role to play, as decarbonization requires a tremendous influx of new capital (as noted in the committee's first report) to achieve net-zero goals alongside equity. With well-defined emission regulations and information, the financial sector is well equipped to channel the necessary resources. To date and although there has been progress in the past few years (TCFD 2022), both regulation and information have been lacking, which could slow the shift of capital flows from carbon-intensive investments to lower-carbon investments. With improved information, forward-looking investors and the expectation of future regulation can accelerate this shift (TCFD 2017).

In its 2021 *Report on Climate-Related Financial Risk*, FSOC identified numerous instances and sources of gaps in data and methodologies—both within member U.S. federal agencies and across them—that would otherwise be useful to these agencies in evaluating climate-related financial risks of regulated entities and financial markets. These data-related challenges include

> Cataloging and analyzing existing data sources, as climate-related data has not been extensively used by financial regulators and investments will be necessary to incorporate and utilize available data. Another set of challenges involves data gaps. For example, current collection of financial data associated with corporate loans may not include important details associated with climate-related risks, such as emissions-related information that may inform transition risks and detailed geographic information on production facilities that could inform exposure of such loans to physical climate risks. A third set of challenges involves combining different types of data (e.g., climate, economic, and financial) from different sources and in different formats. In many cases, data may be difficult to use or combine owing to, for example, inconsistencies, or the lack of definitions, taxonomies, reporting standards, and entity identification that facilitates aggregation and analysis. (FSOC 2021, p. 48)

Disclosing transition risk and climate-related information raises questions regarding how specific disclosures need to be and what information is—and can be—effectively provided. Standardizing and formalizing what disclosures look like raise important issues, particularly because the relevant risks will be different for different types of firms. Some firms, for example, may be more likely to be candidates for rapid decarbonization based on the availability of low-carbon alternatives. Sometimes, the emissions intensity of a firm's products may determine its transition risk; other times, it may be tied to the absolute emission level. Additionally, such things as validating carbon offsets and/or evaluating their permanence, accurately assessing Scope 3 emissions, and accounting for emissions leakage or international outsourcing of emissions (Dai et al. 2021) will be important to meaningful and rigorous disclosures, but all of these issues are difficult to standardize. Nonetheless, standardizing data and methods wherever possible will facilitate comparing the risk exposures of similar firms at the very least, and ideally across broader groupings of enterprises.

Hard data, such as total emissions or the carbon-intensity of products/processes, are more difficult to manipulate than soft data, and providing useful and usable data to inform policy and regulation will be important for avoiding distortions. Hard data are also an essential bedrock for carbon intensity–based performance standards, like the standard proposed in Recommendation 10-9. Standardizing future risk projections to the extent possible would be beneficial, as could standardizing the frequency of disclosure, which could allow for structures such as coupling executive compensation to hitting climate targets (akin to what is currently done with stock prices).

> Finding 11-4: Standardized data and methodologies to measure and report climate risk are a key input to investor decisions about capital allocation and climate risk management. The 2021 FSOC report included numerous recommendations about how to fill these climate-related data and methodological gaps. These recommendations were that FSOC member agencies:
>
> - "Promptly identify and take the appropriate next steps toward ensuring that they have consistent and reliable data to assist in assessing climate-related risks;
> - Use existing authorities to implement appropriate data- and information-sharing arrangements to facilitate the sharing of climate-related data across FSOC members and non-FSOC member agencies to assess climate-related financial risk, consistent with data confidentiality requirements;
> - Coordinate efforts, as appropriate, to address data gaps, including prioritizing data sets and coordinating data acquisition, in order to avoid duplication of effort and facilitate the improvement and coordinated use of data and models across FSOC members;
> - Move expeditiously to develop consistent data standards, definitions, and relevant metrics, where possible and appropriate, to facilitate common

definitions of climate-related data terms, sharing of data, and analysis and aggregation of data; and

- Continue to coordinate with their international regulatory counterparts, bilaterally and through international bodies, as they identify and fill data gaps, address data issues, and develop definitions, data standards, metrics, and tools" (pp. 6–7).

Recommendation 11-4: *Fill Gaps in Federal Financial Risk Data and Information Collection Rules.* **Federal agency decision makers that are members of the Financial Stability Oversight Council (FSOC) should work to implement the recommendations in the 2021 FSOC report to fill climate-related data and methodological gaps and to enhance public climate-related disclosures.**

Recommendation 11-5: *Strengthen Climate Disclosure Rules and Standardize Data and Methods.* **The Securities and Exchange Commission (SEC) should continually strengthen climate disclosure rules within the bounds of its mandate, with particular attention to standardizing data and methods where possible. The Commodity Futures Trading Commission and SEC should develop standardized reporting and tracking of voluntary offset use for firms pursuing net-zero voluntary commitments.**

As the energy transition proceeds and the effects of climate change become more immediate, investors may demand compensation for holding climate risk, and look to hedge this risk by reducing exposure and increasing the required return. There is documented evidence of climate regulatory risks causally affecting bond credit ratings and yield spreads (Seltzer et al. 2022). Attention to the data and methods for reporting and tracking voluntary offset use bears particular mention. Offsets fund projects that reduce emissions in different ways—for example, by replacing carbon-based electricity through funding renewable energy projects or by removing and sequestering carbon through biological or engineering approaches. Government agencies cannot provide certainty where it does not fundamentally exist—for example, about the role of particular kinds of offsets in a future regulatory regime. However, firms should not need to jump through hoops to understand the qualities of offsets they are buying, nor should investors need to do the same regarding the offsets held by firms with whom they invest or provide credit.

Such careful accounting of offset quality would allow different kinds of offsets to be increasingly aligned with different scopes. To achieve net-zero emission, truly permanent offsets will need to be aligned with scope 1 emissions. In advance of that, other types of offset requirements or actions might be aligned with scope 2 and 3 emission, but those will ultimately vanish as net-zero emissions are achieved. Clear disclosure of emission scope and offset qualities would facilitate such an alignment in advance of any policy decisions.

> Finding 11-5: There is currently no good model for estimating the price impacts of decarbonization transition risk, with the best means currently being through estimating differences in how risk is priced across relevant assets. Risk needs to be priced accurately to inform investor decision-making and capital allocation. Additionally, the timing and pace of decarbonization will itself affect climate impacts and future transition risk—for example, with rapid future decarbonization triggered by damages from near-term delays. Uncertainty remains as to whether the energy transition will occur with the rapidity to constitute a systemic risk to the financial system, but this is a topic that regulators are considering and is discussed below.

Financial-Sector Policies Beyond Disclosure

Beyond data collection and mandatory information disclosure, a number of other financial-sector policies have been discussed, proposed, or implemented. These include standardized scoring of climate-friendly activities, regulation of voluntary carbon markets, climate risk monitoring and supervision, capital requirements for banks related to carbon risk, and public ownership and management of fossil assets. The committee briefly discusses the first four topics below, while public ownership and management of fossil assets is discussed in Chapter 12. The first two policies discussed here—standardized scoring of climate-friendly activities and regulation of voluntary carbon markets—address the interest of financial-sector actors seeking to understand the climate change risks and opportunities associated with different economic activities or offsets. The other two discussed here—climate risk monitoring/supervision and capital requirements for banks—focus more on the overall performance of financial markets themselves.

Standardized Scoring of Climate-Friendly Activities

While mandatory information disclosure standardizes the information presented by firms subject to financial regulation, this policy does not attempt to digest that information into any kind of standard investment guidance. Investors themselves must decide how to value each piece of information.

The European Union has taken this a step further with its Taxonomy Regulation:[16]

> The EU taxonomy is a classification system, establishing a list of environmentally sustainable economic activities [that] would provide companies, investors and policymakers with appropriate definitions for which economic activities can be considered environmentally sustainable. In this way, it should create security for investors, protect private investors from greenwashing, help companies to become more climate-friendly, mitigate market fragmentation and help shift investments where they are most needed. (EC n.d.)

Organizations' economic activities are evaluated against six environmental objectives: climate mitigation, climate adaptation, water use and protection, waste (and the circular economy), pollution prevention and control, and biodiversity. Activities are broadly categorized as either eligible or ineligible contributors to these environmental objectives, and then must meet specific technical screening criteria in order to be considered "aligned" with them. In particular, the activity must (1) make a substantial contribution to one of the six objectives, and (2) do no harm to the others (with the technical screening criteria making this explicit). In 2022, firms need to report only the share of their economic activity that is "taxonomy-eligible" with increasing requirements for alignment reporting in 2023 and 2024 (Pettingale et al. 2022).

There is considerable debate about the EU taxonomy (Pacces 2021; Schütze et al. 2020; Zachmann 2022). At its core, the taxonomy is designed to facilitate financial flows toward environmentally sustainable activities, including climate mitigation. However, Zachmann points out that a green premium, which might be 20 basis points, is equivalent to a $1 carbon price—hardly likely to alter investments beyond existing economic incentives. Perhaps more importantly, a binary measure of sustainability may be overly simplistic and create considerable disagreement about assigned scores. While research has estimated how current financing breaks down into aligned and unaligned flows (Alessi et al. 2019), future work will be needed to assess the actual impact of the taxonomy on changing these flows.

Voluntary Carbon-Offset Market Regulation

As noted elsewhere in the committee's report (e.g., Chapter 6 on the electric sector), many private-sector entities have made voluntary commitments to reduce their GHG emissions, often as an element of their ESG commitments. Many firms' commitments include a net GHG emissions target, with the possibility, if not expectation, that the achievement of the goal will rely at least in part on use of carbon offsets. Sales of

---

[16] See https://ec.europa.eu/sustainable-finance-taxonomy/home. South Africa is also adopting a taxonomy, but reporting is not required (National Treasury of South Africa 2022). In the United States, CFTC (2020) suggested a standardized taxonomy for climate risks but not this type of "score" for economic activities.

voluntary offsets approached $2 billion in 2021 (Donofrio et al. 2022) and have been forecast to possibly reach $50 billion by 2030 (Blaufelder et al. 2021).

Voluntary carbon-offset markets have faced skepticism pertaining to the quality of the offsets they are offering. Summarized in a discussion paper from the International Organization of Securities Commissions, vulnerabilities for these markets include credit integrity concerns (including, but not limited to, double-counting, transparency and verification of carbon reduction calculations, and conflicts-of-interest), issues pertaining to market structure (such as issues of legal clarity, data availability, and standardization), as well as the need for responsible, legible communication (IOSCO 2022).

Some have advocated that the U.S. federal government should establish standards for GHG emission offsets used in the voluntary market. Currently, a number of third-party standards have emerged for rating offsets, and several have announced recent efforts to strengthen integrity (Integrity Council for the Voluntary Carbon Market 2022). Fredman and Phillips (2022) discuss the many problems that have arisen among third-party standards and argue that rather than leaving such standard setting to civil society, the federal government could step in to provide official standards. Fredman and Phillips argue that CFTC could and should provide guidance on offset quality and registries while also addressing fraud, brokerage businesses, and derivatives. Recent announcements by CFTC indicate that such guidance may be forthcoming (Ellfeldt 2023).

At the same time, participants in the voluntary market can choose to purchase government-certified offsets (or even allowances) rather than those certified exclusively for voluntary markets (Broekhoff et al. 2019). Government-certified offset standards already exist for multiple regulated emission trading programs, including California's and the European Union's. However, Badgley et al. (2022) discuss similar problems with government-certified offsets. Haya et al. (2020) discuss ways to mitigate the problems of additionality (i.e., ensuring that an offset represents a reduction in emissions that would otherwise not occur), baselines (e.g., for measuring the size of an offset), and perverse incentives (e.g., offset activities that end up directly or indirectly increasing GHG emissions) but conclude that the risk of over-crediting can be reduced but not eliminated.

Firms, investors, policy makers, and other stakeholders may hold different views about the value of precision in addressing such issues, especially in the context of voluntary as opposed to mandatory carbon markets. These views reflect, in essence, the balance between the risk of over-crediting emissions reductions versus the costs of reducing such risks.

Perhaps more to the point, it is not clear what role government regulation should play in the certification of offsets in *voluntary* markets. With the exception of cap-and-trade allowances from a binding jurisdictional cap, government certification does not alter the fundamental trade-off between cost and integrity associated with offset projects that have been recognized for more than a decade (Hall 2007). Rather than attempting to arbitrate this debate, governments could pursue more modest steps by requiring standardized reporting and tracking of offset use, consistent with emerging disclosure requirements. Information that could be required by, say, the federal government might include the activity (or activities) leading to the creation of the offset (e.g., direct air capture, agricultural soils management, reforestation) and location, monitoring/reporting/verification methodologies and third-party certification, tracking and registry information, and cost or price paid.

Climate Risk Monitoring and Supervision

A different arena of potential policy and regulation is related to systemic risks to the financial system posed by both climate impacts and mitigation action. Such risk arises when large, interconnected banks or nonbank financial companies are, in turn, at risk of failing to provide needed services, including the clearing of payments, the provision of liquidity, and the availability of credit. The well-known example of this is the collapse of the market for mortgage-backed securities in 2008 that led to a broader financial crisis and, in turn, the Dodd-Frank Wall Street Reform and Consumer Protection Act, FSOC and other heightened efforts to monitor and supervise the financial sector. Smaller, regional risks are sometimes referred to as "sub-systemic" risks (CFTC 2020).

As noted above, FSOC issued a report in 2021 on climate-related risks to the financial sector. In addition to the findings and recommendations of that report related to collecting and disclosing climate-risk data and information, which were discussed previously in this chapter, FSOC also concluded that financial regulators should assess and mitigate climate-related risks to the stability of the nation's financial institutions and markets.

Importantly, the focus of these recommendations was *not* the feedback of financial sector action to climate change mitigation. Stiroh (2022) explains this distinction in terms of single- and double-materiality, where single-materiality focuses on impacts of firm (e.g., a bank's own) behavior—say, in reducing emissions in its portfolio—on the firm's own risks. Double-materiality considers the impact of the firm's (e.g., the bank's) financial activities on climate change itself (e.g., resulting from its portfolio). For supervisory institutions in the United States, whose mandate is financial stability, the relevant focus is single-materiality.[17]

---

[17] In other countries, such as members of the European Union and the United Kingdom, financial authorities also have secondary mandates to support other government policies.

Many of the FSOC recommendations in this arena revolve around financial regulators' use of scenario analysis and stress testing of financial institutions. In summer 2020, the Network for Greening the Financial System published a guide for such activities (NGFS 2020). This was followed in late 2020 and 2021 by pilots in France, Canada, and England, including a small number of banks (ACPR 2020; BOC 2022; BOE 2021). The European Central Bank completed analysis using existing data (Dunz et al. 2021). A key outcome of the pilot analyses has been the recognized need for development and standardization of methodologies for climate risk assessment and the availability of climate-related data (BOC 2022). In fall 2022, the Federal Reserve Board began a U.S.-based pilot scenario analysis exercise with six of the nation's largest banks (Federal Reserve System 2022a).

> **Recommendation 11-6:** *Implement Financial Stability Oversight Council (FSOC) Recommendations to Ensure the Stability of U.S. Financial Markets.* **Members of FSOC should work to implement the recommendations in their 2021 report to ensure the stability of U.S. financial markets in the face of climate risks. In particular, the Federal Reserve should build on its current pilot efforts to conduct and test scenario analysis to incorporate climate risks in their regular stress testing of large financial institutions.**

Capital Requirements for Banks

A step beyond monitoring and supervision would be to consider changes in regulators' determinations relating to capital requirements for banks. Banks hold capital to absorb unanticipated losses and allow them to continue to operate under conditions of stress. Regulators establish minimum capital requirements based on a bank's loan portfolio to ensure that they can withstand adverse shocks with a high-level of confidence. Gelzinis (2021) suggests a number of changes to the regulatory capital regime in the face of climate risks. Meanwhile, supervisory institutions in different countries are considering the implications of climate risk for their regulatory capital requirements (Holscher et al. 2022).

However, as Holscher et al. (2022) point out, capital requirements are a tool to address unexpected losses; other tools (such as risk-based pricing, loan-loss provisions) make more sense if the problem is higher expected losses. It remains unclear whether climate risks are affected more by increased variability in losses or changes in the mean. Moreover, if the goal is really to facilitate the low-carbon transition, it will be important to consider the mandate of the supervisory institution, noted above.

The preceding discussions have focused on providing information to investors and encouraging federal agencies charged with supervising the financial sector to appropriately consider climate risks. All of these discussions exist within various agencies' mandates to appropriately and equitably protect investors and society. This falls within ordinary prudential actions.

There have been efforts to limit these disclosure and supervisory activities by some sub-national governments. This would inevitably exacerbate climate risks to society through financial market channels and are not consistent with timely and efficient decarbonization efforts.

> Finding 11-6: Federal regulators charged with supervising the financial sector currently have the ability to exercise their responsibilities in ways that inform investors about climate-related risks of firms' activities and that assess financial markets' vulnerability to climate risks.

Table 11-1 summarizes all the recommendations in this chapter regarding how the financial sector and capital markets can support decarbonization across the economy.

SUMMARY OF RECOMMENDATIONS ON ALIGNING THE FINANCIAL SECTOR AND CAPITAL MARKETS WITH THE ENERGY TRANSITION

TABLE 11-1 Summary of Recommendations on Aligning the Financial Sector and Capital Markets with the Energy Transition

Short-Form Recommendation	Actor(s) Responsible for Implementing Recommendation	Sector(s) Addressed by Recommendation	Objective(s) Addressed by Recommendation	Overarching Categories Addressed by Recommendation
11-1: Expand and Extend Funding and Financing Assistance for Actions Benefiting Low-Income and Disadvantaged Households and Communities	Congress and the Environmental Protection Agency	• Buildings • Transportation • Finance • Non-federal actors	• Equity	Ensuring Procedural Equity in Planning and Siting New Infrastructure and Programs Reforming Financial Markets
11-2: Disclose Equity Indicators for Federal Funding of Clean Energy	Office of Management and Budget	• Electricity • Buildings • Transportation • Finance • Non-federal actors	• Equity	Ensuring Procedural Equity in Planning and Siting New Infrastructure and Programs Reforming Financial Markets
11-3: Address Limited Access Faced by Low-Income and Marginalized Households	Treasury Advisory Group on Racial Equity	• Finance	• Equity	Ensuring Procedural Equity in Planning and Siting New Infrastructure and Programs Reforming Financial Markets
11-4: Fill Gaps in Federal Financial Risk Data and Information Collection Rules	Federal agency decision makers that are members of the Financial Stability Oversight Council (FSOC)	• Finance		Rigorous and Transparent Analysis and Reporting for Adaptive Management Reforming Financial Markets

continued

TABLE 11-1 Continued

Short-Form Recommendation	Actor(s) Responsible for Implementing Recommendation	Sector(s) Addressed by Recommendation	Objective(s) Addressed by Recommendation	Overarching Categories Addressed by Recommendation
11-5: Strengthen Climate Disclosure Rules and Standardize Data and Methods	Securities and Exchange Commission and Commodity Futures Trading Commission	• Finance • Non-federal actors		Rigorous and Transparent Analysis and Reporting for Adaptive Management Reforming Financial Markets
11-6: Implement Financial Stability Oversight Council Recommendations to Ensure the Stability of U.S. Financial Markets	FSOC members and the Federal Reserve	• Finance		Reforming Financial Markets

REFERENCES

ACPR (French Prudential Supervision and Resolution Authority). 2020. "Scenarios and Main Assumptions of the ACPR Pilot Climate Exercise." Bank of France. https://acpr.banque-france.fr/en/scenarios-and-main-assumptions-acpr-pilot-climate-exercise.

Alessi, L., S. Battiston, A.S. Melo, and A. Roncoroni. 2019. "The EU Sustainability Taxonomy: A Financial Impact Assessment." European Commission Joint Research Centre. https://data.europa.eu/doi/10.2760/347810.

Atkins, B. 2020. "Demystifying ESG: Its History and Current Status." *Forbes*, June 8. https://www.forbes.com/sites/betsyatkins/2020/06/08/demystifying-esgits-history—current-status.

Aurand, A., D. Emmanuel, M. Clarke, I. Rafi, and D. Yentel. 2022. "The Gap: A Shortage of Affordable Rental Homes." National Low Income Housing Coalition. https://nlihc.org/gap.

Badgley, G., J. Freeman, J.J. Hamman, B. Haya, A.T. Trugman, W.R.L. Anderegg, and D. Cullenward. 2022. "Systematic Over-Crediting in California's Forest Carbon Offsets Program." *Global Change Biology* 28(4):1433–1445. https://doi.org/10.1111/gcb.15943.

Barclays. 2022. "The Transition Mission: 10 ESG Themes for 2022." https://www.cib.barclays/content/dam/barclaysmicrosites/ibpublic/documents/our-insights/ESG10trends2022/Research-2022-ESG-Themes-Road-Survey-Slideshow.pdf.

Berlin, M. 2022. "The CEO Imperative: US Executives Recalibrate Risk Radar." *CEO Magazine North America*. https://ceo-na.com/opinion/the-ceo-imperative-us-executives-recalibrate-risk-radar.

BIL (Bipartisan Infrastructure Law) Summary. 2021. *Bipartisan Infrastructure Investment and Jobs Act Summary: A Road to Stronger Economic Growth.* Maria Cantwell, U.S. Senator for Washington. https://www.cantwell.senate.gov/imo/media/doc/Infrastructure%20Investment%20and%20Jobs%20Act%20-%20Section%20by%20Section%20Summary.pdf.

Blaufelder, C., C. Levy, P. Mannion, and D. Pinner. 2021. "A Blueprint for Scaling Voluntary Carbon Markets to Meet the Climate Challenge." McKinsey. https://www.mckinsey.com/capabilities/sustainability/our-insights/a-blueprint-for-scaling-voluntary-carbon-markets-to-meet-the-climate-challenge.

BOC (Bank of Canada). 2022. "Using Scenario Analysis to Assess Climate Transition Risk: Final Report of the BoC-OSFI Climate Scenario Analysis Pilot." https://www.bankofcanada.ca/wp-content/uploads/2021/11/BoC-OSFI-Using-Scenario-Analysis-to-Assess-Climate-Transition-Risk.pdf.

BOE (Bank of England). 2021. "Guidance for Participants of the 2021 Biennial Exploratory Scenario: Financial Risks from Climate Change." https://www.bankofengland.co.uk/-/media/boe/files/stress-testing/2021/the-2021-biennial-exploratory-scenario-on-the-financial-risks-from-climate-change.pdf.

Broekhoff, D., M. Gillenwater, T. Colbert-Sangree, and P. Cage. 2019. "Securing Climate Benefit: A Guide to Using Carbon Offsets." Stockholm Environment Institute and Greenhouse Gas Management Institute. http://www.offsetguide.org/wp-content/uploads/2020/03/Carbon-Offset-Guide_3122020.pdf.

Broyles, P.A. 1998. "The Impact of Shareholder Activism on Corporate Involvement in South Africa During the Reagan Era." *International Review of Modern Sociology* 28(1):1–19. https://www.jstor.org/stable/41421629.

Carpenter, S., A. Nikolov, S. Norris, and A. Shott. 2021. "2021 CDFI Survey Key Findings." Fed Communities. https://fedcommunities.org/data/2021-cdfi-survey-key-findings.

CEQ (Council on Environmental Quality). 2022a. "Downloads." Climate and Economic Justice Screening Tool. https://screeningtool.geoplatform.gov/en/downloads.

CEQ. 2022b. "Explore the Map." Climate and Economic Justice Screening Tool. https://screeningtool.geoplatform.gov.

CFPB (Consumer Financial Protection Bureau). n.d. "About Us." https://www.consumerfinance.gov/about-us. Accessed March 2, 2023.

CFTC (Commodity Futures Trading Commission). 2020. *Managing Climate Risk in the U.S. Financial System, Report.* Washington, DC. https://www.cftc.gov/sites/default/files/2020-09/9-9-20%20Report%20of%20the%20Subcommittee%20on%20Climate-Related%20Market%20Risk%20-%20Managing%20Climate%20Risk%20in%20the%20U.S.%20Financial%20System%20for%20posting.pdf.

Climate-Related Market Risk Subcommittee. 2020. *Managing Climate Risk in the U.S. Financial System.* Washington, DC: U.S. Commodity Futures Trading Commission.

Dai, R., R. Duan, H. Liang, and L. Ng. 2021. "Outsourcing Climate Change." European Corporate Governance Institute—Finance Working Paper No. 723/2021. https://doi.org/10.2139/ssrn.3765485.

Davidson, C. 2018. "Lack of Access to Financial Services Impedes Economic Mobility." Federal Reserve Bank of Atlanta. https://www.atlantafed.org/economy-matters/community-and-economic-development/2018/10/16/lack-of-access-to-financial-services-impedes-economic-mobility.

Department of the Treasury. 2022. "Fact Sheet: Four Ways the Inflation Reduction Act's Tax Incentives Will Support Building an Equitable Clean Energy Economy," https://home.treasury.gov/system/files/136/Fact-Sheet-IRA-Equitable-Clean-Energy-Economy.pdf.

Department of the Treasury. n.d. "Treasury Advisory Committee on Racial Equity." https://home.treasury.gov/about/offices/equity-hub/TACRE. Accessed January 11, 2023.

Donofrio, S., P. Maguire, C. Daley, C. Calderon, and K. Lin. 2022. "The Art of Integrity: State of the Voluntary Carbon Markets 2022 Q3." Ecosystem Marketplace.

Dunz, N., T. Emambakhsh, T. Hennig, M. Kaijser, C. Kouratzoglou, and C. Salleo. 2021. "ECB's Economy-Wide Climate Stress Test." *ECB Occasional Paper No. 2021/281,* September. https://doi.org/10.2139/ssrn.3929178.

EC (European Commission). n.d. "EU Taxonomy for Sustainable Activities." Sustainable Finance. Accessed January 11, 2023.

Eccles, R., and S. Klimenko. 2019. "The Investor Revolution." *Harvard Business Review.* https://hbr.org/2019/05/the-investor-revolution.

Ellfeldt, A. 2023. "CFTC Chief Sees 'Critical Role' for Agency in Policing Carbon Markets." *Politico,* July 20. https://subscriber.politicopro.com/article/eenews/2023/07/20/cftc-chief-sees-critical-role-for-agency-in-policing-carbon-markets-00107219.

Elsevier. 2022. "Journal of Climate Finance." https://www.journals.elsevier.com/journal-of-climate-finance.

Elwood, D.T., and N.G. Patel. 2018. "Restoring the American Dream: What Would It Take to Dramatically Increase Mobility from Poverty?" U.S. Partnership on Mobility from Poverty. https://www.mobilitypartnership.org/restoring-american-dream.

EPA (Environmental Protection Agency). 2022. "GGRF Request for Information Final." https://www.regulations.gov/document/EPA-HQ-OA-2022-0859-0002.

EPA. 2023a. "Biden-Harris Administration Launches $7 Billion Solar for All Grant Competition for Residential Solar Programs That Lower Solar Costs for Families and Advance Environmental Justice Through Investing in America Agenda." https://www.epa.gov/newsreleases/biden-harris-administration-launches-7-billion-solar-all-grant-competition-fund.

EPA. 2023b. "Biden-Harris Administration Launches $20 Billion in Grant Competitions to Create National Clean Financing Network as Part of Investing in America Agenda." https://www.epa.gov/newsreleases/biden-harris-administration-launches-historic-20-billion-grant-competitions-create.

EPA. 2023c. "EPA Releases Framework for the Implementation of the Greenhouse Gas Reduction Fund as Part of President Biden's Investing in America Agenda." https://www.epa.gov/newsreleases/epa-releases-framework-implementation-greenhouse-gas-reduction-fund-part-president.

Fannie Mae. 2022. "Duty to Serve: Underserved Markets Plan 2022–2024." https://www.fhfa.gov/PolicyProgramsResearch/Programs/Documents/FannieMae2022-24DTSPlan-April2022.pdf.

Federal Reserve System. 2022a. "Financial Stability Report." https://www.federalreserve.gov/publications/files/financial-stability-report-20221104.pdf.

Federal Reserve System. 2022b. "Principles for Climate-Related Financial Risk Management for Large Financial Institutions." 87 FR 75267. Docket No. OP-1793. https://www.federalregister.gov/documents/2022/12/08/2022-26648/principles-for-climate-related-financial-risk-management-for-large-financial-institutions.

Federal Reserve System. 2023a. "Federal Reserve Board Provides Additional Details on How Its Pilot Climate Scenario Analysis Exercise Will Be Conducted and the Information on Risk Management Practices That Will Be Gathered Over the Course of the Exercise." January 17, 2023. https://www.federalreserve.gov/newsevents/pressreleases/other20230117a.htm.

Federal Reserve System. 2023b. "Pilot Climate Scenario Analysis Exercise: Participant Instructions. January 2023. https://www.federalreserve.gov/publications/files/csa-instructions.

Fink, L. 2020. "BlackRock Capital, Letter to CEOs." https://www.blackrock.com/us/individual/larry-fink-ceo-letter.

Fredman, A., and T. Phillips. 2022. "The CFTC Should Raise Standards and Mitigate Fraud in the Carbon Offsets Market." Center for American Progress. https://www.americanprogress.org/article/the-cftc-should-raise-standards-and-mitigate-fraud-in-the-carbon-offsets-market.

FSOC (Financial Stability Oversight Council). 2021. "Report on Climate-Related Financial Risk." https://home.treasury.gov/system/files/261/FSOC-Climate-Report.pdf.

Gelles, D., and H. Tabuchi. 2022. "How an Organized Republican Effort Punishes Companies for Climate Action." *The New York Times*, May 27. https://www.nytimes.com/2022/05/27/climate/republicans-blackrock-climate.html.

Gelzinis, G. 2021. "Addressing Climate-Related Financial Risk Through Bank Capital Requirements." Center for American Progress. https://www.americanprogress.org/article/addressing-climate-related-financial-risk-bank-capital-requirements.

Getter, D.E. 2022. "Community Development Financial Institutions (CDFIs): Overview and Selected Issues." Congressional Research Service. https://crsreports.congress.gov/product/pdf/R/R47217.

Goldstein, M., and M. Farrell. 2022. "BlackRock's Pitch for Socially Conscious Investing Antagonizes All Sides." *The New York Times*, December 23. https://www.nytimes.com/2022/12/23/business/blackrock-esg-investing.html.

Hall, D.S. 2007. "Offsets: Incentivizing Reductions While Managing Uncertainty and Ensuring Integrity." In *Assessing U.S. Climate Policy Options: A Report Summarizing Work at RFF as Part of the Inter-Industry U.S. Climate Policy Forum*. Resources for the Future. https://media.rff.org/documents/CPF_COMPLETE_REPORT.pdf.

Haya, B., D. Cullenward, A.L. Strong, E. Grubert, R. Heilmayr, D.A. Sivas, and M. Wara. 2020. "Managing Uncertainty in Carbon Offsets: Insights from California's Standardized Approach." *Climate Policy* 20(9):1112–1126. https://doi.org/10.1080/14693062.2020.1781035.

Henisz, W., T. Koller, and R. Nuttall. 2019. "Five Ways That ESG Creates Value Getting Your Environmental, Social, and Governance (ESG) Proposition Right Links to Higher Value Creation. Here's Why." *McKinsey Quarterly*, November. https://www.mckinsey.com/capabilities/strategy-and-corporate-finance/our-insights/five-ways-that-esg-creates-value.

Holscher, M., D. Ignell, M. Lewis, and K. Stiroh. 2022. *Climate Change and the Role of Regulatory Capital: A Stylized Framework for Policy Assessment*. Finance and Economics Discussion Series 2022-068. Washington, DC: Board of Governors of the Federal Reserve System. https://doi.org/10.17016/FEDS.2022.068.

Ilhan, E., P. Krueger, Z. Sautner, and L.T. Starks. 2021. "Climate Risk Disclosure and Institutional Investors." Swiss Finance Institute Research Paper No. 19-66. European Corporate Governance Institute—Finance Working Paper No. 661/2020. https://ssrn.com/abstract=3437178.

Integrity Council for the Voluntary Carbon Market. 2022. "The Core Carbon Principles Assessment Framework." https://icvcm.org/assessment-framework.

IOSCO (International Organization of Securities Commissions). 2022. "Voluntary Carbon Markets Discussion Paper." Board of the International Organization of Securities Commissions. Discussion Paper CR/06/22. https://www.iosco.org/library/pubdocs/pdf/IOSCOPD718.pdf.

Jenkins, J.D., E.N. Mayfield, J. Farbes, R. Jones, N. Patankar1, Q. Xu, and G. Schivley. 2022. "Preliminary Report: The Climate and Energy Impacts of the Inflation Reduction Act of 2022." Princeton, NJ: REPEAT Project. https://repeatproject.org/docs/REPEAT_IRA_Prelminary_Report_2022-08-04.pdf, p. 17.

Litterman, B. 2020. "Managing Climate Risk in the U.S. Financial System: Report of the Climate-Related Market Risk Subcommittee." Market Risk Advisory Committee of the U.S. Commodity Futures Trading Commission. https://www.cftc.gov/sites/default/files/2020-09/9-9-20%20Report%20of%20the%20Subcommittee%20on%20Climate-Related%20Market%20Risk%20-%20Managing%20Climate%20Risk%20in%20the%20U.S.%20Financial%20System%20for%20posting.pdf.

Litterman, R. 2020a. "Financial Regulation and Climate Risk Management." Interview with Robert Litterman. Harvard Business School's "Climate Rising" podcast. https://www.hbs.edu/environment/podcast/Pages/podcast-details.aspx?episode=15116819.

Litterman, R. 2020b. "Pricing Climate Risk in the Markets, with Robert Litterman." Resources Radio. https://www.resourcesmag.org/resources-radio/pricing-climate-risk-markets-robert-litterman.

Mills, C.K., and J. Scott. 2022. "Sustainable Affordable Housing: Strategies for Financing an Inclusive Energy Transition." Federal Reserve Bank of New York. https://www.newyorkfed.org/medialibrary/media/outreach-and-education/community-development/fed-affordable-housing-and-energy-transition-final-10-4-22.

NASEM (National Academies of Sciences, Engineering, and Medicine). 2021. *Accelerating Decarbonization of the U.S. Energy System*. Washington, DC: The National Academies Press. https://doi.org/10.17226/25932.

National Treasury of South Africa. 2022. "Media Statement: South Africa's Green Finance Taxonomy Comparison to the EU Taxonomy, to Promote International Green Finance Flows." https://www.treasury.gov.za/comm_media/press/2022/2022111101%20Media%20statement%20-%20Green%20Finance%20Taxonomy%20Treasury.pdf.

Newburger, E. 2023. "Power Reiterates Fed Is Not Going to Become a 'Climate Policy Maker.' " CNBC. https://www.cnbc.com/2023/01/10/powell-reiterates-fed-is-not-going-to-become-a-climate-policymaker.html.

NGFS (Network for Greening the Financial System). 2020. "Guide to Climate Scenario Analysis for Central Banks and Supervisors." https://www.ngfs.net/sites/default/files/medias/documents/ngfs_guide_scenario_analysis_final.pdf.

NYSE (New York Stock Exchange). 2020. "How Should Boards Approach ESG: We Get Perspectives from CERES." https://www.nyse.com/esg/issuer-insights/ceres.

Pacces, A.M. 2021. "Will the EU Taxonomy Regulation Foster Sustainable Corporate Governance?" *Sustainability* 13(21):12316. https://doi.org/10.3390/su132112316.

Pettingale, H., S. de Maupeou, and P. Reilly. 2022. "EU Taxonomy and the Future of Reporting." Harvard Law School Forum on Corporate Governance. https://corpgov.law.harvard.edu/2022/04/04/eu-taxonomy-and-the-future-of-reporting.

PRI (Principles for Responsible Investment). 2021. "Principles for Responsible Investment: An Investor Initiative in Partnership with UNEP Finance Initiative and the UN Global Compact." https://www.unpri.org/download?ac=10948.

Read, T. 2022. "A Sneaky Form of Climate Obstruction Hurts Pension Funds." *The New York Times*, September 17. https://www.nytimes.com/2022/09/17/opinion/climate-change-pension-texas-florida.html.

Schütze, F., J. Stede, M. Blauert, and K. Erdmann. 2020. "EU Taxonomy Increasing Transparency of Sustainable Investments." *DIW Weekly Report*. https://doi.org/10.18723/DIW_DWR:2020-51-1.

SEC (Securities and Exchange Commission). 2022a. "Enhanced Disclosures by Certain Investment Advisers and Investment Companies About Environmental, Social, and Governance Investment Practices, Investment Company Act." *Federal Register* 87:36654–36761. https://www.federalregister.gov/documents/2022/06/17/2022-11718/enhanced-disclosures-by-certain-investment-advisers-and-investment-companies-about-environmental.

SEC. 2022b. "The Enhancement and Standardization of Climate-Related Disclosures for Investors." 17 CFR 210, 229, 232, 239, and 249. https://www.sec.gov/rules/proposed/2022/33-11042.pdf.

SEC. 2022c. "SEC Proposes Rules to Enhance and Standardize Climate-Related Disclosures for Investors." Press Release. https://www.sec.gov/news/press-release/2022-46.

Seltzer, L., L.T. Starks, and Q. Zhu. 2022. *Climate Regulatory Risks and Corporate Bonds*. SSRN Scholarly Paper ID 3563271. Rochester, NY: Social Science Research Network. https://papers.ssrn.com/abstract=3563271.

Shukla, P.R., J. Skea, A. Reisinger, and R. Slade. 2022. "Climate Change 2022: Mitigation of Climate Change." International Panel on Climate Change. https://www.ipcc.ch/report/ar6/wg3/downloads/report/IPCC_AR6_WGIII_FullReport.pdf.

Sorkin, A.R., B. Warner, V. Giang, S. Kessler, S. Gandel, M.J. de la Merced, L. Hirsch, and E. Livni. 2022. "An Anti-E.S.G. Activist Takes on Apple and Disney." *The New York Times*, September 20. https://www.nytimes.com/2022/09/20/business/dealbook/anti-esg-campaign-fund-etf-disney-apple.html.

Stiroh, K.J. 2022. "Climate Change and Double Materiality in a Micro- and Macroprudential Context," Finance and Economics Discussion Series 2022-066. Washington, DC: Board of Governors of the Federal Reserve System. https://doi.org/10.17016/FEDS.2022.066.

TCFD (Task Force on Climate-Related Financial Disclosures). 2017. "Final Report: Recommendations of the Task Force on Climate-Related Financial Disclosures." Financial Stability Board. https://assets.bbhub.io/company/sites/60/2021/10/FINAL-2017-TCFD-Report.pdf.

TCFD. 2022. "2022 Status Report." Financial Stability Board. https://assets.bbhub.io/company/sites/60/2022/10/2022-TCFD-Status-Report.pdf.

UN (United Nations). 2006. "Secretary-General Launches 'Principles for Responsible Investment' Backed by the World's Largest Investors." Press Release, April 27. https://press.un.org/en/2006/sg2111.doc.htm.

Young, D., and M. Reeves. 2020. "The Quest for Sustainable Business Model Innovation." *BCG*. March 10. https://www.bcg.com/publications/2020/quest-sustainable-business-model-innovation.

Zachmann, G. 2022. "Europe's Sustainable Taxonomy Is a Sideshow." *Bruegel* (blog), February 22. https://www.bruegel.org/blog-post/europes-sustainable-taxonomy-sideshow.

Zaidi, A. 2020. "Mandates for Action: Corporate Governance Meets Climate Change." *Stanford Law Review Online*. https://www.stanfordlawreview.org/online/new-mandates-for-action.

Zucker, T., L. Lee, and E. Kim. 2022. "West Virginia v. EPA Casts a Shadow over SEC's Proposed Climate-Related Disclosure Rule." Harvard Law School Forum on Corporate Governance. https://corpgov.law.harvard.edu/2022/08/03/west-virginia-v-epa-casts-a-shadow-over-secs-proposed-climate-related-disclosure-rule.

The Future of Fossil Fuels

ABSTRACT

Today, fossil fuels provide most of the nation's energy supply, and their production, delivery, and use lead to the vast majority of U.S. greenhouse gas (GHG) emissions and ambient air pollution.[1] Most of those CO_2 and ambient air-pollutant emissions result from use of fossil fuels, rather than their production and transportation, with methane leaks during natural gas production, processing, and delivery being the most significant exception. Other chapters of this report address the use of fossil fuels (e.g., in buildings, transport, and power production), while this chapter addresses the fossil fuel industries themselves.

Coal production and employment have declined dramatically over many decades, with economic impacts largely in Appalachian, Midwestern, northern Great Plains, and Rocky Mountain states. In contrast, domestic natural gas and oil production almost doubled over the past 15 years. Looking ahead over the next dozen years, coal use is expected to continue to decline, while oil and natural gas are likely to remain relatively flat and decline somewhat after 2030. Beyond 2030 and even during the current decade, the outlook for fossil fuels is highly uncertain and depends on how world events affecting global energy markets unfold; how fossil fuel prices and production costs vary relative to one another and relative to other energy resources; the development, economics, and deployment of energy technologies and electrification in vehicles, power plants, buildings, and industrial applications; the commercial and political viability of carbon capture; the build-out of energy delivery infrastructure; and other factors.

Further action is necessary to reduce emissions from the production and delivery of fossil fuels and to address transition impacts in regions tied to fossil fuel production; such transitions have been occurring for decades in coal-mining communities and may accelerate after the 2020s in communities where oil and gas are produced. Advanced notice, planning, and preparation will be important to mitigate harms to affected workers and communities.

This chapter overviews developments in fossil fuel markets and policy that have occurred since the committee's first report; assesses the extent to which these policies

---

[1] Ambient air pollution refers to emissions of the criteria air pollutants regulated by the Environmental Protection Agency (EPA) through the National Ambient Air Quality Standards; these pollutants include ground-level ozone, particulate matter, carbon monoxide, lead, sulfur dioxide, and nitrogen dioxide.

and market conditions put the United States on track toward decarbonization by mid-century; discusses what still needs to be done to address barriers and impediments to action in these sectors consistent with a net-zero economy; and provides findings and recommendations. Key recommendations relate to funding and capacity building for fossil fuel communities in transition and for the states where those communities are located; advance notice of facility closures; use of public revenues from taxes and fees collected on fossil fuel extraction and power production in the near term to assist in these transitions; increased analysis and planning to understand the timing and location of communities and regions that will be affected by transitions in infrastructure and extraction; and consideration of GHG emissions in public determinations about proposals for fossil fuel infrastructure additions.

INTRODUCTION

Given the fundamental role that fossil fuel combustion plays in producing GHG emissions, the outlook for production, delivery, and use of fossil fuels is central to the success of decarbonization pathways. In 2022, CO_2 emissions from consumption of oil, natural gas, and coal in the United States were 2,273 million metric tons (MMT), 1,746 MMT, and 935 MMT, respectively (EIA n.d.). Fossil fuels dominate energy consumption across all sectors, as shown in Figure 12-1, accounting for nearly four-fifths of total U.S. energy use as of 2020.[2]

- *Oil:* Approximately one-third of all energy consumption in the United States relies on oil, with the transportation sector almost entirely reliant on petroleum as its energy source. Two-thirds of petroleum use goes to fueling vehicles, and another fourth is used in industrial applications, with the remaining 6 percent used for heating in homes and commercial buildings and power plants providing peaking power (Conte 2021).
- *Natural gas:* Another one-third of U.S. energy use depends on natural gas, with consumption spread across multiple sectors. Thirty-eight percent of natural gas consumption is used to produce power. One-third is consumed

---

[2] Note that Figure 12-1 shows the sources and disposition of energy consumed in the United States. It does not show total U.S. energy production, some of which is exported for use in other countries. Additionally, some domestic fossil fuel consumption comes from imports. For example:
- *Oil:* Historically, the United States has been a net importer of oil and petroleum products, but in recent years the nation has exported approximately the same volumes of oil products as it has imported (EIA 2022a).
- *Natural gas:* The United States has been a net exporter of natural gas since 2017. As of 2021–2022, exports accounted for approximately 10 percent of domestic gas production (EIA 2023a).
- *Coal:* In 2021, the United States exported approximately 14 percent of domestically produced coal (EIA 2022b).

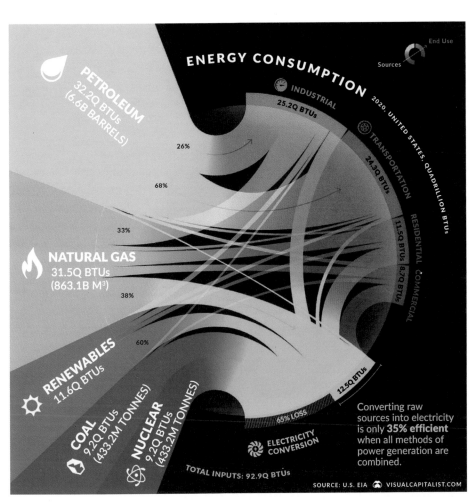

FIGURE 12-1 U.S. energy consumption by energy source and end-use sector. SOURCE: Conte (2021), *Visual Capitalist*.

by industries for energy (e.g., process heating, cogeneration of power and heat) and as a feedstock to produce other products (e.g., fertilizers, chemicals, hydrogen) (Conte 2021). The remaining natural gas consumption in the United States occurs in buildings for heat, hot water, cooking, and other energy services for households and businesses (EIA 2022i).

- *Coal:* Approximately 10 percent of total domestic energy consumption relies on coal. Over 90 percent of coal is used to produce power, with most of the rest going to export markets (Conte 2021).

The outlook for fossil fuel production and use in the United States varies among oil, gas, and coal and by different assumptions about how world events will unfold; how fossil fuel prices and production costs will evolve; the development, economics, and deployment of energy technologies and of strategies to control emissions in vehicles, power plants, buildings, and industrial applications (including, e.g., through carbon capture and storage); the rate at which power lines and other delivery infrastructure are sited and built; and other factors.

These issues were complex and challenging at the start of 2021 when the committee published its first report. At that time and based on then-current law and policy, the Energy Information Administration (EIA) was projecting in its Reference Case that U.S. natural gas production and crude oil production would rise 14 percent and 10 percent, respectively, between 2022 and 2050 (EIA 2021a). The following year, EIA projected in its Reference Case that over that same period (2022–2050), gas production would rise 19 percent and oil production would increase 9 percent (EIA 2022b). EIA's side cases indicated significant variance from these Reference Case outlooks.

The upheaval in global energy systems and the price volatility in oil and gas markets brought about by Russia's war on Ukraine have made these issues even more complex and challenging.[3] Economic-, energy-, and national-security considerations in global markets for fossil fuels (Birol 2022; Bordoff 2022a,b; EFI Foundation 2023; Victor 2021, 2022) add significant uncertainty about future outlooks for coal, oil, and natural gas— all of which are in plentiful supply in the United States. On top of that are challenges, and therefore uncertainties, about nations' abilities to meet their climate commitments (Boehm et al. 2022; Gelles 2023).[4] Additionally, as discussed later in the chapter, recent federal and state actions—including the passage of the Infrastructure Investment and Jobs Act (IIJA) in 2021 and the Inflation Reduction Act (IRA) in 2022—and technology, cost, and market trends could impact U.S. demand for and production of fossil fuels. The most recent EIA Reference Case forecasts that while domestic consumption of natural gas and oil remain relatively flat between 2022 and 2050, domestic production of both fossil fuels will increase over that same period largely as a result of fuel

---

[3] "The global economy is facing significant challenges. Growth has lost momentum, high inflation has broadened out across countries and products, and is proving persistent. Risks are skewed to the downside. Energy supply shortages could push prices higher. Interest rates increases, necessary to curb inflation, heighten financial vulnerabilities. Russia's war in Ukraine is increasing the risks of debt distress in low-income countries and food insecurity. The world is coping with a massive energy price shock. Russia's war of aggression against Ukraine has provoked a massive energy price shock not seen since the 1970s. The increase in energy prices is taking a heavy toll on the world economy" (OECD 2022).

[4] Ahead of COP-27, the Secretary-General of the United Nations said, "the war in Ukraine is putting climate action on the back burner while our planet itself is burning. We even see backsliding in some areas of the private sector—namely, around fossil fuels—while the most dynamic climate actors in the business world continue to be hampered by obsolete regulatory frameworks, red tape and harmful subsidies that send the wrong signals to markets" (Secretary-General of the UN 2022).

exports—with a wide range of outlooks for U.S. production depending on the scenario, ranging, for example, from –40 percent to +52 percent for domestic oil production and –20 percent to +36 percent for domestic natural gas production (reflecting the ranges for low and high gas price scenarios, respectively) (EIA 2023c). Investment in fossil fuels continues apace, even as investment in non-fossil technologies is increasing. IEA's *World Energy Investment 2023* report indicates continued growth in clean energy technology investments, projecting $1.7 trillion USD globally in 2023, compared to about $1 trillion USD for fossil fuels (IEA 2023). Nonetheless, fossil fuel investments are more than twice the level needed in 2030 to meet the IEA's Net-Zero by 2050 Scenario, and oil and gas industry investments in low-carbon technologies remain, on average, at only 5 percent of spending on upstream oil and gas (IEA 2023). Recent analysis indicates that

> phasing out [fossil fuel] transition assets too early may result in wild swings in oil and gas prices, raising energy security and affordability concerns, similar to what has transpired since Russia invaded Ukraine. Oil and gas price volatility can be an additional incentive to accelerate the energy transition, but in the short term it may move the world further away from the path to net zero . . . [and] could result in social and economic pain . . . via a higher cost of living and lower economic growth, and may severely impair the ability of many emerging economies to invest in clean energy sources. (Jain and Palacios 2023)

Currently, fossil fuels provide several functions to the energy system. First, and most obvious, they are primary sources of energy with high energy density. Second, they serve as energy carriers—that is, means of moving energy between one location and another. The other major energy carrier used today is electricity, with the crucial difference that electricity is not a primary source of energy (at least for now, ahead of any role for electricity in producing green hydrogen) but rather a means of moving energy from where it is produced (which could be a fossil fuel power plant, a nuclear plant, or a solar/wind facility) to where it is consumed. Just as electric transmission and distribution lines act as the means of carrying electrical energy, natural gas and oil pipelines act as a means of carrying fossil energy. The third role that fossil fuels currently play is storage. The fossil fuel energy system's underground storage, pipelines, and tanks constitute a massive energy storage system, particularly suited for storing energy for long durations.

As the fossil fuel sector transitions, key uncertainties surround which parts of it will remain intact (presumably supplemented by offsetting carbon capture), which parts of it will be "generalized but replaced," and which parts will be "eliminated and replaced." The "eliminated and replaced" category refers to situations where renewable primary energy sources generate and move electric power, and then use it or store it with batteries or other means of storing electrical energy. "Generalized but replaced" refers to eliminating fossil fuels as a primary energy source but continuing to use renewable fuels as means of carrying and storing energy. These renewable fuels could be biomass-derived or synthetic hydrocarbons produced with renewable energy, which may have compositions very similar to the refined petroleum products in use today (e.g., diesel, gasoline, jet fuel) and therefore provide the same high energy density as fossil fuels. Alternatively, these renewable fuels could be chemical energy carriers (e.g., ammonia or hydrogen) that are quite distinct from current refined petroleum products. Even if fossil fuels are largely eliminated, the mix of energy carriers that society ultimately adopts will have significant implications for regional economic impacts and jobs, as they influence the sectors involved in refining (e.g., synthetic fuel production), midstream transmission and distribution (e.g., pipelines carrying oil, natural gas, hydrogen, ammonia), and energy use. In addition, they influence the extent to which society creates a potentially massive new sector involving carbon capture, utilization, and storage that does not exist today.

> Finding 12-1: In addition to being primary sources of energy, fossil fuels currently also serve two other functions: energy storage and energy carriers. These latter functions would also be served by renewable fuels—that is, chemical energy carriers that are synthesized (rather than extracted as in today's fossil fuels) in a manner that generates no net GHG emissions on a life-cycle basis. Such renewable fuels could be "drop-ins"—that is, synthetic hydrocarbons that are very similar in composition to gasoline, jet fuel, or natural gas currently derived from fossil fuels, but manufactured using biological sources or from hydrogen extracted from water and captured atmospheric CO_2. Alternatively, renewable fuels could consist of chemical energy carriers that currently do not have wide societal usage, such as hydrogen or ammonia.

This chapter examines key considerations about the future of fossil fuels as part of decarbonizing the U.S. economy:

- A brief review of relevant recommendations from the committee's first report;
- An overview of developments in policy and markets that have occurred since that first report was published in early 2021;
- An assessment of the extent to which these policies and market conditions put the United States on track toward decarbonization by midcentury;
- A discussion of what still needs to be done to address barriers and impediments to action in these sectors to put them on a path to a midcentury net-zero economy; and
- Findings and recommendations (included in the relevant sections above). Table 12-5, at the end of the chapter, summarizes all the recommendations that appear in this chapter regarding the future of fossil fuels.

PRIOR COMMITTEE RECOMMENDATIONS RELATED
TO FOSSIL FUEL INDUSTRIES

The committee's first report did not specifically examine what happens to the fossil fuel industries themselves as the nation navigates paths to decarbonization but focused instead on policies in various other sectors that would naturally affect demand for and use of fossil fuels. These policies included:

- An economy-wide price on GHG emissions;
- A clean energy standard for reducing emissions in the power sector;
- Deployment of electric transmission infrastructures to support access to renewables;
- Support for zero- and low-carbon energy technology research, development, demonstration, and deployment (RDD&D);
- Energy efficiency standards and programs to reduce energy use in buildings;
- Performance standards for GHG emissions from motor vehicles;
- Planning and assessment of requirements for a national CO_2 transport network and characterization of geologic storage reservoirs; and
- Low-carbon standards for federal procurements of goods and services (NASEM 2021a).

The committee also made recommendations to address social and economic challenges associated with communities whose economies, social structures, and cultures have been tied to either coal extraction or oil or natural gas production. These included:

- The establishment of a White House Office of Equitable Energy Transitions to report annually on energy equity indicators and triennially on transition impacts and opportunities (among other things);
- The establishment of a National Transition Corporation to ensure coordination and funding for job loss mitigation, critical infrastructure siting and deployment, and equitable access to economic opportunities, and to create public energy equity indicators;
- Funding to support subnational entities' planning for the net-zero transition;
- The creation and funding of 10 regional centers to manage socioeconomic dimensions of the net-zero transition and of a net-zero transition office in each state capital; and
- The funding of local community block grants for planning and to help identify especially at-risk communities (NASEM 2021a).

RECENT DEVELOPMENTS AND EVENTS DIRECTLY AFFECTING FOSSIL FUEL INDUSTRIES

Market Conditions

Since the committee issued its first report, fossil fuel markets have reacted to the combined effects of the global COVID-19 pandemic and Russia's war on Ukraine. But trends in the domestic coal, oil, and natural gas markets were changing even before those events.

Coal Production/Consumption, Employment, and Prices

U.S. coal production and consumption peaked around 2008 (Figure 12-2), largely owing to transitions in the electric industry, where virtually all U.S. coal use occurs (EIA 2020; see Chapter 6). Appalachian coal production had been relatively flat for more than 20 years before that (Tierney 2016), and with a few outlier years, coal-mining employment had declined across American coal-producing regions since the mid-1980s,[5] primarily owing to mechanization. During the past decade, the number of mines in the United States dropped from 1,229 (in 2012) to 512 (in 2021), and production declined 43 percent during that period (EIA 2013, 2022c).

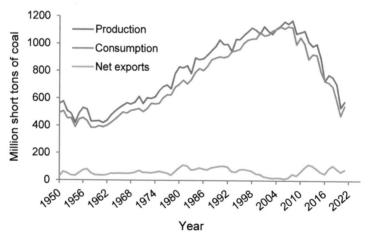

FIGURE 12-2 U.S. coal production (blue), consumption (orange), and net exports (green): annual, 1950–2021. SOURCE: Data from the U.S. Energy Information Administration, June 2022 (EIA 2022c).

---

[5] U.S. coal mining employment in September 2022 (41,000 people) was one-fifth of its levels in 1985 (approximately 178,000) (Federal Reserve Bank of St. Louis 2022).

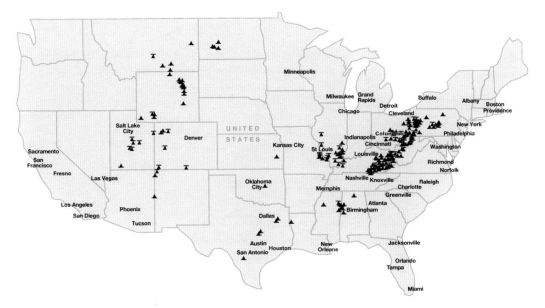

FIGURE 12-3 Coal mines (surface and underground) in the contiguous United States. SOURCE: Data from the U.S. Energy Information Administration, April 2023 (EIA n.d.).

Historically, coal production has occurred primarily east of the Mississippi River in Appalachia and to a lesser extent west of the Mississippi River, especially in the Rockies. Most coal mines (including legacy ones with no current production) are in Appalachia (Figure 12-3), but eastern coal production has been in decline for 3 decades, whereas western coal production increased until it peaked a decade ago (Figure 12-4). Coal mining communities in both the east and west have now experienced economic pressure for many years.

Employment in the coal industry has also been in decline for decades—long before the peak in production. Figure 12-5 shows the long-term trends in coal mining employment, with declines driven primarily by productivity improvements, especially as production shares shifted to the West, where surface-mining techniques required fewer employees per ton of coal produced (Tierney 2016).[6] On top of those productivity improvements, reduction in demand for coal during the past 2 decades further drove down both production and employment.

---

[6] "Coal productivity ranges significantly across production regions, with productivity in the Powder River Basin far exceeding productivity in the Interior and Appalachian regions. U.S. coal mining productivity has increased despite mine closings and decreasing employment and production. Technology and process improvements have contributed to the increase in productivity, but a larger factor is the distribution of productivity across mines. The mines that are first to close during market downturns are often the ones with higher production costs and lower productivity, while more productive mines remain operating, increasing overall productivity" (EIA 2018).

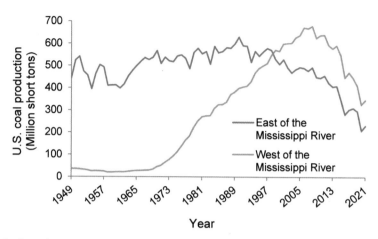

FIGURE 12-4 Coal production east (blue) and west (orange) of the Mississippi River: annual million short tons, 1949–2021. SOURCE: Data from the U.S. Energy Information Administration Report, October 2022 (EIA 2022a).

Coal-related economic activity more broadly—including mining, power production, and/or product manufacturing—occurs in almost every U.S. state, and the economies of nearly a quarter of the states are tied relatively tightly to coal—both as a fuel and as a source of electricity. This is especially true for Wyoming, West Virginia, and Kentucky, all of which experienced a decrease in coal-related activity in the past decade (Table 12-1). West Virginia has many more jobs in coal extraction compared to other

FIGURE 12-5 U.S. coal mining employment (thousand employees): 1990–2022. SOURCE: Data from the Bureau of Labor Statistics (BLS 2023a).

TABLE 12-1 Economic Activity Related to Coal: Ranking of Selected States

	Coal Production: Rank (in terms of % of U.S. production)		Coal-Fired Power Generation: Rank		Coal-Related Employment: Number	
			State's % of U.S. Coal-Fired Generation	State's % Dependence on Coal-Fired Generation	Jobs in Coal Industry	Jobs Related to Coal-Fired Generation
	2012	2021	2021	2021	2021	2021
Wyoming	1 (39%)	1 (41%)	9 (3.6%)	3 (74%)	4,859	936
West Virginia	2 (12%)	2 (14%)	2 (6.6%)	1 (91%)	12,261	1,808
Kentucky	3 (9%)	7 (5%)	5 (5.6%)	4 (71%)	5,425	1,550
Pennsylvania	4 (5%)	3 (7%)	10 (3.3%)	29 (12%)	5,062	1,784
Illinois	5 (5%)	4 (6%)	7 (4.8%)	19 (24%)	2,355	2,068
Texas	6 (4%)	9 (3%)	1 (9.9%)	22 (18%)	2,690	4,689
Indiana	7 (4%)	8 (3%)	4 (6.1%)	6 (58%)	1,924	2,918
Montana	8 (4%)	5 (5%)	29 (1.2%)	9 (44%)	1,161	356
Colorado	9 (3%)	11 (2%)	15 (2.6%)	11 (42%)	1,431	2,467
North Dakota	10 (3%)	6 (5%)	14 (2.7%)	7 (57%)	1,306	869
Ohio	11 (3%)	16 (0.5%)	6 (5.1%)	12 (36%)	937	10,126
New Mexico	12 (2%)	14 (2%)	27 (1.4%)	13 (36%)	931	206
Other States with Relatively High Rank		Utah (10th) (2%), Virginia (12th) (2%), Alabama (13th) (2%)	Missouri (3rd) (6.4%) Michigan (8th) (4.1%) Wisconsin (11th) (3.1%) Alabama (12th) (3%) Utah (13th) (2.9%)	Missouri (2nd) (75%) Utah (5th) (62%) North Dakota (7th) (57%) Nebraska (8th) (50%) Wisconsin (10th) (43%)		

SOURCES: Data for EIA Annual Coal Reports from EIA (2013, 2022a). Data for EIA Electric Generation Data by State from DOE (2022a) and EIA (2023c).

states that ranked relatively high in coal production in 2021. Ohio has many more jobs tied to coal-fired power generation than any other state, even though it ranks sixth in terms of its share of U.S. coal-fired electricity production. Notably, Ohio has more than double the employment of Texas (which, at the first-ranked position, produces 10 percent of all coal-fired electricity in the United States) and employs more people in coal-fired electricity generation than the 2nd-to-5th ranked states (West Virginia, Montana, Indiana, and Kentucky) combined.

Although coal prices were relatively flat over much of the past decade (Bloomberg 2022), power production at coal plants faced stiff competition from favorable prices and power-production economics at gas-fired power plants and the entry of new wind and solar generating units, and thus demand for coal declined. The uptick in coal prices in 2022 has been attributed to increases in demand for coal and energy shortages in Europe and elsewhere in the wake of Russia's invasion of Ukraine (IEA 2022a).

Natural Gas Production/Consumption, Prices, and Employment

In contrast to coal trends in the 21st century, demand for natural gas has remained robust, especially since around 2010, with increases largely driven by power-sector consumption and to a lesser degree by exports of liquefied natural gas (LNG) in the past 5 years (Figure 12-6). As of 2021, electric power accounted for 36 percent of gas

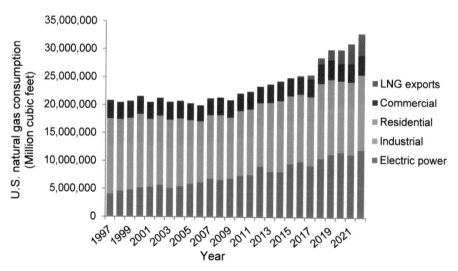

FIGURE 12-6 U.S. consumption of natural gas: 1997–2022 (annual MMcf). SOURCE: Data from the U.S. Energy Information Administration, December 2022 (EIA 2022h,k).

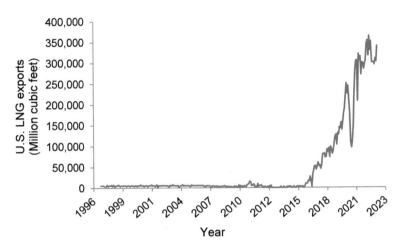

FIGURE 12-7 U.S. exports of LNG: 1997–2022. SOURCE: Data from the U.S. Energy Information Administration, December 2022 (EIA 2022h).

use; industrial use was 27 percent; and residential, commercial, and LNG were 15, 11, and 11 percent, respectively.

Most of this gas comes from domestic resources. Gas imports peaked in 2007, before U.S. production dramatically increased with the application of hydraulic fracturing (EIA 2022j). Domestic gas production has doubled since its recent low point in 2005 (EIA 2022j), with production relatively stable over the COVID-19 pandemic and recent upticks associated with global demand for U.S. LNG (Figures 12-6 and 12-7).

Once highly volatile, natural gas prices have been relatively stable and low over the past 15 years (as production increased since 2007)—again until global demand increased with the war in Ukraine and Russia's cutoff of natural gas supplies to Europe (Figure 12-8).

Also of note is the fundamentally different pricing structure of natural gas relative to fuels like gasoline or crude oil that are liquid at room temperatures and thus easily and inexpensively transported globally. The prices of these fuels reflect conditions in their global markets. By contrast, U.S. natural gas prices tend to be driven by conditions in domestic markets because the fuel (like ethane) is gaseous at room temperature and pressure and has high transportation costs owing to capital-intensive liquefaction and gasification facilities and special-purpose transportation ships. (Figure 12-9 shows the price of several hydrocarbons on an *energy basis*, which reflects a fuel's price relative to

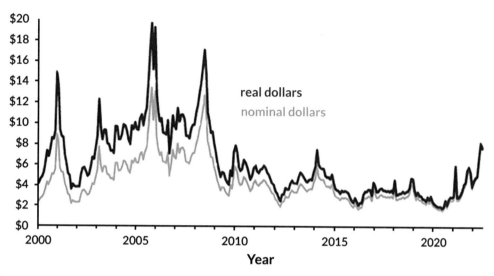

FIGURE 12-8 Monthly average natural gas price, January 2000–June 2022 ($/million BTUs). SOURCE: Data from the U.S. Energy Information Administration and Refinitiv Eikon, July 2022 (EIA 2022p).

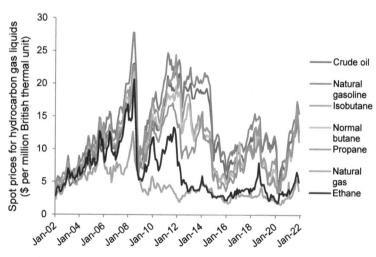

FIGURE 12-9 Price of various oil and natural gas commodities, January 2002–December 2021 ($/million BTUs). SOURCE: Data from the U.S. Energy Information Administration, January 2022 (EIA 2022g).

FIGURE 12-10 Shale gas plays, principal production regions, and pipelines. SOURCE: Adapted from S&P Capital IQ Pro Basemap, annotated with names of key shale gas production regions.

a given amount of energy it provides.) The structure of prices for these different fuels reflects their physical attributes as well as the economics of their markets.

Most of the natural gas resource base and production in the United States occurs in a few regions of the country. Figure 12-10 shows the location of shale gas plays,[7] which are served by an extensive network of high-pressure gas pipelines. In 2021, 79 percent of U.S. gas production was in shale formations (EIA 2022n), with approximately 80 percent of shale gas production occurring in a handful of regions: the Marcellus and Utica (Pennsylvania, West Virginia, Ohio, and New York) and the Permian, Haynesville, Eagle Ford, and Barnett areas of Texas, New Mexico, and Louisiana (Figure 12-11). Figure 12-12 shows the states with the most natural gas production in the past 15 years.

Employment in gas production—typically combined with employment in oil production given the co-location of these fuels in many regions of the United States—has shown the same kind of volatility as gas prices (Figure 12-13). The high prices during much of the 2000s, combined with the decrease in production costs brought about by hydraulic fracturing toward the end of that decade, spurred new gas production, and demand for workers increased in response.

---

[7] A shale gas play is defined as a "set of discovered, undiscovered or possible natural gas accumulations that exhibit similar geological characteristics" (DOE-FECM 2013).

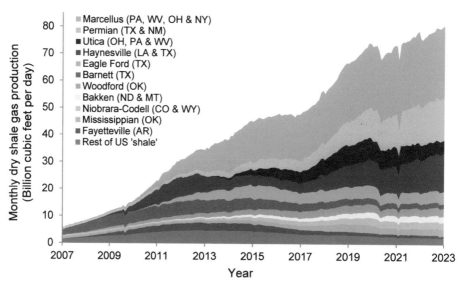

FIGURE 12-11 Monthly dry shale gas production by region, 2007–2022 (Bcf per day). SOURCE: Data from the U.S. Energy Information Administration, December 2022 (EIA 2022d).

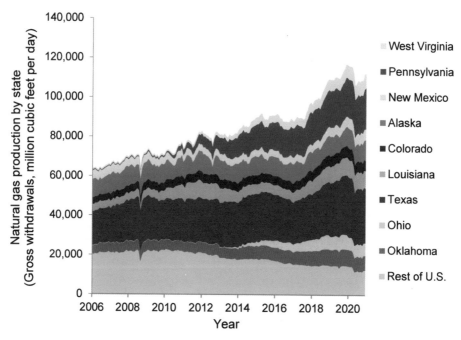

FIGURE 12-12 Monthly natural gas production by state: 2006–2020. SOURCE: Data from the U.S. Energy Information Administration, March 2021 (EIA 2021b).

FIGURE 12-13 Employment in U.S. natural gas and oil extraction: 1990–2022. SOURCE: Data from the Bureau of Labor Statistics (BLS 2023b).

Oil Production

Like natural gas, U.S. oil production has seen growth tied to the application of hydraulic fracturing in basins with unconventional oil more than doubling over the past 15 years (Figure 12-14). (Nationwide employment figures for oil production are included in Figure 12-13.) As of 2021, oil production occurred primarily in the Gulf of Mexico (serviced by firms and people in the Gulf States like Louisiana and Texas), Texas, North Dakota, New Mexico, Oklahoma, Colorado, Alaska, and California (Figure 12-15).

The impact of oil production on these economies has been significant. Most recent estimates of economic impact include the combined effects of oil and gas production and other activities (because many data sources combine the two). For example, as summarized in Table 12-2, a 2021 study by PwC estimated direct economic impacts of oil and gas subsectors in 2019. An RFF analysis found that the upstream and midstream oil and gas sectors had annual public revenues of $34 billion and $8 billion, respectively, on average between 2015 and 2020 (Raimi 2023).

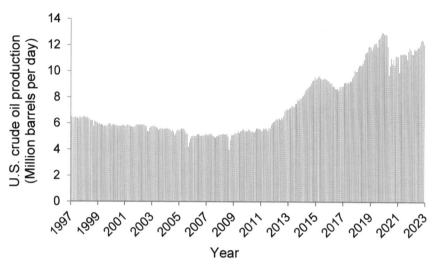

FIGURE 12-14 U.S. production of crude oil: January 1997–July 2022 (million barrels/day). SOURCE: Data from the U.S. Energy Information Administration, December 2022 (EIA 2022n).

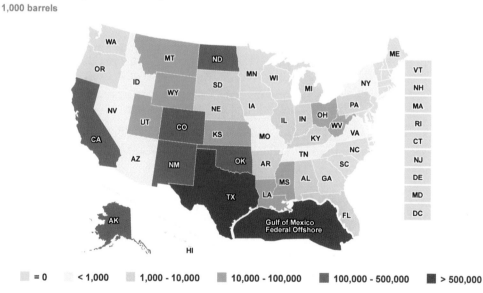

FIGURE 12-15 2021 Crude oil production by state (million barrels). SOURCE: Data from the U.S. Energy Information Administration, *Petroleum Supply Annual*, September 2022 (EIA 2022l).

TABLE 12-2 Direct Impact of Selected Oil and Natural Gas Industry Activities on the U.S. Economy in 2019

NAICS Code	Subsector	Jobs (thousands)[a]	Income from Labor (billions $)[b]	Added Value to Economy (billions $)
211	Oil and gas extraction	507.0	106.6	193.1
213111	Drilling oil and gas wells	68.7	8.2	10.7
213112	Support activities for oil and gas operations	305.6	32.9	43.0
32411	Petroleum refineries	73.1	20.9	136.3

[a] Includes payroll and self-employed jobs, as well as part-time jobs.
[b] Includes wages salaries, benefits, and proprietors' income.
SOURCE: PwC (2021) estimates based on 2019 employment from the U.S. Bureau of Economic Analysis and supplemented by data from the U.S. Bureau of Labor Statics and U.S. Census Bureau and 2019 input–output relationships from the IMPLAN modeling system.

Upstream activities (i.e., oil and gas extraction, drilling of oil and gas wells, and support activities for oil and gas operations) were concentrated most heavily in Texas, Oklahoma, Louisiana, and Colorado (PwC 2021, Table 7).

Trends in Fossil Fuels as of 2023

There are clear trends across the three fossil fuels: coal production continues to decline, while natural gas and oil production have not only avoided declines but in fact have increased in the past decade. U.S. domestic production of oil and natural gas is playing a strategic energy-security role internationally at present, with the United States positioned to rely on domestic production for its own supply and for serving export markets. This is a major change from earlier decades when the nation depended on imports of oil for most of its petroleum use.

Employment

In 2021, direct employment in fossil fuel industries amounted to 743,872, which was 0.5 percent of total employment and 9.5 percent of energy-sector jobs (DOE 2022d).[8] Employment across fossil fuel industries, especially in jobs related to extraction, has been declining.[9] The impacts of such trends are being experienced by communities in many states—in particular in Appalachia, parts of the Rockies, and the south-central parts of the United States—with a higher-than-average percentage of jobs involved in fossil fuels (not counting electricity generation) (Table 12-3). The economies of Alaska, Louisiana, North Dakota, New Mexico, Oklahoma, Texas, West Virginia, and Wyoming currently depend on activities in the fossil fuel sector. However, given the recent

---

[8] The Department of Energy's *United States Energy and Employment Report: 2022* (DOE 2022d, p. 1) defines "energy sector jobs" as "all the professional, construction, utility, operations, and production occupations associated with energy infrastructure, production, and use" and more specifically:

- Electric Power Generation: "jobs across all electric generating technologies. This covers both utility and non-utility generation . . . [and] employment in any firms engaged in the manufacture, operation, and/or maintenance of turbines and other generating equipment, as well as those engaged in the construction and installation of electricity generation plants or other sources of electricity (e.g., solar panels), capital investments, and wholesale parts distribution for all electric generation technologies" (p. 11).
- Fuels: "any work related to fuel extraction, mining, and processing. This includes firms that manufacture machinery that supports oil and gas extraction, as well as coal mining. Agriculture and forestry workers who support fuel production with biodiesels, corn ethanol, and fuel wood are also included . . . [as are jobs for] the production of nuclear fuels for power plants. Jobs in electricity fuel to power vehicles and buildings are reflected in the Electric Power Generation section" (p. 78).
- Motor Vehicles: This includes "the manufacture of new vehicles and parts, construction of manufacturing facilities, and vehicle repair services" (p. 141).
- Energy Efficiency: "employment in the production and installation of products that increase energy efficiency and the provision of services that reduce energy consumption by the end-user. These jobs include building design and contracting services that provide insulation, improve natural lighting, and otherwise reduce energy consumption in residential and commercial areas. Additionally, this sector includes employment in the manufacture of ENERGY STAR labeled products" (p. 130).
- Transmission, Distribution, and Storage: "employment associated with constructing, operating, and maintaining energy infrastructure [including] electric transmission lines, pipeline construction, fuel distribution and transport, and the manufacture of equipment used for electrical transmission. Also included in this sector is employment related to storage technologies such as batteries, pumped storage, compressed air, and other utility-level storage methods. The TDS sector includes both legacy power lines and newer technologies such as microgrids and smart grids" (p. 65).

[9] From 2020 to 2021, for example, and across all jobs in energy "fuels" (including renewable-energy fuels, nuclear fuels, and fossil fuels), the jobs lost were in extraction, which saw a decrease of 46,007 jobs (12 percent). Of these, the majority were in petroleum (31,593, a 6.4 percent decrease), although the greatest percent decrease occurred in coal jobs (a loss of 7,125, or 12 percent) (DOE 2022a).

TABLE 12-3 Jobs in Fossil Fuel Sector: U.S. Totals and Selected States with Above-Average Fossil Fuel Jobs (2021)

	Jobs in Fossil Fuel Industries	Fossil Fuel Jobs as % of Energy-Related Jobs	Fossil Fuel Jobs as % of Total Jobs	Energy Sector in the State with Highest Number of Energy Jobs (2021)
United States	743,872	9.5%	0.5%	Motor Vehicles (2,553,368 jobs) Energy Efficiency (2,164,914 jobs)
Alaska	10,175	41.1%	3.3%	Fuels
Colorado	22,330	15.3%	0.8%	Energy Efficiency
Kansas	9,076	11.7%	0.7%	Motor Vehicles
Louisiana	55,397	37.6%	3.0%	Fuels
Mississippi	6,777	10.1%	0.6%	Motor Vehicles
Montana	4,486	14.5%	0.9%	Transmission, Distribution, Storage
New Mexico	22,161	38.3%	2.8%	Fuels
North Dakota	23,659	46.7%	5.9%	Fuels
Oklahoma	48,258	38.8%	3.1%	Fuels
Pennsylvania	30,859	12.0%	0.5%	Motor Vehicles
Texas	247,604	28.1%	2.0%	Fuels
West Virginia	19,978	27.5%	3.0%	Transmission, Distribution, Storage
Wyoming	16,912	40.6%	6.4%	Fuels

SOURCE: Data from DOE (2022a,d).

increases in production in response to Russia's war on Ukraine, fossil fuel employment was expected to increase in 2022 (DOE 2022d). (Note that for fuels, these employment statistics do not reflect jobs in service sectors associated with these fuels. For example, repair of internal-combustion vehicles using gasoline are reflected in the motor vehicles sector rather than the fuels category.)

As shown in Figure 12-16, a considerable portion of fossil fuel production takes place in areas that are currently considered economically distressed, especially in states close to the Gulf Coast, Appalachia, and parts of the southern Rockies (especially near tribal communities in Arizona and New Mexico).

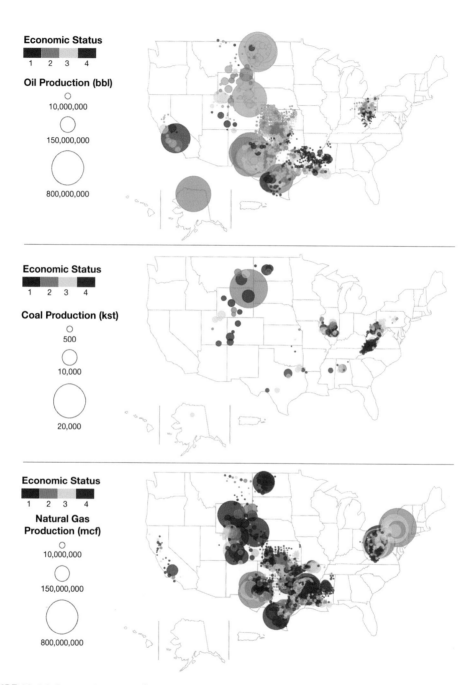

FIGURE 12-16 Economic status of regions with oil (top), coal (middle), and natural gas (bottom) production (2019). Economic status: index from 1 to 4 (4 = distressed) based on 3-year average unemployment rates, per capita market income, and poverty rates at the county level. Circle size reflects the volume of production, and circle color shows economic status, with red indicating the most vulnerable communities. SOURCE: Raimi (2021), https://www.rff.org/documents/3222/WP_21-36.pdf. CC BY-NC-ND 4.0.

Tax Revenues

More than two-thirds of all U.S. states collect revenues via severance or other related taxes on the production of non-renewable natural resources (e.g., oil, natural gas, coal). As shown in Figure 12-17, such taxes provide an important source of revenues to state budgets in several states. In 2019, four states depended on severance tax revenue for more than 5 percent of state and local general revenues: North Dakota (23 percent), Wyoming (8 percent), Alaska (7 percent), and New Mexico (6 percent) (Urban Institute 2019). Five others—Louisiana, Montana, Oklahoma, Texas, and West Virginia—depended on severance taxes for 1–5 percent of state revenue. As severance tax revenues are tied to the production volume and price, fossil fuel severance taxes represent a relatively volatile source of public funds.[10]

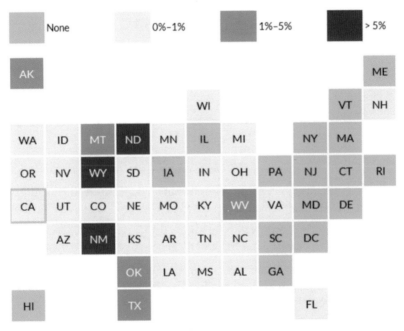

FIGURE 12-17 Severance tax revenue as a percent of state and local general revenues (2020).
NOTE: California does not have a severance tax but levies a fee on in-state oil and gas production.
SOURCES: Courtesy of Urban Institute (2020), with data from U.S. Census Bureau.

_____

[10] For example, Alaska's severance taxes on oil and gas production accounted for 72 percent of state revenues in 2014 and 7 percent in 2019. Severance tax revenues represented the following percentage of public revenues elsewhere: North Dakota (54 percent in 2014, 23 percent in 2019), Wyoming (39 percent in 2014, 8 percent in 2019), and Texas (11 percent in 2014, between 1–5 percent in 2019) (EIA 2015; Urban Institute 2019).

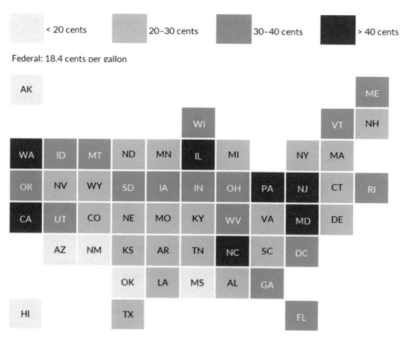

FIGURE 12-18 Motor fuel tax rates by state (as of January 2023). SOURCES: Courtesy of Urban Institute (2023), with data from FTA (2023).

A notable suite of taxes related to fossil fuels are federal and state motor vehicle and highway taxes collected on sales of gasoline and other petroleum products. Since 1993, the federal government has collected 18.4 cents per gallon of gasoline sold and 24.4 cents per gallon of diesel fuel sold. All states and the District of Columbia also collect taxes on fuels, ranging from a high of 57.6 cents per gallon (Pennsylvania) to a low of 8 cents per gallon (Alaska) (FTA 2023) (Figure 12-18). These taxes and fees provide substantial revenues to the federal government (i.e., infusions into the Highway Trust Fund that finances construction on the federal highway and transit systems) and the states. In 2021, federal highway taxes generated approximately $43 billion, and in 2020 (the latest year available), local and state fuel tax revenues amounted to another $52.7 billion (CBO 2022; Tax Policy Center 2022).

Emissions

GHG emissions from fossil fuel combustion are typically allocated to the economic sectors that use the fuels as sources of energy (see Figure 12-1 and the sectoral chapters of this report). However, methane leaks and flaring from oil and natural gas

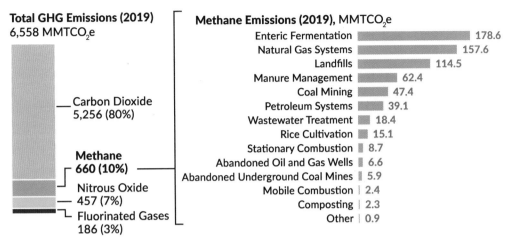

FIGURE 12-19 U.S. GHG emissions and sources of methane emissions (2019). SOURCES: CRS (2022) with data from EPA (2021b).

systems (e.g., production, processing, delivery, abandoned wells) accounted for approximately 3 percent of total U.S. GHG emissions in 2019 (Figure 12-19). Methane emissions, of which fossil fuel systems account for a substantial portion, amount to 10 percent of total U.S. GHG emissions, and methane is a potent GHG, with a 100-year global warming potential (GWP) of 27–30, compared to a reference value of 1 for CO_2 (EPA 2023b).

GHG emissions from coal mining have declined over the past few decades, in line with the decline in overall coal production levels (Figure 12-20). GHG emissions from natural gas and petroleum systems (e.g., production, transmission, distribution, storage) have remained relatively flat over the past 3 decades (Figure 12-21), with reductions in emissions from local gas distribution systems offset by increased emissions from production. As Figures 12-20 and 12-21 illustrate, the magnitude of annual GHG emissions from production, extraction, and transportation of fossil fuels (~300–350 MMT CO_2e) is much less than that from their combustion: 935 MMT CO_2 (coal), 1,746 MMT CO_2 (natural gas), and 2,273 MMT CO_2 (oil) in 2022 (EIA n.d.).

Areas of Appalachia and Louisiana and parts of Gulf Coast Texas with significant fossil fuel production also experience high concentrations of $PM_{2.5}$ (shown in red circles in Figure 12-22 below). $PM_{2.5}$ and other co-pollutants associated with GHG emissions from a fossil-fuel economy are important issues for air pollution in local communities and the environmental justice outcomes of the energy transition. (See Chapters 2 and 3 on equity and health, respectively.)

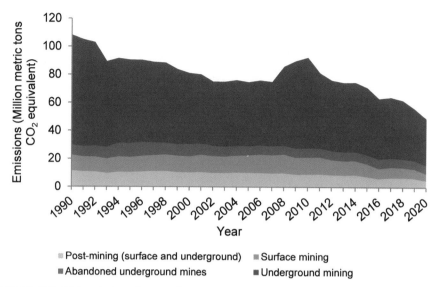

Post-mining (surface and underground)   ■ Surface mining
■ Abandoned underground mines   ■ Underground mining

FIGURE 12-20 U.S. GHG emissions from coal mining: 1990–2020. Not shown are minimal (<0.04 MMT/yr) carbon dioxide emissions from methane flaring. SOURCE: Data from EPA Greenhouse Gas Inventory Data Explorer (EPA n.d.(b)).

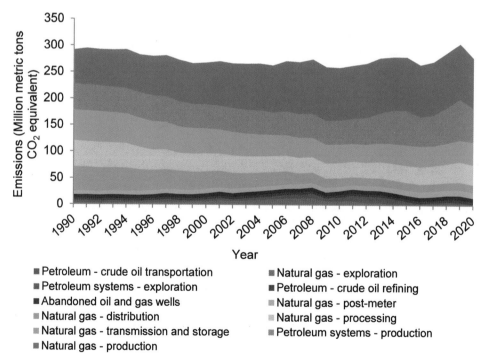

■ Petroleum - crude oil transportation   ■ Natural gas - exploration
■ Petroleum systems - exploration   ■ Petroleum - crude oil refining
■ Abandoned oil and gas wells   ■ Natural gas - post-meter
■ Natural gas - distribution   ■ Natural gas - processing
■ Natural gas - transmission and storage   ■ Petroleum systems - production
■ Natural gas - production

FIGURE 12-21 U.S. GHG emissions from natural gas and petroleum systems: 1990–2020. SOURCE: Data from EPA Greenhouse Gas Inventory Data Explorer (EPA n.d.(b)).

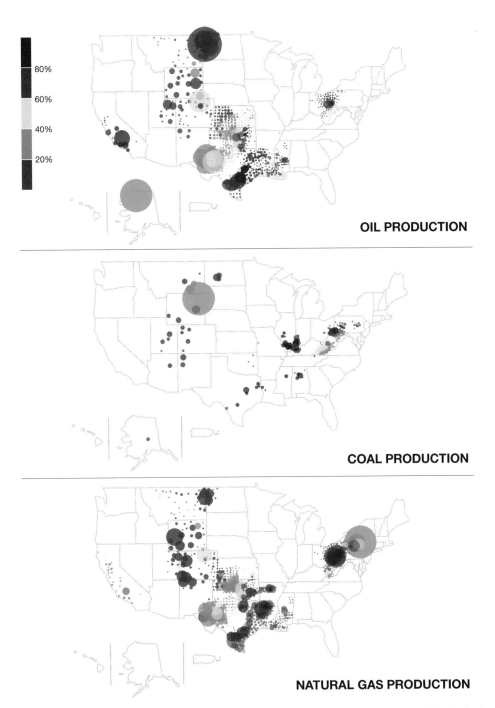

FIGURE 12-22 PM$_{2.5}$ levels in regions with oil, natural gas, and coal production (2019). NOTE: Circle size reflects the relative volumes of production and circle color shows the relative level of PM$_{2.5}$ (with dark red indicating higher levels of particulates). SOURCE: Raimi (2021), https://www.rff.org/publications/data-tools/mapping-vulnerable-communities. CC BY-NC-ND 4.0.

Finding 12-2: U.S. coal production and employment have declined dramatically over many decades, with economic impacts largely in Appalachian, Midwestern, northern Great Plains, and Rocky Mountain states. In contrast, domestic natural gas and oil production have almost doubled over the past 15 years. Most of the GHG and ambient air-pollutant emissions associated with fossil fuels result from their use rather than from their production/extraction and transportation, with methane leaks during natural gas production and processing the most significant exception.

Policy Changes in 2021 and 2022

Since the committee issued its first report in early 2021, federal action on fossil fuel production and transportation has focused on a handful of issues. Notably, even as the United States committed in 2021 to a "Nationally Determined Commitment" under the Paris Accord with a target to "achieve a 50–52 percent reduction from 2005 levels in economy-wide net greenhouse gas pollution in 2030" (White House 2021a), it did not sign on to the "Global Coal to Clean Power Transition Statement" that more than 40 countries signed as part of the 2021 Conference of the Parties (COP26) (Plumer and Friedman 2021).[11] Nor did the United States, or other countries, agree to a phase-out of oil and natural gas as part of the 2022 COP27 meetings (Abnett and Nasralla 2022; Fickling 2022). That said, the United States did take several actions affecting direct emissions from the fossil fuel sector, including signing on to an international agreement with a hundred other countries to limit methane emissions to no more than 2020 levels by 2030 (White House 2021b) and entering into a bilateral agreement with China to cooperate in a number of areas (including on methane emissions) that are critical to encouraging decarbonization and electrification of end uses that currently use fossil fuels (DOS 2021).

Federal Executive Branch Actions Related to Fossil Fuel Industries

Four recent activities by the White House and federal agencies related to the fossil fuel industry and its workers are particularly significant: work on new methane policies by the Environmental Protection Agency (EPA); guidance from the Council on Environmental Quality (CEQ) that GHG emissions be considered in reviews of federal

---

[11] The United States did, however, sign on to the agreement that would end financing of unabated coal, oil, and natural gas facilities in other countries ("Statement on International Public Support for the Clean Energy Transition," 2021).

action under the National Environmental Policy Act (NEPA); policy on reviewing natural gas infrastructure project proposals by the Federal Energy Regulatory Commission (FERC); and formation of and assessments by the Interagency Working Group on Coal and Power Plant Communities and Economic Revitalization.

EPA's Methane Policies

One area of recent executive branch effort focused on the fossil fuel industry involves the regulation of methane emissions from oil and gas production, processing, and transport. In June 2021, President Biden signed a joint resolution of Congress, which, under the Congressional Review Act, repealed a Trump administration rule that removed methane as a pollutant regulated in the oil and gas industries under the Clean Air Act (Traylor and Toley 2021). In November 2021, EPA proposed a policy aimed at reducing methane emissions from new, modified, and reconstructed sources in the oil and natural gas industry. This proposed regulation would establish a comprehensive monitoring program and require companies to fix any leaks discovered at new and existing well sites and compressor stations (i.e., "fugitive emissions"). It would also require states to develop plans to establish emissions standards for facilities within their borders (EPA 2021a). Then in November 2022, EPA proposed supplemental regulations that would strengthen the prior regulatory requirements and complement provisions in the IRA that incentivize implementation of technologies to monitor and mitigate methane emissions (EPA 2022). (See further discussion below.)

CEQ Guidance on GHG Emission Impacts in NEPA Reviews

In January 2023, CEQ published interim guidance that federal agencies consider effects of GHG emissions and climate change in their NEPA reviews of proposed major federal actions (CEQ 2023). Citing concerns about impacts of climate change, and particularly those that could exacerbate environmental injustice, CEQ stated that "NEPA reviews should quantify proposed actions' GHG emissions, place GHG emissions in appropriate context and disclose relevant GHG emissions and relevant climate impacts, and identify alternatives and mitigation measures to avoid or reduce GHG emissions" (p. 1197) It encouraged agencies to pursue actions consistent with national emissions reductions goals (CEQ 2023). CEQ stated its intention to finalize guidance on these issues after receiving and reviewing comments from the public.

FERC Policy on Reviewing Natural Gas Infrastructure Proposals

FERC, which has jurisdiction on whether to approve interstate natural gas pipelines and LNG facilities, issued proposals for how the agency will address (among other things) the need for such facilities, their associated GHG emissions, and their impacts on local communities, with special attention to environmental justice (FERC 2021). First in early 2021 and again in February and March 2022, FERC specifically asked for public comment on whether and how it should incorporate GHG emissions into its environmental reviews and determinations of public need (FERC 2022). Under the current operative 1999 Policy Statement, the commission has approved nearly every pipeline facility proposal and did so upon a showing for "project need" that the pipeline company had an agreement with a party seeking additional pipeline capacity and without other indications of project benefits or costs (including such things as GHG emissions) (Tierney 2019, 2021).

Interagency Working Group on Coal and Power Plant Communities

In January 2021, the Biden administration issued Executive Order (EO) 14008, Tackling the Climate Crisis at Home and Abroad, which among other things established an Interagency Working Group on Coal and Power Plant Communities and Economic Revitalization (IWG). This working group was tasked with coordinating the identification, development, and assessment of economic and social resources to support and revitalize coal, oil and gas, and power plant communities, as well as preparing reports on such efforts (Lawhorn 2022). In April 2021, IWG issued an interim report that identified the places that would be most affected by declines in coal production and use (see Figure 12-23) and recommended actions to address worker and other economic impacts in these communities (IWG 2021).

Although IWG itself has not received congressional appropriations, its website (http://energycommunities.gov) features a clearinghouse of approximately 180 federal funding opportunities (as of May 2023) that is searchable by status, department/agency, funding type, program purpose, and eligible recipients. IWG's interim report made several immediate and near-term recommendations, including (but not limited to):[12]

- Investing in Job-Creating Infrastructure Projects
- Providing Access to Rural Broadband

---

[12] For details, see IWG (2021, pp. 12–16).

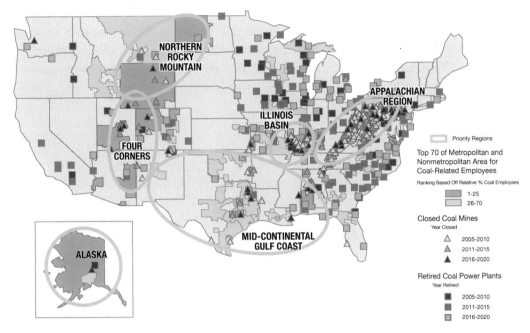

FIGURE 12-23 Adversely impacted coal-dependent regions identified by the Interagency Working Group on Coal and Power Plant Communities and Economic Revitalization (2021). SOURCE: IWG (2021).

- Investing in Technological Innovations
- Funding for Small Businesses
- Investing in Financial Institutions Serving Energy Communities
- Creating Good Jobs by Reclaiming Abandoned Mine Lands
- Financing for Economic Development Aligned Workforce Training
- Funding to Revitalize Brownfields
- Investing in Economic Revitalization in Appalachia

Finding 12-3: The Interagency Working Group on Coal and Power Plant Communities and Economic Revitalization has been helpful for transitions in coal communities, but there is no parallel, coordinated effort under way to address and plan for transitions affecting regions heavily dependent on oil and/or natural gas production.

Recommendation 12-1: *Authorize and Provide Appropriations for State Transition Offices to Address Coal, Oil, and Natural Gas Community Transitions.* **Congress should authorize and provide multi-year appropriations for state transition offices to undertake and implement the recommendations of**

the Interagency Working Group on Coal and Power Plant Communities and Economic Revitalization. Additionally, either this coal-related working group should be expanded to include transitions of oil and natural gas communities or the White House should establish a parallel interagency working group to address such transitions, with congressional funding for the oil and gas transition efforts.

Federal Legislative Action

The IIJA includes provisions to address legacy pollution in and around abandoned coal mines and to plug orphan oil and gas wells. It authorizes $4.7 billion to plug, remediate, and reclaim orphaned wells located on federal, state, tribal, and privately owned lands (IIJA §40601) and $11.3 billion for the Abandoned Mine Reclamation Fund, with increased fees charged for coal production to support abandoned mine reclamation (IIJA §40701).[13] The IIJA also includes provisions for research, development, and demonstration (RD&D) on carbon capture, utilization, and storage (CCUS) technologies that could be used to reduce emissions from facilities relying on fossil energy sources.

In addition to providing support for non-fossil investment, the IRA includes provisions that support development of oil and natural gas on public lands. For example, when the Department of the Interior issues leases for offshore wind in federal waters, the federal government must also offer offshore lease sales for oil and gas development. The IRA increases offshore oil and gas royalty rates and minimum bids associated with onshore oil and gas leases.[14]

The IRA established a methane emissions charge for facilities reporting more than 25,000 tonnes CO_2e per year with methane emissions that exceed a certain threshold; these excess methane emissions are charged at $900 per tonne in 2024, $1,200 per tonne in 2025, and $1,500 per tonne for 2026 and beyond (IRA §60113). It appropriates $1.55 billion (through EPA) in the form of grants, financial incentives, loans, rebates, and other means to support methane emissions monitoring, reporting, and

---

[13] For abandoned mine reclamation fees, Title VII "adjusts the rates of the Abandoned Mine Reclamation Fee to 22.4 cents per ton of coal produced by surface coal mining, 9.6 cents per ton of coal produced by underground mining, and 6.4 cents per ton for lignite coal. This section also extends the fee until 2034" (BIL Summary 2021, p. 81).

[14] The IRA increases the minimum offshore oil and gas royalty rate from 12.5 percent to 16.66 percent and stipulates that the rate be no more than 18.75 percent in the first 10 years after the bill's enactment (IRA §50261). It also increases the minimum for onshore oil and gas leasing from $2 to $10 per acre for the first 10 years after the bill's enactment (IRA §50262).

source plugging; to deploy innovative technologies for methane emissions reduction; and to improve climate resilience, mitigate health impacts, and restore environments in communities affected by oil and gas systems (IRA §60113). Also aimed at reducing GHG emissions, the IRA extends and enhances the tax credit for carbon capture (the so-called 45Q program), adds a tax credit for direct air capture, and includes a "direct pay" provision that creates incentives for non-tax-paying entities to participate in such programs (BPC 2022).

Additionally, the IRA includes two programs to help finance transitions in communities. First, it appropriates $27 billion in funding for the EPA GHG Reduction Fund (discussed in Chapter 11), whose programs may directly or indirectly provide investment benefits in communities adversely affected by the energy transition. Second, it includes a new loan program called the Energy Infrastructure Reinvestment (EIR) Program with $5 billion in appropriations available through September 30, 2026, and a total cap on loans of $250 billion (IRA §50144). It amends Section 1706 of the Energy Policy Act of 2005 so that DOE may "guarantee loans to projects that retool, repower, repurpose, or replace energy infrastructure that has ceased operations, or enable operating energy infrastructure to avoid, reduce, utilize, or sequester air pollutants or anthropogenic emissions of greenhouse gases" (DOE 2022b). DOE's Loan Programs Office provides several examples of eligible project types, including repurposing fossil energy facilities or infrastructure to support clean energy production and adding emissions control technologies, such as carbon capture, to existing energy infrastructure (DOE 2022b).

State Action

Many states have adopted executive or legislative targets to reduce GHG emissions from their economies,[15] including a handful of states with relatively significant economic activity tied to fossil fuels (e.g., coal mining and oil and gas production) (Figure 12-24). Comparing the map in Figure 12-24 with the states listed above in Table 12-1 (states with significant coal mining) and Table 12-2 (states with relatively high levels of employment in fossil fuel industries), "fossil" states committed to GHG emissions reductions are Colorado, Illinois, Montana, New Mexico, and Pennsylvania. Colorado passed legislation committing it to net zero by 2050, while Illinois, Montana, New Mexico, and Pennsylvania have executive branch commitments.

---

[15] As of November 2022, 25 states (as well as the District of Columbia and Puerto Rico) had made some sort of commitment (e.g., economy-wide, power sector, buildings, vehicles, industry, just transition, and/or lands) that is tied directly or indirectly to GHG emissions reductions (UC Berkeley 2023). These include all of the states shown in Figure 12-25, as well as New Hampshire and Hawaii.

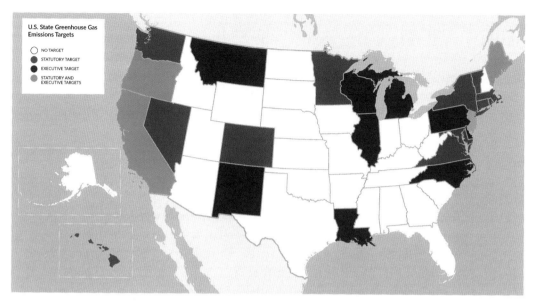

FIGURE 12-24 States with GHG reduction commitments as of 2022. Statutory targets are depicted in blue, executive targets in purple, and both statutory and executive targets in teal. SOURCE: C2ES (2022), © OpenStreetMap contributors, © CARTO.

HOW FAR DO CURRENT/NEW POLICY AND MARKET CONDITIONS GET US?

Several modeling efforts have projected the potential impact of recent federal legislation on production and consumption of fossil fuels, GHG emissions, and economic outcomes as of 2030 and 2035. All found continued demand for oil and natural gas, with a diminishing role for coal.

For example, EIA's assessment (which includes a reference case along with two other bookend cases that rely on high and low assumptions about market participants' "uptake" of financial incentives offered under the IRA) projects that domestic natural gas production will increase through 2050 in all cases, largely owing to increased global demand (EIA 2023d). In the two IRA cases (High Uptake and Low Uptake), natural gas production is estimated to be 1–5 percent lower than in the No IRA case through 2050. Exports of LNG are expected to increase in all cases, and "particularly in the Reference and High Uptake cases because of reduced natural gas demand for power generation and relatively low domestic natural gas prices, which support wider spreads between domestic and international LNG prices" (EIA 2023d, p. 4).

The REPEAT Project analyzed two scenarios of global and domestic demand for oil and gas, with high and low production estimates varying as a function of changes

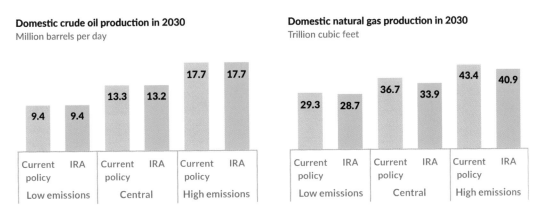

Domestic crude oil production in 2030
Million barrels per day

Domestic natural gas production in 2030
Trillion cubic feet

FIGURE 12-25 Rhodium Group estimates of the impact of the IRA on domestic production of oil and natural gas in 2030. SOURCE: Larsen et al. (2022), *Rhodium Group.*

in domestic demand and product exports (Jenkins et al. 2022b). The REPEAT analysis estimated that the IRA's incentives, combined with provisions in the IIJA, will increase demand for and deployment of CCUS as "a viable economic option for the most heavily emitting industries, such as steel, cement, and refineries, as well as power generation from coal and natural gas" (Jenkins et al. 2022b, p. 17). It projected significant retrofits at existing coal-fired power plants (6 GW) and gas-fired power plants (18 GW) by 2030, resulting in 90 million tons of captured CO_2 at power plants, along with another 110 million tons of CO_2 captured by various industrial sites (Jenkins et al. 2022b). The option to retrofit plants with CCS contributes to continued demand for coal and natural gas, while helping to improve public health outcomes and to reduce employment losses in the oil and gas industries (Jenkins et al. 2022b).

The REPEAT Project also analyzed the implications of the IIJA and IRA for coal consumption under alternative assumptions about the development of electricity transmission infrastructure over the next 12 years. Across all transmission scenarios, coal use is estimated to decline by one-third to one half over that period (Jenkins et al. 2022a).

In its analysis of recent federal legislation, the Rhodium Group observed that the IIJA and IRA will decrease reliance on imported fossil fuels. Rhodium Group's model estimates "the most economical way to meet demand for energy. . . . [T]he clean energy provisions in the IRA drive down demand for petroleum and even more so for natural gas. Domestic production and imports respond accordingly, even though more federal land is available for exploration. In 2030, crude production is effectively flat . . . when comparing the IRA with current policy [which in this analysis includes the IIJA], and gas production declines by 2–7% . . . with the IRA compared to current policy" (Larsen et al. 2022, p. 9; see Figure 12-25).

U.S. liquefied natural gas export projects: existing and under construction (2016–2025)
billion cubic feet per day

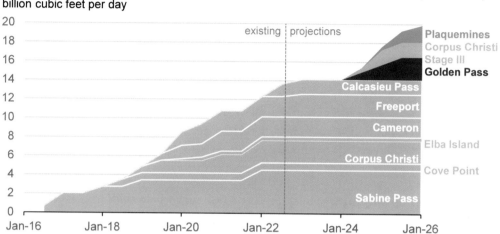

FIGURE 12-26 Existing and projected domestic LNG export facilities as of September 2022. SOURCE: U.S. Energy Information Administration, September 2022 (EIA 2022o).

As a point of reference, in 2021, domestic production of crude oil amounted to 11.2 million barrels per day, and natural gas production was 34.5 trillion cubic feet (EIA 2022j,n). Thus, under the central scenario, the Rhodium Group estimates an increase in domestic oil production by 2030 and a slight decrease in natural gas production by 2030.

The market for exports of natural gas has shown recent strength given global energy market dynamics, including those brought about by Russia's war on Ukraine. Previously approved LNG liquefaction capacity to enable exports of domestic LNG is moving into construction (Figure 12-26), which aligns with expectations in the REPEAT analysis and the Rhodium Group's estimates of continued demand for domestic natural gas production. Estimates show that it takes 3–5 years to build an already-permitted LNG export facility and put it into commercial operation in the United States (Global Energy Monitor 2022). The three projects currently under construction will increase LNG export capacity by more than 40 percent relative to current capacity. The key market pull for expanded U.S. exports of natural gas is the dramatically lower price in the United States relative to global markets (IEA 2022c).

Researchers at the World Resources Institute (WRI) also modeled the impact of recent federal policies on changes in GHG emissions as well as energy-economy employment (Shrestha et al. 2022). The WRI analysis estimated that federal legislation providing tax incentives, infrastructure investments, and targeted spending ("advanced tax credit scenario") would decrease GHG emissions and produce a 5-million net increase

TABLE 12-4 Estimates of Employment Impacts in Two 2035 Policy Scenarios

	2020	2035 Reference Case	2035 Federal Policy Case
Total U.S. energy economy employment	16.584 million	20.807 million	21.677 million
Net change in U.S. energy economy jobs	—	+4.223 million	+5.093 million
Fossil fuel employment	4.268 million	4.115 million	3.214 million
Change in fossil fuel employment	—	−0.153 million	−1.053 million

SOURCE: Data from Shrestha et al. (2022).

in energy sector employment in 2035, which would be 0.87 million more jobs than in a reference case without such policies. However, the analysis also found that there would be employment losses in several areas related to fossil fuels.

The WRI study compared a "reference" scenario without any of the new federal policies, an "advanced tax credit" scenario with new policies similar to those in the IIJA and IRA, and a "net-zero" scenario with additional policies (including sector-specific performance standards such as a clean energy standard and economy-wide net-zero emissions cap). It found that the fossil fuel sector would see significant jobs losses, primarily in petroleum, natural gas, and coal mining and extraction, and that the growth in biofuel and hydrogen employment would not offset these losses (Shrestha et al. 2022). The WRI analysis estimated that the advanced tax credit scenario would lead to a net decrease in fossil fuel jobs and fossil fuel–related electric power jobs of approximately 0.95 million by 2035, relative to the reference scenario (which also was expected to see a job loss relative to 2020). Excerpted estimates of employment impacts in the WRI analysis are shown in Table 12-4.

Economic modeling of recent federal legislation estimates continued reductions in coal production and use, with relatively flat oil and gas production during the next decade. These modeling studies indicate that the legislation will likely put the nation on a path toward decarbonizing the power, building, and transportation sectors, with significant reductions in GHG emissions anticipated over the next decade. Beyond 2030–2035, additional policies, investments, and actions will be required to reach net-zero targets by midcentury. (Discussion of the results of these studies and recommendations related to the sectors that consume fossil fuels can be found in other chapters of this report.)

Finding 12-4: Despite the positive progress in policy change since the committee's first report in early 2021, much more is needed. Federal policy has focused insufficient attention on the recent and anticipated transitions affecting communities tied to fossil fuel extraction and production. A National Transition Corporation (see the committee's first report [NASEM 2021a] and Chapter 1 of this report) offers an institutional framework to fund and guide strategic action on community transition over the next several decades.

CONTINUING IMPEDIMENTS AND GAPS AFFECTING CLIMATE-RELATED ACTION IN THE NATION'S FOSSIL FUEL INDUSTRIES

Compared to business-as-usual outlooks, recent actions in markets and policy will change the nation's fossil fuel production levels, delivery, consumption, and systems. Many changes (notably coal production and use) have been under way for some time, driven by market forces and existing policies. But accelerated change can now be expected because of recent additions to climate and energy policy that will affect not only fossil fuel use, but also all the other systems that support their production, delivery, and transportation within the U.S. economy.

This chapter focuses on gaps in and impediments to new federal policies *directly targeting the fossil fuel industry*, rather than on policies (like the incentives for deployment of alternative technologies) whose indirect impacts will cause most of the expected decline in fossil fuel–related emissions. The primary barriers to intended fossil fuel emissions reductions are the gaps and barriers identified in those other chapters, which might result in either changes in energy-transporting and/or energy-using infrastructure that are too little or too late. If the recent EPA-proposed regulations to control power plant and industrial CO_2 emissions are adopted and implemented,[16] if facility owners take advantage of the IRA's enhanced 45Q financial incentives for implementing carbon capture and sequestration technologies (Kelly and Nelson 2023), and if sufficient renewable and electricity-storage capacity exists in 2030–2035, then coal use might be reduced but not eliminated, while oil and gas consumption will be close to what it is today. Chapter 6 discusses how public pushback could frustrate or prevent the build-out of new transmission capacity or the installation of carbon capture technologies on existing or new power plants, which in turn could impede deployment

---

[16] In May 2023, EPA proposed new regulations to control carbon emissions from existing and new fossil-fueled power plants and large industrial facilities, with emissions standards reflecting the "best system of emission reductions" "based on either 90 percent capture of CO_2 using CCS by 2035, or co-firing of 30 percent by volume low-GHG hydrogen beginning in 2032 and co-firing 96 percent by volume low-GHG hydrogen beginning in 2038" (EPA 2023b).

of renewable capacity. Limited availability of critical minerals necessary for clean energy technologies could also impede renewable deployment. With these constraints, combined with increased electrification, fossil emissions might then actually increase through 2030.

Thus, near-term uncertainty about the size of and activities in the U.S. fossil fuel industry in 2030 is dominated by impediments to policies targeting other sectors, as well as the forces directly affecting global fossil fuel supply, demand, and prices. On the other hand, there is considerable uncertainty about the nature and size of the fossil fuel industry from 2030–2050, which will be strongly shaped by current and future federal policy.

> Finding 12-5: Numerous, significant technological, economic, behavioral, and market uncertainties make it impossible to project with assurance the roles and specific mix of fossil fuels in upcoming decarbonization pathway(s). Nonetheless, most estimates indicate that fossil fuels will continue to provide a significant share of energy during the 2020s, with most of the uncertainty affecting outcomes in the 2030s and beyond. There are at least three key drivers of this uncertainty: (1) policy; (2) the degree to which carbon capture is socially acceptable and cost competitive; and (3) the role of electrification (e.g., as opposed to renewable fuel substitutes for fossil fuels) as means for carrying and storing energy.

Techno-Economic Uncertainties

Uncertainty affects different stakeholders' willingness to accept or reject an ongoing role for fossil fuels in a decarbonized economy. Examples of such uncertainties are multiple and confounding, and affect not only the character of fossil fuels' roles in the economy but also the timing of changes in production, transportation, and use.

Take the future of coal, for example. Demand for coal is tied to transitions in the U.S. power sector, where almost all of U.S. coal is consumed. As explained in Chapter 6, state decarbonization commitments, federal and state policies, and power-market economics that favor renewable power sources and gas-fired generation are expected to decrease coal-fired power generation and demand for domestic coal production dramatically over the next 10–15 years. In the absence of commercially available and economically attractive CCUS technology and the buildout of infrastructure to serve coal-fired power plants outfitted with CCUS, coal plant retirements can be expected to continue over the next decade and a half, perhaps even eliminating coal as a source of generating capacity in the United States by the mid-to-late 2030s.[17] In such circumstances, coal-related employment will continue to decline, putting economic stress on those communities and states whose economies are tied to coal mining and coal-fired power production. On the other hand, many stakeholders' hopes for the success of CCUS may create resistance to walk away in the near term from assets that could have economic and technical value in the future. Some stakeholders involved in the mining sector may even argue that the U.S. coal industry should be a larger player in global markets for coal (where emissions from power plant and industrial uses would not be reflected in U.S. territorial emissions under most modeling approaches, but which would nonetheless be included in global emissions budgets and impact the climate).[18] The presence of critical minerals and rare earth elements (REEs) in coal and coal by-products (see, e.g., DOE 2022c; NETL n.d.; Zhang et al. 2020) could also impact the prospects for coal, as there is great interest in establishing domestic sources of these materials for use in clean energy technologies (DOE 2022c, 2023). Providing an additional revenue stream for coal may lengthen the timeline for coal plant retirements, although processing coal wastes to extract critical minerals and REEs could have the benefit of cleaning up legacy environmental pollution.

Similarly, the outlook for natural gas remains uncertain. The prospects (e.g., timing and cost) for CCUS affect not only the use of coal but also industrial and power-sector demand for natural gas. More broadly, demand for natural gas for power generation varies across regions and the nation as a whole in light of uncertainty about the availability of other forms of flexible generation (e.g., storage, dynamic demand) and long-term on-demand power generation capability (e.g., long-duration storage). Given the current role of natural gas for heating and other end uses in buildings and in industrial applications, the pace of electrification affects how much gas will need to be produced and delivered into various regions of the country.

Uncertainty about the technical and economic ability to repurpose existing fossil-fuel delivery infrastructure (e.g., pipelines) for different energy carriers in the future (e.g., renewable gas, hydrogen) produces another challenge to understanding the

---

[17] Researchers have analyzed the potential to repurpose the sites of recently retired coal-fired power plants. While such proposals might not extend the life of coal as a fossil fuel, they might address transition impacts of communities with economic attachment to such large infrastructure projects (Hansen et al. 2022). The IRA's Energy Infrastructure Reinvestment Program administered by the DOE Loan Programs Office and described above will create incentives for such repurposing of sites.

[18] Note that the United Kingdom recently decided to approve the first coal mine that would open in many decades, with the goal of providing jobs in the United Kingdom to produce steel for domestic and international markets (Ravikumar et al. 2022).

potential roles of such infrastructure in the future. If gas pipelines do not end up being suitable for other uses besides transporting natural gas and the commodity flows on some of those facilities decline (e.g., as a result of electrification of buildings and vehicles, or of decarbonization of the power system), then decisions will ultimately have to be made with regard to continued investment and expenditures to keep those pipelines in safe operating condition and to address any potential stranded costs. Such analyses may be facility-specific, as observed by a National Academies' committee examining market and infrastructure considerations associated with use of CO_2 for economic products and services: "Given the large number of parameters involved, the economic and operational feasibility of repurposing existing natural gas pipelines for transporting CO_2 requires examination on a case-by-case basis using rigorous systems analysis, with techno-economic, safety, environmental, and justice considerations being the overarching factors" (NASEM 2023, p. 100).[19]

Decisions about repurposing and/or decommissioning any existing natural gas pipelines, or about adding new natural gas pipelines, will be made in the first instance by their owners and shippers, but additionally by regulators (e.g., FERC for interstate gas pipelines; state regulatory agencies for intrastate pipelines) in proceedings that will likely become increasingly contentious with different stakeholders' positions on the long-term need for, and safety and economics of, such facilities. These discussions introduce differing perspectives on the potential option value of maintaining infrastructure, for example, and the cost of doing so in a world of declining use of fossil fuel–based natural gas by a system that must remain in safe operations for as long as any consumer demand is served by facilities on that system. (See Chapter 5 on public engagement.)

Because discussions of the future of gas-delivery infrastructure typically do not differentiate between declining use of natural gas derived from fossil fuels and other

---

[19] NASEM (2023) made two potentially relevant recommendations as well:
- Recommendation 5.6. Regulatory authorities in charge of siting infrastructure should account for distributional impacts of CO_2 utilization projects through a process that considers equity and justice for disadvantaged groups, engages impacted communities early and throughout the project planning, and allows for alteration of project design and implementation (p. 125).
- Recommendation 6.1. The U.S. Department of Energy should support its national laboratories, academia, and industry to leverage their competencies in techno-economic and life cycle assessments, as well as integrated systems analysis, to identify the best deployment and investment opportunities from the myriad of utilization options, avoiding those that are technically feasible but not sustainable or economically attractive. These assessments should consider relevant regulatory and policy frameworks and environmental justice impacts, as well as factors that may influence societal acceptance of the technologies (p. 133).

gaseous forms of energy, issues associated with the phase-out of fossil fuels like natural gas do not necessarily mean the phase-out of chemical energy carriers, such as hydrogen, synthetic natural gas, synthetic liquid fuels, and ammonia. The future of these other non-fossil-based energy carriers will influence the future of midstream and downstream delivery infrastructure and systems.

Finding 12-6: Transition impacts on jobs and local communities differ based on whether they are associated with fossil fuel extraction, processing/refining, transportation, or utilization. For example, there may continue to be a significant role for the latter three of these if there is significant uptake of renewable fuels, such as hydrogen or sustainable aviation fuels.

Finding 12-7: If oil and gas production, processing, and delivery volumes decrease in the future, companies in these sectors will continue to need investment to ensure safety and environmental compliance for as long as these facilities are being used during the net-zero transition. With sufficient planning and problem solving, such investments could occur without inadvertently extending the uneconomic lives of the facilities or by maintaining the ability of such facilities to be repurposed for supporting other fuels in a decarbonized economy.

Recommendation 12-2: *Consider Whether Proposed Natural Gas Pipeline Projects Are Needed, Incorporate Greenhouse Gas (GHG) Emissions Impacts into National Environmental Policy Act (NEPA) Analyses, and Require the Use of Depreciation Periods for Pipeline Application Reviews.* **Congress should direct the Federal Energy Regulatory Commission to incorporate the following considerations into the agency's reviews of applications for new interstate natural gas pipeline capacity and liquefied natural gas terminal capacity:**

- a. **Consider whether the proposed project is needed in light of projections of GHG emissions reductions consistent with a net-zero economy by midcentury.**
- b. **Incorporate the GHG emissions impacts associated with production, transportation, and use of the gas that is proposed to be transported by the facility application into NEPA analyses and project need determinations.**
- c. **Require, in decisions regarding the recovery of costs of jurisdictional natural gas facilities, the use of depreciation periods that take into account facility lives consistent with a net-zero economy by midcentury.**

Uncertainty About Behavior in Markets, Social Systems, and Politics

Apart from uncertainty about the technological readiness and costs of various options that affect future demand for fossil fuels, there are other social, political, and consumer preference issues that affect the uptake of equipment, devices, facilities, and infrastructure that relies on other non-fossil fuels.[20]

Events in 2020–2022 have heightened awareness that unexpected global and geopolitical events can disrupt markets for fossil fuels, as occurred in the many months following Russia's invasion of Ukraine, the continuing war there, and the consequences for European reliance on Russian natural gas. These events have already changed the outlook for LNG exports from the United States, as well as the implications for demand for increased gas production domestically.[21] Although European countries have responded with significant modifications in their own energy requirements, uncertainty about these circumstances remains as of the committee's writing.

Demand for natural gas and oil in the United States a decade from now will be greatly affected by the pace and extent to which consumers and businesses purchase electric vehicles (EVs) and that building and industrial plant owners—whether homeowners, landlords, or owners of commercial buildings—convert various end-use equipment (e.g., heating, water heating, cooling) to electricity and otherwise adopt energy efficiency measures. Projections of adoption of non-fossil appliances and conversion of equipment tend to focus on economic considerations. Individual decision makers will make their choices based on economic and other criteria, some of which—like range anxiety over EV driving distances, convenience factors for EV charging, concerns about performance issues associated with heat pumps, uncertainty about the ability of local contractors to install and service electric appliances and/or solar technology systems—are not yet tested or well understood in the literature (see, e.g., IEA 2022b; NASEM 2021b). These factors could render forecasts of electrification to mis-estimate changes in equipment—either overshooting or underestimating what ends up happening.

---

[20] " 'This transition that we need to make to clean energy is complicated and will take time,' Heather Boushey, a member of Mr. Biden's Council of Economic Advisers, said in an interview last year. 'And it requires changing some of the biggest, most expensive things that people buy: a boiler, a car, how they cool their home, what kinds of transportation they drive' " (Tankersley 2023).

[21] The National Academies (NASEM 2023) included a recommendation (4.1) that agencies "with authority over LNG facilities—including the U.S. Department of Energy, the Pipeline and Hazardous Materials Safety Administration, the Federal Energy Regulatory Commission, and the U.S. Coast Guard—should collaborate to assess the possibility of repurposing existing LNG facilities for CO_2 liquefaction, as well as the feasibility of co-designing any future facilities to be able to compress both natural gas and CO_2," (pp. 99–100).

Suppliers of fossil fuels and alternatives may attempt to influence these decisions in one direction or another, making it hard to plan on such things as utilization rates of natural gas pipelines or retail gasoline stations.

In some places, local governments have tried to make certain outcomes more predictable—for instance, by putting in place moratoria on extensions of natural gas service or prohibitions on gas heating systems in new buildings.[22] Some of these have been contested and are subject to legal challenges.[23] Concerns have been raised in Congress to push back on actions that would restrict access to natural gas use in buildings (Shao and Friedman 2023), and as of early 2022, 20 states had enacted "preemptive laws" to foreclose the adoption of such actions by localities in those jurisdictions (Cunningham and Narita 2021).

Utility regulators in various states with carbon-reduction commitments or mandates have begun to focus attention on various issues associated with the future of local gas distribution utilities and their existing customers (Solomon and Tallackson 2022; Zitelman 2022). With electrification of buildings, power systems, and industrial uses, sales of traditional forms of natural gas service will decrease, but customers who remain in need of natural gas service will require access to supply delivered over safely operating local and higher-pressure gas pipelines. Lower sales volumes will naturally lead to revenue erosion for the local utility, which will need to continue to spend money on maintaining and operating the system infrastructure. Fewer sales over an existing asset base with such ongoing costs will have adverse consumer price impacts, all else equal, potentially further reducing demand. Those consumers slower to transition to electrification—which, based on previous examples of shrinking utility customer bases, are likely to be disproportionately lower-income households and people of color—would bear the burden of these higher costs (Davis and Hausman 2022). Local utilities, regulators, consumers, and other stakeholders will increasingly need to wrestle with such difficult challenges, including utility business model considerations.[24]

---

[22] As of February 2022, 74 communities in California have banned natural gas in new buildings (Gable 2021). The City Council of New York City voted to ban natural gas in new buildings (Maldonado 2021).

[23] In Massachusetts, for example, the City of Brookline instituted a moratorium on new gas hook-ups; this action was challenged and found to be unconstitutional as written. Subsequently, the Massachusetts legislature enacted a law in 2021 that included provisions allowing for communities to impose bans on gas in new buildings under certain circumstances. Those provisions are being implemented as of 2023 (Iaconangelo 2023; Van Voorhis 2022).

[24] See, for example, regulatory proceedings related to the future of natural gas in Minnesota (Minnesota Public Utilities Commission 2021, 2022), with illustrative stakeholder views (Drake and Partridge 2021), and in Massachusetts (Massachusetts Department of Public Utilities 2022).

Finding 12-8: Local gas distribution utilities around the country have seen relatively flat sales of natural gas and have dealt with economic, financial, and regulatory considerations relative to these conditions. In parts of the country where electrification of end uses that now use natural gas is occurring or is anticipated to take place in the future, these challenges will become more acute. Many utility regulators and stakeholders in such communities are actively looking at the complex implications for utility business models, investment patterns, service provision, safety, and consumer rate impacts.

Recommendation 12-3: *Require Utilities and Service Providers to Plan for the Transition.* **State regulators should require natural gas distribution utilities and fossil fuel supplier/service providers to plan for transitioning the investment in and operations of their pipeline and service-delivery systems to ensure safe operations and affordable delivery of gas as buildings and other end uses undergo electrification. Such investigations should consider how gas utilities with electric-distribution affiliates might use cost- or revenue-sharing approaches, and/or asset disposition options, to transition their systems safely and affordably.**

Recommendation 12-4: *Consider Adoption of Moratoria on New Gas Lines in Previously Unserved Areas.* **States and communities should consider adopting moratoria on extension of new gas lines into areas previously unserved by natural gas unless or until there is a showing that such extensions will reduce greenhouse gas emissions.**

In places where transition discussions are under way as well as in other places, there are tenacious impediments to action on fossil fuels—many of which are discussed in other chapters. Fundamentally, as described in Chapter 11 and the committee's first report, the absence of a price on carbon produces economic advantages for some, if not many, applications of fossil fuels owing to unpriced and unmitigated CO_2 emissions from power generation using coal and natural gas; consumption of petroleum products in vehicles; use of gas, oil, and propane in buildings; and industrial use of fossil fuels. (See also the chapters on these sectors.)

Public, community, and/or political resistance to or lack of support for build-out of regional infrastructure (e.g., power lines, CO_2 pipelines, sequestration sites, hydrogen facilities) that might otherwise facilitate relatively timely and efficient decarbonization strategies may prolong use of existing fossil infrastructure and production of coal, natural gas, or oil—or accelerate precipitous and destabilizing declines.

Financing and legal instruments related to existing infrastructure may inhibit what might otherwise seem like economic transitions away from use of fossil fuels. For example, the existence of long-term contracts or debt arrangements associated with investments in and/or ownership of existing coal-fired generating assets may impede their retirement, as might the desire to avoid creating stranded costs.

States and communities that depend on fossil-fuel royalty fees, severance taxes, property taxes, or other payments tied to facilities or activities in the fossil-fuel sup-ply chain are particularly vulnerable to decarbonization transitions (Raimi 2023). One state—Illinois—has put in place a legislative requirement that in advance of a closure of investor-owned power plants and coal mining operations, the owner must give advance notice of the anticipated facility closure to local communities and workers.[25] Such advanced notice allows for transition planning and other activities by the state, local community, and workforce.[26] Initiatives from the IWG and some states (e.g., Colorado, see Colorado OJT 2020) offer lessons regarding what kinds of engagement, planning, and support have worked and what types are still missing. The IWG has rec-ommendations related to federal grants on infrastructure projects, provision of access to broadband in rural areas, funding for small businesses and the financial institutions that support them in disadvantaged communities, grants for land reclamation and the revitalization of brownfields, and support for workforce training, which are ripe for implementation in light of the financial resources and technical assistance enabled by the IIJA and IRA (IWG 2021).

Furthermore, the federal government's Highway Trust Fund has been stressed in recent years, in part because of flat collections of federal fuel taxes. There are mul-tiple drivers of such collections, including the fact that the federal gas tax rate has

---

[25] "An owner of an investor-owned electric generating plant or coal mining operation may not order a mass layoff, relocation, or employment loss unless, 2 years before the order takes effect, the employer gives written notice of the order to the following: (1) affected employees and representatives of affected employees; and (2) the Department of Commerce and Economic Opportunity and the chief elected official of each municipal and county government within which the employment loss, relocation, or mass layoff occurs." Also: "An employer which is receiving State or local economic development incentives for doing or continuing to do business in this State may be required to provide additional notice" (Illinois CEJA 2021, pp. 954, 955).

[26] Note that communities that host commercial nuclear power plants have tax bases and local econo-mies tied to the operations of and jobs associated with such facilities. During the operations of such plants, the owners must maintain trust funds to support the costs of decommissioning the plants, with the consum-ers of electricity produced by the plants contributing to those trust funds over the life of the facility (NRC n.d.). In anticipation of closures, the owner must submit decommissioning plans to the Nuclear Regulatory Commission. Technical assistance programs have become available to aid local communities as they transi-tion through the process of plant closures (SmartGrowthAmerica n.d.).

not changed in 3 decades, the increasing efficiency of vehicles (thus requiring lower purchases of petroleum product per mile driven), and the sales of vehicles that run on alternative fuels (e.g., hybrid or all-electric vehicles). These circumstances have eroded the ability of the Highway Trust Fund to pay for the federal share of highway and transit projects (GAO 2022). Other sources of funding—for example, some system that raises revenues based on vehicles' mileage rather than purchases of petroleum products—have been in discussion (GAO 2022; Raimi 2023; TRB and NRC 2015), and the IIJA included a one-time transfer of $118 billion in general funds into the Highway Trust Fund to forestall anticipated insolvency of the fund for 5 more years (CRFB 2022).

> Finding 12-9: Presuming that sales of electric, hybrid, and increasingly fuel-efficient vehicles continue to grow, then revenues from federal highway taxes (and support for the Highway Trust Fund) will decline in future years without some change in policy that affects new and/or revised sources of revenue.

> **Recommendation 12-5:** ***Modify the Design of Taxes on Gasoline, Diesel, and Petroleum Products.*** **Congress and states should replace their current highway user taxes and fees, which are based on gasoline, diesel, and other petroleum products that have been relied on to provide funding for highways and mass transit programs, with alternative funding mechanisms that are consistent with the user-pay principle for highway funding.**

Last, many communities, institutions, investors, and others seek to retain the jobs, profits, tax base and public revenues, service industries, and myriad other economic activities in places where fossil industries currently operate. Some states have attempted, with mixed success, to mandate the continuity of coal-based economies by statute. For example, Montana's legislature passed two laws in 2021 that fined out-of-state companies for refusing to fund long-term coal-fired power plant operations, but those laws were subsequently deemed unconstitutional (IEEFA 2022). Impacts have already been felt in communities where coal is in decline and are anticipated in other communities where fossil fuel extraction transitions are expected to occur in the years ahead. A model for providing federal assistance to communities impacted by fossil fuel transition assistance can be found in the American Rescue Plan's "Coronavirus State and Local Fiscal Recovery Fund," which allows states, localities, and tribal governments to apply for and use funding for a variety of purposes (including replacing lost public sector revenue and responding to the negative economic impacts [of the pandemic] on communities and households, small businesses, and others) (Department of the Treasury 2022).

Finding 12-10: Energy transitions in communities long dominated by fossil fuel industries and related activities are greatly affected by the reality that decisions about facility operations and closures tend to be made by firms that own assets and not by the communities that host them. Firms vary significantly in their sense of responsibilities to assist in these communities' transitions. There are lessons in other technologies (e.g., commercial nuclear reactors), where decommissioning monies have been collected and reserved for use in the period after the plant is in operations, with the idea that the consumers who benefit from the operations of, say, a nuclear plant should pay for its eventual decommissioning and site cleanup and restoration. Such cleanup activities provide economic activities to local communities during transition periods.

Recommendation 12-6: *Require Recipients of Federal Funding to Provide Advance Notice of Facility Closures.* **Congress should require that recipients (e.g., states, localities, companies, non-profits) of funding from federal agencies provide at least 2 years of notice prior to closure of federally funded facilities that would lead to worker layoffs exceeding 99 full-time equivalent employees.**

Finding 12-11: Local and regional economies developed around extraction of fossil fuels will be greatly in need of economic and social support in conjunction with, if not ahead of, these energy transitions. State governments dependent on royalty payments from extraction of fossil fuels will be similarly affected by and in need of fiscal transition strategies. In these locations, planning for public service delivery under changing circumstances is a pressing issue, and one not adequately addressed in employment- or worker-focused assessments.

Recommendation 12-7: *Fund the Decommissioning, Cleanup, and Just Transition for Communities Historically Dependent on Fossil Fuels.* **Congress and state legislatures should ensure the existence of funding in support of decommissioning, cleanup, and just transitions for communities associated with fossil fuel extraction and power production:**

    a. **States that currently receive royalty, severance tax, or revenues associated with extraction of fossil fuels should evaluate—and adjust as needed, the availability of revenue to assist in planning for just transitions in the communities affected by such resource extraction activities. Fossil fuel revenues should not be mixed in with general fund spending but should be invested in permanent funds to ensure long-term** savings, while also providing funding for economic diversification and transition assistance.

b. State agencies with permitting responsibility for oil and gas production within their borders should require the owner/operator of such facilities to maintain adequate financial reserves to cover and ultimately pay for closure and site decommissioning and restoration of the site and for impacts in the host communities.

c. State regulators with jurisdiction over the recovery of a regulated utility's costs related to existing or new fossil power plants should require that the owners of such plants include reserves for the eventual funding of cleanup/decommissioning of the site and its impacts on host communities. State utility regulators should continue to expand opportunities for rate recovery to address remediation/transition costs associated with use of fossil fuels, while protecting the utility bills of low- and moderate-income consumers.

Table 12-5 summarizes all the recommendations in this chapter regarding the future of fossil fuels.

SUMMARY OF RECOMMENDATIONS ON THE FUTURE OF FOSSIL FUELS

TABLE 12-5 Summary of Recommendations on the Future of Fossil Fuels

Short-Form Recommendation	Actor(s) Responsible for Implementing Recommendation	Sector(s) Addressed by Recommendation	Objective(s) Addressed by Recommendation	Overarching Categories Addressed by Recommendation
12-1: Authorize and Provide Appropriations for State Transition Offices to Address Coal, Oil, and Natural Gas Community Transitions	Congress and state transition offices	• Fossil fuels • Non-federal actors	• Equity • Employment	Ensuring Procedural Equity in Planning and Siting New Infrastructure and Programs Ensuring Equity, Justice, Health, and Fairness of Impacts Building the Needed Workforce and Capacity Managing the Future of the Fossil Fuel Sector
12-2: Consider Whether Proposed Natural Gas Pipeline Projects Are Needed, Incorporate Greenhouse Gas Emissions Impacts into National Environmental Policy Act Analyses, and Require the Use of Depreciation Periods for Pipeline Application Reviews	Congress and Federal Energy Regulatory Commission	• Fossil fuels • Transportation	• Greenhouse gas (GHG) reductions	A Broadened Policy Portfolio Managing the Future of the Fossil Fuel Sector

TABLE 12-5 Continued

Short-Form Recommendation	Actor(s) Responsible for Implementing Recommendation	Sector(s) Addressed by Recommendation	Objective(s) Addressed by Recommendation	Overarching Categories Addressed by Recommendation
12-3: Require Utilities and Service Providers to Plan for the Transition	State regulators of natural gas distribution utilities and fossil fuel supplier/ service providers	• Electricity • Buildings • Transportation • Industry • Fossil fuels • Non-federal actors	• GHG reductions • Equity • Health • Public engagement	Ensuring Equity, Justice, Health, and Fairness of Impacts Managing the Future of the Fossil Fuel Sector
12-4: Consider Adoption of Moratoria on New Gas Lines in Previously Unserved Areas	States and communities	• Fossil fuels • Non-federal actors	• GHG reductions	Managing the Future of the Fossil Fuel Sector
12-5: Modify the Design of Taxes on Gasoline, Diesel, and Petroleum Products	Congress and states	• Transportation • Fossil fuels • Non-federal actors	• GHG reductions	Managing the Future of the Fossil Fuel Sector
12-6: Require Recipients of Federal Funding to Provide Advance Notice of Facility Closures	Congress and recipients of federal agency funding	• Fossil fuels • Non-federal actors	• GHG reductions • Equity • Employment • Public engagement	Ensuring Equity, Justice, Health, and Fairness of Impacts Managing the Future of the Fossil Fuel Sector
12-7: Fund the Decommissioning, Cleanup, and Just Transition for Communities Historically Dependent on Fossil Fuels	Congress, state legislatures, state agencies, and state regulators	• Finance • Fossil fuels • Non-federal actors	• GHG reductions • Equity • Health • Public engagement	Ensuring Equity, Justice, Health, and Fairness of Impacts Managing the Future of the Fossil Fuel Sector

REFERENCES

Abnett, K., and S. Nasralla. 2022. "COP27: Energy Crisis Leaves Fossil Fuel Phase-Out Clubs Struggling to Recruit." *Reuters*, November 16. https://www.reuters.com/business/cop/cop27-energy-crisis-leaves-fossil-fuel-phase-out-clubs-struggling-recruit-2022-11-16.

Allen, D. 2016. "Emissions from Oil and Gas Operations in the United States and Their Air Quality Implications." *Journal of the Air & Waste Management Association* 66(6):549–575. https://pubmed.ncbi.nlm.nih.gov/27249104.

Alvarez, R.A., D. Zavala-Araiza, D.R. Lyon, D.T. Allen, Z.R. Barley, A.R. Brandt, K.J. Davis, et al. 2018. "Assessment of Methane Emissions from the U.S. Oil and Gas Supply Chain." *Science* 361(6398):186–188. https://doi.org/10.1126/science.aar7204.

BIL (Bipartisan Infrastructure Law) Summary. 2021. *Bipartisan Infrastructure Investment and Jobs Act Summary: A Road to Stronger Economic Growth.* Maria Cantwell, U.S. Senator for Washington. https://www.cantwell.senate.gov/imo/media/doc/Infrastructure%20Investment%20and%20Jobs%20Act%20-%20Section%20by%20Section%20Summary.pdf.

Birol, F. 2022. "Three Myths About the Global Energy Crisis." *Financial Times*, September 5. https://www.ft.com/content/2c133867-7a89-44d0-9594-cab919492777.

BLS (Bureau of Labor Statistics). 2023a. "All Employees, Thousands, Coal Mining, Seasonally Adjusted." https://data.bls.gov/timeseries/CES1021210001.

BLS. 2023b. "All Employees, Thousands, Oil and Gas Extraction, Seasonally Adjusted." https://data.bls.gov/timeseries/CES1021100001?amp%253bdata_tool=XGtable&output_view=data&include_graphs=true.

Boehm, S., L. Jeffrey, K. Levin, J. Hecke, C. Schumer, C. Fyson, A. Majid, and J. Jaeger. 2022. *State of Climate Action 2022.* Berlin and Cologne, Germany; San Francisco, CA; Washington, DC: Bezos Earth Fund, Climate Action Tracker, Climate Analytics, ClimateWorks Foundation, NewClimate Institute, the United Nations Climate Change High-Level Champions, and World Resources Institute. https://doi.org/10.46830/wrirpt.22.00028.

Bordoff, J. 2022a. "Europe Is Wrong to Blame the U.S. for Its Energy Problems." *The New York Times*, December 2. https://www.nytimes.com/2022/12/02/opinion/europe-ukraine-energy.html.

Bordoff, J. 2022b. "Will Putin's Energy Strategy Backfire?" Interview with Daniel Yergin, September 13. Columbia Energy Exchange. https://www.energypolicy.columbia.edu/will-putin-s-energy-strategy-backfire.

BPC (Bipartisan Policy Center). 2022. "Inflation Reduction Act Summary: Energy and Climate Provisions." https://bipartisanpolicy.org/blog/inflation-reduction-act-summary-energy-climate-provisions.

C2ES (Center for Climate and Energy Solutions). 2022. "U.S. State Greenhouse Gas Emissions Targets." https://www.c2es.org/document/greenhouse-gas-emissions-targets.

CBO (Congressional Budget Office). 2022. "The Budget and Economic Outlook: 2022 to 2032." https://www.cbo.gov/publication/58147#_idTextAnchor264.

CEQ (Council on Environmental Quality). 2023. "National Environmental Policy Act Guidance on Consideration of Greenhouse Gas Emissions and Climate Change." *Federal Register* 88(5):1196–1212. https://www.govinfo.gov/content/pkg/FR-2023-01-09/pdf/2023-00158.pdf.

Colorado OJT (Office of Just Transition). 2020. *Colorado Just Transition Action Plan.* Denver, CO: Colorado Department of Labor and Employment. https://cdle.colorado.gov/sites/cdle/files/documents/Colorado%20Just%20Transition%20Action%20Plan.pdf.

Conte, N. 2021. "This Chart Shows Where America Got Its Energy from in 2020." World Economic Forum. https://www.weforum.org/agenda/2021/07/this-visualization-shows-how-energy-was-consumed-by-the-united-states-in-2020.

CRFB (Committee for a Responsible Federal Budget). 2022. "The Infrastructure Bill's Impact on the Highway Trust Fund." https://www.crfb.org/blogs/infrastructure-bills-impact-highway-trust-fund.

Cunningham, A.M., and K. Narita. 2021. "Gas Interests Threaten Local Authority." *Natural Resources Defense Council* (blog), January 19. Updated February 22, 2022. https://www.nrdc.org/experts/alejandra-mejia/gas-interests-threaten-local-authority-6-states.

Davis, L.W., and C. Hausman. 2022. "Who Will Pay for Legacy Utility Costs?" *Journal of the Association of Environmental and Resource Economists* 9(6):1047–1085. https://doi.org/10.1086/719793.

Department of the Treasury. 2022. "Coronavirus State and Local Fiscal Recovery Funds." https://home.treasury.gov/policy-issues/coronavirus/assistance-for-state-local-and-tribal-governments/state-and-local-fiscal-recovery-funds.

DOE (Department of Energy). 2022a. *Energy Employment by State 2022: United States Energy and Employment Report.* Washington, DC. https://www.energy.gov/sites/default/files/2022-06/USEER%202022%20State%20Report_0.pdf.

DOE. 2022b. "Jigar Shah Explains How Inflation Reduction Act Impacts DOE Loan Program." Department of Energy Loan Program. https://www.energy.gov/lpo/inflation-reduction-act-2022.

DOE. 2022c. *Recovery of Rare Earth Elements and Critical Materials from Coal and Coal Byproducts.* Washington, DC. https://www.energy.gov/sites/default/files/2022-05/Report%20to%20Congress%20on%20Recovery%20of%20Rare%20Earth%20Elements%20and%20Critical%20Minerals%20from%20Coal%20and%20Coal%20By-Products.pdf.

DOE. 2022d. *United States Energy and Employment Report 2022.* Washington, DC. https://www.energy.gov/sites/default/files/2022-06/USEER%202022%20National%20Report_1.pdf.

DOE. 2023. "Biden-Harris Administration Invests $16 Million to Build America's First-of-a-Kind Critical Minerals Production Facility." https://www.energy.gov/articles/biden-harris-administration-invests-16-million-build-americas-first-kind-critical-minerals.

DOE-FECM (Department of Energy Office of Fossil Energy and Carbon Management). 2013. "Shale Gas Glossary." https://www.energy.gov/sites/default/files/2013/04/f0/shale_gas_glossary.pdf.

DOS (Department of State). 2021. "U.S.-China Joint Glasgow Declaration on Enhancing Climate Action in the 2020s." https://www.state.gov/u-s-china-joint-glasgow-declaration-on-enhancing-climate-action-in-the-2020s.

Drake, T., and A. Partridge. 2021. *Decarbonizing Minnesota's Natural Gas End Uses: Stakeholder Process Summary and Consensus Recommendations.* Minneapolis, MN: Great Plains Institute and Center for Energy and Environment. https://e21initiative.org/wp-content/uploads/2021/07/Decarbonizing-NG-End-Uses-Stakeholder-Process-Summary.pdf.

EFI Foundation. 2023. "EFI Foundation Workshop Summary Report: The Role of U.S. Natural Gas Exports in a Low Carbon World." https://energyfuturesinitiative.org/wp-content/uploads/sites/2/2023/04/Role-of-US-Gas-Exports-Work-shop-Report_4-25-2023-2.pdf.

EIA (U.S. Energy Information Administration). 2013. "Annual Coal Report 2012." https://www.eia.gov/coal/annual/archive/05842012.pdf.

EIA. 2015. "Major Fossil Fuel-Producing States Rely Heavily on Severance Taxes." https://www.eia.gov/todayinenergy/detail.php?id=22612.

EIA. 2018. "Average U.S. Coal Mining Productivity Increases as Production Falls." https://www.eia.gov/todayinenergy/detail.php?id=35232.

EIA. 2020. "Almost All U.S. Coal Production Is Consumed for Electric Power." https://www.eia.gov/todayinenergy/detail.php?id=44155.

EIA. 2021a. "Annual U.S. Natural Gas Production Decreased by 1% in 2020." https://www.eia.gov/todayinenergy/detail.php?id=46956.

EIA. 2021b. "Annual Energy Outlook: Table 1. Total Energy Supply, Disposition, and Price Summary." https://www.eia.gov/outlooks/aeo/data/browser/#/?id=1-AEO2021&cases=ref2021&sourcekey=0.

EIA. 2022a. "Annual Coal Report 2021." https://www.eia.gov/coal/annual/pdf/acr.pdf.

EIA. 2022b. "Annual Energy Outlook." https://www.eia.gov/outlooks/archive/aeo22/tables_ref.php, Table 1.

EIA. 2022c. "Coal Explained: Coal Imports and Exports." https://www.eia.gov/energyexplained/coal/imports-and-exports.php.

EIA. 2022d. "Dry Shale Gas Production Estimates by Play." https://www.eia.gov/naturalgas/data.php#production.

EIA. 2022e. "EIA Expects U.S. Petroleum Trade to Shift Toward Net Imports During 2022."

EIA. 2022f. "How Much Shale Gas Is Produced in the United States?" https://www.eia.gov/tools/faqs/faq.php?id=907&t=8.

EIA. 2022g. "Hydrocarbon Gas Liquids Explained: Prices for Hydrocarbon Gas Liquids." https://www.eia.gov/energyexplained/hydrocarbon-gas-liquids/prices-for-hydrocarbon-gas-liquids.php.

EIA. 2022h. "Liquefied U.S. Natural Gas Exports." https://www.eia.gov/dnav/ng/hist/n9133us2m.htm.

EIA. 2022i. "Natural Gas Consumption by End Use." https://www.eia.gov/dnav/ng/NG_CONS_SUM_DCU_NUS_A.htm.

EIA. 2022j. "Natural Gas Explained: Natural Gas Imports and Exports." https://www.eia.gov/energyexplained/natural-gas/imports-and-exports.php.

EIA. 2022k. "Natural Gas Explained: Use of Natural Gas." https://www.eia.gov/energyexplained/natural-gas/use-of-natural-gas.php.

EIA. 2022l. "Oil and Petroleum Products Explained: Where Our Oil Comes From." https://www.eia.gov/energyexplained/oil-and-petroleum-products/where-our-oil-comes-from.php.

EIA. 2022m. "Section 6. Coal." In *Monthly Energy Review*. Washington, DC. https://www.eia.gov/totalenergy/data/monthly/pdf/sec6.pdf.

EIA. 2022n. "U.S. Field Production of Crude Oil." https://www.eia.gov/dnav/pet/hist/LeafHandler.ashx?n=PET&s=MCRFPUS2&f=M.

EIA. 2022o. "U.S. LNG Export Capacity to Grow as Three Additional Projects Begin Construction." https://www.eia.gov/todayinenergy/detail.php?id=53719.

EIA. 2022p. "U.S. Monthly Average Henry Hub Spot Price Nearly Doubled in 12 Months." https://www.eia.gov/todayinenergy/detail.php?id=53039.

EIA. 2023a. "All Energy Infrastructure and Resources." Web Mapping Application. https://atlas.eia.gov/apps/5039a1a01ec34b6bbf0ab4fd57da5eb4/explore.

EIA. 2023b. "Annual Energy Outlook." Outlooks Narrative. https://www.eia.gov/outlooks/aeo/narrative/index.php#Discussion.

EIA. 2023c. "EIA-923 Power Plant Operations Report: Net Generation by State by Type of Producer by Energy Source (EIA-906, EIA-920, and EIA-923), 1990–2021." Historical State Data. https://www.eia.gov/electricity/data/state.

EIA. 2023d. "Issues in Focus: Inflation Reduction Act Cases in the AEO2023." https://www.eia.gov/outlooks/aeo/IIF_IRA.

EIA. 2023e. "Table 5a. U.S. Natural Gas Supply, Consumption, and Inventories." Short-Term Energy Outlook. https://www.eia.gov/outlooks/steo/tables/pdf/5atab.pdf.

EIA. n.d. "Table 11.1 Carbon Dioxide Emissions from Energy Consumption by Source." https://www.eia.gov/totalenergy/data/browser/?tbl=T11.01#/?f=A&start=1973&end=2022&charted=0-1-13. Accessed May 15, 2023.

EPA (Environmental Protection Agency). 2021a. "EPA Proposes New Source Performance Standards Updates, Emissions Guidelines to Reduce Methane and Other Harmful Pollution from the Oil and Natural Gas Industry." https://www.epa.gov/controlling-air-pollution-oil-and-natural-gas-industry/epa-proposes-new-source-performance.

EPA. 2021b. "Inventory of U.S. Greenhouse Gas Emissions and Sinks: 1990–2019." https://www.epa.gov/sites/default/files/2021-04/documents/us-ghg-inventory-2021-main-text.pdf.

EPA. 2022. "EPA's Supplemental Proposal to Reduce Pollution from the Oil and Natural Gas Industry to Fight the Climate Crisis and Protect Public Health: Overview." https://www.epa.gov/system/files/documents/2022-11/Oil%20and%20Gas%20Supplemental.%20Overview%20Fact%20Sheet.pdf.

EPA. 2023a. "Fact Sheet: Greenhouse Gas Standards and Guidelines for Fossil Fuel-Fired Power Plants Proposed Rule." https://www.epa.gov/system/files/documents/2023-05/FS-OVERVIEW-GHG-for%20Power%20Plants%20FINAL%20CLEAN.pdf.

EPA. 2023b. "Understanding Global Warming Potentials." https://www.epa.gov/ghgemissions/understanding-global-warming-potentials.

EPA. n.d.(a). "Greenhouse Gas Inventory Data Explorer: Energy: Coal Mining." https://cfpub.epa.gov/ghgdata/inventoryexplorer/#energy/coalmining/allgas/subcategory/all. Accessed January 13, 2023.

EPA. n.d.(b). "Greenhouse Gas Inventory Data Explorer: Energy: Natural Gas and Petroleum Systems." https://cfpub.epa.gov/ghgdata/inventoryexplorer/#energy/naturalgasandpetroleumsystems/allgas/subcategory/all. Accessed January 13, 2023.

Federal Reserve Bank of St. Louis. 2022. "All Employees, Coal Mining," https://fred.stlouisfed.org/series/CES1021210001. Accessed October 31, 2022.

FERC (Federal Energy Regulatory Commission). 2021. "FERC Revisits Review of Policy Statement on Interstate Natural Gas Pipeline Proposals." https://www.ferc.gov/news-events/news/ferc-revisits-review-policy-statement-interstate-natural-gas-pipeline-proposals.

FERC. 2022. "Order on Draft Policy Statements, 178 FERC ¶ 61,197." Certification of New Interstate Natural Gas Facilities, Docket Nos PL18-1-001, and Consideration of Greenhouse Gas Emissions in Natural Gas Infrastructure Project Reviews, PL21-3-001. https://www.ferc.gov/media/c-1-032422.

Fickling, D. 2022. "The World Will Never Agree to Phase Out Petroleum. And That's OK." *Bloomberg*, November 20. https://www.bloomberg.com/opinion/articles/2022-11-20/cop27-no-agreement-on-petroleum-phase-out-if-emissions-are-falling-that-s-ok?sref=B3uFyqJT&leadSource=uverify%20wall.

FTA (Federation of Tax Administrators). 2023. "State Motor Fuel Tax Rates (January 1, 2023)." https://www.taxadmin.org/assets/docs/Research/Rates/mf.pdf.

Gable, J. 2021. "California's Cities Lead the Way on Pollution-Free Homes and Buildings." *Sierra Club*, July 22. Updated December 20, 2022. https://www.sierraclub.org/articles/2021/07/californias-cities-lead-way-pollution-free-homes-and-buildings.

GAO (Government Accountability Office). 2022. "Highway Trust Fund: Federal Highway Administration Should Develop and Apply Criteria to Assess How Pilot Projects Could Inform Expanded Use of Mileage Fee Systems." https://www.gao.gov/assets/720/718391.pdf.

Gelles, D. 2023. "The War in Ukraine Upended Energy Markets. What Does That Mean for the Climate?" *The New York Times*, January 14. https://www.nytimes.com/2023/01/14/business/energy-environment/davos-energy-climate-ukraine.html?action=click&module=RelatedLinks&pgtype=Article.

Global Energy Monitor. 2022. "How Long Does It Take to Build an LNG Export Terminal in the United States?" https://globalenergymonitor.org/wp-content/uploads/2022/04/GEM-Briefing-LNG-Terminal-Development-Timelines.pdf.

Gorchov Negron, A.M., E.A. Kort, Y. Chen, A. Brandt, M. Smith, G. Plant, A. Ayasse, et al. 2023. "Excess Methane Emissions from Shallow Water Platforms Elevate the Carbon Intensity of US Gulf of Mexico Oil and Gas Production." *Proceedings of the National Academy of Sciences* 120(15):e221527512. https://www.pnas.org/doi/epdf/10.1073/pnas.2215275120.

Hansen, J., W. Jenson, A. Wrobel, N. Stauff, K. Biegel, T. Kim, R. Belles, and F. Omitaomu. 2022. *Investigating Benefits and Challenges of Converting Retiring Coal Plants into Nuclear Plants: Nuclear Fuel Cycle and Supply Chain."* INL/RPT-22-67964. Washington, DC: Department of Energy. https://fuelcycleoptions.inl.gov/SiteAssets/SitePages/Home/C2N2022Report.pdf.

Iaconangelo, D. 2023. "Mass. Unveils Plans to Roll Back Gas in New Buildings." *E&E News: Energywire*, January 9. https://www.eenews.net/articles/mass-unveils-plans-to-roll-back-gas-in-new-buildings.

Illinois CEJA (Climate and Equitable Jobs Act). 2021. Public Act 102-0662, Section 90-60. https://www2.illinois.gov/epa/topics/ceja/Documents/102-0662.pdf.

IEA (International Energy Agency). 2022a. *Coal Market Update—July 2022*. Paris. https://www.iea.org/reports/coal-market-update-july-2022.

IEA. 2022b. *The Future of Heat Pumps*. Paris. https://www.iea.org/reports/the-future-of-heat-pumps.

IEA. 2022c. "Natural Gas Prices in Europe, Asia and the United States, January 2020–February 2022." https://www.iea.org/data-and-statistics/charts/natural-gas-prices-in-europe-asia-and-the-united-states-jan-2020-february-2022.

IEA. 2023. "World Energy Investment 2023." Paris: International Energy Agency. May 2023. https://iea.blob.core.windows.net/assets/8834d3af-af60-4df0-9643-72e2684f7221/WorldEnergyInvestment2023.pdf.

IEEFA (Institute for Energy Economics and Financial Analysis). 2022. "State Laws Propping Up Colstrip Coal-Fired Plant Ruled Unconstitutional." https://ieefa.org/articles/state-laws-propping-colstrip-coal-fired-plant-ruled-unconstitutional.

IWG (Interagency Working Group on Coal and Power Plant Communities and Economic Revitalization). 2021. *Initial Report to the President on Empowering Workers Through Revitalizing Energy Communities*. Washington, DC. https://netl.doe.gov/sites/default/files/2021-04/Initial%20Report%20on%20Energy%20Communities_Apr2021.pdf.

Jain, G., and L. Palacios. 2023. "Investing in Oil and Gas Transition Assets En Route to Net Zero." Center on Global Energy Policy, Columbia University. https://www.energypolicy.columbia.edu/wp-content/uploads/2023/03/TransitionAssets_Commentary_CGEP_030723-2.pdf.

Jenkins, J.D., J. Farbes, R. Jones, N. Patankar, and G. Schivley. 2022a. "Electricity Transmission Is Key to Unlock the Full Potential of the Inflation Reduction Act." REPEAT Project. https://repeatproject.org/docs/REPEAT_IRA_Transmission_2022-09-22.pdf.

Jenkins, J.D., E.N. Mayfield, J. Farbes, R. Jones, N. Patankar, X. Qingyu, and G. Schivley. 2022b. "Preliminary Report: The Climate and Energy Impacts of the Inflation Reduction Act of 2022." REPEAT Project. https://repeatproject.org/docs/REPEAT_IRA_Prelminary_Report_2022-08-12.pdf.

Kelly, C., and S. Nelson. 2023. "Clear Path to a Clean Energy Future 2022." https://static.clearpath.org/2023/02/CPCEF22-2-23.pdf.

Larsen, J., B. King, H. Kolus, N. Dasari, G. Hiltbrand, and W. Herndon. 2022. "A Turning Point for US Climate Progress: Assessing the Climate and Clean Energy Provisions in the Inflation Reduction Act." Rhodium Group. https://rhg.com/research/climate-clean-energy-inflation-reduction-act.

Lawhorn, J.M. 2022. "Inflation Reduction Act Methane Emissions Charge: In Brief." IF12238. Congressional Research Service. https://crsreports.congress.gov/product/pdf/IF/IF12238.

Maldonado, S. 2021. "New York City Banned Gas in New Buildings. Here's What You Need to Know." *The City*, December 15. https://www.thecity.nyc/2021/12/15/22838761/new-york-city-banned-gas-in-new-buildings-what-to-know.

Marchese, A., T. Vaughn, D. Zimmerle, D. Martinez, L. Williams, A. Robinson, A. Mitchell, et al. 2015. "Methane Emissions from United States Natural Gas Gathering and Processing." *Environmental Science and Technology* 49:10718–10727. https://pubs.acs.org/doi/pdf/10.1021/acs.est.5b02275.

Massachusetts Department of Public Utilities. 2022. "Investigation Assessing the Future of Natural Gas in Massachusetts." https://www.mass.gov/info-details/investigation-assessing-the-future-of-natural-gas-in-massachusetts.

Minnesota Public Utilities Commission. 2021. "In the Matter of a Commission Evaluation of Changes to Natural Gas Utility Regulatory and Policy Structures to Meet State Greenhouse Gas Reduction Goals." https://efiling.web.commerce.state.mn.us/edockets/searchDocuments.do?method=showPoup&documentId=%7B808FD37A-0000-C31B-B117-9573E0C381A5%7D&documentTitle=20217-176407-01.

Minnesota Public Utilities Commission. 2022. "The Minnesota Public Utilities Commission Moves Forward Implementing the Natural Gas Innovation Act." https://content.govdelivery.com/bulletins/gd/MNPUBUC-3167ec0?wgt_ref=MNPUBUC_WIDGET_2.

NASEM (National Academies of Sciences, Engineering, and Medicine). 2021a. *Accelerating Decarbonization of the U.S. Energy System*. Washington, DC: The National Academies Press. https://doi.org/10.17226/25932.

NASEM. 2021b. *Assessment of Technologies for Improving Light-Duty Vehicle Fuel Economy: 2025–2035*. Washington, DC: The National Academies Press. https://doi.org/10.17226/26092.

NASEM. 2023. *Carbon Dioxide Utilization Markets and Infrastructure: Status and Opportunities: A First Report*. Washington, DC: The National Academies Press. https://doi.org/10.17226/26703.

NETL (National Energy Technology Laboratory). n.d. "Critical Minerals and Materials." https://netl.doe.gov/resource-sustainability/critical-minerals-and-materials.

NRC (Nuclear Regulatory Commission). n.d. "Backgrounder on Decommissioning of Nuclear Plants." https://www.nrc.gov/reading-rm/doc-collections/fact-sheets/decommissioning.html.

OECD (Organisation for Economic Co-operation and Development). 2022. "Confronting the Crisis: OECD Economic Outlook, November 2022." https://www.oecd.org/economic-outlook/november-2022.

Plumer, B., and L. Friedman. 2021. "Over 40 Countries Pledge at U.N. Climate Summit to End Use of Coal Power." *The New York Times*, November 4. https://www.nytimes.com/2021/11/04/climate/cop26-coal-climate.html.

PwC (PricewaterhouseCoopers). 2021. "Impacts of the Oil and Natural Gas Industry on the US Economy in 2019." Prepared for the American Petroleum Institute. https://www.api.org/-/media/files/policy/american-energy/pwc/api-pwc-economic-impact-report.pdf.

Raimi, D. 2021. "Mapping US Energy Communities." Resources for the Future. https://www.rff.org/publications/data-tools/mapping-vulnerable-communities.

Raimi, D. 2023. "Government Revenue from Fossil Fuels in the US: History and Projections." Written Comments Prepared for the Senate Budget Committee. https://www.budget.senate.gov/imo/media/doc/Mr.%20Daniel%20Raimi%20-%20Testimony%20-%20Senate%20Budget%20Committee5.pdf.

Ramseur, J.L. 2022. "Inflation Reduction Act Methane Emissions Charge: In Brief." R47206. Congressional Research Service. https://crsreports.congress.gov/product/pdf/R/R47206.

Ravikumar, S., M. Mujiva, S. Twidale, W. James, and K. MacLellan. 2022. "Britain Approves First New Coal Mine in Decades Despite Climate Targets." *Reuters*, December 7. https://www.reuters.com/world/uk/britain-approves-first-new-coal-mine-decades-2022-12-07.

Secretary-General of the UN (United Nations). 2022. "Secretary-General's Press Encounter on Pre-COP27." https://www.un.org/sg/en/content/sg/press-encounter/2022-10-03/secretary-generals-press-encounter-pre-cop27.

Shao, E., and L. Friedman. 2023. "Ban Gas Stoves? Just the Idea Gets Some in Washington Boiling." *The New York Times*, January 11. https://www.nytimes.com/2023/01/11/climate/gas-stoves-biden-administration.html?action=click&module=Well&pgtype=Homepage&section=Climate%20and%20Environment.

Shrestha, R., Ji. Neuberger, and D. Saha. 2022. *Federal Policy Building Blocks: To Support a Just and Prosperous New Climate Economy in the United States*. Washington, DC: World Resources Institute. https://files.wri.org/d8/s3fs-public/2022-09/federal-policy-building-blocks-support-just-prosperous-new-climate-economy-united-states_0.pdf.

SmartGrowthAmerica. n.d. "Getting Started with Nuclear Funding: Technical Assistance for Nuclear Communities." https://smartgrowthamerica.org/wp-content/uploads/2021/06/nuclear_funding_resource.pdf.

Solomon, M., and H. Tallackson. 2022. "Transmission, Reliability and Gas System Decarbonization Top of Mind for State Utility Regulators 2022." Utility Dive. https://www.utilitydive.com/news/transmission-reliability-and-gas-system-decarbonization-top-of-mind-for-st/619250.

Tankersley, J. 2023. "Biden Caps Two Years of Action on the Economy, with New Challenges Ahead." *The New York Times*, January 2. https://www.nytimes.com/2023/01/02/business/biden-economy.html.

Tax Policy Center. 2022. "Motor Fuel Tax Revenues (States)." Tax Policy Center of the Urban Institute and Brookings Institution. https://www.taxpolicycenter.org/statistics/motor-fuel-tax-revenue.

Tierney, S.F. 2016. "The U.S. Coal Industry: Challenging Transitions in the 21st Century." White Paper. Analysis Group. https://www.analysisgroup.com/Insights/publishing/the-u-s—coal-industry—challenging-transitions-in-the-21st-century.

Tierney, S.F. 2019. "Time to Move Away from Old Precedents in FERC Pipeline Reviews." *Utility Dive*. https://www.utilitydive.com/news/time-to-move-away-from-old-precedents-in-ferc-pipeline-reviews/567512.

Tierney, S.F. 2021. "Comments on Certification of New Interstate Natural Gas Facilities (May 26, 2021)." https://elibrary.ferc.gov/eLibrary/filelist?accession_number=20210526-5272, p. 2.

Traylor, P., and G. Tolley. 2021. "Congress Repeals Trump-Era Methane Rule for the Oil and Gas Sector: What Happens Now?" Vinson & Elkins. https://www.velaw.com/insights/congress-repeals-trump-era-methane-rule-for-the-oil-gas-sector-what-happens-now.

TRB and NRC (Transportation Research Board and National Research Council). 2015. *Overcoming Barriers to Deployment of Plug-in Electric Vehicles*. Washington, DC: The National Academies Press. https://doi.org/10.17226/21725.

UC (University of California) Berkeley. 2023. "States' Climate Action Map." Berkeley California-China Climate Institute. https://ccci.berkeley.edu/states-climate-action-map.

UN (United Nations) Climate Change. 2021. "Statement on International Public Support for the Clean Energy Transition." UN Climate Change Conference UK 2021 in Partnership with Italy. https://ukcop26.org/statement-on-international-public-support-for-the-clean-energy-transition.

Urban Institute. 2019. "State and Local Backgrounders: Severance Taxes." https://www.urban.org/policy-centers/cross-center-initiatives/state-and-local-finance-initiative/state-and-local-backgrounders/severance-taxes.

Urban Institute. 2023. "State and Local Backgrounders: Motor Fuel Taxes." https://www.urban.org/policy-centers/cross-center-initiatives/state-and-local-finance-initiative/state-and-local-backgrounders/motor-fuel-taxes.

U.S. Climate Alliance. 2021. "U.S. Climate Alliance States Commit to New High-Impact Actions to Achieve Climate Goals and Go Further, Faster, Together." http://www.usclimatealliance.org/publications/2021/11/7/us-climate-alliance-states-commit-to-new-high-impact-actions-to-achieve-climate-goals-and-go-further-faster-together.

Van Voorhis, S. 2022. "Massachusetts Enacts Major Climate Bill with Gas Hookup Bans." *ENR New England*, August 12. https://www.enr.com/articles/54612-massachusetts-enacts-major-climate-bill-with-gas-hookup-bans.

Victor, D. 2021. "Energy Transformations: Technology, Policy, Capital and the Murky Future of Oil and Gas." https://reenergizexom.com/documents/Energy-Transformations-Technology-Policy-Capital-and-the-Murky-Future-of-Oil-and-Gas-March-3-2021.pdf.

Victor, D. 2022. "Opinion: The Vital Role That Energy Markets Play in Russia's War Against Ukraine." *Market Watch*, May 5. https://www.marketwatch.com/story/the-vital-role-energy-markets-play-in-russias-war-against-ukraine-11651768274.

Wade, W. 2022. "U.S. Coal Prices Top $100 a Ton for First Time Since 2008." *Bloomberg News*, April 4. https://www.bloomberg.com/news/articles/2022-04-04/u-s-coal-prices-top-100-a-ton-for-first-time-since-2008.

White House. 2021a. "Fact Sheet: President Biden Sets 2030 Greenhouse Gas Pollution Reduction Target Aimed at Creating Good-Paying Union Jobs and Securing U.S. Leadership on Clean Energy Technologies." https://www.whitehouse.gov/briefing-room/statements-releases/2021/04/22/fact-sheet-president-biden-sets-2030-greenhouse-gas-pollution-reduction-target-aimed-at-creating-good-paying-union-jobs-and-securing-u-s-leadership-on-clean-energy-technologies.

White House. 2021b. "Fact Sheet: President Biden Tackles Methane Emissions, Spurs Innovations, and Supports Sustainable Agriculture to Build a Clean Energy Economy and Create Jobs." https://www.whitehouse.gov/briefing-room/statements-releases/2021/11/02/fact-sheet-president-biden-tackles-methane-emissions-spurs-innovations-and-supports-sustainable-agriculture-to-build-a-clean-energy-economy-and-create-jobs.

Zhang, W., A. Noble, X. Yang, and R. Honaker. 2020. "A Comprehensive Review of Rare Earth Elements Recovery from Coal-Related Materials." *Minerals* 10(5):451. https://doi.org/10.3390/min10050451.

Zitelman, K. 2022. "Potential State Regulatory Pathways to Facilitate Low-Carbon Fuels." National Association of Regulatory Utility Commissioners. https://pubs.naruc.org/pub/895485A7-1866-DAAC-99FB-2F331818510F.

Enhancing and Realizing the Climate Ambitions and Capacities of Subnational Actors: State and Local Government Perspectives

ABSTRACT

Subnational governments—state, county, and local entities—and other non-federal actors play important roles in actions to decarbonize the U.S. economy. Many provisions of recent federal laws will need to be implemented through subnational actors. Some jurisdictions are better prepared and/or more willing than others to engage on these issues. To understand the roles of subnational governments, this chapter explores the current landscape of state and local decarbonization policies, the influences of American federalism, the polarization of climate change as a political issue, and the challenges of uneven national leadership. The chapter then explores how robust, locally relevant, and more flexible federal funding for intergovernmental coordination with subnational agencies with climate-related responsibilities could help lead to more effective decarbonization solutions.

The chapter also examines governance attributes, such as technical capacity, resource availability, and agency coordination, that facilitate the achievement of subnational policy goals. It underscores key provisions of the Infrastructure Investment and Jobs Act (IIJA) and the Inflation Reduction Act (IRA) that depend on subnational action, noting the new or expanded responsibilities that subnational governments will need to take on in order to successfully shepherd the implementation of these laws. Some—but not all—provisions in the IIJA and IRA are flexible and can support capacity-building through strategic and dedicated investment in planning, program development, stakeholder engagement, and staffing at the federal, state, and local levels. The chapter concludes with a call for immediate, reliable, and significant investment in state and local government capacity-building to enable the critical policy, regulatory, and bureaucratic environments needed to deploy climate solutions at scale.

INTRODUCTION

The IRA is the latest and the most significant climate-mitigation-related federal legislation in U.S. history. As noted earlier in the report, the IRA holds the potential to address the climate crisis by modernizing American energy infrastructure and decarbonizing the economy. Along with the IIJA, which directed significant investment toward the nation's energy and transportation infrastructure, the IRA represents a critically important milestone in congressional action and federal investment in U.S. efforts to combat climate change. However, its success is not guaranteed. Non-federal and subnational entities will hold significant influence over whether, when, and how effectively these federal funds are used and implemented.

The many actors in the non-federal ecosystem are described in Figure 13-1: this ecosystem contains a multitude of entities, motivations, and relationships among them, from local, county, state, and tribal governments to private-sector businesses and investors, to civic and community-based organizations, faith groups, political alliances, households and individuals, and others.

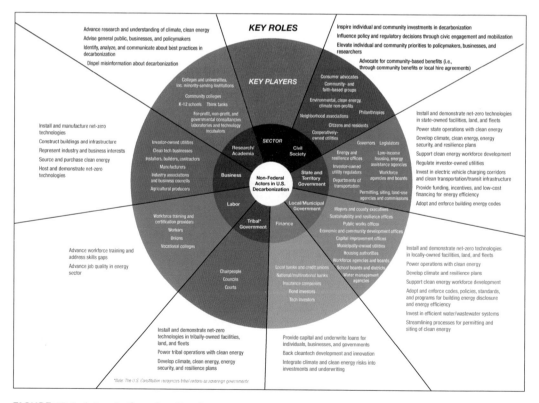

FIGURE 13-1 Actors in the subnational ecosystem.

This chapter will focus specifically on the roles and capacities of state and local governments that exist and operate in relation to the federal government and the other nonfederal stakeholders listed above. While state and local governments do not represent the full subnational ecosystem of climate and clean energy players in the United States, this chapter focuses on them because their roles are foundational and essential to the achievement of the IIJA, the IRA, and net-zero goals. State legislative, executive, and regulatory institutions will play a make-or-break role in the implementation of policies, and the effectiveness of state and local processes for procurement, siting, zoning, infrastructure development, workforce development, and partnership building will be foundational in deploying low-carbon technologies at scale and in an equitable manner.[1]

State and local governments have driven a significant amount of U.S. climate progress over the past decade, sometimes in the face of federal inaction. States, cities, and counties that are committed to climate action currently represent two-thirds of the U.S. population and economy (Zhao et al. 2022). Many continue to adopt their own ambitious, jurisdiction-specific climate, clean energy, energy efficiency, and decarbonization policies, including carbon pricing, clean electricity and renewable portfolio standards, emission limits, zero-emission vehicle deployment, low-carbon fuel standards, buy clean standards, building performance and electrification incentives, and energy codes and standards. In addition to decarbonization policies, a growing number of states and cities are taking steps to address issues related to energy affordability and access, ensure community participation in siting and development of energy infrastructure, and advance environmental justice by identifying and reducing pollution burdens in disadvantaged communities and targeting investments in those communities (Hanus et al. 2023; Ricketts et al. 2020).

Subnational governments have played a critical, complementary, and sometimes contentious role to the federal government by enacting regional and local climate policies and regulations. The diversity and range of climate actions adopted by subnational entities have been key in fostering policy, innovation, and experimentation; driving federal action; and delivering near-term emissions reductions. Policy experimentation at the subnational level often creates the potential for best ideas and practices to spread to other states and localities (horizontal diffusion) and even percolate up to the federal government (vertical diffusion). In fact, action by states, particularly those that are more stringent than the federal policy, has historically spurred the federal government to take new or more robust climate action. For example, California was the first state in the country to adopt appliance efficiency standards in 1976. Other states,

---

[1] For a closer look at the roles of other non-federal actors, such as businesses and civil society, see Kennedy et al. (2021) and Vandenbergh and Gilligan (2017).

including Massachusetts and New York, followed, eventually leading to federal standards in the National Appliance Energy Conservation Act of 1987 (Bianco et al. 2020). More recently, the federal American Innovation and Manufacturing Act of 2020 directed the Environmental Protection Agency (EPA) to phase down the production and consumption of hydrofluorocarbons (HFCs) and built on leadership by several states that have passed laws limiting HFCs in recent years (Rabe 2021). Maine's long-standing investments and programs in cold-climate heat pumps have helped set the stage for other state and federal efforts in the electrification of home heating and cooling (Officer of the Governor Janet T. Mills 2021).

State and local initiatives will continue to be central to decarbonization—although recent legislation is a significant enabler, federal support for decarbonization has not been, nor is it likely to be in the future, a guarantee (Bloomberg 2018; Hultman 2020). Relying on a multitude of diverse policies across subnational governments is an effective although insufficient path, and will remain crucial with or without a national strategy.

Recent federal laws represent the boldest action Congress has taken to address climate change—putting the United States on a possible path to cut greenhouse gas (GHG) emissions by 40 percent below 2005 level by 2030. Yet, the combined forces of the IRA, the IIJA, and CHIPS, complemented with existing subnational decarbonization policies, are by themselves not enough to meet the U.S. Nationally Determined Contribution (NDC) to the Paris Agreement of 50–52 percent emissions reduction by 2030 compared to 2005 levels. Effective state and local implementation of these federal laws and additional federal policies (e.g., tailpipe standards for light-duty vehicles) will be required to make progress in addressing the climate crisis. Analyses show that it is possible to close the approximately 10 percent emissions gap between the NDC and recent federal actions, including the IRA, the IIJA, and CHIPS, with additional federal and subnational actions (Orvis et al. 2022; Zhao et al. 2022). Subnational leadership in climate action is particularly important in that it can lower the barriers and costs for other state and local governments to follow (Peng 2021), especially when exemplified in diverse political, market, and geographic contexts.

The operative word in these analyses is that it is *possible* to close the emissions gap—but by no means a given. For instance, one important action that states can take is to strengthen or adopt clean energy standards in their electricity sector. Yet, currently, fewer than half of the states have laws, executive directives, or voluntary goals committing to 100 percent clean electricity by 2050 or sooner, with Minnesota becoming the latest state to pass legislation requiring electric utilities to use 100 percent clean energy by 2040 (Clean Energy States Alliance 2023). Other critical actions that state and local governments can take include adopting zero-emission vehicle sale targets and mandates, accelerating the retirement of coal-fired power plants, modernizing building energy codes, preparing the clean energy workforce, and implementing fugitive methane leak recovery (Zhao et al. 2022). However, as this chapter explores, the current landscape of climate policies and subnational capacity for decarbonization may be insufficient without federal–subnational coordination, accelerated investments in capacity-building initiatives at the subnational level, and efforts to soften some states' and communities' resistance to climate action by emphasizing the economic, job creation, security, and resilience opportunities that can come from clean energy and climate investments.

In addition to federal action and subnational implementation, private-sector actors are also playing a meaningful role. Further discussion of the importance of private-sector actors in decarbonization efforts is available throughout other chapters in the report; some examples of private-sector impacts on subnational governments' clean energy and climate goals are highlighted in Box 13-1.

The climate actions of state and local governments represent a patchwork quilt of will, capacity, and influence, with deep variations across different policy, regulatory, geographical, and market environments. The evolution of U.S. climate and energy policy is a story of multi-level, multi-nodal governance, decision-making, and action—or, in some circumstances, inaction. This array of climate policies represents a significant risk to the goal of transitioning the entire U.S. economy to net zero by 2050. As Basseches et al. (2022, p. 4) note, despite pockets of advancement and innovation, more consistent and stringent policy coverage is necessary to meet climate mitigation objectives, "necessitating efforts to reduce obstacles to more robust state climate policy activity." Deep decarbonization will remain incomplete so long as subnational actors in any part of the country refuse or struggle to implement and engage their policy makers, regulators, business communities, and public in the complex and difficult tasks of transitioning the U.S. energy system laid out in Chapters 6–12.

To understand the many inconsistencies, obstacles, and opportunities surrounding climate and energy policy, this chapter begins with an exploration of the role of American federalism, climate change as a polarizing political issue, uneven national leadership, regional and geographic differences, and the complexity of solutions. It then examines governance attributes that are critical to achieving subnational public policy goals such as decarbonization, including political messaging and will around clean energy and climate efforts, staffing and resources, technical expertise, autonomy from special interests, and the ability to integrate decarbonization into locally relevant

Public- and Private-Sector Relationships and Interactions

Independent of federal-level policies, state and local governments can influence private-sector investments in decarbonization through mandates and incentives. This includes enacting laws and regulations that require investor-owned electric utilities to use clean energy sources, adjusting zoning and land use to streamline clean energy and climate resilience infrastructure, and setting standards for vehicles and fueling and charging infrastructure, examples of which can be found in Chapters 3, 7, and 9, respectively.

There are several initiatives connecting subnational levels of government and the private sector. Regional efforts include the Northeast and Mid-Atlantic Transportation and Climate Initiative (TCI), the REV-West Initiative, and the Southeast REVI. TCI's original goal was to develop a cap-and-trade program to reduce transportation emissions. Now, TCI's Northeast Electric Vehicle Network is partnered with more than 100 companies, organizations, and jurisdictions to support electric vehicle use. The REV-West Initiative and the Southeast REVI is under the National Association of State Energy Officials (NASEO). Both programs involve collaboration between states and other stakeholders, including companies, to accelerate the adoption of electric vehicle (EV) infrastructure and policy in their respective areas.

Cities can also come together to increase purchasing power and increase EV deployment locally. The EV Purchasing Collaborative was formed in 2017 by Climate Mayors. The collaborative included 31 cities that issued an Electric Vehicle Request for Information. This led to widespread commitments to purchase EVs. Aside from the economic benefit to private automakers, the cultural shift and announcement that many cities want such technology can affect how companies approach their own decarbonization goals.

Other examples stress the potential impact the private sector has in accelerating decarbonization. The Volkswagen Clean Air Act Civil Settlement, for example, led to the construction of EV charging infrastructure in all 50 states. Although this was a reaction to non-compliant practices, it still shows the impact private companies can have on decarbonation if funds are used in particular ways.

Corporate-Led Decarbonization

Many major corporations have made climate pledges to reduce the emissions from their products, supply chains, and daily operations. Chapters 6, 9, and 10 describe some specific examples for the electricity, automotive, and industrial sectors. To make good on such pledges requires decision makers in the energy, technology, and other private sectors (chief executive officers and boards of directors) to allocate significant capital investments and resources required to transition business operations. Such investments include up to $270 billion in domestic clean energy projects and manufacturing over the past year (ACP 2023). As another example, Walmart's Project Gigaton, in partnership with a variety of nonprofit organizations, aims to achieve a billion tons of GHG emissions reductions from Walmart's supply chain by 2030 by targeting energy use, nature, waste, packaging, transportation, and product use and design

BOX 13-1 Continued

(https://www.walmartsustainabilityhub.com/climate/project-gigaton). In addition to commitments by corporate leaders of capital and human resources, transparency and effectiveness remain two critical aspects of such pledges. These aspects often rely on third parties, such as CDP (formerly Carbon Disclosure Project), for public reporting of GHG emissions. In an attempt to validate targets, the Science Based Targets initiative (STBi) was developed to create methods and criteria for evaluating the effectiveness of corporate climate action; only targets that meet strict criteria are approved (WRI 2023).

One method that companies use to lower their net GHG emissions is offsetting emissions. The quality of offsets ranges widely, and several factors must be taken into account, including additionality, baselines, leakage, perverse incentives, durability, emissions factors, do no harm, and scalability. Cames et al. (2016) found that 85 percent of carbon offset projects overestimate their actual impact on net emissions; therefore, transparent reporting and consistent metrics are necessary to track progress. There are several programs that allow companies to report offsets through a database voluntarily, such as Berkeley's Voluntary Registry Offsets Database. For more discussion on voluntary offsets, see Chapter 11.

Corporations have also turned to consulting companies to determine decarbonization strategies. McKinsey Sustainability, for example, "help(s) companies identify decarbonization opportunities that work both environmentally and financially."

Additionally, as the cost of renewable power continues to decrease, a growing number of companies are seeking to locate operations in jurisdictions with low barriers to grid access and interconnection, options third-party renewable energy procurement (such as power purchase agreements), tax exemptions and benefits, strong transmission and distribution infrastructure, and other policies (such as net metering) that enable access to reliable, low-cost energy for their operational needs (Bird et al. 2017). As of the end of 2022, 326 companies contracted 77.4 gigawatts of wind and solar energy across 49 states, the District of Columbia, and Puerto Rico. Of contracted capacity bought by companies, 35 percent is coming from Texas alone (ACP 2022). In this scenario, state and local economic development goals serve as a powerful motivator to integrate diverse and cleaner energy sources onto the grid.

Companies have several approaches they can take to improve decarbonization efforts:

1. Companies can adopt new technologies as they are being developed to expedite their development and deployment. Company interest in new technologies can also include direct funding to research institutions.
2. Companies can make pledges to commit to decarbonization goals publicly. The transportation and electricity sectors, for example, have made pledges indicating the year and emission-level goals (see Chapters 6 and 9). These public declarations can help hold the private sector accountable. Other corporations that sell products to consumers, like Walmart, can also decrease emissions of their day-to-day operations and can make commitments for similar goals. Publishing the specific strategies that companies implement is helpful for others to mimic and for clear communication of what steps are being taken.

continued

3. Collaboration among companies and between companies and other groups is important to take collective action and share knowledge. For example, electric vehicle companies partnering with delivery companies allows for both sectors to contribute to the demand and supply of decarbonization technology. Companies can also collaborate with other subnational groups like local and state governments to increase demand for developed products. Collaboration between subsets of industries, such as construction and developers, is also necessary to show how different groups can benefit, economically and otherwise, from decarbonized technology that may be more expensive up front.

4. Monitoring and reporting emissions is crucial for companies to accurately assess progress toward pledged goals. This can include both internal and external audits of emissions and of specific new technologies and process. Consistent monitoring and public reporting of results can help hold companies accountable to their consumers. If progress is failing to meet interim goals, strategies must be reassessed.

and coherent policy portfolios. Building on this analysis, the chapter recommends strengthening structures that will enhance capacity-building and coordination among subnational actors and the federal government. Next, it highlights key provisions in the IIJA and IRA that will require subnational action and involvement, with a particular focus on the ability of the implementation of these bills to either contribute to more effective state and local policy regimes or, conversely, to deepen existing variations among subnational actors. Last, the chapter concludes with an outlook on how decarbonization policy may continue to play out at the subnational level and among federal, state, and local government actors. Table 13-1, at the end of the chapter, summarizes all the recommendations that appear in this chapter to support subnational actors in policy implementation and advancing decarbonization objectives.

U.S. FEDERALISM AND THE EVOLUTION OF STATE AND LOCAL CLIMATE POLICY

Federalism has been the bedrock of the U.S. system of government for more than two centuries. Over this time, American federalism has evolved to become a mixture of dual powers and responsibilities, and cooperation and conflict between the federal government and the states (Kincaid 2017).

Consistent with the dualism of American federalism, power is divided between the federal government and the states rather than being shared or concentrated at any one level, leading to most policy making as an inherently intergovernmental endeavor.

The U.S. Constitution explicitly recognizes the rights of individual states to function as what Supreme Court Justice Louis Brandeis called "laboratories of democracy," where states can experiment with innovative policies and have the authority to delegate many of their powers to local governments (*New State Ice Co. v. Liebmann* 1932; Tyson and Mendonca 2018). At the next level down, state constitutions create a baseline balance between state and local authority. At one end of this spectrum is state supremacy in the form of "Dillon's rule" (which views local governments as administrative arms of the state, with no inherent lawmaking powers other than what the state expressly grants); on the other, "home rule" (which grants local and municipal governments full capacity to govern affairs within their territorial jurisdiction, subject to state law limitations) (Richardson et al. 2003; Toscano 2018). While 39 states employ Dillon's rule as a starting principle to define local authority, every state constitution devolves some degree of decision making away from the state apparatus, creating unique dynamics and opportunities for local units of government to exercise autonomy.

Over the past decades, the American system of federalism has shifted between cooperation and conflict in continuous evolution among the federal government, states, and localities. In the cooperative variant, federal officials have shown a willingness to negotiate with state and local officials over formulating and implementing policies, and in turn states and localities have engaged the federal government to advance commonly shared goals. Many federal environmental laws—such as the Clean Air Act, the Clean Water Act, and the Safe Drinking Water Act—have operated under the cooperative federalism model, where states have implemented and enforced federal laws while retaining the power to enact policies that are more stringent (Kastorf 2014; Lin 2020). This model of cooperative federalism is by and large still in place today, although the exact scope of federal, state, and sometimes local authority has shifted through the years.

In the uncooperative or conflict model of federalism, relations between the federal and subnational governments are characterized by resistance or direct opposition. In some cases, states and localities initiate new ideas that the federal government is not yet ready to embrace. For instance, many states and cities have moved ahead of the federal government on climate change, gun control, and policies governing democratic processes, such as automatic voter registration (Rose and Goelzhauser 2018). In other cases, some states have advocated for less federal authority and challenged federal government policies to reduce GHGs. The precise nature of the relationship between the federal and lower levels of government varies depending on the political context and whether the federal government's policy priorities and political ideology align with or diverge from those of specific states (Konisky and Woods 2018; Tyson and Mendonca 2018).

As explored below, examining the history and evolution of U.S. climate policy through the lens of federalism helps to explain the current fragmentation in climate policy interest and capacity at the subnational level. This variation both propels and impedes decarbonization efforts to this day.

Early U.S. Climate Policy: Foundational Yet Modest Steps

Since the early 19th century, the migration of settlers from the humid east to the arid west created a need to address adaptation to changing climate conditions in U.S. policy deliberations (Holmes 2015). The issue of climate change due to anthropogenic GHG emissions later emerged in the mainstream public and political consciousness beginning in the mid- to late 20th century. These recent decades saw the establishment of key federal agencies and programs that play an important role in national, state, and local decarbonization efforts—among them are the Department of Energy (DOE), EPA, the Federal Energy Regulatory Commission (FERC), and the White House Council on Environmental Quality. State governments, too, took foundational steps in climate, clean energy, and energy efficiency policy during this early era. Throughout the 1960s and 1970s, state legislatures established Departments of Environmental Protection, Pollution Control Agencies, Air Control Boards, and other entities focused on air and water quality and other natural resource management issues. On the energy front, in response to the energy crisis of the early 1970s, the U.S. State Energy Program created by the Energy Policy and Conservation Act of 1975 (P.L. 94-163) prompted governors from all states and territories to establish State and Territory Energy Offices. These agencies are tasked with convening stakeholders, informing legislators and regulators, and funding and financing energy efficiency and conservation programs (DOE 2023a). In the early 1980s, Iowa became the first state to adopt a renewable portfolio standard (Database of State Incentives for Renewables and Efficiency 2022), and by the end of the 1980s, California passed legislation mandating an inventory of state GHG emissions (Farber 2021).

Importantly, early U.S. climate policy was not marked by the political divisiveness and polarization that characterizes decarbonization efforts today (Worland 2017). Rather, many of the defining and foundational milestones in federal environmental policy were achieved through bipartisan action. Such examples include the Nixon administration's establishment of EPA and the National Oceanic and Atmospheric Administration (NOAA) (Richard Nixon Foundation 2014), widespread bipartisan congressional support for the 1990 Clean Air Act Amendments (S. 1630) (Grassle 2021), and the George H.W. Bush administration's ratification for the United States of the United Nations Framework Convention on Climate Change (Bush 1992).

Rabe (2011, p. 499) argues that very few of these early steps amounted to a "serious policy initiative" but rather sought to recognize the problem of climate change and set a foundational—if imperfect—governance strategy at the federal and state level. Nevertheless, these moves caught the attention of skeptics. Collomb (2014) notes that in addition to the strength of oil and gas interests in sowing doubt and misinformation about climate science, two additional factors became critical in blocking strong climate policy: opposition among small government conservatives and libertarians to regulation, and the potential loss of American prosperity and competitiveness, particularly in relation to emerging economies. "The fear that strong climate action might reduce American competitiveness with rising giants like China is undoubtedly one of the strongest reasons why the Senate refused to ratify the Kyoto Protocol in 1997" (Collomb 2014, p. 8).

Federal Disengagement and the Rise of Subnational Climate Action

State and local efforts to address climate change increased significantly after the United States failed to ratify the Kyoto Protocol and the federal government repeatedly signaled a lack of interest to pursue climate change mitigation strategies (Bromley-Trujillo and Holman 2020). Between roughly 1998 to 2007, some states took unilateral policy steps to reduce GHG emissions by experimenting with cap-and-trade and GHG auction programs,[2] renewable portfolio standards,[3] and a range of other environmental and economic development policy tools. In this period, local governments also organized around climate action. The International Council for Local Environmental Initiatives (ICLEI) created the most extensive network of city climate mitigation action under its Cities for Climate Protection campaign. By 2007, 171 U.S. municipalities had set emission reduction targets and were pursuing GHG reduction strategies (Byrne et al. 2007). Another 435 cities committed to meeting or exceeding the U.S. Kyoto reduction target as part of the U.S. Mayors Climate Protection Agreement, launched in 2005 (Byrne et al. 2007).

While climate and environmental concerns drove some of this activity, in politically conservative states, perceived economic advantage, job creation, and energy security and reliability were likely even more important impetuses (Engel and Barak 2008; Gallagher 2013). Statehouses across the country recognized an opportunity to lessen their dependence on imported energy and expand the market for "home-grown" energy options and locally manufactured/provided goods and services. Similarly, city

---

[2] For example, the Northeast/Mid-Atlantic states that are members of the Regional Greenhouse Gas Initiative (for more information, see https://www.rggi.org) and California.

[3] As of 2007, 25 states had adopted a renewable portfolio standard (Wiser et al. 2008).

officials cited the economic benefits of energy efficiency measures as the primary motivation for actions to reduce GHG emissions (Kousky and Schneider 2003). Addressing health impacts from air pollution, the need for transportation alternatives, and concerns over the livability of their cities were other reasons cited by local governments to justify actions leading to GHG reductions (Betsill 2001).

The framing of the issue as an economic and energy security opportunity rather than as an environmental imperative provided openings for subnational governments on both sides of the political aisle to pursue decarbonization actions, although the most ambitious policy activity at the time tended to be concentrated in regions with Democratic governors and mayors, including the Northeast, Pacific West, and Southwest. Congressional gridlock and federal inaction in this era provided what Rabe describes as "enormous latitude to states to do nothing, pursue a few symbolic steps, enact one or two significant policies, or pursue a far-reaching approach that might position them for regional and national leadership and even global visibility" (Rabe 2011, p. 504).

Federal Reengagement, Mixed Subnational Responses, and Mounting Polarization

Following the Supreme Court's 2007 decision in *Massachusetts v. EPA*[4]—where the Supreme Court ruled that carbon dioxide and GHGs were air pollutants under the Clean Air Act and could be regulated by EPA—and throughout the Obama administration, federal policy proposals explored a variety of different weightings of subnational versus national authority. At one extreme, total preemption of state regulation and policy via a national carbon cap-and-trade regime—an option technically available to Congress through a preemption statute—failed to gain political traction due to a

---

[4] Rabe (2011, pp. 504–505) describes this seminal case: "The first significant indication that the American intergovernmental balance on climate policy might be changing occurred when the U.S. Supreme Court performed the role of intergovernmental umpire. Massachusetts and twelve allied states contended that the federal government's refusal to designate carbon dioxide as an air pollutant under the 1990 Clean Air Act Amendments placed them in danger of such risks as sea level rise owing to climate change. Ten other states took the opposite position, backing the Bush Administration's claim that the federal government lacked statutory authority and those states had no business being in court on such a matter." In 2007, a five-to-four majority of justices ruled in favor of forcing EPA to revisit its refusal to designate carbon dioxide as an air pollutant. "Massachusetts cannot invade Rhode Island to force reductions in greenhouse gas emissions, it cannot negotiate an emissions treaty with China or India, and in some circumstances the exercise of its policy powers to reduce in-state motor-vehicle emissions might well be pre-empted," wrote Associate Justice John Paul Stevens in the majority opinion. "These sovereign prerogatives are now lodged in the Federal Government." This decision represented a federal court response to state pressure to compel federal executive branch action, with potentially far-reaching intergovernmental consequences (Engel 2009; *Massachusetts v. EPA* 2007).

variety of factors, including its encroachment on state policy (Rabe 2011). Less extreme approaches led to some successes in climate policy. These included

1. The renewable fuel standard in the 2007 Energy Independence and Security Act (which created a federal floor for the minimum amount of ethanol blended into gasoline, and a schedule of increase for this minimum, but did not preempt any of the states that had already established their own policies) (Rabe 2011);
2. The 2012 54.5 mile per gallon fuel efficiency standard, in which automotive companies in collaboration with EPA, the Department of Transportation (DOT), and the California Air Resources Board developed the first GHG standards for light duty vehicles (Obama White House 2012); and
3. EPA's granting of a waiver to California in 2013 for its "Advanced Clean Car" regulations (NHTSA 2018), which helped advance what is today a multi-state zero-emissions vehicle program (C2ES 2022).

Similarly, the Obama administration's Clean Power Plan sought to advance a cooperative federalism approach by assigning emissions reduction targets but allowing considerable flexibility to states in how to achieve them (Engel 2015). Even though it drew polarized reactions at the time and was ultimately stayed by the Supreme Court, it is today recognized for having prompted many states to plan for power sector emissions reductions (UCS 2021).

In the years leading up to the enactments of the IIJA and IRA, the landscape of subnational climate policy featured a deepening of state climate positioning, often along partisan political lines. During the Obama and Trump administrations, subnational governments, especially states, pushed back on implementing federal policies owing to political polarization and the associated level of agreement with the current presidential administration. Mounting politicization of climate change in these years (Jaffe 2018) entrenched climate policy activity further in certain states, cities, and regions; emboldened yet others to use their unique subnational authorities to exert pressure in the opposite direction; and left many in between incapacitated to act meaningfully on decarbonization ambitions. While dozens of California cities have moved to ban gas and propane hookups in new construction, 19 other states, collectively representing nearly one-third of residential and commercial natural gas consumption in the United States, have passed legislation preventing localities from doing so (Gleason 2022),[5] even as

---

[5] Localities are not the only targets of such efforts, as political and ideological rifts have also affected the movement of clean energy across state lines. One example is playing out in Millard County, Utah, where, to comply with California's carbon emission standards, the Intermountain Power Agency (IPA) has announced plans to convert its plant, which services the Los Angeles Department of Water and Power, from coal to natural gas by 2025 and eventually be fueled by emission-free hydrogen produced with energy from solar farms under development nearby. In response, the Utah legislature passed a bill stripping IPA of privileges and tax exemptions it has long enjoyed. While this move has not necessarily derailed IPA's decarbonization plans, it is expected to affect its ability to procure low-cost capital as well as its operating revenue (Maffly 2021).

two-thirds of Americans favor using a mix of fossil fuel and renewable energy sources (Tyson 2022).

Conflicting climate-related policies during the Obama and Trump administrations cast a spotlight on the deeply fragmented landscape of subnational climate and decarbonization policies and demonstrated the significant power of committed subnational and non-federal institutions in propelling climate policy even during times of federal stalemate and inaction. Through executive orders and regulations in President Obama's second term, the Obama administration advanced high-profile initiatives that sparked widespread support from some states, localities, and companies and deep political and legal ire from others—including the proposed Clean Power Plan designed to reduce carbon dioxide emissions from the U.S. power sector by 32 percent below 2005 levels by 2030,[6] the rejection of the Keystone XL pipeline, and support for the Paris Agreement (Lavelle 2015).

Following, the Trump administration made moves to reject the previous administration's climate policies—which included replacing the Clean Power Plan with a much weaker Affordable Clean Energy rule, withdrawing from the Paris Agreement, reducing the social cost of carbon, and increasing oil drilling in the Arctic refuge, to name a few (Nuccitelli 2020). Reversing policies from the previous administration, these moves led to legal challenges from 17 states and gave rise to high-profile coalitions that aimed to leverage state, local, and private-sector climate commitments to work toward decarbonization despite federal inaction (Reuters 2017). In the wake of the Trump administration's decision to pull out of the Paris Agreement, a coalition of states, cities, businesses, and universities signed the "We Are Still In" declaration, committing themselves to drive down their GHG emissions consistent with the goals of the Paris Agreement. Since its release in June 2017, more than 3,800 mayors, governors, university presidents, and business leaders—representing more than 155 million Americans and $9 trillion of the U.S. economy—have signed the declaration. Simultaneously, New York City Mayor Michael Bloomberg and California Governor Jerry Brown launched the America's Pledge initiative to aggregate and quantify the actions by these subnational actors to drive down GHG emissions (We Are Still In n.d.). America's Pledge serves as an analytical companion to the We Are Still In movement (We Are Still In and America's Pledge publicly merged in 2021 to form America Is All In). Bottom-up actions by these non-federal actors played a crucial role in keeping the momentum on climate action going during the Trump administration.

---

[6] Owing to changes in policies and the economics of fossil-based electricity generation since its announcement and ultimate Supreme Court stay, the U.S. power sector has already surpassed the GHG targets of the Clean Power Plan, more than 10 years ahead of schedule. In this light, it may feel ironic that within months of its release, 28 states sued Obama's EPA over the plan, and power companies and congressional opponents labeled it "aggressive, impractical, and reckless" (Schaeffer and Pelton 2021).

States and regions doing less to foster clean energy and climate investments risk making their economies vulnerable to the economic transition occurring in the U.S. energy sector (Muro et al. 2019). Resources for the Future finds that regardless of federal climate action through the IIJA and IRA, coal will decline dramatically in the next 2 decades, and impacts will be felt most strongly in localities where public services and local economies continue to rely heavily on revenue and royalties from its production, transportation, processing, and consumption (Raimi et al. 2022).

Similarly, states and regions with poor energy efficiency policies are not only falling behind in helping their industry and businesses capitalize on revenue and productivity gains (Global Alliance for Energy Productivity 2016), but they are also placing economic burdens on their residents, and disproportionately so on disadvantaged communities. Counterintuitively, this holds true especially in regions with the lowest electricity rates in the country: low-income households in Mississippi, South Carolina, Alabama, Georgia, and Arkansas bear the highest energy burdens in the country because of their high levels of consumption owing to low investment in energy efficiency in relation to income (DOE 2018).

On the other hand, jurisdictions that have adopted clean energy and energy efficiency—regardless of partisan positioning or anti-climate positions—are able to embrace tangible opportunities for revenue generation, economic and workforce growth, and technology and business development. For instance, due in large part to public policies and investments, New York's clean economy includes 165,000 workers, is exceeding other industries in growth, and contributes significantly to local, state, and federal revenues through taxes on production and imports (New York State Energy Development Authority 2022). Texas's dominance in wind power and utility-scale solar is reaping significant economic benefits for local communities, and particularly rural landowners—up to $8.8 billion in new tax revenue over the lifetime of the existing fleet (Rhodes 2023), an example highlighting the drawbacks of broadly categorizing states as clean energy actors based solely on their political leaning. In another analysis, a study of the impacts of economy-wide investments in zero-carbon power generation and demand-side technologies in Wisconsin finds up to a 3.0 percent increase in gross state product and 68,000 additional jobs across the utility, manufacturing, and construction sectors by 2050 (Hartvig et al. 2022).

These disconnects reveal some of the costs of climate polarization and gridlock in a federalist system: slower decarbonization, but also missed opportunities for state and local economic development, competitiveness, workforce growth, public health, and resilience in the face of inevitable energy market transitions and rising economic costs of climate change. Subnational governments can have a wide variety of reasons for adopting policies and actions that lead to GHG reductions, and understanding and advocating for those reasons can be an effective way to drive further subnational climate action. At the same time, understanding why some subnational governments are not able or willing to invest, and addressing those concerns, could be key to enhancing adoption of decarbonization policies.

Leading Up to the IIJA and IRA, Subnational Action Falling Short of Decarbonization Goals

Important as they may be for seizing economic development and job creation opportunities, subnational climate ambition and policies remain insufficient and are nowhere near the scale and pace of emissions reduction required across the entire country. An analysis conducted by the Environmental Defense Fund using historical and projected emissions data from Rhodium Group found that even among states that have committed to economy-wide GHG emissions reduction of 26 to 28 percent below 2005 levels by 2025 (in line with the U.S. commitment under the Paris Agreement), the suite of their policies adopted as of 2020—notably, before the passage of the IIJA and IRA—was "not nearly enough" to meet the goals set by the states themselves (Stilson et al. 2020, p. 7). Their remedy called for a significant increase in focus and investment by states to pursue not only "surgical" interventions in clean electricity, vehicle standards, energy efficiency, and electrification, but also comprehensive action with enforceable emissions limits across sectors and equitable outcomes for local communities. Climate scorecards published in 2022 by the Rocky Mountain Institute reveal similar findings for climate "front-runner" states Colorado, California, Illinois, New Jersey, New York, and Washington: none were found to be on track to achieve 50 percent economy-wide emissions reductions by 2030 (RMI 2022). In particular, California, a state well known for its climate policy, will need to triple its historical decarbonization rate in order to meet the target of reducing its economy-wide emissions 40 percent below 1990 levels by 2030. Modeling by Energy Innovation reveals that California's policy commitments, as of 2022, would produce statewide emissions nearly 20 percent above its 2030 target (Busch et al. 2022).

At the local level, analyses prior to the passage of the IIJA and IRA found that even the most populous and well-funded cities had been falling short of emissions targets. A 2020 analysis by Brookings found that fewer than half of the country's 100 largest cities have emissions goals, equating to roughly 6 percent of U.S. emissions in 2017. As of 2020, two-thirds of these cities were lagging in achieving their goals, putting even the 6 percent reduction figure into question (Markolf et al. 2020).

Cities were also found to be vastly undercounting their own emissions (on average by 18 percent), as they lack the tools to measure the emissions they are generating and monitor progress in decarbonization (Gurney et al. 2021). A 2021 study of 167 cities from around the world found that "current inventory methods used by cities significantly vary, making it hard to assess and compare the progress of emission mitigation over time and space" (Wei et al. 2021, p. 2). Accurate data and access to estimations and modeling are critical for making effective decisions. In the absence of accurate emissions data, cities run the risk of not being able to prioritize mitigation solutions, misallocating scarce resources, and failing to course correct. At the same time, research has shown that there are often valid reasons for variability in how cities and communities define and measure sustainability, including GHG emissions. The act of developing locally relevant metrics and inventories is a strong motivator to action and helps ensure that action is consistent with local priorities and values and engages local publics in meaningful ways (Miller 2005, 2007).

The reality that even the best-resourced and most-climate-focused state and local governments had been falling short of meeting self-imposed emissions reductions targets has made the prospects for decarbonizing less populous states, smaller localities, rural and remote communities, and tribal communities especially challenging—even as evidence suggests that many of these places would like to participate in decarbonization and that large majorities of both Republican and Democratic voters support solar and wind expansion (Nicholas Institute for Environmental Policy Solutions 2018; Pew Research Center 2016).

As noted by the Housing Assistance Council, geographic isolation and low levels of economic opportunity have contributed to persistently high poverty for "several predominantly rural regions and populations such as Central Appalachia, the Lower Mississippi Delta, the southern Black Belt, the Colonias region along the U.S.-Mexico border, Native American lands, and migrant and seasonal farmworkers" (Housing Assistance Council 2023). This is often exacerbated by high levels of energy poverty and energy cost burdens in these same communities (Biswas et al. 2022). In this context, even in communities where the will and ambition to reduce emissions are strong, socioeconomic challenges such as lack of access to capital, community capacity, climate threats, poor housing and infrastructure, workforce shortfalls, and the potential disruptions to local (fossil-reliant) economies and workforces are likely to impede efforts (Clean Energy Transition Institute 2023). Even in states and communities with strong climate ambitions, opposition to projects stems from multiple sources: health and safety concerns, institutional and procedural complexities in siting, and fear over diminishing the value (monetary, aesthetic, recreational, and emotional) of the land on which projects are located (Susskind et al. 2022). For more on the equity dimension of the energy transition, see Chapter 2, and for more on the need for federal, subnational, and private-sector investment in public engagement, see Chapter 5.

Today, state and local climate policies continue to be highly variable; Figure 13-2 illustrates the patchwork of subnational ambition and policy making in climate. The hard reality of climate math reveals that there is no likely way to reach U.S. climate targets without achieving significant emission reductions in states with the biggest carbon footprint. Five states alone—Texas, California, Florida, Pennsylvania, and Ohio—account for one-third of total carbon dioxide emissions, while the top 10 emitting states account for 50 percent of the nation's carbon emissions (EIA 2022a,b). Although the United States reduced its energy-related carbon emissions by almost 16 percent between 2011 and 2020, a handful of states, including Idaho, Mississippi, Oregon, and South Dakota, increased their emissions (EIA 2022a,b).

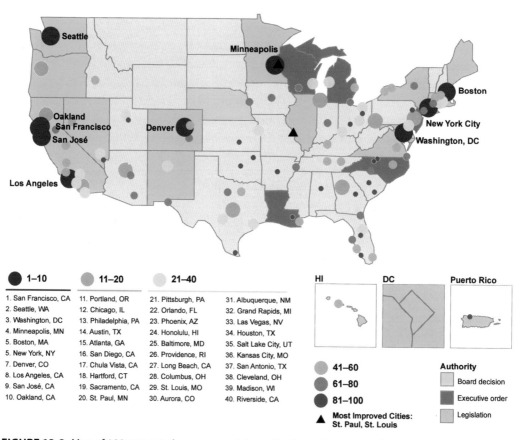

FIGURE 13-2 Map of 100 percent clean energy states with clean city scorecard cities. NOTES: The circles represent the scores for 100 U.S. cities on their efforts to advance clean energy goals. Bubble size is representative of city population. SOURCES: Adapted from City Scorecard Rankings courtesy of Samarripas et al. (2021), *American Council for an Energy-Efficient Economy*, and Map of 100% Clean Energy States courtesy of Clean Energy States Alliance (2023), ©2023 Mapbox ©OpenStreetMap.

Renewable Portfolio Standards (RPSs) have been adopted by many states across the country, but they vary in terms of their stringency (Basseches et al. 2022). At one end, California requires 60 percent and 100 percent of its electricity from renewable energy by 2030 and 2045, respectively. Meanwhile, Ohio—one of the top emitters of carbon dioxide from its power sector—requires that 8.5 percent of electricity sold by the state's utilities be from renewable energy by 2026 (California Energy Commission 2023).[7] In fact, Ohio weakened its RPS from 12.5 percent to 8.5 percent in 2019—even as much of the rest of the Midwest has sought to strengthen their states' clean energy commitment and to reduce the region's reliance on fossil fuels, highlighting the fact that subnational climate policies are not immune to reversal (EIA 2022a,b; Williams 2019). In 2015, West Virginia became the first state to repeal its RPS entirely, even as it was clear the state was running low on easily accessible coal mines, and cheap natural gas was displacing coal generation (Beirne 2015; Bromley-Trujillo and Holman 2020; Tomich 2019).

By changing the fundamental economics of many aspects of the U.S. energy system—notably the power sector—the IRA and IIJA hold the potential to help subnational entities enhance and realize their existing climate goals or, if they have not done so already, begin to take on decarbonization goals and policies. Yet, as this chapter explores further, the question remains as to the willingness and capacity of more subnational entities to enter the climate and clean energy policy arena, and the success of the federal policies in effectively engaging and resonating with a variety of unique state and local priorities and motivations.

A New Era?: The Promise and Complexity of Climate Federalism

The Biden administration policies have heralded a new era of "climate federalism" where federal, state, and local governments collaborate on decarbonization roles and responsibilities (Bianco et al. 2020). Acknowledging the mounting challenges of the global climate crisis, the Biden administration has prioritized progress on climate change more than any previous administration and has also committed to support ambitious climate action by subnational governments (Hale and Hultman 2020).

The urgency and scale of the climate change challenge calls for action and coordination at all levels of government, with each level leveraging its unique areas of strength in the spirit of "climate federalism" (Bianco et al. 2020). There are policy arenas where the federal government can most naturally lead, including in setting national

---

[7] In December 2022, renewable energy including hydro accounted for 38 percent of California's net electricity generation (EIA 2023a). Ohio obtained less than 5 percent of its net electricity generation from renewable energy in December 2022 (EIA 2023b).

emissions targets consistent with science; setting national standards, incentives, and taxes; engaging the international community to advance global climate action; supporting research, development, and demonstration of clean technologies; developing data and reporting tools; and bolstering other areas where federal resources cannot be easily matched at the subnational level. However, planning and implementation of various programs, projects, and policies requires local context, community input, relationships, and expertise that the federal government is not equipped to manage. Most policy areas for addressing the climate crisis could be served by coordination and constructive partnership among federal, state, and local governments. Given the varying capacity and willingness of subnational actors to tackle climate issues, it will be important for the federal government, with its significantly larger financial and technical resources, to look for opportunities to enhance and drive additional subnational climate action and to develop custom-tailored messaging and resources to include more states and localities.

Notwithstanding the significant recent progress made by the federal government and Congress in addressing climate change (Lashof 2023), the promise of climate federalism cannot be taken for granted. While it could be possible to construe the passage of the IIJA and IRA as the start of a new era where federal leadership can solve U.S. climate challenges, the theme of federalism and the need for federal agencies to defer to—and provide support for—subnational roles and authorities persist to this day. As explained below, the success of the IIJA and IRA in achieving deep decarbonization rests heavily on the shoulders of subnational actors. There is no shortage of analyses of what states and localities *should do and must do* to achieve decarbonization; however, the questions of whether they *will* and *can* at sufficient scale and consistency and what the federal government can do to enable most subnational governments to move forward on addressing climate change remain largely unexamined. The next section discusses key factors for effective governance of clean energy, energy efficiency, and climate policy at the subnational level.

> Finding 13-1: State, county, and local governments possess unique policy, regulatory, and financial levers that can inform and accelerate, or conversely stall and impede, efforts to achieve ambitious U.S. decarbonization goals.

> Finding 13-2: The imperative to decarbonize cannot rest solely on the federal government together with traditional subnational climate leaders. More states and localities have to be convinced and equipped to invest in, adopt, and implement climate solutions; in many cases, the most compelling reasoning may not have to do with climate or environmental factors at all, but rather economic, resilience, and energy security motivations. Otherwise, it is unlikely that the United States will achieve its net-zero goals at the scale and urgency needed to address the climate crisis.

Finding 13-3: Because state, county, local, and other non-federal actors play such a critical role in decarbonization, future federal climate action (including Recommendation 1-1 calling for a congressionally mandated GHG budget for the U.S. economy) should meet two critical criteria: (1) it must be developed in meaningful consultation with subnational actors; and (2) it must create a floor—rather than a ceiling—to enable subnational actors to surpass federal climate goals.

Key Dimensions of Effective Subnational Climate Governance

The literature on U.S. state and local actions to address the climate crisis tends to focus on what state[8] and local[9] governments *should* do—including the development of climate and equity action plans, transportation sector policies promoting low-carbon and alternative fuels, building energy codes and disclosures, and electricity sector policies such as renewable portfolio standards, net metering, decoupling, among many other important policies and initiatives that are emphasized in the preceding chapters of this report. However, research exists when it comes to exploring other factors—such as policy coherence, institutional structures, resources, expertise, and bureaucratic insulation from special interests—that are necessary for effective governance of decarbonization, clean energy, and energy efficiency policies at the subnational level. This section will explore these factors, with a particular eye on the investments and coordination that will be needed at both the subnational and

---

[8] Using modeling, Energy Innovation has identified the most impactful policy strategies for states to achieve deep emissions reductions across key sectors: "renewable portfolio standards, and feed-in tariffs; complementary power sector policies, such as utility business model reform; vehicle performance standards; vehicle fuel fees and feebates; electric vehicle policies; urban mobility policies, such as parking restrictions and increased funding for alternative transit modes; building codes and appliance standards; industrial energy efficiency standards; industrial process emission policies; carbon pricing; and [research and development] policies" (Harvey et al. 2018, p. 65).

[9] The World Economic Forum contends that the urban transition is based in the integration of smart energy infrastructure, clean electrification, efficient building standards, and public services (Corvidae et al. 2021). Steps for decarbonizing urban areas include such actions as assessing the energy supply, building energy efficiency profiles to transform existing buildings into efficient and renewable infrastructure, determining dominant modes of mobility, and facilitating zero-waste recovery systems and promoting sustainable consumption (Plastrik and Cleveland 2015). Although local, exurban, rural, and remote regions also face decarbonization priorities and challenges distinct from urban and suburban areas, in more isolated communities, Saha et al. (2021a) recommend key actions such as investments in renewable energy; energy efficiency; transmission, distribution, and storage; environmental remediation of abandoned fossil fuel infrastructure; tree restoration on federal and non-federal lands; and wildfire risk management.

national levels, to navigate IIJA and IRA opportunities effectively. Key dimensions include:

- *Institutional Breadth and Depth*: In analyzing how institutions are structured around climate policy, Dubash (2021, p. 515) emphasizes "the need for an expansive definition of climate institutions, as in many cases new institutions have not been created from the ground up, but have emerged through processes of layering" on existing bodies. Municipal sustainability initiatives, which on average involve seven different units of government,[10] illustrate the breadth of responsibilities and capacities that may be implicated in climate policy making and implementation (Park et al. 2020). Mildenberger (2021, p. S76) highlights both the pros and the cons of this institutional layering. On the one hand, it creates multiple "significant sites of . . . climate policy capacity," diffusing expertise across a multitude of agencies and offices. On the other hand, it can make policy coordination and durability more difficult, as nodes of expertise can be disintegrated from one administration to the next. Recognizing this dynamic, both the Biden administration at the federal level and some subnational governments have espoused a "whole-of-government" approach to climate change, exemplified by such decisions as the administration of Massachusetts Governor Healey's appointment of a cabinet-level Climate Chief tasked with coordinating climate policy across all state agencies and in partnership with local communities[11] (Healey and Driscoll 2023); San Diego's efforts to amend its citywide general plan to encourage sustainable development and GHG emissions reductions (City of San Diego 2023); and New York City Mayor Eric Adams's establishment of the Office of Climate and Environmental Justice, which consolidates key environmental and climate resiliency personnel to promote an integrated approach to the city's climate goals (New York City Government 2022). Such institutional configurations can help centralize coordination, but also empower a broad range of agencies and offices to act and be held accountable for climate goals, and thus potentially

---

[10] "The most common departmental actors include planning, public works, economic and community development, environmental services, municipal utilities, parks and recreation, and the mayor or city manager's office" (Park et al. 2020, p. 436).

[11] While Massachusetts claims to be the first state in the nation to make such an appointment, it is not the first example of governors establishing multi-agency initiatives to address climate change. For example, Louisiana Governor Edwards's Executive Order (EO) 2020-19 appointed a Chief Resilience Officer and directed state agencies to appoint resilience coordinators to serve point on adaptation and resilience initiatives (EO JBE 2020-19). Connecticut Governor Lamont's EO 21-3 required executive branch agencies to report progress on climate mitigation and resilience efforts to the Governor's Council on Climate Change (EO 21-3 2021).

be shielded from political backlash and budget cuts. They can also be especially helpful as states and cities prepare to coordinate unprecedented levels of federal funding, potentially disruptive construction periods, and market and community transitions through the IIJA and IRA (Badlam et al. 2022). After the passage of the IIJA, some states, including Arkansas, California, Pennsylvania, and New Mexico, have designated an infrastructure advisor, agency, or committee tasked with making recommendations to the governor on determining priorities for the billions of dollars in federal infrastructure funding. These steps are designed to help states maximize access to the recent infusion of federal funding, but these kinds of positions and offices need to become the norm for state- and city-level operations in order to provide an integrated, whole-of-government approach to climate leadership.

- *Stable and Adequate Staffing and Access to Resources*: Navigating and realizing decarbonization goals requires deep investments in and support for human capital, which includes the teams, organizations, and processes that advance climate, clean energy, and energy efficiency goals (see also Chapter 5). Factors such as knowledge, motivation, and personal stance on environmental issues can affect the ability of policy makers, implementers, regulators, and other civil servants to advance decarbonization goals. Technical and policy expertise and cultural competencies are especially important for the development of effective and sequential policies,[12] to manage networks of diverse stakeholders, and to address unexpected issues (Lipinski et al. 2021). Such job responsibilities warrant high levels of compensation and benefits, but state and local governments often face hiring and fiscal challenges (Brey 2022). The need for large, high-budget staff and agencies can potentially be offset by access to professional development opportunities (such as networking opportunities and educational conferences that enable staff to understand and apply best practices learned from other jurisdictions' experiences in decarbonization), informational and technical assistance resources, and strategic partnerships with trusted non-governmental partners, such as universities, think-tanks, and others who can provide analytical assistance to governmental entities.
- *Cushioning from Special or Conflicting Interests*: Special interests can pose a significant impediment to climate receptivity and policies (see Chapter 12).

---

[12] The National Renewable Energy Laboratory's "policy-stacking" framework illustrates the complexity and level of expertise needed to navigate effective climate policy at the subnational level. The theory posits that interdependencies among and the sequencing of clean energy policies are important to promoting market certainty, investor confidence, and the likelihood of achieving state policy goals. Successive stages of policy first *prepare*, then *establish*, and last *expand* markets (Krasko and Doris 2012), relying heavily on effective agency planning, expertise, and long-term engagement on clean energy initiatives.

For example, many fossil fuel–dependent states tend to have weaker environmental and climate policies (Basseches et al. 2022) and may export power to neighboring jurisdictions (Popovich and Plumer 2020). Even in states traditionally aligned with the climate movement, special interests can stall or dilute climate bills and regulations (Culhane et al. 2021). Recognizing this challenge, Meckling and Nahm (2022) contend that a cornerstone of state governmental capacity to advance climate policy lies in the ability to mobilize or demobilize interest groups in pursuit of goals, with this ability being especially useful when there are constructive partnerships between the legislative and executive branch. In California's signature 2006 climate policy act, AB 32 Global Warming Solutions Act, the legislature succeeded in limiting the influence of special interest groups by establishing overall emissions targets in statute but delegating contentious policy decisions to an independent regulatory agency, the California Air Resources Board. The board was subsequently able to develop a sweeping, economy-wide plan targeting the transportation and electricity sectors and establishing a statewide carbon trading system. This strategic choice helped to shield climate goals from powerful interests at a pivotal time in California's climate policy trajectory (Meckling and Nahm 2022). Basseches et al. (2022, p. 5) add to this analysis: "when state legislatures delegate significant policy making authority to executive branch agencies, the latter tend to be relatively depoliticized and less susceptible to powerful interest groups. However, the success of administrative delegation is contingent on administrative capacity," emphasizing once more the need for adequate staffing, expertise, and resources at the agency level. In parts of the country where state agencies are limited in their power and authority, broad stakeholder networks that cut across the private and public sectors in support of decarbonization goals are critical to minimizing the ability of any individual interest group to prevent progress.

- *Access to Peer Sharing Networks*: Wiseman (2014) identifies the problem of "regulatory islands" that stem from state and local government failures to share policy insights, experiences, and results, and highlights the importance of shared content in supporting policy experimentation and improvement. Drawing on state clean energy and energy efficiency policy as one positive example, Wiseman notes the positive role federal involvement and support has played in the production of information and peer exchange networks, whether directly through federal agencies or, when state mistrust of federal actors prevents more active involvement, through representative associations, regional organizations, and intermediaries that receive federal support.

Effective governance of decarbonization at the subnational level will require greater investment and focus at all levels of government—federal, state, and local. The potential reward for this subnational capacity building is significant, as state and local authorities and actions will be critical to unlocking the potentially immense environmental and economic development benefits of the IIJA and IRA. However, questions remain as to how to enable state and local government institutions to attain these rewards, particularly in jurisdictions that may lack the history, institutional capacity, and political consensus to navigate the complexities of decarbonization. The following section explores whether and how subnational governments can leverage the IIJA and IRA provisions strategically to build their capacity to advance decarbonization policy.

EXPANDING AND ENHANCING SUBNATIONAL ACTION

Maintaining Policy Coherence by Mainstreaming Net-Zero Goals in Subnational Policy

When presented or perceived as a climate issue, decarbonization may fail to garner sufficient public support and policy attention; yet, when it is tied to immediate and tangible benefits important to local populations, electorates, and economies, the prospects may improve (Li et al. 2023; Tyson et al. 2023; Victor et al. 2017). Kok and de Coninck (2007, p. 588) find that this process of "mainstreaming" climate change concerns into policy domains and priorities that capture locally relevant priorities, goals, and motivators enhances the effectiveness of decarbonization policy by "increasing policy coherence [and] minimizing duplications and contradictory policies." They advise tying decarbonization to adjacent policy motivations such as security of energy supply, air pollution and public health, poverty reduction, agricultural development, and disaster reduction. For more on the energy justice aspect of decarbonization policy, see Chapter 2.

The promise of many clean energy policies is that they can flexibly adapt to different (non-climate) motivators and values that drive political and policy action. In Utah (a net exporter of natural gas, coal, and electricity), for example, former Governor Gary Herbert used local energy production, rural business development, economic competitiveness, and air quality goals to drive efforts in large-scale clean energy storage, electric transportation, and clean energy science, technology, engineering, and mathematics education (Barrett 2019; Citizens for Responsible Energy Solutions 2018; Fazeli 2020; Maffly 2019; Utah Geological Survey 2020). During this term, he also signed into law the 2019 Community Renewable Energy Act to assist communities in achieving 100 percent clean energy by 2030.

Finding 13-4: Decarbonization policy can be tied to a variety of policy motivations and domains, including energy security and resilience, air pollution and public health, economic development and poverty alleviation, and agricultural and rural development, among others. Subnational actors are well positioned to link climate policies to locally relevant priorities and values and to navigate policy environments that may appear to be at odds with a pro-climate agenda.

Implications for Federal–Subnational Coordination and Technical Assistance

Examples like Governor Herbert's, above, illustrate that it is possible to achieve greater GHG reductions, even in areas that may ostensibly appear as unfavorable policy environments for addressing climate change. These examples highlight the possibility that with greater levels of engagement of and deference to state and local champions, authorities, and messengers, federal climate action can resonate in places where it has failed to do so in the past. Hendricks et al. (2020) propose one potential model:

> the federal government can ensure alignment between national mobilization and each state's march toward decarbonization and a just, green economy by establishing federal interagency climate mobilization councils in every state and territory that include all relevant federal agencies operating in that jurisdiction. These councils could be supported by detailed staff from White House offices, for example, from the Council on Environmental Quality.

Federal agencies and offices are well positioned to promote federal–subnational coordination, most notably through strategic partnerships and the delivery of targeted technical assistance resources. For instance, the Joint Office of Energy and Transportation (2022, p. 2) brokered a Memorandum of Understanding (MOU) among DOE, DOT, the National Association of State Energy Officials, and the American Association of State Highway and Transportation Officials to ensure the "strategic, coordinated, efficient, and equitable" investment of electric vehicle charging infrastructure. The MOU identifies activities that will support this coordination, including convenings of national, state, local, tribal, and private-sector actors; the elevation of data, technical, and program assistance needs of states to federal agencies; and communications channels.

Additionally, the Solar Energy Technologies Office has established a States Collaborative as part of its National Community Solar Partnership as a peer exchange space for states interested in accelerating community solar development (DOE n.d.). EPA's State and Local Climate and Energy Program offers targeted, voluntary resources—such as GHG inventory, energy savings, and building energy benchmarking resources—that subnational entities have applied to their own policy making and program design

(EPA 2023c). Similarly, DOE's Office of Cybersecurity, Energy Security, and Emergency Response has made analyses, coordination services, and experts available to help regions and communities navigate energy disruptions and outages through the State, Local, Tribal, and Territorial Program (DOE 2023c).

However, not all federal technical assistance is created equal. In an open letter highlighting the pitfalls of technical assistance that *fails* to respond to on-the-ground realities and needs, the City of Savannah's clean energy program manager Alicia Brown notes: "federally funded technical assistance often provides little in the way of additional expertise" because local sustainability professionals already have topical knowledge [on clean energy and climate solutions] and access to private sector partners (Brown 2023). It also comes "at the expense of directing cold hard cash to organizations that need it."

As Brown's op-ed implies, subnational entities have been excluded from programmatic determinations that directly affect their ability to plan and make use of IIJA and IRA funds—decisions such as the amount of program funding that will be retained for federal agencies to deliver technical assistance; the content, source, and value of the technical assistance delivered to state and local governments; and even the timing of the release of program funds. A 2022 exchange between DOE and the National Association of State Energy Officials (NASEO), representing the state and territory agency recipients of U.S. State Energy Program (SEP) funds, helps illustrate this dynamic. Nearly 9 months after the passage of the IIJA, NASEO called for the immediate release of the $500 million authorized to support state and territory capacity-building and planning:

> We urge DOE to release the entire $500 million to the states and not retain any amount of these critical planning funds for DOE-directed technical assistance or for distribution over a period of years. Opting for state direction over the funds, as Congress intended, enables states to procure assistance from DOE's National Laboratories, local universities and colleges (including minority-serving institutions), and other experts if they so choose, in line with their unique state goals and needs and the statutory requirements of the program. (Terry 2022, p. 1)

In DOE's ensuing program guidance, the allocations to be distributed to states and territories totaled $425,152,000, reserving nearly $75 million (about 15 percent) for overhead and for the newly formed Office of State and Community Energy Programs (SCEP) to oversee existing programs and provide technical assistance in new areas (DOE 2022b). These new areas include transmission and distribution planning; system-wide planning for grid expansion, modernization, and clean energy technology integration; energy security; community energy planning; and clean energy manufacturing—with no acknowledgment that states and localities may already be accessing federal technical assistance on these topics from other parts of DOE, such as the Office of Electricity, Grid Deployment Office, Building Technologies Office, and others. DOE's SCEP began dispersing the IIJA funds for the U.S. State Energy Program in 2023—more than a year after the passage of IIJA—and, as of the writing of this report, has not released any of the $550 million (effectively $430 million after DOE overhead and technical assistance cuts) in Energy Efficiency and Conservation Block Grants authorized for states and localities (DOE 2023b). Both are existing, non-competitive programs with formula allocations that would have been relatively straightforward for DOE procurement and administration to issue.

The failure to prioritize the release of these funds and listen to the needs of states and localities is very likely to hold repercussions for IIJA and IRA implementation. It all but ensures that lesser-resourced states and local governments will be hard-pressed to plan for, staff for, convene stakeholders around, and pursue clean energy funding opportunities competitively.[13] That is, if they choose to do so at all—without the support of planning and capacity-building funds and transparency into federal technical assistance—jurisdictions may be deterred from pursuing funds altogether, which may hit rural, disadvantaged, or lower-income communities hardest, as noted in Chapter 2.

For their part, subnational governments do also need to dedicate resources to communicate and coordinate with federal agencies, and with one another, on IIJA and IRA implementation. Some states, cities, and counties may be well-resourced and driven to interface directly, whether through direct relationships or opportunities, such as Requests for Information, to submit comments and feedback on federal programs and initiatives. For others, active participation in state-, city-, county-, and regional-facing associations and organizations (such as the National Association of Counties, the National Leagues of Cities, the National Association of State Energy Officials, the National Association of Regulatory Utility Commissioners, and others) can provide an efficient throughway to elevate subnational needs and innovations and inform federal technical assistance and processes. While it does not necessarily guarantee outcomes, these organizations offer platforms to deliver messages on behalf of subnational entities

---

[13] Additionally, requirements for local-match can be a deterrent. Analysis from Headwaters Economics reveals that 60 percent of the IIJA's funding for projects designed to help communities prepare for natural disasters requires communities to contribute between 20–30 percent of the project cost. This can put rural communities at a disadvantage. Many lack the resources to both apply for grants and sustain their financial contribution (Headwater Economics 2023).

about the size, scope, and timing of federal programs that they will be responsible for implementing. They can also be critical in relationship-building between federal and subnational agencies, which can heighten accountability and trust on both ends. More systematic coordination can also help federal agencies understand the various ways states and localities are applying federal decarbonization resources on the ground to achieve their unique priorities and goals—whether climate-driven or not—and advance streamlined and equitable processes, particularly for under-resourced civil and public entities.

Finding 13-5: The IIJA and IRA provide federal agencies with an imperative to engage subnational governments and to ensure that federal technical assistance and application processes meet state, county, and local needs.

Recommendation 13-1: *Establish an Ongoing Process to Evaluate and Integrate Feedback into Technical Assistance Processes and Federal Applications.* **The White House Council on Environmental Quality, in cooperation with agencies like the Department of Energy (DOE) and the Environmental Protection Agency, should establish a process to integrate feedback from subnational entities into federal application and technical assistance processes. This process should be relationship-based, iterative, and ongoing, beginning with a national convening through which subnational government entities can elevate concerns and challenges with application, implementation, and technical assistance processes and quality, and continue with a working group dedicated to evaluating and informing federal processes on a semi-annual basis. Observations and recommendations from the convening and working group could be delivered to the Office of Management and Budget, DOE's Office of State and Community Energy Programs, and other relevant offices across the federal agencies, to empower them to adjust processes and resources according to subnational needs and priorities.**

Recommendation 13-2: *Disburse Capacity-Building Funds for State, Local, and Community Recipients Flexibly and Speedily.* **The Department of Energy, Environmental Protection Agency, Department of Transportation, Department of Agriculture, and other federal agencies should be held to strict timelines (with 3–6 months of program passage and funding authorization) to disburse funds to support state, local, and community clean energy and climate planning and capacity building.**

THE IIJA AND IRA AND IMPLEMENTATION AT THE SUBNATIONAL LEVEL

The IIJA provides billions of dollars to modernize the grid, build needed electric transmission, enhance energy system resilience, expand electric vehicle charging infrastructure, and advance building energy efficiency, smart manufacturing, carbon capture and utilization, renewables, and other important energy actions (see Chapter 6). In particular, the IIJA promises an enormous amount of investment into the nation's transportation infrastructure, with surface transportation accounting for $600 billion of the approximately $1 trillion authorized over 5 years (Saha et al. 2022).

The more recent IRA contains an estimated $370 billion in provisions for energy security and climate change, including support for domestic clean energy manufacturing, residential energy efficiency, and electrification rebates, affordable housing grants, environmental and climate justice block grants, and a slew of tax credits expected to ramp up investments in energy efficiency, renewable energy, geothermal, carbon capture and sequestration, and other emissions reductions efforts (see Chapter 6) (White House n.d.).

State and local governments and other subnational actors will play a make-or-break role in the implementation of the IIJA and IRA. All told, approximately $470 billion in the IIJA and as much as $139 billion in the IRA rely on proactive action and investment in state and local governments (Badlam et al. 2022; Elliot and Hettinger 2022). Early implementation moves seem to suggest that at the state level, state energy offices, transportation departments, and/or public utility commissions are likely and well-positioned to lead on coordinating applications, organizing partners, assessing market barriers and opportunities, collecting stakeholder feedback, and implementing project management, oversight, and reporting tools. Although this leadership risks being patchy at the county and local level, Sustainability Offices, Governmental Affairs liaisons, and Public Works Departments are all likely to be in the driver's seat and will need to act effectively to ensure that implementation is successful.

More broadly, the success of IIJA and IRA programs and incentives may also depend on several policy, programmatic, organizational, and regulatory factors that may or may not be in place at the subnational level. For instance, states and localities that have been slow in the past to submit designations to DOT for alternative fuels corridors, or whose electric utilities may not be adequately prepared for vehicle electrification, may have a harder time using their allocation of the $5 billion National Electric Vehicle Infrastructure Formula Program authorized by the IIJA (DOT 2022).

Similarly, if a state has complicated or burdensome electric generator interconnection standards and rules, the effectiveness and value of IRA incentives for grid-connected renewables may be compromised (NREL 2023). As emphasized in Chapter 6, investments in long-range transmission will be another critical success factor in the power sector, but only a few states and localities have invested in strategies to navigate competing interests and accelerate and streamline arduous siting and approval processes (Smith 2021). Further still, analyses suggest that states that participate in regionally organized, wholesale electricity markets can make better use of IRA incentives than those in fragmented systems, as they can support clean energy integration on the grid while preserving reliability and affordability (Lehr and Groves 2022). In the (non-organized) West, two states—Colorado and Nevada—have enacted legislation requiring transmission utilities to join an organized wholesale market by 2030 (Senate Bill 21-072 2021; Senate Bill 448 2021).

As Chapter 7 highlights, states that do not have robust networks in place to ensure that low-income communities can take advantage of investments in energy efficiency upgrades and other clean energy technologies may fall short of meeting key equity goals in these laws.

Further complicating matters, subnational governments have significant discretion on spending in these bills, owing in part to the flexibility built into key provisions. For instance, DOE's program guidance for IIJA §40101(d) "Preventing Outages and Enhancing the Resilience of the Electric Grid" provides $459 million annually over 5 years to state and tribal governments to improve the resilience of the electric grid. While all the eligible projects promote resilience, they vary in terms of their emissions reduction potential, ranging from vegetation management and fire prevention systems to the construction of distributed renewables and battery storage solutions (DOE 2022a).

Similarly, the majority of the IIJA's transportation funding is dedicated to bolstering highway programs, which are for the most part distributed to states via formula funding. A large portion of IIJA funding goes to DOT and flows through state transportation agencies; the increased budget is expected to foster 40 new grant programs, many of which are climate-focused (Alexander et al. 2022). Guidance released by the Federal Highway Administration in December 2021 stressed that the federal government would like to see the influx of highway funding from the IIJA go toward fixing existing roads (over building new highways) and cleaner modes of transportation including public transit and bike lanes (Pollack 2021). However, one transportation-specific analysis concluded that investments funded by the IIJA could have the effect of increasing carbon emissions, depending on how highway funding and other programs are implemented by states, cities, and regional agencies (Georgetown Climate Center 2021).

Finding 13-6: Although the IIJA and IRA represent major investments in U.S. climate efforts, their successful implementation and the likelihood of GHG emissions reductions hinge on policy, market, regulatory, and administrative factors at the subnational level. The likelihood of the bills to advance net-zero goals depends in part on the ability and willingness of states and localities to address complex policy and regulatory obstacles both within their jurisdictions and across federal, regional, state, and local levels of policy making and energy markets. Just as important, their success also hinges on procedural and administrative decisions that subnational actors and government staff will need to make: whether to accept or compete for funds for which they are eligible; how to organize and engage stakeholders in program design and implementation; how to develop program designs and partnerships that achieve GHG reductions equitably and quickly; and how to develop, manage, and complete programs effectively and achieve and report on program impacts and outcomes.

Recommendation 13-3: *Designate an Official or Entity to Track Decarbonization Program Opportunities and Deadlines.* **To ensure that their economies and residents do not lose out on the local development, job, resilience, and other opportunities in the Infrastructure Investment and Jobs Act (IIJA) and the Inflation Reduction Act (IRA), every governor, mayor, and county official should designate an official or entity to track activities, deadlines, and funding opportunities in the IIJA and IRA. This individual or entity should serve as a coordinating body and be empowered to engage other agencies to leverage the collective strength and expertise of the state, county, or city in pursuing funding. In a larger state or city, this coordinating body may be an entire office or a working group of department staff; in smaller jurisdictions, it may start out as a fraction of someone's time, to ensure that critical opportunities to advance state, county, and local economic development and decarbonization goals are not missed.**

As noted previously, there is high variability and inconsistency in subnational climate policy governance. While some states and localities with longer histories of climate, clean energy, and energy efficiency policy leadership may be sufficiently staffed, adequately resourced, and have suitable policies and regulations in place to pursue IIJA and IRA funding, most will likely need to build this capacity and fill gaps as swiftly as possible.

Most of the provisions in the IIJA and IRA flowing to subnational governments seem to place new or expanded responsibilities on recipients, with *de minimis* amounts (typically 10 percent of the allocation) available for use for administrative purposes, and some provisions for planning, data analysis, professional development and education opportunities, and other aspects related to capacity building. Table J-1 in Appendix J highlights key provisions from the IIJA and IRA that engage and implicate state and local governments. Many of these funding opportunities are competitive and opening simultaneously, and therefore will require significant time and attention from applicants even though funding is not guaranteed. Even formula (non-competitive) provisions still require extensive plans to be submitted to federal agencies, reporting on program outcomes, and compliance with sometimes-complex federal rules such as the National Environmental Protection Act, Davis-Bacon, and Buy American and other domestic content requirements. Additionally, while the billions of dollars in IRA clean energy tax incentives will flow directly from the Internal Revenue Service, states and localities will likely need to provide information and answers to their constituents on how to take advantage of them and how they may interact with state and local programs. The fact that subnational governments will need to navigate significant new program development, administrative, program oversight, and stakeholder communications responsibilities, with limited or delayed opportunities to build their own staff and expertise, may deter them from applying for some pots of funding at all—not because of a strategic choice, but owing to strains in capacity. This poses the risk of deepening existing inequities and inconsistencies between climate-leading jurisdictions and state and local governments with less experience and expertise.

On the positive side, the IIJA and IRA do contain a handful of more flexible provisions that can support state and local capacity-building via investments in planning, analysis, program development, stakeholder engagement, and staffing. Key examples for states include $500 million in IIJA funds for the U.S. SEP and a portion of the $550 million in funds under the Energy Efficiency and Conservation Block Grant (EECBG) Program (DOE 2023b). For localities and community-based organizations, these include a portion of the EECBG funds as well as $3 billion in the Environmental and Climate Justice Block Grants from the IRA (EPA 2023a).

These are not insignificant levels of funding. However, when spread across hundreds of state and local jurisdictions and other community-based recipients (not to mention federal agencies' share for administration and technical assistance), it is logical to question whether they will be adequate in positioning subnational actors to make the most of the opportunity, including integrating and braiding disparate funding streams across different types of agencies (environmental, energy, regulatory, etc.) to maximize their impact. Additionally, as previously noted, the formula SEP and EECBG funds authorized in the IIJA have been slow to reach states and localities, placing additional strain on smaller and less-resourced agencies.

The upfront effort it takes for state and local agencies to respond to competitive grant opportunities is significant. To alleviate this burden, the Climate Pollution Reduction Grants (IRA §60114) were designed by Congress as a two-staged program. In the first stage, states and localities may apply to access non-competitive planning grants (up to $3 million per state, $1 million to the 67 most populous metropolitan areas, and $2 million to U.S. territories, and $25 million to tribes) to develop or update climate, energy, or sustainability plans. These planning grants can help position state and local governments to apply for the second, competitive, stage of the program: $4.6 billion focused on implementation (EPA 2023c). While it remains to be seen how the program will play out over the coming years, the non-competitive planning stage structure of CPRG is promising because it does not automatically deter lower-resourced or lower-capacity entities from pursuing funds.

There is a need for increased and more consistent federal and state investment in subnational governance and capacity building, with a particular focus on providing easy-to-access funding and financial opportunities to state agencies and communities that have not typically led in climate, clean energy, energy efficiency, and decarbonization policy.

> Finding 13-7: The IIJA and IRA include many programs and investments that implicate subnational actors, particularly state agencies, as potential recipients of the funding. Most of these programs and funding place responsibility for program oversight, implementation, and reporting in the hands of states and allow for a portion of the funds to be used for administrative, planning, and capacity-building purposes. Both bills' reliance on competitive grant programs likely means that some states and local governments will secure a first-mover advantage in the competition for funding and that others lacking the capacity to apply for these funding programs may lag. A handful of key programs—the U.S. State Energy Program, the Energy Efficiency and Conservation Block Grant Program, the Environmental and Climate Justice Block Grants, and the non-competitive planning stage of EPA's Climate Pollution Reduction Grants—provide sufficiently flexible funding for state and local planning and capacity building. Some subnational governments may rely on these programs to build capacity, understand opportunities to braid disparate funds together for maximal impact, and develop locally relevant messaging that encourages uptake and adoption.

> **Recommendation 13-4: *Structure Competitive Opportunities as Non-Competitive Planning Grants Followed by Competitive Grants.* To support lower-resourced states and localities in accessing funding opportunities,**

federal agencies should structure future competitive opportunities under the Infrastructure Investment and Jobs Act and the Inflation Reduction Act via a two-stage approach: non-competitive planning grants followed by competitive grants. The Environmental Protection Agency's Community Pollution Reduction Grants program offers a model by providing sizable non-competitive planning grants to states and large localities in its first phase, which can help inform state and local efforts to pursue the remaining $4.6 billion in competitive funding in its second phase.

Recommendation 13-5: *Continue to Expand Reliable and Flexible Funding to Subnational Governments.* **Recognizing the central role that subnational actors will play in the Infrastructure Investment and Jobs Act (IIJA) and the Inflation Reduction Act (IRA), Congress should continue to expand reliable, annual, flexible funding to subnational governments for the life of IIJA and IRA programs. To support continued capacity-building, stakeholder engagement, and planning at the subnational level, Congress should expand and continue programs that provide flexible formula funding to states and localities for clean energy and energy efficiency deployment, policy development, and planning, such as the U.S. State Energy Program (SEP) and Energy Efficiency and Conservation Block Grant (EECBG) Program, which have a long-standing history of success. As proven and well-established programs, both SEP and EECBG should be sustained and annually funded at heightened levels ($100 million annually for SEP, roughly double the amount appropriated in recent years, enabling more beneficial state allocation minimums; and $500 million annually for EECBG). For each of these programs, federal contracting officials should reduce the up-front burden of developing written proposals and applications, and instead encourage federal agencies and program officers to interface regularly with recipients and provide customized support as needed.**

CONCLUSION

The formation, evolution, and implementation of decarbonization and clean energy goals depend on responsibilities, investments, and authorities delegated across a multitude of federal, state, local, private-sector, and civic actors. In this regard, deep decarbonization in the United States represents an especially complex goal to navigate, not only for its technical and societal dimensions but also for its need to mobilize action by diffuse governmental, private, and civic sector actors, all with varying priorities and motivations.

State and local decarbonization policies have been critical drivers of emissions reductions, economic and community development, resilience, and security; yet they are also highly variable and insufficient in the face of the climate crisis. This reality underscores the need for greater investment at all levels of government in subnational capacity-building and governance to shepherd the transition to a decarbonized economy—not just in traditionally climate-leading states and cities, but more broadly to include actors with the will, interest, and motivation to decarbonize but who may lack the capacity or support to do so.

The implementation of the IIJA and IRA holds promise as a means of catalyzing significant financial investment in clean energy, energy efficiency, and climate solutions across the United States. While states and localities with sufficient budgets, sophisticated staff, and experience in climate policy and regulatory experimentation may be well positioned to tap into the IIJA and IRA's significant benefits, many others may lack the capacity to navigate these laws and maximize their impact in the context of state and local goals and priorities. Shortfalls in climate-forward state and city emissions reduction achievements demonstrate that it is critical for *more* states, cities, regions, and communities to adopt and implement decarbonization goals. These goals can be aligned with a variety of priorities, including economic development, security, and adaptation, so that all localities can benefit from federal investment, regardless of existing climate policies and politics. Without broader and more stringent efforts, there is a serious risk that the United States will fail to meet its net-zero targets.

Governments at all levels—federal, state, and local—must invest immediately and significantly to build subnational readiness and capacity to meet the historic moment presented by the IIJA and IRA and the challenge posed by deep decarbonization. This will require a thoughtful balance between national leadership and deference to on-the-ground actors. With the IIJA and IRA helping to fundamentally transform the economics and potential reach of decarbonization efforts, subnational action will be more essential than ever to seize the opportunity. For many jurisdictions, taking the first steps—diverting scarce staff time to tracking opportunities, developing project and program ideas, designating lead agencies and individuals, submitting applications, and tying these actions to the immediate and tangible needs and priorities of their constituents can unlock myriad benefits for states, regions, counties, cities, communities, and individuals. Table 13-1 summarizes all the recommendations in this chapter to support subnational actors in policy implementation and advancing decarbonization objectives.

SUMMARY OF RECOMMENDATIONS FOR ENHANCING AND REALIZING THE CLIMATE AMBITIONS AND CAPACITIES OF SUBNATIONAL ACTORS: STATE AND LOCAL GOVERNMENT PERSPECTIVES

TABLE 13-1 Summary of Recommendations for Enhancing and Realizing the Climate Ambitions and Capacities of Subnational Actors: State and Local Government Perspectives

Short-Form Recommendation	Actor(s) Responsible for Implementing Recommendation	Sector(s) Addressed by Recommendation	Objective(s) Addressed by Recommendation	Overarching Categories Addressed by Recommendation
13-1: Establish an Ongoing Process to Integrate Feedback into Federal Application and Technical Assistance Processes	Executive Office of the President	• Non-federal actors	• Equity • Public engagement	Building the Needed Workforce and Capacity
13-2: Disburse Capacity-Building Funds for State, Local, and Community Recipients Flexibly and Speedily	Department of Energy, Environmental Protection Agency, Department of Transportation, Department of Agriculture, and other federal agencies	• Non-federal actors		Building the Needed Workforce and Capacity
13-3: Designate an Official or Entity to Track Decarbonization Program Opportunities and Deadlines	Governors, mayors, and county officials; states, counties, and cities	• Electricity • Buildings • Transportation • Industry • Non-federal actors	• Equity • Public engagement	Building the Needed Workforce and Capacity
13-4: Structure Competitive Opportunities as Non-Competitive Planning Grants Followed by Competitive Grants	Federal agencies	• Electricity • Buildings • Transportation • Industry • Non-federal actors	• Equity	Building the Needed Workforce and Capacity

TABLE 13-1 Continued

Short-Form Recommendation	Actor(s) Responsible for Implementing Recommendation	Sector(s) Addressed by Recommendation	Objective(s) Addressed by Recommendation	Overarching Categories Addressed by Recommendation
13-5: Continue to Expand Reliable and Flexible Funding to Subnational Governments	Congress and federal contracting officials	• Electricity • Buildings • Transportation • Industry • Non-federal actors	• Equity • Employment • Public engagement	Building the Needed Workforce and Capacity

REFERENCES

ACP (American Clean Power Association). 2022. *Clean Energy Powers American Business*. Washington, DC. https://clean-power.org/wp-content/uploads/2023/01/2022_CorporateBuyersReport.pdf.

Alexander, M., H. Argento-McCurdy, A. Barnes, C. Chyung, C. Cleveland, and D. Madson. 2022. *How States Can Use the Bipartisan Infrastructure Law to Enhance Their Climate Action Efforts*. Center for American Progress. https://www.americanprogress.org/article/how-states-can-use-the-bipartisan-infrastructure-law-to-enhance-their-climate-action-efforts.

Badlam, J., A. Bielenberg, S. O'Rourke, A. Kumar, and R. Dunn. 2022. "Impact Officer in Chief: The State Infrastructure Coordinator's Role." *McKinsey & Company Public and Social Sector Insights*, April 20. https://www.mckinsey.com/industries/public-and-social-sector/our-insights/impact-officer-in-chief-the-state-infrastructure-coordinators-role.

Barrett, K. 2019. "Governor Advances Clean Transportation in Utah." *Utah Clean Cities*. https://utahcleancities.org/governor-herbert-signs-11th-annual-alternative-fuels-declaration/#/find/nearest.

Basseches, J.A., R. Bromley-Trujillo, M.T. Boykoff, T. Culhane, G. Hall, N. Healy, D.J. Hess, et al. 2022. "Climate Policy Conflict in the U.S. States: A Critical Review and Way Forward." *Climatic Change* 170(32). https://doi.org/10.1007/s10584-022-03319-w.

Beirne, S. 2015. "West Virginia Sticks with Coal Despite Trends in Favor of Cleaner Energy." *Environmental and Energy Study Institute*.

Bianco, N., F. Litz, D. Saha, T. Clevenger, and D. Lashof. 2020. *New Climate Federalism: Defining Federal, State, and Local Roles in a U.S. Policy to Achieve Decarbonization*. Washington, DC: World Research Institute. https://doi.org/10.46830/wriwp.19.00089.

Bird, L., J. Heeter, E. O'Shaughnessy, B. Speer, C. Volpi, E. Zhou, O. Cook, et al. 2017. *Policies for Enabling Corporate Sourcing of Renewable Energy Internationally: A 21st Century Power Partnership Report*. Task No. DS21.2030. National Renewable Energy Laboratory. https://www.nrel.gov/docs/fy17osti/68149.pdf

Biswas, S., A. Echevarria, N. Irshad, Y. Rivera-Matos, J. Richter, N. Chhetri, M.J. Parmentier, and C.A. Miller. 2022. "Ending the Energy-Poverty Nexus: An Ethical Imperative for Just Transitions." *Science and Engineering Ethics* 28(4):36. https://doi.org/10.1007/s11948-022-00383-4.

Bloomberg Philanthropies. 2018. "Fulfilling America's Pledge: How States, Cities, and Businesses Are Leading the United States to a Low-Carbon Future." www.americaspledge.com.

Brey, J. 2022. "Government Worker Shortages Worsen Crisis Response." Governing the Future of States and Localities. https://www.governing.com/work/government-worker-shortages-worsen-crisis-response.

Bromley-Trujillo, R., and M.R. Holman. 2020. "Climate Change Policymaking in the States: A View at 2020." *Publius: The Journal of Federalism* 50(3):446–472. https://doi.org/10.1093/publius/pjaa008.

Brown, A. 2023. "An Open Letter to the Department of Energy: Enough with the Technical Assistance." *Utility Dive*, January 30. https://www.utilitydive.com/news/an-open-letter-to-the-department-of-energy-enough-with-the-technical-assis/641491.

Bush, G.H.W. 1992. "Statement on Signing the Instrument of Ratification for the United Nations Framework Convention on Climate Change." https://www.presidency.ucsb.edu/node/266987.

Byrne, J., K. Hughes, W. Rickerson, and L. Kurdgelashvili. 2007. "American Policy Conflict in the Greenhouse: Divergent Trends in Federal, Regional, State, and Local Green Energy and Climate Change Policy." *Energy Policy* 35(9):4555–4573. https://doi.org/10.1016/j.enpol.2007.02.028.

California Energy Commission. 2023. "SB 100 Joint Agency Report." https://www.energy.ca.gov/sb100.

Cames, M., R.O. Harthan, J. Fussler, M. Lazarus, C.M. Lee, P. Erickson, and R. Spalding-Fecher. 2016. "How Additional Is the Clean Development Mechanism? Analysis of the Application of Current Tools and Proposed Alternatives." CLIMA/B.3/SERI2013/0026r. Zürich, Switzerland: INFRAS and Seattle, Washington: Stockholm Environment Institute. https://climate.ec.europa.eu/system/files/2017-04/clean_dev_mechanism_en.pdf.

Center for Climate and Energy Solutions. 2022. "U.S. State Clean Vehicle Policies and Incentives." ttps://www.c2es.org/document/us-state-clean-vehicle-policies-and-incentives.

Citizens for Responsible Energy Solutions. 2018. "Utah Governor Gary Herbert and State Legislators Taking a Stand on Clean Energy and the Climate." https://cresenergy.com/utah-governor-gary-hebert-and-state-legislators-taking-a-stand-on-clean-energy-and-the-climate.

City of San Diego. 2023. "Blueprint SD." https://www.sandiego.gov/blueprint-sd.

Clean Energy States Alliance. 2023. "Map and Timelines of 100% Clean Energy States." https://www.cesa.org/projects/100-clean-energy-collaborative/guide/map-and-timelines-of-100-clean-energy-states.

Clean Energy Transition Institute. 2023. "Community-Defined Decarbonization." https://www.cleanenergytransition.org/projects/rural-and-tribal-community-decarbonization/community-defined-decarbonization.

Collomb, J.-D. 2014. "The Ideology of Climate Change Denial in the United States." *European Journal of American Studies* 9(1). https://doi.org/10.4000/ejas.10305.

Corvidae, J., C. dalla Chiesa, R. Denda, J. Hartke, U. Jardfelt, R. Lichtman, G. Long, et al. 2021. *Net Zero Carbon Cities: An Integrated Approach.* Geneva, Switzerland: World Economic Forum.

Culhane, T., G. Hall, and J.T. Roberts. 2021. "Who Delays Climate Action? Interest Groups and Coalitions in State Legislative Struggles in the United States." *Energy Research and Social Science* 79:102114. https://doi.org/10.1016/j.erss.2021.102114.

Database of State Incentives for Renewables and Efficiency. 2022. "Alternative Energy Law (AEL). Database of State Incentives for Renewables and Efficiency." https://programs.dsireusa.org/system/program/detail/265.

DOE (Department of Energy). 2018. *Low-Income Household Energy Burden Varies Among States—Efficiency Can Help in All of Them.* DOE/GO-102018-5122. Washington, DC: EERE Publication and Product Library.

DOE. 2022a. "Bipartisan Infrastructure Law Section 40101(d): Formula Grants to States and Indian Tribes for Preventing Outages and Enhancing the Resilience of the Electric Grid." https://netl.doe.gov/bilhub/grid-resilience/formula-grants.

DOE. 2022b. "State Energy Program IIJA Formula Grant Allocations." https://www.energy.gov/scep/articles/state-energy-program-iija-formula-grant-allocations.

DOE. 2023a. "About the State Energy Program." https://www.energy.gov/scep/about-state-energy-program#history.

DOE. 2023b. "EECBG Program Formula Grant Application Hub." https://www.energy.gov/scep/eecbg-program-formula-grant-application-hub.

DOE. 2023c. "State, Local, Tribal, and Territorial (SLTT) Program." https://www.energy.gov/ceser/state-local-tribal-and-territorial-sltt-program.

DOE. n.d. "Community Solar: About the National Community Solar Partnership." https://www.energy.gov/communitysolar/about-national-community-solar-partnership. Accessed May 15, 2023.

DOT (Department of Transportation). 2022. "President Biden, USDOT and USDOE Announce $5 Billion over Five Years for National EV Charging Network, Made Possible by Bipartisan Infrastructure Law." FHWA 05-22. https://highways.dot.gov/newsroom/president-biden-usdot-and-usdoe-announce-5-billion-over-five-years-national-ev-charging.

Dubash, N.K. 2021. "Varieties of Climate Governance: The Emergence and Functioning of Climate Institutions." *Environmental Politics* 30(Sup1):1–25. https://doi.org/10.1080/09644016.2021.1979775.

EIA (U.S. Energy Information Administration). 2022a. "Energy-Related CO_2 Emission Data Tables." https://www.eia.gov/environment/emissions/state.

EIA. 2022b. "Ohio State Profile and Energy Estimates." Profile Analysis. https://www.eia.gov/state/analysis.php?sid=OH.

EIA. 2023a. "California: State Profile and Energy Estimates." https://www.eia.gov/state/?sid=CA#tabs-4.

EIA. 2023b. "Ohio: State Profile and Energy Estimates." https://www.eia.gov/state/?sid=OH#tabs-4.

Elliot, N.M., and L.A. Hettinger. 2022. "The Inflation Reduction Act: Provisions and Incentives for Local Governments." *Holland and Knight Alert*, October 13. https://www.hklaw.com/en/insights/publications/2022/10/the-inflation-reduction-act-provisions-and-incentives-for-local.

Engel, K.H. 2009. "Whither Subnational Climate Change Initiatives in the Wake of Federal Climate Legislation?" *Publius: The Journal of Federalism* 39(3):432–454.

Engel, K.H. 2015. "EPA's Clean Power Plan: An Emerging New Cooperative Federalism?" *Publius: The Journal of Federalism* 45(3):452–474.

Engel, K.H., and O. Barak. 2008. "Micro-Motives for State and Local Climate Change Initiatives." *Harvard Law and Policy Review* 2:119–137.

EO (Executive Order) JBE 2020-19. 2020. "Coastal Resilience." J.B. Edwards, State of Louisiana Executive Department.

EO 21-3. 2021. N. Lamont, State of Connecticut.

EPA (Environmental Protection Agency). 2023a. "Climate Pollution Reduction Grants." https://www.epa.gov/inflation-reduction-act/climate-pollution-reduction-grants.

EPA. 2023b. "Climate Pollution Reduction Grants Program: Formula Grants for Planning. Program Guidance for States, Municipalities, and Air Pollution Control Agencies." Office of Air and Radiation. https://www.epa.gov/system/files/documents/2023-02/EPA%20CPRG%20Planning%20Grants%20Program%20Guidance%20for%20States-Municipalities-Air%20Agencies%2003-01-2023.pdf.

EPA. 2023c. "Energy Resources for State, Local, and Tribal Governments." https://www.epa.gov/statelocalenergy.

Farber, D. 2021. "State Governmental Leadership in U.S. Climate Policy." *Wilson Center Insight and Analysis*, June 23. https://www.wilsoncenter.org/article/state-governmental-leadership-us-climate-policy.

Fazeli, S. 2020. *States and Cleantech Innovation: An Examination of State Energy Officers' Roles in Clean Energy Technology-Based Economic Development*. Arlington, VA: National Association of State Energy Officials.

Gallagher, K.S. 2013. "Why and How Governments Support Renewable Energy." *Daedalus* 142(1):59–77. https://doi.org/10.1162/DAED_a_00185.

Georgetown Climate Center. 2021. "Issue Brief: Estimating the Greenhouse Gas Impact of Federal Infrastructure Investments in the IIJA." Georgetown Law. https://www.georgetownclimate.org/articles/federal-infrastructure-investment-analysis.html.

Gleason, P. 2022. "Why States Continue to Overrule Local Regulation of Fossil Fuels." *Forbes*, April 19. https://www.forbes.com/sites/patrickgleason/2022/04/19/why-states-continue-to-overrule-local-regulation-of-fossil-fuels.

Global Alliance for Energy Productivity. 2016. *Energy Productivity Playbook: Roadmaps for an Energy Productive Future*. Washington, DC: Alliance to Save Energy.

Grassle, W. 2021. "The Politicization and Polarization of Climate Change." *Claremont McKenna College Theses* 2663. https://scholarship.claremont.edu/cgi/viewcontent.cgi?article=3830&context=cmc_theses.

Gurney, K.R., J. Liang, G. Roest, Y. Song, K. Mueller, and T. Lauvaux. 2021. "Under-reporting of Greenhouse Gas Emissions in U.S. Cities." *Nature Communications* 12(1). https://doi.org/10.1038/s41467-020-20871-0.

Hale, T., and N. Hultman. 2020. "'All In' Climate Diplomacy: How a Biden-Harris Administration Can Leverage City, State, Business, and Community Climate Action." Global Working Paper #147. Brookings Institute.

Hanus, N., J. Barlow, A. Satchwell, and P. Cappers. 2023. "Assessing the Current State of U.S. Energy Equity Regulation and Legislation" GRID Modernization Laboratory Consortium, United States Department of Energy. https://eta-publications.lbl.gov/sites/default/files/equity_db_report_-_v9_-_final.pdf.

Hartvig, Á., A. Vu, J. Stenning, and D. Hodge. 2022. *The Economic Impacts of Decarbonization in Wisconsin*. Northampton, MA: Cambridge Econometrics: Clarity from Complexity.

Harvey, H., R. Orvis, and J. Rissman. 2018. "The Top Policies for Greenhouse Gas Abatement." Pp. 69–120 in *Designing Climate Solutions: A Policy Guide for Low-Carbon Energy*. Washington, DC: Island Press.

Headwater Economics. 2023. "Match Requirements Prevent Rural and Low-Capacity Communities from Accessing Climate Resilience Funding." https://headwaterseconomics.org/equity/match-requirements.

Healey, M., and K. Driscoll. 2023. "Governor Healey Signs Executive Order Creating Massachusetts' First Ever Climate Chief." https://www.mass.gov/news/governor-healey-signs-executive-order-creating-massachusetts-first-ever-climate-chief.

Hendricks, B., R. Gunn-Wright, and S. Ricketts. 2020. "The Greatest Mobilization Since WWII." *Democracy: A Journal of Ideas* 56(Spring).

Housing Assistance Council. 2023. "Persistent Poverty." https://ruralhome.org/our-initiatives/persistent-poverty.

Hultman, N.E., L. Clarke, C. Frisch, K. Kennedy, H. McJeon, T. Cyrs, P. Hansel, et al. 2020. "Fusing Subnational with National Climate Action Is Central to Decarbonization: The Case of the United States." *Nature Communications* 11(5255). https://doi.org/10.1038/s41467-020-18903-w.

Jaffe, C. 2018. "Melting the Polarization Around Climate Change Politics." *Georgetown International Environmental Law Review* 30(3):455–497.

Jenkins, J., E.N. Mayfield, J. Farbes, R. Jones, N. Patankar, Q. Xu, and G. Schivley. 2022. *Preliminary Report: The Climate and Energy Impacts of the Inflation Reduction Act of 2022*. Princeton, NJ: REPEAT Project.

Joint Office of Energy and Transportation. 2022. "*Memorandum of Understanding on Electric Vehicle Infrastructure Deployment*." Edited by American Associate of State Highway and State Transportation Officials, National Association of State Energy Officials and Department of Transportation. https://www.naseo.org/Data/Sites/1/documents/tk-news/naseo-aashto-joet-nevi-mou-signed-final.pdf.

Kastorf, K.G. 2014. "Cooperative Federalism: Is There a Trend Towards Uniform National Standards Under the Clean Air Act?" American Bar Association.

Kennedy, K., W. Jaglom, N. Hultman, E. Bridgwater, R. Mendell, J. O'Neill, R. Gasper, et al. 2021. "Blueprint 2030: An All-In Climate Strategy for Faster, More Durable Emissions Reductions." America Is All In.

Kincaid, J. 2017. "Introduction: The Trump Interlude and the States of American Federalism." *State and Local Government Review* 49(3):156–169.

King, B., J. Larsen, and H. Kolus. 2022. "A Congressional Climate Breakthrough." https://rhg.com/research/inflation-reduction-act.

Kok, M.T.J., and H.C. de Coninck. 2007. "Widening the Scope of Policies to Address Climate Change: Directions for Mainstreaming." *Environmental Science and Policy* 10(7):587–599. https://doi.org/10.1016/j.envsci.2007.07.003.

Konisky, D.M., and N.D. Woods. 2018. "Environmental Federalism and the Trump Presidency: A Preliminary Assessment." *Publius: The Journal of Federalism* 48(3):345–371. https://doi.org/10.1093/publius/pjy009.

Kousky, C., and S.H. Schneider. 2003. "Global Climate Policy: Will Cities Lead the Way?" *Climate Policy* 3(4):359–372.

Krasko, V.A., and E. Doris. 2012. *Strategic Sequencing for State Distributed PV Policies: A Quantitative Analysis of Policy Impacts and Interactions*. National Renewable Energy Laboratory.

Lashof, D. 2023. "Tracking Progress: Climate Action Under the Biden Administration." *World Resources Institute Insights*, January 23.

Lavelle, M. 2015. "In Wake of Court Defeat, Opponents of Obama's Climate Rule Tee Up Seven More Attacks: Battle Over Clean Power Rule Likely to Be Long." *ScienceInsider*. https://www.science.org/content/article/wake-court-defeat-opponents-obama-s-climate-rule-tee-seven-more-attacks.

Lehr, R., and M. Groves. 2022. "Opinion: The West Can Plug into Lower-Cost Electricity with Regional Energy Markets." *Colorado Sun*, October 4.

Li, B., E. Dabla-Norris, and K. Srinivasan. 2023. "Climate Change Support for Climate Action Hinges on Public Understanding of Policy: Novel Survey Shows How Concerned People Are About Climate Change, How They View Mitigation Policies, and What Drives Support for Climate Action." *IMF*. https://www.imf.org/en/Blogs/Articles/2023/02/09/support-for-climate-action-hinges-on-public-understanding-of-policy.

Lin, A. 2020. "Uncooperative Environmental Federalism: State Suits Against the Federal Government in an Age of Political Polarization." *George Washington Law Review* 88(4).

Lipinski, R., W.V. Acker, G. Praninskas, and M.E. Barsi. 2021. "Bureaucrats Must Take a Bigger Role in Fighting Climate Change." *World Bank Blogs*, September 21.

Maffly, B. 2019. "Herbert Unveils Massive Utah Renewable Energy Project, Draws Praise from Trump Official, Protests from Climate Activists." *Salt Lake Tribune*, May 30. https://www.sltrib.com/news/environment/2019/05/30/herbert-unveils-massive.

Maffly, B. 2021. "Utah Lawmakers Punish Intermountain Power, Allege the Utility's Interests No Longer Align with the State." *Salt Lake Tribune*, November 10. https://www.sltrib.com/news/environment/2021/11/11/utah-lawmakers-punish.

Mahajan, M., O. Ashmoore, J. Rissman, R. Orvis, and A. Gopal. 2022. "Modeling the Inflation Reduction Act Using the Energy Policy Simulator." Energy Innovation Policy and Technology. https://energyinnovation.org/publication/modeling-the-inflation-reduction-act-using-the-energy-policy-simulator.

Markolf, S.A., I.M.L. Azevedo, M. Mur, and D.G. Victor. 2020. *Pledges and Progress: Steps Toward Greenhouse Gas Emissions Reductions in the 100 Largest Cities Across the United States.* Washington, DC: Brookings.

Massachusetts v. EPA. 2007. 549 U.S. 947.

Meckling, J., and J. Nahm. 2022. "Strategic State Capacity: How States Counter Opposition to Climate Policy." *Comparative Political Studies* 55(3):493–523. https://doi.org/10.1177/00104140211024308.

Mildenberger, M. 2021. "The Development of Climate Institutions in the United States." *Environmental Politics* 30(Suppl 1):71–92. https://doi.org/10.1080/09644016.2021.1947445.

Miller, C.A. 2005. "New Civic Epistemologies of Quantification: Making Sense of Indicators of Local and Global Sustainability." *Science, Technology, and Human Values* 30(3):403–432.

Miller, C.A. 2007. *Creating Indicators of Sustainability: A Social Approach.* International Institute for Sustainable Development.

Muro, M., D.G. Victor, and J. Whiton. 2019. "How the Geography of Climate Damage Could Make the Politics Less Polarizing." *Brookings.* https://www.brookings.edu/articles/how-the-geography-of-climate-damage-could-make-the-politics-less-polarizing.

New State Ice Co. v. Liebmann. 285 U.S. 262 (1932). https://supreme.justia.com/cases/federal/us/285/262.

New York City Government. 2022. "Mayor Adams' Appointments to Climate Leadership Team Streamlines Multiple City Agencies to Tackle Climate Change." https://www.nyc.gov/office-of-the-mayor/news/053-22/mayor-adams-appointments-climate-leadership-team-streamlines-multiple-city#/0.

New York State Energy Research and Development Authority. 2022. *New York Clean Energy Industry Report.* Albany, NY: New York State Energy Research and Development Authority.

NHTSA (National Highway Traffic Safety Administration). 2018. "Notice of Proposed Rule: The Safer Affordable Fuel-Efficient Vehicles Rule for Model Years 2021–2026 Passenger Cars and Light Trucks." *Federal Register* 83:42986–43500.

Nicholas Institute for Environmental Policy Solutions. 2018. *Rural Attitudes on Climate Change: Lessons from National and Midwest Polling and Focus Groups.* Duke and University of Rhode Island. https://nicholasinstitute.duke.edu/sites/default/files/publications/Rural-Attitudes-on-Climate-Change-Midwest_1.pdf.

NREL (National Renewable Energy Laboratory). 2023. "State, Local, and Tribal Governments: Interconnection Standards." https://www.nrel.gov/state-local-tribal/basics-interconnection-standards.html.

Nuccitelli, D. 2020. "The Trump EPA Is Vastly Underestimating the Cost of Carbon Dioxide Pollution to Society, New Research Finds." *Yale Climate Connections.* https://yaleclimateconnections.org/2020/07/trump-epa-vastly-underestimating-the-cost-of-carbon-dioxide-pollution-to-society-new-research-finds.

Office of Governor Janet T. Mills. 2021. "For Climate Week, Governor Mills Celebrates Maine's Progress Toward Installing 100,000 Heat Pumps by 2025." State of Maine. https://www.maine.gov/governor/mills/news/climate-week-governor-mills-celebrates-maines-progress-toward-installing-100000-heat-pumps.

Orvis, R., A. Gopal, J. Rissman, M. O'Boyle, S. Baldwin, and C. Busch. 2022. *Closing the Emissions Gap Between the IRA and 2030 U.S. NDC: Policies to Meet the Moment.* San Francisco, CA: Energy Innovation Policy and Technology.

Park, A.Y.S., R.M. Krause, and C.V. Hawkins. 2020. "Institutional Mechanisms for Local Sustainability Collaboration: Assessing the Duality of Formal and Informal Mechanisms in Promoting Collaborative Processes." *Journal of Public Administration Research and Theory* 31(2):434–450. https://doi.org/10.1093/jopart/muaa036.

Peng, W., G. Iyer, M. Binsted, J. Marlon, L. Clarke, J.A. Edmonds, and D.G. Victor. 2021. "The Surprisingly Inexpensive Cost of State-Driven Emission Control Strategies." *Nature Climate Change* 11:738–745. https://doi.org/10.1038/s41558-021-01128-0.

Pew Research Center. 2016. "Public Opinion on Renewables and Other Energy Sources [Report]." https://www.pewresearch.org/science/2016/10/04/public-opinion-on-renewables-and-other-energy-sources.

Plastrik, P., and J. Cleveland. 2015. *Framework for Long-Term Deep Carbon Reduction Planning.* Carbon Neutral Cities Alliance.

Pollack, S. 2021. "Information: Policy on Using Bipartisan Infrastructure Law Resources to Build a Better America." https://www.fhwa.dot.gov/bipartisan-infrastructure-law/building_a_better_america-policy_framework.cfm.

Popovich, N., and B. Plumer. 2020. "How Does Your State Make Electricity?" *The New York Times*, October 8. https://www.nytimes.com/interactive/2020/10/28/climate/how-electricity-generation-changed-in-your-state-election.html.

Rabe, B. 2011. "Contested Federalism and American Climate Policy." *Publius: The Journal of Federalism* 41(3):494–521. https://doi.org/10.1093/publius/pjr017.

Rabe, B.G. 2021. "When Climate Policy Works: HFCs and the Case of Short-Lived Climate Pollutants." *Brookings*, June 24. https://www.brookings.edu/blog/fixgov/2021/06/24/when-climate-policy-works-hfcs-and-the-case-of-short-lived-climate-pollutants.

Raimi, D., E. Grubert, J. Higdon, G. Metcalf, S. Pesek, and D. Singh. 2022. *The Fiscal Implications of the US Transition Away from Fossil Fuels*. Washington, DC: Resources for the Future.

Reuters. 2017. "Coalition of 17 States Challenges Trump Over Climate Change Policy." *Guardian*, April 5. https://www.theguardian.com/environment/2017/apr/05/climate-change-legal-challenge-donald-trump.

Rhodes, J.D. 2023. "The Economic Impact of Renewable Energy and Energy Storage in Rural Texas." IdeaSmiths.

Richard Nixon Foundation. 2014. "President Nixon on Establishing the EPA and NOAA." *New Nixon News*, July 9. https://www.nixonfoundation.org/2014/07/president-nixons-message-congress-epa-noaa.

Richardson, J.J., M.Z. Gough, and R. Puentes. 2003. "Is Home Rule the Answer? Clarifying the Influence of Dillon's Rule on Growth Management." Discussion paper prepared for the Brookings Institution Center on Urban and Metropolitan Policy. Virginia Polytechnic Institute and State University.

Ricketts, S., R. Cliffton, L. Oduyeru, and B. Holland. 2020. "States Are Laying a Road Map for Climate Leadership." Center for American Progress.

Rocky Mountain Institute. 2022. "State Climate Scorecards." https://statescorecard.rmi.org.

Rose, S., and G. Goelzhauser. 2018. "The State of American Federalism 2017–2018: Unilateral Executive Action, Regulatory Rollback, and State Resistance." *Publius: The Journal of Federalism* 48(3):319–344. https://doi.org/10.1093/publius/pjy016.

Saha, D., A. Rudee, H. Leslie-Bole, and T. Cyrs. 2021. *The Economic Benefits of the New Climate Economy in Rural America*. Washington, DC: World Resources Institute.

Saha, D., J. Bradbury, J. Kruger, D. Lashof, and F. Litz. 2022. "Beyond Highways: Funding Clean Transportation Through the US Bipartisan Infrastructure Law." *World Resources Institute Insights*, May 6. https://www.wri.org/insights/clean-transportation-us-bipartisan-infrastructure-law.

Samarripas, S., K. Tanabe, A. Dewey, A. Jarrah, B. Jennings, A. Drehobl, H. Bastian, et al. *The 2021 City Clean Energy Scorecard*. Washington, DC: American Council for an Energy-Efficient Economy. https://www.aceee.org/sites/default/files/pdfs/u2107.pdf.

Schaeffer, E., and T. Pelton. 2021. "Greenhouse Gases from Power Plants 2005–2020: Rapid Decline Exceeded Goals of EPA Clean Power Plan." Environmental Integrity Project.

Schneider, S.H., A. Rosencranz, M.D. Mastrandrea, and K. Kuntz-Duriseti. 2009. *Climate Change Science and Policy*. Washington, DC: Island Press. https://stephenschneider.stanford.edu/Books/CCSAP/CCSAP_book.html.

Senate Bill 21-072. 2021. Colorado State Senate.

Senate Bill 448. 2021. 81st Session of Nevada State Senate.

Smith, Jr., W.H. 2021. "Mini Guide on Transmission Siting: State Agency Decision Making." National Association of Regulatory Utility Commissioners Center for Partnerships and Innovation.

Stilson, D., P. Kiely, and R. Zakaria. 2020. *Turning Climate Commitments into Results: Progress on State-Led Climate Action*. Washington, DC: Environmental Defense Fund.

Susskind, L., J. Chun, A. Gant, C. Hodgkins, J. Cohen, and S. Lohmar. 2022. "Sources of Opposition to Renewable Energy Projects in the United States." *Energy Policy* 165:112922. https://doi.org/https://doi.org/10.1016/j.enpol.2022.112922.

Terry, D. 2022. "NASEO Letter to DOE Secretary Granholm on IIJA and IRA Implementation." https://www.naseo.org/Data/Sites/1/documents/tk-news/secretary-granholm-naseo-letter_082422-1.pdf.

Tomich, J. 2019. "Ohio Rolls Back RPS, Boost Nuclear. Here's Why It Matters." *EnergyWire*, July 24. https://www.eenews.net/articles/ohio-rolls-back-rps-boosts-nuclear-heres-why-it-matters.

Toscano, D.J. 2018. "State Preemption and the Fracturing of America." *Harvard Advanced Leadership Initiative Social Impact Review*. https://www.sir.advancedleadership.harvard.edu/articles/state-preemption-and-the-fracturing-of-america.

Tyson, A., C. Funk, and B. Kennedy. 2022. "Americans Largely Favor U.S. Taking Steps to Become Carbon Neutral by 2050." Pew Research Center. https://www.pewresearch.org/science/2022/03/01/americans-largely-favor-u-s-taking-steps-to-become-carbon-neutral-by-2050.

Tyson, A., C. Funk, and B. Kennedy. 2023. "What the Data Says About Americans' Views of Climate Change." Pew Research Center. https://www.pewresearch.org/short-reads/2023/04/18/for-earth-day-key-facts-about-americans-views-of-climate-change-and-renewable-energy.

Tyson, L.D., and L. Mendonca. 2017. "The Progressive Resurgence of Federalism." *Stanford Social Innovation Review* 16(1):C12. https://doi.org/10.48558/YE99-WC16.

Union of Concerned Scientists. 2021. "What Is the Clean Power Plan?" https://www.ucsusa.org/resources/clean-power-plan.

Utah Geological Survey. 2020. "Utah's Energy Landscape." Circular 127. Division of Utah Department of Natural Resources. https://energy.utah.gov/wp-content/uploads/Utahs-Energy-Landscape-5th-Edition.pdf.

Vandenbergh, M., and J.M. Gilligan. 2017. *Beyond Politics: The Private Sector Response to Climate Change.* Cambridge, UK: Cambridge University Press.

Victor, D.G., N. Obradovich, and D.J. Amaya. 2017. "Why the Wiring of Our Brains Makes It Hard to Stop Climate Change." *Brookings Planet Policy*, September 18. https://www.brookings.edu/blog/planetpolicy/2017/09/18/why-the-wiring-of-our-brains-makes-it-hard-to-stop-climate-change.

We Are Still In. n.d. " 'We Are Still In' Declaration." *We Are Still In.* https://www.wearestillin.com/we-are-still-declaration. Accessed May 15, 2023.

Wei, T., J. Wu, and S. Chen. 2021. "Keeping Track of Greenhouse Gas Emission Reduction Progress and Targets in 167 Cities Worldwide." *Frontiers in Sustainable Cities* 3(2021):Article 696381. https://doi.org/10.3389/frsc.2021.696381.

White House. n.d. "Inflation Reduction Act Guidebook." https://www.whitehouse.gov/cleanenergy/inflation-reduction-act-guidebook. Accessed May 15, 2023.

Williams, S. 2019. "Midwest Renewables Surge Forward: But Where Is Ohio?" *NRDC Expert Blog*, March 29. https://www.nrdc.org/bio/samantha-williams/midwest-renewables-surge-forward-where-ohio.

Wiseman, H.J. 2014. "Regulatory Islands." *New York University Law Review* 89(5):1662–1742.

Wiser, R., R. Wiser, G. Barbose, L. Bird, S. Churchill, J. Deyette, and E. Holt. 2008. *Renewable Portfolio Standards in the United States—A Status Report with Data Through 2007.* LBNL-154E. Berkeley, CA: Lawrence Berkeley National Laboratory. https://doi.org/10.2172/927151.

Worland, J. 2017. "Climate Change Used to Be a Bipartisan Issue. Here's What Changed." *Time Magazine*, July 27. https://time.com/4874888/climate-change-politics-history.

World Resources Institute. 2023. "Science Based Targets Initiative (SBTi)." https://www.wri.org/initiatives/science-based-targets.

Zhao, A., S. Kennedy, K. O'Keefe, M. Borreo, K. Clark-Sutton, R. Cui, C. Dahl, et al. 2022. "An 'All-In' Pathway to 2030: Beyond 50 Scenario." Center for Global Sustainability, University of Maryland, America Is All In.

Appendixes

Committee Member
Biographical Information

STEPHEN W. PACALA, *Chair*, is the Frederick D. Petrie Professor of Ecology and Evolutionary Biology at Princeton University. He directs the Carbon Mitigation Initiative, an independent academic research program sponsored by BP and administered by the High Meadows Environmental Institute (HMEI). Dr. Pacala is also a founder and the chair of the board of Climate Central, a nonprofit media organization focusing on climate change. He currently serves on President Joseph R. Biden's Council of Advisors on Science and Technology (PCAST). He chaired the National Academies of Sciences, Engineering, and Medicine's Committee on Carbon Dioxide Removal and Sequestration, which released its report in 2018. His research covers a wide variety of ecological and mathematical topics with an emphasis on interactions among greenhouse gases (GHGs), climate, and the biosphere. Dr. Pacala has an undergraduate degree from Dartmouth College (1978) and a PhD in biology from Stanford University (1982). He serves on the boards of the Environmental Defense Fund and the Hamilton Insurance Group. Among his many honors are the David Starr Jordan Prize and the George Mercer Award of the Ecological Society of America. Dr. Pacala is a member of the National Academy of Sciences and the American Academy of Arts and Sciences.

DANIELLE DEANE-RYAN is a senior fellow at The New School's Tishman Environment and Design Center. She has devoted her career to advancing pioneering strategic, equitable climate crisis solutions across multiple sectors. Previously, she served as the director of Equitable Climate Solutions at the Bezos Earth Fund and the director of the Inclusive Clean Economy Program at the Nathan Cummings Foundation, where she supported collaborations that catalyzed world-leading climate and equity policies. Ms. Deane-Ryan served in the Obama administration as the senior advisor for external affairs and the acting director for stakeholder engagement at the Department of Energy (DOE) Office of Energy Efficiency and Renewable Energy. She is a co-author of the 2019 Clean Energy States Alliance report *Solar with Justice: Strategies for Powering Up Under-Resourced Communities and Growing an Inclusive Solar Market*. Prior to this, Ms. Deane-Ryan was the founding executive director of Green 2.0 and a principal of the Raben Group; she launched the New Constituencies for Environmental Program at the Hewlett Foundation; and she managed the Commission to Engage African

Americans on Energy, Climate Change, and the Environment at the Joint Center for Political and Economic Studies. Ms. Deane-Ryan is on the boards of the Clean Energy States Alliance and Resource Media. She holds an MSc from the London School of Economics in environment and development and a BA from Williams College in political economy with an environmental studies concentration. Williams College awarded its Bicentennial Medal in 2019 to Ms. Deane-Ryan for her contributions to the environmental justice field.

ALEXANDRA "SANDY" FAZELI leads the National Association of State Energy Officials' (NASEO's) policy and program priorities coordination; workforce development; equity, access, and inclusion; and state and local cooperation and coordination on energy, climate, and resilience planning. Ms. Fazeli oversees NASEO's private-sector affiliates program, which connects state energy policy makers, companies, and non-profits, and helps lead the content development of NASEO's major conferences and events. She serves as an adjunct fellow for the Center for Strategic and International Studies Wadhwani Chair in U.S. India Policy Studies and on the City of Minneapolis Community Environmental Advisory Commission and advisory board of the Center for Advanced Energy Studies Energy Policy Institute at Boise State University. She also serves as a non-resident fellow for the Center for Strategic and International Studies. Prior to NASEO, Ms. Fazeli worked on energy efficiency and state policy issues at the Rocky Mountain Institute, the Colorado Energy Office, and the Alliance to Save Energy. She received a BS in foreign service from Georgetown University and an MDP from the University of Denver.

KELLY SIMS GALLAGHER is a professor of energy and environmental policy at The Fletcher School at Tufts University. Dr. Gallagher directs the Climate Policy Lab and the Center for International Environment and Resource Policy at Fletcher. From June 2014 to September 2015, she served in the Obama administration as a senior policy advisor in the White House Office of Science and Technology Policy and as the senior China advisor in the Special Envoy for Climate Change office at the Department of State. Dr. Gallagher is a member of the board of the Belfer Center for Science and International Affairs at Harvard University. She is a member of the executive committee of the Tyler Prize for Environmental Achievement and she also serves on the board of the Energy Foundation. Broadly, she focuses on energy innovation and climate policy. She specializes in how policy spurs the development and deployment of cleaner and more efficient energy technologies, domestically and internationally. She is a member of the Council on Foreign Relations. She is the author of *Titans of the Climate* (2018), *The Global Diffusion of Clean Energy Technologies: Lessons from China* (2014), *China Shifts Gears: Automakers, Oil, Pollution, and Development* (2006), and dozens of other publications. A Truman Scholar, she has an MALD and a PhD in international affairs from The Fletcher School and an AB from Occidental College.

JULIA H. HAGGERTY is an associate professor of geography in the Department of Earth Sciences at Montana State University (MSU), where she holds a joint appointment in the Montana Institute on Ecosystems. She received her BA from Colorado College in liberal arts and her PhD from the University of Colorado in history. An award-winning teacher, Dr. Haggerty teaches courses in human, economic, and energy resource geography at MSU. She also leads the Resources and Communities Research Group in studying the ways rural communities respond to shifting economic and policy trajectories, especially as they involve natural resources. Dr. Haggerty has expertise in diverse rural geographies, including those shaped by energy development, extractive industries, ranching and agriculture, and amenity development and conservation. Partnerships and collaboration with diverse stakeholders are central to her approach. Prior to joining MSU, Dr. Haggerty was a postdoctoral fellow at the University of Otago in New Zealand (2005–2007) and a policy analyst with Headwaters Economics in Bozeman, Montana (2008–2013). She speaks frequently to public audiences about her research and has served on a number of boards and advisory committees from local to international scales.

CHRIS T. HENDRICKSON is the Hamerschlag University Professor of Engineering Emeritus and director of the Traffic 21 Institute at Carnegie Mellon University. Dr. Hendrickson's research, teaching, and consulting are in the general area of engineering planning and management, including transportation systems, design for the environment, system performance, construction project management, finance, and computer applications. Central themes in his work are a systems-wide perspective and a balance of engineering and management considerations. He has co-authored eight books and published numerous articles in the professional literature. He is the editor-in-chief of the American Society of Civil Engineers (ASCE) *Journal of Transportation Engineering Part A (Systems)*. Dr. Hendrickson has been the recipient of the Council of University Transportation Centers Lifetime Achievement Award (2020), the ARTBA Steinburg Award (2019), the Faculty Award of the Carnegie Mellon Alumni Association (2009), the Turner Lecture Award of the ASCE (2002), and the Fenves Systems Research Award from the Institute of Complex Engineering Systems (2002). He is a member of the National Academy of Engineering (NAE) and the chair of the National Academies' Transportation Research Board (TRB) Division Committee, a member of the National Academy of Construction (2014), a fellow of the American Association for the Advancement of Science (AAAS; 2007), a distinguished member of the ASCE (2007), and an emeritus member of the TRB (2004). He earned a bachelor's degree and an MS from Stanford University, an MPhil in economics from Oxford University, and a PhD from the Massachusetts Institute of Technology (MIT).

ADRIENNE HOLLIS[1] leads the National Wildlife Federation's environmental justice team to advance climate justice policy and programs. With more than 20 years of experience in the environmental justice and public health arena as both a toxicologist and an attorney, Dr. Hollis focuses her work on the intersection of public health, environmental justice, and climate change, and on methods for accessing and documenting the health impacts of climate change on communities of color and other traditionally disenfranchised groups. She works to identify priority health concerns related to climate change and other environmental assaults and evaluate climate and energy policy approaches for their ability to effectively address climate change and benefit underserved communities.

JESSE D. JENKINS[2] is an assistant professor and macro-scale energy systems engineer at Princeton University with a joint appointment in the Department of Mechanical and Aerospace Engineering and the Andlinger Center for Energy and Environment. Dr. Jenkins is also an affiliated faculty with the Center for Policy Research in Energy and Environment at the School of Public and International Affairs and an associated faculty at the High Meadows Environmental Institute. At Princeton, Dr. Jenkins leads the Zero-Carbon Energy Systems Research and Optimization Laboratory (ZERO Lab), which focuses on improving and applying optimization-based energy systems models to evaluate and optimize low-carbon energy technologies, guide investment and research in innovative energy technologies, and generate insights to improve energy and climate policy and planning decisions. Dr. Jenkins completed a PhD in engineering systems (2018) and SM in technology and policy (2014) at the Massachusetts Institute of Technology and a BS in computer and information science (2006) at the University of Oregon.

ROXANNE JOHNSON established and currently directs the research department at the BlueGreen Alliance (BGA), a national coalition of labor unions and environmental groups working to build a stronger, fairer economy. In her current role, Ms. Johnson leads BGA's research efforts to understand job creation opportunities in the clean economy. Her team is responsible for conducting manufacturing and policy research in industries such as wind and solar energy, energy efficiency, advanced vehicles, and infrastructure. Her previous work at the Great Plains Institute focused on communicating model results showing potential impacts of energy and transportation policy. Ms. Johnson earned a BS in mathematics and environmental studies from Northland College in Ashland, Wisconsin. She also earned an MS in science, technology, and environmental policy from the Humphrey School of Public Affairs in Minneapolis, Minnesota.

---

[1] Resigned from the committee October 2022.
[2] Resigned from the committee March 2022.

TIMOTHY C. LIEUWEN serves as the executive director of the Strategic Energy Institute at Georgia Technology. He is also a Regents' Professor and the David S. Lewis, Jr. Chair in the School of Aerospace Engineering. He is the founder and chief technology officer of TurbineLogic, an analytics firm working in the gas turbine industry. Dr. Lieuwen is an international authority on gas turbine technologies, both from a research and development perspective and from a field/operational perspective. He has authored or edited four books, including the textbook *Unsteady Combustor Physics*. He has also authored 350 other publications and received 4 patents, all of which are licensed to the gas turbine industry. He is the editor-in-chief of the Aerospace Industries Association (AIAA) Progress book series. He is also the past chair of the Combustion, Fuels, and Emissions Technical Committee of American Society of Mechanical Engineers (ASME) and has served as the associate editor of *Combustion Science and Technology*, *Proceedings of the Combustion Institute*, and the AIAA *Journal of Propulsion and Power*. He is a fellow of ASME and AIAA and a recipient of the AIAA Lawrence Sperry Award, ASME's George Westinghouse Gold Medal, the National Science Foundation (NSF) CAREER Award, and various best paper awards. His board positions include appointment by the Secretary of Energy to the National Petroleum Counsel, the board of governors of Oak Ridge National Laboratory, and board member of the ASME International Gas Turbine Institute. He has also served on a variety of federal review and advisory committees. He holds a PhD in mechanical engineering from the Georgia Institute of Technology. He served on the National Academies' Committee on the Review of NASA Test Flight Capabilities and the Decadal Survey of Aeronautics.

VIVIAN E. LOFTNESS is a university professor and the former head of the School of Architecture at Carnegie Mellon University. Dr. Loftness is an internationally renowned researcher, author, and educator with more than 30 years of focus on environmental design and sustainability, advanced building systems integration, climate and regionalism in architecture, and design for performance in the workplace of the future. She has served on 10 National Academies' committees and the Board on Infrastructure and the Constructed Environment, and she has given 4 congressional testimonies on sustainability. Dr. Loftness is the recipient of the National Educator Honor Award from the American Institute of Architecture Students and the Sacred Tree Award from the U.S. Green Building Council (USGBC). She received her BS and MS in architecture from MIT and served on the national boards of USGBC, the American Institute of Architect (AIA) Committee on the Environment, the Green Building Alliance, Turner Sustainability, and the Global Assurance Group of the World Business Council for Sustainable Development. She is a registered architect and a fellow of the AIA.

CARLOS E. MARTÍN serves as a David M. Rubenstein Fellow at the Brookings Institution's Metropolitan Policy Program and the director of the Remodeling Futures

Program at Harvard University's Joint Center for Housing Studies. Trained as an architect, construction engineer, and historian of technology, Dr. Martín connects the bricks and mortar of housing to social and economic outcomes of occupants, especially at the intersections of environment, energy, and housing with racial equity and income disparity. For more than 20 years, he has led evaluation, research, and policy analysis for federal, state, and civil-sector entities in the fields of energy efficiency, housing construction and design, climate mitigation and adaptation, and energy and environmental justice. Dr. Martín previously led the Urban Institute's Built Environment practice area. He was also part of a core team of researchers looking at the policy and practical methods for assessing equity in energy programs. Before Urban, Dr. Martín was the assistant staff vice president for construction codes and standards at the National Association of Home Builders, the Salt River Project Professor for Energy and the Environment at Arizona State University (ASU), and the coordinator for the Department of Housing and Urban Development's Partnership for Advancing Technology in Housing. He received his BSAD in architecture from MIT and MEng and PhD in civil and environmental engineering from Stanford University.

MICHAEL A. MÉNDEZ is an assistant professor of environmental policy and planning at the University of California (UC), Irvine. He previously was the inaugural James and Mary Pinchot Faculty Fellow in Sustainability Studies and the associate research scientist at the Yale School of the Environment. Dr. Méndez has more than a decade of senior-level experience in the public and private sectors, where he consulted and actively engaged in the policy-making process. This included working for the California State Legislature as a senior consultant, as a lobbyist, and as the vice chair of the Sacramento City Planning Commission. In 2021, California Governor Gavin Newsom appointed Dr. Méndez to the Los Angeles Regional Water Quality Control Board. The board regulates water quality in a region of 11 million people. During his time at UC Irvine and Yale University, he has contributed to state and national research policy initiatives, including serving as an advisor to a California Air Resources Board member, and as a participant of the U.S. Global Change Research Program's workgroup on Climate Vulnerability and Social Science Perspectives. Dr. Méndez is a member of the National Academies' Board on Environmental Change and Society (BECS) and is on the board of directors of the social justice nonprofit Alliance for a Better Community. He also serves as a panel reviewer for the National Academies' Transit Cooperative Research Program (TCRP). Dr. Méndez holds three degrees in environmental planning and policy, including a PhD from UC Berkeley's Department of City and Regional Planning, and a graduate degree from MIT. His research on the intersection of climate change and communities of color has been featured in national publications, including *Urban Land* (published by the Urban Land Institute), the *Natural Resources Defense*

Fund Annual Report, the American Planning Association's *Planning Magazine*, *Green 2.0: Leadership at Work*, *USA Today*, and *Fox Latino News*. His new book *Climate Change from the Streets* (2020) is an urgent and timely story of the contentious politics of incorporating environmental justice into global climate change policy.

CLARK A. MILLER is a professor and the director of the Center for Energy and Society at ASU. He leads sustainability research for the Quantum Energy and Sustainable Solar Technologies Engineering Research Center. He also serves as a member of the steering committee of LightWorks, ASU's university-wide sustainable energy initiative. Dr. Miller's current research focuses on the human and social dimensions of energy transitions, including the social value of distributed renewable energy systems, strategies for addressing poverty and inequality through energy innovation, the organization of urban and regional energy transitions, and the design and governance of solar energy futures. He is an author or editor of eight books, including *The Weight of Light* (2019), *Designing Knowledge* (2018), *The Handbook of Science and Technology Studies* (2016), *The Practices of Global Ethics* (2015), *Science and Democracy* (2015), *Nanotechnology, the Brain, and the Future* (2013), *Arizona's Energy Future* (2011), and *Changing the Atmosphere* (2001). He has published extensively in the fields of energy policy, science and technology policy, the role of science in democratic governance and international relations, the governance of emerging technologies, and the design of knowledge systems for improved decision-making. He holds a PhD in electrical engineering from Cornell University.

JONATHAN A. PATZ is the director of the Global Health Institute at the University of Wisconsin–Madison. He is a professor and the John P. Holton Chair of Health and the Environment with appointments in the Nelson Institute for Environmental Studies and the Department of Population Health Sciences. For 15 years, Dr. Patz served as a lead author for the United Nations Intergovernmental Panel on Climate Change (IPCC)—the organization that shared the 2007 Nobel Peace Prize with Al Gore. He also co-chaired the health expert panel of the *U.S. National Assessment on Climate Change*, a report mandated by the U.S. Congress. In addition to directing the university-wide Global Health Institute, Dr. Patz has faculty appointments in the Nelson Institute, the Center for Sustainability and the Global Environment (SAGE), and the Department of Population Health Sciences. He also directs the NSF-sponsored Certificate on Humans and the Global Environment (CHANGE). Dr. Patz is double board certified, earning medical boards in both occupational/environmental medicine and family medicine, and he received his MD from Case Western Reserve University (1987) and his MPH (1992) from Johns Hopkins University.

KEITH PAUSTIAN is a University Distinguished Professor in the Department of Soil and Crop Sciences and a senior research scientist at the Natural Resource Ecology

Laboratory at Colorado State University. A major focus of his work involves modeling, field measurement, and development of assessment tools for soil carbon sequestration and GHG emissions from soils. Dr. Paustian was the founder of SoilMetrics, which provides modeling software for estimating agricultural GHG emissions, which was acquired by Indigio Ag. He has published more than 380 journal articles and book chapters. Dr. Paustian serves on the Farm and Forest Carbon Solutions Task Force of the Bipartisan Policy Center, the science advisory board of the Rabo Carbon Bank, and on the board of senior advisors of Solutions of the Land. Professional service activities also include coordinating lead author for the IPCC 2006 National Greenhouse Gas Inventory Methods and the IPCC 2003 Good Practice Guidance for Land Use, Land Use Change and Forestry and two National Academies' committees (in 2010–2011 and 2018–2019) related to land use, GHGs, and climate change mitigation. He served as a member of the U.S. Carbon Cycle Science Steering Group, which provides expert input to federal agencies involved in climate and carbon cycle research. He also served on the Voluntary Carbon Standard Steering Committee for Agriculture, Forestry and Other Land Use and on numerous other national and international committees involving climate and carbon cycle research. He is a fellow of the Soil Science Society of America, recipient of the Soil Science Society of America's Outstanding Research Award in 2015, and 2019 winner of the Global Foodshot Groundbreaker Award.

WILLIAM "BILLY" PIZER is a senior fellow and the vice president for research and policy engagement at Resources for the Future. Dr. Pizer was previously the Susan B. King Professor and the senior associate dean for faculty and research at the Sanford School of Public Policy and the faculty fellow at the Nicholas Institute for Environmental Policy Solutions, both at Duke University. His current research examines how we value the future benefits of climate change mitigation, how environmental regulation and climate policy can affect production costs and competitiveness, and how the design of market-based environmental policies can address the needs of different stakeholders. Dr. Pizer has been actively involved in the creation of an environmental program at Duke Kunshan University in China, a collaborative venture between Duke University, Wuhan University, and the city of Kunshan. Before Duke, he was the deputy assistant secretary for environment and energy at the Department of the Treasury from 2008 to 2011, overseeing the Treasury's role in the domestic and international environment and energy agenda of the United States. Prior to that, he was a researcher at Resources for the Future for more than a decade. Dr. Pizer has written more than 50 peer-reviewed publications, books, and articles, and holds a PhD and an MA in economics from Harvard University and a BS in physics from the University of North Carolina at Chapel Hill.

VARUN RAI[3] is a professor in the LBJ School of Public Affairs, where he directs the Energy Systems Transformation Research Group, and is in the Department of Mechanical Engineering. Dr. Rai's interdisciplinary research at the interface of energy systems, behavioral sciences, complex systems, and public policy focuses on enabling a broad diffusion of sustainable energy technologies globally. His research has been presented at several important forums and he serves on the editorial boards of the *Electricity Journal and Energy Research & Social Science*, for which he is also an associate editor. Dr. Rai was a Global Economic Fellow in 2009 and, during 2013–2015, he was a commissioner for the vertically integrated electric utility Austin Energy (~$1.4 billion revenue in 2015). In 2016, the Association for Public Policy Analysis and Management awarded him the David N. Kershaw Award and Prize, which "was established to honor persons who, at under the age of 40, have made a distinguished contribution to the field of public policy analysis and management." Dr. Rai has been the associate dean for research at the LBJ School since 2017. He received his PhD and MS in mechanical engineering from Stanford University and a bachelor's degree in mechanical engineering from the Indian Institute of Technology, Kharagpur.

EDWARD "ED" RIGHTOR is the director of the Center for Clean Energy Innovation (ITIF), which seeks to accelerate the transition of the domestic and global energy systems to low-carbon resources. Prior to joining ITIF, he was the director of the Industrial Program for the American Council for an Energy-Efficient Economy (ACEEE). In that role, Dr. Rightor developed and led the strategic vision for the industrial sector, shaped the research and policy agenda, and convened stakeholders to accelerate energy efficiency and carbon emissions reductions. Prior to joining ACEEE, he held several leadership roles at Dow Chemical during his 31-year career. Through 2017, Dr. Rightor served as the director of strategic projects in Dow's Environmental Technology Center. In this role, he worked with Dow businesses, operations, and corporate groups to reduce air emissions, waste, freshwater intake, and energy use. He also served as the facilitator of Dow's Corporate Water Strategy Team and led teams to establish and pursue Dow's 2025 Sustainability Goals, including the first-ever water goal. Working across global industrial associations, Dr. Rightor spearheaded a roadmap for the chemical industry on paths to reduce energy and GHG emissions. In prior roles, he developed GHG and energy reduction options across Dow's global operations and pursued project funding and implementation. Earlier, he started a new market-facing business in the energy sector, led cross-functional teams to optimize processes (Six Sigma), pioneered technology that led to new materials development, and led

---

[3] Resigned from the committee April 2022.

teams to troubleshoot production challenges. Dr. Rightor earned a doctorate in chemistry from Michigan State University and a BS in chemistry from Marietta College.

PATRICIA "PATY" ROMERO-LANKAO is a sociology professor at the University of Toronto Scarborough and the recipient of a Canada Excellence Research Chair in Sustainability Transitions. Before this, Dr. Romero-Lankao worked at the National Renewable Energy Laboratory's (NREL's) Center for Integrated Mobility Sciences in 2018 as a senior research scientist in a joint appointment with the University of Chicago's Mansueto Institute for Urban Innovation, where she is a research fellow. Previously, she was a senior scientist at the National Center for Atmospheric Research. Throughout her career, she has developed a considerable body of highly regarded sociological and transdisciplinary research resulting in several research grants and some 145 peer-reviewed publications. Her work primarily focuses on crucial intersections among people, energy, mobility, and the built environment in cities around the world. She has developed many innovative methods (e.g., clustering techniques and indices) to examine how inequalities in income, education, and decision-making power across populations relate to the distribution of benefits or negative impacts associated with access to transportation, energy, and related technological innovations (distributional justice). She has also developed tools such as listening sessions and fuzzy cognitive maps to examine the energy and mobility needs of women, elders, the working class, people of color, and other underrepresented groups to inform the understanding and management of these needs (e.g., procedural justice). Dr. Romero-Lankao has extensive experience as a sociologist working across disciplines and at the science–policy interface in the United States, Mexico, and many urban locations internationally. Her leadership of international research has garnered a good deal of recognition—she served as co-leading author in a working group contributing to the Nobel Prize–winning Fourth Assessment Report published by the United Nations' IPCC. She also serves on the editorial board of *Earth's Future* and several other journals and on the steering committee of the U.S. Carbon Cycle Science Program.

DEVASHREE SAHA is a senior associate at WRI United States. In this role, Dr. Saha supports state, city, and federal policy makers as they work to develop and implement policies to reduce GHG emissions and support clean energy. This includes analysis of the economics of climate action; work to develop a new framework for climate federalism that supports and strengthens the partnership between city, state, and federal governments as they work to drive deep emissions reductions; and efforts to advance a fair and equitable transition to a low-carbon economy. Prior to joining WRI, Dr. Saha led the Council of State Government's (CSG's) energy and environmental policy work, where she was responsible for directing research and providing policy analysis and technical assistance to state legislators and executive branch officials. Before joining

CSG, Dr. Saha worked at the Brookings Institution, where her research focused on a wide array of clean energy topics, including examining clean energy innovation trends at the U.S. subnational level, identifying promising clean energy financing mechanisms, and estimating the employment size, nature, and spatial geography of the U.S. clean economy. Earlier in her career, she worked for the National Governors Association, providing governors and their staff with data and guidance on best practices affecting the energy sector. Over her career, she has authored several publications on clean energy that have informed state and city policy making. Dr. Saha holds a PhD in public policy from The University of Texas at Austin and a master's degree in political science from Purdue University.

ESTHER S. TAKEUCHI[4] is a professor at Stony Brook University and a chief scientist at the Brookhaven National Laboratory. Dr. Takeuchi is an energy storage expert who led efforts to invent and refine the lifesaving lithium/silver vanadium oxide (Li/SVO) battery technology, utilized in the majority of today's implantable cardioverter defibrillators (ICDs). Dr. Takeuchi's work was conducted during 22 years at Greatbatch, Inc., a major supplier of pacemaker and ICD batteries. ICD batteries have high energy density with the ability to support intermittent high-power pulses. In addition, they have a long life, are safe, and are durable. In Dr. Takeuchi's innovation, the cathodes employ two metals, silver and vanadium, rather than just one, allowing for more energy. In addition, the Li/SVO chemistry lets the ICD monitor the level of discharge, allowing it to predict end of service in a reliable manner. Today, more than 300,000 ICDs are implanted every year. Dr. Takeuchi received her BA from the University of Pennsylvania and her PhD from The Ohio State University. She joined Greatbatch in 1984, and in 2007, she joined the State University of New York, Buffalo. Dr. Takeuchi is a member of the National Academy of Engineering, has received more than 140 U.S. patents, and is the recipient of the 2008 National Medal of Technology and Innovation.

SUSAN F. TIERNEY, a senior advisor at Analysis Group, is an expert on energy economics, regulation, and policy, particularly in the electric and gas industries. Dr. Tierney consults to businesses, government agencies, foundations, tribes, environmental groups, and other organizations on energy markets, economic and environmental regulation and strategy, and climate-related energy policies. She has participated as an expert in civil litigation cases, regulatory proceedings before state and federal agencies, and business consulting engagements. Previously, she served as the assistant secretary for policy at DOE and was the secretary for environmental affairs in Massachusetts, the commissioner at the Massachusetts Department of Public Utilities, and the executive director of the Massachusetts Energy Facilities Siting Council. She co-authored the

---

[4] Resigned from the committee April 2022.

energy chapter of the *National Climate Assessment* and serves on the boards of the Sloan Foundation, the Coalition for Green Capital, the Barr Foundation, Resources for the Future, and the World Resources Institute. Dr. Tierney taught at the Department of Urban Studies and Planning at MIT and at UC Irvine and has lectured at Harvard University, University of Chicago, Yale University, New York University, Tufts University, Northwestern University, and University of Michigan. She earned her PhD and master's degree in regional planning at Cornell University and her BA at Scripps College.

WILLIAM "REED" WALKER is the Transamerica Professor of Business and Public Policy and Economics at UC Berkeley. His research explores the social costs of environmental externalities such as air pollution and how regulations to limit these externalities contribute to gains and/or losses to society. Dr. Walker is the faculty co-director of UC Berkeley's Opportunity Lab-Climate and Environment Initiative. He is also a research associate at the Energy Institute at Berkeley and a faculty research fellow at the National Bureau of Economic Research. He was a recipient of the Sloan Foundation Research Fellowship and the IZA Young Labor Economist Award. Dr. Walker's work has been supported by the Environmental Protection Agency, NSF, the Robert Wood Johnson Foundation, the Sloan Foundation, and the Smith-Richardson Foundation. He received his PhD in economics from Columbia University.

Disclosure of Unavoidable Conflicts of Interest

The conflict-of-interest policy of the National Academies of Sciences, Engineering, and Medicine (www.nationalacademies.org/coi) prohibits the appointment of an individual to a committee like the one that authored this Consensus Study Report if the individual has a conflict of interest that is relevant to the task to be performed. An exception to this prohibition is permitted only if the National Academies determine that the conflict is unavoidable and the conflict is promptly and publicly disclosed.

When the committee that authored this report was established a determination of whether there was a conflict of interest was made for each committee member given the individual's circumstances and the task being undertaken by the committee. A determination that an individual has a conflict of interest is not an assessment of that individual's actual behavior or character or ability to act objectively despite the conflicting interest.

Michael A. Méndez was determined to have a conflict of interest in relation to his service on the Committee on Accelerating Decarbonization in the United States: Technology, Policy, and Societal Dimensions because of his ownership of stock in Tesla, Inc., an electric vehicle and clean energy company. The National Academies have concluded that the committee must include a member with current experience working at the state and local levels in the policy-making process focusing on connecting climate change and communities of color, and helping to bring local knowledge, culture, and history into policy making to address the complexities of climate change. As his biographical summary makes clear, Dr. Méndez has extensive current experience at the state and local levels, and in linking issues of sustainability, health, and environmental justice into climate change policy. His multifaceted expertise in planning, regulation, legislation, and advocacy uniquely positions him to help the committee evaluate and elucidate the implications of its analysis to impacted communities.

Keith Paustian was determined to have a conflict of interest in relation to his service on the Committee on Accelerating Decarbonization in the United States: Technology, Policy, and Societal Dimensions because he is a paid advisor to Indigo Agriculture, a company that works to build a system for "carbon farming," and was the founder and part owner of Soil Metrics (acquired by Indigo Agriculture in October 2021), which provides modeling software for estimating agricultural greenhouse gas emissions. Dr. Paustian served through June 14, 2022, on the Science Advisory Team at Carbon Direct, which works to expand the development of carbon removal technologies. The National Academies have concluded that the committee must include a member with current experience in and understanding of the mitigation measures for reducing agricultural-sector emissions, their costs, and their overall potential to contribute to emissions reductions. This topic and specific expertise were identified as critical needs after the publication of the first report from this committee. The committee also requires a member with current direct transdisciplinary experience in the modeling, field measurement, and development of assessment tools for soil carbon sequestration and greenhouse gas emissions from soils. As his biographical summary makes clear, Dr. Paustian has extensive current experience in soil organic matter dynamics, carbon and nitrogen cycling in agricultural ecosystems, and assessment of agricultural climate change mitigation strategies.

Edward Rightor was determined to have a conflict of interest in relation to his service on the Committee on Accelerating Decarbonization in the United States: Technology, Policy, and Societal Dimensions because he owns shares in Dow Chemical Company and DuPont. The National Academies have concluded that the committee must include a member with current experience in industrial energy efficiency and reductions in greenhouse gases, waste, and water use to accomplish the tasks for which it was established. The committee also requires current direct experience in business strategy, capital fundraising, and market analysis to drive corporate sustainability programs. As his biographical summary makes clear, Dr. Rightor has extensive current experience providing technical and strategic analyses of sustainability, energy efficiency, and greenhouse gas emissions reduction for manufacturing industries.

Susan F. Tierney was determined to have a conflict of interest in relation to her service on the Committee on Accelerating Decarbonization in the United States: Technology, Policy, and Societal Dimensions because she is currently employed by a consulting company (Analysis Group) that provides analyses of energy markets, clean energy regulatory policy, and resource planning and procurement for a broad range of clients (including grid operators, utility and other energy companies, governments, non-governmental organizations, and energy consumers) in the electric and natural gas industries. The National Academies have concluded that in order for the committee to accomplish the tasks for which it was established, it must include a committee member with current and extensive experience in electric power markets, natural gas markets, federal and state regulations, and utility planning processes. As her biographical summary makes clear, Dr. Tierney has extensive current experience providing technical and market analyses for electricity and gas system policy, planning, and operations.

The National Academies determined that the experience and expertise of the above individuals was needed for the committee to accomplish the task for which it was established. The National Academies could not find other available individuals with the equivalent experience and expertise who did not have a conflict of interest. Therefore, the National Academies concluded that the above conflicts were unavoidable and publicly disclosed them through the National Academies Projects and Activities Repository (NAPAR) (http://webapp.nationalacademies.org/napar).

First Report Policy Recommendations

Table C-1 is a reproduction of the policy table produced to summarize the recommendations to meet net-zero carbon emissions goals in the committee's first report.[1] Along with describing the policy, responsible actor(s), and necessary appropriations, the table also displays icons indicating the importance of each policy to achieving the five technological goals and four socioeconomic goals of the first report, explained in the Key to Icons section at the top of the table.

KEY TO ICONS
DARK GREEN icon indicates that the policy is highest priority and indispensable to achieve the objective. MEDIUM GREEN icon indicates that the policy is important to achieve the objective. LIGHT GREEN icon indicates that the policy would play a supporting role. No icon indicates that the policy would have at most a small positive (or, in some cases, a small negative impact) impact on the objective.

Technological Goals	Socioeconomic Goals
Invest in energy efficiency and productivity	Strengthen the U.S. economy
Electrify energy services in transportation, buildings, and industry	Promote equity and inclusion
Produce carbon-free electricity	Support communities, businesses, and workers
Plan, permit, and build critical infrastructure	Maximize cost-effectiveness
Expand the innovation toolkit	

---

[1] National Academies of Sciences, Engineering, and Medicine, 2021, *Accelerating Decarbonization of the U.S. Energy System*, Washington, DC: The National Academies Press, https://doi.org/10.17226/25932.

TABLE C-1 Summary of Policies Designed to Meet Net-Zero Carbon Emissions Goal and How the Policies Support the Technical and Societal Objectives

Policy	Technological Goals	Socioeconomic Goals	Government Entities	Appropriation	Notes
Establish U.S. commitment to a rapid, just, equitable transition to a net-zero carbon economy.					
U.S. CO_2 and other GHG emissions budget reaching net zero by 2050.			Executive and Congress	$5 million per year.	Budget is central for imposing emissions discipline, although any consequences for missing the target must be implemented through other policies. Funds are primarily for administration of the budget and data collection and management.
Economy-wide price on carbon.			Congress	None. Revenue of $40/t$CO_2$ rising 5% per year, which totals approximately $2 trillion from 2020 to 2030.	Carbon price level not designed to directly achieve net-zero emissions. Additional programs will be necessary to protect the competitiveness of import/export exposed businesses.
Establish 2-year federal National Transition Task Force to assess vulnerability of labor sectors and communities to the transition of the U.S. economy to carbon neutrality.			Congress	$5 million per year.	Task force responsible for design of an ongoing triennial national assessment on transition impacts and opportunities to be conducted by the Office of Equitable Energy Transitions.

continued

TABLE C-1 Continued

Policy	Technological Goals	Socioeconomic Goals	Government Entities	Appropriation	Notes
Establish White House Office of Equitable Energy Transitions: • Establish criteria to ensure equitable and effective energy transition funding. • Sponsor external research to support development and evaluation of equity indicators and public engagement. • Report annually on energy equity indicators and triennially on transition impacts and opportunities.			Congressional appropriation	$25 million per year, rising to $100 million per year starting in 2025.	Federal office establishes targets and monitors and advances progress of federal programs aimed at a just transition.
Establish an independent National Transition Corporation to ensure coordination and funding in the areas of job losses, critical location infrastructure, and equitable access to economic opportunities and wealth, and to create public energy equity indicators.			Congressional appropriation	$20 billion in funding over 10 years.	Primary means to mediate harms that occur during transition, including support for communities that lose a critical employer, support for displaced workers, abandoned site remediation, and opportunities for communities to invest in a wide range of clean energy projects.

Set rules/standards to accelerate the formation of markets for clean energy that work for all.		
Set energy standard for electricity generation, designed to reach 75% zero-emissions electricity by 2030 and decline in emissions intensity to net-zero emissions by 2050.	Congress	None.
Set national standards for light-, medium-, and heavy-duty zero-emissions vehicles, and extend and strengthen stringency of CAFE standards. Light-duty ZEV standard ramps to 50% of sales in 2030; medium- and heavy-duty to 30% of sales in 2030.	Congress	None.
Set manufacturing standards for zero-emissions appliances, including hot water, cooking, and space heating. Department of Energy (DOE) continues to establish appliance minimum efficiency standards. Standard ramps down to achieve close to 100% all-electric in 2050.	Congress	None.

continued

TABLE C-1 Continued

Policy	Technological Goals	Socioeconomic Goals	Government Entities	Appropriation	Notes
Enact three near-term actions on new and existing building energy efficiency, two by DOE/Environmental Protection Agency (EPA)[a] and one by the General Services Administration (GSA).			DOE, GSA	None.	GSA to set a cap on existing and new federal buildings that declines by 3% per year.
Enact five congressional actions to advance clean electricity markets, and to improve their regulation, design, and functioning.[b]			Congress	$8 million per year for Federal Energy Regulatory Commission (FERC) Office of Public Participation and Consumer Advocacy.	Two of these congressional actions involve FERC, and three involve DOE.
Deploy advanced electricity meters for the retail market, and support the ability of state regulators to review proposals for time/location-varying retail electricity prices.			Congressional appropriation for DOE	$4 billion over 10 years.	
Recipients of federal funds and their contractors must meet labor standards, including Davis-Bacon Act prevailing wage requirements; sign Project Labor Agreements (PLAs) where relevant; and negotiate Community Benefits (or Workforce) Agreements (CBAs) where relevant.			Congress.	None.	

744

Action			Lead	Cost	Notes
Report and assess financial and other risks associated with the net-zero transition and climate change by private companies, government agencies, and the Federal Reserve. Private companies receiving federal funds must also report their clean energy research and development (R&D) by category (wind, solar, etc.).			Congress	None.	Risk disclosures to be included in annual SEC reports for private companies. Federal Reserve to use climate-related risks in financial stress tests. Federal agencies to include climate-related risks in all benefit cost analyses. All banks to report on comparative financial investments in all energy sources.
Ensure that Buy America and Buy American provisions are applied and enforced for key materials and products in federally funded projects.			Congress	None.	
Establish an environmental product declaration library to create the accounting and reporting infrastructure to support the development of a comprehensive Buy Clean policy.			Congressional appropriation for EPA and DOE	$5 million per year.	

continued

TABLE C-1 Continued

Policy	Technological Goals	Socioeconomic Goals	Government Entities	Appropriation	Notes
Invest (research, technology, people, and infrastructure) in a U.S. net-zero carbon future.					
Establish a federal Green Bank to finance low- or zero-carbon technology, business creation, and infrastructure.			Congressional authorization and appropriation	Capitalized with $30 billion, plus $3 billion per year until 2030.	Additional requirements include public reporting of both energy equity analyses of investment and leadership diversity of firms receiving funds.
Amend the Federal Power Act and Energy Policy Act by making changes to facilitate needed new transmission infrastructure.[c]			Congress	None.	
Plan, fund, permit, and build additional electrical transmission, including long-distance high-voltage, direct current (HVDC). Require fair public participation measures to ensure meaningful community input.[d]			Congressional authorization and appropriation for DOE and FERC	$25 million per year to DOE for planning; $50 million per year for DOE and FERC to facilitate use of existing rights-of-way; finance build through Green Bank; $10 million per year to DOE for distribution system innovations.	Funds provide support for technical assistance to states, communities, and tribes to enable meaningful participation in regional transmission planning and siting activities. Funds to distribution utilities to invest in automation and control technologies.

Expand EV charging network for interstate highway system.[e]		Congressional directive to Federal Highway Administration (FHWA) and National Institute of Standards and Technology (NIST); congressional appropriations to DOE	$5 billion over 10 years to expand changing infrastructure.	FHWA to expand its "alternative fuels corridor" program. NIST to develop interoperability standards for level 2 and fast chargers. DOE to fund expansion of interstate charging to support long-distance travel and make investments for EV charging for low-income businesses and residential areas.
Expand broadband for rural and low-income customers to support advanced metering.		Congress to authorize and fund rural electric cooperatives and private companies to offer broadband	$0.5 billion for rural electric cooperatives and $1.5 billion for private companies.	10% of investment costs to expand capabilities of smart grid to underserved areas. Grants or loans to rural electric providers and investment tax incentives to companies, both focused on rural and low-income communities.

continued

TABLE C-1 Continued

Policy	Technological Goals	Socioeconomic Goals	Government Entities	Appropriation	Notes
Plan and assess the requirements for national CO_2 transport network, characterize geologic storage reservoirs, and establish permitting rules.[f] Require fair public participation measures to ensure meaningful community input.	⚙ 🗼	〰	Congressional authorization and appropriation to multiple agencies	$50 million to Department of Transportation (DOT) with other agencies involved for 5-year planning plus $50 million for block grants for community and stakeholder engagement. $10 billion during the 2020s to $15 billion total during the 2020s to DOE, United States Geological Survey (USGS), and Department of the Interior (DOI) to characterize reservoirs. Extend 45Q and increase to $70/tCO_2—$2 billion per year.	Modeling studies and other analysis indicate that significant amounts of negative emissions will be needed to meet net-zero emissions. The CO_2 pipeline network is needed even with 100% non-fossil electric power to enable carbon capture at cement and other industrial facilities with direct process emissions of greenhouse gases and to enable capture of CO_2 from biomass or via direct air capture for use in production of carbon-neutral liquid and gaseous fuels.

Establish educational and training programs to train the net-zero workforce, with reporting on diversity of participants and job placement success.[g]			Congressional appropriations to Department of Education, DOE, and NSF	$5 billion per year for GI Bill–like program. $100 million per year for new undergraduate programs. $50 million per year for use-inspired and $375 million per year for other doctoral and postdoctoral fellowships. Eliminate visa restrictions for net-zero students. $7 million over 2020–2025 for the Energy Jobs Strategy Council.	Fields covered include science, engineering, policy, and social sciences, for students researching and innovating in low-carbon technologies, sustainable design, and the energy transition.
Revitalize clean energy manufacturing.[h]			Congressional appropriation and direction of Green Bank and U.S. Export-Import Bank	Manufacturing subsidies for low-carbon products starting at $1 billion per year and phased out over 10 years. No additional appropriation required for loans and loan guarantees from Green Bank and Export-Import Bank.	Export-Import Bank should make available at least $500 million per year in low-carbon product and clean-tech export financing and eliminate support for fossil technology exports.

continued

TABLE C-1 Continued

Policy	Technological Goals	Socioeconomic Goals	Government Entities	Appropriation	Notes
Increase clean energy and net-zero transition RD&D that integrates equity indicators.[i]			Congressional appropriation for and directions to DOE and NSF	DOE clean energy RD&D triples from $6.8 billion per year to $20 billion per year over 10 years. DOE funds studies of policy evaluation at $25 million per year and regional innovation hubs at $10 million per year; DOE- and NSF-funded studies of social dimensions of the transition should be supported by an appropriation of $25 million per year.	Establish criteria for receiving funds on equity analysis, appropriate community input, and leadership diversity of companies applying for public investments. DOE to report on equity impacts and diversity of entities receiving public funds.
Increase funds for low-income households for energy expenses, home electrification, and weatherization.			Congressional appropriation	Increase Weatherization Assistance Program (WAP) funding to $1.2 billion per year from $305 million per year. Direct HHS to increase state's share of LIHEAP funds for home electrification and efficiency.	

Increase electrification of tribal lands.		Congressional appropriation to DOE and U.S. Department of Agriculture (USDA)	$20 million per year for assessment and planning through DOE Office of Indian Energy Policy (DOE-IE) and USDA Rural Utilities Service (USDA-RUS); expand DOE-IE to $200 million per year.	Increase direct financial assistance for the build-out of electricity infrastructure through DOE-IE grant programs.
Assist families, businesses, communities, cities, and states in an equitable transition, ensuring that the disadvantaged and at-risk do not suffer disproportionate burdens.				
Please note that the primary policies targeting fairness, diversity, and inclusion during the transition are the Office of Equitable Energy Transitions and the National Transition Corporation, which are the fourth and fifth policies in this table.				
Establish National Laboratory support to subnational entities for planning and implementation of net-zero transition.		Congressional appropriation	Additional funding to national laboratories' annual funding commencing at the level of $200 million per year, rising to $500 million per year by 2025, and $1 billion per year by 2030.	To establish a coordinated, multi-laboratory capability to provide energy modeling, data, and analytic and technical support to cities, states, and regions to complete a just, equitable, effective, and rapid transition to net zero.
Establish 10 regional centers to manage socioeconomic dimensions of the net-zero transition.[j]		Congressional authorization and appropriations to DOE	$5 million per year for each center; $25 million per year for external research budget to provide data, models, and decision support to the region.	Coordinated by the Office of Equitable Energy Transitions.

continued

TABLE C-1 Continued

Policy	Technological Goals	Socioeconomic Goals	Government Entities	Appropriation	Notes
Establish net-zero transition office in each state capital.			Congressional appropriations	$1 million per year in matching funds for each state.	Coordinate state's effort with federal and regional efforts.
Establish local community block grants for planning and to help identify especially at-risk communities. Greatly improve environmental justice (EJ) mapping and screening tool and reporting to guide investments.			Congressional appropriations to DOE	$1 billion per year in grants administered by regional centers.	Required to qualify for funding from the National Transition Corporation. Block grant funding requires inclusive participation and engagement by historically marginalized and low-income groups.

[a] Direct DOE/EPA to expand its outreach of and support for adoption of benchmarking and transparency standards by state and local government through the expansion of Portfolio Manager. Direct DOE/EPA to further investigate the development of model carbon-neutral standards for new and existing buildings that, in turn, could be adopted by states and local authorities. Policies targeting retrofits of existing buildings will be in the final report.

[b] FERC should work with regional transmission organizations (RTOs) and independent system operators (ISOs) to ensure that markets in all parts of the country are designed to accommodate the shift to 100 percent clean electricity on the relevant timetable. Congress should clarify that the Federal Power Act does not limit the ability of states to use policies (e.g., long-term contracting with zero-carbon resources procured through market-based mechanisms) to support entry of zero-carbon resources into electric utility portfolios and wholesale power markets. Congress should further direct FERC to exercise its rate-making authority over wholesale prices in ways that accommodate state action to shape the timing and character of the transitions in their electric resource mixes. Congress should reauthorize the FERC Office of Public Participation and Consumer Advocacy to provide grants and other assistance to support greater public participation in FERC proceedings. FERC should direct NERC to establish and implement standards to ensure that grid operators have sufficient flexible resources to maintain operational reliability of electric systems. Congress should direct and fund DOE to provide federal grants to support the deployment of advanced meters for retail electricity customers as well as the capabilities of state regulatory agencies and energy offices to review proposals for time/location-varying retail electricity prices, while also ensuring that low-income consumers have access to affordable basic electricity service.

^{c} (1) Establish National Transmission Policy to rely on the high-voltage transmission system to support the nation's (and states') goals to achieve net-zero carbon emissions in the power sector. (2) Authorize and direct FERC to require transmission companies and regional transmission organizations to analyze and plan for economically attractive opportunities to build out the interstate electric system to connect regions that are rich in renewable resources with high-demand regions; this is in addition to the traditional planning goals of reliability and economic efficiency in the electric system. (3) Amend the Energy Policy Act of 2005 to assign to FERC the responsibility to designate any new National Interest Electric Transmission Corridors and to clarify that it is in the national interest for the United States to achieve net-zero climate goals as part of any such designations. (4) Authorize FERC to issue certificates of public need and convenience for interstate transmission lines (along the lines now in place for certification of gas pipelines), with clear direction to FERC that it should consider the location of renewable and other resources to support climate-mitigation objectives, as well as community impacts and state policies as part of the need determination (i.e., in addition to cost and reliability issues) and that FERC should broadly allocate the costs of transmission enhancements designed to expand regional energy systems in support of decarbonizing the electric system.

^{d} (1) Congress should authorize and appropriate funding for DOE to provide support for technical assistance and planning grants to states, communities, and tribal nations to enable meaningful participation in regional transmission planning and siting activities. (2) Congress should authorize and appropriate funding for DOE and FERC to encourage and facilitate use of existing rights of way (e.g., railroad, roads and highways, electric transmission corridors) for expansion of electric transmission systems. (3) Congress should authorize and appropriate funding for DOE to analyze, plan for, and develop workable business model/regulatory structures, and provide financial incentives (through the Green Bank) for development of transmission systems to support development of offshore wind and for development, permitting, and construction of high-voltage transmission lines, including high-voltage direct-current lines.

^{e} (1) Congress should direct the Federal Highway Administration (a) to continue to expand its "alternative fuels corridor" program, which supports planning for EV charging infrastructure on the nation's interstate highways, and (b) to update its assessment of the ability and plans of the private sector to build out the EV charging infrastructure consistent with the pace of EV deployment needed for vehicle electrification anticipated for deep decarbonization, the need for vehicles on interstate highways and in public locations or high-density workplaces, and to identify gaps in funding and financial incentives as needed. In coordination with FHWA, DOE should provide funding for additional EV infrastructure that would cover gaps in interstate charging to support long-distance travel and make investments for EV charging for low-income businesses and residential areas. (2) NIST should develop communications and technology interoperability standards for all EV level 2 and fast charging infrastructure.

continued

f Extend 45Q tax credit for carbon capture, use, and sequestration for projects that begin substantial construction prior to 2030 and make tax credit fully refundable for projects that commence construction prior to December 31, 2022. Set the 45Q subsidy rate for use equal to $35/tCO$_2$ less whatever explicit carbon price is established and the subsidy rate for permanent sequestration to be equal to $70/tCO$_2$ less whatever explicit carbon price is established. A hydrogen pipeline network will ultimately also be needed, but as indicated in Chapter 2 [in first report], the time pressure to build a national hydrogen pipeline network is less severe than for CO$_2$. This is because hydrogen production facilities can be located close to industrial hydrogen consumers, unlike CO$_2$ pipelines, which must terminate in geologic storage reservoirs. Also, hydrogen can be blended into natural gas and transported in existing gas pipelines, and gas pipelines could ultimately be converted to 100% hydrogen.

g (1) Congress should establish a 10-year GI Bill-type program for anyone who wants a vocational, undergraduate, or master's degree related to clean energy, energy efficiency, building electrification, sustainable design, or low-carbon technology. Such a program would ensure that the U.S. workforce transitions along the physical infrastructure of our energy, transportation, and economic systems. (2) Congress should support the creation of innovative new degree programs in community colleges and colleges and universities focused uniquely on the knowledge and skills necessary for a low-carbon economic and energy transformation. (3) Congress should provide funds to create interdisciplinary doctoral and postdoctoral training programs, similar to those funded by the National Institutes of Health (NIH), which place an emphasis on training students to pursue interdisciplinary, use-inspired research in collaboration with external stakeholders that can guide research and put it to use in improving practical actions to support decarbonization and energy justice. (4) Congress should provide support for doctoral and postdoctoral fellowships in science and engineering, policy, and social sciences for students researching and innovating in low-carbon technologies, sustainable design, and energy transitions, with at least 25 fellowships per state to ensure regional equity and build skills and knowledge throughout the United States. (5) The Department of Homeland Security (DHS) should eliminate or ease visa restrictions for international students who want to study climate change and clean energy at the undergraduate and graduate level, where appropriate. (6) Congress should pass the Promoting American Energy Jobs Act of 2019 to reestablish the Energy Jobs Strategy Council under DOE, require energy and employment data collection and analysis, and provide a public report on energy and employment in the United States.

h (1) Congress should establish predictable and broad-based market-formation policies that create demand for low-carbon goods and services, improve access to finance, create performance-based manufacturing incentives, and promote exports. Specifically, Congress should provide manufacturing incentive through loans, loan guarantees, tax credits, grants, and other policy tools to firms that are matched with corresponding performance requirements. Subsidies provided directly to manufacturers must be tied to the meeting of performance metrics, such as production of products with lower embodied carbon or adoption of low-carbon technologies and approaches. Specific items could include expanding the scope of the energy audits in the DOE Better Plants program and expanded technical assistance to focus on energy use and GHG emissions reductions at the 1,500 largest carbon-emitting manufacturing plants; supporting the hiring of industrial plant energy managers by having DOE provide manufacturers with matching funds for 3 years to hire new plant energy managers; enabling the development of agile and resilient domestic supply chains through DOE research, technical assistance, and grants to assist manufacturing facilities in addressing supply chain disruptions resulting from COVID-19 and future crises. (2) Congress should provide loans and loan guarantees to manufacturers to produce low-carbon products, ideally through a Green Bank (see Chapter 4 [in the first report]). (3) Congress should require the U.S. Export-Import Bank to phase out support for fossil fuels and make support for clean energy technologies a top priority with a minimum of $500 million per year. (4) Congress should create a new Assistant Secretary for Carbon Smart Manufacturing and Industry within DOE.

[i] (1) Congress should triple DOE's investments in low- or zero-carbon RD&D over the next 10 years, in part by eliminating investments in fossil-fuel RD&D. These investments should include renewables, efficiency, storage, transmission and distribution (T&D), carbon capture, utilization, and storage (CCUS), advanced nuclear, and negative emissions technologies and increase the agency's funding of large-scale demonstration projects. By eliminating investments in non-carbon capture and storage (CCS) fossil-fuel RD&D, the net increase to the energy RD&D budget will be partially offset. (2) Congress should direct DOE to fund energy innovation policy evaluation studies to determine the extent to which policies implemented (both RD&D investment and market-formation policies) are working. (3) Congress should direct DOE and the National Science Foundation (NSF) to create a joint program to fund studies of the social, economic, ethical, and organizational drivers, dynamics, and outcomes of the transition to a carbon-neutral economy, as well as studies of effective public engagement strategies for strengthening the U.S. social contract for decarbonization. (4) Congress should direct DOE to establish regional innovation hubs where they do not exist or are critically needed using funds appropriated under item 1 above. (5) Congress should direct DOE to enhance public-private partnerships for low-carbon energy.

[j] (1) Congress should coordinate federal agency actions at the regional scale through the deployment of federal agency staff to regional offices. (2) Congress should host a coordinating council of regional governors and mayors that meets annually to establish high-level policy goals for the transition. (3) Congress should establish mechanisms for ensuring the effective participation of low-income communities, communities of color, and other disadvantaged communities in regional dialogue and decision-making about the transition to a carbon-neutral economy. (4) Congress should provide information annually to the White House Office of Equitable Energy Transitions detailing regional progress toward decarbonization goals and benchmarks for equity.

Public Meetings

Public Meeting 1, Virtual
February 8, 2022
Topic: Industrial Decarbonization Options

Public Meeting 2, Virtual
March 7, 2022
Topic: Government Perspectives on Implementing a Just and Equitable Energy Transition

Public Meeting 3, Virtual
March 14, 2022
Topic: Leveraging Financial Systems and Markets—Information, Transition Risk, and Decarbonization

Public Meeting 4, Virtual
March 29, 2022
Topic: Nonprofit Perspectives on Implementing a Just and Equitable Energy Transition

Public Meeting 5, Virtual
April 25–26, 2022
Topic: Perspectives from Subnational Entities on Energy Transitions

Public Meeting 6, Virtual
April 29, 2022
Topic: Leveraging Organizational Capacity for Investment in Deep Decarbonization

Public Meeting 7, Virtual
June 6, 2022
Topics: Soil Carbon Offsets; Research Perspectives on Public Responses to Large-Scale Net-Zero Infrastructure

Public Meeting 8, Virtual
June 13, 2022
Topic: Philanthropic Perspectives on Implementing a Just and Equitable Energy Transition

Public Meeting 9, Virtual
June 28, 2022
Topic: Public Engagement with Distributed Energy Infrastructure

Public Meeting 10, Hybrid
July 26, 2022
In-person participants at Keck Center, Washington, DC
Topic: Pathways to an Equitable and Just Transition Workshop

Public Meeting 11, Virtual
September 9, 2022
Topic: The Role of Manufacturing in Industrial Decarbonization

Public Meeting 12, Virtual
September 30, 2022
Topic: Public Engagement Across the Transmission Development Life Cycle

Public Meeting 13, Virtual
October 3, 2022
Topic: Local Benefits and Compensation Strategies for Deep Decarbonization
Infrastructure

Public Meeting 14, Virtual
October 20–21, 2022
Topic: Perspectives on Priority Actions for the Built Environment and Building
Technologies Research, Development, and Demonstration Needs

Public Meeting 15, Virtual
January 6, 2023
Topic: Transformative Climate Communities Lessons Learned and Best Practices

Decarbonization Technologies and Related Equity and Justice Concerns

Table E-1 describes a broad selection of technologies that may play a role in decarbonization, example equity and justice concerns specific to each technology, and potential methods to mitigate problems and amplify equity and justice.

TABLE E-1 Decarbonization Technologies, Their Description and Role in Decarbonization, Example Equity and Justice Concerns Specific to the Technology, and Potential Equity and Justice Amplifiers and Problem Mitigants

Decarbonization Technology	Technology Description and Role in Decarbonization	Example Equity and Justice Concerns Specific to the Technology	Potential Equity and Justice Amplifiers and Problem Mitigants
Decarbonization technologies in general—apply to all below technologies	*Various technologies that reduce or eliminate emission of greenhouse gases (GHGs) or remove GHGs from the atmosphere.*	*Siting polluting infrastructure in disadvantaged communities.* *Participatory justice.* *Community benefits.* *Workforce opportunities.*	*Develop projects that improve well-being of disadvantaged communities and that engage community members in decision-making about projects that impact them.* *Follow all existing air and water pollution regulations, permits, and other requirements.*
Point source carbon capture (fossil fuel combustion emissions)	Point source carbon capture prevents some or all of the carbon dioxide from being released by a combustion facility, such as a power plant, by capturing the carbon and then using or storing it. The capture may be from the waste gas from combustion (post-combustion), or it may involve transformation of the inputs to remove carbon before combustion and prevent formation of CO_2 (precombustion). This technology may be required to mitigate emissions from some fossil fuel combustion facilities where there is not a zero-emission alternative.	Local air and water emissions from the technologies used to capture the carbon dioxide, such as emissions from the power source, and from amine or other capture chemicals. Continuation of local air and water pollution from the entire fossil fuel life cycle, even though GHG emissions are reduced or eliminated. Opportunity cost: Investing in a nascent technology that allows polluting facilities to continue operation and that may not be implemented to remove GHG emissions at scale.	Implement technologies that capture a greater portion of both GHG and non-GHG air quality emissions, such as processes with extensive gas pretreatment or precombustion capture.

continued

759

TABLE E-1 Continued

Decarbonization Technology	Technology Description and Role in Decarbonization	Example Equity and Justice Concerns Specific to the Technology	Potential Equity and Justice Amplifiers and Problem Mitigants
Point source carbon capture (industrial process emissions)	Carbon capture prevents some or all of the carbon dioxide from being released by an industrial process, such as the chemical reactions that make cement or steel from ores, or that form ethanol from biomass fermentation by capturing the carbon and then using or storing it. This technology may be required to mitigate emissions from some industrial facilities if it is not possible to replace the product or process with a non-emitting substitute.	Local air and water emissions from the technologies used to capture the carbon dioxide, such as emissions from the power source, and from amine or other capture chemicals. Continuation of local air and water pollution from the industrial process or other associated processes, even though GHG emissions are reduced or eliminated.	Implement technologies that capture a greater portion of both GHG and non-GHG air quality emissions, such as processes with extensive gas pretreatment or precombustion capture.
Direct air capture	Direct air capture (DAC) is composed of industrial facilities that process air from the atmosphere to remove some of the CO_2. The CO_2 can then be used or stored. DAC can remove emissions that are already present in the atmosphere.	Local air and water emissions from the technologies used to capture the carbon dioxide, such as emissions from the power source, and from amine or other capture chemicals. Opportunity cost: Local air and water emissions from the GHG emissions that led to the GHG emissions being captured from the atmosphere, if DAC enables the continuation of those processes. Opportunity cost: Investing in a nascent technology that allows polluting facilities to continue operation and that may not be implemented to remove GHG emissions at scale.	Create separate targets for emissions reductions and removals, to ensure that both are pursued concurrently.

Carbon dioxide utilization	Carbon dioxide utilization transforms CO_2 into useful products. It may be used in a net-zero future to provide needed carbon-based products without GHG emissions, or to produce materials that act as long-term carbon storage.	Local pollution from the facilities that transform the carbon dioxide into a product. Opportunity cost: GHG emissions and local pollution from the use of the product created by carbon dioxide utilization, such as combustion of a synthetic fuel.	Mitigate the impacts on GHG and local pollutant emissions from the full life cycle of the carbon dioxide utilization product. Place restrictions on where and when synthetic fuels can be used, to limit exposure of disadvantaged communities to combustion pollutant emissions, such as limiting to use in aviation, rather than on-road or off-road vehicles.
Solar and wind electricity generation	Solar and wind electricity-generating facilities collect energy from the sun or the wind and convert it to electric power. As compared to some other net-zero facilities, they can occupy a large land area.	Siting without community participation. Lack of community benefits. Pollution throughout the life cycle of the generation facilities, including inputs, manufacture, use, and disposal of the generating equipment, particularly waste disposal of solar panels and wind turbines, blades, and towers.	Participatory siting. Development of community benefits, including community ownership of zero-carbon electricity generation. Reuse, recycling, and/or planned disposal of used generating equipment.
Electric transmission	Transmission lines move electric power between areas of high generation to areas of high demand. New technologies for decarbonization will likely require increased electricity use and changes in locations of generation and demand.	Preferential siting in disadvantaged communities. Lack of community benefits. Lack of participatory justice.	Participatory siting. Development of community benefits. When planning electric system investments, consider the benefits of electric systems with fewer transmission requirements, especially those that may have enhanced resiliency, including energy storage.

continued

TABLE E-1 Continued

Decarbonization Technology	Technology Description and Role in Decarbonization	Example Equity and Justice Concerns Specific to the Technology	Potential Equity and Justice Amplifiers and Problem Mitigants
Pipelines	Pipelines move materials such as gaseous and liquid fuels and chemicals between sources and end users. Pipelines may be developed to move CO_2, hydrogen, or synthetic fuels for decarbonization.	Preferential siting in disadvantaged communities. Lack of community benefits. Lack of participatory justice. Safety risks, especially with pipeline leaks or failures. Opportunity cost: Indirectly enabling technologies with pollutant emissions, such as fossil fuel use, to make hydrogen.	Participatory siting. Development of community benefits. Consider the environmental justice benefits of colocation of the producers and users of a commodity, which may prevent the need for pipelines, although it may increase the concentration of polluting facilities into fewer, more greatly impacted communities.
Mining	Some decarbonization technologies will require increased development and use of mineral resources, which will increase mining requirements in some communities, although the mining and other resource extraction requirements for production of coal, oil, and natural gas will decrease.	Local air and water pollution from mining and mineral extraction.	Develop and implement resource extraction technologies that are less polluting for nearby communities. Develop recycling technologies that allow reuse of already mined material, and avoid mining of new, virgin material. Participatory siting. Development of community benefits.

Biomass and biofuels	Biomass and biofuels growth consumes CO_2 from the atmosphere. If all upstream process inputs like fertilizers can be made net-zero emissions, then the carbon in the product made from biomass, like a biofuel, is considered renewable. If the product is combusted or decays, it is net-zero carbon. In some circumstances, a long-lived product can be made, which—if stored for the long term—may result in net-negative carbon.	Local air and water pollution from farming and processing. Opportunity cost: Local air and water pollution from combustion of biofuel products or disposal or decay of other biobased products.	Develop and implement biomass production technologies that are less polluting for nearby communities. Place restrictions on where and when biofuels can be used, to limit exposure of disadvantaged communities to combustion pollutant emissions, such as limiting to use in aviation, rather than on-road or off-road vehicles.
Hydrogen production and use as an energy carrier	Hydrogen is a zero-carbon energy carrier. It can be made from natural gas coupled to carbon capture and storage or can be generated through electrolysis with zero-carbon electricity inputs. It produces no CO_2 when used in a fuel cell or combusted.	Hydrogen combustion produces some local air pollutants like nitrogen oxides. Hydrogen generation, transport, and storage introduce safety concerns for those in very close proximity.	Prioritize hydrogen produced from electrolysis with zero-carbon electricity over production from fossil materials like natural gas with carbon capture. Participatory siting. Development of community benefits. Implement safety mitigants for communities that include hydrogen generation, transportation, and storage or use infrastructure.

continued

TABLE E-1 Continued

Decarbonization Technology	Technology Description and Role in Decarbonization	Example Equity and Justice Concerns Specific to the Technology	Potential Equity and Justice Amplifiers and Problem Mitigants
Nuclear power generation	Nuclear power uses the energy in radioactive materials to power generation of electricity. Nuclear power generation facilities have very low GHG and criteria air pollutant emissions while operating.	Local air and water pollution from uranium mining, milling, and processing, and mining waste disposal. Air, water, and radiation pollution risk from accidents during nuclear power production. Air, water, and radiation pollution risk from processing, storage, transportation, disposal, and long-term management of spent nuclear fuel and other radioactive wastes (low-level, greater than Class C, and high-level).	Significant public engagement at local and national levels. Participatory siting for reactors and fuel cycle facilities, including waste disposal sites. Development of community benefits. Develop and implement resource extraction technologies that are less polluting for nearby communities. Local, state, and federal regulatory processes for reactor and fuel cycle facilities. Mitigation of legacy uranium pollution.

Equity and Justice Scorecard: Inflation Reduction Act Provisions

Through informed assumptions, Table F-1 seeks to quantify the benefits or impacts of federal funding authorized by the Inflation Reduction Act (IRA) on underserved, low-income, or disadvantaged communities. For the purposes of the following table, the committee uses the definition of "disadvantaged communities" provided by the Council for Economic Quality (CEQ): a disadvantaged community is one that is marginalized, underserved, and overburdened by environmental pollution and has other socioeconomic burdens, including low income and high employment (CEQ n.d.).

	Benefit Type	Inclusions
	Direct	• Provisions with mandatory carve-outs targeting for "underserved," "low-income," or "disadvantaged" communities" and tribal nations and communities • Justice-oriented programs that acknowledge and address harms
	Indirect	• Provisions with non-mandatory options for spending on "underserved," "low-income," or "disadvantaged" communities • Provisions with unclear or unknown impacts on disadvantaged communities
	Evaluation and Assessment	• Funding for research, modeling, and monitoring of the impacts of air pollution and climate change on disadvantaged communities • Funding for evaluation of the impact of environmental programs on disadvantaged communities

TABLE F-1 Disadvantaged Community Benefits Scorecard for Select Inflation Reduction Act Provisions

Provision *Description*	Type of Benefit	Percent of Funding to Disadvantaged Communities
§13102—Investment Tax Credit for Energy Property *Provides a tax credit for investment in renewable energy projects. Credit is increased by up to 10% if project is in an energy community.[a] (White House 2023)*		**Unknown** Dependent on whether energy community is also a disadvantaged community
§13103—Increase in Energy Credit for Solar and Wind Facilities Placed in Service in Connection with Low-Income Communities *Provides an additional investment tax credit for small-scale solar and wind facilities in low-income communities. Credit increased by 10% for facilities located in low-income communities or on tribal lands and by 20% for facilities that are part of federally subsidized housing programs or that offer at least 50% of the financial benefits to low-income households. (White House 2023)*		**Unknown** Dependent on location of facility and distribution of benefits produced
§13301—Energy Efficiency Home Improvement Credit *Provides a tax credit for energy-efficiency improvements of residential homes. (White House 2023)*		**Unknown** Dependent on implementation and location of improved homes
§13302—Residential Clean Energy Credit *Provides a tax credit for the purchase of residential clean energy equipment. (White House 2023)*		**Unknown** Dependent on implementation and location of improved homes
§13303—Energy Efficient Commercial Buildings Deduction *Provides a tax deduction for energy efficiency improvements to commercial buildings. (White House 2023)*		**Unknown** Dependent on implementations and location of improved buildings
§13304—Extension, Increase, and Modification of New Energy-Efficient Home Credit *Provides a tax credit for construction of new energy-efficient homes. (White House 2023)*		**Unknown** Dependent implementation and location of new homes built
§13501—Advanced Energy Project Credit *To provide an investment tax credit for qualified advanced energy manufacturing properties, with intentional consideration of projects that create jobs in historically underserved communities. (Jenkins et al. 2022)*		**Unknown** Dependent on implementation and location of manufacturing properties

TABLE F-1 Continued

Provision *Description*	Type of Benefit	Percent of Funding to Disadvantaged Communities
§13701—Clean Electricity Production Tax Credit *Provides a tax credit to produce clean electricity, regardless of the technology used. Credit increases by 10% if project is in an energy community.* (White House 2023)		**Unknown** Dependent on whether energy community is also a disadvantaged community
§13702—Clean Electricity Investment Tax Credit *Provides a tax credit for investments in facilities that generate clean electricity. Credit increases by 10% if facility is in an energy community.* (White House 2023)		**Unknown** Dependent on whether energy community is also a disadvantaged community
§13901—Permanent Extension of Tax Rate to Fund Black Lung Disability Trust Fund *To make permanent the tax rate to pay for the Black Lung Disability Trust Fund.* (Jenkins et al. 2022)		**Unknown** Dependent on implementation and how many of the affected live in disadvantaged communities
§40001—Investing in Coastal Communities and Climate Resilience *To support coastal communities, Indigenous governments, and nonprofit organizations in the conservation, restoration, and protection of marine habitats and resources.* (White House 2023)		**Unknown** Dependent on implementation and on how many coastal communities are disadvantaged communities
§40004—Research and Forecasting for Weather and Climate *To support advancements and improvements in research, observation systems, modeling, forecasting, assessments, and dissemination of information, including climate research.* (White House 2023)		0%
§40005—Computing Capacity and Research for Weather, Oceans, and Climate *To procure additional high-performance computing, data processing capacity, and data management for transaction agreements under the Weather Research and Forecasting Innovation Act.* (White House 2023)		0%

continued

TABLE F-1 Continued

Provision *Description*	Type of Benefit	Percent of Funding to Disadvantaged Communities
§50122—High-Efficiency Electric Home Rebate Program *To award grants to state energy offices and tribal entities to develop and implement a high-efficiency home rebate program.* (White House 2023)		**95%** $4.275 billion in Department of Energy (DOE) grants to state energy offices to fund high-efficiency programs with benefits dependent on implementation
		5% $0.225 billion for grants to Tribes to develop and implement a high-efficiency electric home rebate program
§50123—State-Based Home Energy Efficiency Contractor Training Program *To provide financial assistance to states to develop and implement a training and education program for contractors involved with the installation of home energy efficiency and electrification improvements.* (White House 2023)		**Unknown** Dependent on implementation
§50131—Assistance for Latest Net-Zero Building Energy Code Adoption *To provide grants to states or units of local government to adopt updated building energy codes.* (White House 2023)		**Unknown** Dependent on implementation
§50144—Energy Infrastructure Reinvestment Financing *To guarantee loans to energy infrastructure projects.* (White House 2023)		**Unknown** Dependent on implementation
§50145—Tribal Energy Loan Guarantee Program *To support tribal investments in energy-related projects through direct loans or partial loan guarantees.* (White House 2023)		**100%**
§50231—Domestic Water Supply Projects *To provide domestic water supplies to disadvantaged communities or households that do not have reliable access to domestic water supplies.* (White House 2023)		**100%**

continued

TABLE F-1 Continued

Provision *Description*	Type of Benefit	Percent of Funding to Disadvantaged Communities
§50301—Department of Energy Environmental Reviews *To provide for the development of programmatic environmental documents, development of data or information systems, and the engagement of stakeholders.* (U.S. Congress 2022)	◎	**0%**
§50302—Federal Energy Regulatory Commission Environmental Reviews *To provide for the development of programmatic environmental documents, development of data or information systems, and the engagement of stakeholders.* (U.S. Congress 2022)	◎	**0%**
§50303—Department of the Interior Environmental Reviews *To provide for the development of programmatic environmental documents, development of data or information systems, and the engagement of stakeholders.* (U.S. Congress 2022)	◎	**0%**
§60101—Clean Heavy-Duty Vehicles *To provide funding to offset the costs of replacing heavy-duty commercial vehicles with zero-emission vehicles, deploying necessary infrastructure, and developing and training the workforce.* (White House 2023)	◎	**Unknown** Dependent on implementation
§60102—Grants to Reduce Air Pollution at Ports *To purchase and install zero-emission port equipment and technology and develop climate action plans to address air pollution at ports.* (White House 2023)	◎	**Unknown** Dependent on implementation
§60103—Greenhouse Gas Reduction Fund *To provide competitive grants to mobilize financing for clean energy and climate projects with an emphasis on projects that benefit low-income and disadvantaged communities.* (White House 2023)	◎	**26%** $7 billion to enable low-income and disadvantaged communities to deploy and benefit from zero-emission technology
	◎	**30%** $8 billion reserved for low-income and disadvantaged communities for financial and technical assistance

continued

TABLE F-1 Continued

Provision *Description*	Type of Benefit	Percent of Funding to Disadvantaged Communities
§60104—Diesel Emissions Reductions *To identify and reduce diesel emissions resulting from good movement facilities and vehicles servicing good movement facilities in low-income and disadvantaged communities.* (White House 2023)	◎	**Unknown** Dependent on implementation
§60105—Funding to Address Air Pollution *Part (a). To extend community air monitoring at or near fenceline communities.* *Part (c). To make air quality sensor technology available to low-income and disadvantaged communities.* (White House 2023)	◎	**51%** $0.1175 billion for grants to deploy and maintain fenceline air monitoring, screening air monitors, national air toxic trend stations, and community monitors
	◎	**1%** $0.003 billion for grants to deploy, integrate, and operate air quality sensors in low-income and disadvantaged communities
§60106—Funding to Address Air Pollution at Schools *To provide funding for grants and other activities to monitor and reduce pollution and greenhouse gas emissions in low-income and disadvantaged communities.* (White House 2023)	◎	**100%**
§60107—Low-Emissions Electricity Program *To fund a wide range of activities to encourage low-emissions electricity generation and use through education, technical assistance, and partnerships with consumers, including low-income and disadvantaged communities and local and tribal governments.* (White House 2023)	◎	**20%** $0.017 billion for programs in low-income and disadvantaged communities
	◎	**20%** $0.017 billion for outreach and technical assistance to and partnerships with state, tribal, and local governments
	◎	**0%** $0.001 to assess reductions in greenhouse gas (GHG) emissions that result from changes in domestic electricity generation

TABLE F-1 Continued

Provision *Description*	Type of Benefit	Percent of Funding to Disadvantaged Communities
§60108—Funding for Section 211(O) of the Clean Air Act *To support investments in advanced biofuels and to implement the Renewable Fuel Standard, including the review of transportation fuel impacts on low-income and disadvantaged communities.* (White House 2023)		**Unknown** Dependent on implementation
§60114—Climate Pollution Reduction Grants *To provide grants to states, air pollution control agencies, and Indigenous nations to develop and implement plans for reducing GHG emissions.* (White House 2023)		**Unknown** Dependent on implementation
§60115—Environmental Protection Agency Efficient, Accurate, and Timely Reviews *To provide for the development of environmental data or information systems, stakeholder and community engagement, and the development of geographic information systems and other analysis tools and guidance to improve agency transparency, accountability, and public engagement.* (U.S. Congress 2022)		**0%**
§60201—Environmental and Climate Justice Block Grants *To provide grants and technical assistance to community-based organizations to reduce indoor and outdoor air pollution.* (White House 2023)		**100%**
§60401—Environmental and Climate Data Collection *To improve the availability and use of data to support efforts to address environmental injustice and better protect all communities from the impacts of pollution, to update and improve the Climate and Economic Justice Screening Tool, and to identify ways to improve outcomes for communities with environmental justice concerns.* (White House 2023)		**0%**
§60402—Council on Environmental Quality Efficient and Effective Environmental Reviews *To add staff to support federal agencies and develop tools, guidance, and techniques to increase efficiency and improve community engagement in federal decisions.* (White House 2023)		**0%**

continued

TABLE F-1 Continued

Provision *Description*	Type of Benefit	Percent of Funding to Disadvantaged Communities
§60501—Neighborhood Access and Equity Grant Program *To award competitive grants for projects that improve walkability and safety and provide affordable transportation access and for planning and capacity building activities in disadvantaged or underserved communities.* (White House 2023)	🎯	**62%** $1.9 billion for planning and capacity building in disadvantaged and underserved communities impacted negatively by highways or other transportation facilities
	🎯	**36%** $1.1 billion for projects in economically disadvantaged communities
§80001—Tribal Climate Resilience *Parts (a) and (c). To support climate resilience planning to help sustain tribal ecosystems and natural and cultural resources.* *Part (b). To extend the life of tribal hatcheries and to support hatchery rearing and stocking programs.* (White House 2023)	🎯	**94%** $0.22 billion for tribal climate resilience and adaptation programs
§80003—Tribal Electrification Program *To provide financial and technical assistance to Indigenous communities to increase the number of tribal homes with zero-emission electricity.* (White House 2023)	🎯	**100%**
§80004—Emergency Drought Relief for Tribes *To fund drought relief actions to mitigate impacts for Indigenous communities affected by the Bureau of Reclamation water project.* (White House 2023)	🎯	**100%**

a The IRA defines an "energy community" as (1) a brownfield site, as defined by the Comprehensive Environmental Response, Compensation, and Liability Act of 1980; (2) a metropolitan statistical area or non-metropolitan statistical are that has 0.17 percent or greater direct employment or 25 percent or greater local tax revenues related to the coal, oil, or natural gas industry and has en unemployment rate at or above the national average unemployment rate for the previous year; or (3) a census tract, or directly adjoining census tract, in which a coal mine has closed after 1999 or in which a coal-fired electric generating unit has been retired after 2009 (NETL n.d.).

REFERENCES

Jenkins, J.D., J. Farbes, R. Jones, and E. Mayfield. 2022. "REPEAT Project Section-by-Section Summary of Energy and Climate Policies in the 117th Congress." http://bit.ly/REPEAT-Policies.

NETL (National Energy Technology Laboratory). n.d. "Energy Community Tax Credit Bonus." https://energycommunities.gov/energy-community-tax-credit-bonus.

U.S. Congress. 2022. "H.R.5376—117th Congress (2021–2022): Inflation Reduction Act of 2022." https://www.congress.gov/bill/117th-congress/house-bill/5376.

White House. 2023. *Building a Clean Energy Economy: A Guidebook to the Inflation Reduction Act's Investments in Clean Energy and Climate Action.* Version 2. https://www.whitehouse.gov/wp-content/uploads/2022/12/Inflation-Reduction-Act-Guidebook.pdf.

Disadvantaged Community as Defined by Implementers of Justice40 Covered Programs

Table G-1 displays the various definitions for "disadvantaged community" used across federal agencies with Justice40 covered programs and compares screening tool metrics with those of the Council for Economic Quality's (CEQ's) Climate and Economic Justice Screening Tool (CEJST).

TABLE G-1 Definitions for Disadvantaged Community for Federal Agencies with Justice40 Covered Programs, Comparing Screening Tool Metrics with CEQ's Climate and Economic Justice Screening Tool

Actor	Disadvantaged Community Definition	Metrics That Overlap with CEJST	
CEQ	**Definition:** A community that is marginalized, underserved, and overburdened by pollution and has other socioeconomic burdens. **Screening Methodology:** 30 metrics are grouped into eight burden categories (the parentheses show the number of metrics per category): • Climate change (5). • Energy (2). • Health (4). • Housing (5). • Legacy pollution (5). • Transportation (3). • Water and wastewater (2). • Workforce development (4). CEJST[a] identifies 27,251 census tracts as disadvantaged (33% of the U.S. population) and an additional 1,063 of census tracts are partially disadvantaged communities (White House 2022).	• Abandoned Mine Land • Agriculture Loss Rate • Asthma • Building Loss Rate • Diabetes • Diesel Particulate Matter (PM) Exposure • Energy Cost • Formerly Used Defense Site • Hazardous Waste Facility Proximity • Heart Disease • High School (HS) Education • Historic Underinvestment • Housing Cost • Lack of Green Space • Lack of Indoor Plumbing • Lead Paint	• Linguistic Isolation • Low Income • Low Life Expectancy • Low Median Income • $PM_{2.5}$ in the Air • Population Loss Rate • Poverty • Projected Flood Risk • Projected Wildfire Risk • Risk Management Plan Facility Proximity • Superfund Site Proximity • Traffic Proximity and Volume • Transportation Barriers • Underground Storage Tanks and Releases • Unemployment • Wastewater Discharge
Department of Transportation (DOT)	**Definition:** A historically disadvantaged community is (1) a qualifying census tract; (2) tribal land; or (3) any territory or possession of the United States. **Screening Methodology:** 40 metrics are grouped into five categories of transportation disadvantage (the parentheses show the number of metrics per category): • *Transportation insecurity* occurs when people cannot get to where they need to go (3).	• 200% of Poverty Line • Age of Housing Unit • Annualized Disaster Losses • Anticipated Changes in Extreme Weather • Asthma • Coal Mine Proximity • Diabetes	• Limited English Proficiency • Linguistic Isolation • No High School Diploma • $PM_{2.5}$ Level • Population Poverty • Pre-1980s Housing • Risk Management Sites Proximity

continued

Actor	Disadvantaged Community Definition	Metrics That Overlap with CEJST
	- *Environmental burden* measures air and water pollution from hazardous facilities and the built environment (16). - *Social vulnerability* measures socioeconomic metrics that direct impact quality of life (13). - *Health vulnerability* identifies communities based on adverse health outcomes from exposure to air and water pollution (5). - *Climate and disaster risk burden* measures changes in precipitation, weather, and heat that pose a risk to the transportation system (3). The Equitable Transportation Community[b] Explorer identifies 35% of census tracts as transportation disadvantaged communities (DOT 2023).	- Diesel Particulate Matter Level - Diesel PM Level - Hazardous Sites Proximity - High-Volume Road Proximity - Housing Cost - Housing Cost Burden - Lead Mine Proximity - Traffic Proximity and Volume - Transportation Access - Transportation Cost Burden - Transportation Safety - Treatment and Disposal Facility Proximity - Unemployment
Department of Energy (DOE)	**Definition:** As defined by Young et al. (2021), a disadvantaged community is either (1) a group of individuals living in geographic proximity, such as a census tract, or (2) a geographically dispersed set of individuals who experience common conditions, such as migrant workers or Indigenous people. **Screening Methodology:** 36 burden metrics are grouped into four categories (the parentheses show the number of metrics per category): - Fossil dependence (2). - Energy burden (5). - Environmental and climate hazards (10). - Socioeconomic vulnerabilities (19).	- Climate Hazards - Diesel Particulate Matter Level - Home Age - Household Income - Housing Energy Cost - Housing Plumbing - Linguistic Isolation - National Priorities List Proximity - $PM_{2.5}$ Level - Risk Management Plan Proximity - Traffic Proximity - Unemployment

	Definition	Indicators
	The Disadvantaged Communities Reporter[c] mapping tool identifies 13,581 census tracts as disadvantaged communities (DOE 2023).	• Energy Cost Burden • Environmental Cumulative Impacts • Impacts from Climate Change • Income • Linguistic Isolation • Population Poverty • Tribal Jurisdictions • Unemployment
Department of the Interior (DOI)	**Definition:** A community may be considered disadvantaged based on a combination of burden indicators or based on the community's inclusion in the CEJST (DOI 2022).	See indicator list for CEQ above.
Army Corps of Engineers (USACE) Civil Works Program	**Definition:** USACE uses the CEQ definition of a disadvantaged community and CEJST to implement Justice40 covered programs (Connor 2022). Additional tools will be used for further support, including the Environmental Protection Agency (EPA) EJScreen Tool; the California Communities Environmental Health Screening Tool; New Jersey's Environmental Justice Mapping, Assessment, and Protection Tool; Maryland's Environmental Justice Screen Tool; and North Carolina's Community Mapping System.[d]	
Environmental Protection Agency (EPA)	**Definition:** EPA is developing benefit methodologies to track and report the benefits going toward disadvantaged communities (EPA 2022).	Unknown.

[a] See CEQ (2023).
[b] See DOT (2023).
[c] See Argonne National Laboratory (2022).
[d] See North Carolina Department of Environmental Quality (n.d.).

REFERENCES

Argonne National Laboratory. 2022. "Energy Justice Mapping Tool—Disadvantaged Communities Reporter." https://energyjustice.egs.anl.gov/. Accessed September 1, 2023.

CEQ (White House Council on Environmental Quality). 2023. "Methodology." https://screeningtool.geoplatform.gov/en/methodology#3/33.47/-97.5.

Connor, M.L. 2022. *Implementation of Environmental Justice and the Justice40 Initiative*. Edited by U.S. Army Corps of Engineers. https://api.army.mil/e2/c/downloads/2022/03/22/6ab6eb44/final-interim-implementation-guidance-on-environmental-justice-1.pdf.

DOE (Department of Energy). 2023. "Justice40 Initiative." https://www.energy.gov/diversity/justice40-initiative.

DOI (Department of the Interior). 2022. *Guidance on the Bipartisan Infrastructure Law Abandoned Mine Land Grant Implementation*. https://www.doi.gov/sites/doi.gov/files/bil-aml-guidance.pdf.

DOT (Department of Transportation). 2023. "US DOT Equitable Transportation Community Explorer Methodology." https://experience.arcgis.com/experience/0920984aa80a4362b8778d779b090723/page/Methodology.

EPA (Environmental Protection Agency). 2022. "Justice40 at EPA." https://www.epa.gov/environmentaljustice/justice40-epa.

North Carolina Department of Environmental Quality. n.d. "Community Mapping System Version 1.0." https://ncdenr.maps.arcgis.com/apps/webappviewer/index.html?id=1eb0fbe2bcfb4cccb3cc212af8a0b8c8. Accessed September 1, 2023.

White House. 2022. "Biden-Harris Administration Launches Version 1.0 of Climate and Economic Justice Screening Tool, Key Step in Implementing President Biden's Justice40 Initiative." Press release. https://www.whitehouse.gov/ceq/news-updates/2022/11/22/biden-harris-administration-launches-version-1-0-of-climate-and-economic-justice-screening-tool-key-step-in-implementing-president-bidens-justice40-initiative.

White House. 2023. *Building a Clean Energy Economy: A Guidebook to the Inflation Reduction Act's Investments in Clean Energy and Climate Action*. https://www.whitehouse.gov/wp-content/uploads/2022/12/Inflation-Reduction-Act-Guidebook.pdf.

Young, S.D., B. Mallory, and G. McCarthy. 2021. "Interim Implementation Guidance for the Justice40 Initiative." Memorandum for the Heads of Departments and Agencies. https://www.whitehouse.gov/wp-content/uploads/2021/07/M-21-28.pdf.

Public Health Provisions in the Infrastructure Investment and Jobs Act and the Inflation Reduction Act

Table H-1 displays the wide array of provisions in the Infrastructure Investment and Jobs Act (IIJA) and the Inflation Reduction Act (IRA) with direct or indirect effects on public health, including provisions relating to safety, research, schools, transit, public lands, multimodal transportation, water and energy system safety and accessibility, air pollution and contamination cleanup, public information, and more.

TABLE H-1 Infrastructure Investment and Jobs Act (IIJA) and Inflation Reduction Act (IRA) Provisions Related to Public Health

Provision—Title	Description
IIJA §11109—Surface Transportation Block Grant Program	Promotes state and local transportation decisions and provides flexible funding to address transportation needs (White House 2022). Eligible activities include improving on- and off-road pedestrian and bicycle facilities, environmental mitigation, and creating or improving recreational trails projects.
IIJA §11110—Nationally Significant Freight and Highway Projects	Funds competitive grants for multimodal freight and highway projects to improve the safety, efficiency, and reliability of the transportation in and across rural and urban areas (White House 2022).
IIJA §11111—Highway Safety Improvement Program	Funds safety projects on public roads to save lives and prevent serious injuries (White House 2022). Eligible uses include construction to reduce vehicle speeds and traffic, installation or upgrades for pedestrian and bicyclist traffic control devices, and pedestrian security features designed to slow or stop a motor vehicle as an eligible highway safety improvement project.
IIJA §11112—Federal Lands Transportation Program	Supports critical transportation needs by providing access within national parks, forests, wildlife refuges, and other federal public lands (White House 2022).
IIJA §11114—National Highway Freight Program	Provides funding to states to improve the efficient movement of freight on the National Highway Freight Network (White House 2022).
IIJA §11115—Congestion Mitigation and Air Quality Improvement Program	Provides funds to state and local governments for transportation projects and programs to reduce congestion and improve air quality for areas that do not meet the National Ambient Air Quality Standard for ozone, carbon monoxide, or particulate matter (White House 2022).
IIJA §11119—Safe Routes to Schools	Codifies the Safe Routes to School Program and amends it to apply the program through 12th grade to enable and encourage high school students to walk and bike to school safely (Jenkins et al. 2022).
IIJA §11122—Vulnerable Road User Research	Directs the Federal Highway Administration's Administrator to establish a research plan to prioritize research relating to roadway safety improvements, the impacts of traffic speeds, and tools to evaluate the impact of transportation improvements on projected rates and safety of bicycling and walking (AASHTO 2021).

TABLE H-1 Continued

Provision—Title	Description
IIJA §11124—Consolidation of Programs	Funds transportation safety outreach, training, and education (White House 2022). Eligible activities include Operation Lifesaver, work zone safety grants, and the Public Road Safety Clearinghouse.
IIJA §11127—Nationally Significant Federal Lands and Tribal Projects	Provides funding for the construction and rehabilitation of nationally significant federal land transportation projects and tribal land transportation projects (White House 2022).
IIJA §11130—Public Transit	Supports the construction or installation of traffic signaling and prioritization systems, redesigned intersections that are necessary for the establishment of a bus rapid transit corridor, on-street stations, fare collection systems, information and wayfinding systems, and depots (Jenkins et al. 2022).
IIJA §11132—Rural Surface Transportation Grant Program	Supports projects to improve and expand the surface transportation infrastructure to increase connectivity, improve the safety and reliability of the movement of people and freight, and generate regional economic growth in rural areas (White House 2022).
IIJA §11133—Bicycle Transportation and Pedestrian Walkways	Funds the construction of walkways and bicycle transportation facilities (Jenkins et al. 2022).
IIJA §11201—Metropolitan Planning	Establishes a cooperative and comprehensive framework for making transportation investment decisions in metropolitan areas (White House 2022). Eligible activities include analysis and forecasting of travel demand and system performance, identification and prioritization of improvement needs, and coordination of the planning process and decision-making.
IIJA §11206—Increasing Safe and Accessible Transportation Options	Incentivizes state adoption of complete streets standards and policies, development of a complete streets prioritization plan, active and mass transportation planning, regional and megaregional planning to address travel demand through alternatives to highway travel, or transit-oriented development planning (Jenkins et al. 2022).
IIJA §11304—Intelligent Transportation Systems Program	Fosters innovation in transportation through technology that enhances safety and efficiency while reducing environmental impacts of surface transportation (White House 2022). Eligible uses include research and deployment of tools that facilitate safe, connected, and automated transportation systems.

continued

TABLE H-1 Continued

Provision—Title	Description
IIJA §11402—Reduction of Truck Emissions at Port Facilities	Funds activities to reduce truck emissions at ports, including through port electrification (White House 2022).
IIJA §11403—Carbon Reduction Program	Provides formula grants to states to reduce transportation emissions or develop carbon reduction strategies (White House 2022). Eligible projects include alternative fueling infrastructure, zero-emission construction vehicles, port electrification, energy-efficient traffic lights and streetlights, and bike paths and public transit routes.
IIJA §11404—Congestion Relief Program	Funds integrated and multimodal solutions to reduce congestion and the related environmental costs in the most congested metropolitan areas (White House 2022). Eligible uses include incentive programs that encourage carpooling, non-highway travel, or travel during nonpeak periods and the deployment of integrated congestion management systems or mobility services.
IIJA §11406—Healthy Streets Program	Provides grants to deploy cool pavements and porous pavements and to expand tree cover to mitigate urban heat islands and improve air quality (Jenkins et al. 2022).
IIJA §11509—Reconnecting Communities Pilot Program	Establishes the Reconnecting Communities Pilot Program to remove, retrofit, or mitigate highways and other transportation facilities that create barriers to community connectivity (White House 2022).
IIJA §21201—National Infrastructure Project Assistance (Megaprojects)	Supports large, complex projects that are likely to generate national or regional economic, mobility, or safety benefits (White House 2022). Eligible projects include freight intermodal or rail projects that provide a public benefit, intercity passenger rail projects, and public transportation projects.
IIJA §21202—Local and Regional Project Assistance Grants (RAISE)	Provides funding for grants to state and local entities for projects that will have local and regional impacts (White House 2022). Eligible projects include public transportation projects, surface transportation projects located on tribal land, and other surface transportation infrastructure projects considered necessary to advance the goal of the RAISE program.

TABLE H-1 Continued

Provision—Title	Description
IIJA §26001—Hazardous Materials and Emergency Preparedness Grants	Funds grants for Hazardous Materials Emergency Preparedness, Assistance for Local Emergency Response Training, and Hazardous Materials Instructor Training programs (White House 2022). Eligible uses for the grants include the training of employees in hazardous materials safety.
IIJA §30007—Research, Development, Demonstration, and Deployment Projects	Provides funding to assist projects and activities that advance safe, equitable, and climate-friendly public transportation (White House 2022).
IIJA §30017—State of Good Repair Formula Grants	Funds capital projects to maintain public transportation systems to ensure that transit operates safely, efficiently, and sustainably (White House 2022). Additionally provides funding for technical assistance to support transit providers in enhancing safe, equitable, and climate-friendly public transportation and supports the development of public transportation industry standards.
IIJA §40103—Energy Improvement in Rural or Remote Areas	Provides financial assistance to improve the resilience, safety, and availability of energy (White House 2022). Eligible uses include reducing greenhouse gas (GHG) emissions from energy generation and increasing energy efficiency.
IIJA §40601—Funding to Support Orphan Well Plugging	Supports efforts to establish a program to plug, remediate, and reclaim orphaned wells on federal land (White House 2022). Eligible activities include scientific research on methane emissions associated with orphan oil and gas wells.
IIJA §40209—Advanced Energy Manufacturing and Recycling Grant Program	Funds program for states and Tribes to clean up abandoned coal mine sites and related problems that pose a threat to public health and safety (White House 2022).
IIJA §40701—Abandoned Mine Reclamation Fund	Funds program for states and Tribes to clean up abandoned coal mine sites and related problems that pose a threat to public health and safety (White House 2022).
IIJA §41008—Industrial Emission Demonstration Projects	Funds industrial emissions demonstration projects that test technologies that reduce industrial emissions (White House 2022). Eligible uses include applying principles of sustainable manufacturing to minimize the potential negative environmental impacts while conserving energy and increasing energy efficiency of industrial processes.

continued

TABLE H-1 Continued

Provision—Title	Description
IIJA §41201—Office of Clean Energy Demonstrations	Establishes a new Department of Energy office to oversee and manage demonstration projects and support efforts to commercialize clean energy technologies, reduce costs, and address barriers to widespread deployment (Jenkins et al. 2022).
IIJA §50102—Drinking Water State Revolving Fund	Provides financial assistance to achieve the health protection objectives of the Safe Drinking Water Act (White House 2022). Eligible uses include projects that prioritize serious risks to human health and assist household systems most in need.
IIJA §71101—Clean School Bus Program	Funds the deployment of zero-emission and alternative-fuel school buses (White House 2022).
IIJA Division J, Title III—Water-Related Environmental Infrastructure Assistance	Funds engineering and construction of authorized environmental projects that offer safe water supply, waste disposal, and pollution control to protect human health (White House 2022).
IIJA Division J, Title V—Building Resilient Infrastructure and Communities	Provides financial assistance for various hazard mitigation activities and projects, including projects designed to increase resilience and public safety and to reduce loss of life and infrastructure damage from the effects of climate change (White House 2022).
IIJA Division J, Title VI—Brownfields Projects	Provides technical assistance for brownfield activities that protect human health and the environment (White House 2022). Eligible uses include conducting community engagement and planning, site assessments, and direct site cleanup.
IIJA Division J, Title VI—Legacy Road and Trail Remediation Program	Funds the decommissioning and repairing of roads and trails to mitigate detrimental impacts to public safety and ecosystems or watersheds (White House 2022).
IIJA Division J, Title VI—Pollution Prevention Grants	Funds technical assistance to identify and adopt source reduction practices and technologies that benefit businesses, communities, and local economies (White House 2022). Eligible uses include targeted assistance to businesses for whom lack of information is an impediment.
IIJA Division J, Title VI—Superfund	Funds the cleanup of the nation's most contaminated lands to protect public health and the environment (White House 2022).
IIJA Division J, Title VII—Low-Income Home Energy Assistance Program	Assists eligible low-income households with heating and cooling energy costs, energy crisis assistance, and weatherization and energy-related repairs (White House 2022).

TABLE H-1 Continued

Provision—Title	Description
Inflation Reduction Act (IRA) §13901—Permanent Extension of Tax Rate to Fund Black Lung Disability Trust Fund	Makes permanent the increased coal excise tax rate for funding the Black Lung Disability Trust Fund (CRS 2022).
IRA §30002—Improving Energy Efficiency or Water Efficiency or Climate Resilience of Affordable Housing	Provides funding to the Department of Housing and Urban Development for loans and grants (CRS 2022). The loans and grants must fund projects that address affordable housing and climate change issues. Eligible property includes low-income housing and housing for the elderly or disabled.
IRA §60101—Clean Heavy-Duty Vehicles	Provides incentives to replace eligible medium-duty vehicles (e.g., school buses) and heavy-duty vehicles (e.g., garbage trucks) with zero-emission vehicles (CRS 2022). It also provides funding for a program within the Environmental Protection Agency (EPA) to award grants and rebates for replacing such vehicles with zero-emission vehicles.
IRA §60102—Grants to Reduce Air Pollution at Ports	Provides incentives to reduce air pollution at ports and funding for rebates and grants for carrying out such activities in ports located in areas designated as non-attainment areas under the Clean Air Act (CRS 2022).
IRA §60103—Greenhouse Gas Reduction Fund	Establishes a GHG reduction fund for the deployment and use of zero-emission technologies, including financial and technical assistance to enable low-income and disadvantaged communities to deploy or benefit from zero-emission technologies (CRS 2022).
IRA §60104—Diesel Emissions Reduction	Funds EPA program that gives grants, rebates, and loans to identify and reduce diesel emissions resulting from goods movement facilities as well as vehicles servicing those facilities in low-income and disadvantaged communities (CRS 2022).
IRA §60105—Funding to Address Air Pollution	Provides funding for programs that incentivize activities to deploy, integrate, and maintain methods to monitor air toxins; expand the national ambient air quality monitoring network; deploy, integrate, and operate air quality sensors in low-income and disadvantaged communities; and conduct research and development related to the prevention and control of air pollution (CRS 2022).

continued

TABLE H-1 Continued

Provision—Title	Description
IRA §60106—Funding to Address Air Pollution at Schools	Funds grants and other activities to monitor and reduce GHG emissions and other air pollutants at schools in low-income and disadvantaged communities (CRS 2022). Also provides funding for technical assistance to schools in low-income and disadvantaged communities.
IRA §60107—Low-Emissions Electricity Program	Provides funding for a program that will provide education, technical assistance, and outreach to reduce GHG emissions that result from domestic electricity use (CRS 2022).
IRA §60108—Funding for Section 211(O) of the Clean Air Act	Provides funding to EPA for (1) the development and establishment of tests and protocols regarding the environmental and public health effects of a fuel or fuel additive; (2) the collection and analysis of data to update applicable regulations, guidance, and procedures for determining the amount of GHG emissions from a fuel; (3) the review, analysis, and evaluation of the impacts of all transportation fuels on the public; and (4) supporting investments in advanced biofuels (CRS 2022).
IRA §60109—Funding for Implementation of the American Innovation and Manufacturing Act	Provides funding to EPA to address hydrofluorocarbons (HFCs) through grants for innovative technologies that reclaim or destroy HFCs (CRS 2022). Furthermore, it provides funding for the EPA to deploy new implementation and compliance tools when carrying out the American Innovation Act of 2020.
IRA §60110—Funding for Enforcement Technology and Public Information	Provides funding to update EPA's Integrated Compliance Information System (ICIS) and any associated systems, necessary information technology infrastructure, or public access software tools to ensure access to compliance data and related information (CRS 2022). It also provides funding for grants to states, Tribes, and air pollution control agencies to update their systems to ensure communication with ICIS.
IRA §60111—Greenhouse Gas Corporate Reporting	Provides funding for EPA to support (1) enhanced standardization and transparency of corporate climate action commitments and plans to reduce GHG emissions; (2) enhanced transparency regarding progress toward meeting such commitments and implementing such plans; and (3) progress toward meeting such commitments and implementing such plans (CRS 2022).

TABLE H-1 Continued

Provision—Title	Description
IRA §60112—Environmental Product Declaration Assistance	Provides funding to develop and carry out a program that supports the development, enhanced standardization and transparency, and reporting criteria for environmental product declarations for construction materials and products (CRS 2022). The declarations must include measurements of the GHGs associated with all the relevant stages of production, use, and disposal of the construction materials and products.
IRA §60113—Methane Emissions Reduction Program	Revises the Clean Air Act to create and provide funding for a Methane Emissions Reduction Program and a Methane Emissions Waste Reduction Program (CRS 2022). The Methane Emissions Reduction Program requires EPA to provide financial incentives for the reporting of GHGs, the monitoring of methane, and the reduction of methane emissions. The Methane Emissions Waste Reduction Program requires EPA to impose and collect a charge on methane emissions from a facility.
IRA §60114—Climate Pollution Reduction Grants	Establishes and funds a program that awards grants to states, air pollution control agencies, municipalities, and Tribes for developing and implementing plans to reduce GHG air pollution (CRS 2022).
IRA §60115—Environmental Protection Agency Efficient, Accurate, and Timely Reviews	Funds EPA to provide for the development of efficient, accurate, and timely reviews for permitting and approval processes; environmental data or information systems and geographic information systems; and other analysis tools, techniques, and guidance to improve agency transparency, accountability, and public engagement (CRS 2022).
IRA §60116—Low-Embodied Carbon Labeling for Construction Materials for Transportation Products	Provides funding to develop and carry out a program to identify and label construction materials and products that have substantially lower levels of GHGs associated with all the relevant stages of production, use, and disposal of the materials and products (CRS 2022).
IRA §60201—Environmental and Climate Justice Block Grants	Provides funding to EPA for environmental and climate justice block grants that benefit disadvantaged communities (CRS 2022).

REFERENCES

AASHTO (American Association of State Highway and Transportation Officials). 2021. *AASHTO Comprehensive Analysis of the Bipartisan Infrastructure Bill: Infrastructure Investment and Jobs Act (IIJA)*. Washington, DC: American Association of State Highway and Transportation Officials. https://www.ite.org/pub/?id=0B60D41F-ADD0-33EB-8ECF-15A3CB59D4A4.

CRS (Congressional Research Service). 2022. "Summary: H.R.5376—117th Congress (2021–2022)." https://www.congress.gov/bill/117th-congress/house-bill/5376.

Jenkins, J.D., J. Farbes, R. Jones, and E. Mayfield. 2022. "REPEAT Project Section-by-Section Summary of Energy and Climate Policies in the 117th Congress." http://bit.ly/REPEAT-Policies.

White House. 2022. *Building a Better America: A Guidebook to the Bipartisan Infrastructure Law for State, Local, Tribal, and Territorial Governments, and Other Partners*. Washington, DC. https://www.whitehouse.gov/build/guidebook.

Public Engagement Scorecard: Current Federal Policy Portfolio

Through a thorough analysis of the executive orders (EOs) and major legislative actions taken since 2021, the committee developed Table I-1, a public engagement scorecard of the current federal policy portfolio compared against the public engagement objectives detailed in the committee's first report.

TABLE I-1 Public Engagement Scorecard of the Current Federal Policy Portfolio Compared Against Public Engagement Objectives from the Committee's First Report

Public Engagement Objective	EOs	IIJA	IRA	Other
Prevent Misinformation				
Expand and tighten financial disclosure and transparency requirements.	✗[a]	✗[b]	✗[c]	✗[d]
Enable the cross-flow of information across diverse communities and value systems through new forms of social interaction that provide a foundation basis of trust.	✗	✗[e]	✗	✗
Engage the Public in the Design and Deliberation of Decarbonization Pathways				
Support high-profile regional, bidirectional dialogue and listening sessions that connect national policy making with local communities.	✗[f]	✗	✗	✗[g]
Design strategies that are sensitive and responsive to local and contextual factors, including through the incorporation of the public's perceptions of costs and benefits.	✗	✗[h]	✗	✗
Engage with the public significantly in advance of proposed technological changes.	✗[i]	✗	✗	✗[j]
Support Multifaceted Coordination for Decarbonization Actions				
Engage with younger populations with well-designed public engagement opportunities.	✗	✗	✗	✗
Establish state energy transition offices to support statewide, cross-sectoral coordination.	✗	✗	✗	✗
Enable mayors, governors, and industry leaders to identify, deliberate, and solve cross-border problems and address regional infrastructure needs.	✗	✗	✗	✗[k]
Support local transition planning, community-based action, and community benefits.	✗	✗[l]	✗[m]	✗
Set Standards for Public Participation				
Require a role for representatives of disadvantaged populations in advisory boards and other influential bodies to enable them to participate in meaningful ways.	✗[n]	✗[o]	✗	✗
Invest in comprehensive education and training opportunities focused on energy transitions.	✗	✗[p]	✗[q]	✗[r]
Set and enforce rules for inclusive public participation in the siting of decarbonization infrastructure.	✗[s]	✗	✗[t]	✗[u]
Support local, state, and regional decision-making for transition planning through robust data, modeling, and knowledge infrastructure.	✗[v]	✗[w]	✗[x]	✗[y]

[a] See EO 14030 (2021).
[b] See IIJA §27001 and §11132.

^c See IRA §40003, §60111, §60112, and §60115.

^d The Securities and Exchange Commission's proposed rule on climate-related disclosures (with potential litigation risk due to Supreme Court decision in *West Virginia v. EPA*) (SEC 2022a,b; Uslaner and Horowitz 2022).

^e See IIJA §60102 and §60201; IIJA §11201 and §30002 include the use of social media and other web-based tools to encourage public participation.

^f EO 14008 (2021) established the White House Environmental Justice Advisory Council.

^g See Federal Interagency Thriving Communities Network (DOT 2023).

^h IIJA §13009, §11401, §11509, §24102, §24112, §30002, §30003, §40321, §40806, and §70801 mention the use of public engagement but do not provide additional details about specific approaches.

ⁱ EO 14096 (2023) requires each federal agency to provide opportunities for early and meaningful involvement in the environmental review process by communities with environmental justice concerns potentially affected by a proposed action.

^j The Council on Environmental Quality (CEQ) issued guidance to responsibly develop carbon capture, utilization, and sequestration (CCUS), recognizing the importance of early consultation and meaningful public engagement (White House 2022).

^k See DOT (2023), and CHIPS and Science Act §10621 and §10622.

^l See IIJA §11109, §40552, §40601, and §40701.

^m See IRA §60103, §60114, §60201, and §60501.

ⁿ EO 14091 (2023) requires all federal agencies to conduct proactive engagement with members of underserved communities; identify and develop tools and methods for engagement; create incentives and guidelines for recipients of federal funding to proactively engage with communities; identify funding opportunities for civil society organizations working in and with underserved communities; and address barriers for individuals with disabilities.

^o See IIJA §11509 and §40211. Regarding membership, the former requires "representatives of the community" on its community advisory board but does not specify socioeconomic status or expertise; the latter requires board appointees with expertise in several areas, including diversifying the workforce.

^p See IIJA §40211, §40503, §40511, §40512, §40513, and §25019.

^q See IRA §50123, §60101, and the prevailing wage and apprenticeship requirements in IRA §13101, §13102, §13104, §13105, and §13204.

^r See CHIPS and Science Act §10381.

^s EO 13985 charges the Office of Management and Budget with studying methods for assessing whether agency policies and actions create or exacerbate barriers to full and equal participation by all eligible individuals.

^t See IRA §22004, §50152, §60115, §60402, §60505, and §70007.

^u See DOE (n.d.), White House (2022), and FERC (n.d.).

^v Federal institutions are implementing recommendations from the Interagency Working Group on Equitable Data.

^w See IIJA §40514, §40201, and §40203.

^x See IRA §60401, §50153, and §70005.

^y See CHIPS and Science Act Title I and Title II.

NOTES: Green indicates that the objective was achieved; yellow indicates that progress was made but significant work remains; red indicates that the objective was not included in the policy and the committee will continue to advocate for the objective. EO = Executive Order; IIJA = Infrastructure Investment and Jobs Act; IRA = Inflation Reduction Act.

REFERENCES

DOE (Department of Energy). n.d. "Community Benefit Agreement (CBA) Toolkit." Energy.Gov. Accessed May 16, 2023. https://www.energy.gov/diversity/community-benefit-agreement-cba-toolkit.

DOT (Department of Transportation). 2023. "Federal Interagency Thriving Communities Network." January 3. https://www.transportation.gov/federal-interagency-thriving-communities-network.

EO (Executive Order) 13985. 2021. "Advancing Racial Equity and Support for Underserved Communities Through the Federal Government."

EO 14008. 2021. "Tackling the Climate Crisis at Home and Abroad."

EO 14030. 2021. "Climate-Related Financial Risk."

EO 14091. 2023. "Further Advancing Racial Equity and Support for Underserved Communities Through the Federal Government."

EO 14096. 2023. "Revitalizing Our Nation's Commitment to Environmental Justice for All."

FERC (Federal Energy Regulatory Commission). n.d. "Office of Public Participation (OPP)." https://www.ferc.gov/OPP.

SEC (Securities and Exchange Commission). 2022a. "The Enhancement and Standardization of Climate-Related Disclosures for Investors." 17 CFR 210, 229, 232, 239, and 249. https://www.sec.gov/files/rules/proposed/2022/33-11042.pdf.

SEC. 2022b. "SEC Proposes Rules to Enhance and Standardize Climate-Related Disclosures for Investors." Press release. https://www.§gov/news/press-release/2022-46.

Uslaner, J.D., and W. Horowitz. 2022. "Will the SEC's Proposed Climate Risk Disclosure Rules Survive Supreme Court Scrutiny?" *Practitioner Insights Commentaries* (blog), August 5.

White House. 2022. "CEQ Issues New Guidance to Responsibly Develop Carbon Capture, Utilization, and Sequestration." February 15. https://www.whitehouse.gov/ceq/news-updates/2022/02/15/ceq-issues-new-guidance-to-responsibly-develop-carbon-capture-utilization-and-sequestration.

Select Infrastructure Investment and Jobs Act and Inflation Reduction Act Provisions Implicating Subnational Entities

Table J-1 displays the wide array of provisions in the Infrastructure Investment and Jobs Act (IIJA) and the Inflation Reduction Act (IRA) directing funds, opportunities, and requirements toward subnational entities, and capacity and resource dimensions that funding agencies, applicants, and recipients will need to consider.

TABLE J-1 Infrastructure Investment and Jobs Act (IIJA) and Inflation Reduction Act (IRA) Provisions, Amount of Funding Available, Federal Agency Responsible for Implementation, Opportunities or Requirements for Subnational Entities, and Capacity and Resource Considerations

Section(s)	Name	Amount ($ billion)	Federal Agency	Opportunities/Requirements for Subnational Entities	Capacity and Resource Considerations
Infrastructure Investment and Jobs Act					
40101(c), 40103(b), 40107	Grid Resilience and Innovation Partnerships	10.500	Department of Energy (DOE)	Subnational and tribal agencies are eligible for $3 billion in Smart Grid grants or $5 billion in grid innovation financial assistance (DOE-GD 2023b).	Recipients must compete for funds.
40101(d)	Preventing Outages and Enhancing the Resilience of the Electric Grid	0.459	DOE	State and tribal governments are eligible for grants to undertake strategic planning, community engagement, and investments to improve grid resilience to all hazards (DOE-GD 2022).	
40104, 40108, 40109	U.S. State Energy Program	0.500	DOE	State and Territory Energy Offices receive formula funding for clean energy, energy security, demand response, and collaborative transmission siting planning and activities (DOE-SCEP 2023f).	Allocated on a formula basis to all states and territories.
40314	Regional Clean Hydrogen Hubs	8.000	DOE	Subnational and tribal governments can be potential partners in the hubs, which will include hydrogen producers, potential consumers, connective infrastructure, and communities (DOE-OCED 2023).	Recipients must compete for funds.
40431	Utility Electric Vehicle (EV) Promotion Measures	0.000	Federal Energy Regulatory Commission (FERC)	Each state regulatory authority and each nonregulated utility is required to consider measures to promote greater transportation electrification by amending rates (DOE-EERE 2023).	Entities with existing EV rates are exempt. Legislation does not provide any funding to states.

40502	Energy Efficiency Revolving Loan Fund Capitalization	0.250	DOE	State and Territory Energy Offices can receive grants to establish a revolving loan fund for residential and commercial energy efficiency loans and audits (DOE 2022a).	Distributed to all states based on a formula allocation, with additional funding to "priority states" with high per-capita energy consumption or carbon dioxide emissions.
40503	Energy Auditor Training Grant Program	0.040	DOE	State and Territory Energy Offices can receive grants to support energy auditor training and education (DOE-SCEP 2023b).	Recipients must compete for funds.
40511	Resilient and Efficient Codes Implementation	0.225	DOE	State Energy Offices, Tribal Energy Offices, and partnerships among local code, construction, and energy efficiency entities are eligible for grants supporting building energy codes updates (DOE-EERE 2022).	Recipients must compete for funds.
40541	Energy Improvements at Public School Facilities	0.500	DOE	Local educational agencies can access grants to make clean energy improvements in K–12 public schools, with an emphasis on the highest-needs districts (DOE-SCEP 2023c).	Recipients must compete for funds.
40551	Weatherization Assistance Program	3.500	DOE	State weatherization agencies received expanded formula funding to conduct low-income home energy efficiency and weatherization programs (DOE 2022b).	Up to 15% of a grant may be used for administrative purposes.
40552	Energy Efficiency and Conservation Block Grant Program	0.550	DOE	States and localities receive formula funding for energy efficiency, renewable energy, clean transportation, and financing projects and planning (DOE-SCEP 2023a).	Portion of funds are reserved for competitive program for lesser-populated localities and tribes.

continued

TABLE J-1 Continued

Section(s)	Name	Amount ($ billion)	Federal Agency	Opportunities/Requirements for Subnational Entities	Capacity and Resource Considerations
Division J	National Electric Vehicle Infrastructure Formula Program	5.000	Department of Transportation (DOT)	States, Washington, DC, and Puerto Rico can receive assistance to strategically deploy EV charging infrastructure to establish an interconnected network (DOT 2022).	Funds available as both formula and discretionary funds.
Inflation Reduction Act					
13101, 13102, 13103, 13104, 13105, 13204, 13403, 13404, 13501, 13701, 13702, 13702(h), 13704	Various tax credits		DOT	Tax-exempt organizations, such as state and local governments, are eligible for direct pay of tax incentives (White House 2022).	
22001	Electric Loans for Renewable Energy	1.000	Department of Agriculture (USDA)	Subnational and tribal governments are eligible to receive loans and loan guarantees for the construction of electric distribution, transmission, and generation facilities (White House 2022).	Up to 50% non-federal cost share required but may be waived by Secretary.
22007	Increasing Land Access Program	0.250	USDA	Local and tribal governments, community-development financial institutions, and nonprofit education partners can receive grants for projects that improve land access for underserved farmers, ranchers, and forest landowners (White House 2022).	
23003(a)(2)	Urban and Community Forestry Assistance Program	1.500	USDA	Subnational and tribal agencies can receive grants for tree-planting activities (White House 2022).	Up to 50% non-federal cost share required but may be waived by Secretary.

40001	Investing in Coastal Communities and Climate Resilience	2.600	Department of Commerce	Coastal states, tribal governments, and local governments can receive financial or technical assistance to support coastal resilience (White House 2022).	
40007(a)(1)	Fueling Aviation's Sustainable Transition Through Sustainable Aviation Fuels	0.245	Federal Aviation Administration (FAA)	Subnational and tribal governments can receive funding for projects relating to sustainable aviation fuel (White House 2022).	Between 10%–25% non-federal cost share required.
40007(a)(2)	Fueling Aviation's Sustainable Transition— Technology	0.047	FAA	Subnational and tribal governments can receive funding for projects relating to low-emission aviation technologies (White House 2022).	Between 10%–25% non-federal cost share required.
50121, 50122	Home Energy Rebates—Home Efficiency, Electrification, and Appliance Rebates	8.800	DOE	State Energy Offices can receive grants to develop whole-house and/or high-efficiency electric home rebates. Tribes are eligible for the electric home rebate program (DOE-SCEP 2023d).	Non-federal cost share of 20%–50% required depending on household income level and/or price of appliance. 20% of Electric Home Rebate Program allocations can be used for planning, administration, or technical assistance.
50123	State-Based Home Efficiency Contractor Training Grants	0.200	DOE	States can receive financial assistance to develop and implement training programs for contractors involved in the installation of home energy efficiency and electrification improvements (DOE-SCEP 2023e).	

continued

TABLE J-1 Continued

Section(s)	Name	Amount ($ billion)	Federal Agency	Opportunities/Requirements for Subnational Entities	Capacity and Resource Considerations
50131	Assistance for Latest and Zero Building Energy Code Adoption	1.000	DOE	State and local governments with building code adoption authority can receive grants to adopt updated building energy codes (White House 2022).	
50141	Energy Loan Programs Office	3.600	DOE	Subnational and tribal governments, school districts, housing authorities, and nonprofits are among the eligible recipients to receive loan guarantees for Innovative Clean Energy technologies, including fossil energy, nuclear energy, critical minerals processing, manufacturing, and recycling (White House 2022).	
50144	Energy Infrastructure Reinvestment Financing	5.000	DOE	Subnational governments are anticipated to be among the eligible recipients of loan guarantees for projects that retool, repower, repurpose, or replace energy infrastructure (White House 2022).	
50145	Tribal Energy Loan Guarantee Program	0.075	DOE	Tribal governments and economic development organizations can receive direct loans or partial loan guarantees for a broad range of energy resources, products, and services (White House 2022).	
50152	Siting of Interstate Electricity Transmission Lines	0.760	DOE	Transmission siting authorities or other state, local, or tribal government entities are eligible for grants to facilitate siting of transmission projects and economic development activities in impacted communities (DOE-GD 2023a).	At least 50% non-federal cost share required for grants to siting authorities.

50241	Climate Change Technical Assistance for Territories	0.015	Department of the Interior (DOI)	Territorial governments can receive technical assistance for climate planning, mitigation, adaptation, and resilience (White House 2022).	Funds will support technical assistance, but not direct financial assistance.
60101	Clean Heavy-Duty Vehicles	1.000	Environmental Protection Agency (EPA)	Subnational and tribal governments and school transportation associations can receive grants and rebates to offset the incremental costs of zero-emission vehicles, infrastructure, workforce development and training, and planning and technical assistance (White House 2022).	Recipients must compete for funds.
60103	Greenhouse Gas Reduction Fund	27.000	EPA	States, municipalities, and tribal governments are directly eligible for $7 billion program to provide financial and technical assistance to low-income and disadvantaged communities to deploy zero-emission technologies. Green banks, community development financial institutions, credit unions, housing agencies, and others are eligible for the $20 billion General and Low-Income Assistance Competition (EPA 2023).	Recipients must compete for funds.
60104	Diesel Emissions Reductions	0.060	EPA	Subnational and tribal agencies and port authorities can receive grants, rebates, and loans to reduce diesel emissions resulting from goods movement facilities and services in low-income and disadvantaged communities (White House 2022).	
60105	Funding to Address Air Pollution	0.236	EPA	Subnational and tribal agencies can receive grants and technical assistance across a variety of topic areas: Clean Air Act planning and implementation, mobile source, fenceline air monitoring, multipollutant monitoring, air quality sensors, methane monitoring (White House 2022).	

continued

Section(s)	Name	Amount ($ billion)	Federal Agency	Opportunities/Requirements for Subnational Entities	Capacity and Resource Considerations
60106	Funding to Address Air Pollution at Schools	0.050	EPA	Subnational and tribal governments can receive grants and technical assistance to monitor and reduce pollution and greenhouse gases (GHGs) in schools in low-income and disadvantaged communities (White House 2022).	Recipients must compete for funds.
60109	Implementation of the American Innovation and Manufacturing Act	0.039	EPA	Subnational governments can receive grants to phase down the production and consumption of hydrofluorocarbons (HFCs) (White House 2022).	Recipients must compete for funds.
60113	Methane Emissions Reduction Program	1.550	EPA	Subnational and tribal governments are eligible to receive grants, rebates, or contracts to reduce methane or other GHGs from petroleum and natural gas systems (White House 2022).	
60114	Climate Pollution Reduction Grants	5.000	EPA	Subnational and tribal governments are eligible to receive grants to develop and implement plans for reducing GHG air pollution (White House 2022).	Initial $250 million for planning grants and remaining $4.75 billion for competitive implementation grants.
60201	Environmental and Climate Justice Block Grants	3.000	EPA	Tribal and local governments can partner with community-based nonprofits to receive grants and technical assistance to reduce indoor and outdoor air pollution (White House 2022).	Most recipients must compete for grants.

60501	Neighborhood Access and Equity Grant Program	3.205	DOT	Subnational and tribal governments and special-purpose districts for projects that improve walkability and transportation access (White House 2022).	Recipients must compete for funds. Grants can be used for planning and capacity-building in disadvantaged communities. Non-federal cost of 20% required expect in disadvantaged communities.
60506	Low-Carbon Transportation Materials Program	2.000	DOT	Subnational and tribal governments and special-purpose districts can receive reimbursements or incentives for the use of low-embodied carbon construction materials and products in federally funded highway projects (White House 2022).	
80001	Tribal Climate Resilience	0.235	DOI	Tribes can receive financial assistance to support climate resilience planning, habitat restoration and adaptation, community-directed relocation, fish hatchery operations, and other activities (White House 2022).	
80002	Native Hawaiian Climate Resilience	0.025	DOI	State, local, and Native Hawaiian Community representatives can receive financial assistance to develop and implement a new Native Hawaii Climate Resilience Program (White House 2022).	
80003	Tribal Electrification Program	0.150	DOI	Tribes can receive financial and technical assistance to increase the number of homes with zero-emission electricity (White House 2022).	May also come in the form of direct federal spending.

REFERENCES

DOE (Department of Energy). 2022a. "Energy Efficiency Revolving Loan Fund Capitalization Grant Program (EE RLF) IIJA ALRD." Administrative and Legal Requirements Document (ALDR). https://www.energy.gov/sites/default/files/2022-11/EE-RLF-IIJA-Administrative-and-Legal-Requirements-Document.pdf.

DOE. 2022b. "Weatherization Assistance Program for Low-Income Persons: Application Instructions." https://www.energy.gov/sites/default/files/2022-05/bil-application-instructions_v2.pdf.

DOE-EERE (U.S. Department of Energy Office of Energy Efficiency and Renewable Energy). 2022. "Resilient and Efficient Codes Implementation: Infrastructure Investment and Jobs Act." https://www.energycodes.gov/sites/default/files/2022-07/NECC2022_2_DOE_IIJA_Presentation.pdf.

DOE-EERE. 2023. "Bipartisan Infrastructure Law (Infrastructure Investment and Jobs Act of 2021)." https://afdc.energy.gov/laws/infrastructure-investment-jobs-act.

DOE-GD (U.S. Department of Energy Grid Deployment Office). 2022. "DOE's Implementation Plan for IIJA Section 40101(D)—Formula Grants to States and Indian Tribes for Preventing Outages and Enhancing the Resilience of the Electric Grid." Notice of Intent (No. DE-FOA-0002764) to Issue Formula Grant Administrative and Legal Requirements (ALRD) Announcement No. DE-FOA-0002736. https://netl.doe.gov/sites/default/files/2022-05/IIJA%2040101d%20-%20Notice%20of%20Intent.pdf.

DOE-GD. 2023a. "Frequently Asked Questions on the Transmission Siting and Economic Development Grants Program." https://www.energy.gov/gdo/frequently-asked-questions-transmission-siting-and-economic-development-grants-program.

DOE-GD. 2023b. "Grid Resilience and Innovation Partnerships (GRIP) Program." https://www.energy.gov/gdo/grid-resilience-and-innovation-partnerships-grip-program.

DOE-OCED (Department of Energy Office of Clean Energy Demonstrations). 2023. "OCED Funding Opportunity Exchange." https://oced-exchange.energy.gov/Default.aspx#Foald4e674498-618c-4f1a-9013-1a1ce56e5bd3.

DOE-SCEP (Department of Energy Office of State and Community Energy Programs). 2023a. "EECBG Program Formula Grant Application Hub." https://www.energy.gov/scep/eecbg-program-formula-grant-application-hub.

DOE-SCEP. 2023b. "Energy Auditor Training Grant Program." https://www.energy.gov/scep/energy-auditor-training-grant-program.

DOE-SCEP. 2023c. "Grants for Energy Improvements at Public School Facilities." https://www.energy.gov/scep/grants-energy-improvements-public-school-facilities.

DOE-SCEP. 2023d. "Home Energy Rebate Programs Early Administrative Funds Administrative and Legal Requirements Document (ALRD)." https://www.energy.gov/sites/default/files/2023-03/Home_Energy_Rebates_ALRD.pdf.

DOE-SCEP. 2023e. "State-Based Home Energy Efficiency Contractor Training Grants." https://www.energy.gov/scep/state-based-home-energy-efficiency-contractor-training-grants.

DOE-SCEP. 2023f. "State Energy Program Guidance." https://www.energy.gov/scep/state-energy-program-guidance.

DOT (Department of Transportation). 2022. *Bipartisan Infrastructure Law (BIL): Overview of Highway Provisions*. https://www.fhwa.dot.gov/bipartisan-infrastructure-law/docs/bil_overview_20211122.pdf.

EPA (Environmental Protection Agency). 2023. "EPA Announces Initial Program Design of Greenhouse Gas Reduction Fund." https://www.epa.gov/newsreleases/epa-announces-initial-program-design-greenhouse-gas-reduction-fund.

White House. 2022. *Inflation Reduction Act Guidebook*. https://www.whitehouse.gov/wp-content/uploads/2022/12/Inflation-Reduction-Act-Guidebook.pdf.

Acronyms and Abbreviations

AGC	Associated General Contractors of America
ARP	American Rescue Plan
ARRA	American Recovery and Reinvestment Act
BAS	building automation system
BECCS	biomass energy with carbon capture and storage
BESS	building energy storage system
BEV	battery electric vehicle
BGA	BlueGreen Alliance
BLS	Bureau of Labor Statistics
CAFE	Corporate Average Fuel Economy
CAFO	concentrated animal feeding operation
CARB	California Air Resources Board
CBA	community benefits (or workforce) agreement
CBO	community-based organization
CCS	carbon capture and storage/sequestration
CCU	carbon capture and utilization
CCUS	carbon capture, utilization, and storage
CDC	Centers for Disease Control and Prevention
CDFI	Community Development Financial Institution
CDR	carbon dioxide removal
CEJST	Climate and Economic Justice Screening Tool
CEQ	Council on Environmental Quality
CFTC	Commodity Futures Trading Commission
CHP	combined heat and power
COZ	Climate Opportunity Zone
CZ	commuting zone
DAC	direct air capture
DER	distributed energy resource
DH	district heating
DHS	Department of Homeland Security
DOC	Department of Commerce

DoD	Department of Defense
DOE	Department of Energy
DOI	Department of the Interior
DOT	Department of Transportation
EECBG	Energy Efficiency and Conservation Block Grant
EIA	Energy Information Administration
EJ	environmental justice
EO	Executive Order
EPA	Environmental Protection Agency
ESG	environmental, social, and governance
EUI	energy use intensity
EV	electric vehicle
FCEV	fuel cell electric vehicle
FERC	Federal Energy Regulatory Commission
FIA	Forestry Inventory and Analysis
FOA	funding opportunity announcement
FSOC	Financial Stability Oversight Council
GAO	Government Accountability Office
GEB	grid interactive energy efficient building
GHG	greenhouse gas
GSA	General Services Administration
GWP	global warming potential
HEV	hybrid electric vehicle
HFC	hydrofluorocarbon
HHS	Department of Health and Human Services
HIA	health impact assessment
HUD	Department of Housing and Urban Development
IAC	Industrial Assessment Center
ICE	internal combustion engine
ICEV	internal combustion engine vehicle
IEA	International Energy Agency
IIJA	Infrastructure Investment and Jobs Act
ILO	International Labour Organization
IPCC	Intergovernmental Panel on Climate Change

IRA	Inflation Reduction Act
ISO	independent system operator
ITC	investment tax credit
IWG	Interagency Working Group
JQI	job quality index
LCOE	levelized cost of energy
LDV	light-duty vehicle
LEAP	local energy action plan
LIHEAP	Low-Income Heating Assistance Program
LMI	low and moderate income
LNG	liquefied natural gas
LTAR	Long-Term Agricultural Research
LULUCF	land use, land use change, and forestry
MHD	medium- and heavy-duty
NABTU	North America's Building Trade Unions
NAICS	North American Industry Classification System
NARUC	National Association of Regulatory Utility Commissioners
NASEO	National Association of State Energy Officials
NBCS	Nature-Based Climate Solution
NCTF	National Climate Task Force
NEPA	National Environmental Policy Act
NERC	North American Electric Reliability Corporation
NGO	non-governmental organization
NHTSA	National Highway Traffic Safety Administration
NIST	National Institute of Standards and Technology
NREL	National Renewable Energy Laboratory
NRI	National Resources Inventory
NSF	National Science Foundation
NTC	National Transition Corporation
NTIA	National Telecommunications and Information Administration
OEM	original equipment manufacturer
OMB	Office of Management and Budget
OSHA	Occupational Safety and Health Administration
OSTP	Office of Science and Technology Policy

PEV	plug-in electric vehicle
PHEV	plug-in hybrid electric vehicle
PLA	project labor agreement
PM	particulate matter
PRI	Principles of Responsible Investment
PTC	production tax credit
PUC	public utility commission
PV	photovoltaic
RD&D	research, development, and demonstration
RDD&D	research, development, demonstration, and deployment
RFF	Resources for the Future
RFI	request for information
RTO	regional transmission organization
SEC	Securities and Exchange Commission
STEM	science, technology, engineering, and mathematics
SUV	sport utility vehicle
TCC	Transformative Climate Communities
TCO	total cost of ownership
TCTAC	Thriving Communities Technical Assistance Center
TIP	NSF Directorate for Technology, Innovation and Partnerships
UNFCCC	United Nations Framework Convention on Climate Change
USDA	U.S. Department of Agriculture
USEER	U.S. Energy and Employment Report
USGCRP	U.S. Global Change Research Program
VOC	volatile organic compound
WAP	Weatherization Assistance Program
WHEJAC	White House Environmental Justice Advisory Council
WRI	World Resources Institute
ZEV	zero-emission vehicle